AF389619

Multi-Criteria Decision-Making Techniques for Improvement Sustainability Engineering Processes

Multi-Criteria Decision-Making Techniques for Improvement Sustainability Engineering Processes

Volume 2

Editors

Edmundas Kazimieras Zavadskas
Dragan Pamučar
Željko Stević
Abbas Mardani

MDPI • Basel • Beijing • Wuhan • Barcelona • Belgrade • Manchester • Tokyo • Cluj • Tianjin

Editors
Edmundas Kazimieras
Zavadskas
Vilnius Gediminas Technical
University
Lithuania

Abbas Mardani
Muma College of Business at
University of South Florida
(USF)
Tampa, FL, USA

Dragan Pamučar
University of Defence,
Military academy,
Department of Logistics
Belgrade, Serbia

Željko Stević
Faculty of Transport and Traffic
Engineering,
University of East Sarajevo,
Doboj, Bosnia and Herzegovina
Republic of Srpsk

Editorial Office
MDPI
St. Alban-Anlage 66
4052 Basel, Switzerland

This is a reprint of articles from the Special Issue published online in the open access journal *Symmetry* (ISSN 2073-8994) (available at: https://www.mdpi.com/journal/symmetry/special_issues/Sustainability_Engineering_Processes).

For citation purposes, cite each article independently as indicated on the article page online and as indicated below:

LastName, A.A.; LastName, B.B.; LastName, C.C. Article Title. *Journal Name* **Year**, *Article Number*, Page Range.

Volume 2
ISBN 978-3-03936-792-4 (Hbk)
ISBN 978-3-03936-793-1 (PDF)

Volume 1-2
ISBN 978-3-03936-794-8 (Hbk)
ISBN 978-3-03936-795-5 (PDF)

© 2020 by the authors. Articles in this book are Open Access and distributed under the Creative Commons Attribution (CC BY) license, which allows users to download, copy and build upon published articles, as long as the author and publisher are properly credited, which ensures maximum dissemination and a wider impact of our publications.

The book as a whole is distributed by MDPI under the terms and conditions of the Creative Commons license CC BY-NC-ND.

Contents

About the Editors

Edmundas Kazimieras Zavadskas Ph.D., DSc, is a Professor at the Department of Construction Management and Real Estate, Chief Research Fellow at the Laboratory of Operational Research, Research Institute of Sustainable Construction, Vilnius Gediminas Technical University, Lithuania. He received his Ph.D. in Building Structures (1973) and Dr Sc. (1987) in Building Technology and Management. He is a member of the Lithuanian and several foreign Academies of Sciences, Doctore Honoris Causa from Poznan, Saint Petersburg, and Kiev Universities, and the Honorary International Chair Professor in the National Taipei University of Technology. Awarded by the International Association of Grey System and Uncertain Analysis (GSUA) for his huge input in the field of Grey Systems, Zavadskas has been elected to an Honorary Fellowship of the International Association of Grey System and Uncertain Analysis, a part of IEEE (2016), awarded by "Neutrosophic Science—International Association" for distinguished achievements in neutrosophics, and has been conferred an honorary membership (2016), and awarded the Thomson Reuters certificate as the most highly cited scientist (2014). A highly cited researcher in the field of Cross-Field (2018, 2019), Zavadskas is recognized for exceptional research performance demonstrated by the production of multiple highly cited papers that rank in the top 1% by citations for field and year in the Web of Science. Zavadskas' main research interests include multi-criteria decision-making, operations research, decision support systems, and multiple-criteria optimization in construction technology and management. With over 517 publications in Clarivate Analytics Web of Science, h-index = 65, a number of monographs in Lithuanian, English, German, and Russian, Zavadskas is also Founding Editor-in-Chief of the journals *Technological and Economic Development of Economy* and *Journal of Civil Engineering and Management*, as well as Guest Editor of over 15 Special Issues related to decision-making in engineering and management.

Dragan Pamučar is an Associate Professor at University of Defence in Belgrade, Department of Logistics, Serbia. Dr. Pamucar received a Ph.D. in Applied Mathematics with a specialization in Multi-criteria modeling and soft computing techniques from the University of Defence in Belgrade, Serbia in 2013 and an MSc degree from the Faculty of Transport and Traffic Engineering in Belgrade, 2009. His research interests are in the fields of Computational Intelligence, Multi-Criteria Decision-Making problems, Neuro-Fuzzy systems, Fuzzy, Rough and Intuitionistic Fuzzy Set Theory, and Neutrosophic Theory. Application areas include a wide range of logistics problems. Dr. Pamucar has authored/co-authored over 80 papers published in SCI-indexed international journals including Experts Systems with Applications, Applied Soft Computing, Soft Computing, Computational Intelligence, Computers & industrial Engineering Technical Gazette, Sustainability, Symmetry, Water, Asia-Pacific Journal of Operational Research, Operational Research, Journal of Intelligent and Fuzzy Systems, Land Use Policy, Environmental Impact Assessment Review, International Journal of Physical Sciences, Economic Computation and Economic Cybernetics Studies and Research, and many more. In the last three years, Prof. Pamucar was awarded as top and outstanding reviewer for numerous Elsevier journals, such as Sustainable Production and Consumption, Measurement, Egyptian Informatics Journal, International Journal of Hydrogen Energy, and so on.

Željko Stević is an Assistant Professor at the University of East Sarajevo, Faculty of Transport and Traffic Engineering, Doboj. He received a Ph.D. in Transport and Traffic Engineering from the University of Novi Sad, Faculty of Technical Sciences, in 2018. His interests include logistics, supply chain management, transport, traffic engineering, soft computing, multi-criteria decision-making problems, rough set theory, sustainability, fuzzy set theory, and neutrosophic set theory. He has published over 120 papers in his areas of interest. He has contributed outstanding research to the mentioned fields. In all his research, he has provided very good application studies and practical contributions, solving different problems in transportation, logistics, supply chain management, traffic engineering, the economy, etc. His published studies are very well cited in other research, which can be seen in ResearchGate. He has an h-index of 18 in Google Scholar, 11 in Scopus, and 11 in WoS. Dr. Stević has authored/co-authored papers published in refereed international journals including Applied Soft Computing, Neural Computing and Applications, Sustainability, Symmetry, Engineering Economics, Soft Computing, Transport, Scientometrics, Information, ECECSR, Technical Gazette, SIC, Mathematics (MDPI), and more. He is a member of the Program Committee for specific programs for the Republic of Srpska in Horizon 2020. He is Editor-in-Chief of the journal Operational Research in Engineering Sciences: Theory and Applications. Moreover, he is Guest Editor of the following journals:

- Symmetry (SCI) IF2018=2.143—Special Issue "Multi-Criteria Decision-Making Techniques for Improvement Sustainability Engineering Processes";

- Algorithms (WoS) Special Issue "Algorithms for Multi-Criteria Decision-Making";

- Sustainability (SCI) IF2018=2.592—Special Issue "Operational Research Tools for Solving Sustainable Engineering Problems";

- Logistics—Special Issue "Application of Multi-Criteria Decision-Making Methods for Evaluation in Logistics and Supply Chain". Awards:

- November 2017: the best young researcher of 3rd cycle (doctoral) studies (Festival of Science 2017).

- January 2018: medal of merit for people in the field of education and science.

- September 2019: Top Peer Reviewer in the Global Peer Review Awards 2019 (Publons).

Abbas Mardani, Ph.D., Informetrics Research Group and Faculty of Business Administration; Ton Duc Thang University, Vietnam. Abbas has published more than 130 articles in high-quality journals. He is Editor and Guest Editor of several journals, including the International Journal of Physical Distribution & Logistics Management; Applied Soft Computing, Technological and Economic Development of Economy, Technological Forecasting and Social Change; Computer and Industrial Engineering; Journal of Enterprise Information Management; International Journal of Fuzzy Systems; Symmetry; Energies; Soft Computing Letters, etc. in Elsevier, Springer, Emerald, and MDPI. Abbas' H-indexes in Scopus and Google Scholar are 18 and 22 respectively. His research interests include quality management, sustainable development, fuzzy sets, decision-making, TQM, SCM, sustainability, service quality, and optimization.

Article

Identification of Determinants of the Speed-Reducing Effect of Pedestrian Refuges in Villages Located on a Chosen Regional Road

Alicja Sołowczuk and Dominik Kacprzak *

Department of road and bridge engineering, West Pomeranian University of Technology Szczecin, 71-311 Szczecin, Poland; alicja.solowczuk@zut.edu.pl
* Correspondence: kdim.zut@gmail.com or dominik.kacprzak@zut.edu.pl; Tel.: +48-(091)-449-40-36

Received: 3 April 2019; Accepted: 19 April 2019; Published: 25 April 2019

Abstract: Traffic calming, as a traffic engineering discipline, is becoming an increasingly important aspect of the road engineering process. One of the traffic calming treatments are pedestrian refuges—raised islands located on or at the road centreline. This paper presents factors relevant to the performance of this kind of traffic calming devices retrofitted on the stretches of regional roads in village areas. To this end, speed surveys were carried out before and after the islands in each direction on purposefully chosen test sections. In order to identify the determinants, each test section was characterised by features including the symmetry of the road layout geometry, surrounding features and the existing traffic signs and, last but not least, visibility of the road ahead. The survey data were used by the authors to perform analyses in order to group the speeds at the pedestrian refuges and relate them to specific factors and, finally, identify the determinants of speed reduction. In this way, the authors arrived at a conclusion that the performance of pedestrian refuges depends on a number of factors rather than solely on their geometric parameters. The analyses showed that the pedestrian refuge geometric parameters, features located in its proximity that influence the driver's perception and placement of appropriate marking, can, in combination, result in achieving the desired speed reduction and ensure safety of non-motorised users. These hypotheses were tested on a stretch of a regional road in village area at three points of the process: before upgrading, after installation of pedestrian refuges, and after retrofitting of enhancements.

Keywords: pedestrian refuge; speed reduction; visibility; surrounding environment

---

1. Introduction

Economic growth increases the road traffic and the associated problems are bound to intensify as a result. In the case of villages located on primary routes (further called regional roads) this growth of traffic is more conspicuous, as compared to small or bigger towns, due to accumulation of problems on a relatively short stretch of the road. A shortage of road by-passes in Poland results in the road routes cutting through the centres of settlements, this affecting the quality of life of the local community.

The main factor which, in most cases, has a direct bearing on both the number of road incidents and their severity is the speed of vehicular traffic. Hence, one of the key issues is ensuring safety on the pedestrian crossings in villages. To this end, various traffic calming treatments are installed, positioned both in the entry zones and in the village centre areas. The latter include pedestrian refuges whose primary function is to protect vulnerable road users (VRU). According to the guidelines of [1] the pedestrian refuges should be provided where it is desired to obtain reduction of the 85th percentile speed v_{85} to below 50 km/h.

Incorporation of pedestrian refuges in the design of roads is a most important issue from the traffic safety improvement viewpoint. According to the most recent studies [2] over 30% of accidents involving pedestrians occur on the pedestrian crossings and excessive speed is identified as their cause. The probability of fatality increases with the impact speed, i.e., the speed at which the vehicle hits the pedestrian. Based on the review of the most recent research publications it has been concluded that the pedestrian crossing type has a bearing on the speeds at which it is passed by vehicles, as demonstrated by the speed survey data. The distance from which the driver spots the pedestrian is also most relevant to the theoretical impact speed [3]. Visibility studies confirmed that pedestrian refuges, due to their central location, make the drivers focus their vision on this obstacle, i.e., on the central area of the road. Looking from a distance, the driver can sooner identify the pedestrian crossing and spot a pedestrian about to cross the street. Facing the island ahead, the drivers become more focussed and alert and this increases the distance from which they can spot a pedestrian and reduces the risk of accidents. Moreover, pedestrian refuges create a perceived reduction of the carriageway width, increasing the amount of speed reduction, as compared to conventional pedestrian crossings [4]. Therefore, pedestrian refuges are ranked as one of the best measures to improve traffic safety on pedestrian crossings.

According to the U.S. traffic survey data published in [1], for the road under analysis speed reduction in the range of 13–23% can be expected on the road under analysis after it has been provided with pedestrian refuges. The purpose of this research was to verify if the actual speed reductions achieved in Poland correspond to the U.S. survey results published in [1]. To this end, a number of free-flow and stable-flow speed surveys were carried out in a few villages located on regional roads with pedestrian refuges positioned in the entry zones and in the village centre areas. The upgrading project was carried out in the period 2012–2016 and to date, only isolated road incidents occurred from completion of the works. Only in one village a few incidents were noted in that period, yet none of them involved pedestrians. As such, the scope of this research has been limited to the vehicular speed issue. By selecting recently upgraded stretches of regional roads the authors excluded the effect of deteriorated pavements, lack of footways and other factors that could influence the speeds of travel. This paper presents the results of analyses performed on the survey data.

2. Review of Engineering Requirements Given in Various Design Manuals

The basic engineering requirements to be applied for pedestrian refuges located in Poland are given in [1,5]. The key points concern maintaining the width of the travel lane alongside the pedestrian refuge, as given in the Design Guidelines [6] depending on the level of service of the road and use of symmetric 1:5 to 1:10 tapered hatched markings as the approach end treatments. However, the guidelines [7] are not specific if the lines of P-7b pavement marking should be extended up to the island nose or to the meeting point with the P10 or P11 line marking, (for the meaning and pictures of signs referred to by their acronyms see the table in the Appendix A). However, they give a recommendation the travel lane width should be measured between the centres of the marking lines or between the line centre and face of the curb. Furthermore, guidelines [5] recommend using different taper geometries depending on the pedestrian island position (in the entry zone, central area or in the vicinity of public buildings respectively), with more aggressive design for the village central areas and less aggressive for the entry zones.

The Swedish guidelines [8] in turn do not give detailed geometric parameters for pedestrian refuges. They, however, recommend two different treatments to accompany the pedestrian refuges in lightly trafficked (yet including heavy goods vehicles) and heavily trafficked roads respectively. These are: raising of pavement—to facilitate crossing the road by pedestrians in the first case and installation of post-and-chain barriers as a measure to prevent illegal crossing in the latter case.

In the U.S. guidelines [9] the recommendations concern primarily the island width, which should be in the range of 1.2 to 1.8 m and the lengths of P-21 and P-7b line markings, which should be 30.48 m (100 ft) in built-up areas and 60.96 m (200 ft) in rural areas. Symmetricity of both the island and the hatched markings is required therein. Moreover, P-4 solid line pavement marking is recommended to be placed before the hatched marking over the same length as the taper length. Raised pedestrian crossings are recommended for less busy roads and refuges flush with the road surface are recommended in the case of narrower islands. In guidelines [9] much emphasis is put on installation of raised kerbing and conspicuous markers which is primarily associated with the motorist's perception of closer and more distant parts of the route and outlines of the nearby houses. Visual perception of the road signs and pavement markings by the motorist is covered, for example, in [10]. The issues pertaining to perception of 2D and 3D symmetric images and the effect of this perception on taking decisions by the motorist are covered in [11,12].

Also, the German guidelines [13] pay special attention to the motorist's perception, with the focus put on the visibility of pedestrians to motorists approaching the pedestrian refuge and the need of artificial lighting installed at a height of 3.5–4.5 m to improve visibility when required. The island widths given in the German guidelines [13] are much greater than the values of the U.S. guidelines [9]. A minimum width of 2.0 m is recommended, increased to 2.5–3.0 m if the crossing is intended to be used by cyclists and wheelchair users. The width of travel lanes recommended in [13] to ensure smooth traffic flow is 3.25–3.75 m, depending on the traffic composition. With a greater percentage of heavy vehicles, the lanes should be 3.5 m or 3.75 m wide. With the heavy vehicles, percentage of 1–3% the travel lanes can be 3.25 m wide. In places where an increased number of over-dimensional vehicles involved in seasonal agricultural activities is expected, the geometry of islands should be adjusted accordingly by providing 1.0 m wide overrun strips at the outer carriageway edge, made of irregular cobblestones or fieldstone/flagstone pavers.

In guidelines [13] a lot of attention is paid to the location of pedestrian crossings on the central islands of various shapes installed as traffic calming measures in the entry zones of settlements. In order to warn the motorists of the change of carriageway geometry and enhance understanding the route and layout of the carriageway, the German guidelines recommend highlighting the chicanes by planting of trees or placing street furniture items. The issue of motorist's perception and comprehension of road signs is extensively covered in literature [10–12].

According to the design guidelines of [10], for safety reasons, the central islands used as the village entry treatment should not be combined with pedestrian crossings. This said, local conditions may sometimes require combining these two elements at the village gateway. In these cases, the crossing should be placed on the so-called safe side, i.e., where the drivers are expected to have reduced their speed, and enhanced by trees or street furniture elements.

3. Parameters of the Test Sections

For testing the effect of pedestrian refuges on vehicular speed in their vicinity a few villages were chosen where different refuge islands had been installed: conventional, symmetric about the carriageway centreline and deflecting the path of travel by 1 m on each side (Figure 1a–c), one non-conventional 2.5 m wide pedestrian refuge island on one side of the centreline (Figure 1d) and one asymmetric pedestrian refuge incorporating a 4 m wide island imposing asymmetric lateral shifts by 1 m and 3 m respectively (Figure 1e). One case under analysis was a 2 m wide pedestrian refuge located in the entry zone in place of centre island (Figure 1a).

Most of the analysed 2 m wide pedestrian refuges were accompanied with a 1:5 tapered marking except for one case with the 1:6.5 rate at village centre side and 1:8 rate at the entry zone side (Figure 1c). In another case, a 4 m wide asymmetric pedestrian refuge had 1:15 tapered marking at the side imposing lateral shift by 1 m (Figure 1e). In addition to asymmetric positioning another unusual feature of this pedestrian refuge was an open bus bay positioned tangentially to the travel lane after the island which the drivers took benefit of by accelerating right after passing the pedestrian crossing rather than slowing down below to the upstream speed.

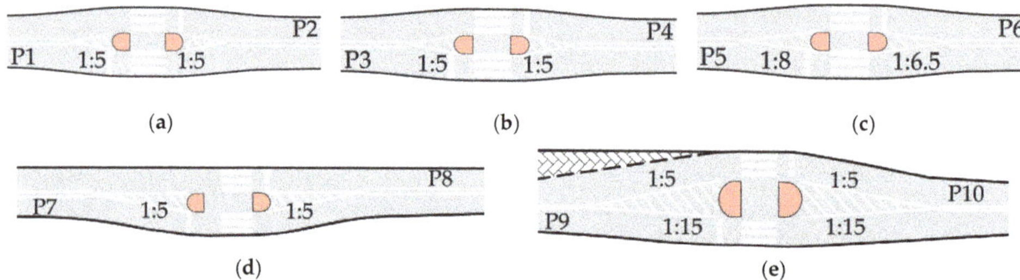

Figure 1. Diagrams of the analysed pedestrian refuges: (**a**) symmetric, 2 m wide (entry zone); (**b**) symmetric, 2 m wide (entry zone; after centre island); (**c**) symmetric, 2 m wide (between two centre islands, both positioned on one side of the centreline); (**d**) 2.5 m wide pedestrian refuge on one side of the centreline (end of entry zone/beginning of the village centre area); (**e**) asymmetric, 4 m wide (village centre, asymmetric shift of the lane alignments by 1 m and 3 m respectively). The test sections P1, P2, … P*n*, are marked on the respective travel lanes before the pedestrian refuges, in this way indicating the traffic direction under analysis.

All of the above-mentioned pedestrian refuges were located in a built-up area marked with D-42 entry signs. The B-33 speed limit sign was not placed in the immediate vicinity of the pedestrian refuge in any of the analysed cases. The daytime speed limit in residential areas in Poland is 50 km/h. The pedestrian refuges chosen for the research feature various geometric parameters, different visibility of the road ahead and traffic control schemes. One of them has a conventional shape, includes a 2 m wide island and approach end treatment with P-21 1:5 tapered markings (Figure 2).

(**a**)   (**b**)

Figure 2. Test sections P1 and P2 with the pedestrian refuge positioned between the end of entry zone and beginning of the village centre area: (**a**) P1—entry zone; (**b**) P2—exit zone.

Two other pedestrian refuges under analysis were located on the stretches surrounded by residential buildings in close proximity (Figure 3) or without any buildings in the surrounding environment (Figure 4). With most of analysed central refuges situated in the settlement areas (Figures 1–10), in two cases they were located in the entry zones, preceded by gateway islands (Figures 3b and 4a). These pedestrian refuges were located quite close to the gateway islands, namely ca. 170 m away. In one case, a very good view was ensured in both directions of travel (Figure 3). The pedestrian refuge included a 2 m wide island and 1:5 tapered markings. In the other case, the gateway island positioned on the approach lane was followed by a horizontal curve (Figure 4b) completely obscuring view on the course of the departure section. The pedestrian refuge presented in Figure 4 has a 2 m wide island and hatched markings with different taper rates of 1:8 on the entry side (Figure 4a) and

1:6.5 on the departure side (Figure 4b), viewing in the direction of the village centre, to accommodate the nearby collector road junction.

(a)   (b)

Figure 3. Test sections P3 and P4, pedestrian refuge placed 170 m after the gateway island symmetrical about the road centreline: (**a**) P3—exit zone; (**b**) P4—entry zone.

(a)   (b)

Figure 4. Test sections P5 and P6, pedestrian refuge placed 170 m after gateway island positioned on one side of the road centreline: (**a**) P5—entry zone; (**b**) P6—exit zone.

One of the analysed refuges was located on one side of the centreline and included a 2.5 m wide island (Figure 5). It was situated between the end of the entry zone and the beginning of the village centre area with nearby buildings spaced 80 m away from the roadway edge. Hatched, 1:5 tapered markings were applied. The buildings were preceded by a bridge lined with a high curb and parapets being visible side obstacles. The bridge was followed by a curve to the right, that reduced the view of the further course of the road (Figure 5a). Viewing in the departure direction, the bridge approach lane was close to the buildings in the village centre area and after the pedestrian refuge the road was surrounded by a forest without any buildings (Figure 5b).

(a) (b)

Figure 5. Test sections P7 and P8 between the end of entry zone and beginning of the village centre area: (**a**) P7—pedestrian refuge on one side of the centreline imposing large lateral shift by 2.5 m; (**b**) P8—departure lane without any imposed lateral shift.

The last case was an unusual, asymmetric pedestrian refuge situated in the village centre area between two bus bays (Figure 6). It included a 4 m wide asymmetric island. In the direction towards the bus bay the travel path is deflected by 1 m and there is a 1:15 tapered marking (Figure 6a). With the nearby positioned open bus bay, the island was designed to impose a 3 m lateral shift and a 1:5 tapered marking was used (Figure 6b). This arrangement resulted in a very sharp lateral deflection of traffic on the approach to the pedestrian refuge and a very convenient departure alignment with possible entering the open bus bay. Very good vision on the road ahead was ensured in both directions of travel.

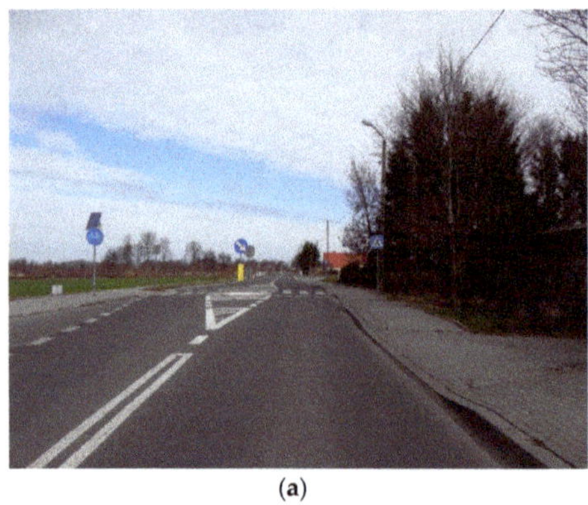

(a) (b)

Figure 6. Test sections P9 and P10 in the village centre area: (**a**) P9—small lateral shift (1 m); (**b**) P10—big lateral shift (3 m).

Table 1 compiles the surrounding environment and land characteristics and visibility conditions on the test sections before the pedestrian refuges and after the pedestrian refuges under analysis.

Table 1. Characteristics of the test sections.

No.	Conditions before Refuge Island		Conditions after Refuge Island		Visibility Conditions [1]
	Surroundings	Buildings	Surroundings	Buildings	
P1	rural area	lack of buildings	forest	lack of buildings	nearby buildings in view
P2	residential area	distant buildings	rural area	lack of buildings	good visibility
P3	forest	lack of buildings	residential area	distant buildings	good visibility
P4	residential area	lack of buildings	forest	lack of buildings	good visibility
P5	residential area	distant buildings	rural area	lack of buildings	good visibility
P6	rural area	lack of buildings	rural area	lack of buildings	240 m sight distance
P7	forest	lack of buildings	residential area	nearby buildings	170 m sight distance
P8	residential area	nearby buildings	forest	lack of buildings	good visibility
P9	residential area	nearby buildings	residential area	nearby buildings	buildings & bus bay in view
P10	residential area	nearby buildings	residential area	lack of buildings	good visibility

[1] Visibility of the road ahead are given in relation to the pedestrian refuge axis.

Wherever "before" appears in this article it designates a location (or locations) upstream of the island viewing in the direction of traffic.

Wherever "after" appears in this article it designates a location (or locations) downstream of the island viewing in the direction of traffic.

4. Study Method

For all the test sections the speed readings were taken between 10:00–15:00 hrs. during weekday, including ca. 70 veh. in free-flow and up to 100 veh. in stable-flow (more congested) conditions. The equipment used both before and after the pedestrian refuges were synchronised SR4 traffic detection devices equipped with automatic speed data logging function (SR4—brand name of the devices used in the survey - Speed Displays Traffic Detection). Additionally, hourly traffic volumes were measured in each case, including determination of the percentages of heavy goods vehicles. The speed data were grouped by direction to calculate the 85th percentile speed v_{85}, average free-flow speed v_{av} and stable-flow speed $v_{av}{}^{pp}$ and also the before/after speed difference Δv and, finally, the speed variation ratio $u_p = \Delta v^{before-after}/v^{before}$ in %. The calculation results are presented in Table 2.

Table 2. Speed distribution parameters, upstream/downstream (before/after) speed difference and speed variation ratio.

No.	Speed before Refuge v^{before}, km/h			Speed after Refuge v^{after}, km/h			Speed Difference Δv, km/h			Speed Variation Ratio u_p, % [1]		
	v_{85}	v_{av}	$v_{av}{}^{pp}$	v_{85}	v_{av}	$v_{av}{}^{pp}$	v_{85}	v_{av}	$v_{av}{}^{pp}$	v_{85}	v_{av}	$v_{av}{}^{pp}$
P1	76.5	65.4	63.8	65.5	58.4	56.8	11.0	7.0	7.0	14	11	11
P2	58.0	51.2	50.8	63.9	56.8	56.8	−5.9	−5.6	−6.1	−10	−11	−12
P3	64.0	55.8	53.9	66.7	56.8	54.7	−2.7	−1.0	−0.9	−4	−2	−2
P4	58.7	51.9	50.7	63.5	54.1	51.3	−4.8	−2.2	−0.6	−8	−4	−1
P5	64.3	55.5	55.5	71.1	60.8	60.8	−6.8	−5.4	−5.4	−10	−10	−10
P6	71.0	62.7	62.7	63.6	58.0	58.0	7.4	4.7	4.7	10	7	7
P7	75.9	67.6	67.4	53.4	48.1	47.2	22.5	19.5	20.3	30	29	30
P8	53.2	47.8	47.5	68.0	59.2	58.4	−14.8	−11.4	−10.9	−28	−24	−23
P9	56.9	52.4	51.7	54.2	47.6	47.1	2.7	4.8	4.6	5	9	9
P10	56.1	44.9	44.3	62.9	53.8	53.5	−6.8	−8.9	−9.2	−12	−20	−21

[1] Speed variation ratio is often used in traffic calming studies, calculated as follows: $u_p = \Delta v^{before-after}/v^{before}$ and given in %.

The measurement data were subjected to statistical inference. After conventional parametric tests, normality of distribution of the respective data sets was assessed with Kolmogorov-Smirnov test and the ranges of results were confirmed with homogeneity tests to remove outliers. The Kolmogorov-Smirnov (K-S) test, χ^2 independence test and, χ^2 median test were carried out for the whole data set comprising the upstream and downstream speeds to check if they belong to one or two different populations. The results of the statistical tests performed on the upstream and downstream speeds are presented in Table 3.

Table 3. Results of statistical tests.

Test Section	Kolmogorov-Smirnov Test λ [1] $H_0: F_1(v^{before}) = F_2(v^{after})$ $H_1: F_1(v^{before}) \neq F_2(v^{after})$		χ^2 Independence Test [2]		χ^2 Median Test [3]	
	Free Flow	**Stable Flow**	**Free Flow**	**Stable Flow**	**Free Flow**	**Stable Flow**
P1	2.20	2.28	**2.61**	4.26	7.03	13.46
P2	1.65	1.73	12.40	13.16	8.36	3.67
P3	**0.70**	**0.73**	0.83	**0.09**	8.45	**0.59**
P4	**0.55**	**0.33**	0.16	**0.45**	10.08	**0.67**
P5	**1.03**	**1.03**	0.55	5.28	12.13	12.13
P6	**1.35**	**1.35**	**2.30**	5.79	6.40	6.40
P7	4.56	5.22	60.42	85.01	77.20	103.91
P8	3.07	3.14	38.47	40.02	35.51	36.21
P9	4.,08	4.51	18.68	32.48	52.05	61.38
P10	**1.13**	1.61	**0.02**	16.60	4.33	**1.60**

[1] For the adopted significance level of $\alpha = 0.05$ the critical value is $\lambda_\alpha = 1.36$. [2] $H_0: P\{V^{before} = v^{before}_i, V^{after} = v^{after}_i\} = P\{V^{before} = v^{before}_i\} P\{V^{after} = v^{after}_i\}$, $H_1: P\{V^{before} = v^{before}_i, V^{after} = v^{after}_i\} \neq P\{V^{before} = v^{before}_i\} P\{V^{after} = v^{after}_i\}$, where: v^{before}_i and v^{after}_i are the respective speed values and, with the adopted significance level of $\alpha = 0.05$ and fourfold table the critical value is: $\chi_\alpha^2 = 3.84$. [3] $H_0: F_1(v^{before}) = F^2(v^{after})$, $H_1: F_1(v^{before}) \neq F^2(v^{after})$, with the adopted significance level of $\alpha = 0.05$ and fourfold table the critical value is: $\chi_\alpha^2 = 3.84$.

The values in boldface in Table 3 are the non-positive results of statistical tests, which do not support rejection of null hypothesis H_0 that the tested features i.e., upstream and downstream speeds belong to the same population.

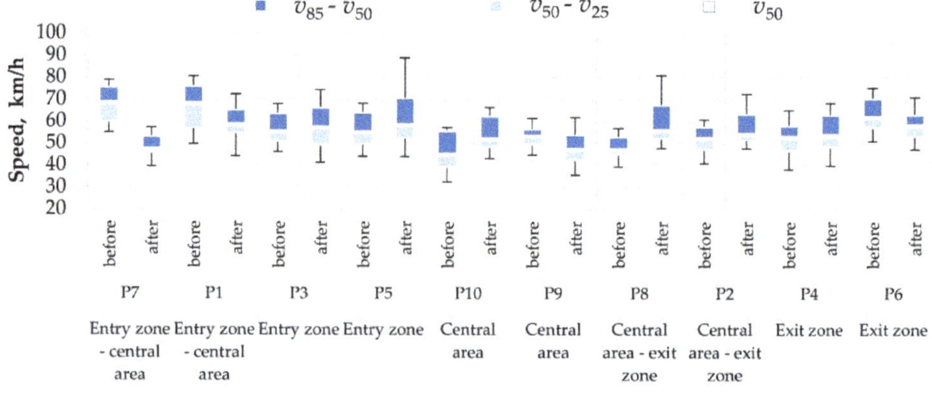

Figure 7. Distribution of speed parameters on the test sections in the range of percentile speeds: v_{85}–v_{50} and v_{50}–v_{25}.

According to the results of statistical tests compiled in Table 3, in some cases we deal with different populations. In the case of the effect of pedestrian refuges on the upstream and downstream speeds the results of statistical tests for free-flow and stable-flow traffic conditions were not always the same. Also, for the same type of traffic, the results of the respective statistical tests were heterogeneous. Such test results can be attributed to various factors influencing the motorists' behaviour in free-flow and stable-flow traffic. In order to enhance perceptions of the results, in Figure 7 they are presented in relation to different speed percentiles and distribution bars in the order of sitting along the road stretch in the village. The analysis of the data presented in Table 3 and in Figure 7 showed that there are a number of factors, in addition to the geometric parameters of the pedestrian refuge, that can have a bearing on the vehicle speeds, including positioning along the road stretch, type of the surrounding environment, distance to the nearest buildings and view of the road ahead. Therefore, a number of different factors will be taken into account in the analysis of the traffic-calming effect of pedestrian refuges using the methodology presented in Figure 8 to determine the probable main determinants of the calculated speeds and their reduction.

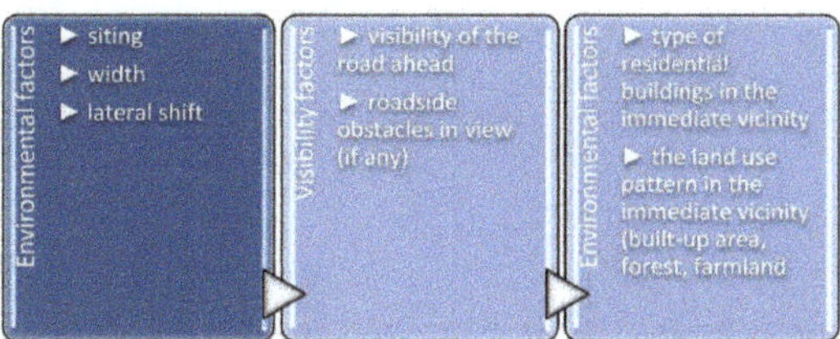

Figure 8. Method of analysis and general and detailed factors under analysis.

5. Results

5.1. Analysis of the Pedestrian Refuge Approach Speeds (Upstream Speeds)

Taking into account the recommendations of [1,5] that retrofitting of pedestrian refuge scan have a speed reducing effect and improve the safety of vulnerable road users on their way across the street, as the first step the authors analysed the speeds measured upstream of the pedestrian refuge v^{before}. The analysis of the values of v^{before} given in Table 2 showed that in no case was the 85th percentile of free-flow speed smaller than 50 km/h, i.e., the speed limit in residential areas. According to the mentioned guidelines [5] the pedestrian refuges should be located where warranted and their geometry should induce reduction of vehicular speed to 50 km/h and less. This requirement is not satisfied by the pedestrian refuges under analysis. However, the average free-flow and stable-flow speeds on the test sections No. P8 and P10 were smaller than 50 km/h.

Taking this into account, the authors analysed also the percentages of vehicles travelling with speeds of ≤50 km/h and > 50 km/h and the results of this analysis are presented in Figure 9. Figure 9 gives also information if the test section is located before or after the pedestrian refuge, i.e., in the entry zone, village centre or exit zone. On the left side of Figure 9, there are test sections situated before the pedestrian refuges positioned in the entry zones and in the village centres and on the right side, there are four test sections located at the border of or within the exit zones. Two test sections—P1 and P7 were located at the border between the entry zone and village centre and, similarly, two sections—P2 and P8 were located at the border of the village centre area and the exit zone and, hence, different designations have been used for them depending on the direction of traffic, before and after the pedestrian refuges in question. The analysis of percentage data showed that only on two test sections, namely P8 and P10 the percentages of speeds below 50 km/h limit before the island locations were ca. 67% or 75%. The approach speeds obtained in the other cases were much higher. With the purpose to establish which determinants could have a predominant effect on the speed measured before the pedestrian refuge or on the percentage of speeds ≤50 km/h Figure 9 presents the relevant characteristics of each pedestrian refuge including its sitting along the road stretch, the surrounding environment, residential buildings and view of the road ahead. For identification of determinants, the test sections in Figure 9 have been put in order of increasing percentage of speeds smaller or equal to 50 km/h in both parts of the chart.

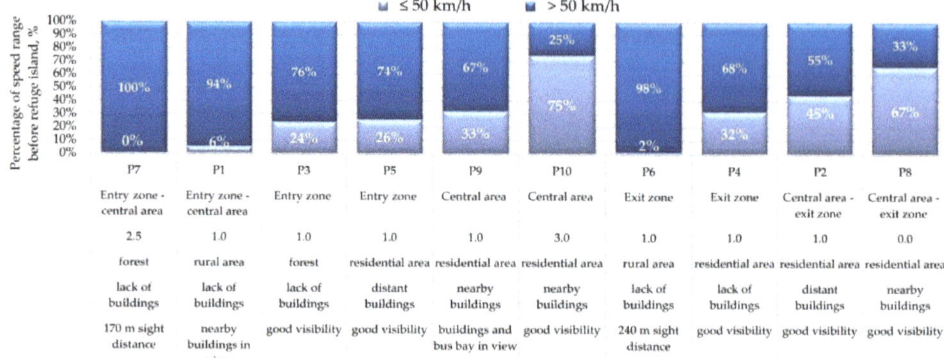

Figure 9. Percentages of the speed ranges before the pedestrian refuges.

Analysing the data presented in Figure 9 above and the features related to the pedestrian refuge siting, surrounding environment and visibility conditions we see that smaller speeds are not directly related to the proximity of residential buildings or view of the road ahead. Only in one case with residential buildings in close proximity where the greatest, 3 m lateral shift in the travelway alignment (P10) was imposed by the pedestrian refuge the approach speeds were the smallest among all the speeds measured before the pedestrian refuges (Table 2), accompanied by the greatest percentage of speeds smaller or equal to 50 km/h (Figure 9). Similar approach speeds and percentage of speeds smaller or equal to 50 km/h were recorded on the test section No. P8 with no lateral shift in the travelway alignment, where the approach section was on a horizontal curve running close to residential buildings. One of the smallest approach speeds was recorded on the test section P7 where 2.5 m lateral shift was imposed, accompanied by the smallest percentage of speeds smaller or equal to 50 km/h. This allows us to state that the width of the pedestrian refuge and the associated lateral shift in the travelway alignment are not the sole determinants of speeds at which the motorists approach the pedestrian refuges.

The field results presented in Table 2 and in Figure 9 were analysed to establish probable determinants that can, in combination, influence the speeds measured before the pedestrian refuges. These determinants, relating to the pedestrian refuges located in the entry zones and in village centre areas are presented in Figure 10a. Therefore, the determinants presented in Figure 10a should be taken into account when siting the pedestrian refuges in the entry zones and this, in combination with appropriate engineering and traffic planning, should result in obtaining approach speeds not exceeding the speed limit in residential areas. Figure 10b, in turn, presents the determinants in relation to the road sections before the pedestrian refuges located in the exit zones.

Summarizing the above analyses we can state that the factors that should be taken into account when siting a pedestrian refuge include both the features of the approach section and the arrangement of residential buildings in the immediate vicinity, as these are the primary factors determining the approach speed and the safety of pedestrians travelling across the street through the crossing in question. Thus, the authors believe that, in light of the above facts, in order to ensure that the approach speeds do not exceed the speed limit in residential areas (≤50 km/h) it is necessary to modify the pedestrian refuges design or thoroughly re-consider the traffic management arrangements. The analyses carried out in this research allow us to conclude that a typical pedestrian refuge imposing 1 m lateral shift on both sides, sited in the exit zone, surrounded by farm fields or open space without residential buildings in close proximity is not an effective measure and does ensure approach speeds smaller or equal to 50 km/h.

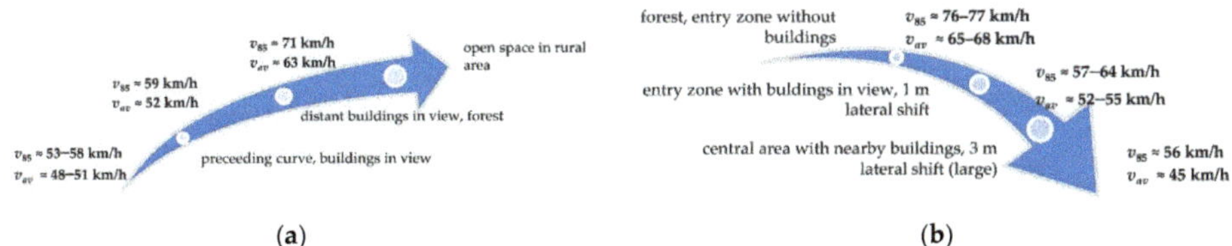

(a)   (b)

Figure 10. Determinants of the approach speed depending on the pedestrian refuge location: (**a**) in the entry zone or in the village centre; (**b**) in the exit zone.

5.2. Comparison of the Pedestrian Refuge Approach Speeds Obtained in This Research with the Results Obtained by Other Researchers

In order to compare the results obtained in this research with the results obtained by other researchers the authors have redrawn the graph relating the speed reduction to the speed value after the central island (Figure 11) published in [14] replacing the shape and width of the central island with the width of pedestrian refuge island. This allowed comparing the results of the research project on German roads published in [14] with the results obtained by the authors in villages located on regional roads in Poland in the vicinity of pedestrian refuges. The results of this comparison are presented in Figure 12. Lateral shift in the travelway alignment was the main factor considered in this case. The research results given in [13] concerned the effect of lateral shift generated by the gateway island on the speed of vehicles after the island, the objective being to achieve reduction of the speed of vehicular traffic in the entry zone by using different types of horizontal deflection measures imposing different lateral shifts. Figure 12 presents the schematic diagrams of the analysed pedestrian refuges accompanied with the typical pavement markings.

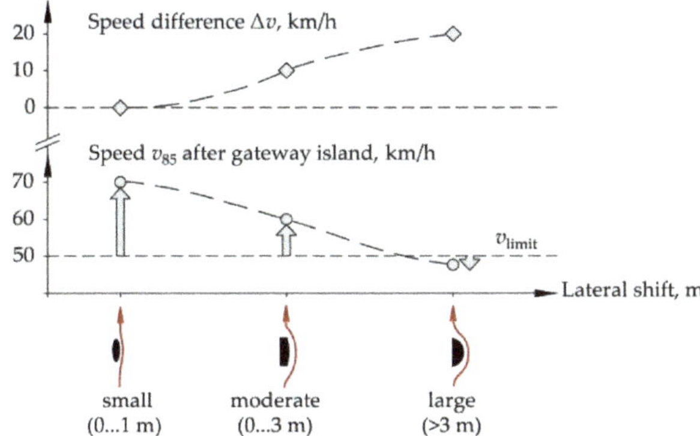

Figure 11. Relationship between the speed parameters Δv_{85} and v_{85} after the gateway island and the lateral shift (redrawn from Wirksamkeit geschwindigkeitsdämpfender Maßnahmen außerorts, 1997 [14], p. 9).

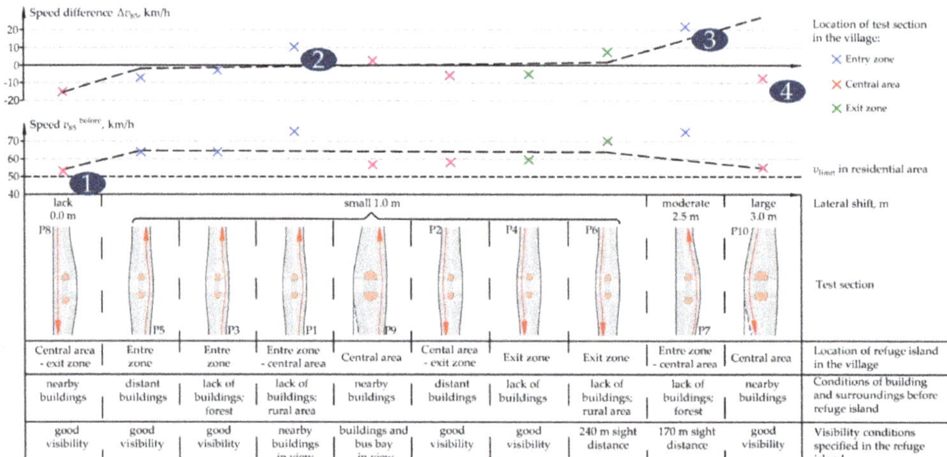

Central area - exit zone	Entre zone	Entre zone	Entre zone - central area	Central area	Cental area - exit zone	Exit zone	Exit zone	Entre zone - central area	Central area	Location of refuge island in the village
nearby buildings	distant buildings	lack of buildings; forest	lack of buildings; rural area	nearby buildings	distant buildings	lack of buildings	lack of buildings; rural area	lack of buildings; forest	nearby buildings	Conditions of building and surroundings before refuge island
good visibility	good visibility	good visibility	nearby buildings in view	buildings and bus bay in view	good visibility	good visibility	240 m sight distance	170 m sight distance	good visibility	Visibility conditions specified in the refuge island

Figure 12. Relationship between the speed parameters: Δv_{85} and v_{85} before the pedestrian refuge and the lateral shift.

Comparing the test results presented in Figure 12 we see that in the entry zones without footways, nearby residential buildings and other such features, pedestrian refuges do not ensure achieving the desired speed reduction, and this irrespective of the approach speed (Figure 3a). Driving with speeds higher than 50 km/h in built-up areas can be explained by good visibility conditions without any buildings or side obstacles in view. The desired speed reduction of $\Delta v_{85} = 10\text{–}20$ km/h is obtained if there are nearby residential buildings, footways, private entries in view (Figure 5a) within a close distance from the pedestrian refuge (up to 100 m).

Unfortunately, appearance of residential buildings in close proximity does not, on its own, have a determining effect on motorists driving at speeds much exceeding the limit of 50 km/h. The authors believe that there are a number of factors relevant to the degree of speed reduction, including, without limitation, the lateral shift, refuge shape (symmetric or adjacent to the road centreline), distance to the nearest residential buildings, visibility conditions before and after the refuge, rather than the lateral shift being one and only factor, as it could be concluded from the research results presented in Figure 11.

Referring to pedestrian refuges located in the entry zones we can conclusively state that they fail to impose the desired reduction of the upstream speed to ≤ 50 km/h irrespective of the speed of the vehicle in the exit zone (Table 2 and Figure 12). Reduction of vehicular speeds in the exit zone was noted only in the case when the sight distance after the pedestrian refuge was ca. 240 m (Figure 4b). In this case speed reduction of $\Delta v_{85} = $ ca. 7 km/h was obtained, yet attention is drawn to the pedestrian refuge approach speeds much exceeding the speed limit in residential areas (50 km/h), this attributed to approach section characteristics i.e., 750 m long straight section of the road surrounded by farm fields devoid of residential buildings in close proximity. These approach conditions are the main determinant of the high speeds before the approach with the value of $v_{85}{}^{before} = $ ca. 71 km/h which is much above the speed limit in residential areas in Poland. Sight distance was, in turn, the main determinant of the recorded speed reduction. In another case, the pedestrian refuge located in the exit zone was preceded by a long straight section (Table 1) of the road with distant, loosely spaced residential buildings (Figure 3b).

The smallest approach speeds were noted on the departure lane with no lateral shift (Figures 5b and 12). However, it must be noted that the approach section followed a winding route with residential buildings in close proximity continuing up to 80 m before the refuge location. The authors believe that the above-described approach conditions were, in this case, the main determinant of the recorded smaller approach speeds rather than the pedestrian refuge installed on the opposite lane. On the other hand, after the pedestrian refuge, the greatest of all increase of speed was noted (Table 2), this attributed to the surrounding environment—a forest devoid of any residential buildings (Figure 5b).

5.3. Summary of the Analyses of Pedestrian Refuge Approach Speeds

Taking this into account, and considering the pedestrian refuge design proposed in [5], the authors are of the opinion that pre-warning devices (thermoplastic strips) or rumble strips should be installed ca. 70 m from the pedestrian refuge axis to alert the motorists of the obstacle ahead. Moreover, the road authority should consider putting up before the pedestrian refuge the B-33 (speed limit) and B-25 (no overtaking) road signs, depending on the needs and the local conditions. The need for the "no overtaking" sign was noted by the authors during the research in the case of pedestrian refuges adjacent to the road centreline where the speeds of travel on the opposing traffic lane (with no lateral shift) much exceeded 80 km/h. These cases were not singular or even rare. The decision whether or not to put up the speed limit sign depends not only on the speeds in the village centre but also on the siting of the pedestrian refuge in relation to the existing access and public buildings layout.

It must be noted that in the cases under analysis the speeds of travel were much above the speed limit in residential areas, except for two cases where free-flow and stable-flow speeds below 50 km/h were obtained (Figure 12). In the first case there was a winding approach section with buildings located in close proximity with no lateral shift (Figure 5b) and in the other case the travelway alignment was shifted by 3 m and the asymmetric design of the island was visible to the motorist approaching it through a straight section (Figure 6b).

5.4. Analysis of the Pedestrian Refuge Departure Speeds

Reduction of vehicular speeds at pedestrian refuges is indirectly related to the issue of obtaining speed reduction also after the island. While the design of pedestrian refuges, as the primary objective, should enforce reduction of vehicular speeds right before and alongside the pedestrian refuge [5], it becomes interesting from the research viewpoint to check their effect on the speed on the analysed stretch of road in the village area. As such, the authors carried out similar analyses for speeds measured after the pedestrian refuges, the results of which are given in Table 2 and in Figure 13, including short descriptions of the relevant factors. Additionally, Figure 13 shows the percentages of motorists not obeying the speed limit in residential areas, as measured after the pedestrian refuges.

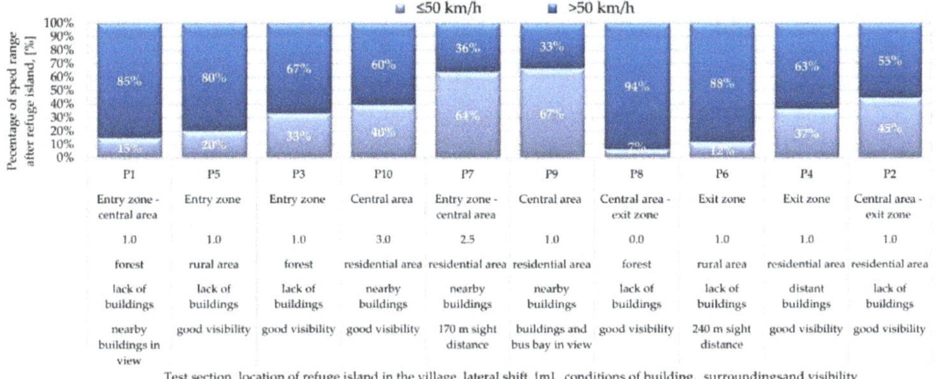

Figure 13. Percentages of a speed ranges after the pedestrian refuges.

The analysis of the data given in Table 2 and in Figure 13 shows that the pedestrian refuges failed to enforce reduction of v_{85} speed to below 50 km/h, i.e., the speed limit in residential areas and, furthermore, the speeds downstream of the pedestrian refuges do not depend on the lateral shift in the travelway alignment. Only in two cases (P7 and P9) the average free-flow and stable-flow speeds were much below the speed limit in residential areas. The determining factor, in this case, was the distance to residential buildings, the type of surrounding environment alongside the road and sight distance after the pedestrian refuge (Figure 14a). However, in the authors' opinion, the most important factor, in this case, is the view of the end treatment of the pedestrian refuge and the lateral shift in relation to the departure lane alignment.

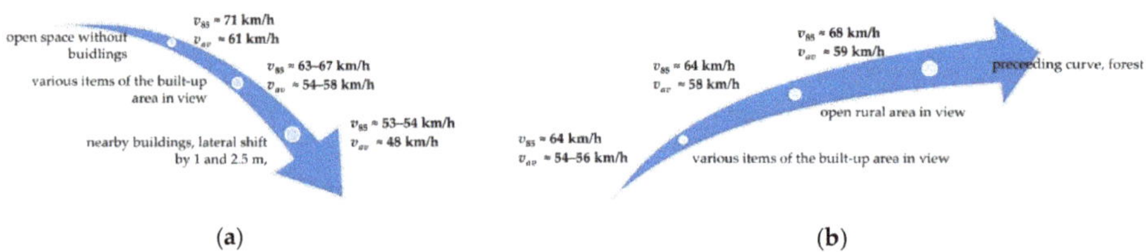

Figure 14. Determinants of the departure speed depending on the pedestrian refuge location: (**a**) in the entry zone or in the village centre; (**b**) in the exit zone.

The speeds recorded on the section after the pedestrian refuge located in the entry zone were much above the speed limit in residential areas. These very high speeds were, in this case, most probably caused by a combination of factors: very long sight distance and nature of the surrounding environment with farm fields, forest and lack of residential buildings in close proximity (Figures 3a and 4a).

In the village centre also the view of the road after the pedestrian refuge, as seen by the driver passing the pedestrian crossing is an important determinant of the speed of travel. If, for example, the pedestrian refuge opens to a wide area of the travel lane and a tangentially positioned open bus bay (Figure 6b) then the drivers, despite a large lateral shift of 3m and the lowest speed recorded before the pedestrian refuges, tend to accelerate seeing end of village at a short distance and lack of any obstacles that would make them slow down.

Conversely, if after passing the pedestrian refuge the motorists sees bridge parapets, residential buildings and also raised kerbing, clearly communicating the end of taper and horizontal deflection (Figure 6a) he/she will maintain the speed at more or less the same level or even slow down in relation to the speed upstream of the pedestrian refuge.

By far the greatest determinant of the speed reduction after the pedestrian refuge is the combination of a large lateral shift, with the end of deflection communicated by a high kerb, and nearby residential buildings or bridge parapets in view (Figure 5a).

Similar to the previous case, the authors established the determinants of the vehicle speeds after the refuges located in the village exit zone (Figure 14b). View of the road ahead, as seen by the driver passing alongside the pedestrian crossing is also highly relevant. If the motorists see bus bays, bike paths, distant buildings and does not see the D-42 built-up area sign he/she will generally drive with more or less the same speed as on the approach to the pedestrian refuge. Conversely, if during passing alongside the pedestrian refuge the driver sees the road ahead surrounded by farm fields (Figure 2b) or by a forest (Figure 5b) with no side obstacles or nearby buildings in view, he/she would accelerate considerably, reaching speeds much greater than on the approach to the pedestrian refuge.

5.5. Comparison of the Pedestrian Refuge Departure Speeds Obtained in this Research with the Results Obtained by Other Researchers

Similar to the analysis presented in 5.2 above (Figure 11), for comparison purposes, the departure speeds after the pedestrian refuges were compared with the results of research given in [14]. The results of this comparison are presented in Figure 15. Analysing the data in Figure 15 we see that in the case of pedestrian refuges located in the village centre with nearby residential buildings (P7, P9) speed reduction can depend on the lateral shift, yet only in combination with other determinants closely associated with the view of the buildings in proximity and the road section ahead. When buildings are located away from the road (sections P1, P4, P5) then the pedestrian refuges cause an increase of speed after the refuge.

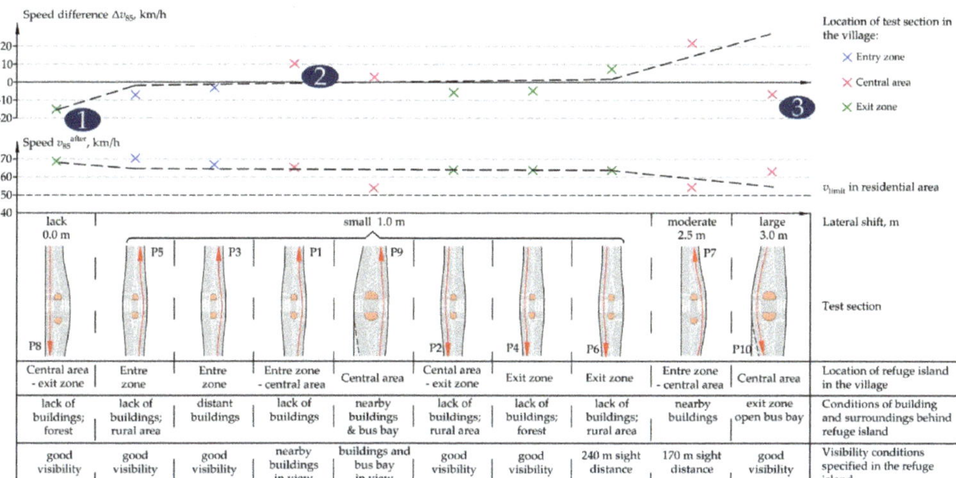

										Location of refuge island in the village
Central area - exit zone	Entre zone	Entre zone	Entre zone - central area	Central area	Cental area - exit zone	Exit zone	Exit zone	Entre zone - central area	Central area	Location of refuge island in the village
lack of buildings; forest	lack of buildings; rural area	distant buildings	lack of buildings	nearby buildings & bus bay	lack of buildings; rural area	lack of buildings; forest	lack of buildings; rural area	nearby buildings	exit zone open bus bay	Conditions of building and surroundings behind refuge island
good visibility	good visibility	good visibility	nearby buildings in view	buildings and bus bay in view	good visibility	good visibility	240 m sight distance	170 m sight distance	good visibility	Visibility conditions specified in the refuge island

Figure 15. Relationship between the speed parameters: Δv_{85} and v_{85} after the pedestrian refuge and the lateral shift.

The only exception to that is the case with an asymmetric pedestrian refuge imposing a 3 m (i.e., large) lateral shift (P10). With the open bus bay positioned tangentially to the departure lane section, this design, allowing the drivers to overrun the bus bay area provokes them to accelerate (Figure 15). This peculiar layout is not in compliance with good engineering practice since locating these two road elements beside each other is in conflict with the principles of traffic calming and protection of vulnerable road users. What can be stated conclusively is that before the pedestrian refuge and alongside the crossing the 85th percentile speed on the approach section and the average speeds on either side were the smallest of all the measured values, which is attributed to large lateral shift. This is to say that although lateral shift by 3 m induced speed reduction, this effect was limited to the section of the road before the pedestrian refuge (Figure 11).

Similar observations can be made when comparing the cases with the pedestrian refuges located in the exit zones (Figure 15). When the pedestrian refuge is followed by a section where the buildings, if any, are spaced away from the road (P3, P6), surrounded by farm fields (P2) or a forest (P8), the increase of speed was noted after the pedestrian refuge. Conversely, with much-reduced sight distance, the pedestrian refuge can contribute to speed reduction, the sight distance being then the determining factor.

5.6. Summary of Analyses of Pedestrian Refuge Departure Speeds

In the analysed cases, the departure speeds and the speed difference across the pedestrian refuge can be influenced by a combination of factors including the lateral shift, sitting along the road stretch, distance to the nearest buildings and visibility conditions.

The authors suggest placing safety barriers, barrier bollards or street furniture items in the vicinity of pedestrian refuges, as per the recommendation of [5]. In accordance with [5] these items should be positioned after the island up to the end of P-21 hatched marking. In the opposite direction these items should be installed on a min. 70 m long section, that is up to the pre-warning device. These measures would probably increase the reduction of the speed value before the pedestrian refuge with the effect maintained on the section past the island.

5.7. Analysis of Speed Variation at the Pedestrian Refuge

The second parameter defining the speed reducing performance of the pedestrian refuges for the approach and departure speeds is the speed reduction alongside the pedestrian refuge. The obtained speed differences are presented in Figure 16, together with a description of developments in the vicinity of the pedestrian refuge and view of the road ahead. From the data presented in Figure 16 it transpires that the main determinants of obtaining the desired speed reduction are the surrounding environment of the pedestrian refuge and view of the road ahead, i.e., accurately communicated information in the driver's field of view. The issue of the driver's perception of the information present in the visual field is studied in [11,12] and the issue of perception of the information given on the new generation traffic signs is covered in [10].

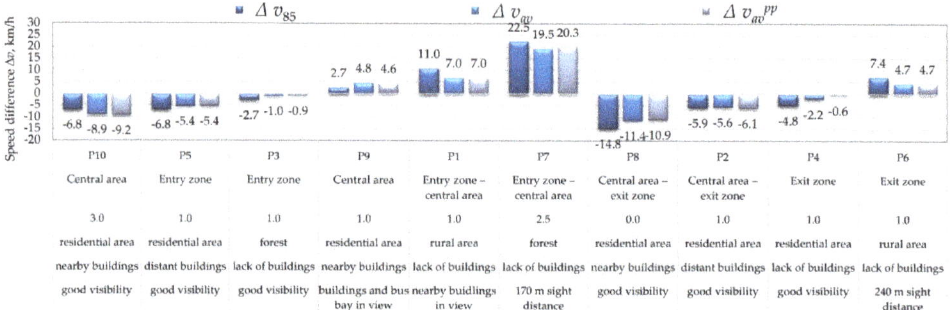

Figure 16. Speed variation on the road section alongside the pedestrian refuge (the sight distance is given in relation to the pedestrian refuge axis).

Lateral shift by 3 m (P10) which was the greatest of all the cases, in combination with good visibility and accessible open bus bay area instead of making the drivers slow down encouraged them to accelerate, both in free-flow and stable-flow conditions. The situation is similar on the two remaining test sections imposing lateral shift by 1 m on both sides with different surrounding features: distant residential buildings—P4 and farm fields in the village centre—P5 where the motorist had a good view of the road ahead.

Speed reducing effect was confirmed on the three other test sections—P1, P7 and P9, with residential buildings in close proximity and different sight distances. The greatest speed reduction was recorded in the case of asymmetric lateral shift in the travelway alignment by 2.5 m, yet this should be attributed to the bridge in view and houses in close proximity of the road rather than to the lateral shift and the pedestrian refuge on one side of the road centreline. The probable determinants of speed variation alongside the pedestrian refuges located in the entry zones or village centres and in the exit zones are presented in Figures 17a and 17b respectively. Similar cases, with the speeds after the pedestrian refuges greater than before were noted on the sections in exit zones (test sections P2, P3 and P8) with good visibility of the road ahead and where the exit section was surrounded by forest rural area or built-up area, yet with buildings located far away from the road. Significant reduction of vehicular speeds was noted only when the motorists approached the fragment of the road giving shorter sight distance (P6).

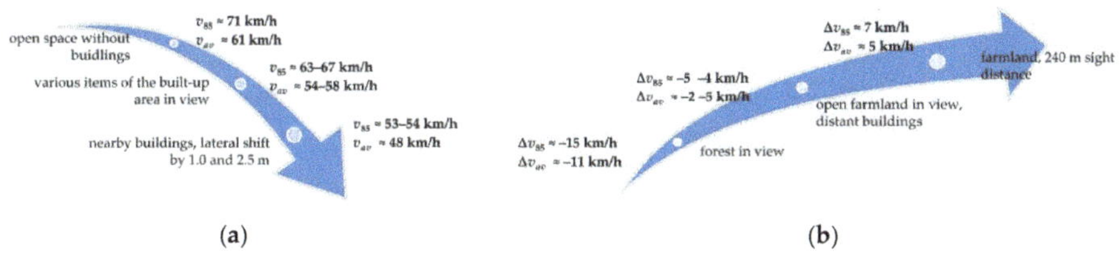

Figure 17. Determinants of the variation of speed at pedestrian refuges located: (**a**) in the entry zone or in the village centre; (**b**) in the exit zone.

Similar relationships and determinants can be found when calculating the speed variation ratio which is a parameter directly related to the above factors. Moreover, the same determinants are confirmed, as presented in Figure 17.

6. Validation of the Results of Analyses with the Speed Data Obtained on a Test Section Including Post-and-Chain Barriers

6.1. Analysis of the Speed Reduction Obtained by Retrofitting Post-and-Chain Barriers along the Regional Road Stretch

In none of the cases described in point No. 5 above did the test section include side obstacles such as post-and-chain barriers, barrier bollards or street furniture items. Taking this into account, in this section, the authors refer the results of the tests carried out in Wrzosowo, originally published in [15]. The measures provided in the entry zone of the analysed village included semi-circular islands on one side of the centreline, imposing lateral shift by 5 m and B-33 speed limit signs imposing 40 km/h speed limit on the road section with residential buildings located in the close proximity. Notwithstanding these measures, the free-flow speeds were higher than the speed limit in residential areas and much higher than the speed limit posted on the B-33 sign. Therefore, post-and-chain barriers were retrofitted alongside the road to protect the pedestrians on the section where the buildings are located close to the road (Figure 18). With the post-and-chain barriers in place, the traffic conditions were re-tested by the authors and the results (Figure 19) confirmed that this has considerably improved the situation on the road, reducing the average free-flow speed to below the speed limit posted on the B-33 sign.

(a) (b)

Figure 18. Stretch of the regional road lined with post-and-chain barriers: (**a**) in the village centre; (**b**) at the pedestrian crossing.

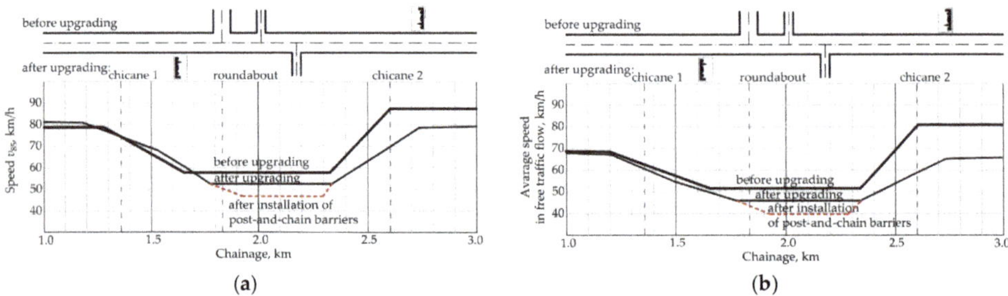

Figure 19. Free-flow speed profiles before and after upgrading of the regional road section in the village area (drawings by D. Kacprzak and A. Sołowczuk [15]): (**a**) v_{85} speed profile; (**b**) v_{av} speed profile.

6.2. Comparison of Speeds Measured before and after Installation of Post-and-Chin Barriers at the Pedestrian Refuge

With the purpose to examine the effect of the post-and-chain barriers installed in the built-up area alongside the section actually occupied by buildings, yet without the speed limit sign B-33, comparative surveys were carried out in the nearby village Wrzosowo Osiedle at the pedestrian refuge located in the village centre area. There the pedestrian refuge generates symmetrical lateral shift by 1.0 m, the same for both directions of traffic, includes 1:15 tapered markings and both the travel lanes are 3.5 wide.

The speeds measured for both directions of traffic are presented in Figures 20 and 21. The analysis of the data given in Figures 20 and 21 showed that after upgrading the speeds exceeded the speed limit in residential areas (50 km/h) in both directions of traffic. On the other hand, installation of post-and-chain barrier-imposed reduction of the average free-flow and stable-flow speeds to below 50 km/h (Figures 21 and 22). Also, the free-flow 85 percentile speed decreased, and decreased much, and although it exceeded the limit of 50 km/h both before and after the pedestrian refuge, the calculated values are by ca. 10 km/h lower than their counterparts before installation of the post-and-chain barriers. This confirms a major effect of the installed post-and-chain barriers in reducing the speeds of traffic. Figure 21 compares the percentages of the analysed speeds before and after installation of post-and-chain barriers on both sides of the road. The percentage of speeds smaller or equal to 50 km/h is much greater after installation of the post-and-chain barriers, as compared to this percentage after upgrading, yet without the barriers in place (see Figure 21).

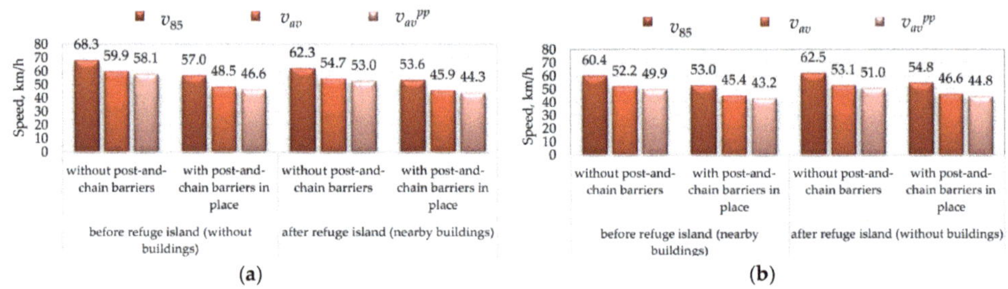

Figure 20. The speed distribution parameters before and after the pedestrian refuges on the test sections with sight distances of: (**a**) 200 m; (**b**) 140 m.

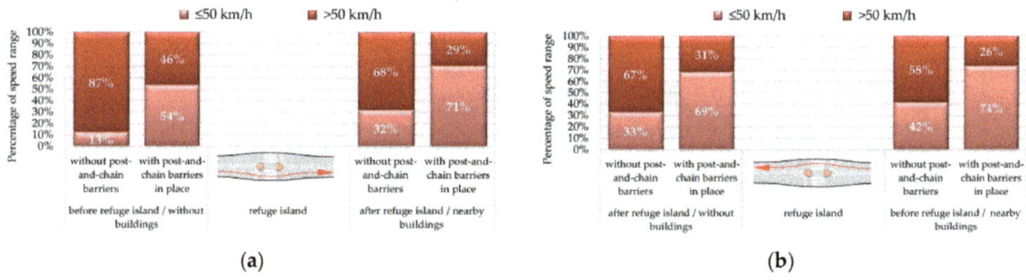

Figure 21. Calculated percentages of ≤50 km/h and >50 km/h speeds on test sections under analysis with sight distances of: (**a**) 200 m; (**b**) 140 m.

The analysis of the data resented in Figure 21 showed that retrofitting of the post-and-chain barriers influenced the drivers' behaviour making them slow down to below the speed limit in residential areas (50 km/h) on the road section concerned. The surrounding environment of the approach section to the pedestrian refuge was found to be also relevant in this case. In the case of approach section surrounded by a forest and post-and-chain barrier limited to a short section preceding the pedestrian refuge, the percentage of speeds ≤50 km/h was smaller before (54%) than after the pedestrian refuge (71%) where the drivers could see the road section lined with post-and-chain barriers over the length of 200 m (Figures 20a and 21a).

A different speed distribution was obtained with the approach section located in the built-up area, yet without buildings in close proximity (Figures 20b and 21b). In the case under analysis, most drivers (up to 74%) approached the pedestrian refuge at speeds smaller or equal to 50 km/h and accelerated after passing the island having in view a road section surrounded by a forest without any buildings (Figure 21b).

6.3. Comparison of Speed Reductions before and after Retrofitting the Post-and-Chain Barriers

Before installation of the post-and-chain barriers a big value of speed reduction was recorded in one case (Figure 22) for one direction of traffic only. The approach section in this case was surrounded by forest groves (Figure 23a) and residential buildings started after the pedestrian refuge, and they were situated not by the footway but spaced away from the road. The drivers were given good visibility with the curve to the right, after which the view was reduced, located ca. 200 m after the pedestrian refuge (Figure 23a). It is probably the combination of the above-mentioned conditions that resulted in such a big reduction in speed alongside the pedestrian refuge. Although installation of post-and-chain barriers brought speed reductions only half as big as previously, it must be underscored that both the upstream and downstream speeds were much smaller than before their installation (Figures 20 and 21).

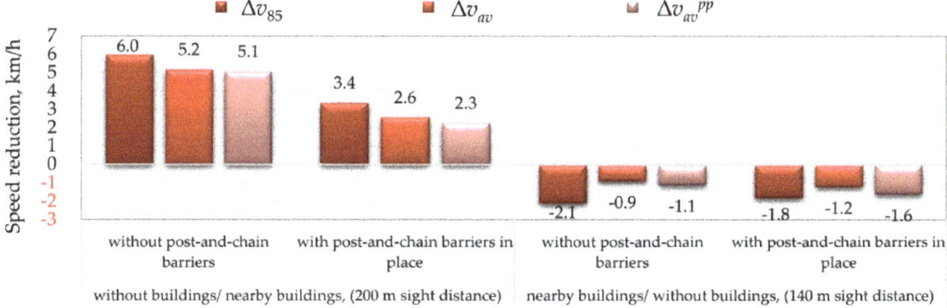

Figure 22. Speed differences on the analysed test sections.

(a)   (b)

Figure 23. Sight distance after the pedestrian refuge: (**a**) 200 m; (**b**) 140 m.

Conversely, on the test sections with the opposite direction of traffic, with shorter sight distances and nearby buildings situated alongside the approach section, where the driver could see the further course of the road surrounded by forest groves no sooner than after passing the pedestrian refuge (Figure 22) the drivers tended to accelerate rather than slow down (Figure 22). The analysis of results shows that speed reduction is obtained by a combined effect of the surrounding environment of the pedestrian refuge approach section and visibility of the surrounding environment of the road section downstream of the pedestrian refuge.

Note, however, that speed difference is purely an analytical parameter, the objective being reduction of free-flow and stable-flow speeds to below the speed limit in residential areas, which was achieved in this case owing to the retrofitted pedestrian refuge and post-and-chain barriers as the primary factors. The authors believe it is a correct approach to consider the speed values as the most important parameter rather than the calculated speed reductions.

7. Discussion and Proposed Engineering Measures

The above results confirm possible speed-reducing effect of pedestrian refuges, although there are a number of relevant factors to be considered before deciding about the refuge layout (symmetric or asymmetric) and width, namely: siting of the pedestrian refuge in the village area, nature of the surrounding environment in the planned location, sight distance, room for widening the carriageway, any bus bays in the vicinity. The obtained speed differences show that the lateral shift is not the sole factor determining the level of speed reduction.

The flow chart for selecting the type and width of the pedestrian refuges on regional roads in villages, based on chosen factors, is presented in Figure 24. The authors recommend, following the German guidelines [13], to consider in the entry zones the environmental factors and use symmetrical central islands combined with pedestrian crossings. In open rural areas, ca. 35 long by 6.25 m wide, symmetric lens-shaped central islands should be used, imposing a large lateral shift in the travelway alignment, passed on a curve with 50 m radius, with min. 3.5 travel lane width, enhanced with barrier bollards and single tree plantings at the pedestrian crossing location (Figure 25a). In forest areas cigar-shaped central islands are recommended, ca. 45 long by 5 m wide, passed on min. 3.5 m wide travel lanes, enhanced with barrier bollards only (Figure 25b). Additional plantings are not required where the surrounding environment is occupied by a forest. According to the Swedish [8] and German [13] guidelines and taking into account the drivers' perception studies reported in [10–12] the authors indicate a need of interactive road signs to be placed in the entry zones to warn the drivers of the central island and pedestrian crossing located ahead of them (activated by motion sensors).

Based on the results of this research, the authors recommend symmetric pedestrian refuge islands, 2 m or 2.5 m wide, the latter to accommodate bicycles, if required, as an appropriate design for the village centre areas. In order to highlight the pedestrian refuge location and influence the driver's perception, it should be accompanied with post-and-chain barriers [5,7] or barrier bollards of various colours and sizes, the latter being a popular measure used in the West European countries. Examples of designs used in Poland are presented in Figures 18 and 23. Figure 26, in turn, shows the proposed design including post-and-chain barriers. According to [5] the post-and-chain barriers should extend ca. 10–15 m upstream of the pedestrian refuge, measuring from the end of the hatched marking and at least to the end of the hatched marking at the other end.

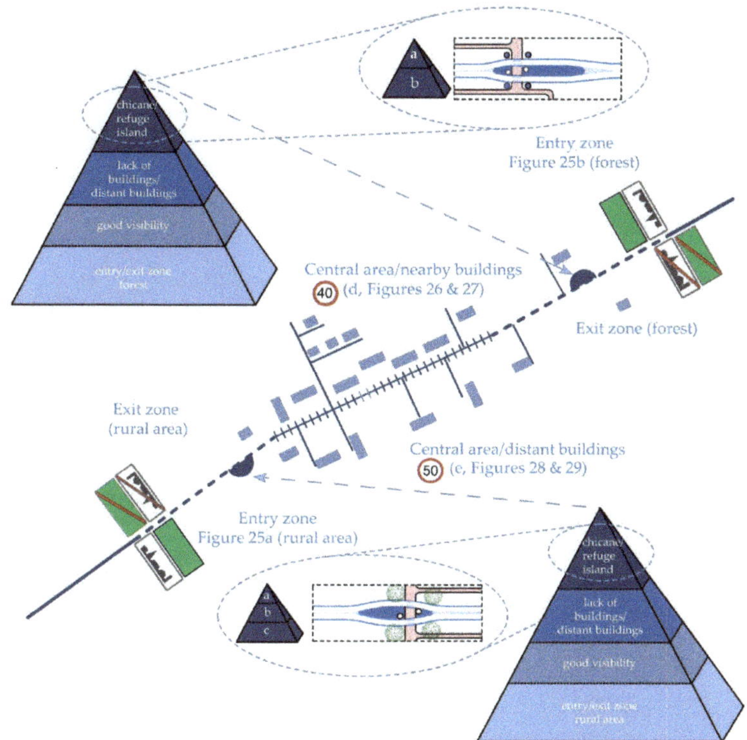

Figure 24. Flow chart for designing pedestrian refuges on sections of regional roads in villages. [a] interactive road signs; [b] barrier bollards; [c] additional plantings; [d] pedestrian refuges with post-and-chain barriers or barrier bollards (Figures 26 and 27); [e] pedestrian refuges combined with kerb buildouts (bulb-outs) or pinchpoints (Figures 28 and 29).

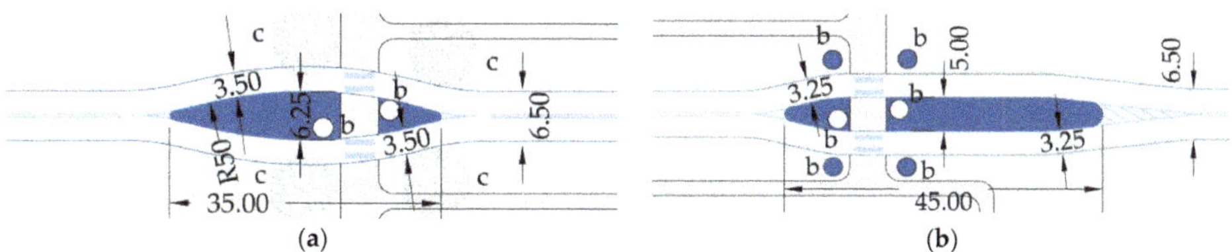

Figure 25. Procedure for designing pedestrian asylums in entry zone: (**a**) rural area; (**b**) forest. [b] barrier bollards U-12c; [c] additional plantings.

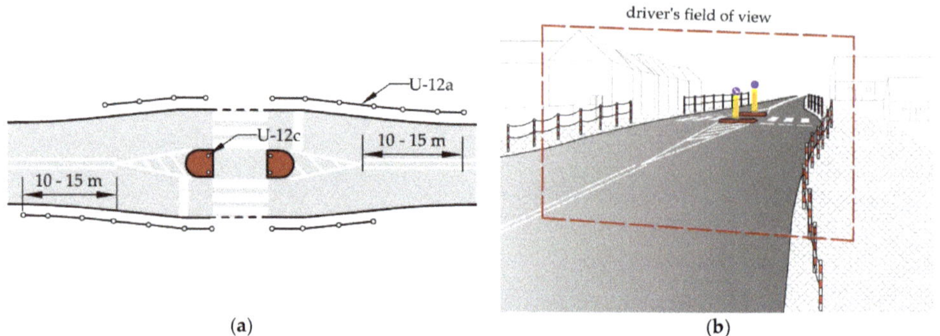

(a)

(b)

Figure 26. Schematic design of pedestrian refuge enhanced by post-and-chain barriers U12-a: (**a**) plan view; (**b**) visualisation of the proposed design (drawings by Agata Misztal).

Figures 26 and 27 present recommendations for village centre area with residential buildings in close proximity, where the pedestrian refuges are enhanced by post-and-chain barriers or barrier bollards to achieve yet smaller speeds and improve the safety of pedestrians right on the crossing.

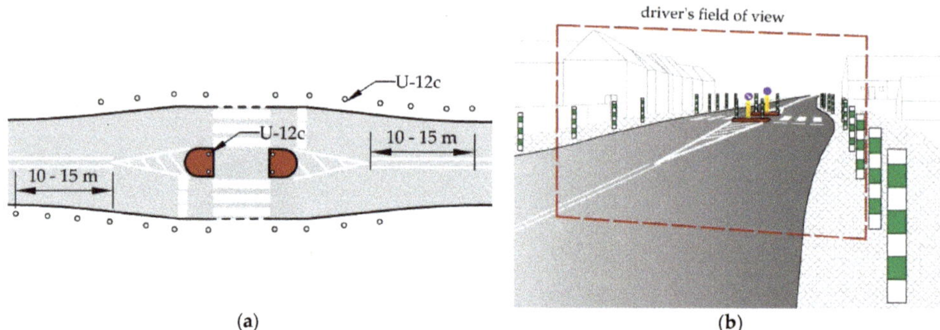

(a)

(b)

Figure 27. Schematic design of pedestrian refuge enhanced by barrier bollards U12-c: (**a**) plan view; (**b**) visualisation of the proposed design (drawings by Agata Misztal).

Based on the results of this research, the authors recommend symmetric pedestrian refuge islands, 2 m or 2.5 m wide, the latter to accommodate bicycles, if required, as an appropriate design for the village centre areas. In order to highlight the pedestrian refuge location and influence the driver's perception, it should be accompanied with post-and-chain barriers [5,7] or barrier bollards of various colours and sizes, the latter being a popular measure used in the West European countries. Examples of designs used in Poland are presented in Figures 18 and 23. Figure 26, in turn, shows the proposed design including post-and-chain barriers. According to [5] the post-and-chain barriers should extend ca. 10-15 m upstream of the pedestrian refuge, measuring from the end of the hatched marking and at least to the end of the hatched marking at the other end.

The travel lane width should be appropriate to the level of service of the road in question. Over the pedestrian crossing length, the travel lane design should not depart from the design on the straight section, the width to comply with the guidelines of [5,6]. According to the recommendations of [5] the width of the travel lane right before and after the pedestrian refuge should be in compliance with the adopted vehicle turning path templates. The angles of hatched marking border lines should follow the recommendations of [6] and the widths of lanes alongside the tapers should provide enough room for the vehicles to pass.

As mentioned earlier, in the West European countries' barrier bollards are used increasingly often at the pedestrian refuges and at the accompanying kerb buildouts. Barrier bollards can be installed both before and after the pedestrian refuge, as shown in Figure 27. They can be installed quicker and at a lower cost than the post-and-chain barriers. Another advantage of barrier bollards is that, in accordance with the guidelines of [7] they can have a decorative form, matching the surroundings and can be finished in colour chosen by the landscape architect. In the West European countries, barrier bollards are fitted with fluorescent strips or LED lighting at the top. The purpose of these treatments is to enhance perception of bollards in limited lighting conditions, during rainfall, at dusk or during the night.

A review of the designs used in West European countries, recommended in the design manuals [16,17] showed that kerb buildouts (bulb-outs) placed symmetrically on either side of the road or pinchpoints are often provided at the pedestrian refuges or gateway islands, as shown in Figures 28 and 29.

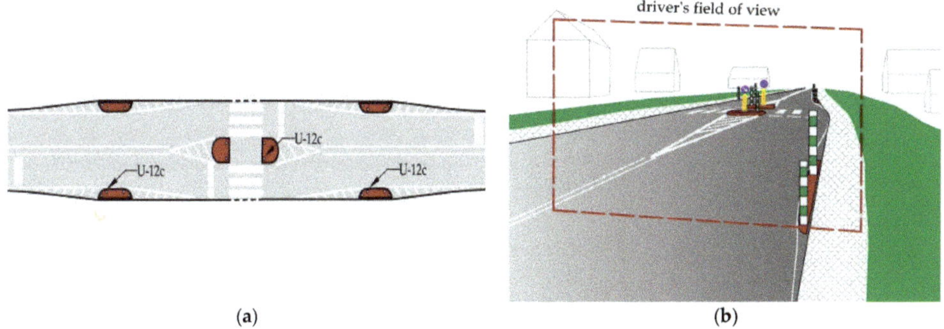

(a)   (b)

Figure 28. Schematic design of pedestrian refuge enhanced by bulb-outs: (**a**) plan view; (**b**) visualisation of the proposed design (drawings by Agata Misztal).

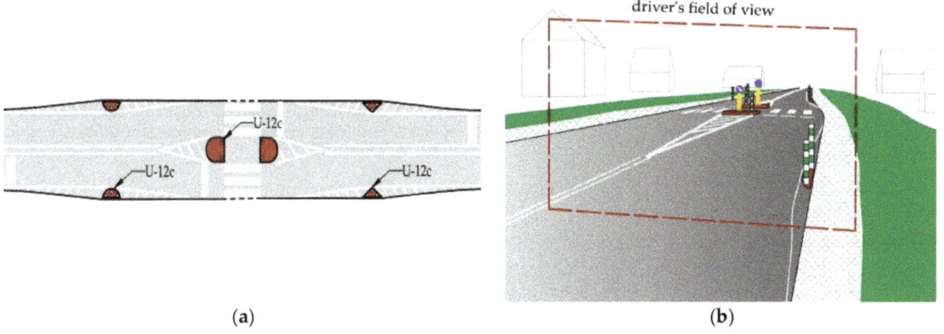

(a)   (b)

Figure 29. Schematic design of pedestrian refuge enhanced by a pinchpoint: (**a**) plan view; (**b**) visualisation of the proposed design (drawings by Agata Misztal).

The side islands can incorporate, as in the West European countries, greenery and four barrier bollards in corners. However, the most important design feature is raised kerbing around the island, provided in order to enhance lateral shift of vehicles and in this way contribute to reducing the speeds to the desired level. Moreover, the side islands should include U-6b object marker signs and signs communicating the horizontal deflection.

The U.S. [16] and U.K. [17] guidelines recommend also using of pinchpoints which are smaller kind of side islands (Figure 29). Due to a smaller size, these islands require a shorter section of widened carriageway and, for this reason, are an option of choice for village centre areas. In this case, they should also be provided with raised kerbing around the perimeter. In the West European countries, planters are placed on them to fit the village centre streetscape. Usually they are enhanced by three barrier bollards placed at the island corners. Although in most cases the islands are triangular in shape, semi-circular islands can also be found. In any case, the islands must incorporate U-6b object markers and bollards. Sometimes, depending on the traffic control scheme, C-10 keep left signs are placed atop the U-6b object markers.

8. Conclusions

The results of this research allow us to conclude that pedestrian refuges imposing symmetric lateral shift by 1 m which are not accompanied by street furniture items have no significant bearing on speed reduction in their vicinity, and this irrespective of their sitting along the stretch of road in the village and geometry of associated pavement markings. Conversely, asymmetric lateral shift in the travelway alignment generated by the refuge island located on one side of the road centreline induces a considerable speed reduction, yet only when the driver sees residential buildings in close proximity of the road.

If a pedestrian refuge located in the village centre area with residential buildings in close proximity of the road is expected to bring down the v_{85} speed to 40 km/h, it must be symmetrical, induce 1 m lateral shift and be enhanced by post-and-chain barriers or barrier bollards. When the desired maximum speed in the village area is 50 km/h and the nearest residential buildings are situated away from the road, the islands should be symmetrical, induce 1m lateral shift and be accompanied by bulb-outs or pinchpoints.

In the village entry zones in open rural areas with scattered residential buildings situated away from the road, pedestrian refuges should be combined with lens-shaped symmetric gateway islands, 6.25 m wide at the widest point. They should be enhanced by planting trees and placing crash-resistant bollards along the footway to influence the drivers' perception and make them alert for pedestrians. On sections surrounded by a forest, cigar-shaped symmetric central islands should be used to save trees by reducing the land take in comparison to lens-shaped islands. These islands must be at least 5 m wide in any case. On sections surround by a forest the pedestrian crossings should be enhanced with barrier bollards. In both cases interactive road signs should be used, coupled with motion sensors, to enhance perception of pedestrians in adverse weather.

Author Contributions: conceptualization, A.S.; methodology, A.S.; formal analysis, A.S. and D.K.; investigation, A.S. and D.K.; resources, A.S. and D.K.; data curation, A.S.; writing—original draft preparation, A.S.; writing—review and editing, A.S. and D.K.; visualization, A.S. and D.K.; supervision, A.S.; project administration, A.S.; funding acquisition, D.K.

Funding: The researches included in the paper were partly financed by grants of WBiA ZUT: DKD Decision No. 517-02-033-6728/17.

Conflicts of Interest: The authors declare no conflict of interest.

Appendix A

No.	Sign	Symbol	Description
1		P-7b	Pavement marking between the travel lane and shoulder (continuous edgeline)

No.	Sign	Symbol	Description
2	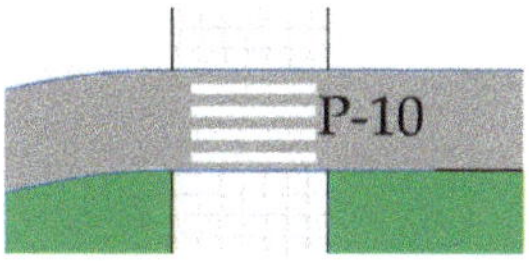	P-10	Pedestrian crossing
3	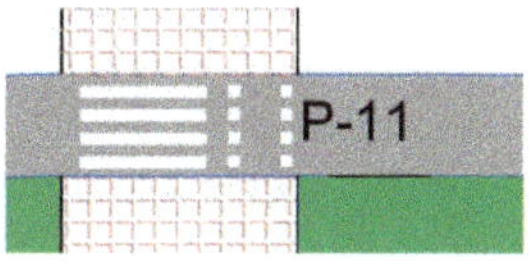	P-11	Cycle crossing
4	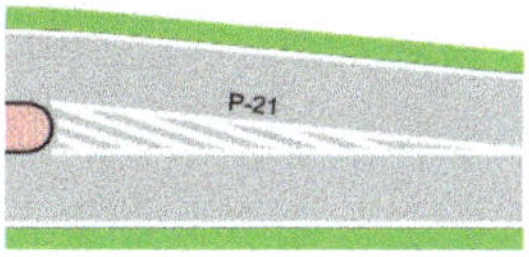	P-21	Tapered hatched area
5		P-4	Double solid line
6		D-42	Built-up area
7		B-33	Speed limit sign
8		B-25	No overtaking sign

No.	Sign	Symbol	Description
9		U-12a	Post-and-chain barrier
10		U-12c	Barrier bollard(s)
11		U-6b	Object marker sign
12		C-10	Keep left sign (for right-hand traffic)
13		C-9	Keep right sign (for right-hand traffic)

References

1. Krystek, R.; Jamroz, K.; Michalski, L.; Banach, B.; Budzyński, M.; Kastner, M.; Kustra, W.; Ryś, A.; Oskarbska, I.; Romanowska, M. *Traffic Calming Engineering on the Roads of the Pomorskie Region in Poland, Part 1: Urban Street Layouts*; GAMBIT Pomorski: Gdańsk, Poland, 2008.
2. Jamroz, K. (Ed.) *Methodology of Systemic Behavioral Studies and Relationships with Drivers*; National Road Safety Council of Poland: Warszawa, Poland; Gdańsk, Poland, 2015.
3. Gaca, S.; Kieć, M. Assessment of Pedestrian Safety Hazards on Designated Pedestrian Crossings in Relation to Vehicle Speeds. *Logistyka* **2014**, *3*, 1844–1853.
4. Vignali, V.; Cuppi, F.; Acerra, E.; Bichicchi, A.; Lantieri, C.; Simone, A.; Costa, M. Effects of Median Refuge Island and Flashing Vertical Sign on Conspicuity and Safety of Unsignalized Crosswalks. *Transp. Res. Part F* **2018**, *60*, 427–439. [CrossRef]
5. Jamroz, K. (Ed.) *Protection of Pedestrians. Manual for Pedestrian Traffic Management Bodies*; National Road Safety Council of Poland: Gdańsk, Kraków; Warszawa, Poland, 2014.
6. Ordinance of the Minister of Transport and Maritime Economy of 2 March 1999. *Published in the Polish Journal of Laws*, Issue No. 43, Item No. 430, as Subsequently Amended; Warszawa, Poland, 1999.
7. Schedules No. 1–4 to the Ordinance of the Minister of Infrastructure of 3 July 2003 on the Engineering Requirements for Road Signs and Signals and Traffic Safety Devices and Placement Considerations. *Published in the Polish Journal of Laws*, Warszawa, Poland, 2003.

8. *Safe Road Design Manual Amendments to the WB Manual, Consulting Services for Safe Road Design*; SweRoad: Stockholm, Sweden, 2011.

9. *Pedestrian Compatible Planning and Design Guidelines*; State of NJ DOT: New York, NY, USA, 1998.

10. Islam, K.T.; Raj, R.G.; Mujtaba, G. Recognition of Traffic Sign Based on Bag-of-Words and Artificial Neural Network. *Symmetry* **2017**, *9*, 138. [CrossRef]

11. Sawada, T.; Li, Y.; Pizlo, Z. Any Pair of 2D Curves Is Consistent with a 3D Symmetric Interpretation. *Symmetry* **2011**, *3*, 365–388. [CrossRef]

12. Jayadevan, V.; Sawada, T.; Delp, E.; Pizlo, Z. Perception of 3D Symmetrical and Nearly Symmetrical Shapes. *Symmetry* **2018**, *10*, 344. [CrossRef]

13. *Directives for the Design of Urban Roads*; RASt 06; Road and Transportation Research Association: Köln, Germany, 2012.

14. *Wirksamkeit Geschwindigkeitsdämpfender Maßnahmen Außerorts*; Hessisches Landesamt für Straßen- und Verkehrswesen: Hessen, Germany, 1997.

15. Kacprzak, D.; Sołowczuk, A. Synergy effect of speed management and development of road vicinity in Wrzosowo. In Proceedings of the Conference Web of Science, World Multidiscipinary Civil Engineering—Architecture—Urban Planning Symposium, Prague, Czech Republic, 18–22 June 2018.

16. *Guidelines for Traffic Calming*; City of Sparks Public Works Traffic Division: Reno, NV, USA, 2007.

17. *Traffic Calming, Local Transport, Note 1, Department for Transport, Department for Regional Development (Northern Ireland), Scottish Executives, Welsh Assembly Government*; TSO: Dublin, UK, 2007.

© 2019 by the authors. Licensee MDPI, Basel, Switzerland. This article is an open access article distributed under the terms and conditions of the Creative Commons Attribution (CC BY) license (http://creativecommons.org/licenses/by/4.0/).

 *symmetry*

Article

Hydrogen Power Plant Site Selection Under Fuzzy Multicriteria Decision-Making (FMCDM) Environment Conditions

Chia-Nan Wang *, Ming-Hsien Hsueh * and Da-Fu Lin *

Department of Industrial Engineering and Management, National Kaohsiung University of Science and Technology, Kaohsiung 80778, Taiwan
* Correspondence: cn.wang@nkust.edu.tw (C.-N.W.); mhhsueh@nkust.edu.tw (M.-H.H.); dafulin01012018@gmail.com (D.-F.L.)

Received: 20 March 2019; Accepted: 17 April 2019; Published: 25 April 2019

Abstract: Fuel and energy are basic resources necessary to meet a country's socioeconomic development needs; further, countries rich in these resources have the best premise for meeting the inputs of an economic system; however, this also poses many political challenges and threats to national security. Vietnam is located in the Southeast Asian monsoon-humid tropical region and has diverse fuel-energy resources such as coal, petroleum, and hydropower, along with renewable energy sources such as solar energy, biomass energy, and geothermal energy. However, the reality of economic development in recent years shows complex fluctuations in fuel and energy usage, i.e., besides the export of coal and crude oil, Vietnam still has imported processed oil products. To overcome this issue, many hydrogen power plants will be built in the future. This is why we propose fuzzy multicriteria decision-making (FMCDM) for hydrogen power plant site selection in this research. All criteria affecting location selection are determined by experts and literature reviews, and the weight of all criteria are defined by a fuzzy analytic hierarchy process (FAHP). The technique for order of preference by similarity to an ideal solution (TOPSIS) is a multicriteria decision analysis method, which is used for ranking potential locations in the final stage. As a result, the decision-making unit, DMU010 (DMU010), has become the optimal solution for building hydrogen power plants in Vietnam. A multicriteria decision-making (MCDM) model for hydrogen power plant site selection in Vietnam under fuzzy environment conditions is a contribution of this study. This research also provides useful tools for other types of renewable energies in Vietnam and other countries.

Keywords: FMCDM; fuzzy logics; fuzzy environment; FAHP; TOPSIS; hydrogen power plant; site selection

---

1. Introduction

Vietnam has the potential to develop its available renewable energy sources. Renewable energy sources that can be exploited and used in practice that have been identified thus far are hydrogen power, wind energy, biomass energy, biogas energy (biogas), biofuel, solar, and geothermal. In the context of Vietnam's energy, demand is increasing. Further, the ability to supply domestic energy resources is limited, while Vietnam's renewable energy potential is huge. A highly effective solution for the present and the future is to consider exploitation of available renewable energy sources for electricity generation, especially hydrogen energy—an idea that is feasible both in technology and economic efficiency and environment.

Today, a new type of renewable energy source that is being exploited is hydrogen (hydrogen, H2). Hydrogen is the highest heat-burning gas of all-natural fuels, e.g., it is used as fuel to launch spacecraft. The important feature of hydrogen is that its molecules do not contain any other chemical elements, e.g., carbon (C), sulfur (S), nitrogen (N), so their combustion product is only water (H2O), which is considered the ideal clean energy resource [1]. Hydrogen is produced from water and solar energy; thus, the collected hydrogen is also called hydrogen by solar energy (solar hydrogen). Planet Earth has a surplus of water and sunshine; therefore, hydrogen, thanks to solar energy, is an endless source of fuel to ensure energy safety for human beings, without fear of exhaustion [1]. Hydrogen is considered to be the renewable energy of the future and has recently received a lot of attention [2]. Many researchers have been involved in dealing with issues to facilitate the introduction of hydrogen into the energy balance [3]. Alternative fuels, including hydrogen-rich fuels, have been studied for use in electricity production [4]. The effect of hydrogen injection as an additional fuel in gas turbine combustion chambers has been assessed [5]. Power plants that utilize hydrogen could potentially have absolutely zero emissions [2].

One of the most important aspects of renewable energy resources is site selection, as decision-makers have to evaluate all quantitative and qualitative factors of the multiple-criteria decision-making (MCDM) process. This is why we proposed a fuzzy MCDM model for hydrogen power plants in this research. MCDM is a decision-making analysis that evaluates multiple (conflicting) criteria as part of the decision-making process. The general process of site selection is shown in Figure 1.

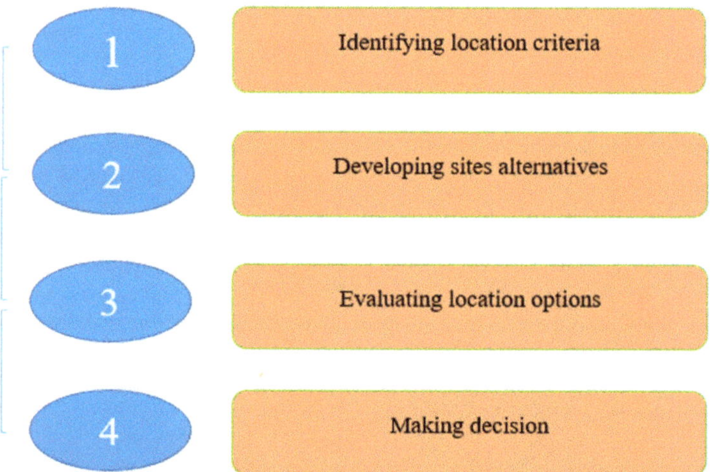

Figure 1. The general process of site selection [6,7].

The primary goal of this work is to propose a fuzzy MCDM model for hydrogen power plant site selection in Vietnam under fuzzy environment conditions. All criteria that affect the location selection are determined by experts and literature reviews, and the weight of all criteria are defined by the fuzzy analytic hierarchy process (FAHP) model. A technique for order of preference by similarity to an ideal solution (TOPSIS) is a multicriteria decision analysis method, which is used for ranking potential locations in the final stage. This research also introduces a useful tool for site selection in other types of renewable energy.

2. Literature Review

Numerous survey studies have recently focused on site selection problems, e.g., Wang et al. [6], who proposed an MCDM model for nuclear power plant (NPP) site selection in Vietnam. In this research, the authors applied fuzzy analytic network process (FANP) and TOPSIS for NPP location selection in Vietnam. Sedady et al. [7] introduced an MCDM model for prioritizing the construction of renewable power plants. The goal of this article is to propose a new MCDM model to define the priority of building renewable power plants considering technical, economic, social, political and environmental factors. Eveloy and Gebreegziabher [8] have considered technical, economic and environmental assessments of projected power-to-gas deployment scenarios on a distributed- to national-scale, as well as their extensions to nuclear-assisted renewable hydrogen.

Biswal and Shukla [9] have developed algorithms for the selection of appropriate locations for the installation of wind turbines. Pamucar et al. [10] combined the use of Geographical Information Systems (GIS) with multi-criteria techniques of Best-Worst method (BWM) and Multi Attributive Ideal-Real Comparative Analysis (MAIRCA) for wind farm location selection. Nicotra et al. [11] presented equivalent small hydropower: A simple method to evaluate energy production by small turbines in collective irrigation systems. Noorollahi [12] proposed multi-criteria decision support system for wind farm site selection using GIS. In this research, the authors considered technical, environmental, economic and geographic standards.

Borah et al. [13], using GIS, developed an analytical framework in which fuzzy logic was used to evaluate suitable sites for turbines for optimum energy output. The factors for a suitable site for energy optimization are environmental, physical and human factors. Öztürk et al. [14] applied GIS for wind turbine location selection in Balikesir, Northwest Turkey. This research identified 12 geographical criteria; the effects and weights of these criteria were defined by considering the relevant literature and field conditions, and various analyses were conducted on these factors with the help of GIS. Mytilinou and Kolios [15] introduced a multi-objective optimization approach applied to offshore wind farm location selection.

Wang et al. [16] proposed an MCDM model for solar power plant location selection in Vietnam. In this research, the authors used fuzzy analytic hierarchy process (FAHP), data envelopment analysis (DEA) to find the best location for building a solar power plant based on both quantitative and qualitative factors. Wang et al. [17] applied an MCDM model for solid waste to energy plant location selection in Vietnam. In this research, the authors applied FANP and TOPSIS for ranking all potential locations. Aktas et al. [18] proposed a hybrid hesitant fuzzy decision-making model for solar power plant location selection. Akkas et al. [19] applied an MCDM model for site selection for a solar power plant in the Central Anatolian Region of Turkey.

All the factors that affect hydrogen power plant location selection are defined by literature reviews and experts, these criteria as shown in Table 1.

Table 1. All criteria affect to location selection.

Main Criteria	Subcriteria	Literature Review
F.1. Social (SOC)	F.1.1. Public acceptance (SOC1) F.1.2. Protection law (SOC) F.1.3. Legal and Regulation compliance (SOC3)	[20] [16] [16,20]
F.2. Environmental (ENV)	F.2.1. Availability of water (EVN1) F.2.2. Water storage (EVN2) F.2.3. Water head (EVN3) F.2.4. Environment affect (EVN4)	[21] [21] [21] [21]
F.3. Technological (TEC)	F.3.1. Distance from major road (TEC1) F.3.2. Distance from power network (TEC2) F.3.3. Potential demand (TEC3)	[22–26] [17,22–27] [16,22]
F.4. Economic (ECO)	F.4.1. Construction cost. (EOC1) F.4.2. Operation and management cost (EOC2) F.4.3. New feeder cost (EOC3)	[16,24,28] [16,20,23] [23]
F.5. Site Characteristics (SIC)	F.5.1. Land use (SIC1) F.5.2. Ecology (SIC2)	[24–26,29–31] [16]

3. Methodology

3.1. Research Development

In this study, we proposed fuzzy MCDM approaches, including an FAHP and TOPSIS approach, for hydrogen power plant site selection. This study has three main stages, as shown in Figure 2.

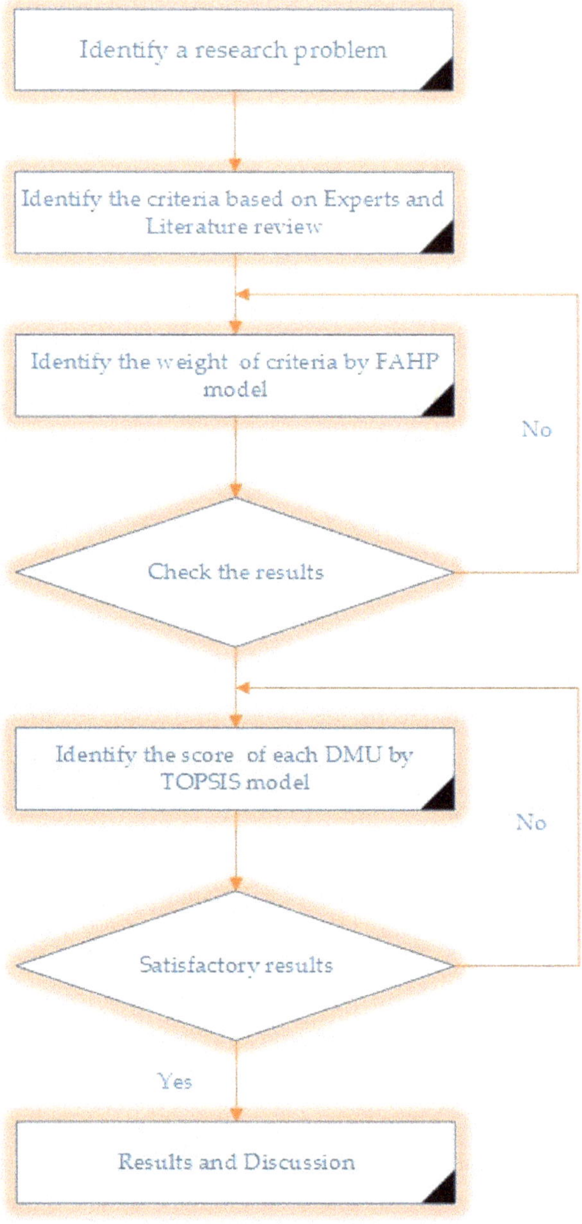

Figure 2. Research methodology.

Step 1: Identify the main criteria and subcriteria. All criteria for hydrogen power plants will be defined through experts and literature reviews.

Step 2: Using the FAHP approach. In this step, an FAHP is used to identify the weight of subcriteria.

Step 3: Ten potential locations can be highly effective for building hydrogen power plants in Vietnam. The TOPSIS approach is used in this stage for ranking all potential locations in this step. The optimal location will be shown based on positive ideal solution (PIS) and negative ideal solution (NIS) value.

3.2. Fuzzy Sets, AHP and TOPSIS Model

3.2.1. Triangular Fuzzy Number (TFN)

The TFN can be defined as (c, d, e) with $c \leq d \leq e$.
The general function of TFN is as follows:

$$\mu\left(\frac{x}{\widetilde{M}}\right) = \begin{cases} 0, & x < d, \\ \frac{x-c}{d-c} & c \leq x \leq d, \\ \frac{e-x}{e-d} & d \leq x \leq e, \\ 0, & x > e, \end{cases} \tag{1}$$

The basic calculations of fuzzy numbers are shown in:

$$\widetilde{M} = (M^{o(y)}, M^{i(y)}) = [c + (d-c)y, \ e + (d-e)y], y \in [0,1] \tag{2}$$

$o(y)$, $i(y)$ indicates both the left side and the right side of a fuzzy number as:

$$(c_1, d_1, e_1) + (c_2, d_2, e_2) = (c_1 + c_2, d_1 + d, e_1 + e_2) \tag{3}$$

$$(c_1, d_1, e_1) - (c_2, d_2, e_2) = (c_1 - c_2, d_1 - d_2, e_1 - e_2)$$

$$(c_1, d_1, e_1) \times (c_2, d_2, e_2) = (c_1 \times c_2, d_1 \times d_2, e_1 \times e_2)$$

$$\frac{(c_1, d_1, e_1)}{(c_2, d_2, e_2)} = (c_1/c, \ d_1/d_2, e_1/e_2)$$

To calculate the priority in the process of pairwise comparisons matrix, that are quantified using a $1 \div 9$ scale.

3.2.2. Analytic Hierarchy Process (AHP) model

AHP is introduced by Saaty [32], this is an MCDM that simplifies complex issues by sorting factors and alternatives in a hierarchical structure by using a pairwise comparison metric [33].

Let $D = \{D_b | b = 1, 2, \ldots, m\}$ be the set of criteria. The pair-wise comparisons metrics on m criteria will be shown in an m × m evaluation matrix, E, in which every element, k_{ab}, is the quotient of weights of the criteria, as shown in (1):

$$E = (k_{ab}), a, b = 1 \tag{4}$$

The relative priorities are given by the Eigenvector (u) corresponding to the largest eigenvector (λ_{max}) as:

$$E_u = \lambda_{max} u \tag{5}$$

The consistency is determined by the relation between the entries of E and Consistency Index (CI):

$$CI = \frac{(\lambda_{max} - m)}{(m - 1)} \tag{6}$$

Consistency Ratio (*CR*) is calculated as the ratio of the *CI* and the Random Consistency Index (*RI*), as shown in (7):

$$CR = \frac{CI}{RI} \tag{7}$$

$CR \leq 0.1$. If the $CR > 0.1$, the evaluation needs to be repeated again for improving consistency.

3.2.3. Technique for Order Preference by Similarity to Ideal Solution (TOPSIS)

Hwang et al. introduced the TOPSIS model [34]. There are five main steps as follows [35].

- Determining TOPSIS requires performance ranking in every option. This can be seen from the formula below:

$$e_{ij} = \frac{X_{ij}}{\sqrt{\sum_{i=1}^{m} X_{ij}^2}} \tag{8}$$

with $i = 1, 2, ..., m$; and $j = 1, 2, ..., n$;
- Calculate the normalized weighted decision matrix.

$$S_{ij} = W_i e_{ij} \tag{9}$$

with $i = 1, 2, ..., m$; and $j = 1, 2, ..., n$;
- Calculate PIS A^+ matrix and NIS A^- matrix.

$$A^+ = s_1^+, s_2^+, \ldots, s_n^+$$
$$A^- = s_1^-, s_2^-, \ldots, s_n^-; \tag{10}$$

where:

$$s_j^+ = \begin{cases} Max\ s_{ij}\ if\ j\ is\ an\ advantage\ factor \\ Min\ s_{ij}\ if\ j\ is\ an\ cost\ factor \end{cases}$$

$$s_j^- = \begin{cases} Max\ s_{ij}\ if\ j\ is\ an\ advantage\ factor \\ Min\ s_{ij}\ if\ j\ is\ an\ cost\ factor \end{cases}$$

- Identifing the gap between the values of each options with (positive ideal solution) PIS matrix and (negative ideal solution) NIS matrix.

Options to PIS.

$$D_i^+ = \sqrt{\sum_{j=1}^{m} \left(s_{ij} - s_i^+\right)^2}; i = 1, 2, \ldots, m \tag{11}$$

Options to NIS.

$$D_i^- = \sqrt{\sum_{j=1}^{m} \left(s_{ij} - s_i^-\right)^2}; i = 1, 2, \ldots, m \tag{12}$$

where D_i^+ is the distance to the PIS for *i* option and D_i^- is the distance to the NIS.

- Calculating the preference value for every alternative (G_i)

$$G_i = \frac{D_i^-}{D_i^- + D_i^+} \quad i = 1, 2, \ldots, m \tag{13}$$

4. Case Study

The energy demand in Vietnam increased by 7.5% annually from 2015 to 2025 because it remains among one of Asia's fastest-growing economies. This study estimates that gas and hydrogen will become the largest sources of energy and will increase to more than 50% of Vietnam's power structure over the next two decades [36]. To support this rapid economic growth, it is necessary that Vietnam continuously supplies new power plant capacity to meet demand. While Vietnam has reserves of oil and coal that provide considerable capacity, traditional hydropower has provided an alternative low-cost base energy source. This study estimates that gas and hydrogen will become the largest energy source by 2015 and will increase to more than 50% in Vietnam's power structure over the next two decades. Vietnam's power generation mix (TWh) is shown in Figure 3.

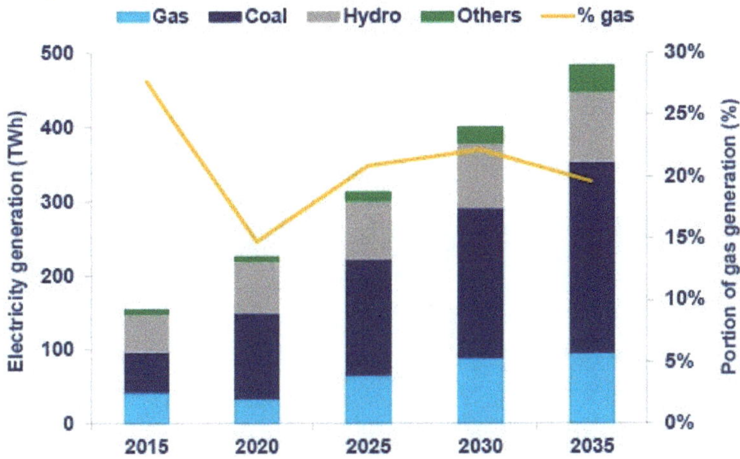

Figure 3. Vietnam power generation mix (TWh). Source: Wook Mackenie.

In recent years, the direction of development of the hydrogen energy industry has changed dramatically. Until now, hydrogen has been mainly used as fuel for automobile transport. It is thought that this will allow a significant reduction in the concentration of toxic substances in the atmosphere—not just carbon dioxide (CO_2) but a combination of toxic wastes when burning carbon-containing fuels. Hydrogen is the simplest and most abundant element on the Earth's surface. Although hydrogen does not exist in nature as monoatomic, by separating hydrogen from other elements, hydrogen can become a perfect energy carrier.

Another advantage is that the process of generating energy does not create any substance other than water. Currently, many available technologies can take advantage of hydrogen to provide energy. The ability to use hydrogen as an efficient fuel source has many advantages, e.g., it not only strengthens national energy safety but also reduces environmental pollution.

An MCDM model based on fuzzy set theory is an effective tool to solve complex selection problems, including many criteria (qualitative and quantitative) with many options [37]. Qualitative standards often have ambiguous characteristics, which are difficult to define correctly, making it difficult to synthesize assessment results according to criteria and decision-making. The fuzzy MCDM method will quantify these criteria, calculate the total scores of the weighted testers of each standard, and help decision-makers obtain a more solid and accurate basis. The assessment of a location is also carried out on such qualitative criteria; thus, the fuzzy MCDM model can be considered as an effective tool to assess site selection. Many studies have applied fuzzy MCDM in the location selection model, e.g., TOPSIS, AHP, ANP, DEA, The Preference Ranking Organization METHod for Enrichment of Evaluations (PROMETHEE). Thus, the authors proposed fuzzy multicriteria decision-making (FMCDM) for hydrogen power plant site selection in this research. All criteria affecting the location selection are determined by experts and literature reviews, and the weight of all criteria are defined by FAHP. The TOPSIS is a multicriteria decision analysis method, which is used for ranking potential locations in the final stage.

Ten potential locations are able to invest in hydrogen power plants, as shown in Table 2 and in Figure 4.

Table 2. List of 10 potential locations.

No.	DMUs	Symbol
1	Can Tho	DMU-001
2	Soc Trang	DMU-002
3	Bac Lieu	DMU-003
4	Hau Giang	DMU-004
5	Tra Vinh	DMU-005
6	Vinh Long	DMU-006
7	Kien Giang	DMU-007
8	Long An	DMU-008
9	An Giang	DMU-009
10	Ca Mau	DMU-010

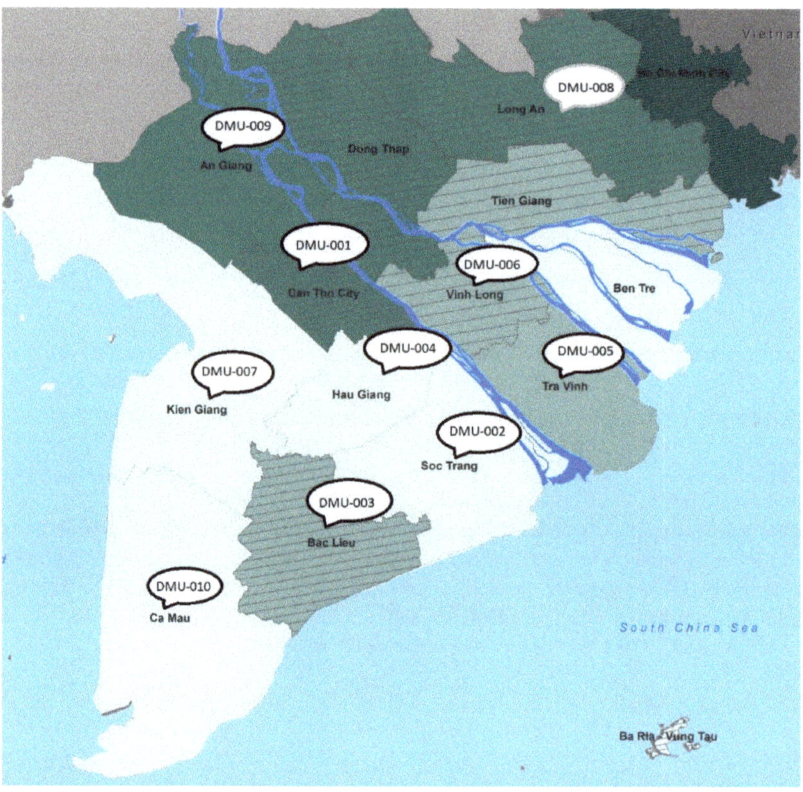

Figure 4. Ten potential locations on the map of Vietnam.

Finding an optimal location is among the most important factors affecting the time at which the project reaches completion. In order to select a good site, the decision-maker must first understand the criteria of site evaluation. Based on experts and literature reviews, the decision-maker must consider social factors, environmental criteria, technological factors, economic criteria and also characteristic factors. The hierarchy of the objectives of this work is shown in Figure 5.

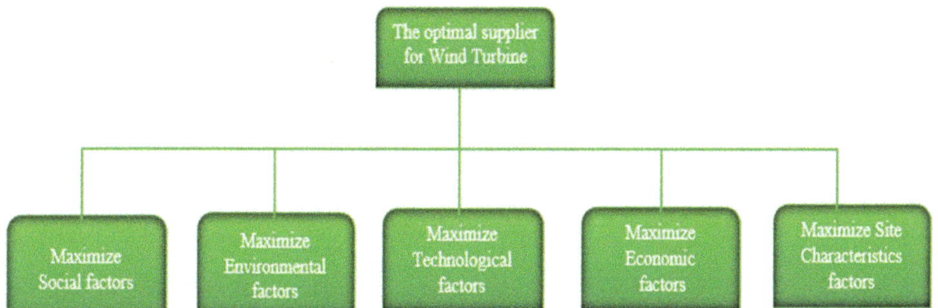

Figure 5. The objectives hierarchy.

The fuzzy comparison matrix of Goal from the FAHP model is shown in Table 3.

Table 3. Fuzzy comparison matrices for GOAL.

Criteria	EIC	ENV	EOC	SOC	TEC
EIC	(1,1,1)	(1/3,1/4,1/5)	(1/3,1/4,1/5)	(1/2,1/3,1/4)	(1,2,3)
ENV	(3,4,5)	(1,1,1)	(1,2,3)	(1,2,3)	(3,4,5)
EOC	(5,4,3)	(1/3,1/2,1)	(1,1,1)	(4,5,6)	(3,4,5)
SOC	(4,3,2)	(1/3,1/2,1)	(1/6,1/5,1/4)	(1,1,1)	(3,4,5)
TEC	(1/3,1/2,1)	(1/5,1/4,1/3)	(1/5,1/4,1/3)	(1/5,1/4,1/3)	(1,1,1)

We have the coefficients $\alpha = 0.5$ and $\beta = 0.5$, and

$$g_{0.5,0.5}(\overline{a_{EIC,TEC}}) = [(0.5 \times 1.5) + (1 - 0.5) \times 2.5] = 2$$
$$f_{0.5}(L_{EIC,TEC}) = (2 - 1) \times 0.5 + 1 = 1.5$$
$$f_{0.5}(U_{EIC,TEC}) = 3 - (3 - 2) \times 0.5 = 2.5$$
$$g_{0.5,0.5}(\overline{a_{TEC,EIC}}) = \tfrac{1}{2}$$

The real number priority when comparing the main criteria pairs is shown in Table 4.

Table 4. Real number priority.

Criteria	EIC	ENV	EOC	SOC	TEC
EIC	1	1/4	1/4	1/3	2
ENV	4	1	2	2	4
EOC	4	1/2	1	5	4
SOC	3	1/2	1/5	1	4
TEC	1/2	1/4	1/4	1/4	1

Calculating the maximum individual value is achieved as follows:

$$M1 = (1 \times 1/4 \times 1/4 \times 1/3 \times 2)^{1/5} = 0.5$$

$$M2 = (4 \times 1 \times 2 \times 2 \times 4)^{1/5} = 2.3$$

$$M3 = (4 \times 1/2 \times 1 \times 5 \times 4)^{1/5} = 2.1$$

$$M4 = (3 \times 1/2 \times 1/5 \times 1 \times 4)^{1/5} = 1.04$$

$$M5 = (1/2 \times 1/4 \times 1/4 \times 1/4 \times 1)^{1/5} = 0.38$$

$$\sum M = 6.32$$

$$\omega_1 = \frac{0.5}{6.32} = 0.08$$

$$\omega_2 = \frac{2.3}{6.32} = 0.36$$

$$\omega_3 = \frac{2.1}{6.32} = 0.33$$

$$\omega_4 = \frac{1.04}{6.32} = 0.16$$

$$\omega_5 = \frac{0.38}{6.32} = 0.06$$

$$\begin{bmatrix} 1 & 1/4 & 1/4 & 1/3 & 2 \\ 4 & 1 & 2 & 2 & 4 \\ 4 & 1/2 & 1 & 5 & 4 \\ 3 & 1/2 & 1/5 & 1 & 4 \\ 1/2 & 1/4 & 1/4 & 1/4 & 1 \end{bmatrix} \times \begin{bmatrix} 0.08 \\ 0.36 \\ 0.33 \\ 0.16 \\ 0.06 \end{bmatrix} = \begin{bmatrix} 0.43 \\ 1.9 \\ 1.87 \\ 0.89 \\ 0.31 \end{bmatrix}$$

$$\begin{bmatrix} 0.43 \\ 1.9 \\ 1.87 \\ 0.89 \\ 0.31 \end{bmatrix} / \begin{bmatrix} 0.08 \\ 0.36 \\ 0.33 \\ 0.16 \\ 0.06 \end{bmatrix} = \begin{bmatrix} 5.4 \\ 5.3 \\ 5.7 \\ 5.56 \\ 5.17 \end{bmatrix}$$

$n = 5$, $\lambda_{\max}$ and CI are calculated as follows:

$$\lambda_{max} = \frac{5.4 + 5.3 + 5.7 + 5.56 + 5.17}{5} = 5.426$$

$$CI = \frac{\lambda_{max} - n}{n - 1} = \frac{5.136 - 5}{5 - 1} = 0.1065$$

To calculate the CR value, we get $RI = 1.12$ with $n = 4$.

$$CR = \frac{CI}{RI} = \frac{0.1065}{1.12} = 0.0951$$

Because $CR = 0.0951$, which is ≤ 0.1, it does not need to be re-evaluated. The weight of all critera are defined by fuzzy AHP are shown in Table 5.

Table 5. The weight of subcriteria.

No.	Criteria	Weight
1	SOC1	0.0199
2	SOC2	0.0593
3	SOC3	0.1329
4	ENV1	0.0290
5	ENV2	0.0478
6	ENV3	0.0681
7	ENV4	0.1293
8	TEC1	0.1106
9	TEC2	0.0392
10	TEC3	0.0208
11	EOC1	0.0396
12	EOC2	0.0227
13	EOC3	0.1037
14	SIC1	0.0354
15	SIC2	0.1418

The normalized matrix and normalized weight matrix, obtained from the TOPSIS model, are shown in Tables 6 and 7.

Table 6. Normalized matrix.

Subcriteria	DMU-1	DMU-2	DMU-3	DMU-4	DMU-5	DMU-6	DMU-7	DMU-8	DMU-9	DMU-10
SOC1	0.2851	0.3258	0.3665	0.3665	0.2851	0.2443	0.3258	0.2851	0.3665	0.2851
SOC2	0.2782	0.3180	0.3180	0.2782	0.3180	0.3577	0.2385	0.3180	0.3577	0.3577
SOC3	0.2825	0.3229	0.3229	0.3229	0.3229	0.2825	0.3632	0.2825	0.2825	0.3632
ENV1	0.3269	0.2860	0.3269	0.3269	0.3677	0.3269	0.3269	0.2860	0.2452	0.3269
ENV2	0.2584	0.3015	0.3015	0.3446	0.3446	0.3015	0.3877	0.3015	0.3015	0.3015
ENV3	0.3269	0.3677	0.3269	0.2860	0.3677	0.2860	0.2860	0.2860	0.2860	0.3269
ENV4	0.3780	0.2940	0.3360	0.2940	0.2940	0.3360	0.2940	0.2940	0.3360	0.2940
TEC1	0.2860	0.3269	0.2860	0.2860	0.3677	0.2860	0.3269	0.2860	0.3269	0.3677
TEC2	0.3671	0.3263	0.3263	0.3263	0.3671	0.2855	0.3263	0.2855	0.2447	0.2855
TEC3	0.3052	0.3052	0.3052	0.3488	0.2616	0.3924	0.3052	0.3052	0.2616	0.3488
ECO1	0.3780	0.2940	0.3360	0.2940	0.2940	0.3360	0.2940	0.2940	0.3360	0.2940
ECO2	0.2860	0.3269	0.2860	0.2860	0.3677	0.2860	0.3269	0.2860	0.3269	0.3677
ECO3	0.3671	0.3263	0.3263	0.3263	0.3671	0.2855	0.3263	0.2855	0.2447	0.2855
SIC1	0.3269	0.2860	0.3269	0.3269	0.3677	0.3269	0.3269	0.2860	0.2452	0.3269
SIC2	0.2584	0.3015	0.3015	0.3446	0.3446	0.3015	0.3877	0.3015	0.3015	0.3015

Table 7. Normalized weight matrix.

Subcriteria	DMU-1	DMU-2	DMU-3	DMU-4	DMU-5	DMU-6	DMU-7	DMU-8	DMU-9	DMU-10
SOC1	0.0261	0.0298	0.0336	0.0336	0.0261	0.0224	0.0298	0.0261	0.0336	0.0261
SOC2	0.0376	0.0430	0.0430	0.0376	0.0430	0.0484	0.0323	0.0430	0.0484	0.0484
SOC3	0.0472	0.0539	0.0539	0.0539	0.0539	0.0472	0.0607	0.0472	0.0472	0.0607
ENV1	0.0257	0.0225	0.0257	0.0257	0.0289	0.0257	0.0257	0.0225	0.0193	0.0257
ENV2	0.0230	0.0268	0.0268	0.0306	0.0306	0.0268	0.0345	0.0268	0.0268	0.0268
ENV3	0.0197	0.0221	0.0197	0.0172	0.0221	0.0172	0.0172	0.0172	0.0172	0.0197
ENV4	0.0367	0.0286	0.0327	0.0286	0.0286	0.0327	0.0286	0.0286	0.0327	0.0286
TEC1	0.0312	0.0357	0.0312	0.0312	0.0401	0.0312	0.0357	0.0312	0.0357	0.0401
TEC2	0.0395	0.0351	0.0351	0.0351	0.0395	0.0307	0.0351	0.0307	0.0263	0.0307
TEC3	0.0197	0.0197	0.0197	0.0225	0.0169	0.0254	0.0197	0.0197	0.0169	0.0225
ECO1	0.0150	0.0116	0.0133	0.0116	0.0116	0.0133	0.0116	0.0116	0.0133	0.0116
ECO2	0.0065	0.0074	0.0065	0.0065	0.0083	0.0065	0.0074	0.0065	0.0074	0.0083
ECO3	0.0381	0.0338	0.0338	0.0338	0.0381	0.0296	0.0338	0.0296	0.0254	0.0296
SIC1	0.0116	0.0101	0.0116	0.0116	0.0130	0.0116	0.0116	0.0101	0.0087	0.0116
SIC2	0.0365	0.0426	0.0426	0.0486	0.0486	0.0426	0.0547	0.0426	0.0426	0.0426

5. Results and Discussion

Owing to its the geographical position, climate and agricultural activities are present in Vietnam; thus, there is a plentiful and quite diverse potential of renewable energy sources, which can be exploited and used as hydrogen power and biomass, wind, solar, geothermal, biofuel, and other new energy sources. In the context of a growing shortage in the domestic energy supply and demand and unpredictable developments, there will certainly be a major impact on supply and demand, along with the prices of traditional energy sources. Thus, exploiting and using renewable energy sources in a considerable and appropriate quantity are urgent requirements.

In this study, the author proposed a fuzzy MCDM model using hybrid FAHP and TOPSIS for site selection of hydrogen power plant projects in Vietnam. Ten potential locations were considered and judged based on five main criteria and 15 subcriteria. All criteria affecting location selection were determined by experts and literature reviews, and the weight of all criteria were defined by FAHP. The TOPSIS was used for ranking potential locations in the final stage. Results are shown in Figures 6 and 7; further, the decision-making unit DMU010 (DMU010) was found to be an optimal solution for building hydrogen power plants in Vietnam.

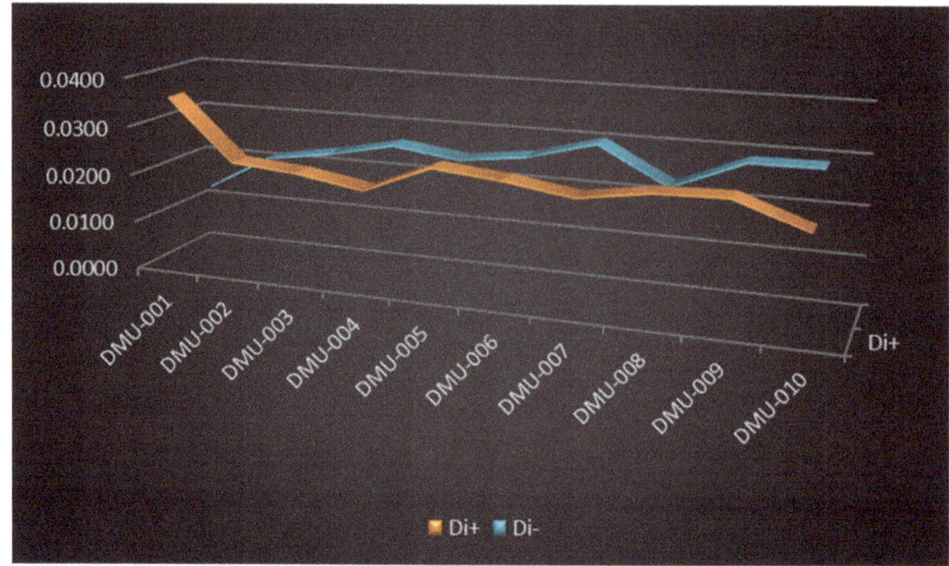

Figure 6. Negative ideal solution (NIS) and positive ideal solution (PIS) value.

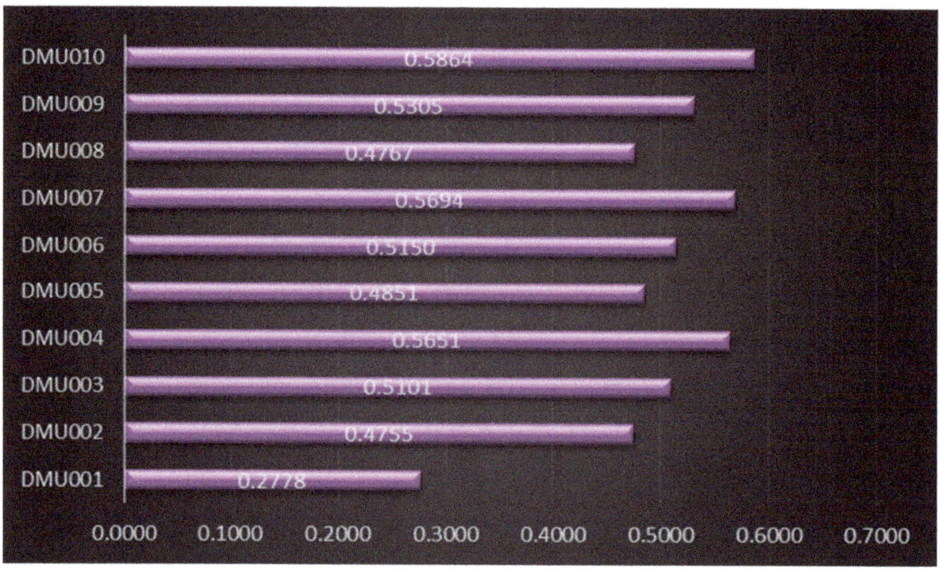

Figure 7. Final ranking score.

6. Conclusions

Energy plays an important role in human life. Industrialization processes have increased the energy demand. Fossil fuels are the main sources of energy for the global economy. However, this fuel is limited and causes environmental problems and climate change; thus, people have found new sources of alternative energy called "renewable energy". Further, this energy source is continuously supplemented by natural processes, including wind power, solar energy, biofuels, hydrogen power, wave energy, and tidal energy, which can be exploited at any time to meet the development needs of the world.

The advantages of Vietnam's natural and climate conditions, such as its coastline of more than 3000 km, rivers and lakes, and tidal energy sources, hydrogen energy, abundant wave energy, and wind energy, have created abundant raw materials for the development of renewable energy. Therefore, the study and access to technologies to maximize and efficiently utilize these energy sources are important tasks for the country toward building sustainable and environmentally friendly energy in the future. Hydrogen is one of the most important energy sources that will increasingly contribute to the world's energy output. Vietnam is an ideal country for investment and expansion of hydrogen power production capacity, thanks to its highly skilled labor force and future development of the energy sector. Site selection is an important issue in renewable energy projects in that a decision-maker must consider qualitative and quantitative factors. Choosing the right location is among the key success factors of renewable energy projects in general and hydrogen power plant projects in particular. Site selection is an MCDM process in which decision-makers have to evaluate qualitative and quantitative factors. Although some studies have applied MCDM approaches for location selection in renewable energy projects, few studies, to the best of our knowledge, have used the MCDM model for hydrogen power plant location selection in fuzzy environment conditions. This is a reason why we proposed an FMCDM model for hydrogen power plant location selection in Vietnam. As a result, decision-making unit DMU010 (DMU010) has become an optimal solution for building hydrogen power plants in Vietnam.

The contribution of this work is to propose an MCDM model for hydrogen power plants site selection in Vietnam under fuzzy environment conditions. The advantages of this proposed model also reside in the evolution of a new approach that is flexible and practical to the decision-maker. This research also provides a useful tool for other types of renewable energies in Vietnam and other countries.

For future study, this FMCDM model can also be used for location selection for other types of renewable energy resources. In addition, different approaches, such as FANP, PROMETHEE, etc., could also be combined for different scenarios.

Author Contributions: Conceptualization, C.-N.W., M.-H.H. and D.-F.L.; Data curation, C.-N.W.; Formal analysis, M.-H.H. and D.-F.L.; Methodology, C.-N.W. and M.-H.H.; Project administration, C.-N.W. and M.-H.H.; Resources, D.-F.L.; Software, D.-F.L.; Writing—original draft, D.-F.L.; Writing—review & editing, C.-N.W. and M.-H.H.

Funding: This research received no external funding.

Conflicts of Interest: The authors declare no conflict of interest.

References

1. Tran Manh Trung. Hydro: nguồn năng lượng mới thay thế dầu - khí trong tương lai. Báo Tuổi Trẻ. 2007. Available online: https://tuoitre.vn/hydro-nguon-nang-luong-moi-thay-the-dau---khi-trong-tuong-lai-220499.htm (accessed on 22 December 2018).
2. Pilavachi, P.A.; Stephanidis, S.D.; Pappas, V.A.; Afgan, N.H. Multi-criteria evaluation of hydrogen and natural gas fuelled power plant technologies. *Appl. Therm. Eng.* **2009**, *29*, 11–12. [CrossRef]
3. Hennicke, P.; Fischedick, M. Towards sustainable energy systems: The related role of hydrogen. *Energy Policy* **2006**, *34*, 1260–1270. [CrossRef]
4. Gökalp, I.; Lebas, E. Alternative fuels for industrial gas turbines (AFTUR). *Appl. Therm. Eng.* **2004**, *24*, 1655–1663. [CrossRef]
5. Juste, G. Hydrogen injection as additional fuel in gas turbine combustor. Evaluation of effects. *Int. J. Hydrog. Energy* **2006**, *31*, 2112–2121. [CrossRef]
6. Wang, C.-N.; Su, C.-C.; Nguyen, V.T. Nuclear Power Plant Location Selection in Vietnam under Fuzzy Environment Conditions. *Symmetry* **2018**, *11*, 548. [CrossRef]
7. Sedady, F.; Beheshtinia, M.A. A novel MCDM model for prioritizing the renewable power plants' construction. *Manag. Environ. Qual. Int. J.* **2019**, *30*, 383–399. [CrossRef]
8. Eveloy, V.; Gebreegziabher, T. A Review of Projected Power-to-Gas Deployment Scenarios. *Energies* **2018**, *11*, 1824. [CrossRef]

9.   Biswal, G.C.; Shukla, S.P. Site Selection for Wind Farm Installation. *Int. J. Innov. Res. Electr. Electron. Instrum. Control. Eng.* **2015**, *3*, 59–61.

10. Pamučar, D.; Gigović, L.; Bajić, Z.; Janošević, M. Location Selection for Wind Farms Using GIS Multi-Criteria Hybrid Model: An Approach Based on Fuzzy and Rough Numbers. *Sustainability* **2017**, *9*, 1315. [CrossRef]

11. Nicotra, A.; Zema, D.A.; D'Agostino, D.; Zimbone, S.M. Equivalent Small Hydro Power: A Simple Method to Evaluate Energy Production by Small Turbines in Collective Irrigation Systems. *Water* **2018**, *10*, 1390. [CrossRef]

12. Noorollahi, Y.; Yousefi, H.; Mohammadi, M. Multi-criteria decision support system for wind farm site selection using GIS. *Sustain. Energy Technol. Assess.* **2016**, *13*, 35–50. [CrossRef]

13. Borah, K.K.; Roy, S.; Harinarayana, T. Optimization in Site Selection of Wind Turbine for Energy Using Fuzzy Logic System and GIS—A Case Study for Gujarat. *Open J. Optim.* **2013**, *2*, 116–122. [CrossRef]

14. Öztürk, B.; Serkendiz, H. Location Selection for Wind Turbines in Balikesir, NW Turkey, Using GIS. *Int. J. Environ. Geoinform.* **2018**, *5*, 284–295. [CrossRef]

15. Mytilinou, V.; Kolio, A.J. A multi-objective optimisation approach applied to offshore wind farm location selection. *J. Ocean. Eng. Mar. Energy* **2018**, *5*, 284–295. [CrossRef]

16. Wang, C.-N.; Nguyen, V.T.; Thai, H.T.N.; Duong, D.H. Multi-Criteria Decision Making (MCDM) Approaches for Solar Power Plant Location Selection in Viet Nam. *Energies* **2018**, *11*, 1504. [CrossRef]

17. Wang, C.-N.; Nguyen, V.T.; Thai, H.T.N.; Duong, D.H. A Hybrid Fuzzy Analysis Network Process (FANP) and the Technique for Order of Preference by Similarity to Ideal Solution (TOPSIS) Approaches for Solid Waste to Energy Plant Location Selection in Vietnam. *Appl. Sci.* **2018**, *8*, 1100. [CrossRef]

18. Aktas, A.; Kabak, M. A Hybrid Hesitant Fuzzy Decision-Making Approach for Evaluating Solar Power Plant Location Sites. *Arabian J. Sci. Eng.* Available online: https://link.springer.com/article/10.1007%2Fs13369-018-3604-5#citeas (accessed on 22 December 2018). [CrossRef]

19. Akkas, O.P.; Erten, M.Y.; Cam, E.; Inanc, N. Optimal Site Selection for a Solar Power Plant in the Central Anatolian Region of Turkey. *Int. J. Photoenergy* **2017**, *2*, 1–13. [CrossRef]

20. Talinli, I.; Topuz, E.; Aydin, E.; Kabakcı, S.B. A Holistic Approach for Wind Farm Site Selection by FAHP. *IntechOpen.* Available online: https://www.researchgate.net/publication/221912736_A_Holistic_Approach_for_Wind_Farm_Site_Selection_by_Using_FAHP (accessed on 22 December 2018). [CrossRef]

21. The Electrical Portal. Site selection of Hydroelectric power plant. Available online: http://www.theelectricalportal.com/2015/08/site-selection-of-hydroelectric-power.html (accessed on 22 December 2018).

22. Ozdemir, S.; Sahin, G. Multi-criteria decision-making in the location selection for a solar PV power plant using AHP. *Measurement* **2018**, *129*, 218–226. [CrossRef]

23. Demirel, T.; Yalcinn, U. Multi-Criteria Wind Power Plant Location Selection using Fuzzy AHP. In Proceedings of the 8th International FLINS Conference, Madrid, Spain, 21–24 September 2008.

24. Uyan, M. GIS-based solar farms site selection using analytic hierarchy process (AHP) in Karapinar region, Konya/Turkey. *Renew. Sustain. Energy Rev.* **2013**, *28*, 7–11. [CrossRef]

25. Carrión, J.A.; Espín Estrella, A.; Aznar Dols, F.; Ridao, A.R. The electricity production capacity of photovoltaic power plants and the selection of solar energy sites in Andalusia (Spain). *Renew Energy* **2008**, *33*, 545–552.

26. Sabo, M.L.; Mariun, N.; Hizam, H.; Mohd Radzi, M.A.; Zakaria, A. Spatial matching of large-scale grid-connected photovoltaic power generation with utility demand in Peninsular Malaysia. *Appl. Energy* **2017**, *191*, 663–688. [CrossRef]

27. Toklu, M.G.; Uygun, Ö. Location Selection for Wind Plant using AHP and Axiomatic Design in Fuzzy Environment. *Period. Eng. Nat. Sci.* **2018**, *6*, 120–128.

28. Kengpol, A.; Rontlaong, P.; Tuominen, M. Design of a decision support system for site selection using fuzzy AHP: A case study of solar power plant in north eastern parts of Thailand. In Proceedings of the PICMET '12: Technology Management for Emerging Technologies, Vancouver, BC, Canada, 29 July–2 August 2012.

29. Idris, R.; Latif, Z.A. GIS multi-criteria for power plant site selection. In Proceedings of the IEEE Control and System Graduate Research Colloquium, Shah Alam, Selangor, Malaysia, 16–17 July 2012.

30. Massimo, A.; Dell'Isola, M.; Frattolillo, A.; Ficco, G. Development of a geographical information system (GIS) for the integration of solar energy in the energy planning of a wide area. *Sustainability* **2014**, *6*, 5730–5744. [CrossRef]

31. Chamanehpour, E.; Akbarpour, A. Site selection of wind power plant using multi-criteria decision-making methods in GIS: A case study. *Comput. Ecology Softw.* **2017**, *7*, 49–64.

32. Saaty, T.L. *How to Make a Decision: The Analytic Hierarchy Process*; Elsevier Science: New York, NY, USA, 1980.

33. Hwang, C.-L.; Yoon, K. *Multiple Attribute Decision Making: Methods and Applications A State-of-the-Art Survey*, 1st ed.; Springer: Berlin/Heidelberg, Germany, 1981.

34. Kusumadewi, S.; Hartati, S.; Harjoko, A.; Wardoyo, R. *Fuzzy Multi-Attribute Decision Making (Fuzzy MADM)*; Graha Ilmu: Yogyakarta, Indonesia, 2006; Volume 14, pp. 78–79.

35. Behzadian, M.; Otaghsara, S.K.; Yazdani, M.; Ignatius, J. A state-of the-art survey of TOPSIS applications. *Exp. Sys. Appli.* **2012**, *39*, 13051–13069. [CrossRef]

36. Ngân, B.T. Vấn đề về năng lượng Việt Nam: khí đốt trong nước và than đá nhập khẩu. Trung tâm Phát triển Sáng tạo Xanh. 2016. Available online: http://greenidvietnam.org.vn/van-de-ve-nang-luong-viet-nam-khi-dot-trong-nuoc-va-than-da-nhap-khau-1.html (accessed on 22 December 2018).

37. Đạt, L.Q.; Phượng, B.H.; Thu, P.; Anh, T.T.L. Xây dựng mô hình ra quyết định đa tiêu chuẩn tích hợp để lựa chọn và phân nhóm nhà cung cấp xanh. *VNU J. Sci. Econ. Bus.* **2017**, *33*, 43–54.

© 2019 by the authors. Licensee MDPI, Basel, Switzerland. This article is an open access article distributed under the terms and conditions of the Creative Commons Attribution (CC BY) license (http://creativecommons.org/licenses/by/4.0/).

 *symmetry*

Article

A New DEA Model for Evaluation of Supply Chains: A Case of Selection and Evaluation of Environmental Efficiency of Suppliers

Evelin Krmac * and Boban Djordjević

Faculty of Maritime Studies and Transport, University of Ljubljana, 6320 Portoroz, Slovenia;
bbn.djordjevic@gmail.com
* Correspondence: evelin.krmac@fpp.uni-lj.si

Received: 31 March 2019; Accepted: 15 April 2019; Published: 18 April 2019

Abstract: Supply Chain Management (SCM) represents an example of a complex multi-stage system. The SCM involves and connects different activities, from customer's orders to received services, all with the aim of satisfying customers. The evaluation of a particular SCM is a complex problem because of the internally linked hierarchical activities and multiple entities. In this paper, the introduction of a non-radial DEA (Data Envelopment Analysis) model for the evaluation of different components of SCM, primarily in terms of sustainability, is the main contribution. However, in order to confirm the novelty and benefits of this new model in the field of SCM, a literature review of past applications of DEA-based models and methods are also presented. The non-radial DEA model was applied for the selection and evaluation of the environmental efficiency of suppliers considering undesirable inputs and outputs resulting in a better ranking of suppliers. Via perturbation of the data used, behavior, as well as the benefits and weaknesses of the introduced model are presented through sensitivity analysis.

Keywords: Supply Chain Management; Data Envelopment Analysis; Non-radial DEA model; Supplier; Efficiency Evaluation; Environment

---

1. Introduction

A prerequisite for providing products and services of high quality at the lowest cost is the effective management of the supply chain (SCM) [1].

The efficiency of the supply chain (SC) is significantly dependent on the coordination both across firms and within firms because each part can influence the SC. When any of the parts lack co-ordination, dramatic effects on the SC can result [2]. Therefore, measuring and monitoring the efficiency of the SC represents one of the most important steps towards its improvement. The DEA method is one of the most often used multi-criteria decision making (MCDM) methods for SC efficiency evaluation. This is why the model used in this paper is based on it.

The DEA method originated from the work of Charnes et al. [3], originally applied in the evaluation of relative efficiency of similar units when there are multiple inputs and outputs. It is one of the most effective approaches in measuring the efficiency of a SC and its components [4]. After the first application of the DEA in the field of SCM, in the literature, various approaches were presented [4]. The main reason for modifications of the DEA lies in the fact that the traditional DEA models cannot be directly employed in the SC evaluation because they consider only inputs and outputs. However, they must be modified in order to be able to include the intermediate products. Moreover, in real applications within the production process, undesirable (bad) outputs can be produced. A good example of such results, pointed out by Mahdiloo et al. [5], are suppliers' carbon emissions. Mahdiloo et al. [5] highlighted that different DEA approaches that consider the undesirable outputs, primarily for the evaluation of the green or sustainable SCM, have been developed and are presented in the literature. However, a DEA model that, besides undesirable outputs, can evaluate efficiency using some undesirable inputs is missing.

Because of the importance and complexity of the SC, as well as the possibility to include undesirable inputs and outputs for the evaluation of different parts of the SC, the aim of this paper is to contribute to the existing literature through the introduction of a non-radial DEA model for efficiency evaluation either of different components of SCM or the whole of SCM. Consequently, the main contribution of the paper is reflected in the introduction of a new DEA model for evaluation of SCM that considers also undesirable outputs. The benefit of the introduced model is related to the possibility of consideration of undesirable inputs as well as outputs simultaneously. With such a model, the evaluation of SCM or their components in terms of sustainability would be possible. In order to check and confirm the novelty of the proposed DEA model, a comprehensive literature review of past applications of the DEA method in SCM and particular areas of SCM is presented. The applicability of the introduced model is presented through the selection and evaluation of the environmental efficiency of suppliers using data taken from Mahdiloo et al. [5]. No matter whether the data were reused from an existing study, the aim was to provide an overview of the behavior of the proposed model and compare it with other models using the same data as Mahdiloo et al. [5]. Because the data were reused from an existing study, testing of data before applying the DEA model was not performed.

With the aim to present the novelty of this paper and to better describe the process of the introduction of our model in the SC efficiency evaluation, a sequence of steps, represented in Figure 1, were performed: (1) systematic literature research; (2) selection of the paper with the most appropriate data set used; (3) the description of the non-radial DEA model itself; (4) the application of the non-radial DEA model using the selected set of data and comparison of the results; and (5) sensitivity analysis of the proposed model.

The following section describes the methodology of the literature review. Section 3 presents the results of the literature review together with the classification of papers according to particular evaluated areas of SCM. Section 4 presents the basics of the non-radial DEA model and its introduction for the evaluation of particular areas of SCM. Within Section 5, the results of the proposed model and sensitivity analysis of the model are presented. Section 6 discusses the methodology and obtained results. Finally, in Section 7, we offer our conclusions, summarizing the literature review, presenting the model, and suggesting future research.

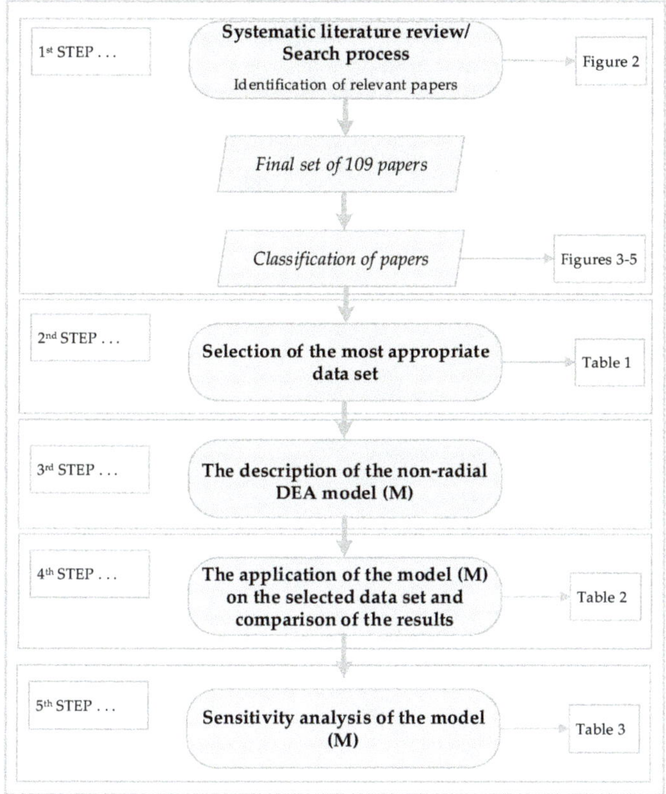

Figure 1. The research process.

2. Previous Research

With the aim of confirmation of the novelty of the introduced non-radial DEA model, an overview of papers related to the application of the DEA method to SCM was performed. During the review, the only paper that reviews the application of DEA in SCM found was that of Soheilirad et al. [4], but it only presents a review of literature published until 2016.

Methodology of Literature Review

The methodology of the literature review has been taken from papers written by Krmac and Djorjević [6] and Djordjević et al. [7]. Accordingly, the literature review was conducted through fundamental guidelines of the systematic literature review. Since ScienceDirect and Scopus represent the two most important (and largest) scientific databases [6], a review of papers published in peer-reviewed journals without limitation on the time period of publishing was performed. However, in order to avoid bias regarding the top journals or the most cited ones, the literature review based on meta-analysis was not conducted.

The review of open-access studies focused on titles, abstracts, and keywords for English-written full-text free-available scientific journal papers, and was performed in December 2018.

The search of databases using keywords such as "Data Envelopment Analysis AND Supply Chain", "Data Envelopment Analysis AND Supply Chain Management", and the variations of both search strings where the abbreviations DEA, SC, and SCM were used, was performed in both databases. Papers that presented an application of the DEA in the SCM field were taken over by first reading the title, keywords, and abstracts. After the initial reading, 222 papers were extracted. Further, after the full-text reading of extracted papers, 109 were selected and considered relevant. Based on the review, the selected papers were classified into main areas such as the evaluation of the SC, the evaluation and the selection of suppliers, as well as the consideration of the SC and the evaluation of suppliers in terms of sustainability. Also, for each of these categories, the application of DEA in combination with other methods was presented. However, those papers that used the DEA method for analyzing more than one aspect or where the DEA was combined with another method or methods were classified as non-categorized. The overall search process is shown in Figure 2.

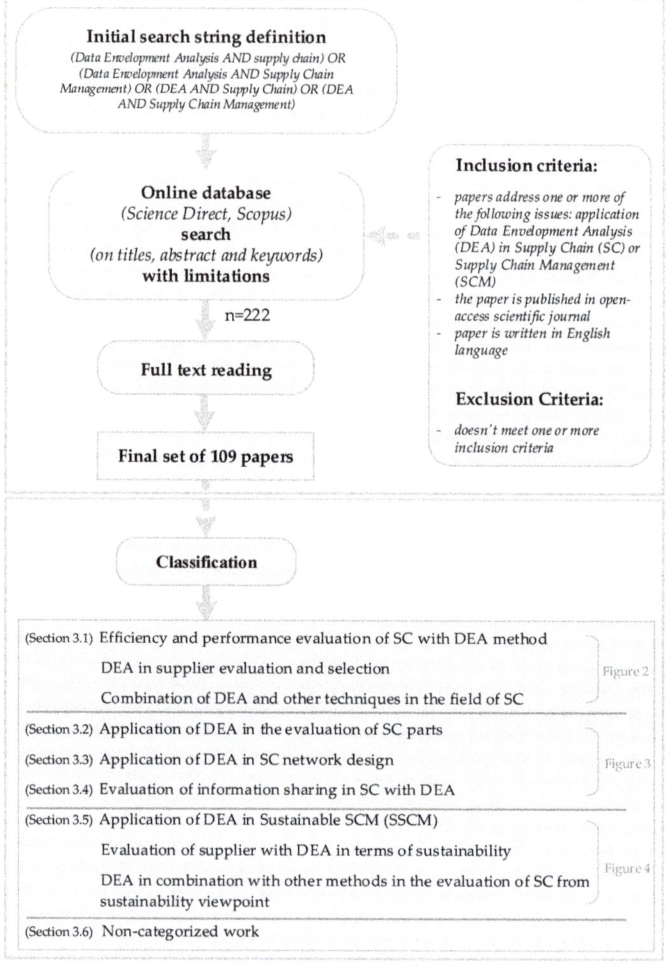

Figure 2. The overall search process.

3. Results of the Literature Review and Classification

In the literature, numerous papers that applied the DEA method in the area of SCM were found. Within this paper, they were classified according to the purpose of the application of DEA and the combinations of DEA with other techniques or approaches. Papers that did not fall into any of the defined categories were classified as "non-categorized works". The overall search process is shown in Figure 2.

3.1. Efficiency and Performance Evaluation of SC with the DEA Method

For the evaluation of SC performances, the DEA method has been extensively employed. Previously, with the DEA method, only the initial inputs and final outputs to measure the efficiency of SCs were used, while intermediate products were ignored. However, the application of the DEA method for measuring the efficiency of the entire SC and all its components at all levels was recognized in ref. [8]. The application of DEA for performance and efficiency evaluation of SCs is summarized in Figure 3.

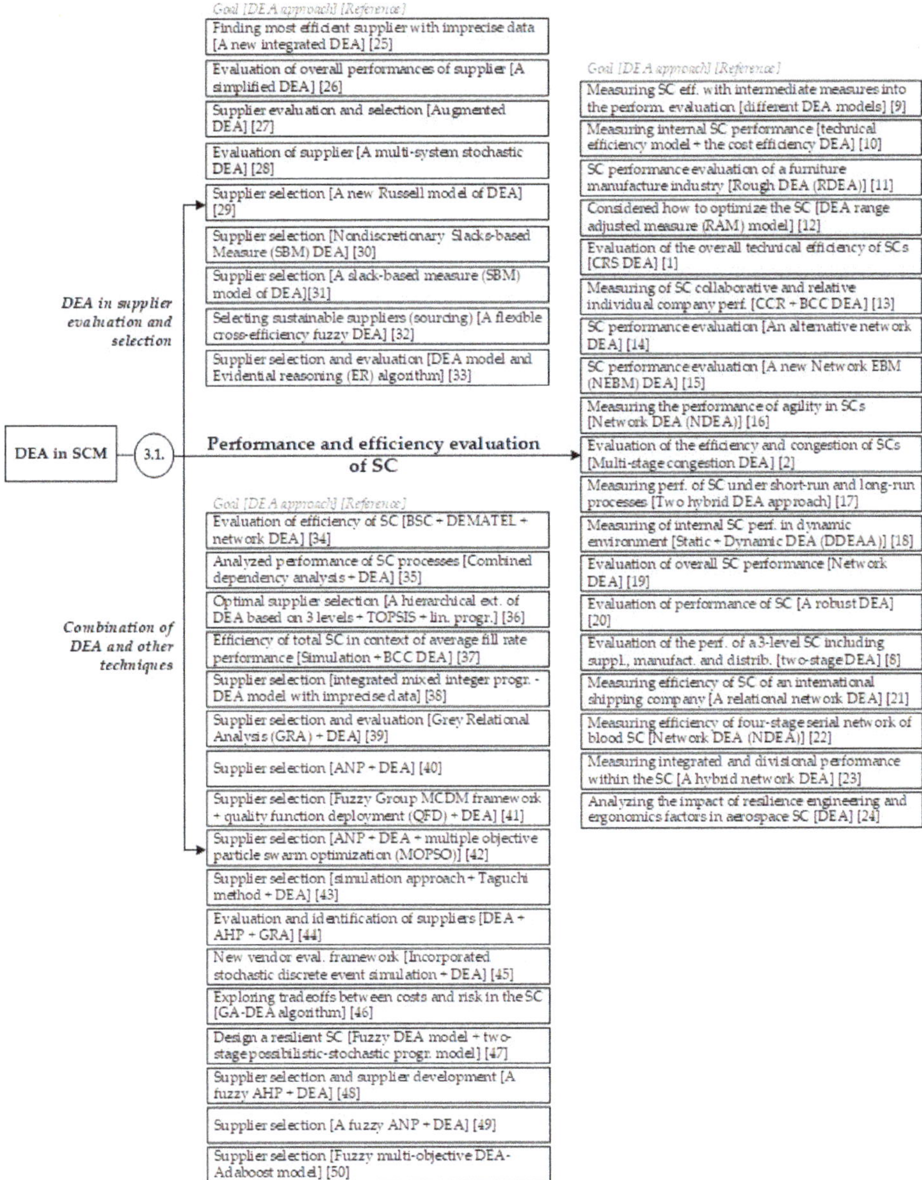

Figure 3. DEA used in performance and efficiency evaluation of SCs [1,2,8–50].

The description of each paper was given true to the short description of the goal followed by a DEA approach in the first square brackets and the reference in the second one. Because the state of SC functioning can be to a large degree linked with the selection of the best suppliers, the papers considering the methods, models and approaches of supplier evaluation and selection were separately classified under the label "DEA in supplier evaluation and selection". Over the years, several techniques such as Analytic Hierarchical Process (AHP), Analytic Network Process (ANP), Linear Programming (LP), Mathematical Programming, Multi-objective Programming, and DEA have been developed to solve the problem efficiently [25]. The papers that considered the evaluation of suppliers using the DEA technique can be also seen in Figure 3.

In order to improve some characteristics of the DEA in the evaluation of SCs and their parts, DEA was also used in combination with other methods. Shafiee et al. [34] created a network DEA for the evaluation of efficiency with a balanced scorecard (BSC) approach where the DEMATEL approach was employed to obtain a network structure of four perspectives of BSC. Many other combinations for the purpose of evaluation of the SC and its different parts were also used in the literature, and they are summarized in Figure 3 as well under the label "Combination of DEA and other techniques".

3.2. Application of DEA in the Evaluation of SC Parts

Papers that applied the DEA method in order to evaluate parts or components of SCs are presented in Figure 4 under the label "DEA evaluation of SC components".

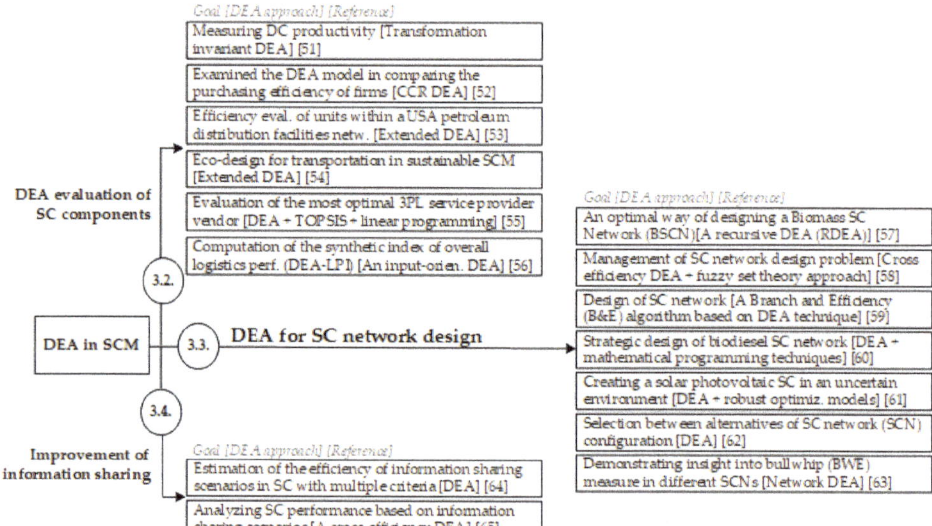

Figure 4. DEA in the evaluation of SC parts, in SC network design, and in the improvement of information sharing among SC stakeholders [51–65].

3.3. Application of DEA in SC Network Design

One of the areas of the SC where the DEA method was applied is the problem of SC network design or selection of optimal network solutions. The group of papers that considers the application of DEA in the SC network field is presented in Figure 4 under the label "DEA for SC network design".

3.4. Evaluation of Information Sharing in SC with DEA

After the previous decade with developed information technology, every firm can now improve its SC strategies with the aim to improve their performance of SCM through well-organized information sharing. In the literature, there are studies like papers written by Chen et al. [64] and Yu et al. [65], that considered the effects of information sharing on the efficiency of SCs (see Figure 4).

3.5. Application of DEA in Sustainable SCM

Recently, in the literature, a lot of papers can be found regarding sustainable SCM (SSCM). Papers that considered SSCM with DEA are summarized in Figure 5.

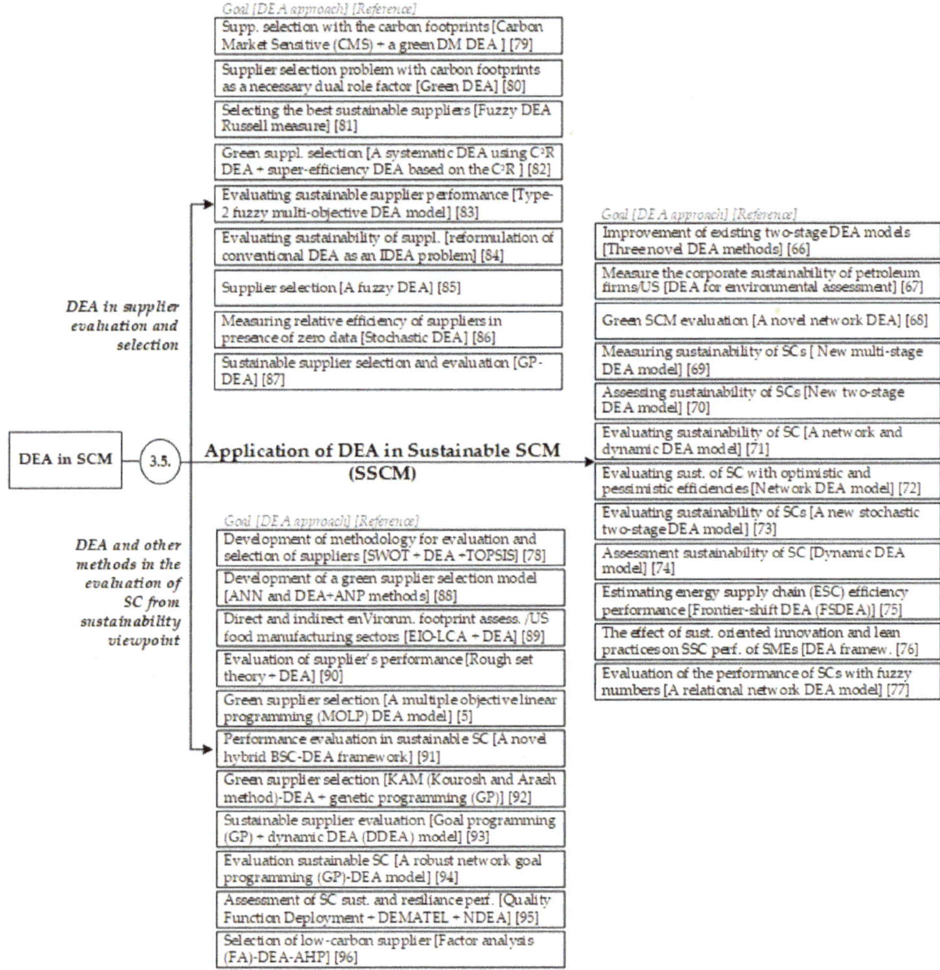

Figure 5. Application of DEA and combinations of DEA and other methods in SSCM [5,66–96].

SSCM is focused on the improvement of economic, social, and environmental performances at the same time. Therefore, sustainable SCM evaluation has become a significant task for each organization. As one of the methods, DEA was recognized as suitable for the evaluation of sustainable SCM [67].

According to ref. [79], in order to develop effective SC, supplier evaluation and selection plays an important role. Within the developed approaches for supplier selection, the main goal was the reduction of SC costs, while environmental criteria were neglected. Nevertheless, the developed environmental criteria should include the comprehensive carbon footprint in the supplier's selection approaches i.e., consideration of the environmental impact of suppliers. Therefore, the authors of ref. [80] proposed an integrated buyer initiated approach for supplier selection that considers two objectives: cost cutting and environmental efficiency. Other DEA approaches in evaluating the sustainability of suppliers are presented in Figure 5 under the label "DEA in supplier evaluation and selection".

With the aim of improving previously used methods, Chen [78] introduced a structured methodology for supplier selection and evaluation based on supply chain integration architecture. Besides this paper, other papers that combine DEA with additional techniques for analysis of different areas of SCs in terms of sustainability can be found in the literature. They are presented in Figure 5 and group under the label "DEA and other methods in the evaluation of SC from a sustainability viewpoint".

3.6. Non-Categorized Works

In order to provide a hybrid method for supplier selection, Sevkli et al. [97] used the DEAHP method—the DEA method embedded into AHP methodology—because DEA still lacks a real application case, in which its implications can be evaluated.

Risk evaluation models that also represent an example of tools for supplier selection between the chance-constrained programming (CCP), multi-objective programming (MOP), and DEA were considered by Wu and Olson [98]. Azadeh and Alem [99] presented three types of models for SC risk and supplier selection under certainty, uncertainty and probabilistic conditions: DEA, Fuzzy DEA, and Chance Constraint DEA. From these studies, it can be seen that DEA has been employed in supplier selection. Further, a new approach also based on DEA, called DEA VaR (value-at-risk), was developed by Wu and Olson [100] for the selection of vendors in enterprise risk management. Visani et al. in ref. [101] used a DEA approach in approximating supplier total cost of ownership. Boudaghi and Saen [102] presented a novel model of data envelopment analysis–discriminant analysis (DEA–DA) to predict group membership of suppliers in a sustainable SC context.

Based on the developed fuzzy network DEA model, Pournader et al. [103] evaluated risk resilience of the overall SC and their individual tiers. The DEA method was also used by Azadeh et al. [104] for analyzing the impact of macro-ergonomic factors in healthcare SC. Further, Amalnick and Saffar [24] conducted an analysis of the impacts of resilience engineering and ergonomic factors in aerospace SC using DEA.

Saranga and Moser [105] presented a comprehensive performance measurement framework using the classical and two-stage Value Chain DEA models for estimating the performance of purchasing and supply management. For measuring the performance of suppliers and manufacturers in SC operations, Amirteimoori and Khoshandam [106] developed a DEA model within their study. A model for performance assessment of an outsourcer's processes in an SC comprised of several internal and external entities was provided by Pournader et al. [107] based on the Slacks-based Measure incorporated into a Hybrid Network DEA. Since a transportation system can be disrupted within the SC, Azadeh et al. [108] designed and simulated an echelon SC in which the preferred scenario was identified using fuzzy DEA.

The DEA method was also used in comparing different aspects. For example, Bayraktar et al. [109] compared SCM and information system practices of small and medium-sized enterprises operating in food products and beverages in Turkey and Bulgaria. Also, by combining the DEA method with other methods, the analysis of SC was conducted. Jalalvand et al. [110] combined DEA and PROMETHEE II as tools to compare SC at the process level, business stage level, and SC level.

4. The Proposal of a Non-Radial DEA Model in SC

Within this part of the paper, the non-radial DEA model M is introduced. The DEA method is a linear programming-based method popularized by Charnes et al. [3] for efficiency evaluation. In the literature, the DEA method has been applied for relative efficiency evaluation of the comparable set of entities, called decision-making units (DMUs) with multiple inputs and outputs, i.e., DMUs that are able to transform multiple inputs to multiple outputs. Based on the obtained results by the application of the DEA method, DMUs are classified as efficient or non-efficient. One of the advantages of the DEA method is that it does not require any prior assumptions on underlying functional relationships between inputs and outputs [7]. The mathematical formulation of the classical input oriented Charnes, Cooper, and Rhodes (CCR) DEA model [111] can be written as:

$$min\theta; s.t \ X\lambda \le \theta x_i, \ Y\lambda \ge y_i, \ \lambda \ge 0,$$

under the assumption that there are n DMUs, m outputs and s inputs, where X and Y represent sets of vectors of inputs and outputs, respectively. θ represents an indicator of technical efficiency where $\theta \in [0,1]$ and indicates how much evaluated entity could potentially reduce its input vector while holding the output constant.

A Brief Description of the Non-Radial DEA Model

As can be seen from the CCR DEA model, as one of the classical DEA models, it is strongly related to, and can be presented through, production theory, in which raw materials and resources are treated as inputs, while products are treated as outputs in the production process [112]. However, in some real applications, the production process may also use undesirable inputs and generate undesirable outputs. However, a method for treating both undesirable inputs and outputs simultaneously in non-radial DEA models has been introduced by Djordjević et al. [7]. One non-radial DEA model was presented by Wu et al. [113] in the field of energy and environmental efficiency. In addition to the advantages of the non-radial DEA model already described, this model was extended in ref. [7] for the proposal of the evaluation of safety at railway-level crossings. Considering the ability of the DEA method in efficiency evaluation and the advantages of the non-radial DEA model, the proposed model M has been chosen for application and evaluation in SC.

The same model could be used in the SC for the evaluation of its different parts/components "as the input" such as the selection of a supplier. Consequently, inputs can be considered as desirable. However, each part of the SC can also represent/include primarily undesirable factors. Therefore, in the paper of ref. [7], in order to allow for the simultaneous reduction of desirable inputs and to obtain an accurate idea regarding the results of efficiency, the non-radial DEA model by authors of ref. [113] was improved in the following way:

$$min \ W_n \frac{1}{N} \sum_{l=1}^{L} \theta_n + W_l \frac{1}{L} \sum_{l=1}^{L} \theta_l + W_j \frac{1}{J} \sum_{j=1}^{J} \theta_j$$

s. t.

$$\sum_{k=1}^{K} \lambda_k x_{nk} \le \theta_n x_{n0}, \ n = 1 \dots N \tag{1}$$

$$\sum_{k=1}^{K} \lambda_k e_{lk} \le \theta_l e_{l0}, \quad l = 1 \dots L \tag{2}$$

$$\sum_{k=1}^{K} \lambda_k y_{mk} \ge y_{m0}, \quad n = 1 \dots M \tag{3}$$

$$\sum_{k=1}^{K} \lambda_k u_{jk} = \theta_j u_{j0}, \ j = 1 \dots J \tag{4}$$

$$\lambda_k \ge 0, \ k = 1 \dots K \tag{5}$$

$$\tag{M}$$

Under the assumption that there are K DMUs, each of them has n desirable and l undesirable inputs in order to produce m desirable and j undesirable outputs, denoted respectively as $x = (x_{1K}, \ldots, x_{nK})$, $e = (e_{1l}, \ldots, x_{LK})$, $y = (y_{mK}, \ldots, y_{mK})$, $u = (u_{1K}, \ldots, u_{JK})$.

This non-radial model M could be projected for efficiency evaluation either of SC components or whole SC. The non-radial model M proportionally decreases the number of undesirable inputs and undesirable outputs as much as possible for the given level of desirable inputs and desirable outputs. The optimal values of unified efficiency are in the interval between 0 and 1. An entity with a higher value of efficiency has better efficiency compared to others. However, if an entity has an objective function equal to 1 it means that the entity is the best, located at the frontier, and could not reduce undesirable input and undesirable output. Such a non-radial model M could therefore be suitable for efficiency evaluation of SC components or a whole SC in terms of sustainability or dimensions of sustainability because it has a relatively strong discriminating power and the capability to expand desirable outputs, simultaneously reducing undesirable outputs. Additionally, unified efficiency can be calculated through decision-maker-specified weights (user-specified weights) assigned to each of these three efficiency scores and depends on the preferences between undesirable inputs utilization and undesirable outputs. However, as with any model, there are some risks related with the application of the non-radial model M. First, because the unified efficiency depends on the selected user-specified weights, the results can be subjective. Consequently, for example, the greater degree of weight for an undesirable output implies the reduction of that output. A second risk of the model M is related to the resultant inaccuracy if not all necessary variables (inputs and outputs) are included. The results of unified efficiency can be inaccurate if the set of data is not comprehensive. The improved non-radial DEA model M in this paper was applied through the evaluation of the environmental efficiency of suppliers. The detailed description of these and other characteristics of the model M can be found in Djordjević et al. [7].

5. Illustration of Application of the Non-Radial Model M—Numerical Example

In this part of the paper, the non-radial DEA model M was applied to the selection and evaluation of the environmental efficiency of suppliers with the aim to present the applicability of the model M within the SC field. Because the new and fresh data was missing, non-radial model M was applied on data used in ref. [5] using the Excel Solver tool. However, the main advantages and purpose of the reuse of data is the comparison of obtained results by different models on the same data. Because the data were reused primarily in order to introduce and present the behavior of the model M, the testing of these data before applying the model M was not performed. The data, inputs and outputs that were used in the paper of Mahdiloo et al. [5] are presented in Table 1. Within the study written by Mahdiloo et al. [5], the number of employees (N1) and energy consumption (L1) were considered as inputs, sales, and return on assets (ROA); and environmental R&D investment were considered as desirable outputs; and carbon dioxide (CO_2) emission as undesirable output. However, for the application of model M, energy consumption was used as an undesirable input.

The basic equation for the evaluation of environmental efficiency of suppliers (EES) of the model M can be written as:

$$\text{EES} = \frac{\text{Desirable Outputs}}{\text{Desirable inputs and Undesirable inputs and outputs}}, \tag{6}$$

where the goal function of the model M can be written as:

$$\min \frac{\sum_{n=1}^{M} \lambda_k y_{mk}}{W_n \frac{1}{N} \sum_{l=1}^{L} \theta_n + W_l \frac{1}{L} \sum_{l=1}^{L} \theta_l + W_j \frac{1}{J} \sum_{j=1}^{J} \theta_j} \tag{7}$$

or, more simply as

$$\min \frac{M1 + M2 + M3}{\frac{1}{3}N_1 + \frac{1}{3}L_1 + \frac{1}{3}J_1}. \tag{8}$$

Table 1. Dataset taken from Mahdiloo et al. [5] for application of model M.

Suppliers	Number of Employees (N1)	Energy Consumption (kWh/year) (L1)	Sales (1000 Korean Won) (M1)	ROA (M2)	Environmental R&D Investment (100,000 Korean Won) (M3)	CO$_2$ (kg) (J1)
1	1112	1267	119,477	0.04046	67	43,562
2	118	968	125,762	0.04499	65	45,000
3	458	1001	58,770	0.02221	57	42,400
4	416	1393	62,989	0.02920	62	43,734
5	413	1586	67,088	0.03269	50	44,890
6	430	1802	72,318	0.03116	36	42,913
7	426	1998	74,626	0.02184	47	39,438
8	452	1824	74,476	0.0348	44	40,078
9	503	1479	79,710	0.03976	47	39,500
10	498	1623	79,384	0.03723	89	45,023
11	192	1322	73,124	0.01269	256	41,324
12	171	831	62,529	0.00385	423	45,000
13	163	913	65,424	0.02776	508	42,400
14	161	893	71,027	0.04847	536	43,734
15	161	903	74,093	0.0514	570	44,890
16	162	778	72,830	0.04356	472	42,913
17	159	710	71,940	0.03932	426	39,438
18	157	695	82,203	0.02599	386	40,078
19	151	637	55,681	0.00001	376	39,500
20	151	781	64,839	0.02742	369	38,570

Comparison of the results of the application of three models, namely Model 2, Model 4, and Model 5, performed by ref. [5], with the results of the use of the introduced non-radial DEA model M on the same data is presented in Table 2.

Table 2. Results of the efficiency from models 2, 4, and 5 presented in Mahdiloo et al. [5] and model M.

Suppliers	Technical Efficiency (Model 2)	Environmental Efficiency (Model 4)	Eco-Efficiency (Model 5)	Model M
1	0.73	0.99	0.99	0.60
2	1.00	1.00	1.00	1.00
3	0.47	0.52	0.52	0.37
4	0.40	0.63	0.63	0.41
5	0.39	0.68	0.68	0.44
6	0.34	0.69	0.69	0.42
7	0.29	0.68	0.68	0.38
8	0.37	0.82	0.82	0.47
9	0.50	0.93	0.93	0.56
10	0.44	0.78	0.78	0.48
11	0.55	0.80	0.80	0.62
12	0.83	0.79	0.83	0.80
13	0.88	0.94	0.94	0.90
14	0.96	0.98	0.98	0.96
15	1.00	1.00	1.00	1.00
16	1.00	0.95	1.00	0.97
17	1.00	0.99	1.00	0.98
18	1.00	1.00	1.00	1.00
19	0.96	0.80	0.96	0.85
20	0.84	0.89	0.89	0.89

Unified efficiency of the model M was obtained based on the same weights for desirable input, undesirable input and output. The same weights, i.e., 1/3, were selected for both desirable and undesirable inputs, as well as undesirable output with the aim to reduce the subjective bias. Using the Excel Solver tool, the results of the non-radial model M were obtained. From Table 2, it can be seen that for Model 2, the suppliers 2, 15, 16, 17, and 18 were rated as the most efficient. Next were 13, 14, 19, and 20. Regarding Model 4, the most efficient suppliers were 2, 15, and 18, while a greater number of suppliers have efficiency near to 1 compared to Model 2. The most efficient suppliers with Model 5 were the same as Model 2. Comparing the results obtained by different models, it can be seen that regarding the environmental efficiency of suppliers, Model M gave similar results to those of Model 4. Hence, suppliers that were rated as efficient within Model 4 were efficient also within Model M. The inefficiency of suppliers obtained by Model 4 was also the same with Model M. The main difference between these two models is the efficiency score. Efficiency scores obtained by Model M for each supplier are lower than efficiency scores obtained by Model 4. The main reason for this lies in the relatively strong discriminating power of model M.

Validation of Non-Radial DEA Model M

The sensitivity analysis of the non-radial DEA model M was performed with the aim to check its validity. It was conducted on data from Mahdiloo et al. [5] as shown in Table 2 using the Excel Solver tool. The main aim of the validation of model M and, therefore, of the sensitivity analysis, was the consideration of the model's behavior. So, the data used has no influence on the sensitivity analysis. The sensitivity analysis was conducted in the same way as in Djordjević et al. [7]. Realization of the sensitivity analysis of non-radial model M is presented in Table 3. The process of sensitivity analysis was conducted based on the certain percentages of perturbation of used data, i.e., 2%, 5% and 10% until the status of at least one supplier was changed [7]. Sensitivity analysis was presented through three cases. In Case 1, both desirable and undesirable inputs, as well as undesirable output, were improved for suppliers with efficiency 1 and worsened for suppliers with efficiency under 1. Within Case 2, perturbation of the data was focused on the increment of undesirable inputs and output for suppliers with efficiency 1 and reduction for those with lower efficiency, while desirable input was fixed. Then, the behavior of model M was checked through Case 3 where desirable outputs were decreased and desirable input increased for suppliers with efficiency 1, and vice versa for inefficient suppliers. For each case, after the data changing model M was solved using Excel Solver. Results of three cases of the sensitivity analysis are presented in Table 3.

Table 3. Results of Sensitivity Analysis of non-radial DEA model M.

Suppliers	Case 1 (C1)			Case 2 (C2)			Case 3 (C3)			Remarks [1]		
	2%	5%	10%	2%	5%	10%	2%	5%	10%	2%	5%	10%
1	0.63	0.67	0.74	0.63	0.66	0.73	0.63	0.67	0.75	C1=C2=C3	C1>C2<C3	C3>C1<C2
2	1.00	1.00	1.00	1.00	1.00	1.00	1.00	1.00	1.00	C1=C2=C3	C1=C2=C3	C1=C2=C3
3	0.39	0.41	0.45	0.39	0.40	0.44	0.39	0.41	0.46	C1=C2=C3	C1>C2<C3	C3>C1>C2
4	0.43	0.45	0.47	0.43	0.44	0.46	0.43	0.45	0.48	C1=C2=C3	C1>C2<C3	C3>C1>C2
5	0.45	0.47	0.49	0.45	0.46	0.48	0.45	0.47	0.50	C1=C2=C3	C1>C2<C3	C3>C1>C2
6	0.43	0.45	0.48	0.43	0.45	0.47	0.43	0.45	0.49	C1=C2=C3	C1=C2=C3	C3>C1>C2
7	0.40	0.42	0.46	0.39	0.41	0.45	0.40	0.43	0.48	C1>C2<C3	C3>C1>C2	C3>C1>C2
8	0.49	0.51	0.53	0.49	0.50	0.53	0.49	0.51	0.54	C1=C2=C3	C1>C2<C3	C1<C3>C2
9	0.58	0.60	0.63	0.58	0.60	0.62	0.58	0.60	0.63	C1=C2=C3	C1=C2=C3	C1>C2<C3
10	0.50	0.52	0.55	0.50	0.51	0.54	0.50	0.52	0.55	C1=C2=C3	C1>C2<C3	C1>C2<C3
11	0.65	0.67	0.69	0.64	0.66	0.68	0.66	0.67	0.71	C3>C1>C2	C1>C2<C3	C3>C1>C2
12	0.83	0.83	0.83	0.82	0.83	0.83	0.83	0.83	0.83	C1>C2<C3	C1=C2=C3	C1=C2=C3
13	0.94	0.95	0.95	0.93	0.95	0.95	0.95	0.95	0.95	C3>C1<C2	C1=C2=C3	C1=C2=C3
14	1.00	1.00	1.00	0.99	1.00	1.00	1.00	1.00	1.00	C1>C2<C3	C1=C2=C3	C1=C2=C3
15	1.00	0.95	0.86	1.00	0.98	0.92	1.00	0.92	0.81	C1=C2=C3	C2>C1>C3	C2>C1>C3
16	1.00	1.00	1.00	0.99	1.00	1.00	1.00	1.00	1.00	C1=C2=C3	C1=C2=C3	C1=C2=C3
17	1.00	1.00	1.00	1.00	1.00	1.00	1.00	1.00	1.00	C1=C2=C3	C1=C2=C3	C1=C2=C3
18	1.00	0.97	0.89	1.00	0.99	0.94	1.00	0.94	0.84	C1=C2=C3	C2>C1>C3	C2>C1>C3
19	0.88	0.88	0.88	0.87	0.88	0.88	0.88	0.88	0.88	C1>C2<C3	C1=C2=C3	C1=C2=C3
20	0.88	0.88	0.89	0.87	0.88	0.89	0.88	0.88	0.89	C1>C2<C3	C1=C2=C3	C1=C2=C3

[1] Remarks: Show the relationships between the results of the efficiency calculated for each supplier and for each data perturbation through Cases 1, 2, and 3.

For most suppliers with efficiency under 1, the score of efficiency was improved (see Table 3). However, for some inefficient suppliers, the efficiency score was not significantly changed, such as, for example, suppliers 12 and 13.

With increments in data for 5% and 10% in Case 1, the suppliers 15 and 18 became efficient, while a score of efficiency was changed gradually for inefficient suppliers. In respect to Case 2, the transformation of results can also be noticed with only a 5% decrement of undesirable input and output for inefficient suppliers, which became efficient, such as suppliers 14 and 16, while the results for some efficient suppliers were transformed to inefficient. In comparison with Case 1, the efficiency and inefficiency of a large number of suppliers were changed. Such information tells us that the results of the non-radial DEA model M probably depend on the undesirable input and output.

For Case 3, results have shown that inefficient suppliers became efficient with 5% decrement/increment. Through comparison with Case 1, it can be concluded that some efficient suppliers became inefficient with a change of desirable outputs and inputs. These results indicated that the suppliers are more sensitive to the data of undesirable output and input (see Case 2). With the aim of further changes from inefficient to efficient or vice versa, it is necessary to apply higher percentages of data perturbations. Meanwhile, it should be pointed out that the efficiency for suppliers such as 2 and 17 are unchanged besides the percentages of data perturbations.

The comparison of the results of three cases is given in Table 3, in the column remarks. It was conducted based on the percentage of data perturbation for each supplier. The results show that the efficiency of a particular supplier was mainly unchanged, i.e., scores of efficiency were the same—in all cases under the 2% of data perturbation. In the case of 5% of data perturbation, the efficiency in Case 2 was lower than those in Case 3 and 1, while it was mainly the same for Cases 1 and 3. Finally, for 10% of data perturbation, the efficiency was mainly lower in Case 2 compared to Cases 3 and 1.

6. Discussion

As can be seen from the literature review, many studies have applied the DEA method for efficiency evaluation in SCM. The main contribution of this research is related to the introduction of the non-radial DEA model for efficiency evaluation of different components of SC. Applicability of the introduced non-radial model M was presented through the evaluation and selection of suppliers using the data from ref. [5]. The proposed tool, i.e., the non-radial model M, is relevant for the selection and evaluation of suppliers. However, it can also be a good tool for considering best practices of all components of SC in terms of sustainability because the model is able to measure efficiency while considering undesirable inputs, as well as undesirable outputs, which appeared in real applications.

Through comparison of the results obtained by the non-radial DEA model M and models developed by Mahdiloo et al. [5] (see Table 2), it can be seen that model M has similar results, i.e., the closest results in terms of efficiency score to Model 2. However, regarding the efficiency and inefficiency of suppliers, results of Model M are the most similar to those of Model 4. Further, the results for each supplier of Model 2 were lower in comparison with other models. The main reason for this is the higher discrimination power of model M.

Based on the results of model M, obtained using the Excel Solver tool, it can be said that this model is more appropriate for efficiency evaluation. First of all, with the model M desirable inputs, undesirable inputs and undesirable outputs are simultaneously reduced. Hence, model M can minimize desirable inputs as well. However, in the case of the application of the non-radial model M without the efficiency score θ_n and weight W_n in terms of desirable inputs, an unreal picture regarding the efficiency can be presented. With model M, the consideration of environmental evaluation and selection of supplier and other components of SC regarding sustainability is more precise, providing better relative efficiency. Further, through the selection of the set of preference weights, the degree of desirability of the adjustment of the input and output levels can be achieved. Therefore, the selection of the weight, for example, for undesirable output, will affect the reduction of that output. Consequently, based on their preferences and the goal of evaluation, decision makers should select weights carefully because the selection of weights can influence the results of model M. In this paper, all weights were selected to be 1/3.

In our case, the environmental efficiency of suppliers was evaluated based on the data taken from Mahdiloo et al. [5]. The suppliers with the greatest relative efficiency were 2, 15 and 18 (see Table 2). Consequently, based on the model M and the results, it can be said that these suppliers for the given level of Sales, ROA and Environmental R&D investment have minimum Energy consumption and CO_2, as well as Number of employees in comparison with other suppliers.

Through the consideration of results of the sensitivity analysis, it can be highlighted that model M is valid. Nevertheless, model M is sensitive to data for a smaller transformation of data that causes reduced stability of the model. The reason lies in the fact that model M has the effect of greater discrimination. Then, the score of efficiency for some suppliers was unchanged, which can be linked to the fact that model M evaluates suppliers, i.e., minimizes inputs for a given level of outputs. Consequently, besides the transformation of data, the score of efficiency for particular suppliers was unchanged as these suppliers have a lower level of desirable outputs in comparison with other suppliers. However, it can be concluded that for a higher percentage of data transformations, the model M can present a picture with higher changes of a score of efficiency. In the case of inaccurate data, the application of model M can present an unrealistic picture regarding the best suppliers. The comparison, given in the column remarks of Table 3, confirms these facts. The comparison was conducted based on the percentage of data perturbation for each supplier. The results show that the efficiency of a particular supplier was mainly unchanged, i.e., scores of efficiency were the same—in all cases under the 2% of data perturbation. In the case of a higher percentage of data perturbation, the efficiency in Case 2 was lower than those in Case 3 and Case 1. The obtained results, therefore, confirm that the behavior and results of the model M can be affected by the accuracy of the data and the selection of inputs and outputs.

Nevertheless, the application of model M with an accurate date can show that the model could be a good tool for supplier evaluation and other parts of SCs in terms of sustainability. Further details of the weaknesses of the model M which can appear during supplier evaluation and selection can be found in Djordjević et al. [7].

7. Conclusions

Through the literature review, various DEA models for evaluation within SCM have been shown. However, just a few of them considered undesirable inputs, which are an inseparable part of real production processes and applications, while consideration of undesirable outputs within DEA models is still missing. Therefore, in order to improve the existing literature, the non-radial DEA model that simultaneously considers undesirable inputs and outputs was proposed. Consequently, within the paper, the introduction and presentation of a new DEA model for the evaluation of different components of SCs, which is the main contribution of this paper, was presented. Applicability of the proposed model was presented through the evaluation of the environmental efficiency of suppliers. In order to confirm the novelty of the introduced non-radial model M within this paper, a comprehensive literature review of the application of the DEA method in SCM was performed. Numerous papers have applied quite a variety of DEA approaches in the field of SCM and its components. These papers were categorized according to the purpose of application of the DEA method. Application of the DEA method in combination with other methods in the field of SCM was presented as a separate category. Finally, papers that used DEA as a part of a developed framework or method, as well as for analyzing two or more aspects of SC, were grouped as non-categorized works. As can be seen, a lot of papers were presented in the evaluation of SC performance and supplier selection in terms of sustainability. Different modifications of the DEA method in SCM are therefore available in the literature. Besides modifications of the DEA model, there are also papers that only considered some inputs and outputs as undesirable factors.

However, it can be concluded that there are not many papers that have considered undesirable factors in/within SCs that can appear in real applications. Some papers that included undesirable factors in the evaluation of SC or different parts of it were focused only on undesirable outputs. Therefore, in this paper, we introduced a DEA model M that besides undesirable outputs can also consider and evaluate undesirable inputs. The proposed new approach of a non-radial DEA model M for the environmental evaluation of suppliers and other components of SC that is able to consider desirable inputs in the goal function, all with appropriate weights, was presented.

With the introduced non-radial model M, a better picture in terms of a score of efficiency can be achieved. Application of model M has been presented based on the data taken from Mahdiloo et al. [5]. Results of the model M, obtained using Excel Solver tool, and results obtained by models applied in Mahdiloo et al. [5], are presented in Table 2. Because the data were reused primarily in order to introduce and present the behavior of model M, testing of these data before applying model M was not performed. From Table 2, it can be seen that for Model 2, the suppliers 2, 15, 16, 17, and 18 were rated as most efficient. For Model 4, the most efficient suppliers were 2, 15, and 18, while the most efficient suppliers within Model 5 were the same as within Model 2. Comparing the results of different models, it can be seen that the model M yielded similar results for environmental efficiency of suppliers as Model 4. The picture regarding the inefficiency of suppliers is the same. The main difference between the model M and other models is in efficiency score, i.e., in the case of the model M, it is lower than in the case of other models. Considering these results, it can be concluded that model M provides more precise results because of the higher discriminatory power.

In order to check the behavior of the model M, using the same data as in Mahdiloo et al. [5], sensitivity analysis was performed using the Excel Solver tool, conducted through three Cases under a certain percentage of data perturbation (2%, 5% and 10%). The results of the sensitivity analysis are presented in Table 3. In Case 1, both desirable and undesirable inputs, as well as undesirable outputs, were improved for suppliers with efficiency 1 and worsened for suppliers with efficiency under 1. Within Case 2, the perturbation of the data was focused on the increment of undesirable inputs and outputs for suppliers with efficiency 1 and on the reduction for those with lower efficiency, while desirable input was fixed. Then, the behavior of model M was checked through Case 3 where desirable outputs were decreased and desirable input increased for suppliers with efficiency 1, and vice versa for inefficient suppliers. Based on the results obtained through the sensitivity analysis, it can be concluded that for most suppliers with efficiency under 1, the score of efficiency was improved. However, for some inefficient suppliers, the efficiency score was not significantly changed; such as, for example, suppliers 12 and 13. Through the comparison of efficiency for a particular supplier, comparing data of different perturbations (Table 3), it can be seen that efficiency was mainly the same in all Cases under the 2% of data perturbation. However, with 5% of data perturbation, the efficiency for each supplier was lower in Case 2 compared to Case 3 and Case 1, while efficiency was mainly the same for Case 3 and Case 1. In addition, regarding 10% of data perturbation, it can be seen that the efficiency of suppliers was mainly lower in Case 2 compared to Case 3 and Case 1.

Model M was taken from Wu et al. [113] and then improved and applied in Djordjević et al. [7]. The comprehensive observation was that efficiency obtained by non-radial DEA model M from ref. [7] was different from the model developed in ref. [113] and that model M gives better efficiency because of the involvement of efficiency score θ_n with weight W_n in the model. The main reason for different results in comparison with results in ref. [113] and in ref. [5] lies in the fact that model M can simultaneously reduce desirable inputs. Through the application of model M for the evaluation of SC, a better picture regarding relative efficiency can be given because the model is more representative and strict. The proposed model M can, therefore, be a good tool for efficiency evaluation of SCs and identification of best practices.

Specifically, the model M is able to measure the efficiency of some components of an SC such as supplier selection and comparison of efficiency between SCs on different (micro and macro) levels over time.

Further, one of the major advantages of the proposed model M are weights that can be assigned to desirable and undesirable inputs and outputs. Through the application of particular weights for inputs and outputs, the level of desirability can be determined, which influences reduction or improvement effects of inputs or outputs. Regarding that, one of the important steps in the non-radial model M should be the careful selection of weights, relying on the preferences of DM and the aim of the evaluation. Based on their degree, they can influence the results of the non-radial DEA model M. Considering the results of the sensitivity analysis of model M presented in Table 3, it can be concluded that the model is valid. However, results of sensitivity analysis also illustrate reduced stability under smaller data transformation.

Bearing in the mind the overview of literature related to the application of the DEA approach in SCM and the introduced non-radial DEA model M, future work can be conducted. For instance, the model M may be applied for the evaluation of components of SC such as supplier selection using experimental or real data. Further, the non-radial DEA model can be also applied to the specific companies within the EU countries or the US. Funding of the best practices among companies and comparisons of between companies or countries can also be realized with model M. In the future, model M can be extended also for the evaluation of whole SC through the inclusion of intermediate variables. Besides the environmental efficiency model M can be applied for measuring other components or whole SC from the perspective of dimensions of sustainability such as economic and social. Moreover, combination and application of DEA with other economic measures such as ROE (Return on Equity) and ROA (Return on Assets) for the purposes of evaluation in SCM in terms of different views of sustainability can be one of the future tasks.

Author Contributions: Both authors contributed equally to this work and have read and approved the final manuscript.

Funding: This research received no external funding.

Conflicts of Interest: The authors declare no conflict of interest.

References

1. Yang, F.; Wu, D.; Liang, L.; Bi, G.; Wu, D.D. Supply chain DEA: Production possibility set and performance evaluation model. *Ann. Oper. Res.* **2011**, *185*, 195–211. [CrossRef]
2. Sharma, M.J.; Yu, S.J. Multi-Stage data envelopment analysis congestion model. *Oper. Res. Int. J.* **2013**, *13*, 399–413. [CrossRef]
3. Charnes, A.; Cooper, W.W.; Rhodes, E. Measuring the efficiency of decision making units. *Eur. J. Oper. Res.* **1978**, *2*, 429–444. [CrossRef]
4. Soheilirad, S.; Govindan, K.; Mardani, A.; Zavadskas, K.E.; Nilashi, M.; Zakuan, N. Application of data envelopment analysis models in supply chain management: A systematic review and meta-analysis. *Ann. Oper. Res.* **2018**, *271*, 915–969. [CrossRef]
5. Mahdiloo, M.; Saen, F.R.; Lee, K.-H. Technical, environmental and eco-efficiency measurement for supplier selection: An extension and application of data envelopment analysis. *Int. J. Prod. Econ.* **2015**, *168*, 279–289. [CrossRef]
6. Krmac, E.; Djordjević, B. An evaluation of train control information systems for sustainable railway using the analytic hierarchy process (AHP) model. *Eur. Transp. Res. Rev.* **2017**, *9*, 35. [CrossRef]
7. Djordjević, B.; Krmac, E.; Mlinarić, T.J. Non-radial DEA model: A new approach to evaluation of safety at railway level crossings. *Saf. Sci.* **2018**, *103*, 234–246.
8. Tavana, M.; Kaviani, M.A.; Caprio, D.D.; Rahpeyma, B. A two-stage data envelopment analysis model for measuring performance in three-level supply chains. *Measurement* **2016**, *78*, 322–333. [CrossRef]
9. Liang, L.; Yang, F.; Cook, W.D.; Zhu, J. DEA models for supply chain efficiency evaluation. *Ann. Oper. Res.* **2006**, *145*, 35–49. [CrossRef]
10. Wong, W.P.; Wong, K.Y. Supply chain performance measurement system using DEA modeling. *Ind. Manag. Data Syst.* **2007**, *107*, 361–381. [CrossRef]
11. Xu, J.; Li, B.; Wu, D. Rough data envelopment analysis and its application to supply chain performance evaluation. *Int. J. Prod. Econ.* **2009**, *122*, 628–638. [CrossRef]

12. Aoki, S.; Naito, A.; Gejima, R.; Inoue, K.; Tsuji, H. Data envelopment analysis for a supply chain. *Artif. Life Robot.* **2010**, *15*, 171–175. [CrossRef]
13. Li, D.-C.; Dai, W.-L. Determining the optimal collaborative benchmarks in a supply chain. *Int. J. Prod. Res.* **2009**, *47*, 4457–4471. [CrossRef]
14. Chen, C.; Yan, H. Network DEA model for supply chain performance evaluation. *Eur. J. Oper. Res.* **2011**, *213*, 147–155. [CrossRef]
15. Tavana, M.; Mirzagoltabar, H.; Mirhedayatian, S.M.; Saen, R.F.; Azadi, M. A new network epsilon-based DEA model for supply chain performance evaluation. *Comput. Ind. Eng.* **2013**, *66*, 501–513. [CrossRef]
16. Khalili-Damghani, K.; Tavana, M. A new fuzzy network data envelopment analysis model for measuring the performance of agility in supply chains. *Int. J. Adv. Manuf. Technol.* **2013**, *69*, 291–318. [CrossRef]
17. Nikfarjam, H.; Rostamy-Malkhalifeh, M.; Mamizadeh-Chatghayeh, S. Measuring supply chain efficiency based on a hybrid approach. *Transp. Res. Part D* **2015**, *39*, 141–150. [CrossRef]
18. Kumar, A.; Mukherjee, K.; Adlakha, A. Dynamic performance assessment of a supply chain process: A case from pharmaceutical supply chain in India. *Bus. Process Manag. J.* **2015**, *21*, 743–770. [CrossRef]
19. Tavassoli, M.; Saen, R.F.; Faramarzi, G.R. Developing network data envelopment analysis model for supply chain performance measurement in the presence of zero data. *Expert Syst.* **2015**, *32*, 381–391. [CrossRef]
20. Khamseh, A.A.; Zahmatkesh, D. Supply chain performance evaluation using robust data envelopment analysis. *Uncertain Supply Chain Manag.* **2015**, *3*, 311–320. [CrossRef]
21. Omrani, H.; Keshavarz, M. A performance evaluation model for supply chain of shipping company in Iran: An application of the relational network DEA. *Marit. Policy Manag.* **2016**, *43*, 121–135. [CrossRef]
22. Moslemi, S.; Mirzazadeh, A. Performance Evaluation of Four-Stage Blood Supply Chain with Feedback Variables Using NDEA Cross-Efficiency and Entropy Measures Under IER Uncertainty. *Numer. Algebra Control Optim.* **2017**, *7*, 379–401. [CrossRef]
23. Huang, C.-W. Assessing the performance of tourism supply chains by using the hybrid network data envelopment analysis model. *Tour. Manag.* **2018**, *65*, 303–316. [CrossRef]
24. Amalnick, M.S.; Saffar, M.M. An integrated approach for supply chain assessment from resilience engineering and ergonomics perspectives. *Uncertain Supply Chain Manag.* **2017**, *5*, 159–168. [CrossRef]
25. Toloo, M.; Nalchigar, S. A new DEA method for supplier selection in presence of both cardinal and ordinal data. *Expert Syst. Appl.* **2011**, *38*, 14726–14731. [CrossRef]
26. Liu, J.; Ding, F.-Y.; Lall, V. Using data envelopment analysis to compare suppliers for supplier selection and performance improvement. *Supply Chain Manag. Int. J.* **2000**, *5*, 143–150. [CrossRef]
27. Wu, T.; Blackhurst, J. Supplier evaluation and selection: An augmented DEA approach. *Int. J. Prod. Res.* **2009**, *47*, 4593–4608. [CrossRef]
28. Wu, D.D. A systematic stochastic efficiency analysis model and application to international supplier performance evaluation. *Expert Syst. Appl.* **2010**, *37*, 6257–6264. [CrossRef]
29. Azadi, M.; Saen, R.F. Supplier Selection using a New Russell Model in the Presence of Undesirable Outputs and Stochastic Data. *J. Appl. Sci.* **2012**, *12*, 336–344.
30. Azadi, M.; Saen, R.F. Developing a Nondisretionary Slacks-based Measure Model for Supplier Selection in the Presence of Stochastic Data. *Res. J. Bus. Manag.* **2012**, *6*, 103–120.
31. Mahdiloo, M.; Noorizadeh, A.; Saen, R.F. Benchmarking suppliers' performance when some factors play the role of both inputs and outputs: A new development to the slacks-based measure of efficiency. *Benchmark. Int. J.* **2014**, *5*, 792–813. [CrossRef]
32. Hatami-Marbini, A.; Agrell, P.J.; Tavana, M.; Khoshnevis, P. A flexible cross-efficiency fuzzy data envelopment analysis model for sustainable sourcing. *J. Clean. Prod.* **2017**, *142*, 2761–2779. [CrossRef]
33. Momeni, M.; Vandchali, R.H. Providing a structured methodology for supplier selection and evaluation for strategic outsourcing. *Int. J. Bus. Perform. Supply Chain Model.* **2017**, *9*, 66–85. [CrossRef]
34. Shafiee, M.; Lotfi, F.H.; Saleh, H. Supply chain performance evaluation with data envelopment analysis and balanced scorecard approach. *Appl. Math. Model.* **2014**, *38*, 5092–5112. [CrossRef]
35. Reiner, G.; Hofmann, P. Efficiency analysis of supply chain processes. *Int. J. Prod. Res.* **2006**, *44*, 5065–5087. [CrossRef]
36. Dotoli, M.; Falagario, M. A hierarchical model for optimal supplier selection in multiple sourcing contexts. *Int. J. Prod. Res.* **2012**, *50*, 2953–2967. [CrossRef]

37. Dev, N.K.; Shankar, R.; Debnath, R.M. Supply chain efficiency: A simulation cum DEA approach. *Int. J. Adv. Manuf. Technol.* **2014**, *72*, 1537–1549. [CrossRef]
38. Toloo, M. Selecting and full ranking suppliers with imprecise data: A new DEA method. *Int. J. Adv. Manuf. Technol.* **2014**, *74*, 1141–1148. [CrossRef]
39. Markabi, M.S.; Sabbagh, M. A Hybrid Method of GRA and DEA for Evaluating and Selecting Efficient Suppliers plus a Novel Ranking Method for Grey Numbers. *J. Ind. Eng. Manag.* **2014**, *7*, 1197–1221.
40. Abdollahi, M.; Arvan, M.; Razmi, J. An integrated approach for supplier portfolio selection: Lean or agile? *Expert Syst. Appl.* **2015**, *42*, 679–690. [CrossRef]
41. Karsak, E.E.; Dursun, M. An integrated supplier selection methodology incorporating QFD and DEA with imprecise data. *Expert Syst. Appl.* **2014**, *41*, 6995–7004. [CrossRef]
42. Che, Z.H.; Chang, Y.F. Integrated methodology for supplier selection: The case of a sphygmomanometer manufacturer in Taiwan. *J. Bus. Econ. Manag.* **2016**, *17*, 17–34. [CrossRef]
43. Azadeh, A.; Zarrin, M.; Salehi, N. Supplier selection in closed loop supply chain by an integrated simulation-Taguchi-DEA approach. *J. Enterp. Inf. Manag.* **2016**, *29*, 302–326. [CrossRef]
44. Prasad, D.G.K.; Subbaiah, V.K.; Prasad, V.M. Supplier evaluation and selection through DEA-AHP-GRA integrated approach—A case study. *Uncertain Supply Chain Manag.* **2017**, *5*, 369–382. [CrossRef]
45. Park, S.; Ok, C.; Ha, C. A stochastic simulation-based holistic evaluation approach with DEA for vendor selection. *Comput. Oper. Res.* **2017**, *100*, 368–378. [CrossRef]
46. Paraizar, M.; Sir, Y.M. A multi-objective approach for supply chain design considering disruptions impacting supply availability and quality. *Comput. Ind. Eng.* **2018**, *121*, 113–130. [CrossRef]
47. Sabouhi, F.; Pishvaee, S.M.; Jabalameli, S.M. Resilient supply chain design under operational and disruption risks considering quantity discount: A case study of pharmaceutical supply chain. *Comput. Ind. Eng.* **2018**, *126*, 657–672. [CrossRef]
48. Diouf, M.; Kwak, C. Fuzzy AHP, DEA, and Managerial Analysis for Supplier Selection and Development; From the Perspective of Open Innovation. *Sustainability* **2018**, *10*, 3779. [CrossRef]
49. Wang, C.-N.; Nguyen, T.V.; Duong, H.D.; Do, T.H. A Hybrid Fuzzy Analytic Network Process (FANP) and Data Envelopment Analysis (DEA) Approach for Supplier Evaluation and Selection in the Rice Supply Chain. *Symmetry* **2018**, *10*, 221. [CrossRef]
50. Cheng, Y.; Peng, J.; Zhou, Z.; Gu, X.; Liu, W. A Hybrid DEA-Adaboost Model in Supplier Selection for Fuzzy Variable and Multiple Objectives. *IFAC PapersOnLine* **2017**, *50*, 12255–12260. [CrossRef]
51. Ross, A.; Droge, C. An integrated benchmarking approach to distribution center performance using DEA modeling. *J. Oper. Manag.* **2002**, *20*, 19–32. [CrossRef]
52. Easton, L.; Murphy, D.J.; Pearson, J.N. Purchasing performance evaluation: With data envelopment analysis. *Eur. J. Purch. Supply Manag.* **2002**, *8*, 123–134. [CrossRef]
53. Ross, A.D.; Droge, C. An analysis of operations efficiency in large-scale distribution systems. *J. Oper. Manag.* **2004**, *15*, 673–688. [CrossRef]
54. Ji, X.; Wu, J.; Zhu, Q. Eco-design of transportation in sustainable supply chain management: A DEA-like method. *Transp. Res. Part D* **2016**, *48*, 451–459. [CrossRef]
55. Haldar, A.; Qamaruddin, U.; Raut, R.; Kamble, S.; Kharat, M.G.; Kamble, S.J. 3PL evaluation and selection using integrated analytical modeling. *J. Model. Manag.* **2017**, *12*, 224–242. [CrossRef]
56. Marti, L.; Martin, J.C.; Puertas, R. A DEA-Logistics Performance Index. *J. Appl. Econ.* **2017**, *20*, 169–192. [CrossRef]
57. Grigoroudis, E.; Petridis, K.; Arabatzis, G. RDEA: A recursive DEA based algorithm for the optimal design of biomass supply chain networks. *Renew. Energy* **2014**, *71*, 113–122. [CrossRef]
58. Dotoli, M.; Epicoco, N.; Falagario, M. A Technique for Supply Chain Network Design under Uncertainty using Cross-Efficiency Fuzzy Data Envelopment Analysis. *IFAC-PapersOnLine* **2015**, *48*, 634–639. [CrossRef]
59. Petridis, K.; Dey, P.K.; Emrouznejad, A. A branch and efficiency algorithm for the optimal design of supply chain networks. *Ann. Oper. Res.* **2017**, *253*, 545–571. [CrossRef]
60. Babazadeh, R.; Razmi, J.; Rabbani, M.; Pishvaee, M.S. An integrated data envelopment analysise mathematical programming approach to strategic biodiesel supply chain network design problem. *J. Clean. Prod.* **2017**, *147*, 694–707. [CrossRef]
61. Dehghani, E.; Jabalameli, M.S.; Jabbarzadeh, A. Robust design and optimization of solar photovoltaic supply chain in an uncertain environment. *Energy Econ.* **2017**, *142*, 139–156. [CrossRef]

62. Pourhejazy, P.; Kwon, O.K.; Chang, Y.-T.; Park, H. Evaluating Resiliency of Supply Chain Network: A Data Envelopment Analysis Approach. *Sustainability* **2017**, *9*, 255. [CrossRef]
63. Goodarzi, M.; Saen, F.R. How to measure bullwhip effect by network data envelopment analysis? *Comput. Ind. Eng.* **2018**, in press. [CrossRef]
64. Chen, M.-C.; Yang, T.; Yen, C.-T. Investigating the value of information sharing in multi-echelon supply chains. *Qual. Quant.* **2007**, *41*, 497–511. [CrossRef]
65. Yu, M.-M.; Ting, S.-C.; Chen, M.-C. Evaluating the cross-efficiency of information sharing in supply chains. *Expert Syst. Appl.* **2010**, *37*, 2891–2897. [CrossRef]
66. Khodakarami, M.; Shabani, A.; Saen, R.F.; Azadi, M. Developing distinctive two-stage data envelopment analysis models: An application in evaluating the sustainability of supply chain management. *Measurement* **2015**, *70*, 62–74. [CrossRef]
67. Sueyoshi, T.; Wang, D. Sustainability development for supply chain management in U.S. petroleum industry by DEA environmental assessment. *Energy Econ.* **2014**, *46*, 360–374. [CrossRef]
68. Mirhedayatian, S.M.; Azadi, M.; Saen, R.F. A novel network data envelopment analysis model for evaluating green supply chain management. *Int. J. Prod. Econ.* **2014**, *147*, 544–554. [CrossRef]
69. Tajbakhsh, A.; Hassini, E. A data envelopment analysis approach to evaluate sustainability in supply chain networks. *J. Clean. Prod.* **2015**, *105*, 74–85. [CrossRef]
70. Izadikhah, M.; Saen, R.F. Evaluating sustainability of supply chains by two-stage range directional measure in the presence of negative data. *Transp. Res. Part D* **2016**, *49*, 110–126. [CrossRef]
71. Kahi, V.S.; Yousefi, S.; Shabanpour, H.; Saen, R.F. How to evaluate sustainability of supply chains? A dynamic network DEA approach. *Ind. Manag. Data Syst.* **2017**, *117*, 1866–1889. [CrossRef]
72. Badiezadeh, T.; Saen, R.F.; Samavati, T. Assessing sustainability of supply chains by double frontier network DEA: A big data approach. *Comput. Oper. Res.* **2017**, *98*, 284–290. [CrossRef]
73. Izadikhah, M.; Saen, R.F. Assessing sustainability of supply chains by chance-constrained two-stage DEA model in the presence of undesirable factors. *Comput. Oper. Res.* **2017**, *100*, 343–367. [CrossRef]
74. Kalantary, M.; Saen, F.R. Assessing sustainability of supply chains: An inverse network dynamic DEA model. *Comput. Ind. Eng.* **2018**, in press. [CrossRef]
75. Zhai, D.; Shang, J.; Yang, F.; Ang, S. Measuring energy supply chains' efficiency with emission trading: A two-stage frontier-shift data envelopment analysis. *J. Clean. Prod.* **2019**, *210*, 1462–1474. [CrossRef]
76. De, D.; Chowdhury, S.; Dey, K.P.; Ghosh, K.S. Impact of Lean and Sustainability oriented innovation on Sustainability performance of Small and Medium Sized Enterprises: A Data Envelopment Analysis-based Framework. *Int. J. Prod. Econ.* **2018**, accepted. [CrossRef]
77. Omrani, H.; Keshavarza, M.; Ghaderib, F.S. Evaluation of supply chain of a shipping company in Iran by a fuzzy relational network data envelopment analysis model. *Sci. Iran. E* **2018**, *25*, 868–890. [CrossRef]
78. Chen, Y.-J. Structured methodology for supplier selection and evaluation in a supply chain. *Inf. Sci.* **2011**, *181*, 1651–1670. [CrossRef]
79. Jain, V.; Kumar, S.; Kumar, A.; Chandra, C. An integrated buyer initiated decision-making process for green greensupplier selection. *J. Manuf. Syst.* **2016**, *41*, 256–265. [CrossRef]
80. Kumar, A.; Jain, V.; Kumar, S. A comprehensive environment friendly approach for supplier selection. *Omega* **2014**, *42*, 109–123. [CrossRef]
81. Azadi, M.; Jafarian, M.; Saen, R.F.; Mirhedayatian, S.M. A new fuzzy DEA model for evaluation of efficiency and effectiveness of supplier sinsustainable supply chain management context. *Comput. Oper. Res.* **2015**, *54*, 274–285. [CrossRef]
82. Shi, P.; Yan, B.; Shi, S.; Ke, C. A decision support system to select suppliers for a sustainable supply chain based on a systematic DEA approach. *Inf. Technol. Manag.* **2015**, *16*, 39–49. [CrossRef]
83. Zhou, X.; Pedrycz, W.; Kuang, Y.; Zhang, Z. Type-2 fuzzy multi-objective DEA model: An application to sustainable supplier evaluation. *Appl. Soft Comput.* **2016**, *46*, 424–440. [CrossRef]
84. Izadikhah, M.; Saen, R.F.; Ahmadi, K. How to assess sustainability of suppliers in volume discount context? A new data envelopment analysis approach. *Transp. Res. Part D* **2017**, *51*, 102–121. [CrossRef]
85. Yu, M.-C.; Su, M.-H. Using Fuzzy DEA for Green Suppliers Selection Considering Carbon Footprints. *Sustainability* **2017**, *9*, 495. [CrossRef]
86. Tavassoli, M.; Saen, F.R. Predicting group membership of sustainable suppliers via data envelopment analysis and discriminant analysis. *Sustain. Prod. Consum.* **2018**, in press. [CrossRef]

87. Ghoushchi, J.S.; Milan, D.M.; Rezaee, J.M. Evaluation and selection of sustainable suppliers in supply chain using new GP-DEA model with imprecise data. *J. Ind. Eng. Int.* **2018**, *14*, 613–625. [CrossRef]
88. Kuo, R.J.; Wang, Y.C.; Tien, F.C. Integration of artificial neural network and MADA methods for green supplier selection. *J. Clean. Prod.* **2010**, *18*, 1161–1170. [CrossRef]
89. Egilmez, G.; Kucukvar, M.; Tatari, O.; Bhutta, M.S. Supply chain sustainability assessment of the U.S. food manufacturing manufacturingsectors: A life cycle-based frontier approach. *Resour. Conserv. Recycl.* **2014**, *82*, 8–20. [CrossRef]
90. Bai, C.; Sarkis, J. Determining and applying sustainable supplier key performance indicators. *Supply Chain Manag. Int. J.* **2014**, *19*, 275–291. [CrossRef]
91. Haghighi, M.S.; Torabi, S.A.; Ghasemi, R. An integrated approach for performance evaluation in sustainable supply chain networks (with a case study). *J. Clean. Prod.* **2016**, *137*, 579–597. [CrossRef]
92. Fallahpour, A.; Olugu, E.U.; Musa, S.N.; Khezrimotlagh, D.; Wong, Y.K. An integrated model for green supplier selection under fuzzy environment: Application of data envelopment analysis and genetic programming approach. *Neural Comput. Appl.* **2016**, *27*, 707–725. [CrossRef]
93. Tavana, M.; Shabanpour, H.; Yousefi, S.; Saen, R.F. A hybrid goal programming and dynamic data envelopment analysis framework for sustainable supplier evaluation. *Neural Comput. Appl.* **2017**, *28*, 3683–3696. [CrossRef]
94. Yousefi, S.; Soltani, R.; Saen, R.F.; Pishvaee, M.S. A robust fuzzy possibilistic programming for a new network GP-DEA model to evaluate sustainable supply chains. *J. Clean. Prod.* **2017**, *166*, 537–549. [CrossRef]
95. Ramezankhani, M.J.; Vahidi, F. Supply chain performance measurement and evaluation: A mixed sustainability and resilience approach. *Comput. Ind. Eng.* **2018**, *126*, 531–548. [CrossRef]
96. He, X.; Zhang, J. Supplier Selection Study under the Respective of Low-Carbon Supply Chain: A Hybrid Evaluation Model Based on FA-DEA-AHP. *Sustainability* **2018**, *10*, 564. [CrossRef]
97. Sevkli, M.; Koh, S.L.; Zaim, S.; Demirbag, M.; Tatoglu, E. An application of data envelopment analytic hierarchy process for supplier selection: A case study of BEKO in Turkey. *Int. J. Prod. Res.* **2007**, *45*, 1973–2003. [CrossRef]
98. Wu, D.; Olson, D.L. Supply chain risk, simulation, and vendor selection. *Int. J. Prod. Econ.* **2008**, *114*, 646–655. [CrossRef]
99. Azadeh, A.; Alem, S.M. A flexible deterministic, stochastic and fuzzy Data Envelopment Analysis approach for supply chain risk and vendor selection problem: Simulation analysis. *Expert Syst. Appl.* **2010**, *37*, 7438–7448. [CrossRef]
100. Wu, D.D.; Olson, D. Enterprise risk management: A DEA VaR approach in vendor selection. *Int. J. Prod. Res.* **2010**, *48*, 4919–4932. [CrossRef]
101. Visani, F.; Barbieri, P.; di Lascio, F.M.; Raffoni, A.; Vigo, D. Supplier's total cost of ownership evaluation: A data envelopment analysis approach. *Omega* **2016**, *61*, 141–154. [CrossRef]
102. Boudaghi, E.; Saen, R.F. Developing a novel model of data envelopment analysis–discriminant analysis for predicting group membership of suppliers in sustainable supply chain. *Comput. Oper. Res.* **2018**, *89*, 348–359. [CrossRef]
103. Pournader, M.; Rotaru, K.; Kach, A.P.; Hajiagha, S.H. An analytical model for system-wide and tier-specific assessment of resilience to supply chain risks. *Supply Chain Manag. Int. J.* **2016**, *21*, 589–609. [CrossRef]
104. Azadeh, A.; Gaeini, Z.; Shabanpour, N. Optimization of healthcare supply chain in context of macro-ergonomics factors by a unique mathematical programming approach. *Appl. Ergon.* **2016**, *55*, 46–55. [CrossRef]
105. Saranga, H.; Moser, R. Performance evaluation of purchasing and supply management using value chain DEA approach. *Eur. J. Oper. Res.* **2010**, *207*, 197–205. [CrossRef]
106. Amirteimoori, A.; Khoshandam, L. A Data Envelopment Analysis Approach to Supply Chain Efficiency. *Adv. Decis. Sci.* **2011**, *2011*, 608324. [CrossRef]
107. Pournader, M.; Kach, A.; Fahimnia, B.; Sarkis, J. Outsourcing performance quality assessment using data envelopment analytics. *Int. J. Prod. Econ.* **2019**, *207*, 173–182. [CrossRef]
108. Azadeh, A.; Atrchin, N.; Salehi, V.; Shojaei, H. Modelling and improvement of supply chain with imprecise transportation delays and resilience factors. *Int. J. Logist. Res. Appl.* **2014**, *17*, 269–282. [CrossRef]
109. Bayraktar, E.; Gunasekaran, A.; Koh, S.L.; Tatoglu, E.; Demirbag, M.; Zaim, S. An efficiency comparison of supply chain management and information systems practices: A study of Turkish and Bulgarian small- and medium-sized enterprises in food products and beverages. *Int. J. Prod. Res.* **2010**, *48*, 425–451. [CrossRef]

110. Jalalvand, F.; Teimoury, E.; Makui, A.; Aryanezhad, M.B.; Jolai, F. A method to compare supply chains of an industry. *Supply Chain Manag. Int. J.* **2011**, *16*, 82–97. [CrossRef]
111. Cooper, W.W.; Seiford, L.M.; Tone, K. *Introduction to Data Envelopment Analysis and Its Use: With DEA-Solver Softwer and References*; Springer: New York, NY, USA, 2006.
112. Djordjević, B.; Krmac, E. *Application of Multicriteria Decision-Making Methods in Railway Engineering: A Case Study of Train Control Information Systems (TCIS), Modern Railway Engineering*; Hessami, A., Ed.; IntechOpen: Rijeka, Croatia, 2018.
113. Wu, J.; Zhu, Q.; Yin, P.; Song, M. Measuring energy and environmental performance for regions in China by using DEA-based Malmquist indices. *Oper. Res. Int. J.* **2015**, *17*, 715–735. [CrossRef]

 © 2019 by the authors. Licensee MDPI, Basel, Switzerland. This article is an open access article distributed under the terms and conditions of the Creative Commons Attribution (CC BY) license (http://creativecommons.org/licenses/by/4.0/).

Article

Modelling Construction Site Cost Index Based on Neural Network Ensembles

Michał Juszczyk * and Agnieszka Leśniak

Cracow University of Technology, Faculty of Civil Engineering, Warszawska 24, 31-155 Cracow, Poland;
alesniak@L3.pk.edu.pl
* Correspondence: mjuszczyk@L3.pk.edu.pl

Received: 11 February 2019; Accepted: 18 March 2019; Published: 20 March 2019

Abstract: Construction site overhead costs are key components of cost estimation in construction projects. The estimates are expected to be accurate, but there is a growing demand to shorten the time necessary to deliver cost estimates. The balancing (symmetry) between time of calculation and satisfaction of reliable estimation was the reason for developing a new model for cost estimation in construction. This paper reports some results from the authors' broad research on the modelling processes in engineering related to estimation of construction costs using artificial intelligence tools. The aim of this work was to develop a model capable of predicting a construction site cost index that would benefit from combining several artificial neural networks into an ensemble. Combining selected neural networks and forming the ensemble-based models compromised their strengths and weaknesses. With the use of data including training patterns collected on the basis of studies of completed construction projects, the authors investigated various types of neural networks in order to select the members of the ensemble. Finally, three models that were assessed in terms of performance and prediction quality were proposed. The results revealed that the developed models based on ensemble averaging and stacked generalisation met the expectations of knowledge generalisation and accuracy of prediction of site overhead cost index. The proposed models offer predictions of cost in an accepted error range and prove to deliver better predictions than those based on single neural networks. The developed tools can be used in the decision-making process regarding construction cost estimation.

Keywords: cost decision making; construction site overhead costs; neural network ensembles; ensemble averaging; stacked generalisation; cost estimation; construction cost management

---

1. Introduction

The success of a construction project is determined by obtaining three fundamental goals of a project—completion within the budget, completion within planned time, and achieving the expected quality of construction works. For the budget issue, cost estimation is a key process. On one hand, the estimates are expected to be accurate; on the other hand, there is a growing demand to shorten the time necessary to deliver cost estimates. These needs justify attempts to employ various tools in fast cost analyses and modelling. The aim of this paper is to present the results of the research on artificial neural networks (ANNs) ensembles as artificial intelligence tools for fast analysis and prediction of site overhead costs. This research is a continuation and extension of previous studies, including prediction of these costs with the use of multilayer perceptron neural networks [1]. It is worth mentioning that mathematical tools—which are constantly being developed—are present in the investigations of a broad variety of problems in the field of construction management and technology. Some interesting examples are applications of fuzzy sets theory and fuzzy logic in construction project risk [2–4], the evaluation of a construction safety management system [5], processes in a construction enterprise [6], the investigation of flow-shop scheduling problems [7], and using multiple criteria decision-making methods for supporting the decision process in construction and building technology [8–10]. There have also been a number of attempts to apply artificial neural networks in the management of construction projects—predicting the completion period of building contracts [11], analysing efficiency and productivity in construction projects [12,13], predicting the maintenance cost of construction equipment [14], supporting bidding decisions [15,16], and facilitating decision making [17–19]. Comprehensive discussion on innovative solutions in the construction industry can be found in Reference [20].

The solutions and models that support cost estimates in construction are explored in many scientific publications. The authors propose a variety of methods, for instance multivariate regression [21], analysis of the selected cost-effectiveness factors [22], a case-based reasoning method [23], fuzzy logic [24], and genetic algorithms [25]. In terms of ANNs, there have been attempts to apply these tools in the field of construction cost management. Some examples are forecasting costs of motorways in different aspects [26], predicting cost deviations in reconstruction, alteration, and rebuilding projects [27], estimating the costs of construction projects [28,29], cost estimates of residential buildings [30], prediction of overhead costs [31,32], cost estimates of buildings' floor structural frames as a higher level of aggregation elements of building information model [33], construction cost of sports fields [34], and shovel capital cost estimation [35].

According to the research presented in Reference [36], the influence of an improper calculation of the overhead costs can create a significant negative financial situation for the contracting company. Generally, the building contractor's overhead costs are divided into two categories—site (project) overhead costs and company's (general) overhead costs [37]. The site (project) overhead costs include items that can be identified with a particular job, but not materials, labour, or production equipment. The company's overhead costs are items that represent the cost of doing business and are often considered fixed expenses that must be paid by the contractor. On the other hand, an overhead cost of a construction project can be defined as a cost that cannot be identified with or charged to a construction project or to a unit of construction production [38]. A new classification of construction companies into competitiveness classes according to the relative value of overhead costs was proposed in Reference [39]. As far as accuracy is concerned, it is more advantageous to calculate both components separately—as is done in Great Britain [40], the US, and Canada [41]. The unstable construction market makes it difficult for construction companies to decide on the optimum level of overhead costs [42].

A number of empirical studies relate to the determination of the project overhead costs. In Reference [43], it is indicated that the method of work is a critical factor affecting the amount spent on project overheads. In Reference [44], the authors pointed that the location of the site could affect a number of project overhead items. In References [31,45,46], research carried out in different countries allowed for the identification of different factors that should be taken into account in site (project) overhead costs.

Studies on construction project overheads and factors that influence their estimates report that it is difficult to determine unambiguously which of the cost components are of the highest importance. Most attention is paid to a detailed calculation of site overheads; however, it is a time-consuming task to take into account all of the possible components of site overhead costs [36].

The aim of the authors' work was to develop a regression model based on the ANNs ensembles, capable of the prediction of site overhead cost index, and, thus, able to support the estimation of site overhead costs in construction projects. An additional research objective was to explore the capabilities of ANNs ensembles in this problem. In the application of ANNs, a very common approach is to select one network to be the core of a developed model. The selection is preceded by a training and performance assessment of numerous networks—compare, e.g., Reference [47]. As an alternative, the employment of a combination of networks i.e., ANNs, offer significant capabilities. Despite their advantages, the ANNs ensembles are rarely reported on for the prediction of widely understood construction costs in research papers.

Site overhead costs can be estimated with the use of preliminaries (compare References [40,41]) —such a method is accurate but time-consuming as all of the cost items must be assessed separately. On the other hand, index methods (compare Reference [36]) allow for quick estimation of site overhead costs, however the accuracy depends on the assumption of the index. The novelty of the approach proposed in this paper relies on the use of knowledge and information from the completed construction projects to train several neural networks, combine them into an ensemble, and assess the site overhead costs on the basis of the predictions produced by the ensemble of neural networks.

The paper content includes an introduction and review of the literature in the above section. Section 2 presents the theoretical background of the problem, and a discussion of the site overhead cost index prediction as a regression problem is presented in Section 3. In Section 4, the authors propose a methodology for the implementation of an ensemble of neural networks (with the use of ensemble averaging and stacked generalisation approaches) for prediction of site overhead cost index, present the results of the studies, and discuss the results. Section 5 includes a summary and conclusions.

2. Background of The Problem, Methods, and Main Assumptions

The development of the proposed model comes down to solving a regression problem and approximation of the true regression function, which is the relationship between the site overhead costs index (as the dependent variable of the model) and a set of selected predictors (as independent variables of the model). The theory of ANNs is widely presented in the literature—for instance, References [47–49]. ANNs, as mathematical tools applied in regression problems, offer an approximation of the true regression function $g(x_j)$ of multiple variables x_j where $j = 1, \ldots ,n$:

$$g(x_j) = f(x_j) + \varepsilon, \tag{1}$$

In the equation (1), function $f(x_j)$, as an approximation of $g(x_j)$, is assumed to be implemented implicitly by a trained single ANN, selected from a number of trained candidate networks, where ε denotes an error of approximation. There are two disadvantages of an approach based on the selection of a single ANN and discarding the rest of the candidate networks [47,48]—the effort required for the training and assessment of the number of candidate networks is wasted. Moreover, the generalisation performance of the chosen network is biased with respect to some part of the input space due to the selection of learning, testing, and validating subsets from the overall number of patterns available for the training process, structure of the network, its parameters, and conditions of training process initialisation. An alternative approach is to combine a number of different ANNs that share common input x_j and form an ensemble (the ANNs may differ in their structures, parameters, and way of training, and the ensemble may even include different kinds of networks). In this paper, the authors consider two alternative approaches that are based on ensembles of neural networks—the first approach is termed as ensemble averaging, and the second one *stacked generalisation*—compare, e.g., References [47,48]. In the next three subsections, the authors systematically present the background of the research and the main assumptions of the model development process.

2.1. Ensemble Averaging

The main assumption for the ensemble averaging approach is that approximation of $g(x_j)$ is done with the use of a linear combination of K-trained ANNs. The formal notation is given by Equation (2):

$$g(x_j) = \frac{1}{K} \sum_{i=1}^{K} f_i(xj) + \varepsilon_i, \tag{2}$$

where $f_i(x_j)$ stands for the approximation and ε_i denotes an error of approximation by i-th neural network for $i = 1, \ldots ,K$. Such a mechanism (compare Reference [48]), which does not involve input signals, where individual outputs of ANNs are combined to produce an overall output, belongs to a class of static structures. The following assumptions can be made [47]—the sum-of-squares error for $f_i(x_j)$ can be given as:

$$E_i^{sos} = \sum \left(\{ g(x_j) - f(x_j) \}^2 \right), \tag{3}$$

where E_i^{sos} corresponds to an integration over x_j, weighted by unconditional density $p(x_j)$:

$$E_i^{sos} \equiv \int \ldots \int \ldots \int \varepsilon_i^2(x_j) p(x_j) dx_1 \ldots dx_j \ldots dx_n, \tag{4}$$

The average error by the networks acting individually can be written as

$$E_{av}^{sos} = \frac{1}{K} \sum_{i=1}^{K} E_i^{sos}. \tag{5}$$

Supposing that the output of the ensemble of networks is the average of outputs of K networks that belong to the ensemble, we have the prediction of the ensemble $f_{ens}(x_j)$:

$$f_{ens}(x_j) = \frac{1}{K} \sum_{i=1}^{K} f_i(x_j). \tag{6}$$

Under the assumption that $\varepsilon_i(x_j)$ are uncorrelated and have zero mean, the relation of the ensemble error to the average error of the networks working separately is:

$$E_{ens}^{sos} = \frac{1}{K^2} \sum_{i=1}^{K} E_i^{sos} = \frac{1}{K} E_{av}^{SOS}. \tag{7}$$

In practice, $\varepsilon_i(x_j)$ are highly correlated and the reduction of the error is much smaller. Typically, some useful reduction of the error is obtained, as the ensemble averaging cannot produce an increase in the expected error:

$$E_{ens}^{sos} \leq E_{av}^{SOS}. \tag{8}$$

The expectation is that differently trained networks converge to different local minima on the error surface, and the overall performance is improved by combining the outputs in some way [47]. The employment of neural networks ensembles may lead to satisfactory results, especially when the number of training patterns is relatively low or the training data is noisy [47,50].

2.2. Stacked Generalisation

The stacked generalisation approach, (compare Reference [47]), is based on combining several trained networks together into a two-level model. The general expectation of such an approach is to improve the generalisation capabilities of the networks acting in isolation. The two-step procedure includes a training set of K level-0 networks, whose outputs are then used to train a level-1 network. One can say that the level-0 networks form an ensemble, and the level-1 network acts as a combiner of the outputs of the networks belonging to the ensemble. The general idea of the approach is presented in Figure 1.

A stacked generalisation-based model combines the outputs of level-0 networks trained with the x_j inputs; the outputs of level-0 networks can be written down as $\hat{y}_i = f_i(x_j)$, with the use of the level-1 network to give the final output. Formally the model can be given as

$$g(x_j) = h(f_i(x_j)) + \varepsilon_{sg}. \tag{9}$$

Consequently, predictions on new data is also a two-step procedure. They are made by presenting new input data to the level-0 networks and computing their outputs, which are then presented to the level-1 network which computes the final output. The general suggestion for the stacked generalisation approach is that the ensemble of level-0 networks should consist of various networks that differ from each other, whilst the level-1 network should have a relatively simple structure [47].

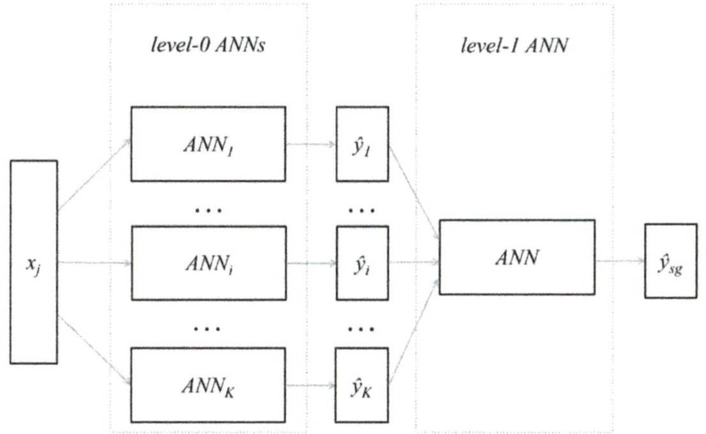

Figure 1. General idea of stacked generalisation approach.

3. Construction Site Overhead Cost Index Prediction as a Regression Analysis Problem—Assumptions for Ensemble Averaging and Stacked Generalisation

The prediction of site overhead cost index by the neural networks ensemble and ensemble averaging approach can be formally given with the following Equations (10) and (11):

$$y = \frac{1}{K}\sum_{i=1}^{K} f_i(xj) + \varepsilon_i, \tag{10}$$

$$\hat{y}_{ens} = \frac{1}{K}\sum_{i=1}^{K} f_i(xj), \tag{11}$$

where y—real life values of site overhead cost index (dependent variable), $\hat{y}_{ens}$—values of y predicted by the ensemble of neural networks, f_i—the i-th mapping function implemented implicitly by the i-th neural network belonging to an ensemble, x_j—dependent variables, input shared by all of the members of the ensemble for $j = 1, \ldots ,m$, ε_i—error of approximation by the i-th member of the ensemble for $i = 1, \ldots , K$.

On the other hand, the prediction by neural networks ensemble and stacked generalisation approach is denoted with Equations (12) and (13):

$$y = h\big(f_i(x_j)\big) + \varepsilon_{sg}, \tag{12}$$

$$\hat{y}_{sg} = h\big(f_i(x_j)\big), \tag{13}$$

where y—as in (11), $\hat{y}_{sg}$—values of y predicted by the stacked generalisation-based two-level model, h—the mapping function implemented implicitly by level-1 neural network, f_i—the i-th mapping function implemented implicitly by i-th level-0 neural network, x_j—as in (11), and ε_{sg}—the error of approximation by the model.

The relationship between the set of selected predictors and the site overhead cost index was investigated by the authors. Eleven independent variables of the model were selected on the basis of studies of literature [28,31,46] and investigations of the number of projects completed in Poland. The training data included samples of real-life values of dependent variables, y, and corresponding vectors of dependent variables, x_j. The value of the dependent variable in the p-th sample ($p = 1, \ldots ,143$) was calculated as follows:

$$y^p = SOC_{ind}{}^p = \frac{SOC^p}{LC^p + MC^p + EC^p + SC^p} \cdot 100\%, \tag{14}$$

where $SOC_{ind}{}^p$—site overhead costs index, SOC^p—site overhead costs observed in reality, LC^p—labour costs observed in reality, MC^p—material costs observed in reality, EC^p—equipment costs observed in reality, and SC^p—subcontractors' costs observed in reality for the p-th observation (sample). Some exemplary data, including cost components present in the Equation (13), in thousands of Euros, and corresponding site overhead cost indexes, are presented in Table 1.

Table 1. Exemplary values of site overhead costs index.

p	SOC^p	LC^p	MC^p	EC^p	SC^p	SOC$_{ind}{}^p$
11	450.00	3828.60	4183.50	336.00	1818.40	4.4%
37	289.00	1693.00	1564.00	85.00	0.00	8.6%
72	812.54	3393.91	2893.45	564.30	5146.69	6.8%
99	217.60	382.36	514.23	48.52	547.43	14.6%

Independent variables of the model were selected on the basis of studies of the literature and investigations of the number of projects completed in Poland. As a result, a set of selected independent variables was proposed; these variables were denoted as x_j, where $j = 1, \ldots ,11$. Three variables brought to the model information about the types of work that were executed in the project were:

- x_1—types of work—general construction works,
- x_2—types of work—installation works,
- x_3—types of work—engineering works.

Another four variables brought to the model information about the construction site location were:

- x_4—construction site location—in the city centre,
- x_5—construction site location—outside the city centre,
- x_6—construction site location—non-urban spaces,
- x_7—distance between the construction site and the company's office.

One of the variables brought to the model information about the duration of construction works was:

- x_8—overall duration of construction works.

Another two variables brought to the model information about the execution of works in winter and about the subcontracted works were:

- x_9—relationship between the amount of works performed in winter to the total amount of works,
- x_{10}—relationship of the amount of works performed by subcontractors to the total amount of works.

The last variable brought to the model information about the main contractor was:

- x_{11}—size and necessary potential of the main contractor.

(When compared to the earlier authors' studies on the problem [1,32], the set of ten independent variables has been expanded. Thorough review of available data, which was collected in the earlier phases of the research, allowed to select an additional variable which brings to the model information about the capabilities of the contractor - namely its size and potential.)

Variables x_1–x_6 were of the nominal type. A binary method of coding was applied in the case of x_1, x_2 and x_3—their values range was 0 or 1. In the case of x_4, x_5 and x_6—a "1 of n" method of coding was applied—the range of values, considered for the three variables altogether, was $1, 0, 0$ or $0, 1, 0$ or $0, 0, 1$.

Variables x_7–x_{10} were of the quantitative type, whereas x_{11} was of the nominal type. A pseudo-fuzzy scaling method of coding was applied for transformation of the original values or information into numerical values into the range 0.1–0.9 in the case of the variables presented in Table 2, but for the variable x_9 the values were scaled into the range 0.0–1.0. The transformation for these variables is presented in Table 2. The rationale for the transformation was to provide a common scale for all of the variables.

Table 2. Transformation of the descriptive values into the numerical values for variables x_7–x_{11}.

Variable	Description	Descriptive Values	Numerical Values
x_7	distance	up to 20 km	0.1
		more than 20 km	0.9
x_8	duration of construction works	up to 6 months	0.1
		between 6 and 12 months	0.5
		more than 12 months	0.9
x_9	share of the amount of works executed in winter	up to 10%	0
		between 10% and 20%	0.1
		between 20% and 40%	0.3
		between 40% and 60%	0.5
		between 60% and 80%	0.7
		between 80% and 90%	0.9
		more than 90%	1
x_{10}	share of subcontractors in the total amount of works	up to 20%	0.1
		between 20% and 50%	0.5
		between 50% and 100%	0.9
x_{11}	size and potential of the main contractor	low	0.1
		average	0.5
		high	0.9

The database that included 143 samples was built on the basis of a survey which was completed by Polish contractors. The survey investigated the factors that influence site overhead costs and the scope and complexity of construction works for completed building projects. The studies of the returned surveys resulted in gathering and ordering data used in the process of ANNs training. Table 3 presents some samples of the training data; exemplary records from the database are given.

Table 3. Exemplary samples of the training data.

p	x_1	x_2	x_3	x_4	x_5	x_6	x_7	x_8	x_9	x_{10}	x_{11}	y
7	1.00	0.00	0.00	0.00	1.00	0.00	0.90	0.50	0.00	0.10	0.50	8.2%
29	1.00	1.00	1.00	0.00	0.00	1.00	0.90	0.10	0.90	0.10	0.90	12.8%
53	1.00	1.00	1.00	0.00	1.00	0.00	0.10	0.50	0.30	0.90	0.50	4.9%
73	1.00	0.00	0.00	1.00	0.00	0.00	0.90	0.50	0.10	0.10	0.10	4.4%
82	1.00	1.00	1.00	0.00	1.00	0.00	0.10	0.10	0.10	0.10	0.10	6.1%
105	1.00	0.00	0.00	0.00	1.00	0.00	0.10	0.10	0.10	0.50	0.10	4.2%
117	0.00	0.00	1.00	0.00	1.00	0.00	0.10	0.10	0.50	0.50	0.50	9.7%

The strategy of the models' development, as well as the assumptions about the training, testing, and performance analysis, are explained in the next section.

4. Models' Development, Results, and Discussion

4.1. Models' Development Strategy

The strategy of the model development included conducting multiple training and testing of a number of different types of single ANNs as candidates to become members of the ensemble, forming the ensemble, and then investigating the two approaches discussed earlier. The strategy is presented schematically in the chart in Figure 2 and then discussed in detail.

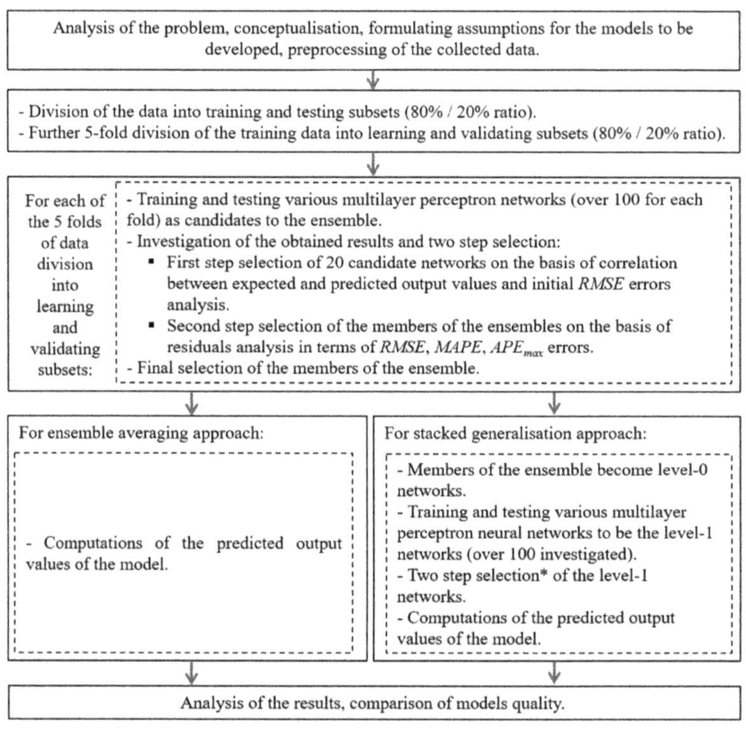

Figure 2. Scheme of the strategy of the models' development.

The whole set of collected data was divided into two main subsets used for training and testing purposes. The testing subset, later referred to as T, was selected carefully to be statistically representative for the whole data collection and included 20% of the samples from the whole set of collected data. The data belonging to this subset did not take part in the training of ANNs and was used for the purposes of examination of single ANNs, as well as the ensemble models built upon the ensemble averaging and stacked generalisation approaches. Samples belonging to the subset T play the role of new cases in prediction performance analysis as well.

The remaining data was used for training i.e., for supervised learning and validating of single ANNs candidates to become members of the ensemble. Later, these subsets are referred to as L and V, respectively, whilst the whole training subset is referred to as $L\&V$. The strategy involved division of the remaining data in the relation $L/V = 80\%/20\%$, repeated five times, so the five folds of data were available for training purposes. Moreover, each of the samples belonging to the $L\&V$ subset took part in supervised learning in four folds and in validating in one fold, so the networks for each fold are trained with data which varies in terms of falling different samples either to the L or V subsets.

Another key assumption was to select one ANN for each of the folds of L and V subsets to become the member of the ensemble. The selection was made on the basis of two-step ANNs' performance analysis and assessment within the sets of networks trained with the use of each fold of L and V subsets. The rationale for such assumption was not only to choose the best networks but also to minimise the risk that the prediction of the model is biased due to the sampling of the L and V subsets. The employed error function and criteria of the trained networks assessment are presented in Table 4. For the purposes of performance assessment and analysis of single ANNs, Pearson's correlation coefficient (15) and error measures (16)–(20) were calculated for the L, V, $L\&V$, and T subsets.

Selection of the ensemble members was preceded by an investigation of a number of various multilayer perceptron (MLP) ANNs with one hidden layer, whose structures included 11 neurons in the input layer, h neurons in the hidden layer, and 1 neuron in the output layer. The choice of the MLP networks relied on their applicability to regression problems (compare References [29,49]).

The networks varied in the number of neurons in the hidden layer (h ranged from 4 to 11), the types of employed activation functions—both in the neurons of the hidden and output layer (sigmoid, hyperbolic tangent, exponential, and linear function) and the initial weights of the neurons—at the beginning of the training process. The Broyden–Fletcher–Goldfarb–Shanno algorithm (BFGS) was used for training individual networks—the details about the algorithm can be found in the literature, e.g., Reference [47]. The choice of the training algorithm was dictated by its availability in the software that were used for neural simulations. As one of the three available algorithms, BFGS offered the fastest performance and best convergence of training and testing processes for the investigated problem. A variety of different combinations of employed activation functions and numbers of neurons in hidden layers that made, altogether, over 100 networks were trained for each of the five folds of L and V subsets.

The first step of selection included an assessment of correlation coefficient between the expected and predicted output and root mean squared error (*RMSE*) values. From the set networks, which fulfilled the conditions of $R_L > 0.90$, $R_V > 0.90$, $R_{L\&V} > 0.90$, and $R_T > 0.90$, the authors initially selected 20 networks for which the differences between $RMSE_L$, $RMSE_V$, $RMSE_{L\&V}$, and $RMSE_T$ were the smallest.

The second step of the selection relied on a thorough review of the initially selected networks for each of the five folds of L and V subsets. The authors carried out a residual analysis, in terms of both measures presented in Table 4, and distributions, dispersions, and values of errors for the samples belonging to the training and testing subsets.

Table 4. Error function and models' performance assessment criteria.

Description	Equation		Used As
sum-of-squares error function	$E_{sos} = \frac{1}{2} \sum\limits_{p \in L} (y^p - \hat{y}^p)^2$	(15)	error function
Pearson's correlation coefficient	$R = \frac{cov(y,\hat{y})}{\sigma_y \sigma_{\hat{y}}}$	(16)	criteria for general assessment of trained ANN's quality (calculated for L, V, $L\&V$, T subsets separately, $cov(y,\hat{y})$—covariance between y and $\hat{y}$, σ_y—standard deviation for y, $\sigma_{\hat{y}}$—standard deviation for $\hat{y}$)
root mean squared error	$RMSE = \sqrt{\frac{1}{c} \sum\limits_{p} (y^p - \hat{y}^p)^2}$	(17)	criteria for assessment of quality and performance of both trained ANNs and developed models based on ensembles (calculated for L, V, $L\&V$, T subsets separately, c stands for L, V, $L\&V$, T subsets cardinality, p stands for index of a training pattern)
mean absolute percentage error	$MAPE = \frac{100\%}{c} \sum\limits_{p} \left| \frac{y^p - \hat{y}^p}{y^p} \right|$	(18)	
absolute percentage error	$APE^p = \left| \frac{y^p - \hat{y}^p}{y^p} \right| \cdot 100\%$	(19)	
maximum of absolute percentage error	$APE_{max} = max \left(\left| \frac{y^p - \hat{y}^p}{y^p} \right| \cdot 100\% \right)$	(20)	

4.2. Results and Discussion

A review and comparison of the network's performance, based on the methodical analysis, allowed for finally choosing five networks—one for each fold of L and V subsets. The five selected networks—later referred to as ANN1, ANN2, ANN3, ANN4, and ANN5—are presented in Table 5.

Table 5. Details of the five networks selected to be the members of the ensemble.

ANN	Structure	Number of Neurons in the Hidden Layer	Hidden Layer Activation Function	Output Layer Activation Function	Number of Training Epochs
ANN1	MLP 11-10-1	10	hyperbolic tangent	hyperbolic tangent	146
ANN2	MLP 11-10-1	10	hyperbolic tangent	hyperbolic tangent	61
ANN3	MLP 11-6-1	6	exponential	linear	109
ANN4	MLP 11-8-1	8	hyperbolic tangent	exponential	67
ANN5	MLP 11-8-1	8	logistic	hyperbolic tangent	73

Table 6 presents the results of training and testing of the five selected networks acting separately. The results in the Table are given according to the criteria presented in Table 4. The results in Table 6 are satisfying, however one can easily see that there are some differences between the performances of the five networks.

Figure 3 presents the scatterplot of the expected and predicted values of SOC_{ind}, points of coordinates $(y^p, \hat{y}^p)$, for the training and testing subsets drawn for the five selected networks acting individually. One can see that, in terms of the criteria shown in Table 4 and according to the results presented in Table 5, the performance of the three networks acting individually was similar and the errors were comparable. However, Figure 3a,b and the analysis of the location and the distribution of the points in the graphs reveal that the predictions for will depended strongly on the choice of a single network acting separately. Although most of the points were distributed along the line of a perfect fit, some points (marked with the ellipses) were placed outside of the cone delimited by percentage errors equal to +25% and −25%.

Table 6. Training results and performance of the selected networks.

ANN	R				RMSE				MAPE			
	Training		Testing		Training		Testing		Training		Testing	
	L	V	L&V	T	L	V	L&V	T	L	V	L&V	T
ANN1	0.9898	0.9662	0.9850	0.9731	0.0112	0.0206	0.01308	0.0181	6.0%	18.3%	8.4%	10.18%
ANN2	0.9861	0.9319	0.9761	0.9729	0.0131	0.0277	0.01643	0.0187	8.7%	14.5%	9.9%	10.19%
ANN3	0.9808	0.9645	0.9778	0.9804	0.0154	0.0198	0.01584	0.0159	10.4%	9.6%	11.3%	12.70%
ANN4	0.9868	0.9555	0.9737	0.9751	0.0128	0.0227	0.01482	0.0175	10.2%	13.1%	10.8%	13.69%
ANN5	0.9855	0.9278	0.9807	0.9881	0.0132	0.0296	0.01724	0.0123	11.1%	17.8%	12.5%	9.15%

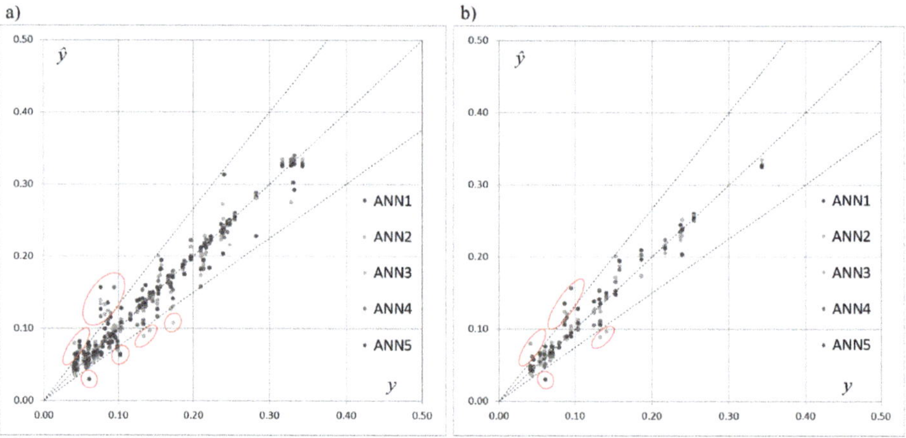

Figure 3. Scatterplots of y and $\hat{y}$ for the five selected neural networks acting separately: (**a**) scatterplot for samples belonging to the training subset, (**b**) scatterplot for samples belonging to the testing subset.

Table 7 presents the maximal values of absolute percentage errors (20) calculated for the five selected ANNs. The values in Table 7 reveal significant errors of predictions, which also justify employment of ensembles of neural networks in the problem.

Table 7. APE_{max} errors obtained for the five selected networks.

ANN	APE_{max}			
	Training			Testing
	L	V	L&V	T
ANN1	65.9%	64.7%	65.9%	76.1%
ANN2	80.7%	45.2%	80.7%	33.4%
ANN3	73.2%	91.8%	91.8%	43.2%
ANN4	50.6%	64.4%	64.4%	70.1%
ANN5	53.3%	79.1%	79.1%	26.6%

The five chosen networks were combined to form the ensemble. The rules presented earlier—Equations (10) and (11)—were employed for implementation of the ensemble averaging approach and the outputs of the model were computed as well as the errors and error measures. This model is later called ENS AV.

To complete the process of model development based on the stacked generalisation approach, the authors investigated a number of artificial neural-network candidates to become the level-1 networks. The investigated networks' structures included five neurons in the input layer (as a consequence of the selection of five ensemble member networks), h neurons in the hidden layer, and one neuron in the output layer. The number of neurons in hidden layer h ranged from one to three, as the structure of the level-1 network was supposed to be simple (compare Section 2.2). The types of employed activation functions and training algorithm were the same as in the case of the training ensemble candidate networks (as presented previously in Section 3). Training patterns that included outputs of the five ensemble member networks as the inputs of level-1 networks, and the accompanying real-life values of SOC_{ind} as the expected outputs, were divided randomly for each investigated network into the learning and validating subset in the proportion L/V = 60%/40%. The investigated networks varied also in the initial weights of the neurons at the beginning of the training process. Altogether, around 100 networks were trained and examined. For the purposes of testing, the authors used the T subset, which was selected in the initial stage of the research (as presented previously in Section 3). The criteria of two-step selection of the level-1 networks were similar as in the case of ensemble candidate networks (as presented previously in Section 3). The final choice of two level-1 networks, namely MLP 5-2-1 and MLP 5-3-1, allowed for the introduction of two alternatively stacked generalisation-based models. The final choice of the two above-mentioned level-1 networks, and further discussion of two alternative models based on stacked generalisation, was due to the comparable quality of these models. These models are later called ENS SG1 and ENS SG2, respectively. The details of the selected level-1 networks are presented in Table 8.

Table 8. Details of the two level-1 networks selected for the stacked generalisation-based models.

Model	Structure	Number of Neurons in the Hidden Layer	Hidden Layer Activation Function	Output Layer Activation Function	Number of Training Epochs
ENS SG1	MLP 5-2-1	2	exponential	linear	40
ENS SG2	MLP 5-3-1	3	exponential	exponential	51

All three proposed models based on the ensemble of networks, namely ENS AV, ENS SG1, and ENS SG2, were assessed in terms of performance and prediction quality. The overall results appear together in Table 9. For the purposes of performance assessment and analysis of ensemble averaging and stacked generalisation-based models, Pearson's correlation coefficient (16) and error measures (17), (18), (19), and (20) were calculated for *L&V* and *T* subsets.

Table 9. Performance measures for the three developed models based on the ensembles of networks.

Model	R		RMSE		MAPE		APE$_{max}$	
	Training	Testing	Training	Testing	Training	Testing	Training	Testing
	L&V	T	L&V	T	L&V	T	L&V	T
ENS AV	0.9869	0.9899	0.0126	0.0112	8.3%	7.1%	42.6%	23.4%
ENS SG1	0.9853	0.9878	0.0135	0.0127	9.6%	9.2%	40.7%	26.7%
ENS SG2	0.9914	0.9922	0.0103	0.0098	7.3%	6.3%	23.2%	18.5%

When the values in Table 9 are collated with values in Tables 5 and 6, the improvements in error measures can be seen easily. The performance of all three models based on the ensembles of networks is better when compared with the performance of the networks acting in isolation. The most evident improvement is achieved for APE_{max}.

Figures 4–6 depict scatterplots of the expected and predicted values of SOC_{ind} for the ENS AV, ENS SG1, and ENS SG2 models. Figures 4–6 present the points of coordinates $(y^p, \hat{y}^p_{ens})$ for the training and testing subsets separately. When compared to Figure 3, these graphs show that combining the five selected ANNs allowed for the compensation of errors made by the ANNs acting in isolation in the case of the ENS AV as well as the ENS SG1 and ENS SG2 models. Although an improvement has been achieved in the case of all three introduced models, one can see that the best performance is provided by ENS SG2, where all of the points are distributed within the cone of acceptable errors. In the case of ENS AV and ENS SG1, there are single points located outside of the cone.

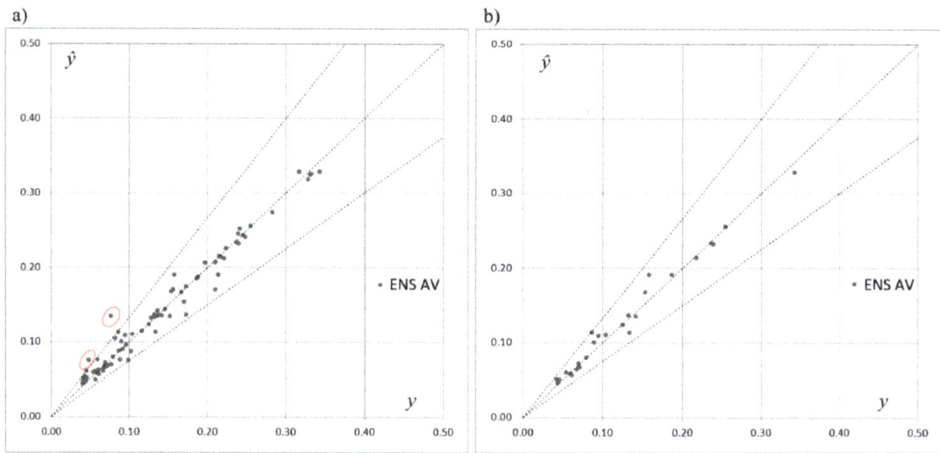

Figure 4. Scatterplot of y and $\hat{y}_{ens}$ for the ensemble, ENS AV, performing ensemble averaging; (**a**) scatterplot for samples belonging to the training subset, (**b**) scatterplot for samples belonging to the testing subset.

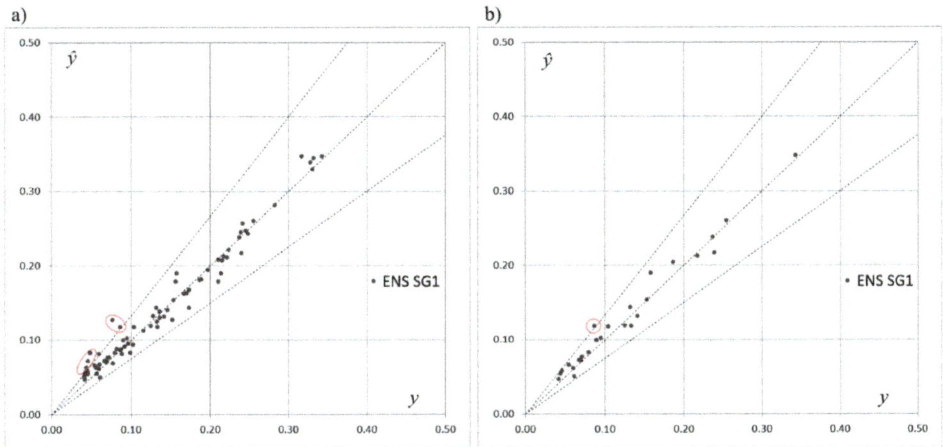

Figure 5. Scatterplot of y and $\hat{y}_{sg}$ for the ensemble, ENS SG1; (**a**) scatterplot for samples belonging to the training subset, (**b**) scatterplot for samples belonging to the testing subset.

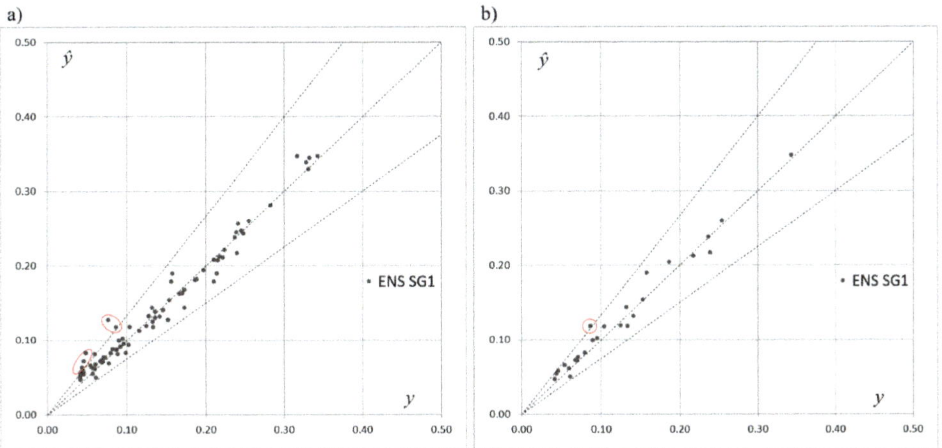

Figure 6. Scatterplot of y and $\hat{y}_{sg}$ for the ensemble, ENS SG2: (**a**) scatterplot for samples belonging to the training subset, (**b**) scatterplot for samples belonging to the testing subset.

Figures 7–9 depict frequencies and distributions of APE^p errors computed for the training and testing subsets for models based on ensembles of networks. The errors have been accumulated and counted in five intervals, whose ranges equalled 5%; one interval accumulated errors greater than 25%:

- interval 1: $0\% \leq APE^p < 5\%$,
- interval 2: $5\% \leq APE^p < 10\%$,
- interval 3: $10\% \leq APE^p < 15\%$,
- interval 4: $15\% \leq APE^p < 20\%$,
- interval 5: $20\% \leq APE^p < 25\%$,
- interval 6: $APE^p \geq 25\%$.

The columns in the Figures 7–9 show the percentage frequencies of the errors that have fallen into one of the intervals. The polylines show the distribution of the errors (cumulative frequencies according to the accepted order of intervals). In Figures 7–9, one can see that, in the case of the ENS

AV and ENS SG1, only a few APE^p errors (19) are greater than 25%, and in the case of ENS SG2, none of them fall into this range. On the contrary, for networks acting separately, the significant number of errors is greater than 25%. These results can be explained through the analysis of the APE^p errors for the networks acting separately. For the networks acting separately (ANN1, ANN2, ANN3, ANN4, ANN5), many of the errors APE^p belonging to the interval 1 were relatively small and close to 0%. On the other hand, these small errors were accompanied by a significant number of errors $APE^p \geq 25\%$, and high values of APE_{max} (compare Table 7). In the case of the ensemble-based models, these errors have been compensated due to the ensemble averaging (ENS AV) or stacked generalisation (ENS SG1, ENS SG2).

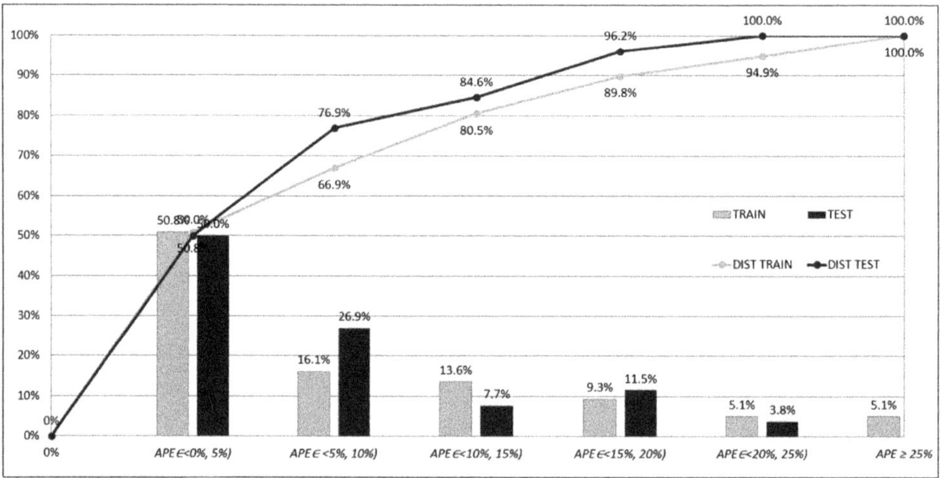

Figure 7. Frequencies and distributions of absolute percentage errors for the ENS AV model computed for the training and testing subsets.

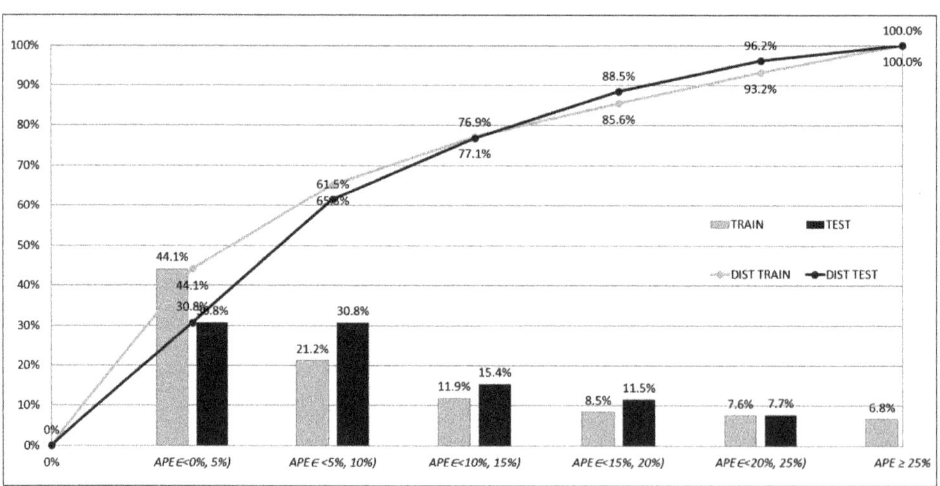

Figure 8. Frequencies and distributions of absolute percentage errors for the ENS SG1 model computed for the training and testing subsets.

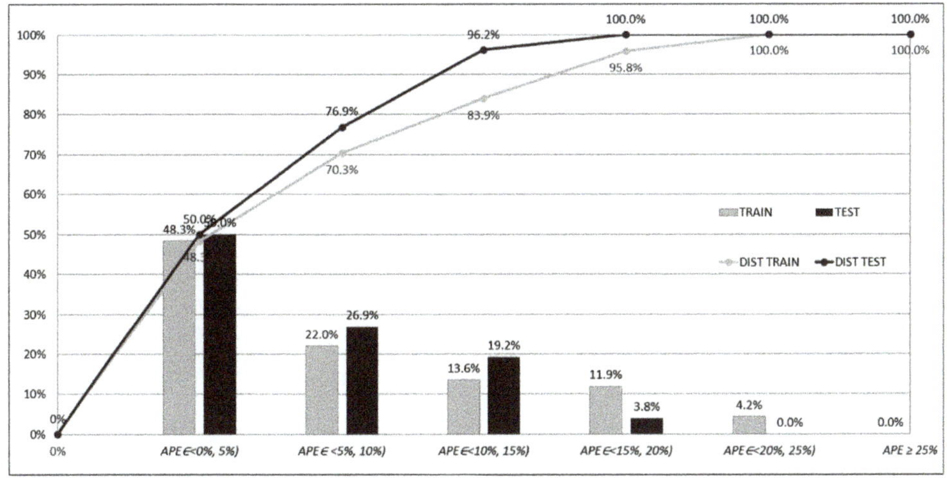

Figure 9. Frequencies and distributions of absolute percentage errors for the ENS SG2 model computed for the training and testing subsets.

The compensation resulted in the collection of most of the prediction errors in the first five intervals. One cost of this compensation is the decrease of the number of small errors, close to 0, in the first interval. The benefit of the compensation, however, is the improvement of the overall prediction performance and better knowledge generalisation. As mentioned previously, one can easily see that the best performance is offered by the ENS SG2 model as there were no errors $APE^p \geq 25\%$.

The analysis of the research results leads to the conclusion that the employment of only one of the five selected networks (as presented in Table 5) to support the prediction of SOC_{ind} would burden the predictions with the choice of a network—this is confirmed by the distribution of points that represent expected and predicted values $(y^p, \hat{y}^p)$ in Figure 3.

On the other hand, combining these five networks to form an ensemble compromises the strengths and weaknesses of the five ANNs—for some data, certain single-acting networks offered good predictions, whilst for others, there were weak predictions. Combining these networks into an ensemble allows for synergy. The decrease in APE_{max}, as well as more stable predictions, are the most beneficial from employment of the ensembles in the models. Furthermore, a risk of errors exceeding the critical level of 25%, in terms of percentage errors is reduced. These benefits have been achieved at some cost, mainly due to compensation of very small and very high errors offered by certain networks acting separately for certain training and testing patterns. However, the compensation of the errors from the ensemble-based models reduces the unwanted oversensitivity of the networks acting separately to certain training patterns.

5. Summary and Conclusions

The authors developed three original models based on ensembles of neural networks aimed at the prediction of site overhead cost index for construction projects. One of the models employed ensemble averaging and two of the models employed stacked generalisation. The developed models are capable of predicting the site overhead cost index with a satisfactory accuracy and, thus, supporting estimates of site overhead costs. In the light of the presented research, the general conclusion is that the employment of the ensemble of neural networks to the models proved to be superior over the approach based on the employment of a single neural network. Moreover, the effort—which is unavoidable in the training, verifying, and selecting number of networks of similar quality—is not wasted. In practical terms, the prediction using the ensemble averaging is simple—it needs an averaging of the outputs of networks belonging to the ensemble. On the other hand, stacked generalisation needs some additional computational effort that includes the training and selection of level-1 networks.

In the proposed approach, the authors employed an ensemble of neural networks to be the core of all three models. All of the proposed models consist of five different MLP networks, chosen from over 500 trained networks (over 100 networks were trained and investigated for each of five folds of training data). The five networks vary in their structure, employed activation functions, and initial conditions of training processes. The performance of the five chosen networks is comparable. However, the predictions made by the networks acting separately are burdened with the conditions of the training processes, sampling of learning and validating, and the specificity of each of the networks. Combining the five networks together leads to improvements in predictions due to compromising the strengths and weaknesses of the five networks. The prediction of the site overhead cost index using the ensemble-based models allowed for compensation of the errors made by the single networks. The predictions based on the three models for which the proposed ensemble is a core are more stable, and the risk of exceeding a critical range of errors is minimised.

The ensembles of neural networks proved their superiority over single neural networks acting in isolation. $MAPE$ testing errors for the five networks acting in isolation ranged between 9.15% and 13.69%, whereas APE_{max} ranged between 26.6% and 76.1%. In the case of the proposed ensembles of networks, both $MAPE$ and APE_{max} errors for testing were lower; values of $MAPE$ ranged between 6.3% and 9.2%, whereas values of APE_{max} ranged between 18.5% and 23.4%. The quality of the ensemble-based model is also visible in the distribution of errors for each of them—more than 90% of the APE testing errors were smaller than 25%.

The three developed models, namely ENS AV (based on the ensemble averaging approach) and ENS SG1 and ENS SG2 (based on stacked generalisation approach), offer comparable prediction quality and performance, however the best results were achieved for ENS SG2. ENS SG2 is a two-level model that employs the five selected ANNs as the level-0 networks and simple ANN as a level-1 network.

The authors' findings, justified by the analysis of the models' performance, is that the developed models are capable of supporting estimation of site overhead costs in building construction projects. In the case of other types of constructions, e.g., bridges, roads, infrastructure, etc., the specificity of the projects must be taken into account and separate models must be developed.

Further research will include studies on the development of models, supporting the cost estimation process on different levels (for certain facilities, construction works, and cost components), that are based on the concept of ensembles of neural networks.

Author Contributions: Conceptualisation, M.J. and A.L.; data curation, A.L.; formal analysis, M.J.; investigation, M.J. and A.L.; methodology, M.J. and A.L.; validation, M.J. and A.L.; visualization, M.J.; writing—original draft, M.J. and A.L.; writing—review and editing, M.J. and A.L.

Funding: No external funding has been received—this research was funded by statutory activities of Cracow University of Technology.

Conflicts of Interest: The authors declare no conflict of interest.

References

1. Leśniak, A.; Juszczyk, M. Prediction of site overhead costs with the use of artificial neural network based model. *Arch. Civ. Mech. Eng.* **2018**, *18*, 973–982. [CrossRef]
2. Gajzler, M.; Zima, K. Evaluation of Planned Construction Projects Using Fuzzy Logic. *Int. J. Civ. Eng.* **2017**, *15*, 641–652. [CrossRef]
3. Leśniak, A.; Plebankiewicz, E. Modeling the decision-making process concerning participation in construction bidding. *J. Manag. Eng.* **2013**, *31*, 04014032. [CrossRef]
4. Skorupka, D. Identification and initial risk assessment of construction projects in Poland. *J. Manag. Eng.* **2008**, *24*, 120–127. [CrossRef]
5. Tam, C.M.; Tong, T.K.; Chiu, G.C.; Fung, I.W. Non-structural fuzzy decision support system for evaluation of construction safety management system. *Int. J. Proj. Manag.* **2002**, *20*, 303–313. [CrossRef]

6.   Hoła, B. Identification and evaluation of processes in a construction enterprise. *Arch. Civ. Mech. Eng.* **2015**, *15*, 419–426. [CrossRef]

7.   Krzemiński, M. KASS v.2.2. Scheduling Software for Construction with Optimization Criteria Description. *Acta Phys. Polon. A* **2016**, *309*, 1439–1442. [CrossRef]

8.   Chatterjee, K.; Zavadskas, E.K.; Tamošaitienė, J.; Adhikary, K.; Kar, S. A hybrid MCDM technique for risk management in construction projects. *Symmetry* **2018**, *10*, 46. [CrossRef]

9.   Tamošaitienė, J.; Zavadskas, E.K.; Šileikaitė, I.; Turskis, Z. A novel hybrid MCDM approach for complicated supply chain management problems in construction. *Procedia Eng.* **2017**, *172*, 1137–1145. [CrossRef]

10. Zavadskas, E.K.; Antucheviciene, J.; Vilutiene, T.; Adeli, H. Sustainable decision-making in civil engineering, construction and building technology. *Sustainability* **2018**, *10*, 14. [CrossRef]

11. Anysz, H.; Zbiciak, A.; Ibadov, N. The Influence of Input Data Standardization Method on Prediction Accuracy of Artificial Neural Networks. *Procedia Eng.* **2016**, *153*, 66–70. [CrossRef]

12. Dikmen, S.U.; Sonmez, M. An Artificial Neural Networks Model for the Estimation of Formwork Labour. *J. Civ. Eng. Manag.* **2011**, *17*, 340–347. [CrossRef]

13. Schabowicz, K.; Hoła, B. Application of artificial neural networks in predicting earthmoving machinery effectiveness ratios. *Arch. Civ. Mech. Eng.* **2008**, *8*, 73–84. [CrossRef]

14. Yip, H.L.; Fan, H.; Chiang, Y.H. Predicting the maintenance cost of construction equipment: Comparison between general regression neural network and Box–Jenkins time series models. *Autom. Constr.* **2014**, *38*, 30–38. [CrossRef]

15. Leśniak, A. Supporting contractors' bidding decision: RBF neural network application. *AIP Conf. Proc.* **2016**, *1738*, 200002. [CrossRef]

16. Wanous, M.; Boussabaine, H.A.; Lewis, J. A neural network bid/no bid model: The case for contractors in Syria. *Constr. Manag. Econ.* **2003**, *21*, 737–744. [CrossRef]

17. Ashraf, M.E. Classifying construction contractors using unsupervised-learning neural networks. *J. Constr. Eng. Manag.* **2006**, *132*, 1242–1253.

18. Mrówczyńska, M. Neural networks and neuro-fuzzy systems applied to the analysis of selected problems of geodesy. *Comput. Assisted Mech. Eng. Sci.* **2011**, *18*, 161–173.

19. Zavadskas, E.K.; Vilutienė, T.; Tamošaitienė, J. Harmonization of cyclical construction processes: A systematic review. *Procedia Eng.* **2017**, *208*, 190–202. [CrossRef]

20. Kapliński, O. Innovative solutions in construction industry. Review of 2016–2018 events and trends. *Eng. Struct. Technol.* **2018**, *10*, 27–33. [CrossRef]

21. Trost, S.M.; Oberlender, G.D. Predicting accuracy of early cost estimates using factor analysis and multivariate regression. *J. Constr. Eng. Manag.* **2003**, *129*, 198–204. [CrossRef]

22. Belniak, S.; Leśniak, A.; Plebankiewicz, E.; Zima, K. The influence of the building shape on the costs of its construction. *J. Financ. Manag. Prop. Constr.* **2013**, *18*, 90–102. [CrossRef]

23. Leśniak, A.; Zima, K. Cost calculation of construction projects including sustainability factors using the Case Based Reasoning (CBR) method. *Sustainability* **2018**, *10*, 1608. [CrossRef]

24. El Sawalhi, N.I. Modelling the parametric construction project cost estimate using fuzzy logic. *Int. J. Emerg. Technol. Adv. Eng.* **2012**, *2*, 2250–2459.

25. Kim, K.J.; Kim, K. Preliminary cost estimation model using case-based reasoning and genetic algorithms. *J. Comput. Civ. Eng.* **2010**, *24*, 499–505. [CrossRef]

26. Wilmot, C.G.; Mei, B. Neural network modeling of highway construction costs. *J. Constr. Eng. Manag.* **2005**, *31*, 765–771. [CrossRef]

27. Attala, M.; Hegazy, T. Predicting cost deviation in reconstruction projects: Artificial neural networks versus regression. *J. Constr. Eng. Manag.* **2003**, *129*, 405–411. [CrossRef]

28. El-Sawalhi, N.I.; Shehatto, O. Neural Network Model for Building Construction Projects Cost Estimating. *J. Constr. Eng. Proj. Manag.* **2014**, *4*, 9–16. [CrossRef]

29. Juszczyk, M. The challenges of nonparametric cost estimation of construction works with the use of artificial intelligence tools. *Procedia Eng.* **2018**, *196*, 415–422. [CrossRef]

30. Juszczyk, M. Application of committees of neural networks for conceptual cost estimation of residential buildings. *AIP Conf. Proc.* **2015**, *1648*, 600008. [CrossRef]

31. El-Sawy, I.Y.; Hosny, H.E.; Razek, M.A. A Neural Network Model for Construction Projects Site Overhead Cost Estimating in Egypt. *Int. J. Comput. Sci. Issues* **2011**, *8*, 273–283.

32. Juszczyk, M.; Leśniak, A. Site Overhead Cost Index Prediction Using RBF Neural Networks. In Proceedings of the 3rd International Conference on Economics and Management (ICEM 2016), Suzhou, China, 2–3 July 2016; DEStech Publications Inc.: Lancaster, CA, USA, 2016; pp. 381–386. [CrossRef]

33. Juszczyk, M. Implementation of the ANNs ensembles in macro-BIM cost estimates of buildings' floor structural frames. *AIP Conf. Proc.* **2018**, *1946*, 020014.

34. Juszczyk, M.; Leśniak, A.; Zima, K. ANN Based Approach for Estimation of Construction Costs of Sports Fields. *Complexity* **2018**, *2018*, 7952434. [CrossRef]

35. Yazdani-Chamzini, A.; Zavadskas, E.K.; Antucheviciene, J.; Bausys, R. A Model for Shovel Capital Cost Estimation, Using a Hybrid Model of Multivariate Regression and Neural Networks. *Symmetry* **2017**, *9*, 298. [CrossRef]

36. Plebankiewicz, E.; Leśniak, A. Overhead costs and profit calculation by Polish contractors. *Technol. Econ. Dev. Econ.* **2013**, *19*, 141–161. [CrossRef]

37. Peurifoy, R.L.; Oberlender, G.D. *Estimating Construction Costs*, 4th ed.; McGraw Hill: New York, NY, USA, 1989.

38. Coombs, W.E.; Palmer, W.J. *Construction Accounting and Financial Management*, 4th ed.; McGraw Hill: New York, NY, USA, 1989.

39. Apanavičienė, R.; Daugėlienė, A. New Classification of Construction Companies: Overhead Costs Aspect. *J. Civ. Eng. Manag.* **2011**, *17*, 457–466. [CrossRef]

40. Chartered Institute of Building. *Project Overheads, in Code of Estimating Practice*, 7th ed.; Wiley-Blackwell: Oxford, UK, 2009.

41. Hegazy, T.; Moselhi, O. Elements of cost estimation: A survey in Canada and United States. *Cost Eng.* **1995**, *37*, 27–30.

42. Assaf, S.A.; Bubshait, A.A.; Atiyah, S.; Al-Shahri, M. The management of construction company overhead costs. *Int. J. Proj. Manag.* **2001**, *19*, 295–303. [CrossRef]

43. Brook, M. *Preliminaries in Estimating and Tendering for Construction Work*; Butterworth-Heinemann: Oxford, UK, 1998.

44. Cooke, B. *Contract Planning and Contractual Procedures*; Macmillan: Basingstoke, UK, 1981.

45. Assaf, S.A.; Bubshait, A.A.; Atiyah, S.; Al-Shahri, M. Project overhead costs in Saudi Arabia. *Cost Eng.* **1999**, *41*, 33–37.

46. Chan, C.T.W. The principal factors affecting construction project overhead expenses: An exploratory factor analysis approach. *Constr. Manag. Econ.* **2012**, *30*, 903–914. [CrossRef]

47. Bishop, P.C. *Neural Networks for Pattern Recognition*; Oxford University Press: Oxford, UK, 1995.

48. Haykin, S. *Neural Networks: A Comprehensive Foundation*; Macmillan Publishing: New York, NY, USA, 1994.

49. Rojas, R. *Neural Networks: A Systematic Introduction*; Springer Science & Business Media: New York, NY, USA, 2013.

50. Krogh, A.; Vedelsby, J. Neural network ensembles, cross validation, and active learning. In *Advances in Neural Information Processing Systems*, 7th ed.; Tesauro, G., Touretzky, D.S., Leen, T.K., Eds.; MIT Press: Cambridge, MA, USA, 1995; pp. 231–238.

 © 2019 by the authors. Licensee MDPI, Basel, Switzerland. This article is an open access article distributed under the terms and conditions of the Creative Commons Attribution (CC BY) license (http://creativecommons.org/licenses/by/4.0/).

Article

Comparative Study of Urban Area Growth: Determining the Key Criteria of Inner Urban Development

Vytautas Palevičius [1], Marija Burinskienė [1], Jurgita Antuchevičienė [2,*] and Jonas Šaparauskas [2]

[1] Department of Roads, Faculty of Environmental Engineering, Vilnius Gediminas Technical University, Saulėtekio al. 11, LT-10223 Vilnius, Lithuania; vytautas.palevicius@vgtu.lt (V.P.); marija.burinskiene@vgtu.lt (M.B.)
[2] Department of Construction Management and Real Estate, Vilnius Gediminas Technical University, Saulėtekio al. 11, LT-10223 Vilnius, Lithuania; jonas.saparauskas@vgtu.lt
* Correspondence: jurgita.antucheviciene@vgtu.lt; Tel.: +370-685-37036

Received: 12 February 2019; Accepted: 16 March 2019; Published: 20 March 2019

Abstract: Urban population is steadily growing worldwide, while the number of people in Eastern Europe is decreasing. These two contradictory trends have outlined the proposal for sustainable solutions to solve civil engineering issues that are aimed at implementing the principles of sustainable development and ensuring a better quality of urban life. When considering the problem that is encountered in Eastern European countries, a multi-criteria model for sustainable urban development has been designed and focused on planning and simulating an inner urban living environment. The suggested model has disclosed the social, economic, environmental, and sustainable components of the infrastructure that are necessary for developing inner urban areas. The components have been adapted and presented in three different size territories covering Lithuanian cities and towns. The applied expert evaluation method has assisted in determining the key criteria that should be considered in order to identify the most important inner areas of urban development. It is expected that this study will extend activities that are performed in the field of improvement of sustainability engineering processes and offer guidelines for other researchers investigating the areas of inner urban development.

Keywords: urban area; sustainable development; urban planning; urban development; urban geographic information system; multi-criteria; criteria weights

---

1. Introduction

Urban population is steadily growing worldwide, while the number of inhabitants in Eastern European countries is decreasing and reallocation between different cities is taking place [1]. In Eastern European countries, due to decline in local production emigration occurs, population decreases, and separate villages disappear in some parts of the territory of the country. Meanwhile, bigger cities are forced to expand their administrative boundaries, thus seeking to manage the unrestrained urban development along the horizontal axis, which has become the largest urban threat of the 21st century. The problem originates from accelerating construction in the suburban area, because a large part of the urban population prefers this location as a place of residence. As a result, the density of urban population decreases, the compactness of cities is diminished, the infrastructure of transport becomes more and more expensive, cars start playing the predominant role, inner urban development is paid less attention, and outer urban development is stimulated.

To avoid the above-mentioned consequences and to find a possibility of balancing (symmetry) needs and their satisfaction, the promotion of inner urban development is necessary, and therefore, the analysis and identification of development trends characteristics of urban areas and based on historical urban plans, maps, general planning, and the main factors that can be considered as the most important expected areas of inner urban development seem to be very important issues. Thus, three Lithuanian different size territories were selected as a research object: Kaunas city, with a population of approximately 300,000 people, Taurage town, with a population of around 20,000 people, and Silale town, with a population of approximately 5000 people.

The main objective of the article is to analyse the variations in three different sized urban territories to define trends in urban development and to identify the main criteria when considering which are the most important areas of inner urban development can be predefined. Thus, an overview of urban growth evolution has been presented. The comparative analysis of the studied urban areas and changes in the population has been carried out. The criteria and groups of criteria having an effect on urban development have been determined. Upon considering the major criteria in inner urban development and applying the expert evaluation method in assessing the significance of multiple criteria, the weights of the criteria have been calculated. Following the identification of the most relevant criteria, recommendations that allow for assessing the possibilities of the growing intensity of urban development have been provided. Additionally, the recommendations allow for weighing the long-term objective of inner urban development, thus increasing the development density up to 100%.

2. Overview of Guidelines and Models for Urban Development

Across Europe, there is a complex mosaic of urban population development [2]. For instance, a decline in the size of urban population and an increase in rural population across Europe have been observed. Central and East Europe are not an exception [3–5]. The ongoing migration of inhabitants to peripheral zones have driven such variations, and therefore suburban areas have grown, their activity has become integrated with the city, and rural areas have been afforded the opportunity to be connected with cities. The development of the urbanized environment and intensive construction processes include former agricultural areas that are located near the city. Good infrastructure, a well-developed transportation system, and lower land prices are important factors having an effect on the rapid spatial distribution of suburbs [6]. An increasing number of construction sites in suburban areas indicate the accelerated growth of low-rise residential buildings in natural, rural, and suburban landscapes, which results in forming the groups of new buildings, frequently urban brownfields, and landfill sites of construction waste [7].

Geoffrey K. F. Tso and Jin Li [8] make a remark that balanced national or regional development has recently gained more and more attention, and a number of countries or regions have implemented strategies to safeguard the sustainable environment. It is assumed that balanced development covers three important areas: finance (economic), social responsibility (social), and ecology (ecologic). These areas are interrelated and affect each other.

Tom Kauko [9] states that sustainable cities exist much longer than discourse on sustainability.

Researchers from the University of Granada [10] assessed sustainable urban planning analysis, thus mainly focusing on urban transport, an increase in energy consumption, green area planning, and waste management planning. The carried out research showed that decision-making institutions, consumers, and residents affected sustainable urban development. According to the scientists from Technical University of Madrid, urban development is indicated by large-scale renovation work [11].

Decision-makers in a field of urban and regional planning in Germany faced new challenges in terms of sustainability, with an emphasis to climate change [12]. High rates of urban sprawl need to be reduced by increased inner-urban development and contribute to the reduction of greenhouse gas emissions at the same time. Hoymann and Goetzke [12] state that strengthening the inner-urban development is particularly effective in terms of reducing built-up and transport area development, which matches the sustainability objective of the German Federal Government for the year 2020.

Plans from private investors often need to be redirected to meet the objectives and constraints of Governmental as well as Municipal authorities, which should also look after the overall sustainability of inner urban development and environmental sustainability above all [13]. According to Gussoni et al. [13], the delayed involvement of a competent department can negatively affect the overall planning procedure and can lead to unsatisfactory outcomes: either the project by the private investors fail or the investor's plans are realized, regardless of "urban sustainability", hampering the development of the whole surrounding area.

Other researchers in articles [14–16] unanimously point out that the current worldwide predominating type of urban development is called urban development, which is accepted as the gradual transformation of the rural environment into the urbanized one, where the outer development of the city and its suburbs absorbs more and more rural and natural areas on the outskirts. Most often, such urban development is described as partly spontaneous, dispersed, of low intensity, and creating the chaotic and fragmented suburbanized landscape. The problems of planning at a higher level, for example, lack of the systemic approach to planning an urban structure at the regional level, and can be highlighted as the main reason for these processes.

The rapid process of urbanization has a negative impact on cities due to the uncontrolled urban expansion. Thus, to successfully manage the problem, urban growth boundaries that are given different names are used. Since the beginning of the 20th century, setting urban boundaries has been used as a tool for urban planning in a number of countries across the world [17]. Urban growth boundaries, or 'green bands', have been accepted as an urban development method that regulates urban growth by defining the area. In the 1950s [18] and 1960s, most urban areas in Japan faced a rapid urban rise, which resulted in the employment of 'green bands'. For instance, the 'yellow line' system [19] in Albania was used for defining 'population centres', highlighting the urban-rural zones. Urban growth boundaries were proposed as one of the first management measures (instruments) for rapid and unrestrained urban growth in the countries such as Saudi Arabia [20], with an annual 6% increase in the urban areas, which resulted in the expansion of urban infrastructure in the most important cities [21]. Urban growth boundaries are also widespread in Canada and the United States of America (USA) as a tool for regional planning [22,23]. For example, the metropolitan zones of Vancouver, Toronto, Ottawa, Waterloo, and Ontario have set urban growth boundaries in order to limit urban growth in the areas of utmost importance and to protect green areas [24]. However, in case of Portland, urban growth boundary was not completely successful [25]. Building new cities in many countries has been intended to have the growth of the urbanized area under control, thus applying the British model based on the garden city idea. However, the effective solution has been adopted developing new cities in Israel, France, and Egypt [26].

R. Giffinger and co-authors [27] analysed intelligent urban models and put emphasis on six axes (aspects) of a smart city: governance, economy, mobility, environment, people, and living conditions. The identified six characteristics were considered to be the relevant group describing the smart city.

A review of scientific literature proposes that the researchers focusing on inner urban development do not analyse the historical development of urban areas. Similarly, scientists use a variety of methods for assessing inner urban development. However, multi-criteria methods, including different components of sustainable development, are insufficiently considered. Thus, it can be concluded that researchers have a narrow focus on the needs of society and, most of all, use their own insights to create the patterns of inner urban development (economic, social, or ecological) in one or several directions, which later becomes meaningless. On these grounds, a challenging task of carrying out historical studies on the growth of urban areas has been approached, which greatly assists in creating a multi-criteria model meeting all the aspects of sustainable development.

3. Research Methodology

Historical cartographic material and the maps of the selected city and towns were used for conducting research on the historical development of urban areas. ArcGIS software displayed the received results. The maps and plans of Kaunas, Taurage and Silale towns were taken from municipal archives and websites.

ArcGis software was applied to analyse the outcomes of variations in urban areas. The carried out analysis covered five stages. At the first stage, cartographic material was prepared—old paper plans and maps were scanned. At the second stage, the attachment of the scanned plans or maps to the coordinate system of Lithuania (LKS94) in ArcGis software was performed. At the third stage, the maximum attachment error was determined and used for making the geometric correction of the attached map. At the fourth stage, the city/town plans attached to the coordinate system of Lithuania (LKS94) were vectorized. At the fifth stage, the comparative analysis of variations in urban areas was carried out. Table 1 presents a more detailed methodology for the survey.

Table 1. Stages of research methodology.

No	Stages	Description
1.		Cartographic material is produced—old paper plans and maps are scanned.
2.		An orthophoto, including LKS94 coordinates and the scanned map, are downloaded into ArcGIS software. 4 check points are set and the scanned map is attached to the orthophoto.
3.		The maximum attachment error is determined and used for making the geometric correction of the attached map.
4.		The attached city plans are vectorized. Vectorization is the creation of the new elements presenting points, lines and areas on the basis of a raster map. This process is analogous to drawing a new map with reference to the outlines of the raster image.
5.		A comparative analysis of variations in urban areas is carried out.

Following the comparative analysis of variations in urban areas, the databases of the analyzed cities were compiled taking into account four groups of indicators:

- Group A—Urban Structure (five criteria),
- Group B—Social Environment (six criteria),
- Group C—Economic Environment (six criteria),
- Group D—Sensitive and Protected Areas (five criteria),
- Group E—Transportation (six criteria), and
- Group F—Land Use (five criteria).

All the above data were used for formulating and selecting the most important criteria and groups of criteria in inner urban development thus identifying similarities and differences in the areas. Schemes were proposed to visually present the collected data (Figures 1–3). To determine and accumulate indicators, the grid system that was suggested by the Lithuanian Department of Statistics according to the data on the general population census of 2011 was employed. According to the findings of the conducted analysis, the main criteria of six groups in inner urban development were proposed and they could be considered for defining the expected most important inner areas of urban development.

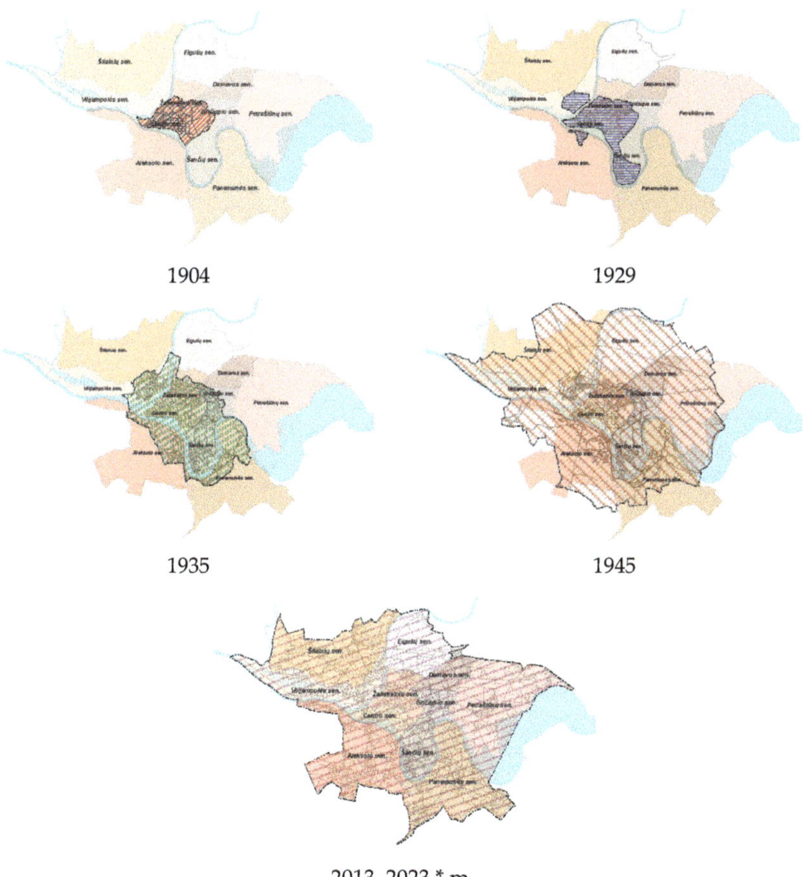

Figure 1. Variations in the area of Kaunas city in terms of the boundaries of neighbourhoods. * Solutions to the general plan of Kaunas city are valid up to 2023.

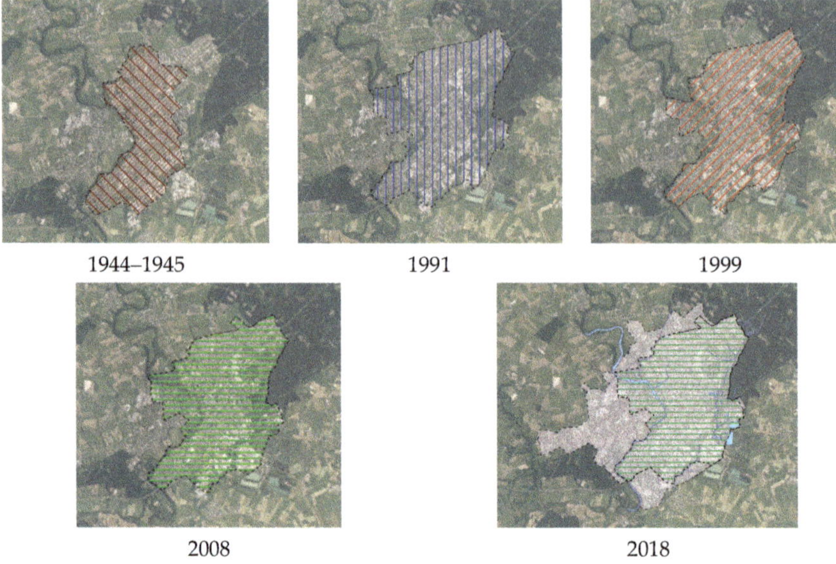

Figure 2. Variations in the development of Taurage town: Grey colour represents the maximum possible urban development according to the general plan of the city of Taurage; hatch green colour represents current situation.

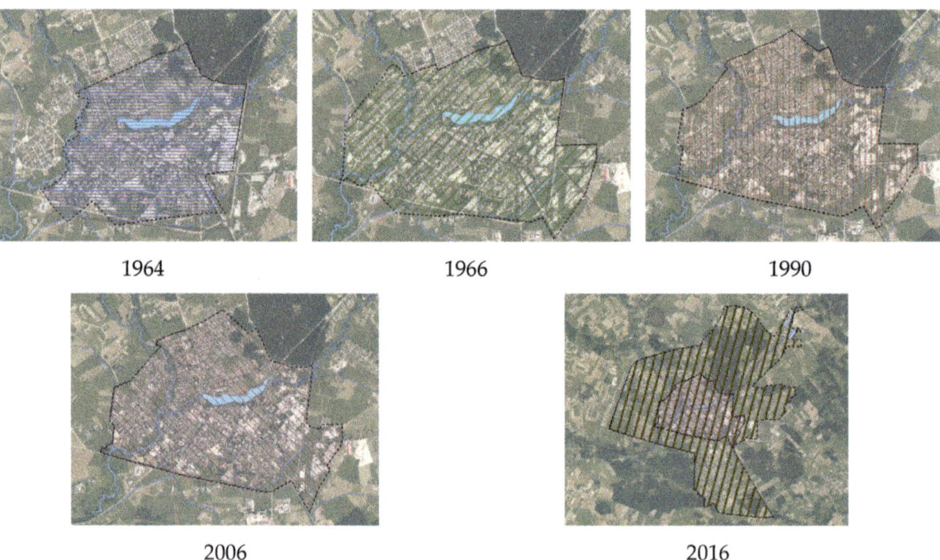

Figure 3. Variations in the development of Silale town.

The next essential and critical procedure is the weighting of criteria, i.e., determining their relative significance to the analysed problem. Both the objective and subjective methods can be used for evaluating the importance of criteria in solving multi-criteria problems [28–30]. The subjective weight approaches reflect judgments of experts, resulting in the acquisition of less rigorous values [31]. The objective weights are achieved by entirely mathematical methods [32]. From a variety of methods, the most useful and practical tools, according to [33], can be considered as a analytical hierarchy process (AHP) [34–36], analytical network process (ANP) [37], superiority and inferiority ranking (SIR) method [38], stepwise weight assessment ratio analysis (SWARA) [39,40], factor relationship (FARE) [41], KEmeny Median Indicator RanksAccordanc (KEMIRA) [42], and the best–worst method (BWM) [43]. A new approach of step-wise weight assessment with symmetric interval type-2 fuzzy sets for determining the subjective weights of criteria has been suggested recently [44]. The recalculation of weights of criteria using the Bayes approach was presented [45].

If a decision is made on the basis of expert evaluation, a degree of the consistency of expert opinions must be assessed. The correlation coefficient can quantify whether the opinions of two experts are consistent; however, if the number of experts exceeds two, then the level of agreement among the ratings of these experts is expressed by the coefficient of concordance [46–50].

Following expert evaluation, the obtained sets t_{ik} are statistically processed. The mean value of evaluating criterion $\bar{t}_i$ is determined by the formula:

$$\bar{t}_i = \frac{\sum\limits_{k=1}^{r} t_{ik}}{r}, k = 1, \ldots, r; i = 1, \ldots n; \tag{1}$$

where t_{ik}—the evaluation of criterion i provided by expert k; r—the number of experts; and, n—the number of criteria.

The relative significance of every indicator q_i is calculated according to the formula:

$$q_i = \frac{\bar{t}_i}{\sum\limits_{i=1}^{n} \bar{t}_i}, i = 1, \ldots, n. \tag{2}$$

The credibility of expert evaluation can be expressed by the concordance coefficient of expert opinions describing the degree of agreement among individual opinions:

$$W = \frac{12S}{r^2 \times (n^3 - n) - r\sum\limits_{k=1}^{r} T_k}, k = 1, \ldots r; \tag{3}$$

$$S = \sum\limits_{i=1}^{n} \left(\sum\limits_{k=1}^{r} t_{ik} - \frac{1}{n}\sum\limits_{i=1}^{n}\sum\limits_{k=1}^{r} t_{ik} \right)^2, k = 1, \ldots, r; i = 1, \ldots n; \tag{4}$$

$$T_k = \sum\limits_{l=1}^{H_l} \left(h_l^3 - h_l \right), k = 1, \ldots, r; l = 1, \ldots h_l. \tag{5}$$

where S is the sum of the squares of deviation from the evaluation results of each indicator; T_k—k the indicator of related ranks; H_l—the number of the groups of equal ranks in k ranking; h_l—the number of equal ranks in the group of related ranks l under the evaluation of expert k; t_{ik}—the rank attributed to indicator i by expert k; r—the number of experts; and, n—the number of evaluated criteria.

The significance of the coefficient of concordance is defined according to the formula:

$$\chi^2 = \frac{12S}{r \times n \times (n+1) - \frac{1}{n-1}\sum\limits_{k=1}^{r} T_k}, k = 1, \ldots r. \tag{6}$$

The overall weights q_i'' can be calculated using Equation (7), which combines the criteria weights in a group q_i and criteria group weights q_j':

$$q_i'' = \frac{q_i q_j'}{\sum\limits_{i=1}^{n} q_i q_j'}, i = 1, \ldots, n; j = 1, \ldots, m. \tag{7}$$

where n—the number of criteria in a group; and, m—the number of criteria groups.

If the value of χ^2 calculated according to formula (6) is higher than χ^2_{tbl}, which depends on the degrees of freedom and the accepted level of significance; the hypothesis regarding the agreement between expert ranks is accepted.

If $\chi^2 < \chi^2_{tbl}$, the positions of experts are considered to be inconsistent.

4. Research Findings

4.1. The Analysed Results of Kaunas City

Kaunas is a city in south-central Lithuania, existing at the confluence of the Neris and Nemunas rivers. It is the second-largest city in Lithuania and the historical centre of Lithuanian economic, academic, and cultural life. In the Russian Empire, it was the capital of the Kaunas Governorate from 1843 to 1915. During the interwar period, it served as the temporary capital of Lithuania, when the current capital Vilnius was part of Poland between 1920 and 1939.

In the period considered, between 1904 and 2023 (solutions to the current general plan are valid up to 2023), both the spatial area of Kaunas city and the number of inhabitants have increased. The population of Kaunas city was steadily growing until 1970, regardless of a sudden drop during war years; however, since 1970, the population has been slightly decreasing (Table 2). The opposite trend is evident while looking at variations in the area. From 1857 to 2013, the area of Kaunas was scattering over, thus occupying new territories closer to the city (Figure 1). Within the analyzed period, the city remained homogeneous and compact, including neighbouring residential areas. Thus, it can be argued that the spatial area of the city is evenly subject to population growth.

Table 2. The final results of the analyzed Kaunas city.

Indicator	1904	1929	1935	1945	2013–2023
Area, km^2	8.78	13.44	39.83	143.69	156.93
	variation	+53.08%	+196.35%	+260.76%	+9.21%
Population	70,920	100,000	155,460	80,000	297,669
	variation	+29,080	+55,460	−75,460	+217,669
Population density, number of people/km^2	8077.45	7440.48	3903.09	556.75	1896.83
	variation	−7.89%	−47.53%	−85.74%	+240.70%
Street length, km	51.48	113.88	249.08	371.71	1016.65
	variation	+62.40	+135.20	+122.63	+644.94
Street density, km/km^2	5.86	8.47	6.26	2.59	6.48
	variation	+44.54%	−26.09%	−58.63%	+150.19%
Street length, per 1000 people, km	0.73	1.14	1.60	4.65	3.42
	variation	+56.16%	+40.35%	+190.63%	−26.45%

Kaunas, Taurage, and Silale cities maps were used in five different periods to evaluate the development of the area of these cities. The maps were randomly selected while considering the precision of historical data found in archives. Our goal was to determine the extent of the urban area varying most of time. The development of Kaunas city took the longest period of time, as changes in the area has covered 120 years. The analysis of the area of Taurage town has been carried out over a period of 74 years, and that of Silale embraced a period of 52 years. The research was aimed at demonstrating the proposed approach to the analysis of old maps, starting from scanning to vectorization on a racist map and a comparison of changes in the analyzed area. The authors suggest that six indicators (the area of the site (km^2), population, population density (people per km^2), street length (km), street density (km/km^2), and length of streets 1000 inhabitants/km.) can perfectly reflect the urban changes in the development of Kaunas city and Taurage and Silale towns.

Figure 1 shows that the urban area covered the entire present Centre neighbourhood and a part of Zaliakalnis and Griciupis neighbourhoods in 1904. A part of the current neighbourhoods of Sanciai, Vilijampole, and Aleksotas joined the area in 1929, a part of Griciupis and Panemune neighbourhoods—in 1935. In 1945, the area of Kaunas was partially formed by all present neighbourhoods of the city.

The analysis of variations in the population of Kaunas city, the existing transportation infrastructure and the expansion of street network infrastructure for the period from 1904 to 2013 (Table 2) shows that the area, population, the general length of streets, and the length of streets per 1000 people were gradually increasing, except for 1945, when the population suddenly dropped to 80,000 and it had an effect on other results, the calculation of which was affected by the population; for example, from 1904 to 1935, the length of streets per 1000 people gradually increased up to 1.60 km, in 1945, the length of streets per 1000 people suddenly increased to 4.65 km, and in 2016, decreased to 3.42 km. The area of Kaunas city has increased by 17.9 times, while the population has grown by 4.2 times. During the analyzed period, Kaunas has become the agglomeration core, together with the settlements that are located along the administrative boundaries of the city and belonging to Kaunas region. Research has shown that Kaunas city has high compactness, reaching 5.92, because of the dense centre in the confluence of Nemunas and Neris rivers and due to apartment blocks build in the northern part of the city forming a large part of the urban area. Kaunas has made an indirect impact on the surrounding areas and encouraged them to develop and enlarge together with the city by expanding its territory. Such variations have significantly reduced the population density in Kaunas city (amounts to 1896.83 people per km^2), which mainly has changed the way of life and vehicles that are used for transportation, because a passenger car has acquired the predominant role.

4.2. The Analysed Results of Taurage Town

Tauragė is an industrial city in Lithuania. It is situated on the Jūra River, close to the border with the Kaliningrad Oblast, and not far from the Baltic Sea coast. Although first mentioned in 1507, Tauragė only received its city charter in 1924. Lithuanian, Swedish, and Danish factories operate in the city. Nowadays, Tauragė is famous for its car markets and adventure park.

The findings of the investigated period from 1944 to 2018 have demonstrated that Taurage town has been expanding steadily by enlarging its area and maintaining a similar shape and the direction of development subject to population changes (Table 3). The population of Taurage town tended to rise until 1989, and it then started decreasing due to a negative natural increase in the population and the negative migration balance. In 2016, a growth in the quantity of people was observed, because the number of town residents was added to that of the citizens of nearby settlements that were planned to be included in the town. The opposite trend is evident when looking at variations in the space of the urban area. From 1653 to 2018, the area of Taurage has gradually expanded. Starting from 2008, according to the general plan of the area of Taurage municipality until 2018, the long-term urban area should increase to 22.33 km^2 (Figure 2) after a decision regarding connecting the surrounding villages to the city. The examined period faced the constantly growing population, which analogically changed the administrative boundary of the town. By 2008, the town was strongly developed in the northeast and southeast directions, and since 2008 southwest and northwest directions have been planned to be developed. The carried out analysis of variations in the area of Taurage town leads to a conclusion that, since 1944, the former elongated city structures situated around railway branches and present Silale Street have not been sufficiently compact, and therefore will become a homogeneous, compact town centre integrated into surrounding areas until 2018 (Figure 2).

Table 3. The final results of the analyzed Taurage town.

Indicator	1944–1945	1991	1999	2008	2018
Area, km^2	7.70	13.90	13.73	14.06	22.33
	variation	+80.52%	−1.22%	+2.40%	+58.82%
Population	10,561	29,996	29,124	26,207	29,003
	variation	+19,435	−872	−2917	+2796
Population density, number of people/km^2	1371.56	2157.99	2121.19	1863.94	1298.84
	variation	+57.34%	−1.71%	−12.13%	−30.32%
Street length, km	41.43	119.50	119.50	175.30	217.48
	variation	+78.07	0.00	+55.80	+42.18
Street density, km/km^2	5.38	8.60	8.70	12.47	9.74
	variation	+59.85%	+1.16%	+43.33%	−21.89%
Street length per 1000 people, km	3.92	3.98	4.10	6.69	7.50
	variation	+1.53%	+3.02%	+63.17%	+12.11%

The analysis of variations in the population of Taurage town and the expansion of street network infrastructure from 1944 until 2018, provided in Table 3, shows that the space of the area, the population before 1991, the length of streets, and that of streets per 1000 people are gradually increasing. The area of the town has expanded by 1.6 times and therefore a growth in the population of Taurage has an effect on the results of the analysis that was conducted in 2018. Since 1991, the population has decreased, and hence population density and people/km^2 have proportionally varied.

The period from 1944 to 2018 ranged from 1371.56 people/km^2 to 2157.99 people/km^2, and it should decrease to 1298.84 people/km^2 in 2018, although the recommended minimum population density in the abovementioned spatial planning norms makes 3000 people/km^2. A decrease in population density results in a reduction in the incomes of the city budget and in the growing costs of maintaining urban infrastructure and public services. Continually decreasing population density indicates that Taurage town needs well-thought-out inner development strategy to the developed territory of town within boundaries.

4.3. The Analysed Results of Silale Town

Silale is a town in Western Lithuania. It is located 30 km north of Taurage. The river Lokysta flows through the town. The town is part of the Samogitian ethnographic region of Lithuania and it was first mentioned in the 16th century. Before the War, a large part of inhabitants were Jewish people. In 1941, the Nazis massacred around 1300 Jews in Silale.

Within the examined period from 1964 to 2016, Silale town developed evenly in all directions, along with variations in the number of the population that fluctuated and faced both an increase and a decrease. For the period from 1989 to 2016, a drop in the population was noticed, which might be affected by tendencies in the aging population, a low birth rate, and the structure of economy, i.e., a reduction in the structural part of agricultural production, migration to metropolitan areas, and foreign countries. The opposite trend is evident when looking at variations in the space of the urban area. From 1964 to 2006, the area of Silale town gradually increased. From 2006, according to the general plan of the town until 2016, the long-term urban area increased to 17.95 km^2 following a decision regarding the connection of surrounding villages to the city (Figure 3). During the period considered, the town formed and became compact, including the surrounding residential areas. Thus, it can be proposed that the urban area was evenly subject to population growth prior to 2006.

The analysis of variations in the population of Silale town and the expansion of the street network infrastructure from 1964 to 2016 are presented in Table 4, showing that the space of the area, the population prior to 1990, the length of streets, and that of streets per 1000 people were gradually increasing. Since 1990, along a drop in the number of people, population density, and the number of people/km^2 have also declined. The findings of the carried out analysis are distorted by the long-term boundary of the area of Silale town presented in the general plan of Silale until 2016 and covering the unified urban structure of Silale, Silai, Balsiai, Vingininkai, and other neighbouring villages. This urban structure is analyzed in the general plan and it is expected to be further developed to give the status of the town, which should strengthen the rank of Silale town in the common system nationwide. The general plan of the area of Silale town until 2016 provides a decreasing number of the population in Silale. Therefore, the situation is intended to be compensated by incorporating surrounding villages into the urban area, thus extending administrative boundaries (approximately 5.2 times) and increasing urban population. Such a decision should obviously increase the space of the urbanized area, however this is a dubious need from a demographic viewpoint of Silalė town and the whole of Lithuania. Prior to 2006, Silale maintained a homogeneous and compact structure with neighbouring surrounding villages having very strong links with the city center. With reference to geospatial data on the results of the population and housing census in the Republic of Lithuania (2011), grids were used for calculating the population of the surrounding settlements that were planned to be incorporated into the city. According to the data collected in 2011, the population of Silale town increased by 3480 inhabitants; nevertheless, it should be taken into account that, from 2011 to the beginning of 2016, the population in Silale decreased by 1.59%—from 5486 to 5400 inhabitants, and therefore the population of the incorporated territories was reduced by 1.59% to 3387 people. The analysis of urban development that was scheduled in the general plan of the city by 2016 points out the opposite trend: the city expanded its area, the population increased by 3387 people, but the population density considerably decreased (from 2006 to 2016, even 71.30%) and the urban structure changed and became less compact than before.

Table 4. The final results of the analyzed Silale town.

Indicator	1964	1966	1990	2006	2016
Area, km^2	2.13	3.18	3.50	3.48	17.95
	variation	+49.30%	+10.06%	−0.57%	+415.80%
Population	2995	2995	6308	5935	8787
	variation	0	+3313	−373	+2852
Population density, number of people/km^2	1406.10	941.82	1802.29	1705.46	489.53
	variation	−33.02%	+91.36%	−5.37%	−71.30%
Street length, km	12.41	15.67	35.91	39.16	103.67
	variation	+3.26	+20.24	+3.25	+64.51
Street density, km/km^2	5.83	4.93	10.26	11.42	0.17
	variation	−15.44%	+108.11%	+11.31%	−98.51%
Street length per 1000 people, km	4.14	5.23	5.69	6.36	11.80
	variation	+26.33%	+8.80%	+11.78%	+85.53%

5. Expert Evaluation of Criteria in Inner Urban Development

Cities are constantly expanding externally, regardless of the ever-decreasing population. The analysis of territorial variations in Kaunas city, as well as in Taurage and Silale towns, has showed that, according to the latest general plans of Silale and Taurage, the outer development of these cities is scheduled, although the areas still have room to expand inside. There is a risk, because this option allows for the expansion of urban areas and carrying out construction works in the large area, thus providing perfect conditions for settling small groups of urban areas that are situated around the densely developed towns of Taurage and Silale. These examples show that, in order to manage urban and outer development, the main criteria in inner development must be analyzed, so that the towns could seize all the inside opportunities for expansion. Figure 4 presents a scheme for the formulated and selected key groups of criteria in urban development. The criteria affecting inner urban development have been formulated and divided into six groups: Group A—Urban Structure (five criteria), Group B—Social Environment (six criteria), Group C—Economic Environment (six criteria), Group D—Sensitive and Protected Areas (five criteria), Group E—Transportation (six criteria), and Group F—Land Use (five criteria).

This study applies for expert evaluation in turn to assess the relative importance of criteria in inner urban development. The experts are selected professionals in spatial planning and transportation systems and those employed in state institutions and all having a Master's or PhD degree or work experience exceeding 10 years.

The core of the expert evaluation method is determining the relative importance of the criteria under consideration in planning sustainable urban development employing the quantitative assessment of expert opinions and processing the evaluation results. Ten experts, with reference to their knowledge and experience, were asked to rank (i.e., to assign scores based on the scoring scale) inner urban development criteria that are presented in the questionnaire and provided in Figure 4. According to the above-presented technique (Equations (1)–(7)), the summarized expert opinion was accepted as the result of the solution to the problem.

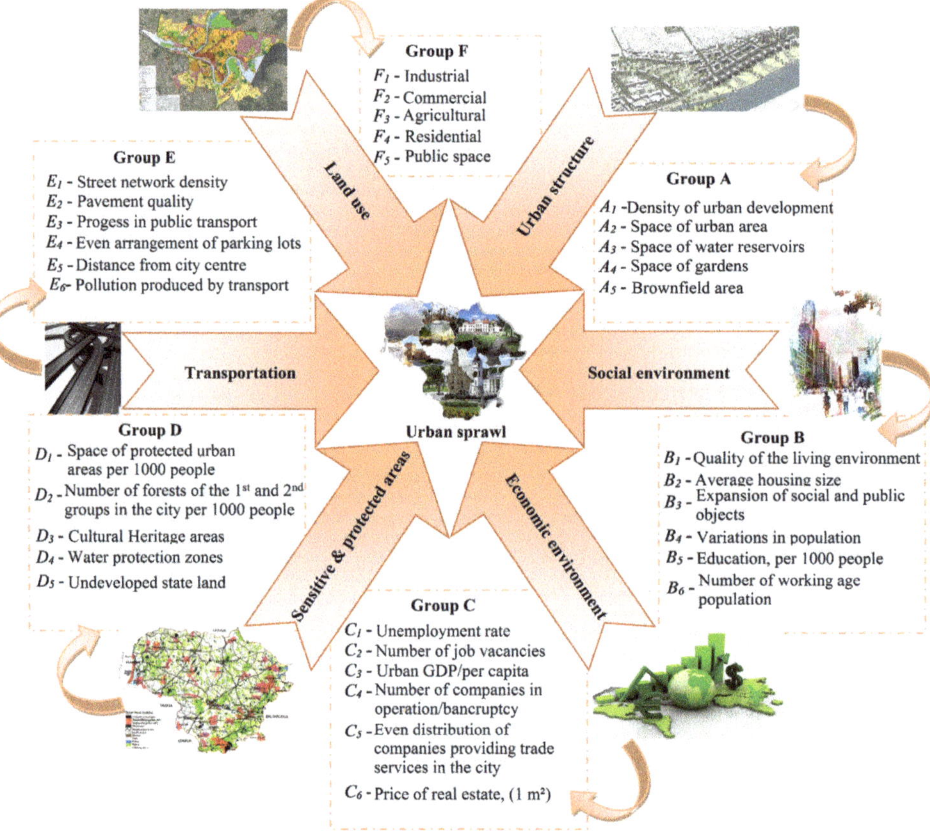

Figure 4. Scheme for the groups of criteria in inner urban development.

Tables 5 and 6 provide the criteria evaluated and ranked (arranged in order of importance) by the experts in separate groups and the importance of the groups of criteria.

Table 5. Ranking the results of the groups of criteria.

Experts	Groups of Criteria	A	B	C	D	E	F
	Expert 1	6	1	4	2	3	5
	Expert 2	6	5	4	2	3	1
	Expert 3	6	4	5	2	1	3
	Expert 4	6	4	6	1	3	2
	Expert 5	6	2	1	3	5	4
	Expert 6	5	4	3	1	6	2
	Expert 7	6	3	4	1	5	2
	Expert 8	6	2	1	4	3	5
	Expert 9	6	5	3	1	4	2
	Expert 10	6	3	2	1	5	4
Relative Significance q_i		0.28	0.16	0.16	0.09	0.18	0.14
Rank		1	3–4	3–4	6	2	5
Coefficient of Concordance W		0.518					
Significance of the Coefficient χ^2		25.91					
Agreement among Expert Positions $\chi^2 > \chi^2_{tbl}$		25.91 > 15.09 Agreement among expert positions is sufficient when reliability $p = 0.01$					

Table 6. The results of ranking criteria in the groups.

Groups of Criteria	Criteria	Unit of Measurement	Weight in the Group	Coefficient of Concordance and Its Significance	Rank in the Group
A	A_1	%	0.31		1
	A_2	km^2	0.21	0.484	2
	A_3	km^2	0.11	19.36 > 13.28 ($p = 0.01$)	5
	A_4	km^2	0.17		4
	A_5	km^2	0.20		3
B	B_1	grade	0.26		1
	B_2	m^2	0.14		4–5
	B_3	hectares	0.20	0.351	2
	B_4	%	0.16	17.54 > 15.09 ($p = 0.01$)	3
	B_5	per 1000 people	0.11		6
	B_6	number	0.14		4–5
C	C_1	%	0.16		3–4
	C_2	number	0,16		3–4
	C_3	GDP per capita	0.14	0.402	5
	C_4	number	0.09	20.08 > 15.09 ($p = 0.01$)	6
	C_5	%	0.18		2
	C_6	m^2	0.26		1
D	D_1	per 1000 people, km	0.27		1
	D_2	per 1000 people, km	0.20	0.460	3
	D_3	hectares	0.25	18.40 > 13.28 ($p = 0.01$)	2
	D_4	hectares	0.19		4
	D_5	hectares	0.09		5
E	E_1	km/km^2	0.22		1–3
	E_2	grade	0.11		5
	E_3	%	0.22	0.502	1–3
	E_4	%	0.14	25.09 > 15.09 ($p = 0.01$)	4
	E_5	km	0.22		1–3
	E_6	grade	0.09		6

Table 6. *Cont.*

Groups of Criteria	Criteria	Unit of Measurement	Weight in the Group	Coefficient of Concordance and Its Significance	Rank in the Group
	F_1	hectares	0.15		4
	F_2	hectares	0.26		2
F	F_3	hectares	0.09	0.724 $28.96 > 13.28\ (p = 0.01)$	5
	F_4	hectares	0.31		1
	F_5	hectares	0.19		3

The most important group of criteria according to experts' opinion is Group A—Urban Structure. It is an obvious leader, with relative significance $q_i = 0.28$. The next follow Group E—Transportation, Group B—Social Environment, and then Group C—Economic Environment.

The overall weights of criteria are calculated using Equation (7), which combines criteria weights in a group and criteria group weights. Table 7 presents the results.

The most important criterion according to experts with $q_i'' = 0.09$ is density of urban development, the next follow space of urban area and brownfield area, with $q_i'' = 0.06$ each (Table 7).

Table 7. The results of overall ranking of criteria.

Criteria	Overall Weight of the Criterion	Overall Rank	Criteria	Overall Weight of the Criterion	Overall Rank	Criteria	Overall Weight of the Criterion	Overall Rank
A_1	0.0877	1	B_1	0.0411	20	C_1	0.0260	7
A_2	0.0579	2	B_2	0.0221	27	C_2	0.0260	21–24
A_3	0.0317	12	B_3	0.0312	21–24	C_3	0.0222	13
A_4	0.0467	4	B_4	0.0259	28–29	C_4	0.0145	18
A_5	0.0560	3	B_5	0.0175	33	C_5	0.0291	28–29
			B_6	0.0221		C_6	0.0421	21–24
D_1	0.0246	20	E_1	0.0394	10	F_1	0.0205	25
D_2	0.0180	27	E_2	0.0197	26	F_2	0.0364	11
D_3	0.0222	21–24	E_3	0.0403	8–9	F_3	0.0121	32
D_4	0.0174	28–29	E_4	0.0249	19	F_4	0.0439	5
D_5	0.0078	33	E_5	0.0403	8–9	F_5	0.0271	15
			E_6	0.0154	30			

6. Discussion

Expert evaluation has shown that the criteria having the most powerful impact on urban development can be found in each of the groups.

To ensure the sustainable development of the area, the urban structure and the density of the urban development, $q_i = 0.31$ each (Table 6) must be primarily evaluated. These are the most important criteria in determining how the city should be further developed to make it more convenient to live selecting an appropriate layout of residential areas to ensure the quality of the living environment. The latter criterion, in particular, determines the price of real estate, which is a complex one, reflecting the level of the quality of life. This criterion includes all income, embracing that of shadow economy in order to obtain highest quality private housing that meets the needs of the family.

The space of protected areas ($q_i = 0.27$) remains important for urban life (Table 6), which shows that inhabitants seek to have a sustainable city with a sufficient number of green areas that should assure the proximity of recreational areas to residential buildings.

Another group of criteria describing the urban transportation system is very important to the urban population. The equal importance of criteria ($q_i = 0.22$) illustrates that the distance from the city centre, the density of the street network, and progress in public transport are equally important for the population (Table 6). All of these criteria describe the balance of urban population, because the closer you can get to the city centre, the more comfortable and faster you can reach attraction objects of the city. Street density and progress in public transport determine daily travel time of the population.

The results of the survey indicate that inner city development is proposed to be given priority when considering the development of the urban area. This shows that the group of indicators for the urban transport system accepts the distance from the city centre and the level of progress in public transport as extremely relevant and, in all cases, the latter will be higher in the places of dense street networks and urban development, because public transport works efficiently only in the areas with a population density of more than 92 people/ha.

The greatest weights of criteria are obtained in all different domains and they show that urban development is a multifaceted process that is determined by composite criteria, which lead to consistency or symmetry in urban development.

As a result of the density decrease of urban population, the compactness of cities is diminished, the infrastructure of transport becomes more and more expensive, cars start playing the predominant role, it is payed less attention to inner urban development, and outer urban development is stimulated.

While planning without taking into account the most important urban indicators (density and intensity of build-up territory) and increasing the number of additional areas increases the cost of maintaining urban infrastructure and it is typical of all the cities analysed, especially in the most recent (2006–2008) Master plans.

As for further studies, it is possible to elaborate on the extent of external development, depending on the size of the urban area, the type of a building, and the means to manage urban development, thus ensuring the quality of life in the compactly developed city, i.e., possibility of balancing (symmetry) of needs and their satisfaction.

7. Conclusions

According to foreign experience, urban development as the biggest issue of the 21st century can be managed by determining the urban growth limits beyond which construction is prohibited or restricted. Another way is using the resources of inner urban development and increasing the density of urban development, thus simultaneously promoting the compactness of applying brownfields and undeveloped territories inside the city borders.

While taking into account the evolution of urban development in Lithuania and abroad, and due to the permanently decreasing population, the cities and towns of Lithuania should focus on inner urban development, although the historical cartographic material of Kaunas city (1904–2023) and Taurage (1944–2018) and Silale (1964–2016) towns, as well as the analysis of variations in urban boundaries, have demonstrated that the design of the initial general plans for cities for the period 2006–2008 overestimated the privatization of land ownership and unduly expanded urban administrative boundaries, despite the fact that the territories had been developed sufficiently evenly. Nevertheless, if we compare the development of cities area, we can see that only the last Master Plan proposed large territorial development, which is due to restoration of private property.

The multi-criteria development model that is proposed by the authors allows for using it in any country or city where the suggested methodology may assist in interviewing urbanists, real estate developers, and planning experts to identify the most important criteria that, when considering the opinion of experts, have a profound effect on urban development.

In the case of Lithuanian cities and towns, the key criteria embrace urban structure, the density of urban development (0.31), which is the aspect defining it, space for residential areas (0.31), and the space of protected areas within the city per 1000 people (0.27). For assessing the effect of infrastructure, transportation plays a crucial role and it can be characterized by the density of the street network, the level of progress in public transport, and the distance from the city centre to developing areas (all above mentioned criteria take weight equal to 0.22). As a result, the price of real estate and the quality of the living environment (weight 0.26) are mentioned.

The results of the study could be incorporated in further steps of the research—ranking areas of urban development that are based on identified criteria and applying multi-criteria decision-making (MCDM) methods.

Author Contributions: All authors contributed equally to this work. V.P., M.B., J.A. and J.Š. designed the research, collected and analyzed the data and the obtained results, performed the development of the paper. All the authors have read and approved the final manuscript.

Funding: This research received no external funding.

Conflicts of Interest: The authors declare no conflict of interest.

References

1. Wolff, M.; Wiechmann, T. Urban growth and decline: Europe's shrinking cities in a comparative perspective 1990–2010. *Eur. Urban Reg. Stud.* **2018**, *25*, 122–139. [CrossRef]
2. Kabisch, N.; Haase, D.; Haase, A. Urban Population Development in Europe, 1991-2008: The Examples of Poland and the UK. *Int. J. Urban Reg. Res.* **2012**, *36*, 1326–1348. [CrossRef]
3. Ubarevičiene, R.; Burneika, D.; Kriaučiunas, E. The sprawl of Vilnius city—establishment of growing urban region. *Ann. Geogr.* **2010**, *43–44*, 96–107.
4. Sporna, T.; Kantor-Pietraga, I.; Krzysztofik, R. Trajectories of depopulation and urban shrinkage in the Katowice Conurbation, Poland. *Espaces-Popul. Soc.* **2015**, *3*, 6102. [CrossRef]
5. Sporna, T. The suburbanisation process in a depopulation context in the Katowice conurbation, Poland. *Environ. Socio-Econ. Stud.* **2018**, *6*, 57–72. [CrossRef]
6. Petersen, J.P. The Linkage of Urban and Energy Planning for Sustainable Cities: The Case of Denmark and Germany. *Int. J. Environ. Chem. Ecol. Geol. Geophys. Eng.* **2016**, *10*, 439–442.
7. Burinskienė, M.; Bielinskas, V.; Podviezko, A.; Gurskienė, V.; Maliene, V. Evaluating the Significance of Criteria Contributing to Decision-Making on Brownfield Land Redevelopment Strategies in Urban Areas. *Sustainability* **2017**, *9*, 759. [CrossRef]
8. Tso, G.K.F.; Li, J. Using Index to Measure and Monitor Progress on Sustainable Development. In *Sustainable Development—Education, Business and Management—Architecture and Building Construction—Agriculture and Food Security*; Ghenai, C., Ed.; InTech: Rijeka, Croatia, 2012; pp. 175–196. ISBN 978-953-51-0116-1.
9. Kauko, T. Sustainable Development of the Built Environment: The Role of the Residential/Housing Sector. In *Sustainable Development—Education, Business and Management—Architecture and Building Construction—Agriculture and Food Security*; Ghenai, C., Ed.; InTech: Rijeka, Croatia, 2012; pp. 161–174. ISBN 978-953-51-0116-1.
10. Martos, A.; Pacheco-Torres, R.; Ordóñez, J.; Jadraque-Gago, E. Towards successful environmental performance of sustainable cities: Intervening sectors. A review. *Renew. Sustain. Energy Rev.* **2016**, *57*, 479–495. [CrossRef]
11. Pombo, O.; Rivela, B.; Neila, J. The challenge of sustainable building renovation: Assessment of current criteria and future outlook. *J. Clean. Prod.* **2016**, *123*, 88–100. [CrossRef]
12. Hoymann, J.; Goetzke, R. Simulation and Evaluation of Urban Growth for Germany Including Climate Change Mitigation and Adaptation Measures. *Isprs Int. J. Geo-Inf.* **2016**, *5*, 101. [CrossRef]
13. Gussoni, A.; De Salvia, A.; Parolin, M.; Piana, A. An integrated approach for the sustainable inner development of brownfields. In *Proceedings of the 10th International UFZ-Deltares/TNO Conference on Soil-Water Systems, Milan, Italy, 3–6 June 2008*; CONSOIL 2008: Theme F—Sustainable & Risk Based Land Management; F&U Confirm: Leipzig, Germany, 2008; pp. 159–167.
14. Ewing, R.; Hamidi, S.; Grace, J.B.; Wei, Y.D. Does urban sprawl hold down upward mobility? *Landsc. Urban Plan.* **2016**, *148*, 80–88. [CrossRef]
15. Arundel, R.; Ronald, R. The role of urban form in sustainability of community: The case of Amsterdam. *Environ. Plan. B Urban Anal. City Sci.* **2017**, *44*, 33–53. [CrossRef]
16. De Vos, J.; Van Acker, V.; Witlox, F. Urban sprawl: Neighbourhood dissatisfaction and urban preferences. Some evidence from Flanders. *Urban Geogr.* **2016**, *37*, 839–862. [CrossRef]
17. Zhu, C.; Ji, P.; Li, S. Effects of urban green belts on the air temperature, humidity and air quality. *J. Environ. Eng. Landsc. Manag.* **2017**, *25*, 39–55. [CrossRef]
18. Nakagawa, K. Municipal sizes and municipal restructuring in Japan. *Lett. Spat. Resour. Sci.* **2016**, *9*, 27–41. [CrossRef]
19. Tsenkova, S. *Housing Change in East and Central Europe: Integration or Fragmentation?* Routledge: London, UK, 2017.

20. Alqurashi, A.F.; Kumar, L. Spatiotemporal patterns of urban change and associated environmental impacts in five Saudi Arabian cities: A case study using remote sensing data. *Habitat Int.* **2016**, *58*, 75–88. [CrossRef]
21. Mubarak, F. Urban growth boundary policy and residential suburbanization: Riyadh, Saudi Arabia. *Habitat Int.* **2004**, *28*, 567–591. [CrossRef]
22. Smith, B.E.; Haid, S. The rural-urban connection: Growing together in Great Vancouver. *Plan Can.* **2004**, *44*, 36–39.
23. Nelson, A.C.; Moore, T. Assessing urban-growth management—The case of Portland, Oregon, the USAs largest urban-growth boundary. *Land Use Policy* **1993**, *10*, 293–302. [CrossRef]
24. Tayyebi, A.; Pijanowski, B.C.; Tayyebi, A.H. An urban growth boundary model using neural networks, GIS and radial parameterization: An application to Tehran, Iran. *Landsc. Urban Plan.* **2011**, *100*, 35–44. [CrossRef]
25. Jun, M.-J. The Effects of Portland's Urban Growth Boundary on Urban Development Patterns and Commuting. *Urban Stud.* **2004**, *41*, 1333–1348. [CrossRef]
26. AbuSada, J.; Thawaba, S. Multi criteria analysis for locating sustainable suburban centers: A case study from Ramallah Governorate, Palestine. *Cities* **2011**, *28*, 381–393. [CrossRef]
27. Giffinger, R.; Fertner, C.; Kramar, H.; Meijers, E. *City-Ranking of European Medium-Sized Cities*; Centre of Regional Science: Vienna, UT, USA, 2007; pp. 1–12.
28. Zavadskas, E.K.; Podvezko, V. Integrated determination of objective criteria weights in MCDM. *Int. J. Inf. Technol. Decis. Mak.* **2016**, *12*, 267–283. [CrossRef]
29. Zavadskas, E.K.; Cavallaro, F.; Podvezko, V.; Ubarte, I.; Kaklauskas, A. MCDM Assessment of a Healthy and Safe Built Environment According to Sustainable Development Principles: A Practical Neighborhood Approach in Vilnius. *Sustainability* **2017**, *9*, 702. [CrossRef]
30. Cavallaro, F.; Zavadskas, E.K.; Raslanas, S. Evaluation of Combined Heat and Power (CHP) Systems Using Fuzzy Shannon Entropy and Fuzzy TOPSIS. *Sustainability* **2016**, *8*, 556. [CrossRef]
31. Yazdani, M.; Hashemkhani Zolfani, S.; Zavadskas, E.K. New integration of MCDM methods and QFD in the selection of green suppliers. *J. Bus. Econ. Manag.* **2016**, *17*, 1097–1113. [CrossRef]
32. Ustinovichius, L. Methods of determining objective, subjective and integrated weights of attributes. *Int. J. Manag. Decis. Mak.* **2007**, *8*, 540–554. [CrossRef]
33. Hashemkhani Zolfani, S.; Yazdani, M.; Zavadskas, E.K. An extended stepwise weight assessment ratio analysis (SWARA) method for improving criteria prioritization process. *Soft Comput.* **2018**, *22*, 7399–7405. [CrossRef]
34. Saaty, T.L. How to make a decision: The analytic hierarchy process. *Eur. J. Oper. Res.* **1990**, *48*, 9–26. [CrossRef]
35. Ramanathan, R.; Ramanathan, U. A qualitative perspective to deriving weights from pairwise comparison matrices. *Omega* **2010**, *38*, 228–232. [CrossRef]
36. Deng, X.; Hu, Y.; Deng, Y.; Mahadevan, S. Supplier selection using AHP methodology extended by D numbers. *Expert Syst. Appl.* **2014**, *41*, 156–167. [CrossRef]
37. Saaty, T.L. *Decision Making with Dependence and Feedback: The Analytic Network Process*; RWS Publications: Pittsburgh, PA, USA, 2001.
38. Xu, X. The SIR method: A superiority and inferiority ranking method for multiple criteria decision making. *Eur. J. Oper. Res.* **2001**, *131*, 587–602. [CrossRef]
39. Keršulienė, V.; Zavadskas, E.K.; Turskis, Z. Selection of rational dispute resolution method by applying new step-wise weight assessment ratio analysis (SWARA). *J. Bus Econ. Manag.* **2010**, *11*, 243–258. [CrossRef]
40. Ghorshi Nezhad, M.R.; Hashemkhani Zolfani, S.; Moztarzadeh, F.; Zavadskas, E.K.; Bahrami, M. Planning the priority of high tech industries based on SWARA-WASPAS methodology: The case of the nanotechnology industry in Iran. *Econ. Res. Ekon. Istraž.* **2015**, *28*, 1111–1137. [CrossRef]
41. Ginevicius, R. A new determining method for the criteria weights in multi-criteria evaluation. *Int. J. Inf. Technol. Decis. Mak.* **2011**, *10*, 1067–1095. [CrossRef]
42. Krylovas, A.; Zavadskas, E.K.; Kosareva, N.; Dadelo, S. New KEMIRA method for determining criteria priority and weights in solving MCDM problem. *Int. J. Inf. Technol. Decis. Mak.* **2014**, *13*, 1119–1134. [CrossRef]
43. Rezaei, J. Best-worst multi-criteria decision-making method. *Omega* **2015**, *53*, 49–57. [CrossRef]

44. Keshavarz-Ghorabaee, M.; Amiri, M.; Zavadskas, E.K.; Turskis, Z.; Antucheviciene, J. An Extended Step-Wise Weight Assessment Ratio Analysis with Symmetric Interval Type-2 Fuzzy Sets for Determining the Subjective Weights of Criteria in Multi-Criteria Decision-Making Problems. *Symmetry* **2018**, *10*, 91. [CrossRef]
45. Vinogradova, I.; Podvezko, V.; Zavadskas, E.K. The Recalculation of the Weights of Criteria in MCDM Methods Using the Bayes Approach. *Symmetry* **2018**, *10*, 205. [CrossRef]
46. Kendall, M. *Rank Correlation Methods*, 4th ed.; Griffin: London, UK, 1970; p. 272.
47. Kendall, M.G.; Gibbons, J.D. *Rank Correlation Methods*, 5th ed.; Oxford University Press: New York, NY, USA, 1990.
48. Zavadskas, E.K.; Peldschus, F.; Kaklauskas, A. *Multiple Criteria Evaluation of Projects in Construction*; Technika: Vilnius, Lithuania, 1994.
49. Zavadskas, E.K.; Vilutiene, T. A multiple criteria evaluation of multi-family apartment block's maintenance contractors: I—Model for maintenance contractor evaluation and the determination of its selection criteria. *Build. Environ.* **2006**, *41*, 621–632. [CrossRef]
50. Jakimavicius, M.; Burinskiene, M.; Gusaroviene, M.; Podviezko, A. Assessing multiple criteria for rapid bus routes in the public transport system in Vilnius. *Public Transp.* **2016**, *8*, 365–385. [CrossRef]

© 2019 by the authors. Licensee MDPI, Basel, Switzerland. This article is an open access article distributed under the terms and conditions of the Creative Commons Attribution (CC BY) license (http://creativecommons.org/licenses/by/4.0/).

Article

Using Stochastic Decision Networks to Assess Costs and Completion Times of Refurbishment Work in Construction

Grzegorz Śladowski *, Bartłomiej Szewczyk, Bartłomiej Sroka and
Elżbieta Radziszewska-Zielina

Faculty of Civil Engineering, Institute of Construction Management, Cracow University of Technology,
31-155 Cracow, Poland; bszewczyk@L3.pk.edu.pl (B.S.); bsroka@L3.pk.edu.pl (B.S.);
eradzisz@L3.pk.edu.pl (E.R.-Z.)
* Correspondence: gsladowski@L3.pk.edu.pl; Tel.: +48-12-628-23

Received: 5 March 2019; Accepted: 14 March 2019; Published: 19 March 2019

Abstract: According to the concept of sustainable development, the process of extending the life-cycle of existing buildings (including historical ones) through their restoration not only generates benefits from their layer use but also—and primarily—constitutes a chance for their substance to survive for use by future generations. Building restoration projects are usually difficult to plan, primarily due to the limited amount of information on the technical condition of existing structures and their historical substance, which often makes the scope of renovation works difficult to determine. At the stage of planning such a project, it is, therefore, reasonable to consider various scenarios of its implementation, the occurrence of which can be both random and can be generated by the decision-maker. Unfortunately, in practice, the right tools for planning such projects are not used, which in effect generates problems associated with underestimating their completion time and costs. In subject literature, there are proposals of the use of stochastic and decision networks to assess the course of various projects that are characterised by having indeterminate structures. However, these networks are limited to modelling tasks that either occur purely randomly or are fully generated by decision-makers. There are no studies that enable the modelling and optimisation of the structure of a project while taking into consideration both the random and decision-based nature of carrying it out. In the article, the authors proposed a stochastic decision network that enables the correct modelling of projects with a multi-variant structure of being carried out. For the purpose of analysing the network model, elements of mathematical programming were used to determine optimal decisions (in terms of expected costs and completion times of carrying out a project) that control the structure of the project being modelled. The entirety of the authors' proposal was backed by a calculation experiment on an example of a refurbishment construction project, which confirmed the application potential of the proposed approach.

Keywords: decision network; stochastic network; graphical evaluation and review technique (GERT); mathematical programming; building refurbishment project planning

---

1. Introduction

One of the missions of contemporary civilisation is the protection of cultural heritage by preventing the decay of its elements, the appropriate conservation, development and propagation of its values. Historical structures are an essential element of historical heritage and in the reality of today can only survive if they are considered to fulfil some useful function by society [1,2]. In the Treaty of Lisbon [3], we can find numerous references to the subject of the relationship between the concept of sustainable development and historical heritage. The preservation of a historical structure and restoring its utilitarian value is associated with the process of its restoration. The character and scope of restoration work primarily depends on its type of architectural and structural system, the state of the preservation of its historical substance, the technical condition of the building, including the quality of its physical and chemical properties and the mechanisms of the materials used in its construction, as well as the suitability of these materials for reuse in the existing structure.

Determining the estimated values of the time and cost of restoring a historical structure is a specific process that is difficult to carry out due to the following factors:

- The complicated course of design work and the unpredictable process of obtaining legal opinions and approvals of proposed refurbishment solutions, the lack of approval of design solutions by architectural conservation authorities.
- Difficulty in determining the scope of work (a high probability of additional or replacement types of work) due to the limited identification of the structure of an existing building, e.g., the technical condition of foundations uncovered while performing construction work can require different forms of reinforcing them.
- In the case of historical buildings, it is possible that discoveries will be made during construction work, resulting in additional archaeological digs, as well as more construction and conservation work [4], e.g., the replacement of existing plasters on a historical structure can result in the discovery of polychromes underneath.

The above-mentioned problems cause attempts at estimating the time and cost of carrying out such projects to be affected by a high level of imprecision. An analysis of the results of surveys that had been conducted among developers, designers and contractors from Poland, Lithuania and Slovakia who possess experience in such projects confirmed the occurrence of the problem of the low accuracy of assessments of the costs of project completion [5]. The respondents highlighted the necessity of including the possibility of performing additional and replacement works as a part of the plan in cost assessments, which will, in turn, reduce the deviation of the actual expenditure incurred from their planned volumes [5]. Therefore, it became necessary to develop an appropriate planning approach to the initial (e.g., at the stage of the restoration work feasibility study) assessment of the costs of such projects, in which different scenarios of carrying them out will be taken into account.

So-called task networks are some of the typically used tools in the planning and controlling of projects. These networks are based on graphs and are a form of the graphical representation of a project's plan, with a defined structure of dependencies between project tasks. Regardless of the type of these networks, it is possible to distinguish elements common to them, such as tasks, events, dependencies between tasks, as well as parameters (time, cost and other types of parameters) characteristic of the projects modelled using these networks.

One of the criteria of the classification of task networks is the division based on their logical structure, which can either be determinate or non-determinate. Networks with a determinate structure (so-called deterministic networks) are suitable only for modelling one scenario of the carrying out of a planned project, for which tasks and the technological and organisational relations between them are clearly defined.

Many methods of the analysis of task networks with a non-determinate logical structure were published in subject literature, with the most well-known methods being: CPM (Critical Path Method) and its probabilistic version, PERT (Program Evaluation and Review Technique). Both methods were initially developed and are still being expanded by [6–10]. However, the determinate logical structure of such networks, enabling only the modelling of one scenario of the carrying out of a planned project, is not suitable for the planning and assessment of the results of the projects being discussed in the article—the restoration of buildings—the planning of which should take into consideration alternative scenarios of carrying them out. Experience has shown that using traditional networks with a determinate logical structure to model these highly specific construction projects is a planning error, which means that these projects are often overestimated or underestimated in relation to their actual costs and completion times [11].

Networks with a non-determinate logical structure are the appropriate tool for planning such projects because their structure makes it possible to take into consideration variations in the tasks being performed during the carrying out of a project. If the variants of a planned project are carried out randomly, then the network describing such a process will be called a stochastic network. If, however, the carrying out of the aforementioned variants of the planned project is a result of the decision-maker's preferences, the network describing such a process will be called a decision network. In the following sections, the authors of this article have provided a brief characteristic of alternative-stochastic and alternative-decision networks, in order to later propose an innovative approach integrating the capabilities of both networks for planning such highly specific projects like the restoration of buildings.

1.1. Stochastic Networks—Literature Review

Eisner [12] first proposed the concept of the network planning of projects with a non-determinate, random structure, by developing the GAN convention (Generalized Activity Network) with a vertex topology which enabled the generation of variant dependencies in the structure of a network. Pritsker [13] developed the GERT (Graphical Evaluation and Review Technique) method that was later further developed by Whitehouse [14], along with its simulation variant, GERTS (Graphical Evaluation and Review Technique Simulation). Both approaches have become the basic methods of analysing stochastic networks with deterministic or probabilistic parameters. It is worth mentioning that the analysis of stochastic networks is still being developed mainly in the context of introducing fuzzy data into the model [11,15–19], based on fuzzy logic presented by Zadeh [20].

In subject literature, we can find examples of using stochastic networks to manage supply chains [21] for business process modelling [22], for planning aircraft production processes [23] and for the analysis of parallel processes of creating new products [24].

In the field of construction projects, Kosecki [25] provided examples of the use of stochastic networks to plan the refurbishment of historical buildings, Hougui [26] carried out an analysis of the process of building a hydroelectric power plant, Pena-Mora and Li [27] applied a stochastic network to the dynamic planning and control of the design and construction of building structures, Gao et al. [28] used them to develop a risk management plan for preparatory stages of large construction projects while Wang et al. [29] used such an approach for quality management in the construction of a concrete dam. Radziszewszka-Zielina et al. [11] used a fuzzy stochastic network (the time, cost and task completion probability parameters are type-2 fuzzy sets) for planning refurbishment works on road surfaces and the reconstruction of a historical retaining wall.

1.2. Decision Networks—Literature Review

Ignasiak [30], who first developed decision networks, introduced a definition of the structure of such a network that allowed the modelling of variants of a planned project and the selection (in an algorhythmic manner based on discrete programming) of the optimal variant within the adopted criterion of minimising the costs of the carrying out of said project. Śladowski [31] extended the structure of the above-mentioned network through additional logical forms of vertices, which enabled a more flexible approach to modelling the network dependencies of the planned project. In addition, the abovementioned work presents a practical example of the use of these networks in the analysis of technological and organisational variants of the construction of reinforced concrete foundations of a building.

The DCPM (Decision Critical Path Method), which was developed independently of the methods mentioned above, extended the basic method of network analysis, namely CPM, and enabled time and cost analysis by reducing networks with a non-determinate structure through the addition of so-called decision vertices. The DCPM method is still being developed by Zhang et al. [32] and Zhang et al. [33]. Recently, Ibadov [34], Ibadov and Kulejewski [35], proposed a network model with a fuzzy decision node for planning construction projects that feature the analysis of uncertain parameters.

1.3. Research Aims

The networks described above enable the modelling of variants of a planned project, which is generated in either a random or decision-based manner. However, it should be highlighted that the implementation of the considered scenarios of the carrying out of projects like the restoration of buildings can have a character that is both random and decision-based. Therefore, the use of stochastic networks to model such projects, in which alternative tasks in a network are treated only as random events, proves to be a rather limited approach. In subject literature one can find proposals (Więckowski [36]) of a generalized NNM (Numerical Network Modelling) network model of construction projects in which, in addition to variants of tasks carried out in a random manner, the author introduces vertices to the network, which he called decision vertices. However, this concept is associated with the defining of additional constraints in the structure of the network, which, as a consequence, has nothing to do with the decision-based nature of the implementation of the modelled variants of the carrying out of planned construction projects. Therefore, what is needed is a tool that will combine the stochastic and decision-based nature of such projects in order to properly model and optimise their plan in terms of the expected costs needed for carrying them out.

The purpose of this article is to develop:

1. An innovative approach to planning the restoration of buildings, allowing for the consideration of various completion scenarios, the occurrence of which can be both random and decision-based. The following will be required for this goal to be achieved:

   a. Defining the stochastic-decision structure of the network model containing an appropriate topology of vertices with deterministic, stochastic and decision emitters (as a directed, non-cyclical graph, with one initial vertex and numerous final end vertices).

   b. Developing a nonlinear one-and multi-criteria binary programming model for the purpose of optimising (in the time-cost aspect) various scenarios of the carrying out of the project that is being modelled by the network.

   c. As a result, the decision-maker, by specifying their preferences as to the planned result of a project and their risk aversion, will receive an optimal (in terms of expected time and costs) scenario of carrying out their project. In the case of a multi-criteria analysis (time and cost), the decision-maker will also be able to specify different values of weights for the expected time and cost of the planned project in the goal function. As a part of the results, the decision-maker will also obtain information on the type of technical solutions or the manner of carrying out the work that should be included in the plan of the optimal scenario of carrying out the project.

2. Developing a digital application of the approach and performing a calculation experiment within which the effectiveness of the stochastic decision network will be demonstrated in relation to the traditional approach.

2. Method

A verbal description of the process that is a building's restoration project can be transformed into a description made by using a stochastic decision network, the structure of which will enable the modelling and analysis of various scenarios of the planned project.

2.1. Definition of the Structure of a Stochastic Decision Network

The given graph $G = [Y, U, P]$ is directed, where: Y is any finite set of elements, U is a non-empty two-unit relationship $U \subset Y \times Y$, $|Y| \geq 2$ specifies the cardinality of set Y and P is a definite function on set U that takes on values $0 < p_{ij} \leq 1$. Elements $y \in Y$ will be called vertices while $\langle y_i, y_j \rangle \in U$ where $i \prec j$ will be called ordered pairs (the arcs of the graph). The graph G fulfills the following conditions: it is consistent, acyclic and there is exactly one starting vertex and at least one endpoint vertex in the graph. Vertex $y \in Y$ in this network represents an event and, on the one hand, determines the achievement of a certain state or a goal achieved by specific subsets of actions symbolised by arcs $\langle y_i, y_j \rangle \in U$, while on the other it conditions completion for other definite subsets of arcs. Based on the above definition, vertices are divided into two groups:

- **Receivers**, defining the conditions for achieving a given state (receiver activation);
- **Emitters**, specifying the conditions for the carrying out of specific arcs that originate from it (Table 1).

Table 1. Graphical representation of the logical forms of receivers and emitters, as well as their reception and emission conditions in the stochastic decision network.

The Name of the Receiver/Emitter	Graphical Representation of the Form of Logical Reception and Emission of Activities (Arches)	Conditions for the Reception and Emission of Activities (arcs) within the Structure of the Network
Receiver "AND"		The "AND" receiver of vertex y will be activated if and only if all the actions of $u_1 \ldots u_n$ entering it will be completed.
Receiver "inclusive-or IOR"		The receiver "or" of vertex y will be activated if and only if at least one of the actions $u_1 \ldots u_n$ entering it will be completed.
Emitter "deterministic"		The deterministic emitter of vertex y enables the carrying out of all actions $u_1 \ldots u_n$ from it, provided that the vertex has been activated.
Emitter "stochastic"		The stochastic emitter of vertex y allows the performance of only one of the actions $u_1 \ldots u_n$ from it with a certain probability, provided that the node has been activated. At the same time, the condition $$\sum_{j \in \Gamma_i} p_{ij} = 1$$ must be met where: Γ_i is a set of direct successors.
Emitter "decision"		The y-vertex decision emitter only allows one of the actions $u_1 \ldots u_n$ that are outbound from it, provided that the vertex has been activated. The decision maker determines which task/action will be carried out.

Nodes in the model are created by connecting a receptor with an emitter. Connecting different emitter types of receptors defined in Table 1 makes it possible to obtain six different types of vertices, which are presented in Table 2.

The non-determinate logical structure of the considered stochastic decision network contains certain possible structures (possible sub-networks) that represent the scenarios of the carrying out of the process that is being modelled. A possible structure (possible sub-network) in a stochastic decision network is a sub-network based on a directed graph $G^* = [Y^*, U^*, P^*]$, where, $Y^* \subset Y$, $U^* \subset U$ are non-empty, and must meet the following conditions: it is consistent and acyclic. The starting point of the graph $G = [Y, U, P]$ of a stochastic decision network is also the starting point of the graph $G^* = [Y^*, U^*, P^*]$, Graph $G^* = [Y^*, U^*, P^*]$, contains at least one endpoint belonging to the $G = [Y, U, P]$ graph of the stochastic decision network. If a node with a deterministic or stochastic emitter in the $G = [Y, U, P]$ graph of a stochastic decision network belongs to graph $G^* = [Y^*, U^*, P^*]$, then all of the direct successors of this vertex belong to it as well. If the vertex with the decision emitter in graph $G = [Y, U, P]$ of the stochastic decision network belongs to graph $G^* = [Y^*, U^*, P^*]$, then one and only one of the direct successors of each of these vertices belongs to it as well. Therefore, any possible structure (possible sub-network) of a stochastic decision network can be determined by means of a vector characterized in the following manner:

$$\{\lambda_{ij}\}, \text{ where } \lambda_{ij} = \begin{cases} 0 \; dla \; \langle y_i, y_j \rangle \notin U^* \\ 1 \; dla \; \langle y_i, y_j \rangle \in U^* \end{cases} \tag{1}$$

Table 2. The six possible two-element network combinations.

Emitter \ Receiver	AND	Inclusive-Or IOR
deterministic	◯	◯
stochastic	◯	◇
decision	◖	◁

2.2. Optimisation Model

The possible structures defined above (possible sub-networks) constitute a set of possible scenarios of the carrying out of the project that is being modelled by the network.

In order to choose the best variant of carrying out the project in terms of time and cost, the authors proposed binary programming optimisation models making it possible to determine:

- The shortest expected completion time of the scenario of the planned project;
- The lowest expected cost of the carrying out of the scenario of the planned project;
- The shortest completion time and lowest cost of the carrying out of the scenario of the planned project.

Based on the obtained results, the decision-maker, based on their risk aversion, may select decisions that specify the final choice of the possible structure of the planned project. In the case of multi-criteria (time and cost) analysis, in addition to risk aversion, the decision-maker may determine weights for the expected time and costs of the completion of the planned project.

Below is a mathematical presentation of the abovementioned optimisation models.

Symbols concerning network structure:

s—starting vertex, $s \in Y$
k—end vertex, $s \in Y$
r—any other vertex, $r \in Y$
D—set of permissible solutions
Γ_r—set of direct successors r
Γ_r^-—collection of direct predecessors of r
$\pi^+ r$—out-degree (number of actions (arcs) exiting a vertex r)
$\pi^- r$—in-degree (number of actions (arcs) entering a vertex r)
λ_{ij}—binary decision variable determining the existence of the i-j action

Symbols concerning network parameters:

p_{ij}—probability of i-j action

α_{ij}—accumulated probability of the occurrence of i-j action (taking into account the probability of the occurrence of preceding tasks)

δ_r—probability of the occurrence of vertex r

c_{ij}—the cost of action i-j

d_{ir}—duration of action i-j w_t, w_c—weights for the criteria of time and cost, respectively

δ_{k_req}—required probability of occurrence of the final node k

T_r—expected completion time of vertex r

C_r—expected cost of completing vertex r

GOAL FUNCTION—EXPECTED TIME

$$F_d = T_k \rightarrow min \tag{2}$$

Goal Function—Expected Cost

$$F_c = C_k \rightarrow min \tag{3}$$

TWO-CRITERIA GOAL FUNCTION (meta-criterion function)

$$F_{cd} = w_t \frac{T_k}{\max_{\{\lambda_{ij}\} \in D} (T_k)} + w_c \frac{C_k}{\max_{\{\lambda_{ij}\} \in D} (C_k)} \rightarrow min \tag{4}$$

$$w_t, w_c \geq 0 \text{ and } w_t + w_c = 1 \tag{5}$$

CONSTRAINT CONDITIONS CONCERNING POSSIBLE STRUCTURES

$$\lambda_{ij} = BIN \tag{6}$$

$$\delta_1 = 1 \tag{7}$$

$$\alpha_{ij} = \lambda_{ij} \cdot \delta_i \cdot p_{ij} \tag{8}$$

$$\delta_j = \sum_{\Gamma_j^-} \alpha_{ij}, \text{ for receiver "or"} \tag{9}$$

$$\delta_j = \prod_{\Gamma_j^-} \alpha_{ij}, \text{ for receiver "and"} \tag{10}$$

$$\delta_k \geq \delta_{k\_req} \tag{11}$$

δ_{k_req} changed in the range from 0 to 1 with a step, e.g., 0.1. for $s \in Y$ (**"deterministic"** emitter) start node [30]

$$\sum_{r \in \Gamma s} \lambda_{sr} = \pi^+ s \tag{12}$$

for $r \in Y$ (**"and"** receiver, **"deterministic"** emitter) [31]

$$\pi^+ r \lambda_{ir} - \sum_{j \in \Gamma r} \lambda_{rj} = 0 \text{ where } i \in \Gamma_r^- \tag{13}$$

for $r \in Y$ (**"and"** receiver, **" decision"** emitter) [31]

$$\sum_{i \in \Gamma_r^-} \lambda_{ir} - \sum_{j \in \Gamma r} \pi^- r \lambda_{rj} = 0 \tag{14}$$

for r $\in$ Y (**"or"** receiver, **"deterministic"** emitter)

$$\pi^+ r\lambda_{ir} - \sum_{j\in\Gamma r} \lambda_{rj} \leq 0 \; dla \; i \in \Gamma_r^- \tag{15}$$

$$\sum_{j\in\Gamma r} \alpha_{rj} - \sum_{i\in\Gamma_r^-} \pi^+ r \, \alpha_{ir} \leq 0 \tag{16}$$

for r $\in$ Y (**"or"** receiver, **"decision"** emitter)

$$\sum_{j\in\Gamma r} \alpha_{rj} - \sum_{i\in\Gamma_r^-} \pi^+ r\alpha_{ir} \leq 0 \tag{17}$$

$$\sum_{i\in\Gamma_r^-} \lambda_{ir} - \sum_{j\in\Gamma r} \pi^- r\lambda_{rj} \leq 0 \tag{18}$$

$$\sum_{j\in\Gamma r} \lambda_{rj} \leq 1 \tag{19}$$

CONSTRAINT CONDITIONS CONCERNING TIME ANALYSES

$$T_1 = 0 \tag{20}$$

for r $\in$ Y (**"and"** receiver)

$$T_r \geq (T_i + t_{ir})\lambda_{ir} \tag{21}$$

for r $\in$ Y (**"or"** receiver)

$$T_r \cdot \sum_{i\in\Gamma_r^-} \alpha_{ir} = \sum_{i\in\Gamma_r^-} (T_i + t_{ir})\alpha_{ir} \tag{22}$$

CONSTRAINT CONDITIONS CONCERNING COST ANALYSES

$$C_1 = 0 \tag{23}$$

for r $\in$ Y (**"and"** receiver)

$$C_r = \sum_{i\in\Gamma_r^-} \left(\frac{C_i}{\pi^+ i_c} + c_{ir} \right) \lambda_{ir} \tag{24}$$

$$\pi^+ i_c = \begin{cases} \pi^+ i \; for \; deterministic \; emitter \\ 1 \; for \; nondeterministic \; emitter \end{cases}$$

for r $\in$ Y (**„or"** receiver)

$$C_r \cdot \sum_{i\in\Gamma_r^-} \alpha_{ir} = \sum_{i\in\Gamma_r^-} (C_i + c_{ir})\alpha_{ir} \tag{25}$$

3. Calculation Experiment

In order to perform the operational verification of the mathematical optimisation model defined in Section 3.2, the authors used an example from literature on the planning the renovation of the foundations of a building using an exemplary stochastic network [25]. The example selected by the authors is relatively simple (academic) which is meant to further simplify the presentation of how the optimisation model works.

3.1. Construction of a Stochastic Decision Network Model

The structure of the network model proposed in [25] takes into account various variants of the course of the renovation of a building's foundations, generated in a random manner based on statistical data collected by the author (Figure 1). For the presented network model to supplement it, the authors of this article introduced estimated values of completion time and costs of individual tasks.

The stochastic nature of the network model causes tasks that are, for example, the effect of assessing the technical condition of the foundations after their excavation, referred to as "serious damage" (action 3,4) and as "minor damage" (action 3–5), to have a random character.

However, in this model it is also possible to distinguish actions related to, for example, the choice of a procedure in the case of a weak subbase in the repair of damaged foundations (actions: 7,8, 7–11, 7–14), which should not be generated randomly because their nature is clearly decision-based.

Therefore, in order to solve the problem specified above, the authors of this article proposed replacing specific stochastic emitters of the network with decision emitters (Figure 2) which undoubtedly makes the network model a better representation of reality and introduces a wider range of possible analysis results. The choice, for example, of how to address a weak subbase in foundation repair will depend on the decision-maker and their risk aversion in the context of the probability of reaching the final vertex (successful refurbishment of the foundations—vertex 17—or abandoning the renovation and dismantlement—vertex 11). Because there are several decision emitters in the model presented in the example, the result will contain not one optimal decision, but a sequence of decisions to be taken by the decision-maker in the context of the preferences they expect.

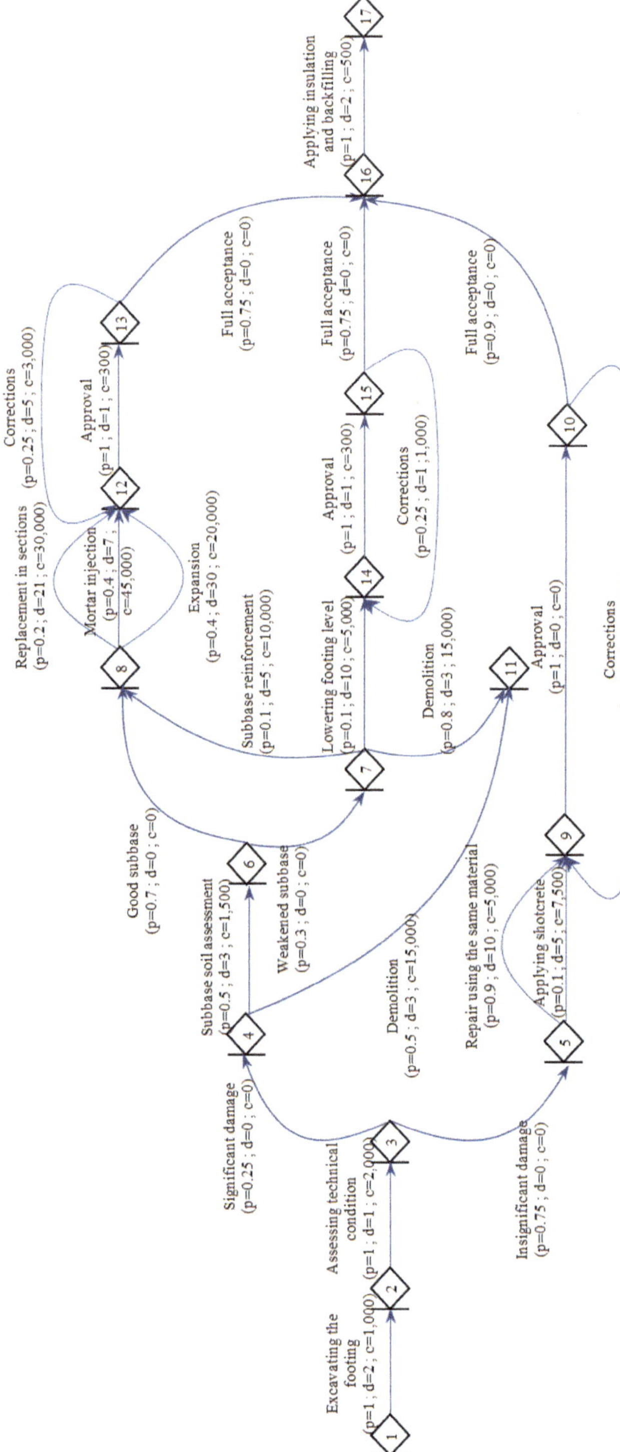

Figure 1. A stochastic network model for the renovation of a building's foundations. Values in brackets signify: "p" probability of occurrence of activities, "d" duration and "c" cost of their implementation, source: based on [25].

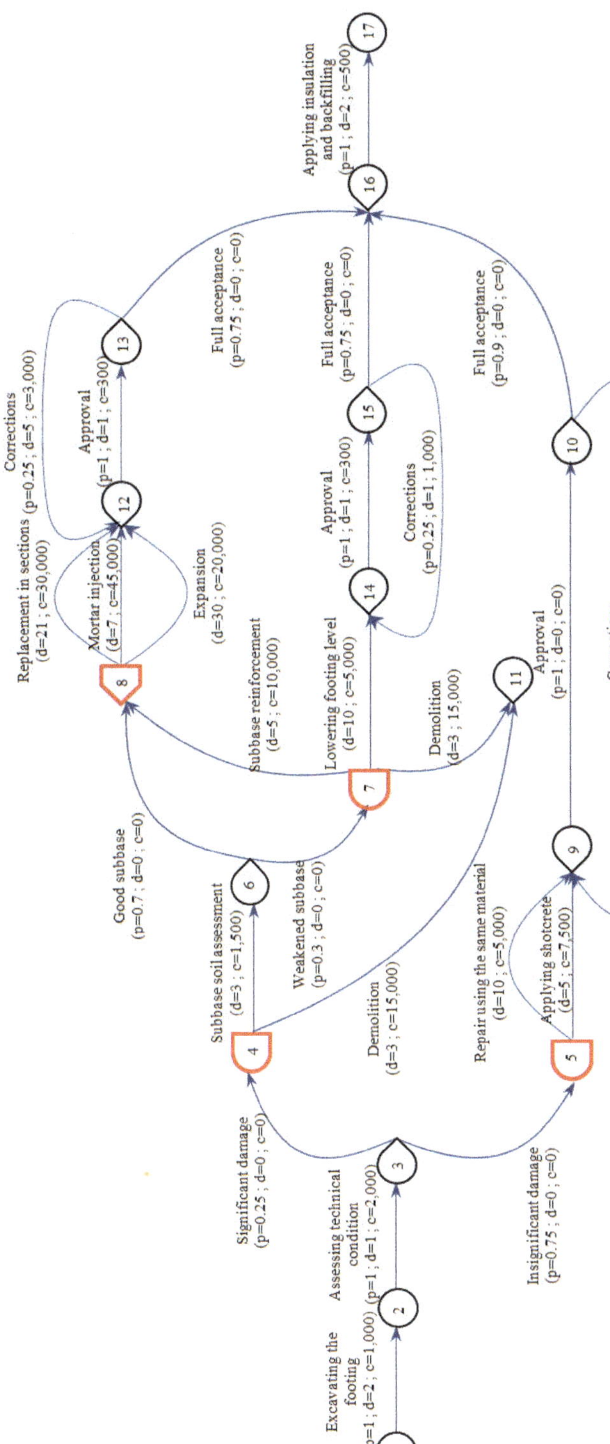

Figure 2. Stochastic decision network model for the renovation of a building's foundations with introduced vertices with decision emitters. The values in brackets signify, respectively: "p" probability of occurrence of activities, "d" duration and "c" cost of carrying them out.

3.2. Analysis of the Network Model

The input data related to the structure of the stochastic decision network of the project being modelled, as well as the values of parameters related to it (the probability of the occurrence of tasks, and the time and cost of performing them) were introduced using a computer application developed by the authors. The application was written in the Python programming language. The user can easily enter data in the form of a vertex matrix (along with determining the types of the vertices) and arcs matrix (along with defining parameters, such as time, cost, probability) into the program.

The application, based on a user-defined structure of the stochastic decision network of the modelled project, consisting of the vertices and arcs of the graph together with their loads (parameters), automatically generates a mathematical optimisation model. In order to simplify the calculations, the introduction of input data was preceded by a reduction of cycles in the considered stochastic decision network, using the well-known method of graph reduction for this purpose. The analysis of the mathematical model of the defined stochastic decision network of a building's foundations was carried out separately for the aspects of: completion time minimisation, cost minimisation and (two-criteria) minimisation of both completion time and costs of the project with the sample values set by the authors, with scales equal to 0.5. The following methods were used for the purpose of the abovementioned analysis:

- Brute force;
- APOPT solver (for Advanced Process OPTIMIZER) is a software package for solving large-scale optimisation problems (http://apopt.com/). The program is used to solve linear problems (LP), square (QP), non-linear (NLP) and mixed problems (MIP, MILP, MINLP). The APOPT solver was used with the APMonitor service [37].

Table 3 presents a set of solutions that are possible for the analysed project. Figures 3–8 present the results of optimisation obtained by means of a brute force analysis at different levels of risk aversion of the decision-maker. The numbers of possible solutions constituting the solutions obtained have been marked on the charts. Figures 9 and 10 also present the results of two-criteria optimisation for different combinations of criteria weights.

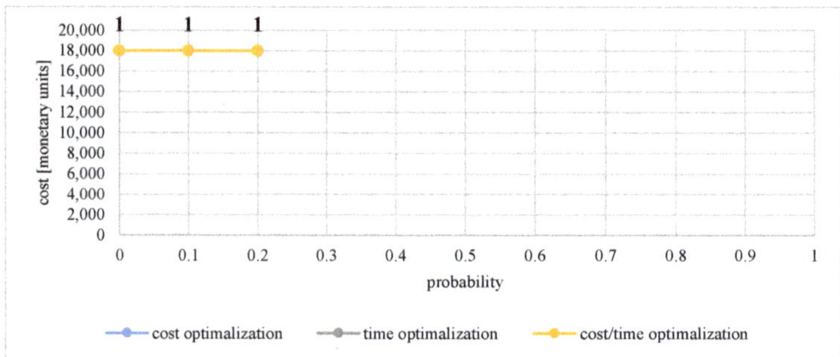

Figure 3. The expected cost of carrying out the 11th vertex at different probability levels for different types of optimisation.

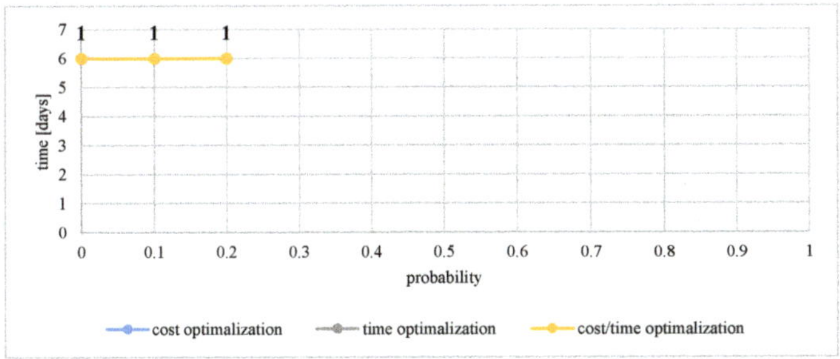

Figure 4. The expected completion time of the 11th vertex at different probability levels for different types of optimisation.

Table 3. The D set of possible solutions for the stochastic decision network of the refurbishment of the foundations of a building.

$\{\lambda_{ij}\}$ i-j	1	2	3	4	5	6	7	8	9	10	11	12	13	14	15	16	17	18	19	20
1–2	1	1	1	1	1	1	1	1	1	1	1	1	1	1	1	1	1	1	1	1
2–3	1	1	1	1	1	1	1	1	1	1	1	1	1	1	1	1	1	1	1	1
3–4	1	1	1	1	1	1	1	1	1	1	1	1	1	1	1	1	1	1	1	1
3–5	1	1	1	1	1	1	1	1	1	1	1	1	1	1	1	1	1	1	1	1
4–6	0	0	1	1	1	1	1	1	1	1	1	1	1	1	1	1	1	1	1	1
4–11	1	1	0	0	0	0	0	0	0	0	0	0	0	0	0	0	0	0	0	0
5–9	1	0	1	1	1	1	1	1	1	1	1	0	0	0	0	0	0	0	0	0
5–9	0	1	0	0	0	0	0	0	0	0	0	1	1	1	1	1	1	1	1	1
6–7	0	0	1	1	1	1	1	1	1	1	1	1	1	1	1	1	1	1	1	1
6–8	0	0	1	1	1	1	1	1	1	1	1	1	1	1	1	1	1	1	1	1
7–8	0	0	1	1	1	0	0	0	0	0	0	1	1	1	0	0	0	0	0	0
7–11	0	0	0	0	0	1	1	1	0	0	0	0	0	0	1	1	1	0	0	0
7–14	0	0	0	0	0	0	0	0	1	1	1	0	0	0	0	0	0	1	1	1
8–12	0	0	1	0	0	1	0	0	1	0	0	1	0	0	1	0	0	1	0	0
8–12	0	0	0	1	0	0	1	0	0	1	0	0	1	0	0	1	0	0	1	0
8–12	0	0	0	0	1	0	0	1	0	0	1	0	0	1	0	0	1	0	0	1
9–10	1	1	1	1	1	1	1	1	1	1	1	1	1	1	1	1	1	1	1	1
10–16	1	1	1	1	1	1	1	1	1	1	1	1	1	1	1	1	1	1	1	1
12–13	0	0	1	1	1	1	1	1	1	1	1	1	1	1	1	1	1	1	1	1
13–16	0	0	1	1	1	1	1	1	1	1	1	1	1	1	1	1	1	1	1	1
14–15	0	0	0	0	0	0	0	0	1	1	1	0	0	0	0	0	0	1	1	1
15–16	0	0	0	0	0	0	0	0	1	1	1	0	0	0	0	0	0	1	1	1
16–17	1	1	1	1	1	1	1	1	1	1	1	1	1	1	1	1	1	1	1	1

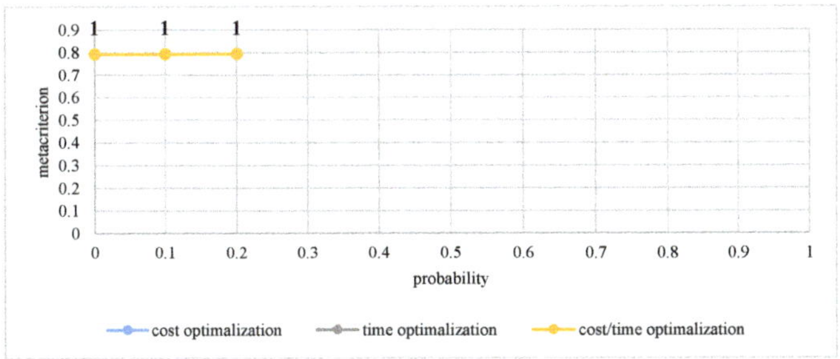

Figure 5. Value of the meta-criterion of carrying out vertex 11 at different probability levels for different types of optimisation.

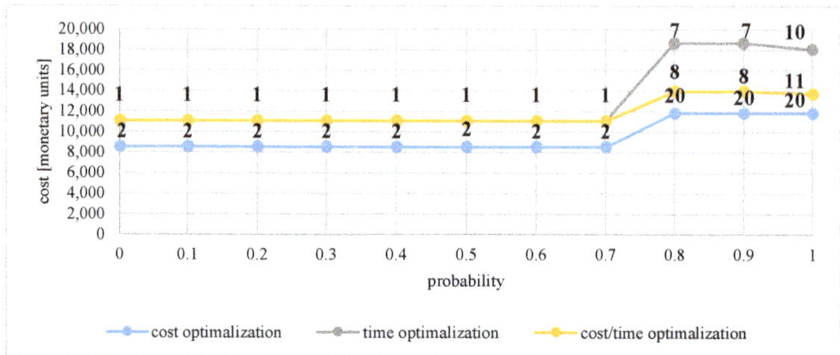

Figure 6. The expected cost of the carrying out of vertex 17 at different probability levels for different types of optimisation.

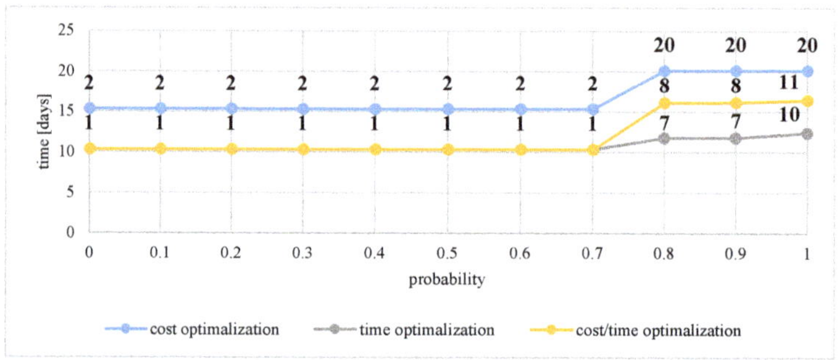

Figure 7. The expected completion time of vertex 17 at different probability levels for different types of optimisation.

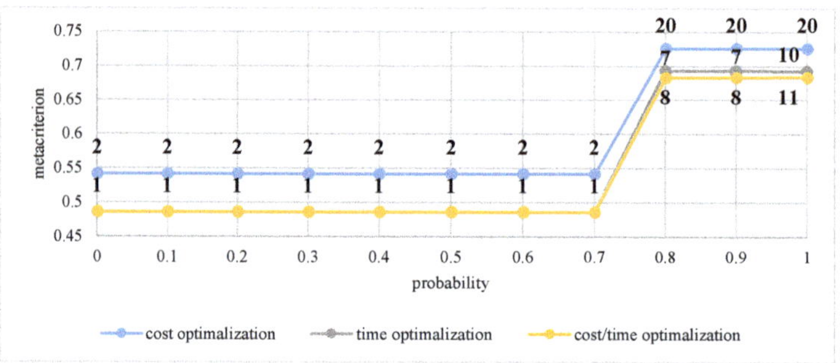

Figure 8. Value of the meta-criterion for the carrying out of vertex 17 at different probability levels for different types of optimisation.

In Table 4, the authors compared the results obtained with the brute force method with the results obtained using the APOPT solver for different levels of probability of reaching end vertices No. 11 and No. 17. The problem of finding a solution using the APOPT solver was noted for several probability values. However, in these cases, the program returned a result if the probability value given by the authors was slightly greater than its threshold value (e.g., instead of 0.3, the value was set to 0.3001). The proposed APOPT solver uses an active-set algorithm. Initially, this algorithm searches for a permissible solution. A small change in task parameter values can cause a permissible solution to either be found or not. This problem results from the fact that the optimisation under consideration is from the Mixed Integer Nonlinear Programming (MINLP) class. So far, no methods capable of effectively solving this class of problems for large-sized tasks have been developed. However, despite what has been stated above, the computer application in question is very practical because it makes it possible to obtain results faster than through the use of the brute force method. In Table 4, solutions obtained using the APOPT solver that differed from solutions obtained using the brute force method (or for which APOPT did not find a solution) were highlighted in colour

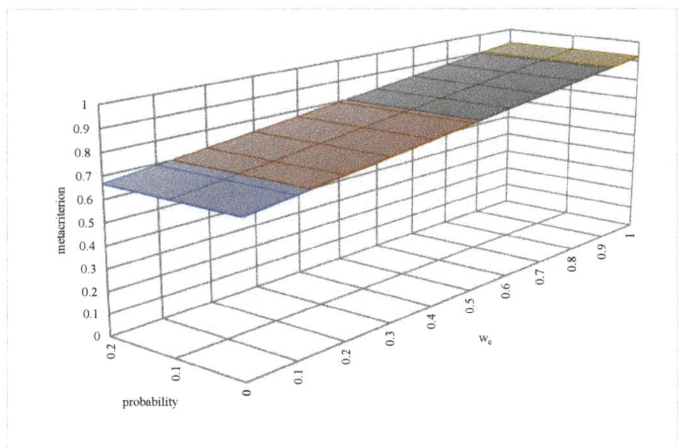

Figure 9. Two-criteria optimisation results for the completion of vertex 11 at different probability levels and different combinations of criterion weights.

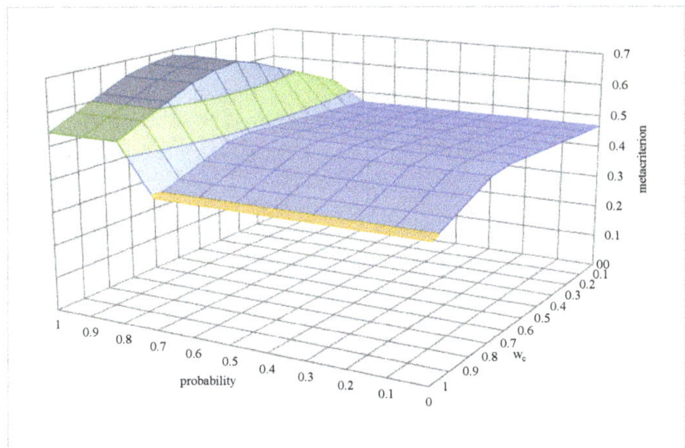

Figure 10. Two-criteria optimisation results for the completion of vertex 17 at different probability levels and different combinations of criteria weights.

Table 4. Comparison of the results obtained using the brute force method with the results obtained with APOPT solver for different levels of probability of reaching the vertices No. 11 and No. 17. (* Solutions were found by increasing probability by a small margin).

	Probability	Results of Cost Optimisation at Different Probability Levels		Results of Time Optimisation at Different Probability Levels		The Results of Two-Criteria Optimisation at Different Probability Levels	
		APOPT	Brute Force	APOPT	Brute Force	APOPT	Brute Force
Vertex 11	>0	18,000.00	18,000.00	6.00	6.00	0.794872	0.794872
	≥0.1	18,000.00	18,000.00	solution not found	6.00	0.794872	0.794872
	≥0.2	18,000.00	18,000.00	6.00	6.00	0.794872	0.794872
Vertex 17	>0	8555.56	8555.56	10.33	10.33	0.485979	0.485979
	≥0.1	8555.56	8555.56	10.33	10.33	0.485979	0.485979
	≥0.2	8555.56	8555.56	10.33	10.33	0.485979	0.485979
	≥0.3	solution not found *	8555.56	10.33	10.33	0.485979	0.485979
	≥0.4	8555.56	8555.56	solution not found *	10.33	0.485979	0.485979
	≥0.5	8555.56	8555.56	10.33	10.33	0.485979	0.485979
	≥0.6	solution not found *	8555.56	10.33	10.33	0.485979	0.485979
	≥0.7	8555.56	8555.56	10.33	10.33	0.485979	0.485979
	≥0.8	solution not found *	11,841.67	12.37	11.78	0.683898	0.683426
	≥0.9	11,841.67	11,841.67	12.37	11.78	0.683898	0.683426
	=1	11,841.67	11,841.67	12.37	12.37	0.683898	0.683898

4. Analysis of Results

When analysing the results, it can be observed that in the case of the 11th vertex, regardless of the type of optimisation used, the optimal solution is a possible solution $\{\lambda_{ij}\}_1$ (the graphs overlap). It is also impossible to achieve this with a probability of 0.3 or higher. As for vertex 17, the same results are obtained in the case of time and cost optimisation for a probability value of up to 0.7 (also a possible solution $\{\lambda_{ij}\}_1$). However, in the case of probabilities higher than 0.7, one could observe discrepancies in the solutions that can be seen when using different types of optimisation.

The decision-maker acquires information about decisions from the obtained results (represented by graph arcs originating at decision emitters) and that should be included in the project's final plan, taking into account the personal risk aversion and preferences as to the significance of expected times and costs of the final results of the planned project. These decisions define/determine the optimal plan of the carrying out of the project under consideration, in terms of the expected value of time and costs, i.e., the optimal possible sub-network. For example, let the decision-maker's risk aversion correspond to a minimum probability of to reach vertex No. 17 being 0.8. Let their preference regarding the validity of the expected costs be 0.3 and the expected time—0.7. Therefore, with the assumptions above, the carrying out of the renovation of foundations (vertex No. 17) within an optimal time and at an optimal cost is determined by the possible solution $\{\lambda_{ij}\}_7$, for which the optimal metacriterion function value is 0.629 (expected cost = PLN 18688.29 and expected time = 11.78 days, with a probability = 0.925).

Therefore, the decision-maker receives information that, at the planning stage of the project, he or she should plan the analysis of the subbase's soil (action: 4–6) and choose solutions for repairing foundations like applying shotcrete (action: 5–9) and traditional repair (action: 8–12), activities that are optimal in the context of minimising the time and costs of the renovation of foundations. However, the decision-maker should take into account the fact that for such a plan, there is a relatively small, 0.075 probability of reaching vertex no. 11 (instead of vertex no. 17). This can take place in the event of serious damage to the foundations (action 3–4) and weakened soil (action 6,7), which will result in the decision to commence with dismantlement (action 7–11). The value of the metacriteria function for the 11th vertex will then be 1.0 (expected cost = 19,500 PLN and expected time = 9 days) (Figure 11).

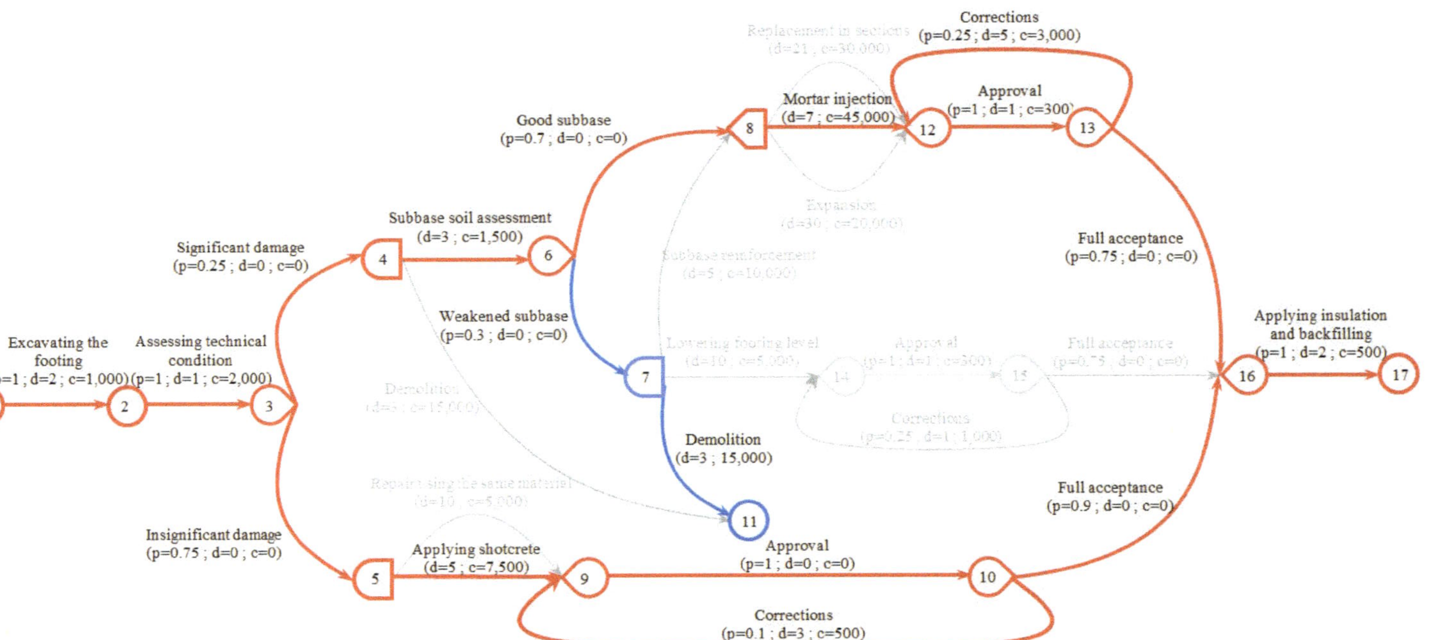

Figure 11. The sub-network $\{\lambda_{ij}\}_7$, that defines/determines the plan of carrying out the renovation of foundations (reaching vertex no. 17) with a probability equal to min. 0.8 with the minimum expected time and cost for the weighted preferences of the decision-maker at a level of 0.3 for the expected costs and 0.7 for the expected project completion time.

Finally, as mentioned in Section 3.1, the use of decision emitters in the structure of the considered stochastic network (Figure 1) is justified due to the decision character of alternatives emerging from vertices 4, 5, 7 and 8. Consequently, this approach allows one to generate an optimal (in terms of time and cost) possible sub-network, whose expected values of the time and cost of carrying out each vertex are smaller than the corresponding values in the case of a classical stochastic network (Table 5). The resulting values for the stochastic network were obtained by using the graph reduction method. Graph reduction methods are still being intensively developed [38].

Table 5. Comparison of the values of expected completion times and costs of carrying out the final vertices for specific probabilities of carrying them out in the case of the stochastic network model and the stochastic decision network of the planned project of renovating foundations.

		Stochastic Network		Stochastic Decision Network		
No. of the Final Vertex	Probability of Being Carried Out	Expected Cost of Completion [Monetary Units]	Expected Completion Time [Days]	The optimal Expected Completion Cost [Monetary Units]	The Optimal Expected Completion Time [Days]	Metacriterion for the Weights: $w_t, w_c = 0.5$
11	0.155	21290.32	9.58	18,000.00 for sub-networks: $\left\{\lambda_{ij}\right\}_1$	6.00 for sub-networks: $\left\{\lambda_{ij}\right\}_1$	0.795 for sub-networks: $\left\{\lambda_{ij}\right\}_1$
17	0.845	11891.62	14.85	11,841.00 for sub-networks: $\left\{\lambda_{ij}\right\}_{20}$	11.78 for sub-networks: $\left\{\lambda_{ij}\right\}_7$	0.683 for sub-networks: $\left\{\lambda_{ij}\right\}_8$

5. Conclusions and Discussion

The restoration of a historical building is a chance to extend its life-cycle, which, according to the concept of sustainable development, will make it possible to preserve its substance and cultural values for future generations.

Determining the estimated completion time and cost of the restoration work to be done on a historical structure is a highly distinct and difficult process, as it requires that various scenarios of completing such projects are taken into account.

The planning of construction projects with a non-determinate course, such as the renovation of buildings and structures, requires the use of appropriate tools that enable the effective modelling of the scenarios being considered, as well as the estimation of the time and costs of carrying them out. Analysis of individual alternatives of the carrying out the project can be expanded to include experiments conducted in order to design various solutions. For instance, in the article [39] it was demonstrated that an experimental propagation of errors can affect the prediction of pedestrian suspended bridge fluttering, which is associated with the cost of their construction. In subject literature, there are many proposals of the use of stochastic networks for this purpose, which are based on the assumption that variant scenarios for the carrying out of such projects are generated in a random manner. However, in practice, the carrying out of the considered scenarios of such projects, such as the restoration of historical structures, may also have a decision-based character. The choice of the type of technical solution or the way of carrying out the work should not be treated as a random event, but as a decision option for the decision-maker to consider. The network planning and analysis methods proposed in literature do not allow for the modelling of construction projects with a non-determinate course, including both a random and decision-based character of carrying them out.

In the work, the authors proposed an innovative approach in the form of a stochastic decision network, enabling the integration of both the random and decision-based character of the planned project. The authors defined the structure of the network model by introducing an appropriate vertex topology and defined conditions for possible structures (possible sub-networks) generated by vertices with decision emitters. For the purpose of optimisation (in terms of time and cost) of the project plan being modelled by the stochastic decision network, the authors developed non-linear models of one and multi-criteria binary programming. The authors also developed a computer application written in the Python programming language, which allows for the easy and quick input of data about the structure and parameters of stochastic networks in order to generate an optimisation model for the user. To analyse the model, the authors used two methods: brute force and APOPT solver and then compared the obtained results. The authors are aware of the limitations and imperfections of the APOPT solver. In the future, it is planned to use metaheuristic methods to optimise the network in question, such as simulated annealing, genetic algorithms or Monte Carlo tree search.

As a result, the approach developed by the authors enables the decision-maker—by specifying their preferences as to the outcome of the planned project (reaching the selected final vertex in the network) and risk aversion—to obtain the optimal (in terms of expected time and costs) solution (sub-network). In the case of a multi-criteria analysis (of time and cost), the decision-maker may also define different values of weights for the expected time and costs of the planned project in the goal function. As a part of the results, the decision-maker will get information on the type of technical solutions or the manner of carrying out the work that should be included in the plan of the optimal scenario of carrying out the project.

In order to confirm the effectiveness of the aforementioned approach, the authors, based on the conducted calculation experiment, demonstrated the advantage of the stochastic decision network over the classical stochastic network in the case of projects whose implementation structure has both a random and decision-based character.

In practice, the developed approach can be useful at the stage of developing a project's feasibility study and can be used to analyse various scenarios of carrying it out in terms of probability and the assessment of its cost and completion time. It should be noted that the optimisation of alternative technical solutions can also be useful during the stage of the design of a new building or planning construction and renovation work.

The example presented in the article features a small number of calculations as it was used to demonstrate how the method works. In the future, the method should be tested on more complicated construction project network models. Furthermore, as part of further research, the potential of stochastic decision networks should be expanded by developing a functional time-cost relation at the level of individual activities in the network. In addition, the time, cost or the amount of resources needed to carry out a given task does not necessarily have to be a deterministic value. For many projects, e.g., the renovations of buildings, these values may be random variables of a given probability distribution or, in the absence of empirical data, may take on fuzzy values, for instance.

In addition, the proposed structure of the stochastic and decision network can, in the future, become a basis for the use of random growing mechanisms, based on the illustrative problems [40–42].

Author Contributions: Conceptualization: G.Ś.; methodology: G.Ś., B.S. (Bartłomiej Szewczyk), B.S. (Bartłomiej Sroka); software: B.S. (Bartłomiej Sroka); validation: B.S. (Bartłomiej Szewczyk); formal analysis: G.Ś., B.S. (Bartłomiej Szewczyk), B.S. (Bartłomiej Sroka); investigation, resources, data curation, writing—original draft preparation, writing—review and editing, visualization, supervision, project administration, G.Ś., B.S. (Bartłomiej Szewczyk), B.S. (Bartłomiej Sroka), E.R.-Z.; funding acquisition: E.R.-Z.

Funding: This research received no external funding.

Conflicts of Interest: The authors declare no conflict of interest.

References

1. Zavadskas, E.K.; Antucheviciene, J. Development of an indicator model and ranking of sustainable revitalization alternatives of derelict property: A Lithuanian case study. *Sustain. Dev.* **2006**, *14*, 287–299. [CrossRef]

2. Zavadskas, E.K.; Antucheviciene, J. Multiple criteria evaluation of rural building's regeneration alternatives. *Build. Environ.* **2007**, *42*, 436–451. [CrossRef]

3. European Union Treaty of Lisbon Amending the Treaty on European Union and the Treaty establishing the European Community. 2007. Available online: http://data.europa.eu/eli/treaty/lis/sign (accessed on 28 January 2019).
4. Śladowski, G.; Radziszewska-Zielina, E. Description of technological and organisational problems in construction works using the example of restoration of the outer courtyard on Wawel Hill. *Tech. Trans.* **2015**, *2*, 117–125.
5. Radziszewska-Zielina, E.; Kutut, V.; Mesároš, P.; Sroka, B.; Szewczyk, B.; Śladowski, G. Problems of assessing the duration and cost of restoration projects on the example of Poland, Slovakia and Lithuania. *Sci. Rev. Eng. Environ. Sci.* **2018**, *27*, 3–8. [CrossRef]
6. Lootsma, F.A. Theory and Methodology stochastic and fuzzy PERT. *Eur. J. Oper. Res.* **1989**, *143*, 174–183. [CrossRef]
7. Chanas, S.; Zieliński, P. Critical path analysis in the network with fuzzy activity times. *Fuzzy Sets Syst.* **2001**, *122*, 195–204. [CrossRef]
8. Lin, F.T.; Yao, J.S. Fuzzy critical path method based on signed-distance ranking and statistical confidence-interval estimates. *J. Supercomput.* **2003**, *24*, 305–325. [CrossRef]
9. Chen, C.-T.; Huang, S.-F. Applying fuzzy method for measuring criticality in project network. *Inf. Sci.* **2007**, *177*, 2448–2458. [CrossRef]
10. Hajdu, M. Effects of the application of activity calendars on the distribution of project duration in PERT networks. *Autom. Constr.* **2013**, *35*, 397–404. [CrossRef]
11. Radziszewska-Zielina, E.; Śladowski, G.; Sibielak, M. Planning the reconstruction of a historical building by using a fuzzy stochastic network. *Autom. Constr.* **2017**, *84*, 242–257. [CrossRef]
12. Eisner, H. A Generalized Network Approach to the Planning and Scheduling of a Research Project. *Oper. Res.* **1962**, *10*, 115–125. [CrossRef]
13. Pritsker, A.A.B. *GERT: Graphical Evaluation and Review Technique*; Rand Corporation: Santa Monica, CA, USA, 1966.
14. Whitehouse, G.E. *Systems Analysis and Design Using Network Techniques*; Prentice-Hall: Englewood Cliffs, NJ, USA, 1973.
15. Cheng, C.-H. Fuzzy consecutive-k-out-of-n:F system reliability. *Microelectron. Reliab.* **1994**, *34*, 1909–1922. [CrossRef]
16. Cheng, C.-H. Fuzzy repairable reliability based on fuzzy gert. *Microelectron. Reliab.* **1996**, *36*, 1557–1563. [CrossRef]
17. Gavareshki, M.K. New fuzzy GERT method for research projects scheduling. In Proceedings of the 2004 IEEE International Engineering Management Conference, Singapore, 18–21 October 2004; Volume 2, pp. 820–824.
18. Hashemin, S.S. Fuzzy completion time for alternative stochastic networks. *J. Ind. Eng. Int.* **2010**, *6*, 17–22.
19. Norouzi, G.; Heydari, M.; Noori, S.; Bagherpour, M. Developing a mathematical model for scheduling and determining success probability of research projects considering complex-fuzzy networks. *J. Appl. Math.* **2015**. [CrossRef]
20. Zadeh, L.A. Fuzzy sets. *Inf. Control* **1965**, *8*, 338–354. [CrossRef]
21. Zanddizari, M. Set confidence interval for customer order cycle time in supply chain using GERT method. In Proceedings of the 17th IASTED international Conference on Modelling and Simulation, Montreal, QC, Canada, 24–26 May 2006; pp. 213–218.
22. Aytulun, S.K.; Guneri, A.F. Business process modelling with stochastic networks. *Int. J. Prod. Res.* **2008**, *46*, 2743–2764. [CrossRef]
23. Zhang, Z.; Ye, W.; Xia, Z. Research on A GERT Based Risk Transfer Model in Complex Product Development Process. *J. Grey Syst.* **2015**, *27*, 99–111.
24. Nelson, R.G.; Azaron, A.; Aref, S. The use of a GERT based method to model concurrent product development processes. *Eur. J. Oper. Res.* **2016**, *250*, 566–578. [CrossRef]
25. Kosecki, A. Use of stochastic network models in managing the renovation and rehabilitation projects. In *Integrated Approaches to the Design and Management of Buildings Reconstruction: International University Textbook*; EuroScientia: Wambeek, Belgium, 2014; pp. 52–60.
26. Hougui, Z. Study on stochastic model for the construction of concrete longitudinal cofferdam in the Three Gorges Project. *J. Hydraul. Eng.* **1997**, *8*, 57–60.
27. Pena-Mora, F.; Li, M. Dynamic planning and control methodology for design/build fast-track construction projects. *J. Constr. Eng. Manag.* **2001**, *127*, 1–17. [CrossRef]

28. Gao, X.; Shen, Y.; Lin, J. Schedule risk management at early stages of large construction projects based on the GERT model. In Proceedings of the IEEE Conference Anthology, Shenyang, China, 7–9 November 2014; pp. 1–4.

29. Wang, B.; Li, Z.; Gao, J.; Vaso, H. Critical Quality Source Diagnosis for Dam Concrete Construction Based on Quality Gain-loss Function. *J. Eng. Sci. Technol. Rev.* **2014**, *7*, 137–151. [CrossRef]

30. Ignasiak, E. Optymalne struktury projektów w sieciach alternatywno-decyzyjnych (Optimal project structures in alternative-decision networks). *Prz. Stat. (Stat. Rev.)* **1975**, *22*, 323–333.

31. Śladowski, G. Modeling technological and organizational variants of building projects. *Tech. Trans.* **2014**, *5*, 261–269.

32. Zhang, L.H.; Liu, X.; Zhong, G. Bound Duration Theorem and DCPM Network Reduction. In Proceedings of the 2009 International Conference on Electronic Commerce and Business Intelligence, Beijing, China, 6–7 June 2009; pp. 230–233.

33. Zhang, L.; Meng, X.; Cui, L. New Method for DCPM by Substituting Critical Path. In Proceedings of the 2011 International Conference on Management and Service Science, Wuhan, China, 12–14 August 2011; pp. 1–4.

34. Ibadov, N. The Alternative Net Model with the Fuzzy Decision Node for the Construction Projects Planning. *Arch. Civ. Eng.* **2018**, *64*, 3–20. [CrossRef]

35. Ibadov, N.; Kulejewski, J. Construction projects planning using network model with the fuzzy decision node. *Int. J. Environ. Sci. Technol.* **2019**, 1–8. [CrossRef]

36. Więckowski, A. Principles of the NNM method applied in the analysis of process realisation. *Autom. Constr.* **2002**, *11*, 409–420. [CrossRef]

37. Hedengren, J.D.; Shishavan, R.A.; Powell, K.M.; Edgar, T.F. Nonlinear modeling, estimation and predictive control in APMonitor. *Comput. Chem. Eng.* **2014**, *70*, 133–148. [CrossRef]

38. Shang, Y. Subgraph Robustness of Complex Networks Under Attacks. *IEEE Trans. Syst. Man Cybern. Syst.* **2017**, *99*, 1–12. [CrossRef]

39. Rizzo, F.; Caracoglia, L.; Montelpare, S. Predicting the flutter speed of a pedestrian suspension bridge through examination of laboratory experimental errors. *Eng. Struct.* **2018**, *172*, 589–613. [CrossRef]

40. Shang, Y. Distinct Clusterings and Characteristic Path Lengths in Dynamic Small-World Networks with Identical Limit Degree Distribution. *J. Stat. Phys.* **2012**, *149*, 505–518. [CrossRef]

41. Shang, Y. Geometric Assortative Growth Model for Small-World Networks. *Sci. World J.* **2014**, *2014*, 1–8. [CrossRef] [PubMed]

42. Shang, Y. On the likelihood of forests. *Phys. A Stat. Mech. Its Appl.* **2016**, *456*, 157–166. [CrossRef]

© 2019 by the authors. Licensee MDPI, Basel, Switzerland. This article is an open access article distributed under the terms and conditions of the Creative Commons Attribution (CC BY) license (http://creativecommons.org/licenses/by/4.0/).

Article

An Extension of the CODAS Approach Using Interval-Valued Intuitionistic Fuzzy Set for Sustainable Material Selection in Construction Projects with Incomplete Weight Information

Jagannath Roy [1], Sujit Das [2], Samarjit Kar [1] and Dragan Pamučar [3,*]

[1]   National Institute of Technology Durgapur, Durgapur 713209, India; jaga.math23@gmail.com (J.R.); dr.samarjitkar@gmail.com (S.K.)
[2]   National Institute of Technology Warangal, Warangal 506004, India; sujit.das@nitw.ac.in
[3]   Department of Logistics, Military academy, University of Defence in Belgrade, Pavla Jurisica Sturma 33, 11000 Belgrade, Serbia
*   Correspondence: dpamucar@gmail.com

Received: 21 February 2019; Accepted: 15 March 2019; Published: 18 March 2019

Abstract: Optimal selection of sustainable materials in construction projects can benefit several stakeholders in their respective industries with the triple bottom line (TBL) framework in a broader perspective of greater business value. Multiple criteria of social, environmental, and economic aspects should be essentially accounted for the optimal selection of materials involving the significant group of experts to avoid project failures. This paper proposes an evaluation framework for solving multi criteria decision making (MCDM) problems with incomplete weight information by extending the combinative distance assessment (CODAS) method with interval-valued intuitionistic fuzzy numbers. To compute the unknown weights of the evaluation criteria, this paper presents an optimization model based on the interval-valued intuitionistic fuzzy distance measure. In this study, we emphasize the importance of individual decision makers. To illustrate the proposed approach, an example of material selection in automotive parts industry is presented followed by a real case study of brick selection in sustainable building construction projects. The comparative study indicates the advantages of the proposed approach in comparison with the some relevant approaches. A sensitivity analysis of the proposed IVIF-CODAS method has been performed by changing the criteria weights, where the results show a high degree of stability.

Keywords: material selection; multiple criteria decision making; IVIFN; CODAS; sustainability

---

1. Introduction

Selecting the right material to develop a particular product is essential for each of the organizations to survive in the competitive business sectors. One suitable material can substantially minimize the production cost and maximizes the profit. It is also needed to improve the product performance and customer satisfaction. Due to the presence of varieties types of almost similar kind of materials, the task of material selection has become one of the most challenging tasks in the real life environment [1–3]. Thus, the material selection process is essential in many practical industrial problems like automotive parts manufacturing, selection of robots and forklifts, designing the femoral component of cemented total hip/knee replacement, building constructions, etc. Hence, in the last few decades, several researchers have focused on this domain. This paper deals with the material selection problem with sustainability perspectives for a construction company.

The material selection problem (MSP) in the presence of imprecise and incomplete information is regarded as an important area of research where researchers try to solve ill-structured complicated real circumstances. Researchers have applied classical fuzzy set theory (CFST) to explore and solve the MSP under uncertain environment where they mainly considered the individual performance of every material. Due to the complicated and ill-structured characteristics and several practical contexts of MSP, it is necessary to adopt/use newer advanced techniques which can flexibly handle uncertain data while assessing the performance of alternative materials. In the recent years, many researchers have proposed novel methods for solving MSP under uncertainty, such as, two interval type 2 fuzzy TOPSIS method [4], interval 2-tuple linguistic model [5], and soft computing tool based on fuzzy grey relations [6]. However, only a few researchers [7] have used interval-valued intuitionistic fuzzy sets (IVIFSs) in MSP under incomplete weight information. The IVIFS theory [8] is more practical and reliable approach than CFST for handling imprecision and fuzziness in DMs' judgments in real world decision making problems. IVIFS considers both the degree of membership and nonmembership value of a component in a given set and they can take interval values rather than exact numbers. When DMs often find it difficult to evaluate the uncertain material performance with just a single valued number using CFST, IVIFS offers more flexibility and DMs get an extra degree of freedom to evaluate the same. Hence, it has become considerably important to explore MSP under uncertainty with more efficient and appropriate mathematical approaches that use the IVIFSs for better handling of imprecise information. Due to its inherent flexible nature, recently many authors have contributed on IVIFS and applied it in decision problems. Some significant contributions of IVIFS in decision making problems are narrated below. Cheng [9] proposed a decision making method to select hotel locations using IVIFS, where the authors used IVIFS to denote the values and weights of the attributes. Wen et al. [10] introduced a new method to aggregate the IVIFNs when they are distributed all over a region. For that purpose the authors studied the interval-valued intuitionistic fuzzy definite integral and based on that they proposed interval-valued intuitionistic fuzzy definite integral (IVIFDI) operator. Rashid et al. [11] studied interval-valued knowledge measure for the IVIFSs by extending the knowledge measure of IFSs and proposed interval-valued information entropy measure for IVIFSs. Wang and Chen [12] proposed a decision making method using IVIFSs, linear programming (LP), and the extended TOPSIS method, where they used LP methodology to obtain optimal weights of attributes. Gupta et al. [13] combined extended TOPSIS and LP method to propose a multi-attribute group decision making (MAGDM) method in the context if IVIFS. Interval-valued intuitionistic fuzzy matrix (IVIFM) was used by Das et al. [14] to propose a decision making approach, where the authors assigned confident weights to the experts and then presented a decision making algorithm [15].

MSP has been recognized as an MCDM problem [7] and one has to take care of several criteria (sometimes referred as objectives) systematically, such as enactment, cost, use, mobility, transportation, availability, disposal, environmental norms, maintenance, etc., while seeking the best material [1–3]. The applicability of MSP in the context of MCDM to design a product/service has been explored in the literature survey segment of this paper. The availability of numerous MCDM tools has helped the DMs to identify and select the optimal choice for their MSP (refer to Section 2.2). The CODAS method is a new evaluation tool, recently proposed by Ghorabaee [16], has been proved to be efficient to deal with MCDM problems. Due to its inherent characteristics, the CODAS method possesses a systematic and simple computation procedure which is logically sound to represent the underlying principle of real life decision making problems. Thus, it has become significant to incorporate CODAS method in MSP for evaluating and ranking the alternative materials under incomplete criteria weight information and IVIFNs context.

The above-mentioned deliberations motivate us to extend the CODAS method for material selection problems (MSPs) under uncertainty. Then the extended CODAS method is used to develop a novel MCDM framework using IVIFNs under difficult situations where criteria weight are partial known or totally unknown. The projected MSP solution procedure is able to replicate both subjective judgments and objective information in practical engineering, manufacturing, and industrial problems in the context of IVIF. Finally, one illustrative example is given to examine the proposed IVIF-CODAS method for MSP followed by a real case study which is demonstrated for a sustainable building construction company. The outcomes of the proposed research framework can help DMs (engineers, manufacturers, and designers) to have an effective decision for complicated MSPs in uncertainty.

The rest of the paper is structured as follows. Section 2 offers a literature review of MSP and sustainability perspectives along with IVIFSs and MCDM. Section 3 discusses the preliminaries on IVIFSs. Next, the extended CODAS method with IVIFNs and incomplete weight information is presented in Section 4. Two illustrative examples are discussed in Section 5 to show the usefulness of the projected evaluation framework. Section 6 explains the outcomes through result comparison and sensitivity analysis while Section 7 concludes the current research work.

2. Literature survey

2.1. Material Selection in Construction Industry and Sustainability

The construction industry has been revealed as the fastest growing industry throughout the world due to the constant increase in urban population. This fastest growth has influenced the society for better economic and social movement and simultaneously has triggered the environmental pollution factors. Some researchers [17–19] investigated that the energy used ratio in recent infrastructures is six times more than that of the older ones, especially in United Arab Emirates (UAE). Furthermore, it was documented (https://ccap.org/assets/Success-Stories-in-Building-Energy-Efficiency_CCAP.pdf) that the construction industry consumes 40% of the world's energy. These factors had inspired many countries to consider eco-friendly infrastructures such as buildings which gives more importance to sustainable construction rather than economic concerns [3]. Since sustainable construction has a direct impact on environment, economy, and society, the construction agencies are continuously trying to adopt it in their work cultures mainly in the form of sustainable design, structure, and material selection. This kind of changed scenario has drawn the focus of many researchers and consequently, several research works have been carried out on sustainable construction. As a pioneer of this concept, Kibert [20] stated that "Sustainable construction is the creation and responsible management of a healthy built environment based on resource efficient and ecological principles". Among many other sustainable construction factors, sustainable material selection imparts a key role and directly effects building sustainability. Radhi [21] performed an experiment on UAE construction and found the impact of UAE construction on global warming. Elchalakani and Elgaali [22] combined the effects of recycled aggregate and recycled water and discussed the strength and durability of recycled concrete. The authors prepared a moderate strength concrete using recycled water and recycled aggregate obtained from construction wastes. Al-Hajj and Hamani [23] investigated the existing studies regarding the sources of waste and suggested some measures to reduce it. The authors noticed that the lack of awareness and poor design were the main causes of material waste. Despite the need to explore MSP in construction projects under sustainability perspectives, only a few research papers [18,24,25] are found in the literature. A more extensive literature survey can be found in [18,23]. As per our knowledge, no researcher has explored the MSP with sustainability norms in the Indian scenario. Hence, this paper attempts to fill this gap and enrich the literature of MSP in sustainable construction projects.

2.2. MSP and Various MCDMs

To find the optimal choice of MSP in construction and design engineering, one can consider diverse methodologies, such as MCDM techniques, statistical approaches, artificial intelligence, mathematical programming, and hybrid methods. Among them, MCDM tools are most widely used for MSP, since they can easily and successfully solve the evaluation problems that are complex and have multiple conflicting objectives/criteria. For example, Bahraminasab and Jahan [26] applied a comprehensive VIKOR (VIsekriterijumska optimizacija i KOmpromisno Resenje) method to find the best material for a femoral component of total knee replacement. A new framework was developed by Jahan et al. [27] for weighting of criteria in MSP. Chauhan and Vaish [28] proposed a hybrid evaluation model including entropy, VIKOR, and TOPSIS (Technique for order preference by similarity to ideal solution) methods to select the soft and hard magnetic materials. Interval 2-tuple linguistic VIKOR [3], multi-objective optimization on the basis of ratio analysis [29] (MOORA), complex proportional assessment [30] (COPRAS) are also successfully used in MSP. In recent years, Anojkumar et al. [31] used fuzzy AHP (Analytic Hierarchy Process), VIKOR and TOPSIS for MSP in the sugar industry. Furthermore, interval type 2 fuzzy TOPSIS [4], Multi-Attributive Border Approximation area Comparison [7] (MABAC), hybrid MCDM framework with DEMATEL, ANP and TOPSIS models [18], neutrosophic MULTIMOORA [32], and grey-correlation-based hybrid MCDM method [33] are notable contributions in MSP in recent times. All the authors except Xue et al. [7] have used either fuzzy sets or crisp sets.

Although many material selection methods are available in the literature, still there is a need to explore the MCDM techniques by incorporating IVIFS theory. IVIFNs remove the limitations of CFST and offer a more rational and computational flexibility to address uncertainty and ambiguity in data. IVIFS theory is receiving much interest from researchers and has been effectively used in a diverse domain of real life problems. This motives us to extend the CODAS method with IVIFNs and apply it to two real problems with incomplete criteria weight information. In this article, the readers will find a systematic and comprehensive research framework for MSP, which is capable of discoursing with the subsequent research questions: (1) what are the sustainability indicators for MSPs in construction projects in India? (2) How to set priorities of these indicators in the evaluation process? (3) Which should be the optimal choice for a sustainable alternative (here, brick) for building construction?

3. Preliminaries

This section reviews the related ideas. Interval-valued intuitionistic fuzzy set (IVIFS) [8] is a generalization of intuitionistic fuzzy set (IFS). Compared to IFS, where the values of membership and non-membership functions are exact numbers, an IVIFS has the characteristic that those values are in intervals [0, 1] instead of exact numbers. Hence, IVIFS is more suitable in uncertain situations.

Definition 1. *Let X be a universal set. An IVIFS A in X is expressed as*

$$A = \left\{ \left\langle x, \left[\mu_A^l(x), \mu_A^r(x) \right], \left[v_A^l(x), v_A^r(x) \right] \right\rangle \middle| x \in X \right\} \tag{1}$$

where $\left[\mu_A^l(x), \mu_A^r(x) \right] \in [0,1]$ *and* $\left[v_A^l(x), v_A^r(x) \right] \in [0,1]$ *are respectively the interval-valued degrees of membership and non-membership of an element* $x \in X$ *to A, and the sum of upper bounds of these two interval-valued degrees is not greater than 1,* $0 \leq \mu_A^r(x) + v_A^r(x) \leq 1$. *If* $\mu_A^l(x) = \mu_A^r(x)$ *and* $v_A^l(x) = v_A^r(x), \forall x \in X$, *then the IVIFS* $A = \left\{ \left\langle x, \left[\mu_A^l(x), \mu_A^r(x) \right], \left[v_A^l(x), v_A^r(x) \right] \right\rangle \vee x \in X \right\}$ *is reduced to IFS, denoted by* $A = \{ x, \mu_A(x), v_A(x)] \rangle \vee x \in X \}$, *where* $\mu_A(x) = [\mu_A^l(x), \mu_A^r(x)]$ *and* $v_A(x) = \left[v_A^l(x), v_A^r(x) \right]$. *Hence, Atanassov's IFS can be considered as a special case of IVIFS.*

For a fixed $x \in X$, *an object* $\left(\left[\mu_A^l(x), \mu_A^r(x) \right], \left[v_A^l(x), v_A^r(x) \right] \right)$ *is called interval-valued intuitionistic fuzzy number (IVIFN). Let* $\beta = ([p,q], [r,s])$ *be an IVIFN, where* $0 \leq p \leq q \leq 1, 0 \leq r \leq s \leq 1$ *and* $q + s \leq 1$. *Then the score function [34,35] S of* β *is defined as* $S(\beta) = (p - r) + (q - s)/2$, *where* $S(\beta) \in [0,1]$. *The accuracy function [34,35] H of* β *is defined as* $H(\beta) = (p + r) + (q + s)/2$, *where* $H(\beta) \in [0,1]$.

Xu and Jian [35] compared two IVIFNs using score and accuracy functions which is defined below.

Let $\beta_1 = ([p_1, q_1], [r_1, s_1])$ *and* $\beta_2 = ([p_2, q_2], [r_2, s_2])$ *be two IVIFNs, then*

1. *If* $S(\beta_1) < S(\beta_2)$, *then* $\beta_1 < \beta_2$;
2. *If* $S(\beta_1) = S(\beta_2)$, *then*

   - *If* $H(\beta_1) = H(\beta_2)$, *then* $\beta_1 = \beta_2$;
   - *If* $H(\beta_1) < H(\beta_2)$, *then* $\beta_1 < \beta_2$;

Xu and Chen [36] proposed a similarity measure between two IVIFNs $\beta_1 = ([p_1, q_1], [r_1, s_1])$ and $\beta_2 = ([p_2, q_2], [r_2, s_2])$ defined as

$$S(\beta_1, \beta_2) = \frac{1}{4}(|p_1 - p_2| + |q_1 - q_2| + |r_1 - r_2| + |s_1 - s_2|) \tag{2}$$

Let $\beta = ([p,q],[r,s])$, $\beta_1 = ([p_1,q_1],[r_1,s_1])$, $\beta_2 = ([p_2,q_2],[r_2,s_2])$ be three IVIFNs and $\lambda > 0$. Some of their basic operational laws [34,37] are given below.

(1) $1 - \beta_1 = \beta_1{}^c = ([r_1,s_1],[p_1,q_1])$
(2) $\beta_1 \cap \beta_2 = ([min(p_1,p_2), min(q_1,q_2)], [max(r_1,r_2), max(s_1,s_2)])$
(3) $\beta_1 \cup \beta_2 = ([max(p_1,p_2), max(q_1,q_2)], [min(r_1,r_2), min(s_1,s_2)])$
(4) $\beta_1 + \beta_2 = ([p_1 + p_2 - p_1p_2, q_1 + q_2 - q_1q_2], [r_1r_2, s_1s_2])$
(5) $\beta_1 \cdot \alpha_2 = ([p_1p_2, q_1q_2], [r_1 + r_2 - r_1r_2, s_1 + s_2 - s_1s_2])$
(6) $\lambda\beta = ([1 - (1 - p)^\lambda, 1 - (1 - q)^\lambda], [r^\lambda, s^\lambda])$
(7) $\beta^\lambda = ([p^\lambda, q^\lambda], [1 - (1 - r)^\lambda, 1 - (1 - s)^\lambda])$

Definition 2. *Let $\beta_j = ([p_j, q_j],[r_j, s_j])$ $(j = 1,2, \dots ,n)$ be a collection of IVIFNs, and $w = (w_1, w_2, \dots , w_n)^T$ be their associated weight vector, with $0 \leq w_j \leq 1$ and $\sum_{j=1}^{n} w_j = 1$, then he interval valued intuitionistic fuzzy weighted geometric (IVIFWG) operator is defined as*

$$\text{IVIFWG}(\beta_1, \beta_2, \dots, \beta_n) = \prod_{j=1}^{n} \beta_j^{w_j} = \left(\left[\prod_{j=1}^{n} p_j^{w_j}, \prod_{j=1}^{n} q_j^{w_j} \right], \left[\prod_{j=1}^{n}(1 - r_j)^{w_j}, \prod_{j=1}^{n}(1 - s_j)^{w_j} \right] \right) \tag{3}$$

Definition 3. *According to Park et al. [38] the distance measures between two IVIFSs are defined as follows:*
The Hamming distance $d_H(\beta_1, \beta_2)$ and Euclidean distance $d_E(\beta_1, \beta_2)$ for the IVIFNs $\beta_1 = ([p_1, q_1], [r_1, s_1])$ and $\beta_2 = ([p_2, q_2], [r_2, s_2])$ are computed as:

$$d_H(\beta_1, \beta_2) = \frac{1}{4}(|p_1 - p_2| + |q_1 - q_2| + |r_1 - r_2| + |s_1 - s_2|) \tag{4}$$

$$d_E(\beta_1, \beta_2) = \sqrt{\frac{1}{4}\left[(p_1 - p_2)^2 + (q_1 - q_2)^2 + (r_1 - r_2)^2 + (s_1 - s_2)^2\right]} \tag{5}$$

Definition 4. *Let $\widetilde{A}_1 = \left[\left(\left[p_j^{(1)}, q_j^{(1)}\right], \left[r_j^{(1)}, s_j^{(1)}\right]\right)\right]_{1 \times n}$ and $\widetilde{A}_2 = \left[\left(\left[p_j^{(2)}, q_j^{(2)}\right], \left[r_j^{(2)}, s_j^{(2)}\right]\right)\right]_{1 \times n}$ be two IVIFNs in the universe $X = \{x_1, x_2, \dots , x_n\}$, then the distance measure between β_1 and β_2 is defined as follows:*

$$D\left(\widetilde{A}_1, \widetilde{A}_2\right) = \left[\frac{1}{4n} \sum_{j=1}^{n} \left\{ \left|p_j^{(1)} - p_j^{(2)}\right|^\lambda + \left|q_j^{(1)} - q_j^{(2)}\right|^\lambda + \left|r_j^{(1)} - r_j^{(2)}\right|^\lambda + \left|s_j^{(1)} - s_j^{(2)}\right|^\lambda \right\}\right]^{1/\lambda} \tag{6}$$

Particularly, if $\lambda = 1$, then Equation (6) becomes the Hamming distance:

$$d_H\left(\widetilde{A}_1, \widetilde{A}_2\right) = \frac{1}{4n} \sum_{j=1}^{n} \left\{ \left|p_j^{(1)} - p_j^{(2)}\right| + \left|q_j^{(1)} - q_j^{(2)}\right| + \left|r_j^{(1)} - r_j^{(2)}\right| + \left|s_j^{(1)} - s_j^{(2)}\right| \right\} \tag{7}$$

If $\lambda = 2$, then Equation (6) is degenerated to the Euclidean distance:

$$d_H\left(\widetilde{A}_1, \widetilde{A}_2\right) = \sqrt{\frac{1}{4n} \sum_{j=1}^{n} \left\{ \left|p_j^{(1)} - p_j^{(2)}\right|^2 + \left|q_j^{(1)} - q_j^{(2)}\right|^2 + \left|r_j^{(1)} - r_j^{(2)}\right|^2 + \left|s_j^{(1)} - s_j^{(2)}\right|^2 \right\}} \tag{8}$$

4. Proposed CODAS Method Using IVIFNs

This section presents an extension of the CODAS method based on IVIFS to deal with MCDM problems. Let D = {$d_1, d_2, \ldots, d_l$} be the group of decision makers, C = {$c_1, c_2, \ldots, c_n$} be the set of criteria, and A = {$a_1, a_2, \ldots, a_m$} be the set of alternatives. The group of experts/decision makers D = {$d_1, d_2, \ldots, d_l$} provide their opinions regarding the criteria C = {$c_1, c_2, \ldots, c_n$} corresponding to each alternative A = {$a_1, a_2, \ldots, a_m$} using linguistic terms which are presented by IVIFNs. In this algorithm, we consider that significance of individual decision makers are different and the weights of the decision makers are expressed using fuzzy membership grades. We also consider that opinions of individual decision makers about the importance of various criteria are different. A flow chart of the proposed approach is given in Figure 1.

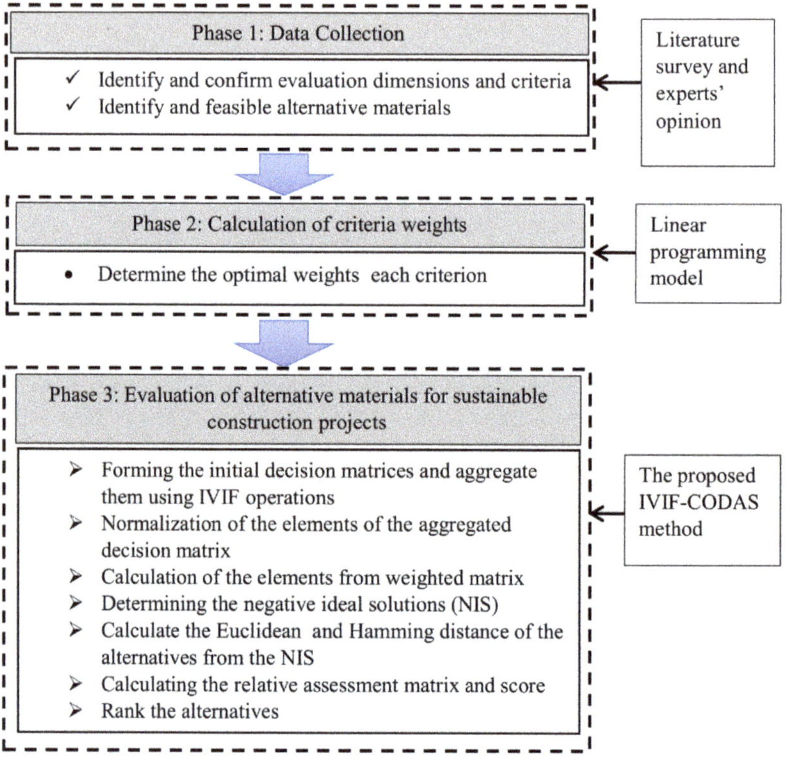

Figure 1. Flow diagram of the proposed IVIF-CODAS.

A step wise illustration of the proposed approach is given below.

Step 1. Opinion of each expert is expressed using decision matrix given below.

$$
X^k = \begin{pmatrix} x_{11}^k & x_{12}^k & \cdots & x_{1n}^k \\ x_{21}^k & x_{22}^k & \cdots & x_{2n}^k \\ \cdots & \cdots & \cdots & \cdots \\ x_{m1}^k & x_{m2}^k & \cdots & x_{mn}^k \end{pmatrix} \quad (k = 1, 2, \ldots, l) \tag{9}
$$

Here x_{ij}^k denotes the evaluating value of *i*th (i $\in$ {1, 2, $\ldots$, m}) alternative with respect to *j*th (j $\in$ {1, 2, $\ldots$, n}) criterion and *k*th (k $\in$ {1, 2, $\ldots$, l}) decision maker which is expressed as IVIFNs.

Step 2. Interval-valued intuitionistic fuzzy weighted geometric (IVIFWG) aggregation operator is used to aggregate the opinion of individual decision makers. The aggregated/collective decision matrix is formed as

$$X = \begin{pmatrix} x_{11} & x_{12} & \cdots & x_{1n} \\ x_{21} & x_{22} & \cdots & x_{2n} \\ \cdots & \cdots & \cdots & \cdots \\ x_{m1} & x_{m2} & \cdots & x_{mn} \end{pmatrix} \tag{10}$$

where $x_{ij} = \text{IVIFWG}\left(x_{ij}^1, x_{ij}^2, \ldots, x_{ij}^l\right)$ and l be the number of decision makers.

Step 3. Calculate the weights of evaluation criteria

Let $w = (w_1, w_2, \ldots, w_n)^T$ be the weight vector of the criteria C_j, $j = 1, 2, \ldots, n$, where $w_j > 0$ ($\forall j$) and $\sum_{j=1}^{n} w_j = 1$. The known criteria weights are divided into five basic ranking forms [39–41] for $i \neq j$ as given below:

(1) A weak ranking: $H_1 = \{w_i \geq w_j\}$;

(2) A strict ranking: $H_2 = \left\{w_i - w_j \geq \beta_j \big| \beta_j \rangle 0\right\}$;

(3) A ranking of differences: $H_3 = \{w_i - w_j \geq w_l - w_l | j \neq k \neq l\}$;

(4) A ranking with multiples: $H_4 = \left\{w_i \geq \beta_j w_j \big| 0 \leq \beta_j \leq 1\right\}$;

(5) An interval form: $H_5 = \left\{\beta_i \geq w_i \leq \beta_j + \epsilon_i \big| 0 \leq \beta_i \leq \beta_i + \epsilon_i\right\}$;

For simplicity, let H denote the set of criteria weight information given by DMs and $H = H_1 \cup H_2 \cup H_3 \cup H_4 \cup H_5$. This approach uses the known criteria weight information to define the weights of evaluation criteria.

In MSP, the significance of a criterion is determined by evaluating the performance values of the alternatives for that criterion. When the performance values of the alternatives differ a little for a particular criterion, then that criterion is considered to be less significant for choosing the best material. Similarly, when the performance values of the alternatives differ much for a particular criterion, then that criterion is considered to be much significant for choosing the best material. This observation leads to assign less weight to the less significant criteria and more weight to the much more significant criteria [42]. A criterion is said to have no significance in the material selection process when the performance values of the alternatives are same for that criteria [43].

When the criteria weight information is partially known, this paper presents an optimization model using the IVIF distance measure to compute the evaluating criteria weights. Below, we define the distance between the alternative A_i and other alternatives corresponding to the criterion C_j.

$$D_{ij} = \frac{1}{m-1} \sum_{g=1, g \neq i}^{m} d_H\left(x_{ij}, x_{gj}\right); i = 1, 2, \cdots, m; j = 1, 2, \cdots, n. \tag{11}$$

The overall distance measures of all the alternatives for the criterion C_j is presented as:

$$D_j = \frac{1}{m-1} \sum_{i=1}^{m} \sum_{g=1, g \neq i}^{m} d_H\left(x_{ij}, x_{gj}\right), j = 1, 2, \cdots, n. \tag{12}$$

Next the weighted distance function is formulated as given below.

$$D(w) = \sum_{j=1}^{n} D_j w_j = \sum_{j=1}^{n} \sum_{i=1}^{m} D_{ij} w_j = \frac{1}{m-1} \sum_{j=1}^{n} \sum_{i=1}^{m} \sum_{g=1, g \neq i}^{m} d_H\left(x_{ij}, x_{gj}\right) w_j \tag{13}$$

Hence, a suitable weight vector of criteria $w = (w_1, w_2, \ldots, w_n)^T$ is needed to maximize $D(w)$, and therefore, we can present the optimization model defined below:

$$(M-1) = \begin{cases} \text{Maximize } D(w) = \frac{1}{m-1} \sum_{j=1}^{n} \sum_{i=1}^{m} \sum_{g=1, \, g\neq i}^{m} d_H\left(x_{ij}, x_{gj}\right) w_j \\ \text{subject to } w \in H, \; \sum_{j=1}^{n} w_j = 1, \quad w_j \geq 0, \quad j = 1, 2, \ldots, n \end{cases} \tag{14}$$

The optimal solution w^* is obtained by solving the model $(M-1)$. We use the optimal solution w^* as the weight vector for the evaluation criteria.

In another case, when the information concerning criteria weights is totally unknown, we can develop another model for optimization to find the optimal weights of criteria:

$$(M-2) = \begin{cases} \text{Maximize } D(w) = \frac{1}{m-1} \sum_{j=1}^{n} \sum_{i=1}^{m} \sum_{g=1, \, g\neq i}^{m} d_H\left(x_{ij}, x_{gj}\right) w_j \\ \text{subject to } \quad \sum_{j=1}^{n} w_j = 1, \quad w_j \geq 0, \quad j = 1, 2, \ldots, n \end{cases} \tag{15}$$

Lagrange's method is used to solve the preceding Model (15) and the corresponding optimal solutions are normalized to determine the criteria weight vector.

$$w_j = \frac{\sum_{i=1}^{m} \sum_{g=1, g\neq i}^{m} d_H\left(x_{ij}, x_{gj}\right) w_j}{\sum_{j=1}^{n} \sum_{i=1}^{m} \sum_{g=1, g\neq i}^{m} d_H\left(x_{ij}, x_{gj}\right) w_j} \tag{16}$$

Step 4. The collective decision matrix is normalized by determining the highest IVIFN under each criterion for all the alternatives and then performing the division operation of between the highest IVIFN and the corresponding IVIFN as given below.

$$\widetilde{N} = \left[\tilde{n}_{ij}\right]_{m \times n} \tag{17}$$

where

$$\tilde{n}_{ij} = \begin{cases} \dfrac{x_{ij}}{\max\limits_{1 \leq i \leq m}\left(x_{ij}\right)}; & \text{if } j \in B \\[4mm] 1 - \dfrac{x_{ij}}{\max\limits_{1 \leq i \leq m}\left(x_{ij}\right)}; & \text{if } j \in C \end{cases} \tag{18}$$

In the above operations, B and C respectively represent the sets of benefit and cost criteria, and $\tilde{n}_{ij}$ denote the normalized performance values in terms of IVIFNs.

Step 5. The weighted normalized decision matrix is determined by performing product operation between the aggregated criteria weights and the normalized performance values of the criteria corresponding to the alternatives, which is give below. It is noted that both the aggregated criteria weights and the evaluating values are expressed as IVIFNs.

$$\widetilde{R} = \left[\tilde{r}_{ij}\right]_{m \times n} \tag{19}$$

where

$$\tilde{r}_{ij} = w_j \tilde{n}_{ij} \tag{20}$$

Here the weight of *j*th criterion is denoted by w_j.

Step 6. The interval-valued intuitionistic fuzzy negative ideal solution is computed as

$$\widetilde{NS} = \left[\tilde{ns}_{ij}\right]_{m \times n} \tag{21}$$

where $\tilde{ns}_{ij} = \min\limits_{1 \leq i \leq m}\{\tilde{r}_{ij}\}$. Here $\tilde{ns}_{ij}$ is an IVIFN for each criterion $j \in \{1, 2, \ldots, n\}$.

Step 7. IVIFN-based Hamming (HD) and Euclidean distances (ED) of the alternatives $i \in \{1, 2, \ldots, m\}$ from the interval-valued intuitionistic fuzzy negative ideal solution $(\widetilde{NS})$ are computed as

$$ED = \sum_{j=1}^{n} d_E\left(\tilde{r}_{ij}, \, \tilde{ns}_{ij}\right) \tag{22}$$

$$HD = \sum_{j=1}^{n} d_H\left(\tilde{r}_{ij}, \, \tilde{ns}_{ij}\right) \tag{23}$$

Step 8. The relative assessment matrix (RA) is computed as

$$RA = \left[p_{is}\right]_{m \times m} \tag{24}$$

where

$$p_{is} = (ED_i - HD_s) + \{\psi(ED_i - ED_s) \times (HD_i - HD_s)\} \tag{25}$$

where $i, s \in \{1, 2, \ldots, m\}$ and ψ is a threshold function defined below.

$$\psi(t) = \left\{ \begin{array}{ll} 1; & \text{if } |t| \geq \theta \\ 0; & \text{if } |t| < \theta \end{array} \right. \tag{26}$$

Decision maker can set the threshold parameter (θ). Here we consider $\theta = 0.02$.
Step 9. The assessment score (AS_i) of each alternative is computed, which is given below.

$$AS_i = \sum_{s=1}^{m} p_{is} \quad (\forall i \in \{1, 2, \ldots, m\}) \tag{27}$$

5. Application of the IVIF–CODAS in MSPs

5.1. Illustrative Example

Here, a MSP is illustrated to show the applicability and effectiveness of the proposed IVIF-CODAS approach. We have considered an automotive parts factory in India [7], where the factory want to find the best material for the automotive instrument panel. In the selection process, a group of five experts/decision makers (DM1, DM2, DM3, DM4, DM5) evaluates four materials/alternatives (A1, A2, A3, A4) based on the values of eight evaluation criteria (C1, C2, C3, C4, C5, C6, C7, C8) and finds the suitable alternative. The structure of the problem in given in Figure 2. The linguistic assessments (using Table 1) of the alternatives given by the decision makers are shown in Table 2. This example assumes that the group of five decision makers and the set of eight criteria have their individual weights/importance. The weights of decision makers and criteria are respectively expressed using fuzzy membership grades and IVIFNs.

Table 1. Linguistic terms and corresponding IVIFNs for evaluating materials.

Linguistic Terms	Corresponding IVIFNs
Very High (VH)	([0.9,1], [0,0])
High (H)	([0.8,0.8], [0.1,0.1])
Medium High (MH)	([0.6,0.7], [0.2,0.3])
Medium (M)	([0.5,0.5], [0.4,0.5])
Medium Low (ML)	([0.3,0.4], [0.5,0.6])
Low (L)	([0.2,0.2], [0.7,0.7])
Very Low (VL)	([0,0.1], [0.8,0.9])

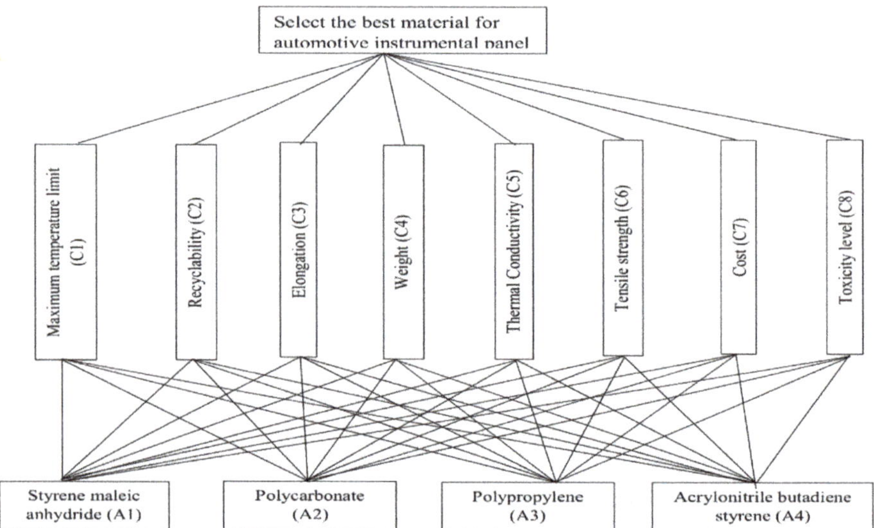

Figure 2. Structure of material selection in automotive industry.

Table 2. Decision matrix with linguistic ratings.

		Max	Max	Max	Max	Max	Max	Min	Min
		C1	C2	C3	C4	C5	C6	C7	C8
DM1	A1	MH	H	MH	MH	H	MH	MH	MH
	A2	M	MH	H	H	H	MH	H	MH
	A3	VH	H	VH	H	H	VH	H	VH
	A4	H	VH	H	H	VH	H	VH	H
DM2	A1	MH	H	MH	MH	MH	H	MH	MH
	A2	MH	M	MH	M	MH	M	M	M
	A3	H	VH	H	VH	H	VH	H	H
	A4	H	H	VH	H	VH	H	VH	VH
DM3	A1	MH	H	MH	H	MH	MH	H	MH
	A2	MH	MH	H	H	H	H	MH	H
	A3	VH	VH	H	H	VH	VH	H	H
	A4	H	VH	H	H	H	VH	H	H
DM4	A1	H	MH	H	MH	MH	H	MH	H
	A2	MH	MH	M	MH	M	MH	M	MH
	A3	H	VH	H	VH	H	H	VH	H
	A4	H	H	VH	H	H	H	VH	VH
DM5	A1	MH	H	MH	H	MH	H	MH	MH
	A2	H	MH	H	MH	H	MH	M	M
	A3	VH	H	VH	H	VH	H	VH	H
	A4	H	VH	H	H	H	VH	H	M

- Assessment of decision makers regarding the alternatives on each criteria are shown using linguistic terms which are given in Table 2, where the IVIFNs corresponding to each linguistic terms are given in Table 1.
- The collective decision matrix is shown in Table 3 which is computed by aggregating the opinions of the decision makers (DM1, DM2, ... , DM5).
- The consensus making model [39] is used to determine the weights of the evaluation criteria. In our study, the criteria weights are unknown while the weights of experts ($\sum_{k=1}^{5} \lambda_k = 1$) are

known in advance. Hence, applying model $(M-1)$ and assuming that the criteria weights are partially known as follows: $H = \{w_1 \leq 0.08, w_1 = w_2, 0.10 \leq w_3 \leq 0.20, w_4 + w_2 \geq 0.3, 0.13 \leq w_5 \leq 0.20, 0.12 \leq w_6 \leq 0.17, 0.12 \leq w_7 \leq 0.16, w_8 = w_6, w_5 - w_6 \geq 0.05, w_5 - w_7 \geq 0.03, w_j \geq 0, j = 1, 2, \ldots, 8, \sum_{j=1}^{8} w_j = 1\}$. To compute the criteria priorities, Equations (16)–(18) and model $(M-1)$ are used to develop the linear programming model given below:

$$\begin{cases} \max D(w) = 0.450w_1 + 0.590w_2 + 0.550w_3 + 0.421w_4 \\ \quad + 0.550w_5 + 0.530w_6 + 0.640w_7 + 0.605w_8 \\ \quad \text{subject to to } w \in H \end{cases}$$

- The final criteria weights (w^*) are obtained by solving the above model and they are represented by the following weight vector $w^* = (0.050, 0.050, 0.100, 0.250, 0.170, 0.120, 0.140, 0.120)^T$.
- Weighted normalized decision matrix, shown in Table 4, is computed by finding the normalized decision matrix and then combining the criteria weights obtained in the previous step with the normalized decision matrix.
- Interval-valued intuitionistic fuzzy negative ideal solution for each of the criteria is shown in Table 5.
- Hamming distance and Euclidean distance of the alternatives from the Interval-valued intuitionistic fuzzy negative ideal solution is shown in Table 6.
- Relative assessment matrix, assessment scores, and rank of the alternatives are given in Table 7.
- Table 8 shows the ranking orders by the proposed method along with four other existing methods for result comparisons.

Table 3. Collective decision matrix.

	C1	C2	C3	C4
A1	([0.636, 0.719], [0.181, 0.264])	([0.755, 0.779], [0.121, 0.144])	([0.636, 0.719], [0.181, 0.264])	([0.645, 0.724], [0.176, 0.255])
A2	([0.668, 0.716], [0.204, 0.284])	([0.687, 0.753], [0.169, 0.247])	([0.775, 0.787], [0.136, 0.171])	([0.782, 0.800], [0.124, 0.157])
A3	([0.931, 0.956], [0.021, 0.021])	([0.943, 0.978], [0.010, 0.010])	([0.915, 0.925], [0.036, 0.036])	([0.926, 0.946], [0.026, 0.026])
A4	([0.925, 0.925], [0.036, 0.036])	([0.947, 0.967], [0.016, 0.016])	([0.941, 0.956], [0.021, 0.021])	([0.925, 0.925], [0.036, 0.036])
	C5	C6	C7	C8
A1	([0.654, 0.729], [0.171, 0.245])	([0.713, 0.758], [0.141, 0.186])	([0.618, 0.709], [0.191, 0.282])	([0.636, 0.719], [0.181, 0.264])
A2	([0.775, 0.787], [0.136, 0.171])	([0.727, 0.774], [0.149, 0.208])	([0.718, 0.741], [0.180, 0.237])	([0.727, 0.774], [0.149, 0.208])
A3	([0.921, 0.935], [0.031, 0.031])	([0.954, 1.000], [0.000, 0.000])	([0.904, 0.904], [0.046, 0.046])	([0.915, 0.925], [0.036, 0.036])
A4	([0.964, 1.000], [0.000, 0.000])	([0.925, 0.925], [0.036, 0.036])	([0.964, 1.000], [0.000, 0.000])	([0.941, 0.956], [0.021, 0.021])

Table 4. Weighted normalized decision matrix.

	C1	C2	C3	C4
A1	([0.042,0.049], [0.028,0.035])	([0.056,0.056], [0.014, 0.021])	([0.090,0.105], [0.030, 0.045])	([0.042,0.049], [0.014, 0.021])
A2	([0.035,0.035], [0.028, 0.035])	([0.042,0.049], [0.014, 0.021])	([0.120,0.120], [0.030, 0.045])	([0.056,0.056], [0.014, 0.021])
A3	([0.063,0.070], [0.028, 0.035])	([0.056,0.056], [0.014, 0.021])	([0.135,0.150], [0.030, 0.045])	([0.056,0.056], [0.014, 0.021])
A4	([0.056,0.056], [0.028, 0.035])	([0.063,0.070], [0.014, 0.021])	([0.120,0.120], [0.030, 0.045])	([0.056,0.056], [0.014, 0.021])

	C5	C6	C7	C8
A1	([0.144,0.144], [0.018, 0.018])	([0.087,0.102], [0.029, 0.044])	([0.102,0.119], [0.034, 0.051])	([0.087,0.102], [0.029, 0.044])
A2	([0.144,0.144], [0.018, 0.018])	([0.087,0.102], [0.029, 0.044])	([0.102,0.119], [0.017, 0.017])	([0.087,0.102], [0.029, 0.044])
A3	([0.144,0.144], [0.018, 0.018])	([0.087,0.102], [0.000, 0.000])	([0.102,0.119], [0.017, 0.017])	([0.087,0.102], [0.000, 0.000])
A4	([0.162,0.180], [0.018, 0.018])	([0.087,0.102], [0.015, 0.015])	([0.102,0.119], [0.000, 0.000])	([0.087,0.102], [0.015, 0.015])

Table 5. Negative ideal solutions.

Criteria	IVIF Negative Ideal Solutions
C1	([0.1407 0.1703], [0.7286 0.7692)]
C2	([0.1520 0.1791], [0.7167 0.7572)]
C3	([0.4668 0.5526], [0.2885 0.3857)]
C4	([0.1427 0.1722], [0.7191 0.7598)]
C5	([0.5886 0.7290], [0.1710 0.2450)]
C6	([0.5158 0.5787], [0.2650 0.3472)]
C7	([0.5401 0.6705], [0.2120 0.3007)]
C8	([0.4668 0.5526], [0.2885 0.3857)]

Table 6. Euclidean and Hamming distance matrices.

	A1	A2	A3	A4
ED	0.0182	0.1899	0.767	0.8317
HD	0.0117	0.1264	0.5361	0.5824

Table 7. Assessment matrix and scores.

	Relative Assessment Matrix				Appraisal Scores	Ranking
	A1	A2	A3	A4		
A1	0	−0.2864	−1.2731	−1.3842	−2.9437	4
A2	0.2864	0	−0.9867	−1.0977	−1.798	3
A3	1.2731	0.9867	0	−0.1111	2.1487	2
A4	1.3842	1.0977	0.1111	0	2.593	1

Table 8. Comparison with other models.

MCDM Methods	Ranking Order
Classical CODAS	A4 > A3 > A2 > A1
Fuzzy CODAS	A4 > A3 > A2 > A1
IVIF-VIKOR	A3 > A4 > A2 > A1
IVIF-TOPSIS	A4 > A3 > A2 > A1
The proposed IVIF-CODAS	A4 > A3 > A2 > A1

5.2. A Real Case of MSP in Sustainable Construction Projects

This section presents a sustainable material selection problem for construction projects using the proposed IVIF-CODAS approach. Applicability and usefulness of the proposed model are validated through a real case study of a construction company. Initially, we sent our research proposals to 10 national builders, who are associated with construction, materials supply, and providing services mainly in India. Three out of ten companies provided positive responses to our proposal and our research team accomplished necessary groundwork on these three construction companies which are located in the "Durgapur-Asansol" division of West Bengal. We have had selected *"Maha Prabhu Buiders: A national company (name changed)"* for implementing our proposed research framework. The company has several operating units all over the country. The primary goal of this study is to provide wide-ranging results by incorporating the perceptions of all actors in the construction field since this research work is completely based on the decision makers' (refer to Table 9) judgments. Therefore, we integrated some major perspectives: clients, the Construction Company, manufacturers, consultants, and material suppliers.

Table 9. Description of experts.

Decision Makers	Expertise
DM1	Head of establishing standards and techniques with 21 years of work experience
DM2	Health, Safety and Environment (HSE) management employee and the head of operations evaluation with 20 years of work experience
DM3	Expert supervisor of construction project implementation with 21 years of work experience
DM4	Project manager with 17 years of work experience
DM5	Financial manager with 18 years of work experience

The main reason for selecting *"Maha Prabhu Buiders"* over other two volunteering companies is because they arrange for all kind of services such as construction, consultants and architects, material suppliers, and so on. The central public works department (CPWD-2014) of India has been promoting the norms and guidelines of sustainable construction in India to increase the awareness in public, private sectors, and most importantly among the habitants. Following to the CPWD (2014) guidelines, this particular company has been striving in sustainable construction projects and most of all other activities. The general manager discussed with our team and informed the new trend in clients. The trend is not limited to embracing the concept of sustainable design of buildings but also in materials (bricks, mortar, concrete, cement, plasters, etc.). Hence, they recognized our research framework, and acknowledged the results that will classify which sustainable materials are most preferable. The consultant will endorse these results, and as a product manufacturer, the company can advertise that specific material with its sustainability indicating tags. On acceptance of the research proposal, we arranged a three-day workshop for the managing directors of the company, architects, and aforementioned clients who encourage sustainability issues in designing buildings as well as selecting materials, engineers and skilled professionals from the company. The problem goal and structure were explained to all the participants and, after several rounds of discussion, the materials selected to be evaluated are burnt clay bricks. Bricks are used substantially in almost every type of construction projects including bridges, housing, firms, hospitals, etc.

As soon as the usefulness of the proposed framework was revealed, our research team began the preliminary investigation on the sustainable alternatives of burnt clay bricks. We found there are three popular alternative bricks which are usually used in sustainable construction projects. The detailed description and characteristics of alternatives bricks can be found in Dhanjode et al. [44] and Mahendran et al. [45].

For evaluating the alternative bricks for sustainable practices, the team of experts ($DM1$, $DM2$, $DM3$, $DM4$, $DM5$) conducts the performance testing to decide the most suitable brick. Next, the sustainable indicators are considered as evaluation criteria which were recommended by the team of experts of the company and approved with the literature. In this concern the proposed framework is applied to find the most sustainable brick(s) based on the considered criteria of sustainable practices in construction industry. To accomplish the purpose of this assignment, a three-phase methodology was used: data collection, data aggregation using IVIFWG operator, and evaluation of bricks and selecting the most suitable brick(s) using extended IVIF-CODAS method. The decision structure of the problem is depicted in Figure 3 in which the goal of this project is located in the top level, followed by the most cited factors and criteria of sustainability. Final level of the hierarchy deals with evaluation of the alternative bricks as sustainable material for building construction.

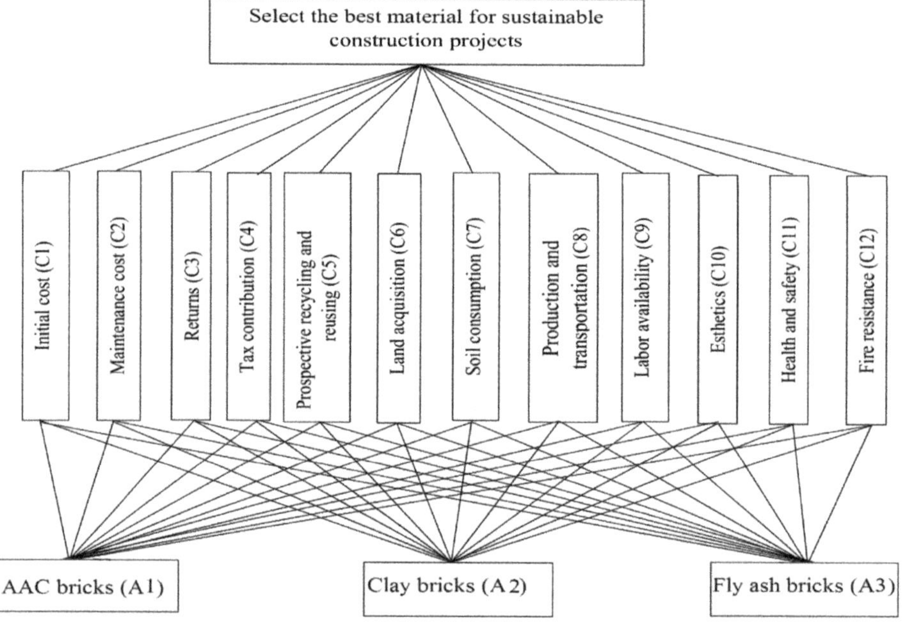

Figure 3. Structure of sustainable material selection problem in construction industry.

In the selection process, a group of five experts $(DM1, DM2, DM3, DM4, DM5)$ evaluates three sustainable materials/bricks $(A1, A2, A3)$ based on the values of twelve evaluation criteria $(C1, C2, C3, C4, C5, C6, C7, C8, C9, C10, C11, C12)$ and finds the suitable material. Similar to the previous case study, the preferences of the decision makers about the materials against the criteria are given in linguistic terms as shown in Table 1. This example also assumes that the group of five decision makers and the set of twelve criteria have their individual weights/importance. The weights of decision makers and criteria are respectively expressed using fuzzy membership grades and IVIFNs. The structure of the sustainable material selection problem is given below in Figure 3.

Step 1. *Data collection*

After finalizing the hierarchical structure (Figure 3) of brick selection problem the required data are collected for conducting the evaluation process. For evaluating the sustainable material for construction projects, sustainability parameters are regarded as criteria in this study. In search of the most suitable sustainability parameters, we go through the present literatures available in reliable journals such as Taylor & Francis, Elsevier, MDPI, and Springer. Articles with environmental, economical, and social sustainability indicators and sustainable material selection and evaluation in the context of the construction and manufacturing projects were selected for this research work. After discussion with the relevant professionals, some criteria were selected which are given in Table 10. Next, each decision maker opined on the importance of individual criterion. Finally, the alternatives (list of bricks) were designed, which was already completed in previous steps.

Table 10. Objectives and criteria.

Dimension	Criteria	Description
Economical	Initial cost (C1)	Cost considered for purchasing/manufacturing of materials
	Maintenance cost (C2)	Cost considered for maintaining in its lifetime
	Disposal cost (C3)	Cost considered for material disposal
	Tax contribution (C4)	Tax regarding the materials
Environmental	Raw material extraction (C5)	It is necessary to manufacture the final material
	Land acquisition (C6)	Land required for the material construction
	Soil consumption (C7)	It is required by the material at the time of manufacturing and operation
	Production and transportation (C8)	Comfortable transportation and production is important
Social	Fire resistance (C9)	Necessary arrangements to resist fire
	Esthetics (C10)	Looking of the material
	Use of local material (C11)	To develop society, more use of local material is needed
	Labor availability (C12)	Quality labor is vital for production

Step 2. *Calculation of criteria weights*

Similar to previous example, according to model $(M-1)$ and Equation (14), the weight vector of the three objectives/dimensions (economic, environmental, and social) is calculated as $(w_{EC}, w_{EN}, w_{SO})^T = (0.4268, 0.3568, 0.2164)^T$. Likewise, the local and global criteria weights are computed and presented in Table 11.

Table 11. Aggregated weights of sustainable material selection indicators.

Dimension	Criteria	Local Weights	Global Weights	Rank
EC	-	0.4268	-	-
	C1	0.3487	0.1488	2
	C2	0.1857	0.0793	6
	C3	0.4031	0.1720	1
	C4	0.0625	0.0267	11
EN	-	0.3568	-	-
	C5	0.3838	0.1369	3
	C6	0.3396	0.1212	4
	C7	0.0143	0.0051	12
	C8	0.2623	0.0936	5
SO	-	0.2164	-	-
	C9	0.2387	0.0517	9
	C10	0.2180	0.0472	10
	C11	0.3000	0.0649	7
	C12	0.2434	0.0527	8

Step 3. *Evaluating the best sustainable material using IVIF-CODAS method*

The IVIF-CODAS method is used to prioritize the optimal choice for sustainable material selection problem. We follow the step-by-step calculation mechanism is discussed in Section 4 and phase 3 in Figure 1. The mechanism includes fuzzy aggregation of individual judgments followed by normalization. The criteria weights that are obtained using Equations (11)–(16) in the second phase are lodged via Equation (20) to the normalized data yielding weighted normalized decision matrix.

Next, we find the negative ideal solutions (NIS) are defined in Equation (21), which are used as criteria reference points to obtain optimal solution. The Euclidean (ED) and Hamming (HD) distances between the interval-valued intuitionistic fuzzy negative ideal solution and criteria functions in the weighted normalized decision matrix(R) are calculated using Equation (19). Finally, we determine the relative assessment matrix (RA) which is computed according to Equation (24). Now, assessment score of each alternative material is calculated taking the column sum of the RA matrix. Ultimately, the alternative/material with highest assessment score is the most desirable alternative in the material selection problem. In our problem, we find the "Fly ash bricks ($A3$)" is the optimal choice for sustainable construction projects in India.

6. Result Discussion

The goal of this section is to analyze the results of the proposed framework in order to validate its rationality and practicality. For exploring the most significant dimensions/aspects of construction projects and to avoid complexity, the sustainability indicators (criteria) were classified into 3 major aspects of sustainability. To keep it simple, we assume all three dimensions should be equally important. On contrast, each individual criterion has its own importance as a sustainable indicator for evaluating sustainable materials in construction industry. Table 11 demonstrates economic ($C1, C2, C3, C4$), environmental ($C5, C6, C7, C8$), and social ($C9, C10, C11, C12$) aspects including the corresponding criteria within them. From Table 11, it is clear that "Prospective recycling and reusing" ($C5$), "Returns" ($C3$), and "Use of local material" ($C11$) are top three important criteria among the twelve sustainable indicators considered here. However, usually there are conflicting concerns on priorities of the economic and environmental factors [18]. Hence, it becomes difficult to adjust them under such conflicts. However, in the viewpoint of sustainable construction in India, our findings show that "Returns" ($C3$) is the most important criterion in the economic dimension. It is quite reasonable for a company to devote its main effort to get higher returns. This fact is supported by the results we obtained by IVIFWG operator. In this cluster, the descending order of criteria is $C3 > C1 > C2 > C4$.

For the second dimension, "Prospective recycling and reusing" ($C5$) occupies the most significant sustainable indicator among the four environmental factors. In this group, the descending order of criteria is as follows: $C5 > C6 > C8 > C7$. Now, to justify such order of criteria importance, some of the works in the existing literature are cited. For green practices the 3R (reduce, recycle and reuse) policy [46,47] is the most sustainable strategy for providing the promotion of environmental performance and development. Finally, "health and safety" ($C11$) grips the top rank in the social dimension. Construction companies in a nation like India deals with more issues than other countries, since a large number of accidental cases as well as safety-related difficulties happen in construction. Table 11 reflects this fact that "Health and safety" ($C11$) and "Fire resistance" ($C12$) come first rather thinking much of "Labor availability" ($C9$), and "Esthetics" ($C10$) in real life situations.

After finalizing the criteria weights, we aim to evaluate and select the most suitable brick(s) as alternative material for sustainable construction. The linguistic assessments of the three alternatives supplied by the five decision makers are shown in Table 12. The decision matrix in Table 12 is transformed to the aggregated IVIF decision matrix (Table 13) by using Equation (10). Next, the average decision matrix is normalized with the help of Equation (18) and after plugging the criteria weights the weighted normalized decision matrix (refer to Table 14) is computed. Interval-valued intuitionistic fuzzy negative ideal solution for each of the criteria is shown in Table 15. Hamming distance and Euclidean distance of the alternatives from the Interval-valued intuitionistic fuzzy negative ideal solution is shown in Table 16. Finally, Table 17 tells the overall performance scores of all the sustainable alternatives of brick, in which "Fly ash bricks" ($A3$) bears the top rank with a high assessment score of 2.158. In addition, "AAC bricks" ($A3$) and "Clay Bricks" ($A2$) have got second and third ranks, with assessment scores -0.976 and -1.182, respectively. The final ranking order of alternative bricks for sustainable construction is: $A3 > A1 > A2$. Additionally, Table 18 shows the ranking orders by the proposed method along with four other existing methods for result comparisons.

Table 12. Decision matrix with linguistic ratings.

		C1	C2	C3	C4	C5	C6	C7	C8	C9	C10	C11	C12
DM1	A1	MH	MH	MH	MH	MH	MH	M	MH	MH	H	H	H
	A2	H	H	H	MH	H	H	VH	MH	M	M	MH	M
	A3	H	MH	H	MH	H	H	H	H	MH	H	H	H
DM2	A1	VH	MH	MH	VH	MH	MH	M	MH	VH	H	H	H
	A2	H	VH	H	MH	H	VH	VH	MH	M	M	VH	VH
	A3	H	MH	VH	MH	VH	H	VH	VH	MH	VH	H	H
DM3	A1	MH	MH	MH	MH	MH	MH	M	MH	MH	MH	H	H
	A2	H	MH	H	MH	MH	H	MH	MH	M	M	MH	M
	A3	H	MH	MH	MH	H	H	H	H	MH	H	H	MH
DM4	A1	M	MH	MH	MH	MH	M	M	M	MH	H	H	H
	A2	H	M	H	M	M	H	M	MH	M	M	MH	M
	A3	H	MH	M	MH	H	H	H	H	M	H	M	H
DM5	A1	VH	MH	MH	VH	MH	MH	M	MH	VH	H	H	H
	A2	H	VH	H	MH	H	VH	VH	MH	M	M	VH	VH
	A3	H	MH	VH	MH	VH	H	VH	VH	MH	VH	H	H

Table 13. Aggregated IVIF decision matrix.

	C1	C2	C3	C4
A1	([0.673, 0.754], [0.174, 0.264])	([0.600, 0.700], [0.200, 0.144])	([0.600, 0.700], [0.200, 0.264])	([0.691, 0.793], [0.125, 0.255])
A2	([0.800, 0.800], [0.100, 0.343])	([0.733, 0.785], [0.145, 0.356])	([0.800, 0.800], [0.100, 0.249])	([0.584, 0.666], [0.245, 0.288])
A3	([0.800, 0.800], [0.100, 0.046])	([0.600, 0.700], [0.200, 0.046])	([0.733, 0.785], [0.145, 0.056])	([0.600, 0.700], [0.200, 0.056])

	C5	C6	C7	C8
A1	([0.600, 0.700], [0.200, 0.245])	([0.584, 0.666], [0.245, 0.186])	([0.500, 0.500], [0.400, 0.282])	([0.584, 0.666], [0.245, 0.264])
A2	([0.704, 0.726], [0.180, 0.249])	([0.834, 0.865], [0.061, 0.340])	([0.760, 0.839], [0.117, 0.383])	([0.600, 0.700], [0.200, 0.373])
A3	([0.834, 0.865], [0.061, 0.076])	([0.800, 0.800], [0.100, 0.036])	([0.834, 0.865], [0.061, 0.066])	([0.834, 0.865], [0.061, 0.071])

	C9	C10	C11	C12
A1	([0.691, 0.793], [0.125, 0.264])	([0.755, 0.779], [0.111, 0.264])	([0.800, 0.800], [0.100, 0.264])	([0.800, 0.800], [0.100, 0.264])
A2	([0.500, 0.500], [0.400, 0.373])	([0.500, 0.500], [0.400, 0.373])	([0.691, 0.793], [0.125, 0.373])	([0.614, 0.637], [0.264, 0.373])
A3	([0.584, 0.666], [0.245, 0.071])	([0.854, 0.904], [0.051, 0.071])	([0.746, 0.746], [0.170, 0.071])	([0.755, 0.779], [0.111, 0.071])

On comparison with Clay bricks (*A2*), both of the Fly ash bricks (*A3*) and AAC bricks (*A1*), which consume less energy and have better thermal insulation, are found to be stronger than conventional clay bricks. In addition to this, although Dhanjode et al. [44] argued that the Fly ash bricks and AAC brick save 82 and 95 carbon tax respectively compared to clay bricks to its environmental properties. However, in Indian perspectives, only a few research works are found on AAC bricks while AAC bricks have been used for the construction of several buildings. Global unavailability of the AAC bricks often arises serious concern since it increases transportation and inventory costs. Moreover, Fly ash bricks are more cost and energy saving alternative choice than other ones. Mahendran et al. [45] asserted that Fly ash brick is the most favorable choice among the building blocks in perspectives of strength, heating load, framed and load bearing buildings. The above findings have selected Fly ash brick to be the most sustainable construction material for use in this case study. When we shared our results to the relevant builders, they appreciated our task and adopted this material in their own construction business. The builders also discussed with their clients about the benefits of using this material and suggested them to use it.

Table 14. Weighted normalized decision matrix.

	C1	C2	C3	C4
A1	([0.149, 0.180], [0.718, 0.763])	([0.133, 0.167], [0.727, 0.724])	([0.448, 0.557], [0.294, 0.376])	([0.129, 0.159], [0.702, 0.760])
A2	([0.149, 0.180], [0.693, 0.788])	([0.133, 0.167], [0.708, 0.793])	([0.448, 0.557], [0.205, 0.363])	([0.129, 0.159], [0.743, 0.771])
A3	([0.149, 0.180], [0.693, 0.693])	([0.133, 0.167], [0.727, 0.693])	([0.448, 0.557], [0.245, 0.200])	([0.129, 0.159], [0.727, 0.696])
	C5	C6	C7	C8
A1	([0.540, 0.700], [0.200, 0.245])	([0.422, 0.508], [0.348, 0.329])	([0.437, 0.473], [0.416, 0.301])	([0.429, 0.512], [0.344, 0.477])
A2	([0.540, 0.700], [0.180, 0.249])	([0.422, 0.508], [0.189, 0.456])	([0.437, 0.473], [0.140, 0.399])	([0.440, 0.538], [0.344, 0.477])
A3	([0.540, 0.700], [0.061, 0.076])	([0.422, 0.508], [0.222, 0.206])	([0.437, 0.473], [0.085, 0.090])	([0.612, 0.665], [0.344, 0.477])
	C9	C10	C11	C12
A1	([0.166, 0.217], [0.773, 0.786])	([0.181, 0.213], [0.773, 0.786])	([0.554, 0.598], [0.295, 0.502])	([0.192, 0.218], [0.722, 0.786])
A2	([0.120, 0.137], [0.773, 0.786])	([0.120, 0.137], [0.773, 0.786])	([0.479, 0.593], [0.295, 0.502])	([0.147, 0.174], [0.722, 0.786])
A3	([0.140, 0.182], [0.773, 0.786])	([0.205, 0.247], [0.773, 0.786])	([0.517, 0.558], [0.295, 0.502])	([0.181, 0.213], [0.722, 0.786])

Table 15. Negative ideal solutions.

NIS	IVIF Negative Ideal Solutions
C1	([0.149, 0.180], [0.718, 0.788])
C2	([0.133, 0.167], [0.727, 0.793])
C3	([0.448, 0.557], [0.294, 0.376])
C4	([0.129, 0.159], [0.743, 0.771])
C5	([0.540, 0.700], [0.200, 0.249])
C6	([0.422, 0.508], [0.348, 0.456])
C7	([0.437, 0.473], [0.416, 0.399])
C8	([0.429, 0.512], [0.344, 0.477])
C9	([0.120, 0.137], [0.773, 0.786])
C10	([0.120, 0.137], [0.773, 0.786])
C11	([0.479, 0.558], [0.295, 0.502])
C12	([0.147, 0.174], [0.722, 0.786])

Table 16. Euclidean and Hamming distance matrices.

	A1	A2	A3
ED	0.3519	0.3258	0.9647
HD	0.2107	0.1682	0.6426

Table 17. Assessment matrix and scores.

	Relative Assessment Matrix			Appraisal Scores	Ranking
	A_1	A_2	A_3		
A_1	0.000	0.069	−1.045	−0.976	2
A_2	−0.069	0.000	−1.113	−1.182	3
A_3	1.045	1.113	0.000	2.158	1

Table 18. Comparison with other models.

MCDM Methods	Ranking Order
Classical CODAS	$A_3 > A_2 > A_1$
Fuzzy CODAS	$A_3 > A_1 > A_2$
IVIF-VIKOR	$A_3 > A_2 > A_1$
IVIF-TOPSIS	$A_3 > A_1 > A_2$
The proposed IVIF-CODAS	$A_3 > A_1 > A_2$

6.1. Comparisons

In this section, we perform the necessary comparative analysis with a set of a few existing approaches to prove the practicality and efficiency of the proposed IVIF-CODAS method. The CODAS [16], fuzzy CODAS [48], IVIF-TOPSIS [49], IVIF-VIKOR [50] methods were modified using fuzzy and IVIF numbers. The reason behind choosing these methods is their stability and reliability in producing satisfactory solutions/results. It is essential to measure the reliability of the results attained by the proposed model for crosschecking the optimal results/alternatives. For such action in MCDM problems the most common measure is the comparison of the results produced by other stable and reliable methods [16,51].

Now, the classical CODAS needs crisp numbers as inputs. Therefore, the crisp numbers corresponding to their linguistic ratings are used to perform the CODAS algorithm. Here, the order of ranking is similar to that of the original study. On contrast, fuzzy CODAS can adopt linguistic ratings as triangular/trapezoidal fuzzy numbers. The outcome of fuzzy CODAS is also same as the ranking produced by IVIF-CODAS. Finally, IVIF-TOPSIS and IVIF-MABAC are used to solve the same material selection problem. Table 18 shows that $A3$ (Fly ash bricks) occupies the first ranking in all the methods except in IVIF-VIKOR. There is a ranking swapping between $A3$ (Fly ash bricks) and $A1$ (AAC bricks).

From the comparative analysis, we can summarize that the results are harmonious to each other. Below we give some advantages of the proposed IVIF-CODAS method.

(1) Crisp ratings is used to evaluate the classical CODAS method but this ratings often fail in real life scenario as real life problems are much uncertain. For example, the construction company may consider some criteria as highly important and to signify the importance, the corresponding rating scale need to be more flexible. This "highly important" term can be preferably expressed using as an IVIF number ($[8, 10], [0, 0]$) rather than a single crisp number 9. However, in this paper, we use IVIF numbers to assess the alternative bricks and criteria importance since DMs can flexibly express their opinions using IVIF numbers.

(2) Compared with fuzzy CODAS, the proposed IVIF-CODAS has an advantage. Grattan-Guinness [52] argued that it is a difficult task for decision makers to represent linguistic ratings in the form of a single membership degree in classical fuzzy set theory. In response, Atanassov [53] introduced IFS as an extension of fuzzy sets. In IFS, hesitation margin is introduced as a new concept and the sum of membership and non-membership degree may be less than one. However, both in fuzzy set and IFS, the membership values are exact and crisp in nature. To present the membership and non-membership values are in intervals, Atanassov and Gargov [8] extended IFS in IVIFS. Thus, in group decision making problems the extended IVIF-CODAS method offers a better treatment in handling uncertainty in the decision making process.

(3) As it is difficult to show the applicability and trustworthiness of a newly proposed method, hence it is necessary to assess it in solving several MCDM problems. Wang et al. [54] asserted a comparison which is only way to apprehend the validity of newly proposed MCDM model (here, IVIF-CODAS). To justify any proposed approach, one has to compare it with several related approaches for the same problem. Accordingly, we have presented two illustrative examples and found encouraging results that show the similarity of IVIF-CODAS to other methods. One can

consider this to be one of the advantages of the novel approach that is reckoned to be applicable irrespective of its case studies.

6.2. Sensitivity Analysis

It is well known that the results of MCDM methods can be highly influenced by the weight coefficients of the evaluation criteria. Hence, a sensitivity analysis is discussed in the next two paragraphs, where we compute the final ranking of the alternative bricks by changing the criteria weights. Small variations in the rank of alternatives are noticed due to the small variations in the weight coefficients. Hence, the results of MCDM methods are analyzed by their sensitivity to these changes [55,56]. Below we present a sensitivity analysis using 10 different scenarios (Table 19). We apply the expression given in Equation (28) for changing the criteria weights.

$$w_j^{new} = w_j^{old} \pm \alpha w_j^{old} \tag{28}$$

Table 19. Ranking for different scenarios.

Alternatives	Scenario 1	Scenario 2	Scenario 3	Scenario 4	Scenario 5	Scenario 6	Scenario 7	Scenario 8	Scenario 9	Scenario 10
A_1	2	2	2	2	1	2	2	2	2	1
>A_2	3	3	3	3	3	3	3	3	3	3
>A_3	1	1	1	1	2	1	1	1	1	2

Here α is the percentage of change of w_j^{old}. In this work, the sustainable material selection and criteria weights evaluation are performed by human inputs. The robustness testing of the final ranking has been conducted by considering the changed weights of the criteria. Small changes in criteria weights of the alternatives $A1, A2$ and $A3$ have a negligible effect in the final ranking in the brick selection problem. The result of the performed sensitivity analyses enforce the proposal that Fly ash bricks ($A3$) have the highest priority followed by AAC bricks ($A1$) and Clay bricks ($A2$). The computed ranking order (Figure 4) $A3$ (Fly ash bricks) $A1$ (AAC bricks) $A2$ (Clay bricks) can be followed in eight out of 10 scenarios. Thus, $A3$ (Fly ash bricks) has been ranked as maximum number of scenarios except in scenario 5 and 10. In these two cases, there are noticeable changes (increase and or decrease) in priorities of criteria set which obviously affect the final ranking as $A1$ (AAC bricks) $A3$ (Fly ash bricks) $A2$ (Clay bricks).

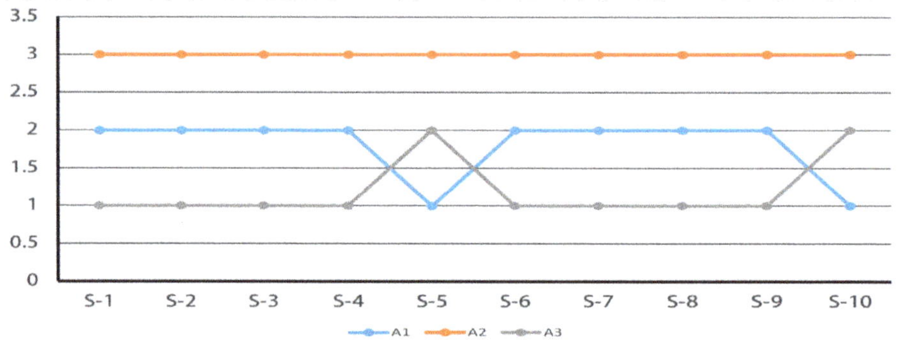

Figure 4. Different ranking positions due to sensitivity analysis.

We find the ranking is consistent unless some noticeable changes are made in the criteria weights. The robustness of the sustainable materials has been shown by sensitivity analysis (Table 19). $A3$ (Fly ash bricks) or *A*1 (AAC bricks) occupies top rank in all cases and may be selected as best by the decision makers. The sensitivity analysis becomes meaningful to assess the bricks as sustainable building blocks of construction projects.

7. Conclusions

Sustainable material selection as well as suitable material selection in uncertainty for construction industry is mandatory to sustain in the competitive market and for eco-friendly environment. Since material selection is one kind of MCDM problem, an efficient MCDM method is needed to cope with the market challenges. In this study, the classical CODAS method is extended with IVIFNs known as IVIF-CODAS for comprehensive, rational, and sensible decision making and especially to handle the uncertain material selection problem. Since human reasoning capability is inherently inexact in nature, we have used linguistic terms to present the opinions of decision makers. IVIFN has been used in this study, since it uses intervals rather than exact numbers to represent the membership and non-membership functions. The comparative study shows that the proposed method is consistent and efficient in respect to the other existing methods. This ranking method will enable clients, the Construction Company, manufacturers, consultants, and material suppliers to understand central public works department norms in India and keep at parity with global standards.

Although this research article delivers numerous valuable implications, a few of the project implications are mentioned here. In any industry it is highly important to keep a balance between the cost parameters (material cost and its operating cost, etc.); hence, managers or engineers face difficulty in selecting the finest material only on the basis of its sustainability benefit. Also, the incorporation of TBL (environment, economy, and society) factors is equally difficult and needs a typical framework in material selection problem. This research work offers such a framework that aids the engineers and experts to decide the most suitable material under sustainability constraints especially in the Indian construction industries. Most research papers in the literature mainly focused on environmental concerns in the construction sector. In comparison to them this paper assists both scientific and societal contributions which contributes towards social progress along with the eco-friendly set-ups. Additionally, this study functions as a reference for researchers and practitioners working in the same field in an Indian context. As a whole, this paper helps and inspires architects, engineers, and other construction managers to respond and to decide the best sustainable material within their viable business set-up.

In future, IVIF-CODAS method can be used to solve other complex MCDM problems, like market segment evaluation and selection, project selection, supplier selection, etc.

Author Contributions: The individual contribution and responsibilities of the authors were as follows: J.R., S.D. and S.K. designed the research, collected and analyzed the data and the obtained results, performed the development of the paper. D.P. provided extensive advice throughout the study, regarding the research design, methodology, findings and revised the manuscript. All the authors have read and approved the final manuscript.

Funding: This work was supported by the Department of Science and Technology, India under the INSPIRE fellowship programme [grant no. DST/INSPIRE Fellowship/2013/544].

Acknowledgments: We wish to express our deepest appreciation to the editors and the anonymous reviewers.

Conflicts of Interest: The authors declare no conflict of interest.

References

1. Jahan, A.; Edwards, K.L. A State-of-the-Art Survey on the Influence of Normalization Techniques in Ranking: Improving the Materials Selection Process in Engineering Design. *Mater. Des.* **2015**, *65*, 335–342. [CrossRef]
2. Yazdani, M.; Payam, A.F. A Comparative Study on Material Selection of Microelectromechanical Systems Electrostatic Actuators Using Ashby, VIKOR and TOPSIS. *Mater. Des.* **2015**, *65*, 328–334. [CrossRef]
3. Petković, D.; Madić, M.; Radovanović, M.; Gečevska, V. Application Of The Performance Selection Index Method For Solving Machining MCDM Problems. *FU Mech. Eng.* **2017**, *15*, 15–97. [CrossRef]

4. Liao, T.W. Two Interval Type 2 Fuzzy TOPSIS Material Selection Methods. *Mater. Des.* **2015**, *88*, 1088–1099. [CrossRef]

5. Liu, H.-C.; Liu, L.; Wu, J. Material Selection Using an Interval 2-Tuple Linguistic VIKOR Method Considering Subjective and Objective Weights. *Mater. Des. 1980–2015* **2013**, *52*, 158–167. [CrossRef]

6. Mousavi, S.M.; Vahdani, B.; Tavakkoli-Moghaddam, R.; Tajik, N. Soft Computing Based on a Fuzzy Grey Group Compromise Solution Approach with an Application to the Selection Problem of Material Handling Equipment. *Int. J. Comput. Integr. Manuf.* **2014**, *27*, 547–569. [CrossRef]

7. Xue, Y.-X.; You, J.-X.; Lai, X.-D.; Liu, H.-C. An Interval-Valued Intuitionistic Fuzzy MABAC Approach for Material Selection with Incomplete Weight Information. *Appl. Soft Comput.* **2016**, *38*, 703–713. [CrossRef]

8. Atanassov, K.; Gargov, G. Interval Valued Intuitionistic Fuzzy Sets. *Fuzzy Sets Syst.* **1989**, *31*, 343–349. [CrossRef]

9. Cheng, S.-H. Autocratic Multiattribute Group Decision Making for Hotel Location Selection Based on Interval-Valued Intuitionistic Fuzzy Sets. *Inf. Sci.* **2018**, *427*, 77–87. [CrossRef]

10. Wen, M.; Zhao, H.; Xu, Z.; Lei, Q. Definite Integrals for Aggregating Continuous Interval-Valued Intuitionistic Fuzzy Information. *Appl. Soft Comput.* **2018**, *70*, 875–895. [CrossRef]

11. Rashid, T.; Faizi, S.; Zafar, S. Distance Based Entropy Measure of Interval-Valued Intuitionistic Fuzzy Sets and Its Application in Multicriteria Decision Making. *Adv. Fuzzy Syst.* **2018**, *2018*, 1–10. [CrossRef]

12. Wang, C.-Y.; Chen, S.-M. Multiple Attribute Decision Making Based on Interval-Valued Intuitionistic Fuzzy Sets, Linear Programming Methodology, and the Extended TOPSIS Method. *Inf. Sci.* **2017**, *397–398*, 155–167. [CrossRef]

13. Gupta, P.; Mehlawat, M.K.; Grover, N.; Pedrycz, W. Multi-Attribute Group Decision Making Based on Extended TOPSIS Method under Interval-Valued Intuitionistic Fuzzy Environment. *Appl. Soft Comput.* **2018**, *69*, 554–567. [CrossRef]

14. Das, S.; Kar, S.; Pal, T. Group Decision Making Using Interval-Valued Intuitionistic Fuzzy Soft Matrix and Confident Weight of Experts. *J. Artif. Intell. Soft Comput. Res.* **2014**, *4*, 57–77. [CrossRef]

15. Das, S.; Kar, M.B.; Pal, T.; Kar, S. Multiple Attribute Group Decision Making Using Interval-Valued Intuitionistic Fuzzy Soft Matrix. In Proceedings of the 2014 IEEE International Conference on Fuzzy Systems (FUZZ-IEEE), Beijing, China, 7–11 July 2014; pp. 2222–2229. [CrossRef]

16. Stanujkić, D.; Karabašević, D. An extension of the WASPAS method for decision-making problems with intuitionistic fuzzy numbers: A case of website evaluation. *Oper. Res. Eng. Sci. Theor. Appl.* **2018**, *1*, 29–39. [CrossRef]

17. Gharzeldeen, M.N.; Beheiry, S.M. Investigating the Use of Green Design Parameters in UAE Construction Projects. *Int. J. Sustain. Eng.* **2015**, *8*, 93–101. [CrossRef]

18. Popovic, M.; Kuzmanovic, M.; Savic, G. A comparative empirical study of Analytic Hierarchy Process and Conjoint analysis: Literature review. *Decis. Mak. Appl. Manag. Eng.* **2018**, *1*, 1–153. [CrossRef]

19. Medineckiene, M.; Zavadskas, E.K.; Björk, F.; Turskis, Z. Multi-Criteria Decision-Making System for Sustainable Building Assessment/Certification. *Arch. Civ. Mech. Eng.* **2015**, *15*, 11–18. [CrossRef]

20. Kibert, C.J. *Sustainable Construction: Proceedings of the First International Conference of CIB TG 16, November 6-9, 1994, Tampa, Florida, USA*; University of Florida Center: Gainesville, FL, USA, 1994.

21. Radhi, H. On the Effect of Global Warming and the UAE Built Environment. In *Global Warming*; InTech: Vienna, Austria, 2010.

22. Elchalakani, M.; Elgaali, E. Sustainable Concrete Made of Construction and Demolition Wastes Using Recycled Wastewater in the UAE. *J. Adv. Concr. Technol.* **2012**, *10*, 110–125. [CrossRef]

23. Al-Hajj, A.; Hamani, K. Material Waste in the UAE Construction Industry: Main Causes and Minimization Practices. *Archit. Eng. Des. Manag.* **2011**, *7*, 221–235. [CrossRef]

24. Ljungberg, L.Y. Materials Selection and Design for Development of Sustainable Products. *Mater. Des.* **2007**, *28*, 466–479. [CrossRef]

25. Florez, L.; Castro-Lacouture, D. Optimization Model for Sustainable Materials Selection Using Objective and Subjective Factors. *Mater. Des.* **2013**, *46*, 310–321. [CrossRef]

26. Bahraminasab, M.; Jahan, A. Material Selection for Femoral Component of Total Knee Replacement Using Comprehensive VIKOR. *Mater. Des.* **2011**, *32*, 4471–4477. [CrossRef]

27. Jahan, A.; Mustapha, F.; Sapuan, S.M.; Ismail, M.Y.; Bahraminasab, M. A Framework for Weighting of Criteria in Ranking Stage of Material Selection Process. *Int. J. Adv. Manuf. Technol.* **2012**, *58*, 411–420. [CrossRef]

28. Chauhan, A.; Vaish, R. Magnetic Material Selection Using Multiple Attribute Decision Making Approach. *Mater. Des. 1980–2015* **2012**, *36*, 1–5. [CrossRef]

29. Karande, P.; Chakraborty, S. Application of Multi-Objective Optimization on the Basis of Ratio Analysis (MOORA) Method for Materials Selection. *Mater. Des.* **2012**, *37*, 317–324. [CrossRef]

30. Maity, S.R.; Chatterjee, P.; Chakraborty, S. Cutting Tool Material Selection Using Grey Complex Proportional Assessment Method. *Mater. Des. 1980–2015* **2012**, *36*, 372–378. [CrossRef]

31. Anojkumar, L.; Ilangkumaran, M.; Vignesh, M. A Decision Making Methodology for Material Selection in Sugar Industry Using Hybrid MCDM Techniques. *Int. J. Mater. Prod. Technol.* **2015**, *51*, 102–126. [CrossRef]

32. Zavadskas, E.K.; Bausys, R.; Juodagalviene, B.; Garnyte-Sapranaviciene, I. Model for Residential House Element and Material Selection by Neutrosophic MULTIMOORA Method. *Eng. Appl. Artif. Intell.* **2017**, *64*, 315–324. [CrossRef]

33. Tian, G.; Zhang, H.; Feng, Y.; Wang, D.; Peng, Y.; Jia, H. Green Decoration Materials Selection under Interior Environment Characteristics: A Grey-Correlation Based Hybrid MCDM Method. *Renew. Sustain. Energy Rev.* **2018**, *81*, 682–692. [CrossRef]

34. Xu, Z.S. Methods for Aggregating Interval-Valued Intuitionistic Fuzzy Information and Their Application to Decision Making. *Control Decis.* **2007**, *22*, 215–219. [CrossRef]

35. Xu, Z.-S.; Jian, C. Approach to Group Decision Making Based on Interval-Valued Intuitionistic Judgment Matrices. *Syst. Eng.-Theory Pract.* **2007**, *27*, 126–133. [CrossRef]

36. Xu, Z.S.; Chen, J. An Overview of Distance and Similarity Measures of Intuitionistic Fuzzy Sets. *Int. J. Uncertain. Fuzziness Knowl.-Based Syst.* **2008**, *16*, 529–555. [CrossRef]

37. Xu, Z. A Deviation-Based Approach to Intuitionistic Fuzzy Multiple Attribute Group Decision Making. *Group Decis. Negot.* **2010**, *19*, 57–76. [CrossRef]

38. Park, J.H.; Lim, K.M.; Park, J.S.; Kwun, Y.C. Distances between Interval-Valued Intuitionistic Fuzzy Sets. *J. Phys. Conf. Ser.* **2008**, *96*, 012089. [CrossRef]

39. Zhang, X.; Xu, Z. Soft Computing Based on Maximizing Consensus and Fuzzy TOPSIS Approach to Interval-Valued Intuitionistic Fuzzy Group Decision Making. *Appl. Soft Comput.* **2015**, *26*, 42–56. [CrossRef]

40. Hashemi, H.; Bazargan, J.; Mousavi, S.M.; Vahdani, B. An Extended Compromise Ratio Model with an Application to Reservoir Flood Control Operation under an Interval-Valued Intuitionistic Fuzzy Environment. *Appl. Math. Model.* **2014**, *38*, 3495–3511. [CrossRef]

41. Chen, T.-Y. Interval-Valued Intuitionistic Fuzzy QUALIFLEX Method with a Likelihood-Based Comparison Approach for Multiple Criteria Decision Analysis. *Inf. Sci.* **2014**, *261*, 149–169. [CrossRef]

42. Pamučar, D.; Lukovac, V.; Božanić, D.; Komazec, N. Multi-criteria FUCOM-MAIRCA model for the evaluation of level crossings: Case study in the Republic of Serbia. *Oper. Res. Eng. Sci. Theor. Appl.* **2018**, *1*, 108–129. [CrossRef]

43. Hassanpour, M. Evaluation of Iranian Wood and Cellulose Industries. *Decis. Mak. Appl. Manag. Eng.* **2019**, *2*, 13–34. [CrossRef]

44. Dhanjode, C.S.; Ralegaonkar, R.V.; Dakwale, V.A. Design and Development of Sustainable Construction Strategy for Residential Buildings: A Case Study for Composite Climate. *Int. J. Sustain. Constr. Eng. Technol.* **2013**, *4*, 12–20.

45. Mahendran, K.; Sivaram, T.; Shahulhameed, M.; Logaraja, R. A Comparative Study on Various Building Blocks as an Alternative to Conventional Bricks. In Proceedings of the International Conference on Emering Trends in Engineering and Management Research, Nashik, Maharastra, 23 March 2016; pp. 1097–1109.

46. Benton, R. Reduce, Reuse, Recycle . . . and Refuse. *J. Macromarket.* **2015**, *35*, 111–122. [CrossRef]

47. Hamid, Z.A.; Anuar Mohamad Kamar, K. Aspects of Off-site Manufacturing Application towards Sustainable Construction in Malaysia. *Constr. Innov.* **2012**, *12*, 4–10. [CrossRef]

48. Pamucar, D.; Bozanic, D.; Lukovac, V.; Komazec, N. Normalized weighted geometric bonferroni mean operator of interval rough numbers—Application in interval rough DEMATEL-COPRAS. *FU Mech. Eng.* **2018**, *16*, 171–191. [CrossRef]

49. Ashtiani, B.; Haghighirad, F.; Makui, A.; Montazer, G.A. Extension of Fuzzy TOPSIS Method Based on Interval-Valued Fuzzy Sets. *Appl. Soft Comput.* **2009**, *9*, 457–461. [CrossRef]

50. Vahdani, B.; Hadipour, H.; Sadaghiani, J.S.; Amiri, M. Extension of VIKOR Method Based on Interval-Valued Fuzzy Sets. *Int. J. Adv. Manuf. Technol.* **2010**, *47*, 1231–1239. [CrossRef]

51. Pamučar, D.; Petrović, I.; Ćirović, G. Modification of the Best–Worst and MABAC Methods: A Novel Approach Based on Interval-Valued Fuzzy-Rough Numbers. *Expert Syst. Appl.* **2018**, *91*, 89–106. [CrossRef]
52. Grattan-Guinness, I. Fuzzy Membership Mapped onto Intervals and Many-Valued Quantities. *Z. Für Math. Log. Grundlagen Math.* **1976**, *22*, 149–160. [CrossRef]
53. Atanassov, K.T. Intuitionistic Fuzzy Sets. *Fuzzy Sets Syst.* **1986**, *20*, 87–96. [CrossRef]
54. Wang, P.; Li, Y.; Wang, Y.-H.; Zhu, Z.-Q. A New Method Based on TOPSIS and Response Surface Method for MCDM Problems with Interval Numbers. *Math. Probl. Eng.* **2015**, *2015*, 1–11. [CrossRef]
55. Debnath, A.; Roy, J.; Kar, S.; Zavadskas, E.; Antucheviciene, J. A Hybrid MCDM Approach for Strategic Project Portfolio Selection of Agro By-Products. *Sustainability* **2017**, *9*, 1302. [CrossRef]
56. Chatterjee, P.; Mondal, S.; Boral, S.; Banerjee, A.; Chakraborty, S. A novel hybrid method for non-traditional machining process selection using factor relationship and Multi-Attributive Border Approximation Method. *FU Mech. Eng.* **2017**, *15*, 15–439. [CrossRef]

© 2019 by the authors. Licensee MDPI, Basel, Switzerland. This article is an open access article distributed under the terms and conditions of the Creative Commons Attribution (CC BY) license (http://creativecommons.org/licenses/by/4.0/).

Article

A New Hybrid MCDM Model: Sustainable Supplier Selection in a Construction Company

Bojan Matić [1], Stanislav Jovanović [1], Dillip Kumar Das [2], Edmundas Kazimieras Zavadskas [3], Željko Stević [4,*], Siniša Sremac [1] and Milan Marinković [1]

[1] Faculty of Technical Sciences, University of Novi Sad, Trg Dositeja Obradovica 6, 21000 Novi Sad, Serbia; bojanm@uns.ac.rs (B.M.); stasha@uns.ac.rs (S.J.); sremacs@uns.ac.rs (S.S.); milan.marinkovic@uns.ac.rs (M.M.)

[2] Department of Civil Engineering, Central University of Technology, Free State, Bloemfontein 9300, South Africa; ddas@cut.ac.za

[3] Institute of Sustainable Construction, Faculty of Civil Engineering, Vilnius Gediminas Technical University, Sauletekio al. 11, LT-10223 Vilnius, Lithuania; edmundas.zavadskas@vgtu.lt

[4] Faculty of Transport and Traffic Engineering Doboj, University of East Sarajevo, Vojvode Mišića 52, 74000 Doboj, Bosnia and Herzegovina

[*] Correspondence: zeljkostevic88@yahoo.com or zeljko.stevic@sf.ues.rs.ba

Received: 21 February 2019; Accepted: 3 March 2019; Published: 8 March 2019

Abstract: Sustainable development is one of the most important preconditions for preserving resources and balanced functioning of a complete supply chain in different areas. Taking into account the complexity of sustainable development and a supply chain, different decisions have to be made day-to-day, requiring the consideration of different parameters. One of the most important decisions in a sustainable supply chain is the selection of a sustainable supplier and, often the applied methodology is multi-criteria decision-making (MCDM). In this paper, a new hybrid MCDM model for evaluating and selecting suppliers in a sustainable supply chain for a construction company has been developed. The evaluation and selection of suppliers have been carried out on the basis of 21 criteria that belong to all aspects of sustainability. The determination of the weight values of criteria has been performed applying the full consistency method (FUCOM), while a new rough complex proportional assessment (COPRAS) method has been developed to evaluate the alternatives. The rough Dombi aggregator has been used for averaging in group decision-making while evaluating the significance of criteria and assessing the alternatives. The obtained results have been checked and confirmed using a sensitivity analysis that implies a four-phase procedure. In the first phase, the change of criteria weight was performed, while, in the second phase, rough additive ratio assessment (ARAS), rough weighted aggregated sum product assessment (WASPAS), rough simple additive weighting (SAW), and rough multi-attributive border approximation area comparison (MABAC) have been applied. The third phase involves changing the parameter ρ in the modeling of rough Dombi aggregator, and the fourth phase includes the calculation of Spearman's correlation coefficient (SCC) that shows a high correlation of ranks.

Keywords: sustainability; supplier selection; construction; FUCOM; rough COPRAS; rough Dombi aggregator

---

1. Introduction

Sustainable engineering implies the execution of all processes and activities respecting all aspects of sustainability: economic, social, and environmental aspects. In addition, it is necessary to take into account the interactions and symmetry between them. This is confirmed by Hutchins et al. [1] according to whom it is necessary to define and understand the relationships that exist among aspects of sustainability and how they influence each other. In the last two decades, according to Vanalle et al. [2], companies around the world have been showing increasing concern about the impact of their operations on the environment, which arises as a result of pressure by legal regulations, customers, and competitors. Taking this into account, construction companies operate under great pressure due to their potentially negative impact on the environment and a complete, sustainable supply chain. In line with sustainability—that has become inevitability—and urgent need, supply chains are also changing, and their focus is no longer just on rationalizing costs, but also on environmental concerns. On this basis, sustainable supply chain management (SSCM) and green supply chain management (GSCM) have been established. The SSCM concept, according to Sen et al. [3], is an integrated approach that links economic and social thinking together with environmental awareness in traditional supply chain management. SSCM is based on the idea that in addition to constant monitoring of economic values, companies must consider environmental and social aspects, too. This implies that, in order to achieve sustainability, companies should solve environmental issues together with meeting social standards at all levels of supply chain [4], and at the same time, achieving certain economic effects. In order to achieve the effects of SSCM, according to Rabbani et al. [5], a large number of individual participants in a supply chain, starting from suppliers to top managers, have to take into account sustainable aspects. At the very beginning of the sustainability concept, according to Singh and Trivedi [6], the focus was mainly on environmental issues, and much less on social aspects, as it was thought that by managing and reducing negative impacts on the environment, companies would achieve competitive advantages. Nowadays, it is a different situation and, therefore, the evaluation and selection of suppliers based on an equal number of criteria by all aspects of sustainability has been performed in this paper.

This paper has several interrelated objectives. The first aim of this research refers to the development and detailed description of the algorithm of a new rough complex proportional assessment (COPRAS) method. The second aim that appears as a causal link to the previous one refers to the development of a new hybrid model which implies the integration of full consistency method (FUCOM), rough Dombi aggregator, and rough COPRAS method. The third aim of the paper is to popularize the FUCOM method, which contributes to the objective determination of weight criteria values, as well as to popularize the application of MCDM methods in integration with rough numbers.

After the introductory part, which explains the aims and motivation for this research, the paper consists of five more sections. The second section presents a two-phase procedure for reviewing the situation in the field. A review of MCDM methods in sustainable civil engineering and a review of MCDM methods for sustainable supplier selection are presented. The third part includes the developed methodology of this paper. At the beginning of the section, the process of research is presented with the contributions and advantages of this paper. Then, the FUCOM method is briefly explained in the first part, while the algorithm of the developed rough COPRAS method is elaborated and explained, in detail, in the second part. The fourth section describes a complete procedure for the selection of a sustainable supplier in a construction company. A detailed calculation for each step of the developed methodology is presented in order to make the model much more understandable to readers. The fifth section is a sensitivity analysis and discussion. The sensitivity analysis implies the already described four-phase procedure, followed by a discussion of the results obtained. In the sixth section, concluding observations with paper contributions and guidelines for future research are provided.

2. Literature Review

2.1. Review of MCDM Methods in Sustainable Civil Engineering

Formal decision-making methods can be used to help improve the overall sustainability of industries and organizations [7]. According to Zavadskas et al. [8], as sustainable development is becoming more relevant, more and more articles are being published related to sustainability in the field of construction. According to same authors, sustainable decision-making in civil engineering, construction, and building technology can be supported by fundamental scientific achievements and multi-criteria decision-making (MCDM) theories that, according to Mardani et al. [9], are widely used. In the field of construction, increasing attention is being paid to energy efficiency and smart buildings, and therefore it is necessary to go towards sustainability in the design and construction of facilities and infrastructure.

Construction is an area that interacts enormously with the natural environment. A large percentage of raw materials are obtained from the earth, and in their treatment, processing, and the construction of buildings, certain environmental pollution is inevitable. Lombera and Rojo [10] use the Spanish MIVES (in English, integrated value model for sustainable assessment) methodology to define criteria for the sustainability of industrial buildings and to select the optimum solution with regard to them. A similar study is presented in [11], where authors also use the MIVES method but in combination with Monte Carlo simulation, in order to assess the sustainability of concrete structures. De la Fuente et al. [12] also apply the MIVES methodology together with the analytic hierarchy process (AHP) method in order to reduce subjective human impact on the selection of sewage pipe material. The MIVES methodology is also used in [13] in assessing the sustainability of alternatives—the types of concrete and their reinforcement for application in tunnels. The problem of monitoring, repairing, and returning to the function of steel bridge structures is a major challenge for engineers, especially because it is necessary to make key decisions, and wrongly made decisions can be very costly. In order to exclude subjectivity in selecting alternatives, Rashidi et al. [14] presented the decision support system (DSS), within which the simplified AHP (S-AHP) method was used. S-AHP combines simple multi-attribute rating technique (SMART) and AHP method. The aim is to help engineers in planning the safety, functionality, and sustainability of steel bridge structures. Jia et al. [15] present a framework for the selection of bridge construction between the ABC (Accelerated Bridge Construction) method and conventional alternatives, using the technique for order of preference by similarity to ideal solution (TOPSIS) and fuzzy TOPSIS methods.

Formisano and Mazzolani [16] present a new procedure for the selection of the optimum solution for seismic retrofitting of existing buildings which involves the application of three MCDM methods: TOPSIS, elimination and choice expressing reality (ELECTRE), and VlšeKriterijumska Optimizacija i Kompromisno Rešenje (VIKOR). Terracciano et al. [17] selected cold-formed thin-walled steel structures for vertical reinforcement and energy retrofitting systems of existing masonry constructions using TOPSIS method. Šiožinytė et al. [18] apply the AHP and TOPSIS grey MCDM methods to select an optimum solution for modernizing traditional buildings. Khoshnava et al. [19] apply MCDM methods to select energy efficient, ecological, recyclable materials for building, with respect to the three pillars of sustainability. In order to evaluate 23 criteria in the selection of materials, they use the decision-making trial and evaluation laboratory (DEMATEL) hybrid MCDM method together with the fuzzy analytic network process (FANP). Akadiri et al. [20] use fuzzy extended AHP (FEAHP) in order to select sustainable building materials. In [21], the ANP method is used to select an environmentally friendly method for the construction of a highway, since it can have a great impact on the environment. Most systems for evaluating the sustainability of facilities take into account only the environmental aspect and the environmental impact. However, it is necessary to take into account all three basic principles of sustainability, and thus Raslanas et al. [22], in their work, develop a system for evaluating the sustainability of recreational facilities using the AHP method. MCDM tool, according to Kumar et al. [23], is becoming popular in the field of energy planning due to the flexibility it provides to the decision-makers to take decisions while considering all the criteria and objectives simultaneously. MULTIMOORA and TOPSIS are used in [24] for sustainable decision-making in the energy planning. The authors have concluded that hydro and solar power systems were identified as the most sustainable. A study performed in [25] deals with developing a sustainability assessment framework for assessing technologies for the treatment of urban sewage sludge based on the logarithmic fuzzy preference programming-based fuzzy analytic hierarchy process (LFPPFAHP) and extension theory. Salabun et al. [26] developed an MCDM model with

COMET method for offshore wind farm localization. This method is also used in [27] for sustainable manufacturing and for solving the problem of the sustainable ammonium nitrate transport in [28].

2.2. Review of MCDM Methods for Sustainable Supplier Selection

The selection of suppliers is a constant process that requires the consideration of a certain number of criteria needed to make a decision on the selection of the most suitable suppliers [29–31]. According to Yazdani et al. [32], supplier evaluation and selection is a significant strategic decision for reducing operating costs and improving organizational competitiveness to develop business opportunities. Therefore, it is necessary to pay special attention to the selection of suppliers, including all aspects of sustainability.

The supplier selection, according to many authors, is one of the most demanding problems of sustainable supply chain management [33]. Fuzzy approach in combination with TOPSIS method is applied in [34] for assessing the sustainable performance of suppliers. In order to select suppliers in terms of sustainability, Dai and Blackhurst [35] present an integrated approach based on AHP and the quality function deployment (QFD) method. For the sustainable supplier selection, Azadnia et al. [36] propose an integrated approach that, in addition to the Fuzzy AHP method, is based on multi-objective mathematical programming, as well as on rule-based weighted fuzzy method. In [37], the assessment of sustainable supply chain management and the selection of suppliers are performed using grey theory in combination with the DEMATEL method, while Luthra et al. [38] present an integrated approach consisting of a combination of AHP and VIKOR method based on 22 criteria for all three aspects of sustainability. Sustainable supplier selection of raw materials in order to achieve sustainable development of the company is performed in [39], based on the fuzzy entropy–TOPSIS method. Hsu et al. [40] present a hybrid approach based on several MCDM methods in order to select suppliers in terms of carbon emissions. The evaluation of the supplier performance in the field of electronic industry in order to implement green supply chains is a topic of research in [41]. The authors use rough DEMATEL–ANP (R′AMATEL) in combination with rough multi-attribute ideal real comparative analysis (R′MAIRCA) method. Liu et al. [42] select the suppliers of fresh products using best worst method (BWM) and multi-objective optimization on the basis of the ratio analysis (MULTIMOORA) method. Kusi-Sarpong et al. [43] present a framework for ranking and selecting the criteria for sustainable innovations in supply chain management based on the BWM method. A quantitative assessment of the performance of a sustainable supply chain is presented in [44] based on fuzzy entropy and fuzzy Multi-Attribute Utility Theory (MAUT) methods. Das and Shaw [45] propose a model based on AHP and fuzzy TOPSIS method for selecting a sustainable supply chain, taking into account carbon emissions and various social factors. Luthra et al. [46] propose the application of Delphi and fuzzy DEMATEL methods for identifying and evaluating guidelines for the application information and communication technologies in sustainable initiatives in supply chains. In [47], a framework that identifies sustainable processes in supply chains for individual industries in India is presented. The ranking of industry branches is carried out using six fuzzy MCDM methods. Liou et al. [48] are proposed hybrid model consists of DEMATEL, ANP, and COPRAS-G methods for improving green supply chain management. They have used 12 criteria for supplier selection in the electronic industry, and provided a systemic analytical model for the improvement of parts of the supply chain management.

3. Methods

Figure 1 presents the methodology used in this paper, which consists of four phases:

- I—initial research and data collection
- II—developing methodology
- III—sustainable supplier selection
- IV—sensitivity analysis

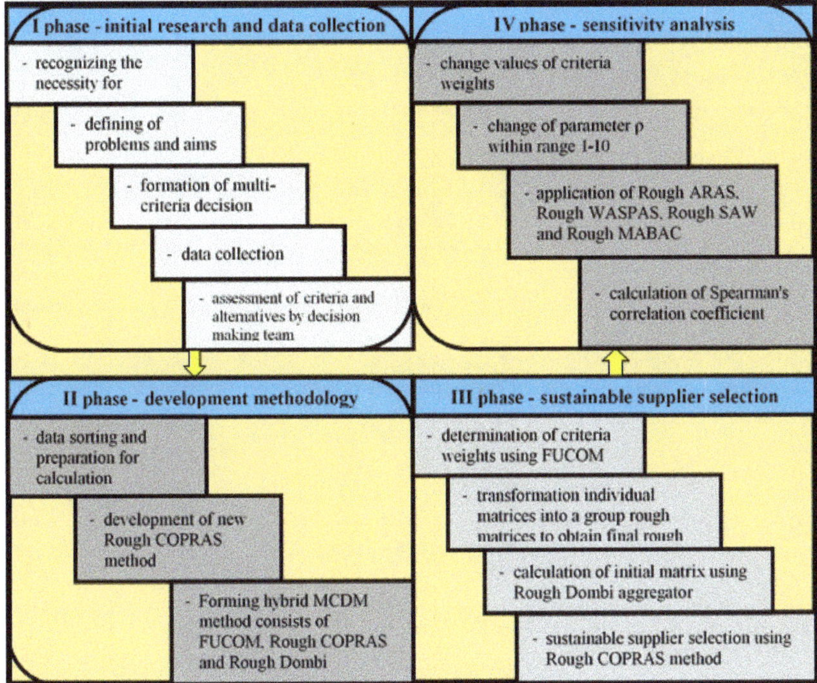

Figure 1. Research flow with proposed model.

The first phase consists of five steps. First, recognition of the need for this research and definition of the problems and aims of the research are performed in the first two steps, and the MCDM model is formed in the third step. After forming the model and defining all elements, the criteria, the alternatives, and the team of experts, the processes of data collection begins, which is the fourth step of the first phase. In the last step, the evaluation of the mutual significance of the criteria and evaluation of the alternatives by the formed team of experts are carried out. The second phase consists of three steps, where the first one is collected data sorting and preparation for their insertion into the developed model. The second step is the development and detailed description of the rough COPRAS method, and the creation of a hybrid MCDM model in the last step of this phase.

The third phase provides a detailed calculation for the evaluation and selection of a sustainable supplier, which consists of four steps. First, the determination of the criteria values using the FUCOM method is carried out, and then the transformation of the obtained values into rough numbers, in order to perform averaging using the rough Dombi aggregator and obtain the final values of the criteria. Subsequently, in the third step, the rough Dombi aggregator is again used to obtain an initial rough matrix, in order to make a decision on the selection of a sustainable supplier using the rough COPRAS method in the fourth step. The final phase is the sensitivity analysis already explained in the previous section.

We have decided to extent the COPRAS method with rough numbers from following reasons. Rough set theory is vague, subjective, and imprecise, while the COPRAS method, according to Mulliner et al. [49], allows for both benefit and cost criteria to be incorporated with one analysis without difficulty or question. The main advantage of COPRAS method compared with other MCDM methods is to be able to show utility degree. Also, COPRAS method has a simple procedure to use.

The following is a brief summary of the FUCOM method algorithm in the first part and the detailed algorithm description of rough COPRAS method in the second part.

3.1. Full Consistency Method (FUCOM)

The FUCOM method has been developed by Pamučar et al. [50] for determining the weights of criteria. It is a new method that, according to authors, represents a better method than AHP (analytical hierarchy process) and BWM (best worst method). So far, it has been applied in studies [51–54]. It consists of the following three steps:

Step 1. In the first step, the criteria from the predefined set of evaluation criteria $C = \{C_1, C_2, \ldots, C_n\}$ are ranked. The ranking is performed according to the significance of the criteria, i.e., starting from the criterion which is expected to have the highest weight coefficient to the criterion of the least significance.

$$C_{j(1)} > C_{j(2)} > \ldots > C_{j(k)} \tag{1}$$

Step 2. In the second step, a comparison of the ranked criteria is carried out and the comparative priority $(\varphi_{k/(k+1)}, k = 1, 2, \ldots, n$, where k represents the rank of the criteria) of the evaluation criteria is determined.

$$\Phi = \left(\varphi_{1/2}, \varphi_{2/3}, \ldots, \varphi_{k/(k+1)} \right) \tag{2}$$

Step 3. In the third step, the final values of the weight coefficients of the evaluation criteria $(w_1, w_2, \ldots, w_n)^T$ are calculated. The final values of the weight coefficients should satisfy the two conditions:

(1) that the ratio of the weight coefficients is equal to the comparative priority among the observed criteria $(\varphi_{k/(k+1)})$ defined in Step 2, i.e., that the following condition is met:

$$\frac{w_k}{w_{k+1}} = \varphi_{k/(k+1)}. \tag{3}$$

(2) In addition to Condition (3), the final values of the weight coefficients should satisfy the condition of mathematical transitivity, i.e., that $\varphi_{k/(k+1)} \otimes \varphi_{(k+1)/(k+2)} = \varphi_{k/(k+2)}$.

Since $\varphi_{k/(k+1)} = \frac{w_k}{w_{k+1}}$ and $\varphi_{(k+1)/(k+2)} = \frac{w_{k+1}}{w_{k+2}}$, $\frac{w_k}{w_{k+1}} \otimes \frac{w_{k+1}}{w_{k+2}} = \frac{w_k}{w_{k+2}}$ is obtained.

Thus, another condition that the final values of the weight coefficients of the evaluation criteria need to meet is obtained, namely

$$\frac{w_k}{w_{k+2}} = \varphi_{k/(k+1)} \otimes \varphi_{(k+1)/(k+2)}. \tag{4}$$

Based on the defined settings, the final model for determining the final values of the weight coefficients of the evaluation criteria can be defined.

$$
\begin{aligned}
&\min \chi \quad s.t. \\
&\left| \frac{w_{j(k)}}{w_{j(k+1)}} - \varphi_{k/(k+1)} \right| \leq \chi, \quad \forall j \\
&\left| \frac{w_{j(k)}}{w_{j(k+2)}} - \varphi_{k/(k+1)} \otimes \varphi_{(k+1)/(k+2)} \right| \leq \chi, \quad \forall j \\
&\sum_{j=1}^{n} w_j = 1, \quad \forall j \\
&w_j \geq 0, \quad \forall j\ .
\end{aligned}
\tag{5}
$$

3.2. A Novel Rough COPRAS Method

The COPRAS method was expanded with rough numbers as part of a sensitivity analysis in the research [55]. So far, a complete algorithm that can enrich the theoretical field of multi-criteria decision-making has not been demonstrated. From this aspect, the algorithm presented below represents a significant contribution to the literature that addresses the problems of multi-criteria decision-making. It should be pointed out that the COPRAS method with interval rough numbers has been developed in [56], which differs from the proposed algorithm in this paper.

Rough COPRAS consists of the following steps:

Step 1: Forming a multi-criteria model. In the initial step, it is necessary to create a multi-criteria model with all necessary elements. Create a set of n alternatives that will be evaluated based on m criteria assessed by e experts.

Step 2: Forming an initial matrix for group decision-making (6). In this step, it is necessary to transform the individual matrices formed by experts' evaluations into an initial group rough matrix. In order to achieve this, it is necessary to apply basic operations with rough numbers.

$$
X = \begin{array}{c} A_1 \\ A_2 \\ \cdots \\ A_n \end{array} \begin{bmatrix} \overset{C_1 \quad\quad C_2 \quad \cdots \quad C_m}{RN(x_{11})} & RN(x_{12}) & \cdots & RN(x_{1n}) \\ RN(x_{21}) & RN(x_{22}) & \cdots & RN(x_{2n}) \\ \cdots & \cdots & \cdots & \cdots \\ RN(x_{m1}) & RN(x_{m2}) & \cdots & RN(x_{mm}) \end{bmatrix}_{m \times n}, \tag{6}
$$

where $RN(x_{ij})$ is an estimated value of the ith alternative in relation to the jth criterion, n is the number of alternatives, and m is the number of criteria.

Step 3: Normalization of the initial rough decision-making matrix applying the linear normalization procedure (7).

$$
r_{ij} = \frac{x_{ij}^L; x_{ij}^U}{\sum x_{ij}^L; x_{ij}^U} = \left[\frac{x_{ij}^L}{\sum x_{ij}^U}, \frac{x_{ij}^U}{\sum x_{ij}^L} \right] \tag{7}
$$

Step 4: Forming a weighted normalized matrix using the following Formula (8):

$$
D = \left[d_{ij}^L; d_{ij}^U \right] = \left[w_j^L \times r_{ij}^L; w_j^U \times r_{ij}^U \right], \tag{8}
$$

where $\left[r_{ij}^L; r_{ij}^U \right]$ is the normalized rough value of the ith alternative in relation to the jth criterion and w_j is the weight or significance of the jth criterion.

Step 5: In this step, it is necessary to calculate the sum of the weighted normalized values for both types of criteria, for benefit criteria using Equation (9):

$$
S_{+i} = \left[s_{ij}^{+L}; s_{ij}^{+U} \right]_{1 \times n}, \tag{9}
$$

and for cost criteria using Equation (10):

$$
S_{-i} = \left[s_{ij}^{-L}; s_{ij}^{-U} \right]_{1 \times n}. \tag{10}
$$

Step 6: Determining the inverse summarized matrix for cost criteria (11):

$$
(S_i^-)^{-1} = \left[\frac{1}{s_{ij}^{-U}}; \frac{1}{s_{ij}^{-L}} \right]. \tag{11}
$$

Step 7: Determining the sum of the matrix for cost criteria (12) and the sum of its inverse matrix (13) so that two matrices 1×1 are obtained:

$$
\left(\overline{S_i^-} \right) = \sum \left[s_i^{-L}; s_i^{-U} \right]_{1 \times 1}, \tag{12}
$$

$$\left(\overline{S_i^-}\right)^{-1} = \sum \left[\frac{1}{s_{ij}^{-U}}; \frac{1}{s_{ij}^{-L}}\right]_{1\times 1}.$$

(13)

Step 8: Determining the relative significance for each alternative. The relative weight Q_i for the ith alternative is calculated applying Equation (14):

$$Q_i = S_{+i} \times \frac{\left(\overline{S_i^-}\right)}{S_{-i} \times \left(\overline{S_i^-}\right)^{-1}}.$$

(14)

Step 9: Determining the priorities of alternatives. The priority in comparing the alternatives is identified on the basis of their relative weight, where the alternative with a higher relative weight value is given a higher priority or a rank, and the alternative with a maximum value represents the most acceptable alternative.

$$A^* = \left\{ A_i | \max_i Q_i \right\}.$$

(15)

4. Case Study

Sustainable supplier selection in the construction company was carried out on the basis of 21 criteria shown and explained in Table 1: economic, social, and environmental criteria. Each of these main criteria consists of seven subcriteria. The set of criteria used in this study was selected according to relevant literature, and based on interviews with authorized and managerial persons in the construction company. The first subcriterion that belongs to economic criteria C_{11} (costs/prices) and the sixth subcriterion (consumption of resources) are the cost criteria, while the others are the benefit criteria.

Table 1. Criteria for sustainable supplier selection.

Id	Criteria	References
C_1		**Economic criteria**
C_{11}	Costs/prices	The final cost to purchase a unit of raw or semi-finished product
C_{12}	Quality	Quality is the degree to which a set of product characteristics meets customer requirements
C_{13}	Delivery	The capability of transporting goods from a source location to a predefined destination
C_{14}	Flexibility	Demand that can be profitably sustained, and time or cost required to add new products to existing production operations
C_{15}	Technology capability	The sum of all the knowledge of an enterprise in support of technological innovation
C_{16}	Financial ability	The capital needed to maintain normal business activities for an enterprise during a certain period of time
C_{17}	Partnership relations	Determining the willingness to establish long-term and close business relations with suppliers to jointly develop the market

Table 1. *Cont.*

Id	Criteria	References
C_2		**Social criteria**
C_{21}	Reputation	Reputation marks the general opinion of the supplier, which relates to his reputation
C_{22}	Safety and health at work	Concerned with the safety, health, and welfare of people at work
C_{23}	Employees' rights	A group of legal rights and claimed human rights having to do with labor relations between workers and their employers
C_{24}	Local community influence	Neighboring relations between the company and the local government, the community, and all residents, representing the public image of the organization
C_{25}	Training of employees	The process of enhancing the skills, capabilities, and knowledge of employees for a particular job
C_{26}	Respect of rights and policies	Enterprises comply with all laws and regulations of the country, assume legal obligations, and promote good social public morals
C_{27}	Disclosing information	Providing information to stakeholders about the materials used, carbon emissions, toxins released during production, etc.
C_3		**Environmental criteria**
C_{31}	Green image	The identity that consumers prioritize environmental conservation and sustainable business practices
C_{32}	Environmental protection management system	A system that comprehensively evaluates the internal and external environmental performance of an organization
C_{33}	Pollution control	The control of pollutants that are released into air, water, or soil
C_{34}	Green products	Environmentally conscious products which are pollution-free, resource-saving, or renewable and recyclable
C_{35}	ECO design	An approach to designing products with special consideration for the environmental impacts of the product during its whole lifecycle
C_{36}	Consumption of resources	The use of non-renewable or, less often, renewable resources
C_{37}	Green competences	The capacity to balance the containment relationships between economic and environmental performance

In this study, the team of five experts took part in the process of determination of weight coefficients of criteria and assessment of alternatives. Experts with a minimum of six years' experience in civil engineering were chosen. After interviewing the experts, the collected data were processed, and the aggregation of expert opinion was obtained. The collecting of data was carried out in the period from November 2018 until January 2019.

4.1. Determining Criteria Weights Using the FUCOM Method

In the following section, a detailed overview is provided of determining weight coefficients of the first-level criteria.

Step 1. In the first step, the decision-makers (DMs) ranked the criteria: DM_1: $C_1 > C_3 > C_2$; DM_2: $C_1 > C_2 > C_3$; DM_3: $C_1 > C_3 > C_2$; DM_4: $C_3 > C_1 > C_2$; and DM_5: $C_1 > C_3 > C_2$.

Step 2. In the second step, the decision-makers compared, in pairs, the ranked criteria from step 1. The comparison is made according to the first-ranked criterion, based on the above scale $[1,7]$. This is how the importance of the criteria is obtained ($\varpi_{C_{j(k)}}$) for all the criteria ranked in step 1 (Table 2).

Table 2. Significance of criteria.

DM_1			
Criteria	C_1	C_3	C_2
Significance ($\omega_{C_{j(k)}}$)	1	2.2	2.8
DM_2			
Criteria	C_1	C_2	C_3
Significance ($\omega_{C_{j(k)}}$)	1	2.7	2.7
DM_3			
Criteria	C_1	C_3	C_2
Significance ($\omega_{C_{j(k)}}$)	1	3.1	3.4
DM_4			
Criteria	C_3	C_1	C_2
Significance ($\omega_{C_{j(k)}}$)	1	1.7	2
DM_5			
Criteria	C_1	C_3	C_2
Significance ($\omega_{C_{j(k)}}$)	1	1.6	1.9

Based on the obtained significance of criteria, comparative significance values of criteria for each expert are calculated as follows:

$$DM_1 : \varphi_{C_1/C_3} = 2.2/1 = 2.2, \ \varphi_{C_3/C_2} = 2.8/2.2 = 1.27;$$

$$DM_2 : \varphi_{C_1/C_2} = 2.7/1 = 2.7, \ \varphi_{C_2/C_3} = 2.7/2.7 = 1;$$

$$DM_3 : \varphi_{C_1/C_3} = 3.1/1 = 3.1, \ \varphi_{C_3/C_2} = 3.4/3.1 = 1.10;$$

$$DM_4 : \varphi_{C_3/C_1} = 1.7/1 = 1.7, \ \varphi_{C_1/C_2} = 2/1.7 = 1.18;$$

$$DM_5 : \varphi_{C_1/C_3} = 1.6/1 = 1.6, \ \varphi_{C_3/C_2} = 1.9/1.6 = 1.19.$$

Step 3. Final values of weight coefficient should satisfy two conditions:

(1) Final values of weight coefficient should satisfy the condition where

$$DM_1 : w_1/w_3 = 2.2, \ w_3/w_2 = 1.27;$$

$$DM_2 : w_1/w_2 = 2.7, \ w_2/w_3 = 1;$$

$$DM_3 : w_1/w_3 = 3.1, \ w_3/w_2 = 1.10;$$

$$DM_4 : w_3/w_1 = 1.7, \ w_1/w_2 = 1.18;$$

$$DM_5 : w_1/w_3 = 1.6, \ w_3/w_2 = 1.19.$$

(2) In addition to the defined relations, final values of weight coefficients should satisfy also the condition of mathematical transitivity, $w_1/w_2 = 2.2 \cdot 1.27 = 2.8$, $w_1/w_3 = 2.7 \cdot 1 = 2.7$, $w_1/w_2 = 3.1 \cdot 1.10 = 3.4$, $w_3/w_2 = 1.7 \cdot 1.18 = 2$ and $w_1/w_2 = 1.6 \cdot 1.19 = 1.9$.

By applying Expression (5), the models for determining weight coefficients of the first-level criteria for each decision-maker can be defined:

DM_1 (First level)
$$\min \chi$$
$$s.t. \begin{cases} \left| \frac{w_1}{w_3} - 2.2 \right| = \chi, \quad \left| \frac{w_3}{w_2} - 1.27 \right| = \chi, \\ \left| \frac{w_1}{w_2} - 2.8 \right|, \\ \sum_{j=1}^{3} w_j = 1, \quad w_j \geq 0, \quad \forall j \end{cases}$$

DM_2 (First level)
$$\min \chi$$
$$s.t. \begin{cases} \left| \frac{w_1}{w_2} - 2.7 \right| = \chi, \quad \left| \frac{w_2}{w_3} - 1 \right| = \chi, \\ \left| \frac{w_1}{w_3} - 2.7 \right|, \\ \sum_{j=1}^{3} w_j = 1, \quad w_j \geq 0, \quad \forall j \end{cases}$$

DM_3 (First level)
$$\min \chi$$
$$s.t. \begin{cases} \left| \frac{w_1}{w_3} - 3.1 \right| = \chi, \quad \left| \frac{w_3}{w_2} - 1.10 \right| = \chi, \\ \left| \frac{w_1}{w_2} - 3.4 \right|, \\ \sum_{j=1}^{3} w_j = 1, \quad w_j \geq 0, \quad \forall j \end{cases}$$

DM_4 (First level)
$$\min \chi$$
$$s.t. \begin{cases} \left| \frac{w_3}{w_1} - 1.7 \right| = \chi, \quad \left| \frac{w_1}{w_2} - 1.18 \right| = \chi, \\ \left| \frac{w_3}{w_2} - 2 \right|, \\ \sum_{j=1}^{3} w_j = 1, \quad w_j \geq 0, \quad \forall j \end{cases}$$

DM_5 (First level)
$$\min \chi$$
$$s.t. \begin{cases} \left| \frac{w_1}{w_3} - 1.6 \right| = \chi, \quad \left| \frac{w_3}{w_2} - 1.19 \right| = \chi, \\ \left| \frac{w_1}{w_2} - 1.9 \right|, \\ \sum_{j=1}^{3} w_j = 1, \quad w_j \geq 0, \quad \forall j \end{cases}$$

By solving the models presented, the values of weight coefficients for the first-level criteria for every decision-maker are obtained, as shown in Table 3.

Table 3. Values of weight coefficients for the first level of decision-making according to each DM.

Id	DM_1	DM_2	DM_3	DM_4	DM_5	Final Rough Values
C_1	0.552	0.574	0.618	0.282	0.465	[0.394, 0.566]
C_2	0.197	0.213	0.182	0.239	0.244	[0.199, 0.230]
C_3	0.251	0.213	0.200	0.479	0.291	[0.228, 0.350]
DFC	0.0013	0.0000	0.0020	0.0015	0.0011	-

The final values shown in the last column of Table 3 are obtained by rough operations and the rough Dombi aggregator. First, the transformation of individual matrices into a group rough matrix is performed as follows:

$$\tilde{c}_1 = \{0.552, 0.574, 0.618, 0.282, 0.465\},$$

$$\underline{Lim}(0.552) = \frac{1}{3}(0.552 + 0.282 + 0.465) = 0.433, \quad \overline{Lim}(0.552) = \frac{1}{3}(0.552 + 0.574 + 0.618) = 0.581,$$

$$\underline{Lim}(0.574) = \frac{1}{4}(0.552 + 0.574 + 0.282 + 0.465) = 0.468, \quad \overline{Lim}(0.574) = \frac{1}{2}(0.574 + 0.618) = 0.596,$$

$$\underline{Lim}(0.618) = \frac{1}{5}(0.552 + 0.574 + 0.618 + 0.282 + 0.465) = 0.498, \quad \overline{Lim}(0.618) = 0.618,$$

$$\underline{Lim}(0.282) = 0.282, \quad \overline{Lim}(0.282) = \frac{1}{5}(0.552 + 0.574 + 0.618 + 0.282 + 0.465) = 0.498,$$

$$\underline{Lim}(0.465) = \frac{1}{2}(0.282 + 0.465) = 0.373, \quad \overline{Lim}(0.465) = \frac{1}{4}(0.552 + 0.574 + 0.618 + 0.465) = 0.552,$$

$$RN(c_1^1) = [0.433, 0.581]; \ RN(c_1^2) = [0.468, 0.596]; \ RN(c_1^3) = [0.498, 0.618];$$
$$RN(c_1^4) = [0.282, 0.498]; \ RN(c_1^5) = [0.373, 0.552].$$

Subsequently, the rough Dombi aggregator is applied and final rough values of the criteria at the first decision-making level are obtained. The aggregation is performed as follows.

After the transformation has been completed, five rough matrices, to which the operations of the rough Dombi aggregator is applied, are obtained. As mentioned in the previous part of the paper, the research has involved five experts who are assigned the same weight values of 0.200. Based on the presented values, Expression (8) from [56], and assuming that $\rho = 1$ is at the position of C_1, the aggregation of values is performed as follows:

$$RNDWGA(c_1) = \begin{cases} \underline{Lim}(c_1) = \dfrac{\sum_{j=1}^{5} \underline{Lim}(\varphi_j)}{1 + \left\{ \sum_{j=1}^{5} w_j \left(\dfrac{1 - f\left(\underline{Lim}(\varphi_j)\right)}{f\left(\underline{Lim}(\varphi_j)\right)} \right)^{\rho} \right\}^{1/\rho}} = \dfrac{2.052}{1 + \left(0.200 \times \left(\frac{1-0.210}{0.210}\right) + 0.200 \times \left(\frac{1-0.229}{0.229}\right) + \ldots + 0.200 \times \left(\frac{1-0.180}{0.180}\right)\right)} = 0.394 \\[2em] \overline{Lim}(c_1) = \dfrac{\sum_{j=1}^{5} \overline{Lim}(\varphi_j)}{1 + \left\{ \sum_{j=1}^{5} w_j \left(\dfrac{1 - f\left(\overline{Lim}(\varphi_j)\right)}{f\left(\overline{Lim}(\varphi_j)\right)} \right)^{\rho} \right\}^{1/\rho}} = \dfrac{2.848}{1 + \left(0.200 \times \left(\frac{1-0.204}{0.204}\right) + 0.200 \times \left(\frac{1-0.211}{0.211}\right) + \ldots + 0.200 \times \left(\frac{1-0.193}{0.193}\right)\right)} = 0.567 \end{cases}$$

Similarly, the decision-makers have ranked the criteria of the second level and the significance of criteria is obtained (Table 4).

Table 4. The ranking and significance of the second-level criteria for a group of economic factors.

DM₁						
Economic factors	C_{11}	C_{12}	C_{13}	C_{16}	C_{14}	C_{17}	C_{15}
$\omega_{C_{j(k)}}$	1	1.25	1.6	2.1	2.2	2.8	3.6

DM₂						
Economic factors	C_{12}	C_{11}	C_{16}	C_{13}	C_{14}	C_{17}	C_{15}
$\omega_{C_{j(k)}}$	1	1.1	2.1	2.3	3.1	3.3	3.8

DM₃						
Economic factors	C_{11}	C_{12}	C_{13}	C_{16}	C_{14}	C_{15}	C_{17}
$\omega_{C_{j(k)}}$	1	1	1.5	1.9	2.6	3.3	3.5

DM₄						
Economic factors	C_{12}	C_{11}	C_{13}	C_{14}	C_{16}	C_{17}	C_{15}
$\omega_{C_{j(k)}}$	1	1.05	1.9	2.3	2.4	2.8	3.7

DM₅						
Economic factors	C_{12}	C_{16}	C_{11}	C_{13}	C_{14}	C_{15}	C_{17}
$\omega_{C_{j(k)}}$	1	1.3	1.35	1.45	2	2.3	2.4

Based on the calculation, in the same way as with the criteria on the first level of decision-making, the calculation for the second decision-making level is made, and the values are shown in Table 5 for a group of economic criteria, in Tables 6 and 7 for a group of social criteria, and in Tables 8 and 9 for a group of environmental criteria.

Table 5. Values of weight coefficients for the second decision-making level according to each decision-maker for a group of economic criteria.

Id	DM_1	DM_2	DM_3	DM_4	DM_5	**Final Rough Values**
C_{11}	0.250	0.245	0.242	0.241	0.163	[0.203, 0.244]
C_{12}	0.200	0.269	0.242	0.252	0.219	[0.219, 0.253]
C_{13}	0.157	0.118	0.161	0.133	0.152	[0.131, 0.154]
C_{14}	0.114	0.087	0.093	0.110	0.110	[0.095, 0.109]
C_{15}	0.070	0.071	0.073	0.068	0.096	[0.071, 0.083]
C_{16}	0.119	0.128	0.127	0.106	0.169	[0.117, 0.145]
C_{17}	0.089	0.082	0.063	0.091	0.092	[0.074, 0.088]
DFC	0.0045	0.0005	0.0013	0.0017	0.0015	-

Table 6. The ranking and significance of the second-level criteria for a group of social factors.

DM_1							
Social factors	C_{22}	C_{21}	C_{26}	C_{23}	C_{27}	C_{24}	C_{25}
$\omega_{C_{j(k)}}$	1	1.3	1.5	1.9	2.3	2.4	3

DM_2							
Social factors	C_{22}	C_{21}	C_{26}	C_{24}	C_{23}	C_{27}	C_{25}
$\omega_{C_{j(k)}}$	1	1.15	1.8	2.1	2.6	2.85	3.2

DM_3							
Social factors	C_{21}	C_{22}	C_{26}	C_{23}	C_{24}	C_{27}	C_{25}
$\omega_{C_{j(k)}}$	1	1	1.4	2.2	2.3	2.9	3.8

DM_4							
Social factors	C_{21}	C_{22}	C_{26}	C_{23}	C_{27}	C_{24}	C_{25}
$\omega_{C_{j(k)}}$	1	1.2	1.55	2	2	2.1	2.7

DM_5							
Social factors	C_{22}	C_{21}	C_{26}	C_{23}	C_{27}	C_{24}	C_{25}
$\omega_{C_{j(k)}}$	1	1.15	1.7	1.8	2.35	2.65	2.95

Table 7. Values of weight coefficients for the second decision-making level according to each decision-maker for a group of social criteria.

Id	DM_1	DM_2	DM_3	DM_4	DM_5	**Final Rough Values**
C_{21}	0.185	0.220	0.237	0.233	0.210	[0.204, 0.231]
C_{22}	0.241	0.253	0.237	0.189	0.241	[0.216, 0.244]
C_{23}	0.127	0.097	0.108	0.115	0.134	[0.107, 0.127]
C_{24}	0.101	0.121	0.104	0.112	0.090	[0.100, 0.114]
C_{25}	0.081	0.079	0.063	0.086	0.081	[0.074, 0.081]
C_{26}	0.161	0.141	0.169	0.149	0.142	[0.144, 0.161]
C_{27}	0.105	0.089	0.082	0.115	0.102	[0.090, 0.107]
DFC	0.0016	0.0022	0.0001	0.020	0.0013	-

Table 8. The ranking and significance of the second-level criteria for a group of environmental factors.

				DM_1			
Environmental factors	C_{33}	C_{35}	C_{32}	C_{31}	C_{34}	C_{36}	C_{37}
$\omega_{C_{j(k)}}$	1	1.2	1.3	1.75	2.2	2.7	3.6
				DM_2			
Environmental factors	C_{33}	C_{35}	C_{31}	C_{32}	C_{36}	C_{34}	C_{37}
$\omega_{C_{j(k)}}$	1	1.1	1.2	2.3	2.4	3.15	3.45
				DM_3			
Environmental factors	C_{33}	C_{35}	C_{32}	C_{31}	C_{34}	C_{36}	C_{37}
$\omega_{C_{j(k)}}$	1	1.4	1.7	2.1	2.2	2.7	2.8
				DM_4			
Environmental factors	C_{35}	C_{33}	C_{32}	C_{34}	C_{36}	C_{31}	C_{37}
$\omega_{C_{j(k)}}$	1	1.3	1.4	1.8	2.2	2.3	2.6
				DM_5			
Environmental factors	C_{33}	C_{35}	C_{32}	C_{31}	C_{34}	C_{36}	C_{37}
$\omega_{C_{j(k)}}$	1	1.2	1.5	1.6	2.4	2.7	3.2

Table 9. Values of weight coefficients for the second decision-making level according to each decision-maker for a group of environmental criteria.

Id	DM_1	DM_2	DM_3	DM_4	DM_5	Final Rough Values
C_{31}	0.134	0.198	0.121	0.101	0.147	[0.119, 0.161]
C_{32}	0.180	0.103	0.148	0.165	0.158	[0.130, 0.165]
C_{33}	0.234	0.238	0.252	0.178	0.237	[0.210, 0.242]
C_{34}	0.106	0.076	0.115	0.128	0.098	[0.090, 0.115]
C_{35}	0.195	0.216	0.180	0.231	0.197	[0.192, 0.214]
C_{36}	0.086	0.099	0.094	0.105	0.088	[0.089, 0.099]
C_{37}	0.065	0.069	0.090	0.091	0.074	[0.071, 0.086]
DFC	0.0012	0.0014	0.0017	0.0020	0.0014	-

Based on the significance of the groups of criteria (economic, social, and environmental) and applying Equation (5), the models for each decision-maker are formed. Solving these models, we obtain the values of weight coefficients per decision-makers (Table 10).

Table 10. Final values of criteria.

Main criteria	Values of Main Criteria	Subcriteria	Values of Subcriteria	Final Weights
		w_{11}	[0.203, 0.244]	[0.080, 0.138]
		w_{12}	[0.219, 0.253]	[0.086, 0.144]
		w_{13}	[0.131, 0.154]	[0.052, 0.087]
Economic	[0.203, 0.244]	w_{14}	[0.095, 0.109]	[0.037, 0.062]
		w_{15}	[0.071, 0.083]	[0.028, 0.047]
		w_{16}	[0.117, 0.145]	[0.046, 0.082]
		w_{17}	[0.074, 0.088]	[0.029, 0.050]

Table 10. *Cont.*

Main criteria	Values of Main Criteria	Subcriteria	Values of Subcriteria	Final Weights
		w_{21}	[0.204, 0.231]	[0.041, 0.053]
		w_{22}	[0.216, 0.244]	[0.043, 0.056]
		w_{23}	[0.107, 0.127]	[0.021, 0.029]
Social	[0.199, 0.230]	w_{24}	[0.100, 0.114]	[0.020, 0.026]
		w_{25}	[0.074, 0.081]	[0.015, 0.019]
		w_{26}	[0.144, 0.161]	[0.029, 0.037]
		w_{27}	[0.090, 0.107]	[0.018, 0.025]
		w_{31}	[0.119, 0.161]	[0.027, 0.057]
		w_{32}	[0.130, 0.165]	[0.030, 0.058]
		w_{33}	[0.210, 0.242]	[0.048, 0.085]
Environmental	[0.228, 0.350]	w_{34}	[0.090, 0.115]	[0.021, 0.040]
		w_{35}	[0.192, 0.214]	[0.044, 0.075]
		w_{36}	[0.089, 0.099]	[0.020, 0.035]
		w_{37}	[0.071, 0.086]	[0.016, 0.030]

The final values of weight coefficients by all criteria are obtained by multiplying the weight coefficients of the main criteria with the subcriteria of the group to which they belong. As can be seen from Table 10, the most important criteria belong to the group of economic and then environmental criteria, which is understandable with regard to the area of existence of the company in which the research has been carried out.

4.2. Ranking Alternatives Using a New Rough COPRAS Method

Table 11 presents the evaluation of alternatives according to all criteria based on the linguistic scale 1–7. In evaluating the alternatives, five decision-makers participated, whose expertise has already been described in the previous section.

Table 11. Comparison of alternatives by five decision-makers.

Id	A_1					A_2					A_3					A_4					A_5				
	E_1	E_2	E_3	E_4	E_5	E_1	E_2	E_3	E_4	E_5	E_1	E_2	E_3	E_4	E_5	E_1	E_2	E_3	E_4	E_5	E_1	E_2	E_3	E_4	E_5
C_{11}	3	2	3	2	3	4	3	2	3	2	3	3	1	2	2	5	4	3	4	4	6	3	3	4	3
C_{12}	4	4	5	4	4	5	6	4	4	4	5	7	6	5	6	4	4	5	4	4	4	3	6	3	5
C_{13}	5	5	4	4	4	3	5	5	3	3	5	5	6	5	5	3	4	4	4	4	6	4	4	4	4
C_{14}	4	3	3	3	4	5	6	7	5	7	4	6	7	6	7	7	5	6	6	5	6	5	7	5	6
C_{15}	4	6	4	5	3	5	6	5	5	5	5	6	6	5	6	6	5	5	6	5	7	5	5	5	5
C_{16}	6	6	7	6	6	5	6	7	4	7	5	5	6	5	6	4	4	5	4	4	4	4	5	4	7
C_{17}	5	4	4	3	5	5	7	6	5	6	5	6	6	6	6	6	4	6	6	4	7	4	6	4	6
C_{21}	5	4	4	4	5	4	7	7	4	7	4	6	5	6	5	4	4	5	4	6	4	4	5	4	5
C_{22}	5	3	3	3	5	5	6	6	4	7	5	6	5	7	5	4	4	6	4	4	4	4	5	4	7
C_{23}	4	3	4	3	4	4	5	6	4	6	5	6	6	6	5	5	4	5	5	4	5	4	5	4	5
C_{24}	4	4	4	4	4	5	5	6	5	6	4	5	6	5	6	6	4	5	5	4	5	4	5	4	5
C_{25}	5	3	4	3	5	6	6	6	7	6	6	6	6	6	6	5	4	4	7	6	5	6	4	5	4
C_{26}	5	2	2	3	5	5	5	6	5	7	4	5	6	5	6	4	4	4	4	4	5	4	4	4	3
C_{27}	5	4	5	5	5	5	5	6	5	6	6	5	7	6	7	4	4	5	4	4	4	4	5	4	5
C_{31}	5	3	4	3	5	4	6	7	6	7	6	5	7	5	7	5	5	6	3	5	5	5	6	7	6
C_{32}	5	6	5	3	5	5	6	6	5	6	7	5	7	4	5	4	4	5	4	4	4	4	5	4	5
C_{33}	4	3	5	3	4	6	6	7	6	7	7	5	7	5	7	6	5	5	4	5	4	5	5	5	5
C_{34}	3	3	5	2	3	5	5	6	5	7	6	6	6	4	6	4	3	5	4	5	4	3	5	3	5
C_{35}	4	4	5	4	4	5	6	6	5	6	6	6	6	6	6	4	4	5	4	4	4	4	5	3	5
C_{36}	3	5	3	4	3	6	6	5	7	5	5	5	6	5	4	3	4	4	3	4	4	4	4	4	4
C_{37}	5	3	4	4	5	5	5	6	5	5	7	5	6	5	4	6	4	5	6	4	6	4	5	4	5

In order to be able to apply the developed methodology, the transformation of individual matrices into a group rough matrix is performed first. An example of calculating the value of the third alternative according to criterion C_{11} is given below:

$$\tilde{c}_{11} = \{3,3,1,2,2\},$$

$$\underline{Lim}(3) = \tfrac{1}{5}(3+3+1+2+2) = 2.2, \ \overline{Lim}(3) = 3, \ \underline{Lim}(1) = 1, \ \overline{Lim}(1) = \tfrac{1}{5}(3+3+1+2+2) = 2.2,$$
$$\underline{Lim}(2) = \tfrac{1}{3}(1+2+2) = 1.67, \ \overline{Lim}(2) = \tfrac{1}{4}(3+3+2+2) = 2.5$$

$$RN\left(A_3^1\right) = RN\left(A_3^2\right) = [2.2,3]; \ RN\left(A_3^3\right) = [1,2.2]; \ RN\left(A_3^4\right) = RN\left(A_3^5\right) = [1.67,2.5].$$

Subsequently, the rough Dombi aggregator is applied and the final rough values of alternatives are obtained. The aggregation on the same example of the third alternative for criterion C_{11} is carried out as follows:

$$RNDWGA(A_3) = \begin{cases} \underline{Lim}(A_3) = \dfrac{\sum_{j=1}^{5} \underline{Lim}(\varphi_j)}{1+\left\{\sum_{j=1}^{5} w_j \left(\dfrac{1-f\left(\underline{Lim}(\varphi_j)\right)}{f\left(\underline{Lim}(\varphi_j)\right)}\right)^{\rho}\right\}^{1/\rho}} = \dfrac{8.74}{1+\left(0.200\times\left(\frac{1-0.252}{0.252}\right)+0.200\times\left(\frac{1-0.252}{0.252}\right)+...+0.200\times\left(\frac{1-0.191}{0.191}\right)\right)} = 1.609 \\[4ex] \overline{Lim}(A_3) = \dfrac{\sum_{j=1}^{5} \overline{Lim}(\varphi_j)}{1+\left\{\sum_{j=1}^{5} w_j \left(\dfrac{1-f\left(\overline{Lim}(\varphi_j)\right)}{f\left(\overline{Lim}(\varphi_j)\right)}\right)^{\rho}\right\}^{1/\rho}} = \dfrac{13.2}{1+\left(0.200\times\left(\frac{1-0.227}{0.227}\right)+0.200\times\left(\frac{1-0.227}{0.227}\right)+...+0.200\times\left(\frac{1-0.189}{0.189}\right)\right)} = 2.603 \end{cases}$$

In the same way, other values for all alternatives are obtained according to all the criteria, which creates the initial aggregated matrix shown in Table 12.

Table 12. Initial aggregated rough matrix.

Id	C_{11}		C_{12}		C_{21}		C_{22}		C_{23}		C_{31}	
A_1	2.321	2.826	4.038	4.339	4.151	4.622	3.276	4.204	3.333	3.830	3.408	4.488
A_2	2.318	3.195	4.157	4.997	4.915	6.465	4.846	6.219	4.421	5.496	5.216	6.578
A_3	1.609	2.603	5.342	6.222	4.704	5.623	5.160	6.007	5.344	5.833	5.429	6.502
A_4	3.614	4.326	4.038	4.339	4.157	4.997	4.074	4.648	4.038	4.339	4.138	5.283
A_5	3.183	4.325	3.397	4.880	4.151	4.622	4.189	5.350	3.691	4.619	5.342	6.222
Σ	13.05	17.27	20.97	24.78	22.08	26.33	21.54	26.43	20.83	24.12	23.53	29.07

The summarized values for each criterion, which are necessary for the application of normalization in the next step, are shown in the last row of Table 12. Applying Equation (7), the normalized value for the third alternative according to criterion C_{11} will be

$$r_{31} = \left[\frac{1.609}{17.27}; \frac{2.603}{13.05}\right] = [0.09, 0.20].$$

The last row of Table 13 presents the values of the criteria obtained by applying the FUCOM method, which are necessary to create a weighted normalized matrix.

Table 13. Normalized rough matrix.

Id	C_{11}		C_{12}		C_{21}		C_{22}		C_{23}		C_{31}	
A_1	0.13	0.22	0.16	0.21	0.16	0.21	0.12	0.20	0.14	0.18	0.12	0.19
A_2	0.13	0.24	0.17	0.24	0.19	0.29	0.18	0.29	0.18	0.26	0.18	0.28
A_3	0.09	0.20	0.22	0.30	0.18	0.25	0.20	0.28	0.22	0.28	0.19	0.28
A_4	0.21	0.33	0.16	0.21	0.16	0.23	0.15	0.22	0.17	0.21	0.14	0.22
A_5	0.18	0.33	0.14	0.23	0.16	0.21	0.16	0.25	0.15	0.22	0.18	0.26
w_j	0.080	0.138	0.086	0.144	0.041	0.053	0.043	0.056	0.021	0.029	0.027	0.057

The fourth step is the weighting of normalized rough matrix (Table 14) by multiplying all the values of the normalized matrix with the weights of the criteria by applying Equation (8).

$$d_{31} = [0.09 \times 0.08; 0.02 \times 0.138] = [0.007, 0.028]$$

Table 14. Weighted normalized rough matrix.

Id	C_{11}		C_{12}		C_{21}		C_{22}		C_{23}		C_{31}	
A_1	0.13	0.22	0.16	0.21	0.16	0.21	0.12	0.20	0.14	0.18	0.12	0.19
A_2	0.13	0.24	0.17	0.24	0.19	0.29	0.18	0.29	0.18	0.26	0.18	0.28
A_3	0.09	0.20	0.22	0.30	0.18	0.25	0.20	0.28	0.22	0.28	0.19	0.28
A_4	0.21	0.33	0.16	0.21	0.16	0.23	0.15	0.22	0.17	0.21	0.14	0.22
A_5	0.18	0.33	0.14	0.23	0.16	0.21	0.16	0.25	0.15	0.22	0.18	0.26

The next step is to summarize the values of the alternatives depending on the type of criteria, and two matrices are obtained. The first matrix refers to the sum of the values of alternatives according to the benefit group of criteria, while another one refers to cost criteria. In this research, cost criteria are C_{11} and C_{36}.

The matrix for alternatives according to benefit criteria is

$$S_i^+ = \begin{bmatrix} 0.096, & 0.215 \\ 0.122, & 0.274 \\ 0.131, & 0.285 \\ 0.105, & 0.226 \\ 0.104, & 0.240 \end{bmatrix}.$$

An example of the calculation for the third alternative is

$$S_3^{L+} = \begin{bmatrix} 0.019 + 0.011 + 0.007 + 0.005 + 0.008 + 0.006 + 0.007 + 0.008 + 0.005 + 0.004 \\ +0.003 + 0.006 + 0.004 + 0.005 + 0.005 + 0.010 + 0.004 + 0.010 + 0.003 \end{bmatrix} = [0.131],$$

$$S_3^{U+} = \begin{bmatrix} 0.043 + 0.023 + 0.017 + 0.011 + 0.019 + 0.012 + 0.014 + 0.016 + 0.008 + 0.007 \\ +0.005 + 0.010 + 0.007 + 0.016 + 0.016 + 0.023 + 0.012 + 0.019 + 0.007 \end{bmatrix} = [0.285].$$

The matrix for alternatives according to cost criteria is

$$S_i^- = \begin{bmatrix} 0.014, & 0.037 \\ 0.015, & 0.044 \\ 0.011, & 0.037 \\ 0.020, & 0.052 \\ 0.018, & 0.053 \end{bmatrix}.$$

An example of calculation is as follows:

$$S_3^{L-} = [0.007 + 0.004] = [0.011],$$

$$S_3^{U-} = [0.028 + 0.009] = [0.037].$$

After that, it is necessary to calculate the inverse values of the matrix S_i^- by applying Equation (11), which is

$$\left(S_i^-\right)^{-1} = \begin{bmatrix} 27.289, & 74.026 \\ 22.553, & 64.948 \\ 27.353, & 87.059 \\ 19.137, & 50.864 \\ 19.035, & 54.824 \end{bmatrix}$$

In the next step, it is first necessary to calculate the sum by column for cost criteria applying Equation (12), and the following values are obtained:

$$\left(\overline{S_i^-}\right) = \sum \left[s_i^{-L}; s_i^{-U}\right] = [0.078, 0.222],$$

and then applying Equation (13) to calculate the sum for the inverse matrix, which will be

$$\left(\overline{S_i^-}\right)^{-1} = \sum \left[\frac{1}{s_{ij}^{-U}}; \frac{1}{s_{ij}^{-L}}\right] = [115.367, 331.722].$$

In the next step, it is necessary to determine the relative significance for each alternative. The relative weight Q_i by alternatives is

$$Q_i = \begin{bmatrix} 0.102, & 0.357 \\ 0.127, & 0.399 \\ 0.137, & 0.453 \\ 0.109, & 0.324 \\ 0.108, & 0.346 \end{bmatrix}.$$

The ith alternative is calculated using Equation (14). An example of the calculation for the third alternative is

$$Q_3^L = S_{3+}^L + \frac{\left(S_3^{-L}\right)}{S_{3-}^U \times \left(S_3^{-U}\right)^{-1}} = 0.131 + \frac{0.078}{0.037 \times 331.722} = 0.137,$$

$$Q_3^U = S_{3+}^U + \frac{\left(S_3^{-U}\right)}{S_{3-}^L \times \left(S_3^{-L}\right)^{-1}} = 0.285 + \frac{0.222}{0.011 \times 115.367} = 0.453.$$

In the last step, the alternatives are ranked from the highest to the lowest value, and the results are as follows: $A_3 > A_2 > A_1 > A_5 > A_4$.

5. Sensitivity Analysis and Discussion

The sensitivity analysis has been performed throughout four phases, the first of which involves the creation of nine scenarios where the weights of criteria are modeled. The second phase involves the application of different methods, that is, comparative analysis, while the third phase implies the change of the parameter ρ into the values of 1–10. The fourth phase includes the application of Spearman's correlation coefficient for the ranks of alternatives throughout the first two phases.

Figure 2 presents the ranks of alternatives throughout nine scenarios. The first scenario implies that all criteria are equally important, while in the second one, the six most important criteria (C_{11}, C_{12}, C_{13}, C_{16}, C_{22}, C_{33}) are reduced by 4%, and others are increased by 2%. In the third set, the six most important criteria are eliminated, and in the fourth one, the most important criteria are increased by 4%, while the rest are reduced by 2%. The fifth scenario involves the elimination of seven least important criteria (C_{23}, C_{24}, C_{25}, C_{27}, C_{34}, C_{36}, and C_{37}). In the sixth set, the criteria that belong to the economic group are reduced by 4%, while the criteria of the social group are proportionally increased. The values of environmental criteria remain unchanged. The seventh set implies a reverse situation from the aspect of economic and social criteria in relation to the sixth set. In the eighth scenario, decision-making is based only on economic criteria, and in the ninth scenario, only on environmental criteria.

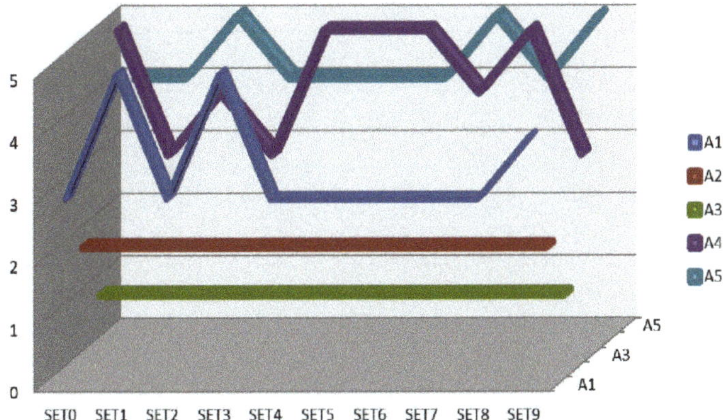

Figure 2. Sensitivity analysis by changing the weight values of criteria.

The ranks of alternatives do not change in the fourth, fifth, sixth, and eighth criteria, which implies that the most important criteria play a very important role in the decision-making process in this research. This is confirmed by the fact that there are significant changes in the rankings in the first and third sets when all the criteria are equal, i.e., when the six most important ones are eliminated. In other scenarios there are no significant changes. It is important to emphasize that the two alternatives that represent the best solution, A_3 and A_2, do not change ranks in any scenario, which implies that they are insensitive to the changes in the significance of the criteria.

Figure 3 shows the comparison of the proposed model with other approaches developed recently: rough WASPAS [57], rough MABAC [58], rough SAW [59], and rough ARAS [60].

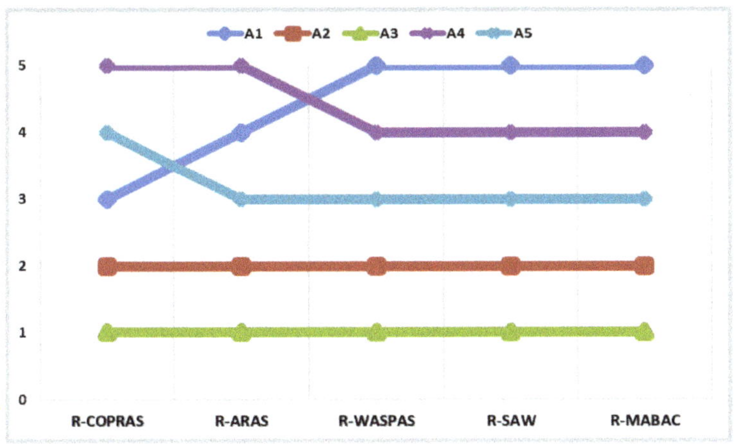

Figure 3. Comparison of the results of the developed model with other methods.

Observing the results obtained by other methods, the stability of the two best alternatives do not come into question, since they continue to take the first two positions using all the methods. The highest similarities of ranks obtained with rough COPRAS have the alternatives obtained with rough

ARAS, where only the first and fifth alternatives change their positions. Slightly bigger changes in ranks are found with other methods.

Table 15 presents the part of the sensitivity analysis that relates to the change of parameter ρ.

Table 15. Ranks of alternatives depending on the change of parameter ρ.

Parameter ρ	Q_i	Rank
$\rho = 1$	$Q_1 = 0.141$; $Q_2 = 0.148$; $Q_3 = 0.171$; $Q_4 = 0.130$; $Q_5 = 0.133$	$A_3 > A_2 > A_1 > A_5 > A_4$
$\rho = 2$	$Q_1 = 0.141$; $Q_2 = 0.149$; $Q_3 = 0.174$; $Q_4 = 0.130$; $Q_5 = 0.133$	$A_3 > A_2 > A_1 > A_5 > A_4$
$\rho = 3$	$Q_1 = 0.142$; $Q_2 = 0.150$; $Q_3 = 0.178$; $Q_4 = 0.131$; $Q_5 = 0.133$	$A_3 > A_2 > A_1 > A_5 > A_4$
$\rho = 4$	$Q_1 = 0.143$; $Q_2 = 0.150$; $Q_3 = 0.182$; $Q_4 = 0.132$; $Q_5 = 0.134$	$A_3 > A_2 > A_1 > A_5 > A_4$
$\rho = 5$	$Q_1 = 0.144$; $Q_2 = 0.151$; $Q_3 = 0.185$; $Q_4 = 0.132$; $Q_5 = 0.134$	$A_3 > A_2 > A_1 > A_5 > A_4$
$\rho = 6$	$Q_1 = 0.144$; $Q_2 = 0.151$; $Q_3 = 0.187$; $Q_4 = 0.133$; $Q_5 = 0.134$	$A_3 > A_2 > A_1 > A_5 > A_4$
$\rho = 7$	$Q_1 = 0.145$; $Q_2 = 0.152$; $Q_3 = 0.189$; $Q_4 = 0.134$; $Q_5 = 0.134$	$A_3 > A_2 > A_1 > A_5 = A_4$
$\rho = 8$	$Q_1 = 0.146$; $Q_2 = 0.152$; $Q_3 = 0.191$; $Q_4 = 0.134$; $Q_5 = 0.134$	$A_3 > A_2 > A_1 > A_5 = A_4$
$\rho = 9$	$Q_1 = 0.146$; $Q_2 = 0.153$; $Q_3 = 0.192$; $Q_4 = 0.135$; $Q_5 = 0.134$	$A_3 > A_2 > A_1 > A_5 > A_4$
$\rho = 10$	$Q_1 = 0.147$; $Q_2 = 0.153$; $Q_3 = 0.194$; $Q_4 = 0.135$; $Q_5 = 0.135$	$A_3 > A_2 > A_1 > A_5 = A_4$

Changing the parameter ρ does not change significantly the initial results obtained. For the parameters $\rho = 1$–6, the same ranks are obtained as with the hybrid FUCOM–rough COPRAS model. The only changes in ranks are for parameters $\rho = 7$, $\rho = 8$, and $\rho = 10$ when the fourth and fifth alternative belongs to the same rank, and when $\rho = 9$, the fourth and fifth alternative change their positions while others remain unchanged. Based on the overall sensitivity analysis with the change of parameter ρ, it can be concluded that the model is not sensitive to these changes.

At the end of the sensitivity analysis, the calculated Spearman's correlation coefficient for the first two phases is given (Table 16). For the third phase, calculation is not performed, since it is obvious that there is almost a complete correlation and, as already mentioned, the change of this parameter does not significantly affect the ranking of the alternatives.

Table 16. Spearman's correlation coefficient for the first two phases of sensitivity analysis.

First Phase									
SET	SET1	SET2	SET3	SET4	SET5	SET6	SET7	SET8	SET9	Average
SET0	0.600	0.900	0.600	1.000	1.000	1.000	0.900	1.000	0.700	0.856

Second Phase					
Methods	R-COPRAS	R-ARAS	R-WASPAS	R-SAW	R-MABAC
R-COPRAS	1.000	0.900	0.700	0.700	0.700
R-ARAS		1.000	0.900	0.900	0.900
R-WASPAS			1.000	1.000	1.000
R-SAW				1.000	1.000
R-MABAC					1.000
Overall average					0.920

Concerning the first phase of the sensitivity analysis in which the weights of the criteria change in sets, it can be seen that the model is sensitive to their changes. The initial set has a full correlation with four sets (4, 5, 6, and 8), while the smallest correlation SCC = 0.600 is with the first and third set, in which the ranks of two alternatives change for a total of three positions. In the second and seventh sets, the two last alternatives change positions between each other, so SCC = 0.900 with the initial set. In the ninth set, there is a change in the rank of three alternatives with SCC = 0.700. The total average value of SCC is 0.856, which represents a high correlation of ranks, regardless of the changes mentioned.

In the second phase, it can be observed that rough COPRAS has the highest correlation with the rough ARAS method of 0.900, while with other methods, rough WASPAS, rough SAW, and rough MABAC, SCC = 0.700. Taking this into account, it is concluded that rough WASPAS, rough SAW, and rough MABAC have a complete correlation, which ultimately implies that the average SCC = 0.920, which is a very high correlation of ranks.

6. Conclusions

This paper has proposed a new hybrid model that integrates FUCOM with the rough COPRAS method using the rough Dombi aggregator. This is the first time in the literature that this kind of model has been applied, that integrates the positive aspects of FUCOM method, rough set theory, rough Dombi aggregator for group decision-making, and the COPRAS method, which is one of the main contributions of this paper. In addition, the detailed and demonstrated algorithm of the rough COPRAS method also contributes to the overall field of multi-criteria decision-making.

Based on the 21 criteria of sustainability, a total of five suppliers in a construction company were considered, where it was concluded that the third and second suppliers are the best solutions regardless of any change in the model. This has been proven throughout a comprehensive sensitivity analysis in which different scenarios—with a change in the weight of criteria—were formed. The two mentioned alternatives are not sensitive to any changes in the values of the criteria. In addition, neither the change of parameter ρ, which is an integral part of the rough Dombi aggregator, affects the rankings of the third and second supplier, which has been confirmed by comparison with other approaches. The best solution in this model is completely insensitive, i.e., stable, while the ranks of other alternatives vary depending on the method of modeling the sensitivity analysis.

The developed model can be useful in other areas of engineering, but also when making real life decisions, since it adequately treats uncertainties by applying the theory of rough sets and subjectivity by applying the FUCOM method. Thus, it is possible to make more accurate and valid decisions that can have a huge impact on a sustainable supply chain. Future research related to this study will address the development and application of a similar model with the FUCOM method and an uncertainty theory, e.g., grey theory.

Author Contributions: Each author has participated and contributed sufficiently to take public responsibility for appropriate portions of the content. Conceptualization, S.S. and B.M.; methodology, Ž.S., S.S. and E.K.Z.; validation, D.K.D., and S.J.; formal analysis, Ž.S. and E.K.Z.; investigation, S.J. and M.M.; writing—original draft preparation, M.M.; writing—review and editing, E.K.Z. and D.K.D.; supervision, B.M.; project administration, B.M.

Funding: This research received no external funding.

Acknowledgments: The authors acknowledge the support of research projects TR 36017, funded by the Ministry of Science and Technological Development of Serbia.

Conflicts of Interest: The authors declare no conflict of interest.

References

1. Hutchins, M.J.; Sutherland, J.W. An exploration of measures of social sustainability and their application to supply chain decisions. *J. Clean. Prod.* **2008**, *16*, 1688–1698. [CrossRef]
2. Vanalle, R.M.; Ganga, G.M.D.; Godinho Filho, M.; Lucato, W.C. Green supply chain management: An investigation of pressures, practices, and performance within the Brazilian automotive supply chain. *J. Clean. Prod.* **2017**, *151*, 250–259. [CrossRef]
3. Sen, D.K.; Datta, S.; Mahapatra, S.S. Sustainable supplier selection in intuitionistic fuzzy environment: A decision-making perspective. *Benchmarking Int. J.* **2018**, *25*, 545–574. [CrossRef]
4. Seuring, S. A review of modeling approaches for sustainable supply chain management. *Decis. Support Syst.* **2013**, *54*, 1513–1520. [CrossRef]
5. Rabbani, M.; Foroozesh, N.; Mousavi, S.M.; Farrokhi-Asl, H. Sustainable supplier selection by a new decision model based on interval-valued fuzzy sets and possibilistic statistical reference point systems under uncertainty. *Int. J. Syst. Sci. Oper. Logist.* **2017**, 1–17. [CrossRef]
6. Singh, A.; Trivedi, A. Sustainable green supply chain management: trends and current practices. *Compet. Rev.* **2016**, *26*, 265–288. [CrossRef]

7. Zavadskas, E.K.; Govindan, K.; Antucheviciene, J.; Turskis, Z. Hybrid multiple criteria decision-making methods: A review of applications for sustainability issues. *Econ. Res. Ekon. Istraživanja* **2016**, *29*, 857–887. [CrossRef]

8. Zavadskas, E.; Antucheviciene, J.; Vilutiene, T.; Adeli, H. Sustainable decision-making in civil engineering, construction and building technology. *Sustainability* **2018**, *10*, 14. [CrossRef]

9. Mardani, A.; Jusoh, A.; Halicka, K.; Ejdys, J.; Magruk, A.; Ahmad, U.N. Determining the utility in management by using multi-criteria decision support tools: a review. *Econ. Res. Ekon. Istraživanja* **2018**, 1–51. [CrossRef]

10. Lombera, J.T.S.; Rojo, J.C. Industrial building design stage based on a system approach to their environmental sustainability. *Constr. Build. Mater.* **2010**, *24*, 438–447. [CrossRef]

11. Del Cano, A.; Gomez, D.; de la Cruz, M.P. Uncertainty analysis in the sustainable design of concrete structures: A probabilistic method. *Constr. Build. Mater.* **2012**, *37*, 865–873. [CrossRef]

12. De la Fuente, A.; Pons, O.; Josa, A.; Aguado, A. Multi-Criteria Decision Making in the sustainability assessment of sewerage pipe systems. *J. Clean. Prod.* **2016**, *112*, 4762–4770. [CrossRef]

13. De la Fuente, A.; Blanco, A.; Armengou, J.; Aguado, A. Sustainability based-approach to determine the concrete type and reinforcement configuration of TBM tunnels linings. Case study: Extension line to Barcelona Airport T1. *Tunn. Undergr. Space Technol.* **2017**, *61*, 179–188. [CrossRef]

14. Rashidi, M.; Ghodrat, M.; Samali, B.; Kendall, B.; Zhang, C.W. Remedial Modelling of Steel Bridges through Application of Analytical Hierarchy Process (AHP). *Appl. Sci.* **2017**, *7*. [CrossRef]

15. Jia, J.M.; Ibrahim, M.; Hadi, M.; Orabi, W.; Xiao, Y. Multi-Criteria Evaluation Framework in Selection of Accelerated Bridge Construction (ABC) Method. *Sustainability* **2018**, *10*. [CrossRef]

16. Formisano, A.; Mazzolani, F.M. On the selection by MCDM methods of the optimal system for seismic retrofitting and vertical addition of existing buildings. *Comput. Struct.* **2015**, *159*, 1–13. [CrossRef]

17. Terracciano, G.; Di Lorenzo, G.; Formisano, A.; Landolfo, R. Cold-formed thin-walled steel structures as vertical addition and energetic retrofitting systems of existing masonry buildings. *Eur. J. Environ. Civ. Eng.* **2015**, *19*, 850–866. [CrossRef]

18. Siozinyte, E.; Antucheviciene, J.; Kutut, V. Upgrading the old vernacular building to contemporary norms: multiple criteria approach. *J. Civ. Eng. Manag.* **2014**, *20*, 291–298. [CrossRef]

19. Khoshnava, S.M.; Rostami, R.; Valipour, A.; Ismail, M.; Rahmat, A.R. Rank of green building material criteria based on the three pillars of sustainability using the hybrid multi criteria decision making method. *J. Clean. Prod.* **2018**, *173*, 82–99. [CrossRef]

20. Akadiri, P.O.; Olomolaiye, P.O.; Chinyio, E.A. Multi-criteria evaluation model for the selection of sustainable materials for building projects. *Autom. Constr.* **2013**, *30*, 113–125. [CrossRef]

21. Ozcan-Deniz, G.; Zhu, Y.M. A multi-objective decision-support model for selecting environmentally conscious highway construction methods. *J. Civ. Eng. Manag.* **2015**, *21*, 733–747. [CrossRef]

22. Raslanas, S.; Kliukas, R.; Stasiukynas, A. Sustainability assessment for recreational buildings. *Civ. Eng. Environ. Syst.* **2016**, *33*, 286–312. [CrossRef]

23. Kumar, A.; Sah, B.; Singh, A.R.; Deng, Y.; He, X.; Kumar, P.; Bansal, R.C. A review of multi criteria decision making (MCDM) towards sustainable renewable energy development. *Renew. Sustain. Energy Rev.* **2017**, *69*, 596–609. [CrossRef]

24. Streimikiene, D.; Balezentis, T.; Krisciukaitiené, I.; Balezentis, A. Prioritizing sustainable electricity production technologies: MCDM approach. *Renew. Sustain. Energy Rev.* **2012**, *16*, 3302–3311. [CrossRef]

25. An, D.; Xi, B.; Ren, J.; Ren, X.; Zhang, W.; Wang, Y.; Dong, L. Multi-criteria sustainability assessment of urban sludge treatment technologies: Method and case study. *Resour. Conserv. Recycl.* **2018**, *128*, 546–554. [CrossRef]

26. Sałabun, W.; Wątróbski, J.; Piegat, A. Identification of a multi-criteria model of location assessment for renewable energy sources. In Proceedings of the International Conference on Artificial Intelligence and Soft Computing, Zakopane, Poland, 12–16 June 2016; Springer: Cham, Switzerland, 2016; pp. 321–332.

27. Watróbski, J.; Sałabun, W. The characteristic objects method: A new intelligent decision support tool for sustainable manufacturing. In Proceedings of the International Conference on Sustainable Design and Manufacturing, Heraklion, Greece, 4–6 April 2016; pp. 349–359.

28. Wątróbski, J.; Sałabun, W.; Karczmarczyk, A.; Wolski, W. Sustainable decision-making using the COMET method: An empirical study of the ammonium nitrate transport management. In Proceedings of the 2017 Federated Conference on Computer Science and Information Systems (FedCSIS), Prague, Czech Republic, 3–6 September 2017; pp. 949–958.

29. Stević, Ž.; Vasiljević, M.; Puška, A.; Tanackov, I.; Junevičius, R.; Vesković, S. Evaluation of suppliers under uncertainty: a multiphase approach based on fuzzy AHP and fuzzy EDAS. *Transport* **2019**, *34*, 52–66. [CrossRef]

30. Badi, I.; Ballem, M. Supplier selection using the rough BWM-MAIRCA model: A case study in pharmaceutical supplying in Libya. *Decis. Mak. Appl. Manag. Eng.* **2018**, *1*, 16–33. [CrossRef]

31. Badi, I.; Abdulshahed, A.; Shetwan, A. A case study of supplier selection for steelmaking company in Libya by using Combinative Distance-based ASsessemnt (CODAS) model. *Decis. Mak. Appl. Manag. Eng.* **2018**, *1*, 1–12. [CrossRef]

32. Yazdani, M.; Chatterjee, P.; Zavadskas, E.K.; Zolfani, S.H. Integrated QFD-MCDM framework for green supplier selection. *J. Clean. Prod.* **2017**, *142*, 3728–3740. [CrossRef]

33. Keshavarz Ghorabaee, M.; Amiri, M.; Zavadskas, E.K.; Antucheviciene, J. Supplier evaluation and selection in fuzzy environments: A review of MADM approaches. *Econ. Res. Ekon. Istraživanja* **2017**, *30*, 1073–1118. [CrossRef]

34. Govindan, K.; Khodaverdi, R.; Jafarian, A. A fuzzy multi criteria approach for measuring sustainability performance of a supplier based on triple bottom line approach. *J. Clean. Prod.* **2013**, *47*, 345–354. [CrossRef]

35. Dai, J.; Blackhurst, J. A four-phase AHP-QFD approach for supplier assessment: a sustainability perspective. *Int. J. Prod. Res.* **2012**, *50*, 5474–5490. [CrossRef]

36. Azadnia, A.H.; Saman, M.Z.M.; Wong, K.Y. Sustainable supplier selection and order lot-sizing: an integrated multi-objective decision-making process. *Int. J. Prod. Res.* **2015**, *53*, 383–408. [CrossRef]

37. Su, C.M.; Horng, D.J.; Tseng, M.L.; Chiu, A.S.F.; Wu, K.J.; Chen, H.P. Improving sustainable supply chain management using a novel hierarchical grey-DEMATEL approach. *J. Clean. Prod.* **2016**, *134*, 469–481. [CrossRef]

38. Luthra, S.; Govindan, K.; Kannan, D.; Mangla, S.K.; Garg, C.P. An integrated framework for sustainable supplier selection and evaluation in supply chains. *J. Clean. Prod.* **2017**, *140*, 1686–1698. [CrossRef]

39. Zhao, H.R.; Guo, S. Selecting Green Supplier of Thermal Power Equipment by Using a Hybrid MCDM Method for Sustainability. *Sustainability* **2014**, *6*, 217–235. [CrossRef]

40. Hsu, C.W.; Kuo, T.C.; Shyu, G.S.; Chen, P.S. Low Carbon Supplier Selection in the Hotel Industry. *Sustainability* **2014**, *6*, 2658–2684. [CrossRef]

41. Chatterjee, K.; Pamucar, D.; Zavadskas, E.K. Evaluating the performance of suppliers based on using the R'AMATEL-MAIRCA method for green supply chain implementation in electronics industry. *J. Clean. Prod.* **2018**, *184*, 101–129. [CrossRef]

42. Liu, A.J.; Xiao, Y.X.; Ji, X.H.; Wang, K.; Tsai, S.B.; Lu, H.; Cheng, J.S.; Lai, X.J.; Wang, J.T. A Novel Two-Stage Integrated Model for Supplier Selection of Green Fresh Product. *Sustainability* **2018**, *10*. [CrossRef]

43. Kusi-Sarpong, S.; Gupta, H.; Sarkis, J. A supply chain sustainability innovation framework and evaluation methodology. *Int. J. Prod. Res.* **2018**, *6*, 1–9. [CrossRef]

44. Erol, I.; Sencer, S.; Sari, R. A new fuzzy multi-criteria framework for measuring sustainability performance of a supply chain. *Ecol. Econ.* **2011**, *70*, 1088–1100. [CrossRef]

45. Das, R.; Shaw, K. Uncertain supply chain network design considering carbon footprint and social factors using two-stage approach. *Clean Technol. Environ. Policy* **2017**, *19*, 2491–2519. [CrossRef]

46. Luthra, S.; Mangla, S.K.; Chan, F.T.S.; Venkatesh, V.G. Evaluating the Drivers to Information and Communication Technology for Effective Sustainability Initiatives in Supply Chains. *Int. J. Inf. Technol. Decis. Mak.* **2018**, *17*, 311–338. [CrossRef]

47. Padhi, S.S.; Pati, R.K.; Rajeev, A. Framework for selecting sustainable supply chain processes and industries using an integrated approach. *J. Clean. Prod.* **2018**, *184*, 969–984. [CrossRef]

48. Liou, J.J.; Tamošaitienė, J.; Zavadskas, E.K.; Tzeng, G.H. New hybrid COPRAS-G MADM Model for improving and selecting suppliers in green supply chain management. *Int. J. Prod. Res.* **2016**, *54*, 114–134. [CrossRef]

49. Mulliner, E.; Malys, N.; Maliene, V. Comparative analysis of MCDM methods for the assessment of sustainable housing affordability. *Omega* **2016**, *59*, 146–156. [CrossRef]

50. Pamučar, D.; Stević, Ž.; Sremac, S. A new model for determining weight coefficients of criteria in mcdm models: Full consistency method (fucom). *Symmetry* **2018**, *10*, 393. [CrossRef]

51. Nunić, Z.B. Evaluation and selection of Manufacturer PVC carpentry using FUCOM-MABAC model. *Oper. Res. Eng. Sci. Theory Appl.* **2018**, *1*, 13–28. [CrossRef]

52. Zavadskas, E.K.; Nunić, Z.; Stjepanović, Ž.; Prentkovskis, O. A novel rough range of value method (R-ROV) for selecting automatically guided vehicles (AGVs). *Stud. Inform. Control* **2018**, *27*, 385–394. [CrossRef]

53. Pamučar, D.; Lukovac, V.; Božanić, D.; Komazec, N. Multi-criteria FUCOM-MAIRCA model for the evaluation of level crossings: case study in the Republic of Serbia. *Oper. Res. Eng. Sci. Theory Appl.* **2018**, *1*, 108–129. [CrossRef]

54. Prentkovskis, O.; Erceg, Ž.; Stević, Ž.; Tanackov, I.; Vasiljević, M.; Gavranović, M. A New Methodology for Improving Service Quality Measurement: Delphi-FUCOM-SERVQUAL Model. *Symmetry* **2018**, *10*, 757. [CrossRef]

55. Stević, Ž.; Pamučar, D.; Vasiljević, M.; Stojić, G.; Korica, S. Novel integrated multi-criteria model for supplier selection: Case study construction company. *Symmetry* **2017**, *9*, 279. [CrossRef]

56. Pamučar, D.; Božanić, D.; Lukovac, V.; Komazec, N. Normalized weighted geometric bonferroni mean operator of interval rough numbers-application in interval rough DEMATEL-COPRAS model. *Facta Univ. Ser. Mech. Eng.* **2018**, *16*. [CrossRef]

57. Stojić, G.; Stević, Ž.; Antuchevičienė, J.; Pamučar, D.; Vasiljević, M. A Novel Rough WASPAS Approach for Supplier Selection in a Company Manufacturing PVC Carpentry Products. *Information* **2018**, *9*, 121. [CrossRef]

58. Roy, J.; Chatterjee, K.; Bandhopadhyay, A.; Kar, S. Evaluation and selection of Medical Tourism sites: A rough AHP based MABAC approach. *arXiv* **2016**.

59. Stević, Ž.; Pamučar, D.; Zavadskas, E.K.; Ćirović, G.; Prentkovskis, O. The Selection of Wagons for the Internal Transport of a Logistics Company: A Novel Approach Based on Rough BWM and Rough SAW Methods. *Symmetry* **2017**, *9*, 264. [CrossRef]

60. Radović, D.; Stević, Ž.; Pamučar, D.; Zavadskas, E.; Badi, I.; Antuchevičiene, J.; Turskis, Z. Measuring performance in transportation companies in developing countries: a novel rough ARAS model. *Symmetry* **2018**, *10*, 434. [CrossRef]

© 2019 by the authors. Licensee MDPI, Basel, Switzerland. This article is an open access article distributed under the terms and conditions of the Creative Commons Attribution (CC BY) license (http://creativecommons.org/licenses/by/4.0/).

Article

A Hybrid Fuzzy Group Multi-Criteria Assessment of Structural Solutions of the Symmetric Frame Alternatives

Zenonas Turskis [1],*, Kęstutis Urbonas [2] and Alfonsas Daniūnas [2]

[1] Institute of Sustainable Construction, Faculty of Civil Engineering, Vilnius Gediminas Technical University, Saulėtekio Ave. 11, LT-10223 Vilnius, Lithuania

[2] Department of Steel and Composite Structures, Faculty of Civil Engineering, Vilnius Gediminas Technical University, Saulėtekio Ave. 11, LT-10223 Vilnius, Lithuania; kestutis.urbonas@vgtu.lt (K.U.); alfonsas.daniunas@vgtu.lt (A.D.)

* Correspondence: zenonas.turskis@vgtu.lt

Received: 10 January 2019; Accepted: 13 February 2019; Published: 19 February 2019

Abstract: Structural designers that design buildings use different criteria to select the frames' materiality and structural solutions. Very often, the primary test is the cost of construction. Sometimes, solutions are determined by the terms of structure, architectural preferences, technological needs, fire safety requirements, environmental conditions, exploitation costs over the life of the building, ecological aspects, and experience, etc. This paper proposes an approach for analyzing the structural elements of buildings taking into account the impact on the environment using jointly incorporating subjective and objective aspects. The objective to combine the most important criteria into a single unit and carry out the overall assessment could be done by giving each variable a weighted value and perform a so-called multi-criteria analysis. This article shows the efficiency of the structural solution of the one-story building. The case study presents an investigation and comparison of five possible symmetrical structural solutions by multi-criteria assessment methods: The analysis of three steel frameworks differs majorly due to the beam-column characteristics, as well as precast RC frame structures case and combined steel beams and RC columns frame option. Possible solutions must meet all the essential requirements of the building, including mechanical resistance and stability. The obtained results show a broad assessment of the structural solutions of the building.

Keywords: symmetry; sustainability; ecological impact; structural solution; steel frame; load-bearing structure; Multi-Criteria Decision-Making (MCDM); Multi-Criteria Decision-Making; ARAS-F; MULT-F

---

1. Introduction

Environmental restoration, revival and recovery are vital principles for sustainable development. There is a symmetrical balance when all the parts of the objects are well balanced [1]. Correct, logical and rational construction projects are reliable and sound products that for a long time have met the critical architectural, quality and design requirements; safety; price; and influence, and are expected to have a lower long-term impact on the environment [2]. Designers make their final decisions according to several requirements. Scientists have proposed many strategies to improve the profitability of the construction industry and apply sustainable construction methods [3]. The evolution of architecture has highlighted the advantages of the principle of symmetry [4]. It has emerged as artifacts, buildings and artificial environments [5]. It affects such building conditions as structural efficiency, attractive structures, economic production, and functional or aesthetic requirements. Geometric symmetry means symmetry in the plane, and structural symmetry says that the centers of mass and resistance are at the same point. The ideal shape is the most straightforward: round. Besides, large L-shaped, zigzag-shaped or large wing structures are undesirable in hazardous areas. It includes compliance with standardization requirements, production of repeat elements and mass production that reduces production costs [6]. Some architecture and basic principles (targeted planning and symmetric arrangement, vertical support elements and symmetrical structural elements) enhance the load on structural elements [7]. Symmetry is an essential element of architecture that reflects the balance between building construction and loading. Symmetric simple geometry structures are safer, more efficient and more predictable than asymmetric structures. The asymmetric building is the weakest when there is a dynamic cross-force due to involved displacement associated with the base shear. Therefore, symmetrical shapes are preferred but not asymmetric. Proper design efficiency contributed to the appropriate arrangement of vertical bearing elements, as well as uniform and balanced openings distribution. Therefore, symmetry and regularity are generally reliable [8].

The selection process of the fundamental building system process shows the trade-off among different options. Decision-makers could use multi-criteria decision-making (MCDM) methods such as AHP, TOPSIS, and COPRAS [9], as well as PROMETHEE families [10] and others to determine the best choice. Decisions made in complex contexts need these methods for practical solutions. The fact that construction materials contribute to sustainable building management has been proven by many studies [11,12]. Energy consumption and (CO_2) emissions are the two critical indicators of sustainability in the construction industry. The primary source of adverse environmental impact on the life cycle of buildings is energy consumption at the stage of long-term building use [13]. The building sector accounts for about 40% of global primary energy consumption. When the service-life regarding the structural safety or serviceability of a deteriorating building does not meet the original target, the options for life cycle maintenance strategies need to be changed. The selection typically depends on costs and execution time [14] being redefined and developed throughout the early design stages. The structure produced by using only environmentally friendly materials is not necessary for a sustainable building. It is required to optimize the selection of materials for greater sustainability [15]. However, adequately selected materials and technologies, suppliers and contractors significantly improve the performance of the building [16–18].

Steel industry and research have been struggling to improve its environmental performance [15] and enable satisfying clients' needs, providing high structural quality and performance while pursuing sustainability [19]. Cold-formed structures have one of the top load capacity-to-weight ratios among the common structural components and bring economic, social and environmental benefits by decreasing raw materials consumption, with lighter foundations preserving the soil and its movements, the economy in handling and transportations and reduces labor fatigue. Furthermore, such structures present a rival structural behavior under seismic loads [20]. Bitarafan specified the suitability of cold-formed steel structures for naturally damaged regions, studying the more suitable construction techniques [21]. Steel is the world's most used and recycled metal. The iron and steel industry is known to be the most significant energy consuming manufacturing sector, consuming 5% of the world's total energy consumption and emitting about 6% of the entire anthropogenic CO_2 [22]. While different materials can only be down-cycled, steel can be recycled countless times keeping its properties and quality (multi-cycling) [23]. Using scrap, the production of steel through EAF (electric arc furnace) instead of through BOF (basic oxygen furnace) can reduce about 32.14% up to 40.32% of the CO_2 emissions per ton of steel. According to Junichiro [24], the energy consumption through EAF is about 10.2 GJ per steel ton whereas through BOF it is 32.9 GJ/t. These values are in the range presented by Flues et al. [25]. Also, recent data from World Steel Association [26] shows that to recycling 1 ton of steel spares more than 1.4 t of iron ore, 1.4 t of CO_2 emissions, 120 of limestone, 740 kg of coal, and two-thirds of the amount of energy spent in the steel production process.

The production of cement and cement-based composites is not an environmentally friendly process. Therefore, evaluating the environmental stress produced by concrete structures during the different phases of life is a fundamental design requirement [27]. Stakeholders estimate it either by the global energy consumed or, equivalently, by the CO_2 released during the entire cycle of life. Concrete is predominantly utilized in buildings and infrastructure worldwide by using ordinary Portland cement (OPC) as a binder. In recent years, the annual world cement production has grown from 1.0 billion tons to approximately 1.7 billion tons, which is enough to produce 1 m^3 of concrete per person.

As a result, the cement industry is commonly regarded as being in a period of high growth. However, the sector has been confronted since the late 1990s by the need to reduce its environmental load, including carbon dioxide (CO_2) emissions. Some estimates suggest that the amount of CO_2 emitted from the global output of OPC may be as high as 7% of the total global CO_2 emissions. Furthermore, the production of OPC involves severe collateral environmental impacts, such as environmental pollution caused by dust and the enormous energy consumption required from having a plasticity temperature over 1300 °C. For these reasons, the cement industry has been challenged in the past 10 years to effectively reduce and control CO_2 emissions effectively.

Nevertheless, progress and innovation in materials and construction processes felt towards these goals. As stated by Burgan and Sansom [28], the path towards sustainability reflects acting in the three main impact areas—environment, society and economy. The sustainability of development is in line with the needs of the current generation in designing, managing and navigating change, ensuring that future generations can meet their needs [29]. A sustainable society considers two critical issues: environment and safety [30]. Development of a sustainable product includes dematerialization, recycling, and compelling design considerations. Ecological design, reduced use of energy, and focus on utility instead of ownership are also important issues in this concept. Adopting sustainable construction practices involves integrating all of the principles of sustainable construction (SC) into the construction project's life cycle plan, with every stakeholder having responsibility for carrying out sustainability practices. Every stakeholder specifically contributes when improving sustainability, while owners play an essential role in requiring other stakeholders to adopt SC practices [31,32]. Stakeholders need to shape the products and the building's life cycle [33]. The owners' subsequent decision-making and practices are more likely to promote the start of projects, and the real driving force for SC can come from the owners [33]. From the economic perspective, steel solutions enable less construction and operational costs, reducing the life cycle costs, being also a cost- and time-efficient solution [14]. On the environmental front, lightweight structures represent a decrease in raw materials' consumption, allow lighter foundations, preserve the soil, and reduce its movement; also, steel is entirely recyclable. The steel structures can be easily re-used or adapted to new functionalities. Structural designers use different criteria to select structural solutions [34]. Designers significantly impact the sustainable performance of a building by selecting the proper materials [35].

The common practice in design processes is an economic assessment or classic approach using the reliability theory and risk management of several of the most possible structural solutions [36]. As a result, stakeholders select the option with the lowest price, while omitting non-economic factors (ecological, social, metaphysical (feelings), and cultural aspects) [37–39]. The selection of suitable materials for building design is essential [40]. Designers rarely use multi-criteria analysis to solve complicated problems. Besides, unresolved issues regarding the subjective qualitative measurements and criteria weights are present in problem solution models [41].

2. Multi-Criteria Decision-Making

An integral part of contemporary human activities is choosing the most efficient solutions and justifying the selected alternatives and judgments of selected justifying procedures. Muckler and Seven [42] pointed out that all objective measurement involves subjective judgments. Firstly, developers of plans select which problems must be solved and which ones not. Humans implement almost all decisions of civil engineers in practice. Humans necessarily filled all measurement in science and technology is filled with subjective elements, whether in selecting measures or in collecting, analyzing or interpreting data. In Kant's view, all knowledge begins with human experience and is concurrent with the experience. The need for qualitative multi-criteria evaluation caused it—information contents by the inexact scale of measurement. It is consecutive. The main problem, however, is dealing with qualitative information. Many methods, qualitative data consider as pseudo-metric data, but officially forbid it as a way to consider qualitative details. Qualitative multi-criteria methods, in general, have to be survivable from the classification of the actual data. The lack of information in a multi-criteria analysis may emerge from two sources: (1) an imprecise definition of alternatives, evaluation criteria and preferences (or preference scenarios); and (2) an inaccurate measurement of the effects of other options on evaluation criteria (the 80s called impact matrix) and preference weights. One symmetry description is to say that it is the result of a balanced proportion harmony.

It is worth noting that besides the methodological developments, there are a large number of successful applications of MCDM methods to real-world problems that have made MCDM a domain of great interest, both for academics and for industry practitioners [43]. The increasing complexity of the rapidly evolving business, engineering, science, and technology environments entails making the right decisions when considering environmental, market and economic considerations. The stages of a typical MCDM procedure in civil engineering are shown in Figure 1. Often, different MCDM techniques do not lead to the same results. Multi-criteria utility models are models designed to obtain the utility of items or alternatives that can be evaluated on more than one criterion.

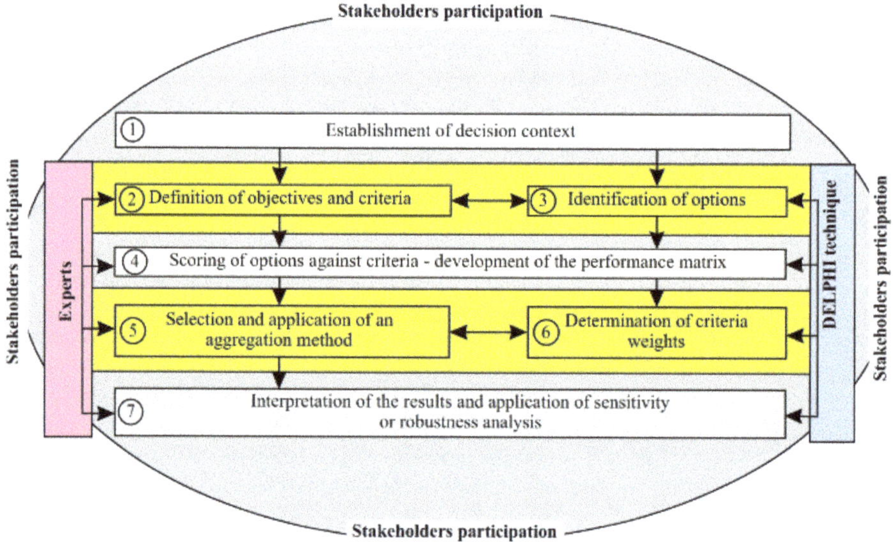

Figure 1. A procedure of MCDM (multi-criteria decision-making) in civil engineering.

2.1. Available MCDM Methods for Problem-Solving

In the last three decades, scientists have developed dozens of MCDM methods [44] that have been applied to address various issues in civil engineering [45–47] and used for sustainability problems [48]. A large variety of different problems emerging in civil engineering projects can adequately be addressed using the MCDM methodology and its related techniques [49]. They differ concerning how they combine the data. MCDM methods are broadly classified into two classes: discrete MCDM or MADM and continuous MODM (Multi-Objective Decision Making) optimization methods [50]. They all require the definition of options and criteria, and most of them demand a measure (e.g., weights) for assessing the relative significance of the criteria.

2.2. A Hybrid MCDM Model for Problem Solution

This work uses a hybrid method, i.e., a combination of fuzzy ARAS, fuzzy form of the multiplicative utility function, and DHP (Figure 2). Belton and Stewart [51] point to the need for such an integrated approach; applying hybrid methods or multiple different techniques simultaneously to the same decision problem might well serve the purpose from the behavioral and educational point of view. Researchers and practitioners had and lately increasingly supported the use of hybrid methods. Such approaches most frequently use two or more MCDM methods or a combination of the MCDM methods and other decision support approaches.

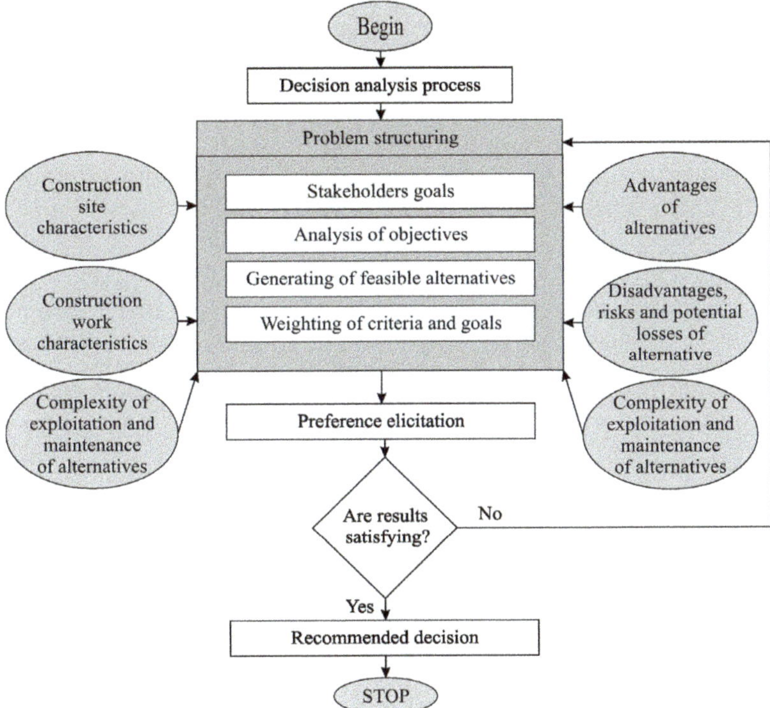

Figure 2. Structure of the hybrid decision support system, based on the ARAS-F, fuzzy multiplicative utility method and Delphi Hierarchy Process.

The integrated approach consists of three phases. In phase 1, a teamwork approach is taken to formulate the ideas for decision-making. In this stage, decision-makers define the alternatives. In the first step, a team of five selected experts selects the methods to evaluate and score options, determine communication procedures for the evaluation process and create an evaluation timeline. The most popular methods are rated according to Saaty's and Ergu's [52] 16 criteria. Later, three different MCDM methods are selected to solve the problem: The Analytic Hierarchy Process (AHP) method to determine criteria weights, the ARAS-F, and the fuzzy product model (modified by the paper's authors) for multi-criteria assessment of feasible alternatives.

In the second step, a set of evaluation criteria using the nominal group technique Delphi are established.

In the third step, a set of cardinal criteria weights to determine the relative importance of evaluation categories (based on the AHP) is provided.

Finally, the overall multi-attribute utility performance score for each feasible alternative is determined.

2.2.1. Fuzzy Number

Various types of membership functions are used. In this study, fuzzy triangular numbers (TFN) are used. A fuzzy set is a class of objects with a continuum of membership grades. A membership function, which assigns to each object a degree of membership ranging between zero and one, characterizes the set. A fuzzy set A defined in space X is a set of pairs:

$$A = \{(x, \mu_A(x)), x \in X\}, \forall x \in X, \tag{1}$$

where fuzzy set A is characterized by its membership function $\mu_A : X \to [0; 1]$ $\mu_A : X \to 0$, which associates with each element $x \in X$, a real number $\mu_A(x) \in [0; 1]$. The value $\mu_A(x)$ at x represents the grade of membership of x in A and is interpreted as the membership degree to which x belongs to A. So the closer the value $\mu_A(x)$ is to one, the more x belongs to A, $x \in 0$.

A crisp or ordinary subset A of X can also be seen as a fuzzy set in X with a membership function as its characteristic function, i.e.,

$$\mu_A(x) = \begin{cases} 1, & x \in A; \\ 0, & x \notin A. \end{cases} \tag{2}$$

The set X, specifically as a universe of discourse, can be written as $\subseteq X$. Sometimes a fuzzy set A in x is denoted by listing the ordered pairs $(x, \mu_a(x))$, where the elements with zero degrees are usually not listed. Thus, fuzzy set A in X can be represented as $A = \{(x, \mu_A(x))\}$, where $x \in X$ and $\mu_A : X \to [0; 1]$.

When the universe of discourse is discrete and finite with cardinality n, that is $X = \{x_1, x_2, \ldots, x_n\}$ $X = x_1, x_2, \ldots, x_n$, the fuzzy set A can be represented as

$$A = \sum_{i=1}^{n} \frac{\mu_A(x_i)}{x_i} = \frac{\mu_A(x_1)}{x_1} + \frac{\mu_A(x_2)}{x_2} + \cdots + \frac{\mu_A(x_n)}{x_n}, \tag{3}$$

When the universe of discourse X is an interval of real numbers, fuzzy set A is expressed as

$$A = \int_X \frac{\mu_A(x)}{x}. \tag{4}$$

A fuzzy number is a triangular fuzzy number (α, β, γ) if three parameters $(\alpha < \beta < \gamma)$ fully describes its membership function.

$$\mu_A(x) = \begin{cases} \frac{1}{\beta-\alpha}x - \frac{\alpha}{\beta-\alpha}, & \text{if } x \in [\alpha, \beta] ; \\ \frac{1}{\beta-\gamma}x - \frac{\alpha}{\beta-\gamma}, & \text{if } x \in [\beta, \gamma] ; \\ 0, & \text{otherwise}. \end{cases} \tag{5}$$

2.2.2. Defuzzification

A defuzzification process is applied to obtain a crisp output. Defuzzification is the production of a quantifiable result in fuzzy logic, given fuzzy sets and corresponding membership grades. Scientists proposed different defuzzification techniques. The triangular membership function is the most typical (Figure 3). Laarhoven and Pedrycz [53] defined the basic operations of fuzzy triangular numbers $\tilde{n}_1$ and $\tilde{n}_2$ as follows:

$$\tilde{n}_1 \oplus \tilde{n}_2 = (n_{1\alpha} + n_{2\alpha}, n_{1\beta} + n_{2\beta}, n_{1\gamma} + n_{2\gamma}), \quad \text{addition} \tag{6}$$

$$\tilde{n}_1(-)\tilde{n}_2 = \left(n_{1\alpha} - n_{2\alpha}, n_{1\beta} - n_{2\beta}, n_{1\gamma} - n_{2\gamma}\right), \quad \text{subtraction} \tag{7}$$

$$\tilde{n}_1 \otimes \tilde{n}_2 = \left(n_{1\alpha} \times n_{2\alpha}, n_{1\beta} \times n_{2\beta}, n_{1\gamma} \times n_{2\gamma}\right), \quad \text{multiplication} \tag{8}$$

$$\tilde{n}_1(\div)\tilde{n}_2 = \left(\frac{n_{1\alpha}}{n_{2\gamma}}, \frac{n_{1\beta}}{n_{2\beta}}, \frac{n_{1\gamma}}{n_{2\alpha}}\right), \quad \text{division} \tag{9}$$

$$k\tilde{n}_1 = \left(kn_{1\alpha}, kn_{1\beta}, kn_{1\gamma}\right), \quad \text{multiplication by constant} \tag{10}$$

$$(\tilde{n}_1)^{-1} = \left(\frac{1}{n_{1\gamma}}, \frac{1}{n_{1\beta}}, \frac{1}{n_{1\alpha}}\right), \quad \text{inverse} \tag{11}$$

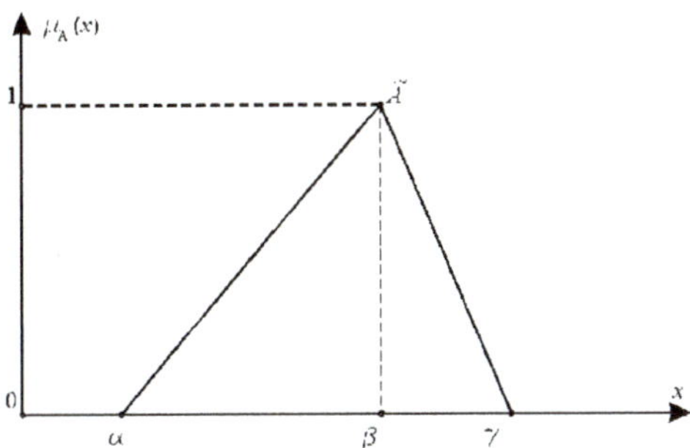

Figure 3. Triangular membership function.

2.3. The DHP (Delphi Hierarchy Process)

To strengthen the AHP method, Khorramshahgol and Moustakis [54] suggested a new DHP (Delphi Hierarchy Process) technique, which incorporates the Delphi method to collect expert judgments (Figure 4). It makes use of the advantages of AHP in determining the weights of criteria and the simplicity of fuzzy ARAS or fuzzy form of the multiplicative utility function for ranking alternatives.

Most human solutions are compromises between competing goals. They refer to the assessment of multiple alternatives [55]. A fundamental problem of decision theory is to choose the proper approach to derive weights for a set of criteria according to significance. Relative importance is usually judged according to several criteria [56]. A variety of methods are proposed for eliciting weights. There is no best way to set criteria weights. The ratio method, swing method and tradeoff method (called pricing out) [57] are general weight elicitation procedures applied in engineering researches. Schoemaker and Waid [55] compared five fundamentally different approaches for determining such weights.

The review of past works shows that AHP is the most common MCDM method used to solve civil engineering multi-attribute decision-making problems. The oldest reference dates from 1972 [58]). Later, Saaty [56] described the technique. An analysis process using the AHP is shown in Figure 4. The AHP modeling process is based on four principles: structuring the decision problem, measurement technique (the impact of the elements of the hierarchy is assessed through paired comparisons done separately in reference to each of the aspects of the level immediately above), data collection, determination of normalized weights, and synthesizing to find a solution to the problem [59].

For the establishment of a pair-wise comparison matrix A; Let $C_1, C_2, \ldots, C_n$ denote the set of elements, while a_{ij} represents a quantified judgment on a pair of elements C_i and C_j. The relative importance of the two aspects is rated using a scale (Table 1). These scales yield an $n \times n$ matrix A as $C_1, C_2, \ldots, C_n$ where $a_{ij} = 1$ and $a_{ij} = 1/a_{ij}, i, j = 1, 2, \ldots, n$. In matrix A, the problem becomes one of assigning to the n elements $C_1, C_2, \ldots, C_n$ a set of numerical weights $w_1, w_2, \ldots, w_n$ that reflect the recorded judgments.

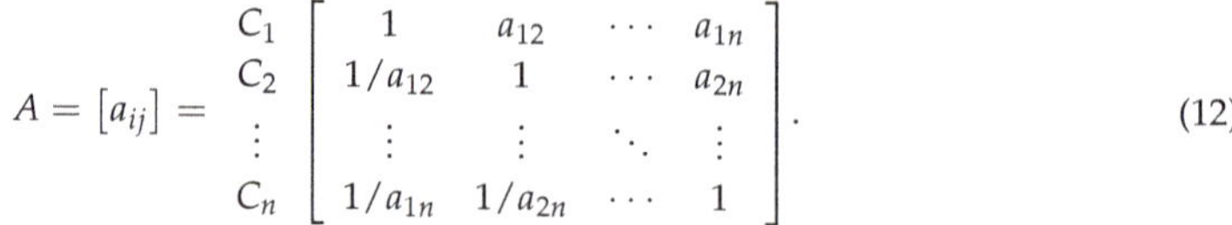

$$A = [a_{ij}] = \begin{array}{c} C_1 \\ C_2 \\ \vdots \\ C_n \end{array} \begin{bmatrix} 1 & a_{12} & \cdots & a_{1n} \\ 1/a_{12} & 1 & \cdots & a_{2n} \\ \vdots & \vdots & \ddots & \vdots \\ 1/a_{1n} & 1/a_{2n} & \cdots & 1 \end{bmatrix}. \tag{12}$$

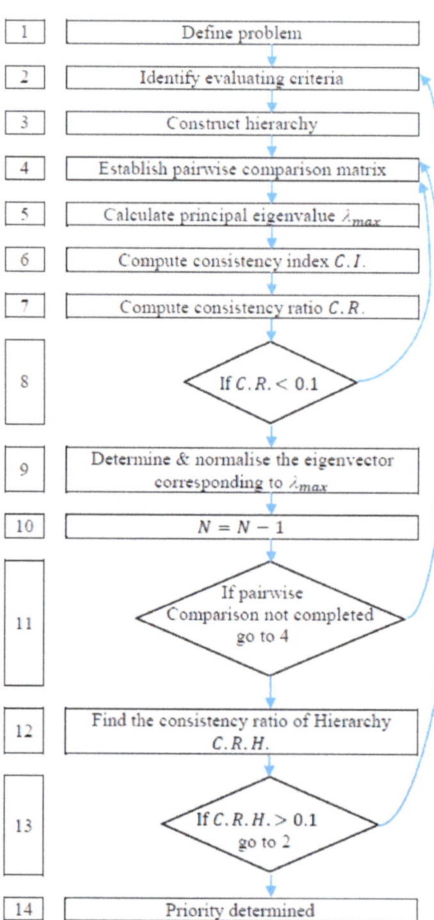

Figure 4. The process of the AHP.

Saaty recommends a nine-level dominance scale, which Saaty describes by appealing to Miller's [60] magical number seven plus two (Table 1) [56]. There are $n(n-1)/2$ judgments required to develop an $n \times n$ judgment matrix since reciprocals are automatically assigned in each pair-wise comparison.

According to Saaty, the largest eigenvalue $\lambda_{\max}$ would be

$$\lambda_{\max} = \sum_{j=1}^{n} a_{ij}(w_j/w_i). \tag{13}$$

Saaty proposed utilizing the consistency index (*CI*) and consistency ratio (*CR*) to verify the consistency of the comparison matrix. *CI* and *CR* are defined as follows:

$$CI = \frac{(\lambda_{max}-1)}{(n-1)}, \quad CR = \frac{CI}{RI}.$$

Here, the *RI* (Table 2) represents the average consistency index over numerous random elements of the same order reciprocal matrices.

2.3.1. Fuzzy Group Criterion Weight is Determined as Follows:

After the criteria weights from the AHP are established, the synthesizing of the ratio judgments is performed.

Suppose that $\widetilde{W} = \left[\widetilde{w}_1, \widetilde{w}_n\right] = \left[\widetilde{w}_j\right]$ are fuzzy group weights for n criteria and $\widetilde{w}_j$ is the fuzzy triangular number

$$\widetilde{w}_j = \left(w_{jl}, w_{jm}, w_{ju}\right), \tag{14}$$

where $w_{jl} = \min_k y_{jk}$, $j = \overline{1,n}$, $k = \overline{1,p}$ is the minimum possible value, $w_{j\alpha} = \left(\prod_{k=1}^{p} y_{jk}\right)^{\frac{1}{p}}$, $j = \overline{1,n}$, $k = \overline{1,p}$ is the most probable value and $w_{j\beta} = \max_k y_{jk}$, $j = \overline{1,n}$, $k = \overline{1,p}$ is the maximal possible value of j criterion.

$$w_{j\gamma} = \max_k y_{jk}, \quad j = \overline{1,n}, \quad k = \overline{1,p}.$$

Table 1. Saaty's original nine-point scale of relative importance for pairwise comparison.

Meaning	Diagonal Elements $i = j$	C_i and C_j Are Equally Important	C_i Is Weakly More Important Than C_j	C_i Is Strongly More Important Than C_j	C_i Is Demonstratively More Important Than C_j	C_i is Absolutely More Important Than C_j	Compromise between Two Judgments	If Element C_j Dominates over Element C_i
a(i,j)	1	1	3	5	7	9	2, 4, 6, 8	a(i,j) = 1/a(j,i)

Table 2. Random consistency indices for the different number of criteria (*n*).

n	1	2	3	4	5	6	7	8	9	10
RI	0	0	0.58	0.9	1.12	1.24	1.32	1.41	1.45	1.49

The $CR \leq 0.1$ indicates that the matrix reached consistency.

2.3.2. Additive Ratio Assessment Method (ARAS) with Fuzzy Criteria Values (ARAS-F)

The MCDM ARAS method [61,62] with a fuzzy criteria values method [63] was selected to solve the problem. At the first stage, a problem is represented by the fuzzy decision-making matrix of preferences for m reasonable alternatives rated on n criteria:

$$\widetilde{X} = \begin{bmatrix} \widetilde{x}_{01} & \cdots & \widetilde{x}_{0j} & \cdots & \widetilde{x}_{0n} \\ \vdots & \ddots & \vdots & \ddots & \vdots \\ \widetilde{x}_{i1} & \cdots & \widetilde{x}_{ij} & \cdots & \widetilde{x}_{in} \\ \vdots & \ddots & \vdots & \ddots & \vdots \\ \widetilde{x}_{m1} & \cdots & \widetilde{x}_{mj} & \cdots & \widetilde{x}_{mn} \end{bmatrix}; \quad i = \overline{0,m}; \quad j = \overline{1,n}, \tag{15}$$

where $\widetilde{x}_{ij}$—fuzzy value represents the performance value of the i alternative regarding the j criterion, $\widetilde{x}_{0j}$—the optimal value of j criterion. A tilde "~" is placed above a symbol if the symbol represents a fuzzy set.

If the optimal value of j criterion is unknown, then

$$\begin{aligned}
\widetilde{x}_{0j} &= \max_i \widetilde{x}_{ij}, \quad \text{if } \max_i \ \widetilde{x}_{ij} \text{ is preferable, and} \\
\widetilde{x}_{0j} &= \min_i \widetilde{x}_{ij}^*, \quad \text{if } \min_i \ \widetilde{x}_{ij}^* \text{ is preferable.}
\end{aligned} \tag{16}$$

At the next stage, dimensionless criteria values $\widetilde{\widetilde{x}}_{ij}$ from the matching criteria are calculated and expressed in the normalized decision-making matrix $\overline{\widetilde{X}}$:

$$\overline{\widetilde{X}} = \begin{bmatrix}
\widetilde{\widetilde{x}}_{01} & \cdots & \widetilde{\widetilde{x}}_{0j} & \cdots & \widetilde{\widetilde{x}}_{0n} \\
\vdots & \ddots & \vdots & \ddots & \vdots \\
\widetilde{\widetilde{x}}_{i1} & \cdots & \widetilde{\widetilde{x}}_{ij} & \cdots & \widetilde{\widetilde{x}}_{in} \\
\vdots & \ddots & \vdots & \ddots & \vdots \\
\widetilde{\widetilde{x}}_{m1} & \cdots & \widetilde{\widetilde{x}}_{mj} & \cdots & \widetilde{\widetilde{x}}_{mn}
\end{bmatrix} ; \tag{17}$$

$$i = \overline{0, m}; \quad j = \overline{1, n} \ .$$

The criteria, whose preferred values are maximal, are standardized as follows:

$$\widetilde{\widetilde{x}}_{ij} = \frac{\widetilde{x}_{ij}}{\sum\limits_{i=0}^{m} \widetilde{x}_{ij}}. \tag{18}$$

The cost type criteria (preferred values are minimal) are normalized as follows:

$$\widetilde{\widetilde{x}}_{ij} = \frac{\frac{1}{\widetilde{x}_{ij}}}{\sum\limits_{i=0}^{m} \frac{1}{\widetilde{x}_{ij}}}. \tag{19}$$

The third stage, normalized-weighted matrix $\widehat{\widetilde{X}}$ is defined. The sum of the weights w_j is limited as follows:

$$\sum_{j=0}^{n} w_j = 1. \tag{20}$$

$$\widehat{\widetilde{X}} = \begin{bmatrix}
\widehat{\widetilde{x}}_{01} & \cdots & \widehat{\widetilde{x}}_{0j} & \cdots & \widehat{\widetilde{x}}_{0n} \\
\vdots & \ddots & \vdots & \ddots & \vdots \\
\widehat{\widetilde{x}}_{i1} & \cdots & \widehat{\widetilde{x}}_{ij} & \cdots & \widehat{\widetilde{x}}_{in} \\
\vdots & \ddots & \vdots & \ddots & \vdots \\
\widehat{\widetilde{x}}_{m1} & \cdots & \widehat{\widetilde{x}}_{mj} & \cdots & \widehat{\widetilde{x}}_{mn}
\end{bmatrix} ; \tag{21}$$

$$i = \overline{0, m}; \quad j = \overline{1, n} \ .$$

Normalized-weighted values of all the criteria are calculated as follows:

$$\widehat{\widetilde{x}}_{ij} = \widetilde{\widetilde{x}}_{ij} \widetilde{w}_j; \quad i = \overline{0, m}, \tag{22}$$

where w_j is the weight (importance) of the j criterion and $\overline{x}_{ij}$ is the normalized rating of the j criterion. The following task determines the values of the optimality function:

$$\widetilde{S}_i = \sum_{j=1}^{n} \widehat{\widetilde{x}}_{ij}; \quad i = \overline{0, m}, \tag{23}$$

where $\widetilde{S}_i$ is the value of the optimality function of the i selection.

The most significant value is the best, and the last value is the worst.

The result of fuzzy decision making for each alternative is a fuzzy number $\widetilde{S}_i$. There are several methods for defuzzification. The center-of-area is the most practical and easy to apply for this reason:

$$S_i = \frac{1}{3}\left(S_{1\alpha} + S_{i\beta} + S_{i\gamma}\right). \tag{24}$$

The equation used for the calculation of the utility degree K_i of an alternative A_i is given below:

$$K_i = \frac{S_i}{S_0}; \quad i = \overline{0,m}, \tag{25}$$

where S_i and S_0 are the optimal criterion values, obtained from Equation (24).

2.3.3. The fuzzy Multiplicative Utility Function

The transitive decomposable model was introduced by Krantz et al. [64] as a natural generalization of the model. It amounts to replacing the addition operation by a general function that is increasing in each of its arguments.

The criteria values x_{ij}^* with favorable minimal values are transformed as follows:

$$\widetilde{x}_{ij} = \frac{1}{\widetilde{x}_{ij}^*}. \tag{26}$$

The optimality function values are calculated as follows:

$$U_i = \prod_{j=1}^{n} \widetilde{x}_{ij} = \left(\prod_{j=1}^{n}\left(\widetilde{x}_{ij\alpha}\widetilde{x}_{ij\beta}\widetilde{x}_{ij\gamma}\right)\right)^{\frac{1}{3}} \quad i = \overline{0,m}, \quad j = \overline{0,n}. \tag{27}$$

The equation used for the calculation of the utility degree K_i of an alternative A_i is given below:

$$K_i = \frac{U_i}{U_0}; \quad i = \overline{0,m}, \tag{28}$$

where U_i and U_0 are the optimal criterion values, obtained from Equation (27).

2.3.4. Integrated Utility Function

An integrated utility value of a considered alternative is calculated as given below:

$$D_i = 0.5\left[(K_{Ai}K_{Ui})^{0.5} + 0.5(K_{Ai} + K_{Ui})\right]; \quad i = \overline{0,m}, \tag{29}$$

where K_{Ai} are K_i values obtained from Equations (25) and (27) respectively.

3. Description of the Problem

The authors investigated a real case study as an object for multi-criteria assessment. The dimensions of the building are $78 \times 9 \times 3.5$ (h) m. A structure consists of 14 transverse frames 9×3.5 (h) m (see Figure 5) set in step 6 meters. There are three identical buildings at the site. Therefore, 42 frames are needed.

Only structural elements of the frames are included in the analysis. Bracings and other secondary elements are disregarded in the investigation. The objective of the research is a selection and multi-criteria assessment of structural solutions. The building design must satisfy the essential requirements of the building. Five possible structural solutions are investigated. There are three cases of steel frames (Case 1, Case 2 and Case 3); those significant differences are due to the initial rotational stiffness of the beam-to-column joints [65–67], one frame from the precast RC columns and beam (Case 4), and one frame from the precast RC columns and steel beam (Case 5).

As the span of the frame is only 9 meters, the beams are used as the main roof load-bearing structural elements in all cases. Steel grade S355 in all cases is used for the steel elements. All the structural calculations are performed according to the rules of design codes. Fulfillment of the ultimate limit state and serviceability limit state checked. The lists of the materials were compounded for all cases.

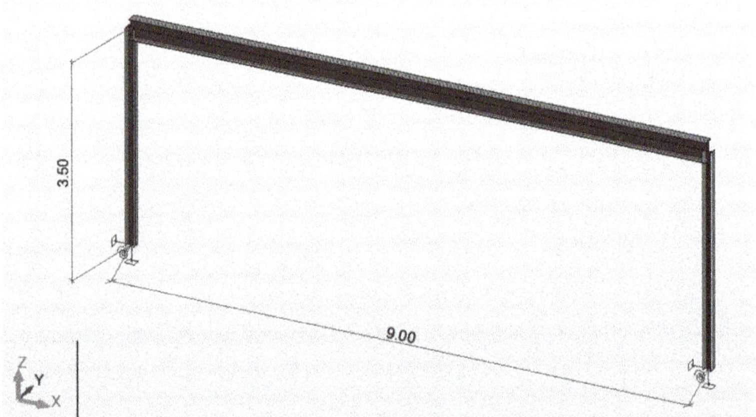

Figure 5. Analyzed symmetric frame.

Case 1. Beam cross-section of the IPE300 and columns cross-sections of the HEA160 profiles. For beam-to-column, connection six M20 bolts of 10.9 grade and 20-mm thickness end-plate were used. Horizontal stiffeners 8 mm of thickness were used as well. For column-to-foundation, joint four M20 bolts of 10.9 grade and 25-mm thickness were base plate used. The joints of the frame are shown in Figure 6. The initial rotational stiffness of beam-to-column joints was 14270 kNm/rad. For column-to-foundation joints, the initial rotational stiffness was 15500 kNm/rad. The element and fittings fulfill the requirements of limit states.

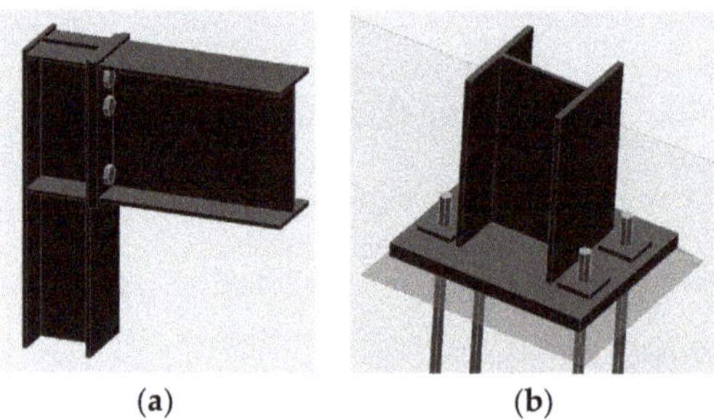

(a)   (b)

Figure 6. Bolted beam-to-column with (**a**) transverse stiffeners and column-to-foundation (**b**) joints.

Case 2. Beam cross-section of the IPE330 and columns cross-sections of the HEA140 profiles. For beam-to-column, connection six M20 bolts of 10.9 grade and 16-mm thickness end-plate used. No additional stiffeners were used. For column-to-foundation, joint four M20 bolts of 10.9 grade and 25-mm thickness base plate were used. The joints of the frame are shown in Figure 7. The initial rotational stiffness of beam-to-column joints was 9730 kNm/rad. For column-to-foundation joints, the initial rotational stiffness was 9300 kNm/rad. In comparison to Case 1, these joints are more flexible, because they have no stiffeners in the column for the beam-to-column joint and the column has a smaller cross-section. The element and fittings fulfill the requirements of limit states.

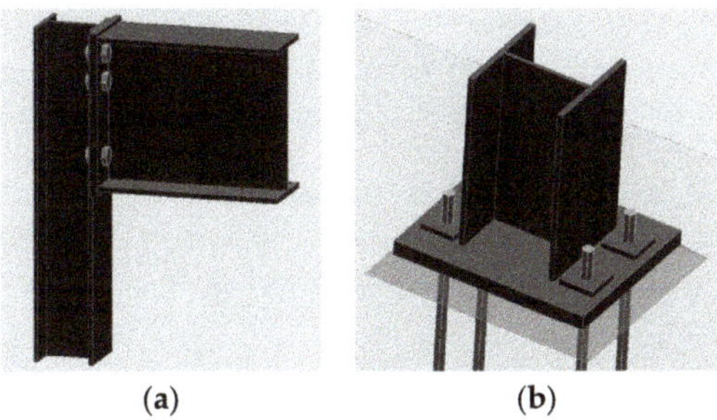

Figure 7. Unstiffened bolted beam-to-column (**a**) and column-to-foundation (**b**) joints.

Case 3. Beam cross-section of the IPE360 and columns cross-sections of the HEA140 profiles. The beam of the column joints is pinned. No bending moments in beam-to-column connections exist. For column-to-foundation, joint four M20 bolts of 10.9 grade and 25-mm thickness base plate were used. The joints of the frame are shown in Figure 8. The initial column-to-foundation joints was 10,800 kNm/rad. The beam on the columns was supported, that means no bending moments in the ends of the beam and the top of the columns appeared. The element and joints fulfill the requirements of limit states.

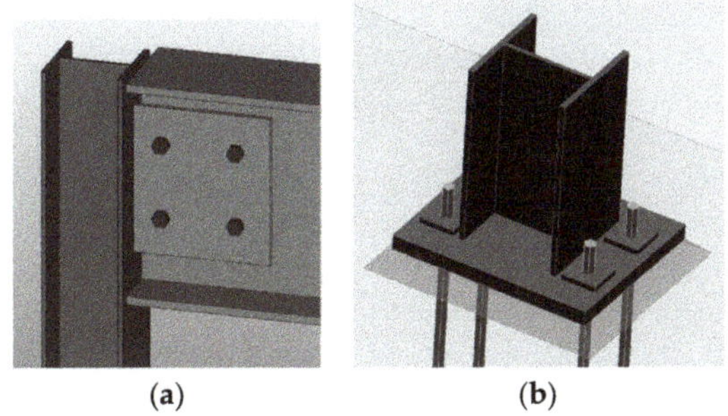

Figure 8. Pinned beam-to-column (**a**) and column-to-foundation (**b**) joints.

Case 4. In this case, the frame was made from precast concrete beam and columns. The columns on the foundations had fixed rigidity. The beam to the columns was connected as pinned. The columns were designed from 250×250 mm squares and beams from 250×400 (h) rectangular cross-sections. The joints of the frame are shown in Figure 9.

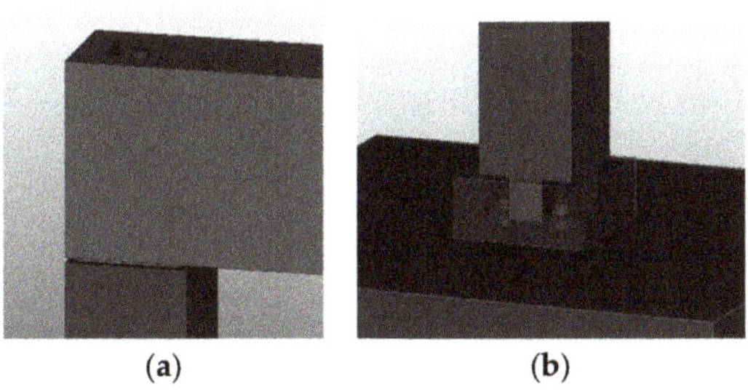

Figure 9. Pinned beam-to-column (**a**) and column-to-foundation (**b**) joints.

The column-to-foundation joints are rigid. This joint was designed using the system of anchor bolts and column shoes for fixing columns to the support rigidly (Peikko certificate). The beam on the columns was supported directly. Columns were reinforced using regular unstrained reinforcement, and beams were reinforced using unstrained and pre-stressed reinforcement as well. The structures of the frame fulfill the requirements of limit states.

Case 5. In this case, the elements of the frame were made from different materials. The columns were designed as precast RC elements. The beams are made of steel. The columns on the foundations had fixed rigidity. The beam on the columns was connected as pinned. The columns were designed by a 250×250 mm square cross-section. Beam cross-section of the IPE360 profile. The joints of the frame are shown in Figure 10.

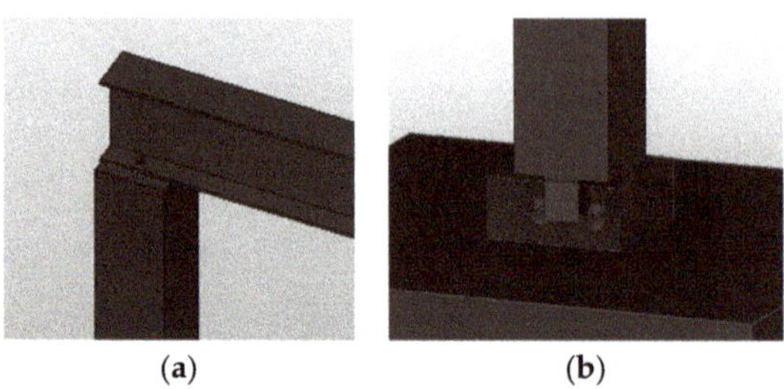

(a)   (b)

Figure 10. Pinned beam-to-column (**a**) and column-to-foundation (**b**) joints.

The column-to-foundation joints are as rigid as in Case 4. The columns are reinforced using regular unstrained reinforcement. The steel beam on the columns was supported directly. The structures of the frame fulfill the requirements of limit states.

4. Problem Solution

Criteria set and weights of criteria were determined by applying DHP. A group of experts was formed, consisting of three civil engineers, one architect and one economist. It aimed to determine feasible alternatives, criteria set for assessment and criteria weights (Table 3).

Table 3. Criteria set determined for the case study, based on the DHP.

Abbreviation	Criterion Name
x_1	Costs, €
x_2	Impact on the environment
x_3	Installment time, hours
x_4	Weight, tons
x_5	Consumption of steel, tons
x_6	Consumption of concrete, m^3

Five experts (E1, ... , E5) assessed the importance of the criteria according to the original Saaty's nine-point scale as shown in Table 4. The fuzzy criteria weights are determined by using the AHP method according to all experts' opinions (Table 5).

Table 4. The opinion of experts regarding criteria importance according to Saaty's nine-point scale.

The Opinion of Expert E_1						
	x_1	x_2	x_3	x_4	x_5	x_6
x_1	1.00	3.00	5.00	6.00	7.00	8.00
x_2	0.33	1.00	3.00	4.00	5.00	6.00
x_3	0.20	0.33	1.00	2.00	3.00	4.00
x_4	0.17	0.25	0.50	1.00	2.00	3.00
x_5	0.14	0.20	0.33	0.50	1.00	2.00
x_6	0.13	0.17	0.25	0.33	0.50	1.00

The Opinion of Expert E_2
$\vdots$

The Opinion of Expert E_5						
	x_1	x_2	x_3	x_4	x_5	x_6
x_1	1.00	2.00	5.00	6.00	7.00	8.00
x_2	0.50	1.00	2.00	4.00	5.00	7.00
x_3	0.20	0.50	1.00	4.00	5.00	3.00
x_4	0.17	0.25	0.25	1.00	2.00	3.00
x_5	0.14	0.20	0.20	0.50	1.00	2.00
x_6	0.13	0.14	0.33	0.33	0.50	1.00

Table 5. Criteria values for frame alternatives.

Criteria	Criteria Weights Determined by Expert					w_j		
	E_1	E_2	E_3	E_4	E_5	α	γ	β
x_1	0.464	0.406	0.412	0.422	0.437	0.406	0.428	0.464
x_2	0.249	0.274	0.276	0.276	0.257	0.249	0.266	0.276
x_3	0.121	0.159	0.156	0.142	0.152	0.121	0.145	0.159
x_4	0.079	0.090	0.086	0.080	0.071	0.071	0.081	0.090
x_5	0.052	0.042	0.041	0.046	0.048	0.041	0.046	0.052
x_6	0.035	0.030	0.029	0.034	0.036	0.029	0.033	0.036
CI	0.050	0.126	0.138	0.045	0.055			
RI			1.24					
CR	0.041	0.101	0.111	0.037	0.045			

Tables 6 and 7 present the considered alternatives representing the frames. A multi-criteria evaluation was performed using the ARAS-F and fuzzy multiplicative utility function (MULT-F) values. The ranks (R) assigned in the assessment for the case study are shown in Table 8. Beside this, in Table 8, the integrated assessment of alternatives is presented.

The calculations made with the ARAS-F and the MULT-F gave different results. According to the ARAS-F method, the alternative A4 representing frame, which was made from precast concrete beam and columns, was best suited, while the alternative A_1, representing the beam cross-section of the IPE300 and columns cross-sections of the HEA160 profiles was ranked second. According to the MULT-F method, the alternative A_1, representing the beam cross-section of the IPE300 and columns cross-sections of the HEA160 profiles, was best suited; while the alternative A_4, representing the frame, was made from precast concrete beam and columns representing the beam cross-section of the IPE300 and columns cross-sections of the HEA160 profiles was ranked fourth. The alternative A_1 was the best suited according to the integrated assessment of alternative performances.

Table 6. Criteria values for frame alternatives: initial DMM.

A_i	x_1			x_2			x_3			x_4			x_5			x_6		
	α	γ	β	α	γ	β	α	γ	β	α	γ	β	α	γ	β	α	γ	β
w_j	0.406	0.428	0.464	0.249	0.266	0.276	0.121	0.145	0.159	0.071	0.081	0.090	0.041	0.046	0.052	0.029	0.033	0.036
Opt.	min	min	min	min	min	min	min	min	min	min	min	min	min	min	min	min	min	min
A_1	33,300	33,300	33,300	74,195	81,614.7	89,776	227	272	326	27.2	27.2	27.2	27.2	27.2	27.2	0.1	0.1	0.1
A_2	34,100	34,100	34,100	75,832	83,414.7	91,756	232	278	334	27.8	27.8	27.8	27.8	27.8	27.8	0.1	0.1	0.1
A_3	38,000	38,000	38,000	84,559	93,014.7	102,316	258	310	372	31	31	31	31	31	31	0.1	0.1	0.1
A_4	29,900	29,900	29,900	34,405	37,845	41,630	328	394	473	141	141	141	6	6	6	135	135	135
A_5	37,200	37,200	37,200	71,107	78,218.1	86,040	303	363	436	66.3	66.3	66.3	24	24	24	42.3	42.3	42.3
A_0	24,917	24,917	24,917	28,670	31,538	34,691	189	227	272	23	23	23	5	5	5	0.08	0.08	0.08

Table 7. Criteria values for frame alternatives: normalized-weighted DMM.

A_i	x_1			x_2			x_3			x_4			x_5			x_6		
	α	γ	β	α	γ	β	α	γ	β	α	γ	β	α	γ	β	α	γ	β
Opt	max	max	max	max	max	max	max	max	max	max	max	max	max	max	max	max	max	max
A_1	0.066	0.069	0.075	0.024	0.031	0.039	0.015	0.026	0.042	0.015	0.017	0.019	0.003	0.003	0.004	0.007	0.008	0.009
A_2	0.064	0.067	0.073	0.023	0.030	0.038	0.015	0.026	0.041	0.015	0.017	0.019	0.003	0.003	0.004	0.007	0.008	0.009
A_3	0.057	0.060	0.066	0.021	0.027	0.034	0.013	0.023	0.037	0.013	0.015	0.017	0.003	0.003	0.003	0.007	0.008	0.009
A_4	0.073	0.077	0.083	0.051	0.066	0.083	0.011	0.018	0.029	0.003	0.003	0.004	0.013	0.015	0.017	0.000	0.000	0.000
A_5	0.059	0.062	0.067	0.025	0.032	0.040	0.011	0.020	0.031	0.006	0.007	0.008	0.003	0.004	0.004	0.000	0.000	0.000
A_0	0.088	0.092	0.100	0.062	0.080	0.100	0.018	0.032	0.050	0.018	0.021	0.023	0.016	0.018	0.020	0.008	0.009	0.010

Table 8. The integrated solution results: ranks of alternatives.

A_i	Solution Methods							
	ARAS-F			MULT-F			Integrated	
	S	K	R	U	K	R	K	R
A_1	0.157	0.616	2	0.0170	0.560	1	0.587	1
A_2	0.154	0.602	3	0.0167	0.549	2	0.576	2
A_3	0.139	0.543	4	0.0152	0.502	3	0.522	3
A_4	0.182	0.714	1	0.0054	0.179	4	0.402	4
A_5	0.127	0.496	5	0.0051	0.169	5	0.311	5
A_0	0.255	1.000		0.0303				

5. Results and Conclusions

In today's business environment, decision-making is a difficult and time-consuming process involving many criteria. In most cases, these criteria have imprecise and vague values and are challenging. In this study, the integration of the DHP, Fuzzy ARAS, and fuzzy multiplicative multi-criteria utility function shows the significant advantage in data mining for processing uncertain information in effective alternative evaluations.

Five possible solutions to the framed structure were analyzed. Three cases were designed using structural steel elements. One example with precast RC elements and one case mixed RC columns and steel beam. For all cases, lists of materials were combined and multi-criteria analysis performed.

Evaluating construction solutions for the implementation is a complex task which requires proper consideration of the technique and engineering management.

To overcome this problem, the model developed was based on the DHP, fuzzy ARAS, and fuzzy multiplicative utility function value to collect and analyze the judgments of experts for the selected criteria and potential alternatives.

In this study, the MCDM model considered six criteria for evaluating as follows: x_1—Costs, x_2—Impact on the environment, x_3—Installment time, x_4—Weight (tons), x_5—Consumption of steel, x_6—Consumption of concrete. The criteria set listed from the most important in decreasing importance order. The problem solution using the ARAS-F method result shows that the best method is to use alternative A_4 and after that A_1 alternative. The worst alternative is A_5 and the second worst is A_3. The difference among scores of the best and worst alternatives is 44%. The MULT-F method shows that the best alternative is A_1 alternative, the A_2 alternative ranks as the second best, while A_5 is the worst alternative, and A_4 is the second worst alternative. The alternatives rank is as follows: $A_4 \succ A_1 \succ A_2 \succ A_3 \succ A_5$. The MULT-F method is not sensitive to the criteria weights values, and the ratio of the best and worst scores is 3.25. The alternatives rank as follows $A_1 \succ A_2 \succ A_3 \succ A_4 \succ A_5$. We offer to rank alternatives according to integrated utility values of investigated alternatives. In this case, the best alternative is the first, and the worst one is the fifth alternative. The ratio of the best alternative score to the worst alternative score equals 89 percent. The final ranking of alternatives is as follows: $A_1 \succ A_2 \succ A_3 \succ A_4 \succ A_5$.

The proposed hybrid assessment approach has significantly reduced the required number of experts' judgments.

Multi-criteria assessment has shown that the alternative A1 is the best suited according to the integrated evaluation of alternative performances and constructors should implement it in practice.

The fuzzy ANP (Analytic Network Process) could be developed and implemented for future research work, based on fuzzy linguistic preference relations or its hybrid approaches with many different methods such as fuzzy the PROMETHEE, fuzzy ELECTRE, fuzzy VIKOR, fuzzy SAW, fuzzy ARAS, and fuzzy TOPSIS.

This paper shows that the hybrid approach presented here is useful in the evaluation of alternatives in a significant number of decision-making problems.

Author Contributions: Conceptualization, A.D., Z.T. and K.U.; methodology, Z.T., A.D.; software, Z.T.; validation, Z.T., A.D. and K.U.; formal analysis, Z.T.; investigation, A.D. and K.U.; resources, A.D.; data curation, K.U.; original draft preparation, A.D. and Z.T.; writing—review and editing, Z.T., and K.U.; supervision, A.D.

Funding: This research received no external funding.

Conflicts of Interest: The authors declare no conflict of interest.

References

1. Hashemkhani Zolfani, S.; Zavadskas, E.K.; Turskis, Z. Design of products with both International and Local perspectives based on Yin-Yang balance theory and SWARA method. *Econ. Res. -Ekon. Istraživanja* **2013**, *26*, 153–166. [CrossRef]
2. Yepes, V.; García-Segura, T.; Moreno-Jiménez, J.M. A cognitive approach for the multi-objective optimization of RC structural problems. *Arch. Civ. Mech. Eng.* **2015**, *15*, 1024–1036. [CrossRef]
3. Peldschus, F.; Zavadskas, E.K.; Turskis, Z.; Tamošaitienė, J. Sustainable assessment of construction site by applying game theory. *Inz. Ekon. -Eng. Econ.* **2010**, *21*, 223–237.
4. Sousa, J.P.; Xavier, J.P. Symmetry-based generative design and fabrication: A teaching experiment. *Autom. Constr.* **2015**, *51*, 113–123. [CrossRef]
5. Salunkhe, A.U. Behavior of RCC In-Filled Buildings with Different Configurations of Plan under Seismic Force. In Proceedings of the 13th World Conference on Earthquake Engineering, Vancouver, BC, Canada, 1–6 August 2004.
6. Jaganathan, S.; Nesan, L.J.; Ibrahim, R.; Mohammad, A.H. Integrated design approach for improving architectural forms in industrialized building systems. *Front. Archit. Res.* **2013**, *2*, 377–386. [CrossRef]
7. Pachta, V.; Papayianni, I. The study of the historic buildings of Eclecticism in Thessaloniki under the prism of sustainability. *Procedia Environ. Sci.* **2017**, *38*, 283–289. [CrossRef]
8. Banginwar, R.S.; Vyawahare, M.R.; Modani, P.O. Effect of plans configurations on the seismic behaviour of the structure by response spectrum method. *Int. J. Eng. Res. Appl. (IJERA)* **2012**, *2*, 1439–1443.
9. Yin, H.; Xiao, Y.; Wen, G.; Qing, Qi.; Deng, Y. Multiobjective optimization for foam-filled multi-cell thin-walled structures under lateral impact. *Thin-Walled Struct.* **2015**, *94*, 1–12. [CrossRef]
10. Balali, V.; Zahraie, B.; Roozbahani, A. A Comparison of AHP and PROMETHEE Family Decision Making Methods for Selection of Building Structural System. *Am. J. Civ. Eng. Archit.* **2014**, *2*, 149–159. [CrossRef]

11. Bakens, W. Sustainable Building and Construction: Contributions by International Organisations. In *Smart and Sustainable Built Environments*; Blackwell Publication: Hoboken, NJ, USA, 2005; pp. 275–288.
12. Giama, E.; Papadopoulos, A.M. Assessment tools for the environmental evaluation of concrete, plaster and brick elements production. *J. Clean. Prod.* **2015**, *99*, 75–85. [CrossRef]
13. Pajchrowski, G.; Noskowiak, A.; Lewandowska, A.; Strykowski, W. Materials composition or energy characteristic—What is more important in environmental life cycle of buildings? *Build. Environ.* **2014**, *72*, 15–27. [CrossRef]
14. Fiorino, L.; Iuorio, O.; Macillo, V.; Landolfo, R. Performance-based design of sheathed CFS buildings in seismic area. *Thin-Walled Struct.* **2012**, *61*, 248–257. [CrossRef]
15. Weisenberger, G. Sustainability and the structural engineers. *Pract. Period. Struct. Des. Constr.* **2011**, *16*, 146–150. [CrossRef]
16. Aretoulis, G.; Kalfakakou, G.; Striagka, F. Construction material supplier selection under multiple criteria. *Oper. Res.* **2010**, *10*, 209–230. [CrossRef]
17. Zavadskas, E.K.; Kaklauskas, A.; Turskis, Z.; Kalibatas, D. An approach to multi-attribute assessment of indoor environment before and after refurbishment of dwellings. *J. Environ. Eng. Landsc. Manag.* **2009**, *17*, 5–11. [CrossRef]
18. Zavadskas, E.K.; Turskis, Z.; Antucheviciene, J. Selecting a contractor by using a novel method for multiple attribute analysis: Weighted Aggregated Sum Product Assessment with grey values (WASPAS-G). *Stud. Inform. Control* **2015**, *24*, 141–150. [CrossRef]
19. Zavadskas, E.K.; Vilutienė, T.; Turskis, Z.; Šaparauskas, J. Multi-criteria analysis of Projects' performance in construction. *Arch. Civ. Mech. Eng.* **2014**, *14*, 114–121. [CrossRef]
20. Dubina, D. Behavior and performance of cold-formed steel-framed houses under seismic action. *J. Constr. Steel Res.* **2008**, *64*, 896–913. [CrossRef]
21. Bitarafan, M.; Zolfani, H.S.; Arefi, S.L.; Zavadskas, E.K. Evaluating the construction methods of cold-formed steel structures in reconstructing the areas damaged in natural crises, using the methods AHP and COPRAS-G. *Arch. Civ. Mech. Eng.* **2012**, *12*, 360–367. [CrossRef]
22. Quader, M.A.; Ahmed, S.; Ghazilla, R.A.R.; Ahmed, S.; Dahari, M. Evaluation of criteria for CO_2 capture and storage in the iron and steel industry using the 2-tuple DEMATEL technique. *J. Clean. Prod.* **2015**, *120*, 1–14. [CrossRef]
23. Gervásio, H. *A Sustentabilidade do Aço e das Estruturas Metálicas, Construmetal 2008*; Congresso Latino-Americano da Construção Metálica: São Paulo, Brazil, 2008.
24. Junichiro, O.; Keigo, A.; Toshimasa, T.; Miyuki, N.; Kenichi, W.; Fuminori, S. International comparisons of energy efficiency in power, steel, and cement industries. *Energy Policy* **2012**, *44*, 118–129.
25. Flues, F.; Rübbelke, D.; Vögele, S. *Energy Efficiency and Industrial Output: The Case of the Iron and Steel Industry*. FEEM Working Paper No. 96. 2013. Available online: https://www.feem.it/m/publications_pages/201311261233104NDL2013-096.pdf (accessed on 18 February 2019).
26. World Steel Association. *Fact Sheet—Raw Materials*; World Steel Association: Brussels, Belgium, 2014.
27. Fantilli, A.P.; Chiaia, B. Eco-mechanical performances of cement-based materials: An application to self-consolidating concrete. *Constr. Build. Mater.* **2013**, *40*, 189–196. [CrossRef]
28. Burgan, B.A.; Sansom, M.R. Sustainable steel construction. *J. Constr. Steel Res.* **2006**, *62*, 1178–1183. [CrossRef]
29. World Commission on Environment and Development (WCED). *Our Common Future (The Brundtland Report)*; Oxford University Press: New York, NY, USA, 1987; p. 43.
30. Yagi, K.; Halada, K. Materials development for a sustainable society. *Mater. Des.* **2001**, *22*, 143–146. [CrossRef]
31. Shen, L.Y.; Song, S.C.; Hao, J.L.; Tam, V.W.Y. Collaboration among project participants towards sustainable construction e a Hong Kong study. *Open Constr. Build. Technol. J.* **2008**, *2*, 59–68. [CrossRef]
32. Abidin, N.Z.; Pasquire, C.L. Revolutionize value management: A mode towards sustainability. *Int. J. Proj. Manag.* **2007**, *25*, 275–282. [CrossRef]
33. Shen, L.; Tam, V.W.Y.; Tam, L.; Ji, Y. Project feasibility study: The key to successful implementation of sustainable and socially responsible construction management practice. *J. Clean. Prod.* **2010**, *18*, 254–259. [CrossRef]
34. Šaparauskas, J.; Zavadskas, E.K.; Turskis, Z. Selection of façade's alternatives of commercial and public buildings based on multiple criteria. *Int. J. Strateg. Prop. Manag.* **2011**, *15*, 189–203. [CrossRef]

35. Florez, L.; Castro-Lacouture, D. Optimization model for sustainable materials selection using objective and subjective factors. *Mater. Des.* **2013**, *46*, 310–321. [CrossRef]

36. Turskis, Z.; Gajzler, M.; Dziadosz, A. Reliability, risk management, and contingency of construction processes and projects. *J. Civ. Eng. Manag.* **2012**, *18*, 290–298. [CrossRef]

37. Van Kesteren, I.E.H. Product designers' information needs in materials selection. *Mater. Des.* **2008**, *29*, 133–145. [CrossRef]

38. Medineckiene, M.; Zavadskas, E.K.; Björk, F.; Turskis, Z. Multi-criteria decision-making system for sustainable building assessment/certification. *Arch. Civ. Mech. Eng.* **2015**, *15*, 11–18. [CrossRef]

39. Štreimikienė, D.; Šliogerienė, J.; Turskis, Z. Multi-criteria analysis of electricity generation technologies in Lithuania. *Renew. Energy* **2016**, *85*, 148–156. [CrossRef]

40. Takano, A.; Hughes, M.; Winter, S. A multidisciplinary approach to sustainable building material selection: A case study in a Finnish context. *Build. Environ.* **2014**, *82*, 526–535. [CrossRef]

41. Iwaro, J.; Mwasha, A.; Williams, R.G.; Zico, R. An Integrated Criteria Weighting Framework for the sustainable performance assessment and design of building envelope. *Renew. Sustain. Energy Rev.* **2014**, *29*, 417–434. [CrossRef]

42. Muckler, F.A.; Seven, S.A. Selecting performance measures: "Objective" versus "subjective" measurement. *Hum. Factors* **1992**, *34*, 441–455. [CrossRef]

43. Stefanoiu, D.; Borne, P.; Popescu, D.; Filip, F.G.; El Kamel, A. *Optimization in Engineering Sciences. Approximate and Metaheuristic Methods*; John Wiley & Sons: Hoboken, NJ, USA, 2014.

44. Zavadskas, E.K.; Turskis, Z. Multiple criteria decision making (MCDM) methods in economics: An overview. *Technol. Econ. Dev. Econ.* **2011**, *17*, 397–427. [CrossRef]

45. Zavadskas, E.K.; Antucheviciene, J.; Turskis, Z.; Adeli, H. Hybrid multiple-criteria decision-making methods: A review of applications in engineering. *Sci. Iran. Trans. Civ. Eng.* **2016**, *23*, 1–20.

46. Turskis, Z.; Zavadskas, E.K.; Antucheviciene, J.; Kosareva, N. A hybrid model based on fuzzy AHP and fuzzy WASPAS for construction site selection. *Int. J. Comput. Commun. Control* **2015**, *10*, 113–128. [CrossRef]

47. Zavadskas, E.K.; Turskis, Z.; Bagočius, V. Multi-criteria selection of a deep-water port in the Eastern Baltic Sea. *Appl. Soft Comput.* **2015**, *26*, 180–192. [CrossRef]

48. Zavadskas, E.K.; Govindan, K.; Antucheviciene, J.; Turskis, Z. Hybrid multiple criteria decision-making methods: A review of applications for sustainability issues. *Econ. Res. -Ekon. Istraživanja* **2016**, *29*, 857–887. [CrossRef]

49. Keršulienė, V.; Turskis, Z. A hybrid linguistic fuzzy multiple criteria group selection of a chief accounting officer. *J. Bus. Econ. Manag.* **2014**, *15*, 232–252.

50. Zavadskas, E.K.; Turskis, Z.; Kildienė, S. State of art surveys of overviews on MCDM/MADM methods. *Technol. Econ. Dev. Econ.* **2014**, *20*, 165–179. [CrossRef]

51. Belton, V.; Stewart, T.J. *Multiple Criteria Decision Analysis: An Integrated Approach*; Kluwer Academic Publishers: Dordrecht, The Netherlands, 2001.

52. Saaty, T.L.; Ergu, D. When is a Decision-Making Method Trustworthy? Criteria for Evaluating Multi-Criteria Decision-Making Methods. *Int. J. Inf. Technol. Decis. Mak.* **2015**, *14*, 1171–1187. [CrossRef]

53. Van Laarhoven, P.J.M.; Pedrycz, W. A fuzzy extension of Saaty's priority theory. *Fuzzy Sets Syst.* **1983**, *11*, 229–241. [CrossRef]

54. Khorramshahgnol, R.; Moustakis, V.S. Delphic hierarchy process (DHP): A methodology for priority setting derived from the Delphi method and Analytical Hierarchy Process. *Eur. J. Oper. Res.* **1988**, *37*, 347–354. [CrossRef]

55. Schoemaker, P.J.H.; Waid, C.C. An experimental comparison of different approaches to determining weights in Additive Utility Models. *Manag. Sci.* **1982**, *28*, 182–196. [CrossRef]

56. Saaty, T.L. A scaling method for priorities in hierarchical structures. *J. Math. Psychol.* **1977**, *15*, 234–281. [CrossRef]

57. Keeney, R.L.; Raiffa, H. *Decisions with Multiple Objectives*; Wiley: New York, NY, USA, 1977.

58. Saaty, T.L. *An Eigenvalue Allocation Model for Prioritization and Planning*; Working Paper; University of Pennsylvania, Energy Management and Policy Center: Philadelphia, PA, USA, 1972.

59. Saaty, T.L. *The Analytic Hierarchy Process*; McGraw-Hill: New York, NY, USA, 1980.

60. Miller, G.A. The magical number seven, plus or minus two: Some limits in our capacity for processing information. *Psychol. Rev.* **1956**, *63*, 81–97. [CrossRef]

61. Zavadskas, E.K.; Turskis, Z. A new additive ratio assessment (ARAS) method in multicriteria decision-making. *Technol. Econ. Dev. Econ.* **2010**, *16*, 159–172. [CrossRef]

62. Turskis, Z.; Zavadskas, E.K. A novel method for multiple criteria analysis: Grey additive ratio assessment (ARAS-G) method. *Informatica* **2010**, *21*, 597–610.

63. Turskis, Z.; Zavadskas, E.K. A new fuzzy additive ratio assessment method (ARAS-F). Case study: The analysis of fuzzy multiple criteria in order to select the logistic centers location. *Transport* **2010**, *25*, 423–432. [CrossRef]

64. Krantz, D.H.; Luce, R.D.; Suppes, P.; Tversky, A. *Foundations of Measurement: Additive and Polynomial Representations*; Academic Press: New York, NY, USA, 1971; Volume 1.

65. Daniūnas, A.; Urbonas, K. Analysis of the steel frames with semirigid beam-to-beam and beam-to-column knee joints under bending and axial forces. *Eng. Struct.* **2008**, *30*, 3114–3118. [CrossRef]

66. Daniūnas, A.; Urbonas, K. Influence of the semi-rigid bolted joints on the frame behavior. *J. Civ. Eng. Manag.* **2010**, *16*, 237–241. [CrossRef]

67. Gečys, T.; Daniūnas, A.; Bader, T.K.; Wagner, L.; Eberhardsteiner, J. 3D finite element analysis and experimental investigations of a new type of timber beam-to-beam connection. *Eng. Struct.* **2015**, *86*, 134–145. [CrossRef]

© 2019 by the authors. Licensee MDPI, Basel, Switzerland. This article is an open access article distributed under the terms and conditions of the Creative Commons Attribution (CC BY) license (http://creativecommons.org/licenses/by/4.0/).

Article

Application of Fuzzy Analytical Network Process (ANP) and VIKOR for the Assessment of Green Agility Critical Success Factors in Dairy Companies

Ahmad Bathaei [1], Abbas Mardani [1], Tomas Baležentis [2,*], Siti Rahmah Awang [1], Dalia Streimikiene [2], Goh Chin Fei [1] and Norhayati Zakuan [1]

[1] Azman Hashim International Business School, Universiti Teknologi Malaysia (UTM), Skudai 81310, Johor, Malaysia; ahmadbathaei@gmail.com (A.B.); mabbas3@live.utm.my (A.M.); sitirahmah@utm.my (S.R.A.); gcfei@utm.my (G.C.F.); norhayatimz@utm.my (N.Z.)

[2] Lithuanian Institute of Agrarian Economics, V. Kudirkos g. 18, 03105 Vilnius, Lithuania; dalia@mail.lei.lt

* Correspondence: tomas@laei.lt; Tel.: +370-5262-2085

Received: 14 December 2018; Accepted: 12 February 2019; Published: 16 February 2019

Abstract: Manufacturing companies are facing rapid and unanticipated changes in their business environment. Most of these companies need to find new strategies to remain competitive in the market. Therefore, the main purpose of this study is to integrate the Fuzzy Analytical Network Process (ANP) and VIKOR methods to evaluate the green agile factors and sub-factors in the dairy companies in Iran. To find the green agile factors and sub-factors, this study used the expert's opinions and literature review. Data is collected from four dairy companies. The results of this study showed that the most important green agility factors are: trust-based relationship with suppliers, flexible production capacity, versatile workers, compliance with quality standards for a new product, and workers' willingness to learn. In addition, the results indicated that the green agility organization is one of the strategies that help companies to stay in the market. To validate the results, this study used four methods, including TOPSIS, ARAS, EDAS, and MABAC. The necessity of a reaction to the increasing customer choices, environmental concerns, and competitiveness among manufacturers across the globe has engaged the industry to embrace innovative manufacturing strategies.

Keywords: Supply Chain Management; Multi-Criteria Decision-Making; Green Supplier; Sustainability

---

1. Introduction

In recent years, firms have been obliged to make changes to their business process because of market transformations and technology innovations. Cost, quality, timeliness, and even flexibility are progressively becoming order qualifiers, hence pushing firms to devise businesses gravitating around innovativeness, responsibility, and customer intimacy [1]. Golpîra et al. [2] believe that it is important for companies to obtain a balance between economic interests and environmental protection, especially because of altered consumers' behavior toward green products and services.

Lee et al. [3] reported that green business accepts the environmental principles and respects the environment, which improves the quality of life for customers and protects existing resources. Green business operations involve reducing, reusing, recycling, reworking, returning, and remanufacturing [4]. Green marketing is focused on developing and marketing the products and services that can satisfy the customers' needs while taking environmental sustainability into account [5]. In addition, firms can focus on developing new and clean products. If products are overpriced or produced with lower quality or fail to consider the environmental benefits, the customers will not be attracted to them, and this will affect the firm's overall performance [6].

In recent years, environmental management issues have become more and more important to both public and private organizations [7]. Improvement of image, profitability, levels of emissions or customer satisfaction are several reasons for organizations to further consider environmental issues. The companies need to minimize adverse environmental impacts and waste of resources and raw materials during the procedure from the very beginning to the final stage and disposal of products [3]. Jovanović et al. [8] reported that organizations are in need of continuous change and development and require the implementation of various strategies. Agility is the normal evolution of flexibility over time which can help modern business organizations to remain in the competitive markets [9].

Implementation of agile methods can help companies to tailor the services and products to dynamic markets [10]. An agile organization is not only compatible with the business environment and ready for these anticipated changes but also qualified enough to sense the changes and respond to them in a quick and effective manner [11]. Nowadays, agility is a necessity for ensuring a competitive advantage and surviving. Customers demand the best products at a better price, less time, more customized, and in the desired value. This brings some problems for companies attempting to increase their market share. These companies are encountering a dynamic and unpredictable environment. Thus, agility and agility assessment of systems have been recognized as a necessary step for competing in a highly turbulent environment. Agile methods can help firms to have the best reaction to these challenges [12].

Agile manufacturing environment should be implemented in a consistent and systematic manner. Agile companies must be innovative, highly responsive, constantly experimenting in order to improve existing products and processes, and striving for less variability and greater capability [13]. Manufacturing practices for managing agility include enterprise integration, shared database, multimedia information network, product and process modeling, intelligent process control, virtual factory, design automation, super-computing, product data standards, paperless transactions via Electronic Data Interchange (EDI), high-speed information highway, etc. [14]. Ip et al. [15] suggested that the order of introduction of agility on shop floor should be adopting cellular layout followed by the reduction in a number of setups, paying attention to integrated quality, preventive maintenance, production control, inventory control, and finally improving relations with suppliers.

One of the goals of agile manufacturing is to produce customized products in a short time at low cost [16]. Another goal of all agile methods is to deliver products quickly and to adapt to changes in the process, product, and environment [17]. Jayatilleke and Lia [18] suggest that a wide variety of organizational settings have accepted the agile methods. Some methods are suitable for certain organizational environments while for a smaller organization, agile development is suitable.

Agile development designs new business models to enhance competitiveness and urges the need for a new organization model [19]. To win the competition in the global manufacturing environment, cooperation and collaboration among enterprises have played a key role in recent years [9]. Some factors affecting the environmental concerns include reduced response time to the customers, need to reach world-class score-cards, and coexistence with international competitors. These are the crucial factors in regards to the market needs [20].

Despite all these benefits, dairy companies all over the world suffered a dramatic decline in sale during the last decades. Iran, as the biggest dairy producer in the Middle East, producing 1.5 million tons of milk per year [21], is not an exception. The selected companies for this study (i.e., Kaleh, Haraz, Gela, and Saleh) are located in the same province (Mazandaran) and play an important role in the dairy market in Iran. They deliver their products to all parts of the country and also export them to countries, such as Afghanistan, Iraq, etc. These four companies employ several methods to introduce their products to the market, but nowadays they face problems to respond to the customers' demand and attract them to their brands. These companies produce a variety of dairy products and distribute them in a competitive market. Thus, these companies need to identify factors, metrics, and measures of green agility in order to satisfy the demands of the market. Many researchers [22,23] have discussed certain green production practices, such as green manufacturing, raw material reduction, and environmental design. Through the process of green production, the quality and variety of products must be taken into consideration [24]. The previous research has discussed the factors that can improve the green agility in the companies. However, there is a lack of research regarding the investigation of these factors in Iranian dairy companies. Therefore, the current study tries to fill this gap by identifying the key factors and sub-factors and providing a method for their measurement. Moreover, a comparative analysis of the green agility levels of these companies is provided.

2. Literature Review

The main factors determined in the previous studies are investigated in this paper. They are divided into five main categories: Market and customer agility, Technology agility, Production agility, Management agility, and Workforce agility.

2.1. The Market and Customer

Aravindraj and Vinodh [25] refer to this factor as an agile manufacturing capability. The customers' demand is increasing; thus, complex and dynamic actions are required to give appropriate services to customers. They expect to receive the products within a shorter period. Manufacturing organizations are attempting to be agile to produce a variety of products within a short period in a cost-effective manner. Heinonen and Strandvik [26] noted that providers see customers as targets to be activated and controlled, and the main concern for manufacturers is to differentiate themselves from their direct competitors. Peng et al. [27] believe that customers' behavior is changing. These dramatic changes to the fundamental characteristics of markets and business environments have precipitated an interest in creating new models for organizations. Based on researchers' opinion, markets need green agile products that can support the customers' demand. Customers pay attention to products that are perfectly suited to the environment.

2.2. Technology

Ji Sun et al. [28] noted that in such a dynamic environment, companies face increasing competition, including severe competence gaps, which presents a fundamental threat to their competitiveness and mere existence. In general, technology appropriation is greatly dependent upon and shaped by the surrounding environment, including social and economic forces beyond managerial intent. Matikiti et al. [29] believe that considering the social nature of technology and environment—when technology implies the relationship between social actors, and environment represents opportunities and constraints that can potentially promote or thwart an individual's goal attainments—we can understand that there is an interaction between technology and environment in reality. This is because how individuals use appropriate technology for their own goals is dependent on whether or not other social actors can create opportunities and/or remove constraints [27].

Mergel et al. [30] found that agile software development approaches involve creating, testing, and improving technology products incrementally in short, iterative sprints. The goal is to increase response to changes or mistakes discovered in the development processes. The overall project is broken down into small modules and short sprint cycles. Many of these agile principles have also made it into the agile development manifesto. Developing the technology approach also involves creating, testing, and improving technology products incrementally in short, iterative sprints. Based on Hausman and Johnston research [31], to gain a competitive advantage, it is crucial to develop innovations and technologies during the recession period. The literature on innovation notes that firms' innovative capacity depends greatly on external competitive pressures. Dai et al. [32] believe that to compete and succeed in stable markets, there is a need for different resources and innovative strategies. The high-tech entrepreneurial ventures have responded to the economic crisis through investments in product innovation and expansion into international markets. However, low-tech industries have to face additional difficulties in managing R&D projects during a crisis, as they require greater internal organizational capabilities to fit to rapidly changing external environments. It is important to have innovation activities in the long term to endure competitive pressures. In addition, firms need to develop strong internal capabilities to support their strategic objectives and survive during economic downturn conditions [33]. To answer the customers' demand, it is important to have facilities that can help us produce the products as soon as possible.

2.3. Production

Aqlan et al. [34] noted that dynamic changes in market demands and companies' strategies require the flexible introduction of new products and implementation of continuous improvement to internal processes in order to cope with the changes. One of the improving changes is consolidating production lines, especially when demand decreases and companies' strategies change.

Hasan et al. [35] believe that the traditional production layout is facing challenges as the product demands become smaller and shorter in lead-times. Lead times, set up times, work in process, quality, machine utilization, and employee job satisfaction are related to the production agility. Agile manufacturing is focused on these factors based on customers' demand [36]. According to emerging economy firms' (EEFs) opinion, to update the ideas about suppliers and move towards producing low-cost products, companies should focus on process development in a way to improve the products through moving from the state of mere imitation to innovation. They have gained momentum in many industries and have competed with developed firms not only within their own emerging markets but also within more developed ones. Successful EEFs are growing faster than their counterparts from developed markets and have been identified as global challenges [37]. This factor, which plays an important role in agile paradigms, affects key issues, such as modeling, producing, and delivering the products to customers.

2.4. Management

Famiyeh et al. [38] believe that Environment Management Practice (EMP) is a tool for an organization to manage the impacts of its activities on the environment. It provides a structured approach to planning and implementing environmental protection measures. EMP monitors environmental performance similar to the way a financial management system monitors expenditure and income, which enables an organization to regularly check its financial performance. Additionally, EMP integrates environmental management into a company's daily operations, long-term planning, and other quality management systems. Rathi [39] found that EMP is also one of the tools an organization can use to implement environmental policy. It illustrates an extension of the core principles of total quality programs to manage the environment. In other words, EMP can be described as the systematic application of business management to environmental issues [40]. It is important to understand the organizations' adoption behavior and identify the determinants of innovation by distinguishing the types of innovation. As adopting EMP involves implementing new or modified processes, techniques or systems to reduce environment damages, the adoption behavior can be considered as a technical innovation process [40].

Agile innovation management describes a set of project management and software development processes, adjusted procurement procedures combined with HR policies and organizational and managerial approaches in a way to support innovative digital service delivery in government. Innovation in government software development happens using an agile software development approach adopted from the private sector [30]. The firm's structure affects the firm's conduct, hence influencing its performance. Originally, most researchers took the approach of studying the structure of the industry and its direct links with the performance achieved [41]. It is important to test how political connections affect accounting quality, as stakeholders rely on corporate disclosure to improve their decision-making quality [42]. Managers make the decision about the company's strategy and can integrate agility into their decision-making processes.

2.5. Workforce

Agility of the workforce is broader than its flexibility, and it addresses a more strategic level. Workforce agility adds issues, such as motivation, attitude, behavior, and abilities, to human factors [43]. The Pitafi et al. [44] indicated that the agility shows an employee's ability of percipience and capability of responding to external changes, which requires the acquisition, interpretation, and utilization of relevant information. These information-processing procedures have an influence on employees in the workplace. Agility of employees reflects their ability to deal with environmental uncertainty through sensing and responding to external changes [45]. To build such agility, employees need to have sufficient sources of information and capabilities for processing such information. Specifically, agility contains the component of promptly sensing external changes, which requires employees to acquire a variety of information from multiple resources [44]. The workforce is one of the organizational parts in manufacturing without which no product can be produced. Therefore, this factor plays a key role in any company. Literature consists of studies conducted to find out agile factors. Table 1 shows the relevant factors and drivers.

Table 1. A review of research on the green agile factors.

Factors	Name of Sub-Factors	Reference
Market and customer	Matching customer feedback with products	Elgammal et al. [46]
	Flexible business	Ravichandran [47]
	Customer satisfaction rate of new product	Mourtzis et al. [48]
	Fast production and introducing the new product on time	Pinna et al., Morgan et al., Lo et al. [49–51]
	Respond quickly to competitors	Dikert et al. [52]
Technology	Diversity of equipment, technology, and operational workstations	Meneses et al. [53]
	Level of company's information excellence	Zraková et al. [54]
	Integration of technology and information	Zraková et al. [54]
	Network and information utilization rate for employees	Ravichandran [47]
	Applying the new communication media	Carr et al. [55]
Production	Modular design	Elgammal et al. [46]
	Flexible production capacity	Ravichandran, Chan et al., Queiroz et al. [47,56,57]
	Relationship based on trust with suppliers	Elgammal et al., Ravichandran [46,47]
	Decreasing non-added value costs	Rungi and Del Prete [58]
	Focusing on the costs of the system and identifying the activities that can add value	Rasnacis and Berzisa [59]
	To invest in the latest techniques, models, and design method	França et al. [60]
	Fixed manufacturing costs based on customer product pricing	Liu et al. [61]
	Short production development cycle	Paschek et al., Bondar et al. [62,63]
	Material transfer speed	Elgammal et al. [46]
	Creativity in products	Ravichandran [47]
	Product quality throughout the product longevity	Elgammal et al., Rasnacis and Berzisa [46,59]
	Resource optimization	Elgammal et al. [46]
	Cope with the change	Klein and Reinhart [64]
	Regarding quality standards in the production of new products	He and Yu [65]
Management	Hierarchy organizational chart beds	Pitafi et al. [43]
	Delegating management	Garwood and Poole [66]
	Management's interest in full automation	Karpinsky et al., Mossalam and Arafa [67,68]
	Management's interest in delivering new models	Mossalam and Arafa [69]
	Promoting a culture of transformation and modernization	Khoshlahn and Ardabili [70]
Workforce	New and existing employees' enthusiasm toward learning and training	Rathi et al. [38]
	Teamwork	Rasnacis and Berzisa, Hilt et al. [59,71]
	Institutionalizing staff design proposals	Zan et al. [72]
	Multi-skilled and flexible staff	Elgammal et al. [46]
	Collaboration interface	Valipour Khatir et al. [73]
	Creativity	Aqlan et al. [33]

3. Methods and Data

Data collection is important in any research since data leads to information. The more complete the information, the more correct and error-free results will be obtained. For this reason, the information about the factors and sub-factors are all collected from credible sources. Figure 1 shows the process of methodology. The research model is developed based on 21 experts' opinions. To collect the data from the selected companies, three separate questionnaires were prepared and filled by the participating specialists of the companies to detect the degree of factors' importance. The first questionnaire was used to determine the sub-factors. The experts then ranked the sub-factors from one to nine (lowest to highest importance), as shown in Table 2. 17 sub-factors from 35 sub-factors were finally selected and assessed. In the second questionnaire, paired comparisons were made between factors and sub-factors since the research was to compare fuzzy criteria. Nine-hour fuzzy spectrum was used that was preferably the same, interstitial, less preferred, in between, very little, intermediate, very high priority, and very little in between. In the third questionnaire, the companies were evaluated based on the green agility and ranked by VIKOR. A spectrum of seven language variables, such as very weak, weak, weak to moderate, moderate, relatively good, good, and very good, was used for the evaluation of factors and sub-factors within the company. Tables 2 and 3 show the equivalent fuzzy numbers, while Table 4 shows the experts' opinions about the important factors.

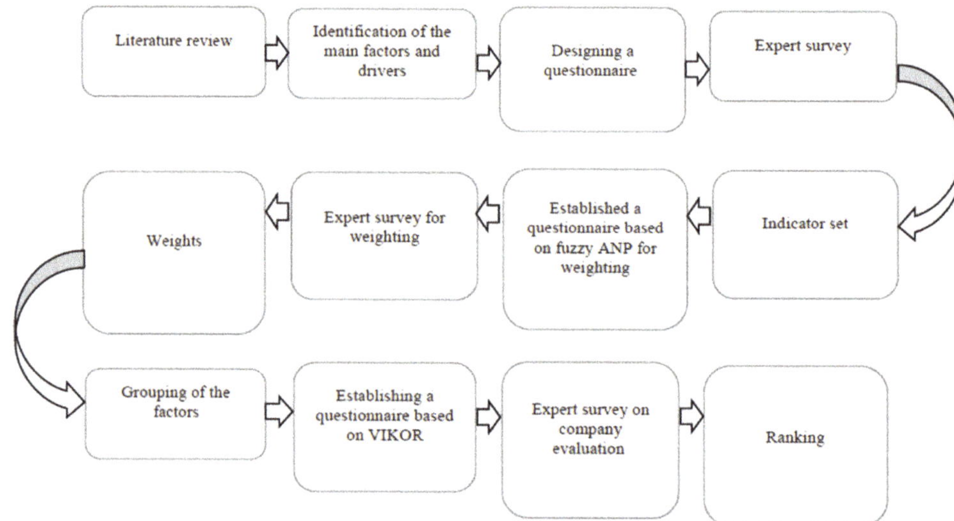

Figure 1. Research framework.

Table 2. Linguistic variables for the Fuzzy ANP and the corresponding triangular fuzzy numbers.

Linguistic Variables	Fuzzy Number	Inverse Fuzzy Number
Same preferences	(1,1,1)	(1,1,1)
Intermediary	(1,2,3)	(1/3,1/2,1)
A little preferred	(2,3,4)	(1/4,1/3,1/2)
Intermediary	(3,4,5)	(1/5,1/4,1/3)
Equally Preferred	(4,5,6)	(1/6,1/5,1/4)
Intermediary	(5,6,7)	(1/7,1/6,1/5)
Preferred a lot	(6,7,8)	(1/8,1/7,1/6)
Intermediary	(7,8,9)	(1/9,1/8,1/7)
Completely Preferred	(9,9,9)	(1/9,1/9,1/9)

Table 3. Triangular fuzzy numbers for evaluation of the alternatives.

Linguistic Variable	Fuzzy Equivalent
Very weak	(0,0,1)
Weak	(0,1,3)
Weak to moderate	(1,3,5)
Moderate	(3,5,7)
Almost good	(5,7,9)
Good	(7,9,10)
Very good	(9,10,10)

In this paper, the Fuzzy Analytical Network Process (ANP) and VIKOR methods are used to find and evaluate the green agile factors and sub-factors. The ANP is a generalization of the Analytic Hierarchy Process (AHP), popularly known as AHP. AHP is a theory of prioritization that derives relative scales of absolute numbers known as 'priorities' from judgments expressed numerically on an absolute fundamental scale [74]. The ANP framework has three basic features, which are useful in multi-criteria decision-making problems: (1) modeling the system's complexity, (2) measuring on a ratio scale, and (3) synthesizing. The local priorities in ANP are established in the same manner as they are in AHP using pairwise comparisons and judgments. However, the supermatrix approach, popularly known as the ANP approach, is becoming an attractive tool to understand more of the complex decision problem as it overcomes the limitation of the AHP's linear hierarchy structure [75].

The aim of Fuzzy ANP is to capture the 'fuzziness' or the vagueness-type uncertainties in the evaluation of remedial countermeasures, particularly, at the initial phase of remediation planning. Due to the complexity and uncertainty involved, as well as the inherent subjective nature of human judgments, it is sometimes unrealistic and infeasible to acquire exact judgments in pairwise comparisons. It is more natural or easier to provide verbal judgments when giving a subjective assessment. Based on the concept of fuzzy logic and the VIKOR method, the proposed VIKOR method has been developed to provide a rational, systematic process to discover the best solution and a compromise solution that can be used to resolve a fuzzy multi-criteria decision-making problem. The proposed VIKOR allows decision-makers to specify the preferred solutions for a given decision problem in real organizational settings [76]. The calculations were carried out by MATLAB and Excel software.

Table 4. Experts' opinion on the importance of factors.

Code	Factor	Sub-Factor	Average
A1		New and existing employees' enthusiasm toward learning and training	7.058
A2	Workforce	Teamwork	7.025
A3		Multi-skilled and flexible Staff	7.05
B1		Matching customer feedback on products	7.066
B2	Market and customer	Flexible business	7.116
B3		Customer satisfaction rate of new product	7.258
B4		Respond quickly to competitors	7
C1		Diversity of equipment, technology, and operational workstations	7.15
C2	Technology	Level of company's information system excellence	7.025
C3		Integration of technology and information	7.025
D1		Management's interest in full automation	7.025
D2	Management	Management's interest to deliver new models	7.225
D3		Promoting a culture of transformation and modernization	7.041
E1		Flexible production capacity	7.433
E2	Production	Relationship based on trust with suppliers	7.308
E3		Innovation in products	7.55
E4		Quality standards in the production of new products	7.041

3.1. Fuzzy ANP

Saaty [74] introduced the ANP technique in 1996. In this study, this technique was combined with the fuzzy approach. In this research, the triangular fuzzy numbers were used (see Figure 2). Table 4 shows the experts' opinions. In the next paragraph, the Fuzzy ANP steps are shown. Tables 5 and 6 show the super initial matrix and super normalized matrix.

Table 5. Initial supermatrix.

-	Goal	Workforce	Market and Customer	Technology	Management	Production	A1	A2	A3	B1	B2	B3	B4	C1	C2	C3	D1	D2	D3	E1	E2	E3	E4
Goal	0	0	0	0	0	0	0	0	0	0	0	0	0	0	0	0	0	0	0	0	0	0	0
Workforce	0.47	0	0.58	0.23	0.14	0.05	0	0	0	0	0	0	0	0	0	0	0	0	0	0	0	0	0
Market and customer	0.15	0.49	0	0.29	0.17	0.05	0	0	0	0	0	0	0	0	0	0	0	0	0	0	0	0	0
Technology	0.05	0.59	0.23	0	0.13	0.05	0	0	0	0	0	0	0	0	0	0	0	0	0	0	0	0	0
Management	0.25	0.56	0.26	0.14	0	0.05	0	0	0	0	0	0	0	0	0	0	0	0	0	0	0	0	0
Production	0.08	0.55	0.25	0.14	0.06	0	0	0	0	0	0	0	0	0	0	0	0	0	0	0	0	0	0
A1	0	0.25	0	0	0	0	0	0.9	0.1	0.07	0.14	0.21	0.58	0.13	0.73	0.14	0.12	0.27	0.61	0.14	0.15	0.33	0.39
A2	0	0.1	0	0	0	0	0.8	0	0.2	0.56	0.17	0.19	0.07	0.09	0.42	0.48	0.66	0.26	0.08	0.09	0.57	0.16	0.18
A3	0	0.65	0	0	0	0	0.8	0.2	0	0.55	0.09	0.08	0.27	0.09	0.66	0.25	0.11	0.44	0.44	0.22	0.59	0.09	0.09
B1	0	0	0.06	0	0	0	0.08	0.27	0.65	0	0.47	0.08	0.44	0.27	0.11	0.62	0.6	0.2	0.2	0.06	0.54	0.11	0.29
B2	0	0	0.63	0	0	0	0.08	0.27	0.65	0.07	0	0.67	0.26	0.44	0.11	0.44	0.71	0.14	0.14	0.26	0.59	0.06	0.09
B3	0	0	0.25	0	0	0	0	0	0	0.21	0.71	0	0.08	0.56	0.13	0.31	0.54	0.35	0.11	0.26	0.58	0.05	0.11
B4	0	0	0.07	0	0	0	0.08	0.27	0.65	0.72	0.09	0.19	0	0.59	0.3	0.12	0.46	0.32	0.22	0.11	0.51	0.05	0.32
C1	0	0	0	0.65	0	0	0.08	0.27	0.65	0.37	0.16	0.16	0.23	0	0.13	0.88	0.23	0.64	0.14	0.05	0.59	0.17	0.2
C2	0	0	0	0.08	0	0	0.65	0.08	0.27	0	0	0	0	0.2	0	0.8	0.09	0.35	0.56	0.34	0.3	0.12	0.24
C3	0	0	0	0.27	0	0	0	0	0	0.61	0.22	0.12	0.06	0.89	0.11	0	0.67	0.09	0.23	0.1	0.12	0.31	0.48
D1	0	0	0	0	0.27	0	0.67	0.24	0.08	0.37	0.24	0.16	0.23	0.08	0.25	0.67	0	0.83	0.17	0.31	1.14	0.1	0.09
D2	0	0	0	0	0.66	0	0.21	0.71	0.08	0.03	0.3	0.33	0.34	0.64	0.08	0.28	0.88	0	0.13	0.23	0.61	0.05	0.12
D3	0	0	0	0	0.07	0	0.24	0.09	0.67	0.51	0.26	0.17	0.07	0.57	0.32	0.11	0.88	0.13	0	0.6	0.22	0.05	0.12
E1	0	0	0	0	0	0.19	0.1	0.25	0.65	0.04	0.6	0.26	0.1	0.62	0.1	0.28	0.71	0.22	0.08	0	0.24	0.12	0.64
E2	0	0	0	0	0	0.52	0.06	0.71	0.24	0.11	0.24	0.05	0.61	0.59	0.08	0.33	0.72	0.21	0.07	0.61	0	0.1	0.29
E3	0	0	0	0	0	0.08	0.12	0.27	0.61	0.26	0.13	0.04	0.56	0.24	0.1	0.66	0.3	0.09	0.61	0.11	0.25	0	0.65
E4	0	0	0	0	0	0.21	0.71	0.24	0.06	0.45	0.27	0.07	0.22	0.17	0.75	0.08	0.66	0.07	0.27	0.09	0.25	0.66	0
SUM	1	2.2	1.32	0.79	0.5	0.19	4.52	4.22	4.26	4.93	4.17	2.77	4.14	6.17	4.38	6.45	8.32	4.63	4.05	3.57	7.26	2.53	4.27

Table 6. Normalized supermatrix.

-	Goal	Workforce	Market and Customer	Technology	Management	Production	A1	A2	A3	B1	B2	B3	B4	C1	C2	C3	D1	D2	D3	E1	E2	E3	E4
Goal	0	0	0	0	0	0	0	0	0	0	0	0	0	0	0	0	0	0	0	0	0	0	0
Workforce	0.47	0	0.25	0.13	0.09	0.04	0	0	0	0	0	0	0	0	0	0	0	0	0	0	0	0	0
Market and customer	0.15	0.15	0	0.16	0.11	0.04	0	0	0	0	0	0	0	0	0	0	0	0	0	0	0	0	0
Technology	0.05	0.19	0.1	0	0.09	0.04	0	0	0	0	0	0	0	0	0	0	0	0	0	0	0	0	0
Management	0.25	0.18	0.11	0.08	0	0.04	0	0	0	0	0	0	0	0	0	0	0	0	0	0	0	0	0
Production	0.08	0.17	0.11	0.08	0.04	0	0	0	0	0	0	0	0	0	0	0	0	0	0	0	0	0	0
A1	0	0.08	0	0	0	0	0	0.19	0.02	0.01	0.03	0.08	0.14	0.02	0.17	0.02	0.01	0.06	0.15	0.04	0.02	0.13	0.09
A2	0	0.03	0	0	0	0	0.17	0	0.04	0.11	0.04	0.07	0.02	0.02	0.1	0.08	0.08	0.06	0.02	0.03	0.08	0.06	0.04
A3	0	0.2	0	0	0	0	0.17	0.04	0	0.11	0.02	0.03	0.07	0.01	0.15	0.04	0.01	0.1	0.11	0.06	0.08	0.04	0.02
B1	0	0	0.02	0	0	0	0.02	0.06	0.12	0	0.12	0.03	0.11	0.04	0.03	0.1	0.07	0.04	0.05	0.02	0.08	0.04	0.07
B2	0	0	0.27	0	0	0	0.02	0.06	0.12	0.02	0	0.24	0.06	0.07	0.03	0.07	0.09	0.03	0.04	0.07	0.08	0.03	0.02
B3	0	0	0.11	0	0	0	0	0	0	0.04	0.17	0	0.02	0.09	0.03	0.05	0.07	0.08	0.03	0.07	0.08	0.02	0.03
B4	0	0	0.03	0	0	0	0.02	0.06	0.12	0.15	0.02	0.07	0	0.1	0.07	0.02	0.06	0.07	0.05	0.03	0.07	0.02	0.08
C1	0	0	0	0.36	0	0	0.02	0.06	0.12	0.08	0.04	0.06	0.06	0	0.03	0.14	0.03	0.14	0.03	0.01	0.08	0.07	0.05
C2	0	0	0	0.05	0	0	0.14	0.02	0.05	0	0	0	0	0.03	0	0.12	0.01	0.08	0.14	0.09	0.04	0.05	0.06
C3	0	0	0	0.15	0	0	0	0	0	0.12	0.05	0.04	0.01	0.14	0.03	0	0.08	0.02	0.06	0.03	0.02	0.12	0.11
D1	0	0	0	0	0.18	0	0.14	0.05	0.02	0.08	0.06	0.06	0.06	0.01	0.06	0.1	0	0.18	0.04	0.09	0.16	0.04	0.02
D2	0	0	0	0	0.44	0	0.05	0.15	0.02	0.01	0.07	0.12	0.08	0.1	0.02	0.04	0.11	0	0.03	0.06	0.08	0.02	0.03
D3	0	0	0	0	0.05	0	0.05	0.02	0.12	0.1	0.06	0.06	0.02	0.09	0.07	0.02	0.11	0.03	0	0.17	0.03	0.02	0.03
E1	0	0	0	0	0	0.16	0.02	0.05	0.12	0.01	0.15	0.09	0.02	0.1	0.02	0.04	0.09	0.05	0.02	0	0.03	0.05	0.15
E2	0	0	0	0	0	0.44	0.01	0.15	0.04	0.02	0.06	0.02	0.15	0.1	0.02	0.05	0.09	0.05	0.02	0.17	0	0.04	0.07
E3	0	0	0	0	0	0.07	0.03	0.06	0.11	0.05	0.03	0.02	0.14	0.04	0.02	0.1	0.04	0.02	0.15	0.03	0.03	0	0.15
E4	0	0	0	0	0	0.18	0.15	0.05	0.01	0.09	0.07	0.02	0.05	0.03	0.17	0.01	0.08	0.02	0.07	0.03	0.04	0.26	0
SUM	1	1	1	1	1	1	1	1	1	1	1	1	1	1	1	1	1	1	1	1	1	1	1

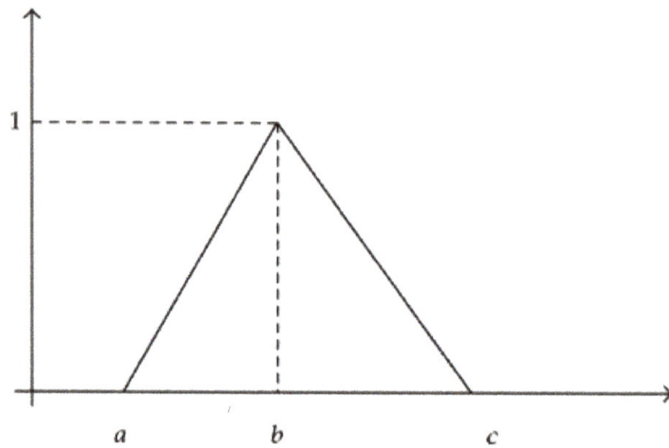

Figure 2. Fuzzy triangular number.

The procedure for implementing the Fuzzy ANP method is as follows:

This method was employed to calculate the priority weights from fuzzy comparison matrices. Chang's method [77] is relatively simpler than other kinds of the Fuzzy AHP method. The steps of Chang's extent analysis method are provided below. Let $X = \{x_1, x_2, \ldots, x_n\}$ be an object set and $U = \{u_1, u_2, \ldots, u_m\}$ be a goal set. According to the method of Chang's extent analysis, each object is taken, and an extended analysis for each goal (g_i) is performed [77]. Thus, m, extent analysis values for each object, can be obtained with the following signs:

$$M_{g_i}^1, \ M_{g_i}^2, \ \ldots, \ M_{g_i}^m, \ i = 1, 2, \ldots, n, \tag{1}$$

where $M_{g_i}^j$ ($j = 1, 2, \ldots, m$) are triangular fuzzy numbers.

Step 1: The value of fuzzy synthetic extent with respect to the i-th objective is defined as:

$$S_i = \sum_{j=1}^{m} M_{g_i}^j \otimes \left[\sum_{i=1}^{n} \sum_{j=1}^{m} M_{g_i}^j \right] \tag{2}$$

To obtain $\sum_{j=1}^{m} M_{g_i}^j$, the fuzzy addition operation of m extent analysis values for a particular matrix is performed as:

$$\sum_{j=1}^{m} M_{g_i}^j = \left(\sum_{j=1}^{m} l_j, \sum_{j=1}^{m} m_j, \sum_{j=1}^{m} u_j \right) \tag{3}$$

To obtain $\left[\sum_{i=1}^{n} \sum_{j=1}^{m} M_{g_i}^j \right]^{-1}$, the fuzzy addition operation of $M_{g_i}^j$ ($j = 1, 2, \ldots, m$) values are processed as:

$$\sum_{i=1}^{n} \sum_{j=1}^{m} M_{g_i}^j = \left(\sum_{i=1}^{n} l_i, \sum_{i=1}^{n} m_i, \sum_{i=1}^{n} u_i \right) \tag{4}$$

and then the inverse of the vector in Eqution (7) is obtained as:

$$\left[\sum_{i=1}^{n} \sum_{j=1}^{m} M_{g_i}^j \right]^{-1} = \left(\frac{1}{\sum_{i=1}^{n} u_i}, \frac{1}{\sum_{i=1}^{n} m_i}, \frac{1}{\sum_{i=1}^{n} l_i} \right) \tag{5}$$

Step 2: The degree of probability of $M_2 = (l_2, m_2, u_2) \geq M_1 = (l_1, m_1, u_1)$ is defined as:

$$V(M_2 \geq M_1) = \sup_{y \geq x}\left[\min(\mu_{M_1}(x), \mu_{M_2}(y))\right] \tag{6}$$

and can be equivalently expressed as follows:

$$V(M_2 \geq M_1) = hgt(M_1 \cap M_2) = \mu_{M_2}(d) = \begin{cases} 1 & \text{if } m_2 \geq m_1 \\ 0 & \text{if } l_1 \geq u_2 \\ \dfrac{l_1 - u_2}{(m_2 - u_2) - (m_1 - l_1)} & \text{otherwise} \end{cases} \tag{7}$$

where d is the ordinate of the highest intersection point D. To compare M_1 and M_2, we need both the values of $V(M) \geq M_2)$ and $V(M_2 \geq M_1)$.

Step 3: For the degree probability of a convex fuzzy number to be greater than k convex fuzzy numbers, M_i ($i = 1, 2, \ldots, k$) can be obtained as:

$$V(M \geq M_1, M_2, \ldots, M_K) = V\left[(M \geq M_1) \text{ and } (M \geq M_2) \text{ and } \ldots \text{ and } (M \geq M_K)\right] = \min V(M \geq M_i), \quad i = 1, 2, \ldots, k, \tag{8}$$

Assume that $d'(A_i) = \min V(S_i \geq S_K)$ for $k = 1, 2, \ldots, n, k \neq i$. Then, the weight vector is given by:

$$W' = (d'(A_1), d'(A_2), \ldots, d'(A_n))^T \tag{9}$$

where A_i are n elements.

$$d(A_i) = \frac{d'(A_i)}{\sum\limits_{i=1}^{n} d'(A_i)}$$

Step 4: The normalized weight vector elements are:

$$W = (d(A_1), d(A_2), \ldots, d(A_n))^T \tag{10}$$

where W contains crisp numbers.

3.2. The VIKOR Method

In this method, at first, surveys are conducted on selected companies' experts, and after examining the sub-factors in their company, they are informed about the desirability of sub-factors in their company. Figure 3 shows the steps involved.

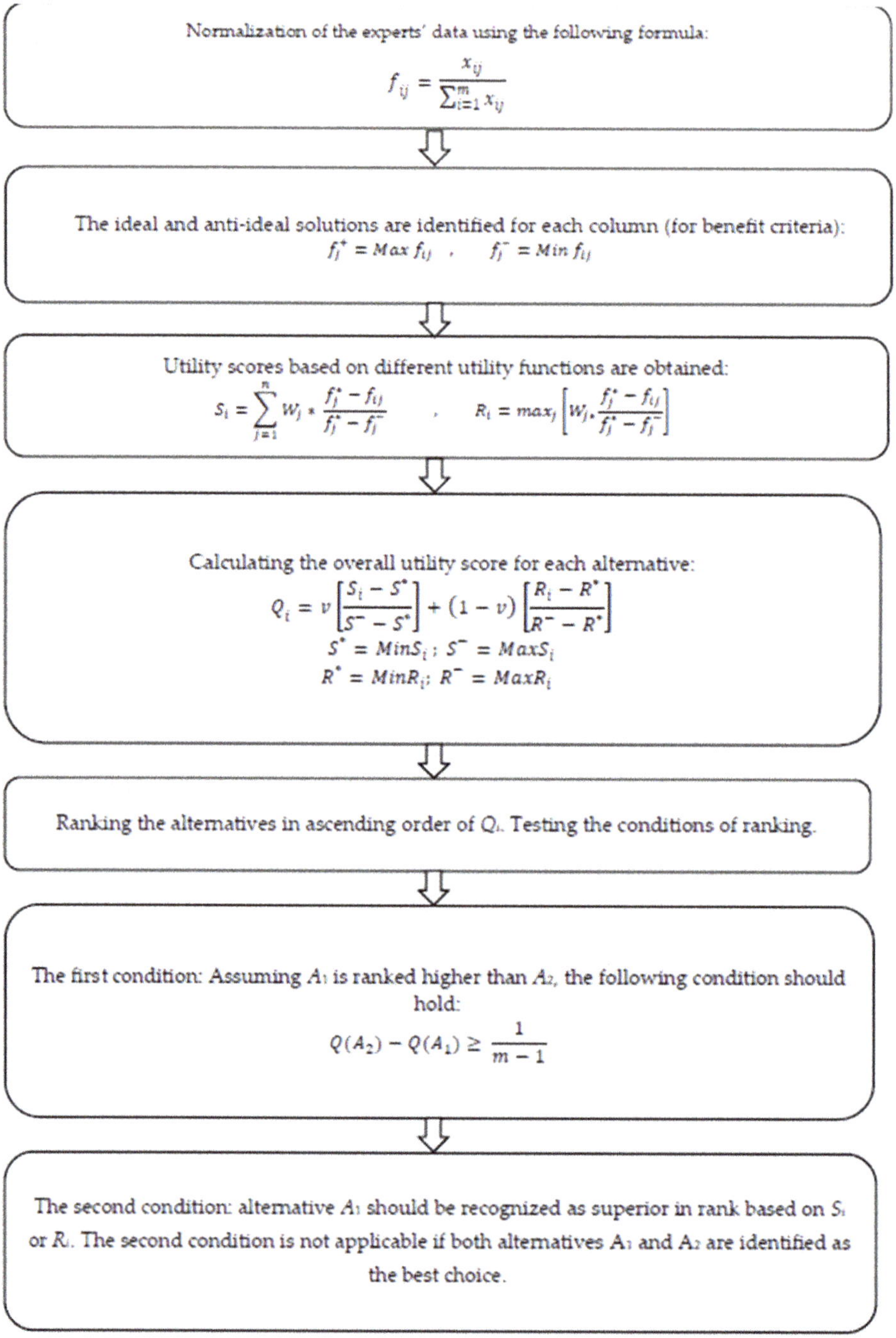

Normalization of the experts' data using the following formula:

$$f_{ij} = \frac{x_{ij}}{\sum_{i=1}^{m} x_{ij}}$$

The ideal and anti-ideal solutions are identified for each column (for benefit criteria):

$$f_j^* = Max\, f_{ij} \quad , \quad f_j^- = Min\, f_{ij}$$

Utility scores based on different utility functions are obtained:

$$S_i = \sum_{j=1}^{n} W_j * \frac{f_j^* - f_{ij}}{f_j^* - f_j^-} \quad , \quad R_i = max_j \left[W_j . \frac{f_j^* - f_{ij}}{f_j^* - f_j^-} \right]$$

Calculating the overall utility score for each alternative:

$$Q_i = v \left[\frac{S_i - S^*}{S^- - S^*} \right] + (1 - v) \left[\frac{R_i - R^*}{R^- - R^*} \right]$$
$$S^* = Min S_i \,;\, S^- = Max S_i$$
$$R^* = Min R_i \,;\, R^- = Max R_i$$

Ranking the alternatives in ascending order of Q_i. Testing the conditions of ranking.

The first condition: Assuming A_1 is ranked higher than A_2, the following condition should hold:

$$Q(A_2) - Q(A_1) \geq \frac{1}{m - 1}$$

The second condition: alternative A_1 should be recognized as superior in rank based on S_i or R_i. The second condition is not applicable if both alternatives A_1 and A_2 are identified as the best choice.

Figure 3. The procedure of VIKOR method.

4. Results

The analysis began with the weight setting. Table 4 shows the experts' opinion about the importance of the selected factors. They prioritized the factors provided in the questionnaire on a scale from 0 to 10. Then, based on the Delphi method, we selected the factors that have got more than seven points on average. Thus, the 17 factors were included in the final list. In the next step, the factors need to be weighted using Fuzzy ANP (see Tables 5 and 6).

After checking the consistency ratio (CR), the weights were transferred to the initial supermatrix (Table 5). However, to use the data, the supermatrix should be normalized. Table 6 shows the normalized supermatrix. The sub-factors' weights are presented in Table 7. These weights show that sub-factors are important (to varying extent) in the decision process. These weights are then used for VIKOR-based decision-making.

Table 7. The weights of sub-factors.

Factors	Code	Sub-Factors	Weight
Workforce	A1	Multi-skilled and flexible Staff	0.068
Workforce	A2	Teamwork	0.0593
Workforce	A3	New and existing employees' enthusiasm toward learning and training	0.0627
Market and customer	B1	Flexible Business	0.0578
Market and customer	B2	Matching customer feedback on products	0.0581
Market and customer	B3	Customer satisfaction rate of new product	0.0453
Market and customer	B4	Respond quickly to competitors	0.0579
Technology	C1	Level of company's information excellence	0.0572
Technology	C2	Diversity of equipment, technology, and operational workstations	0.0497
Technology	C3	Integration of technology and information	0.05
Management	D1	Promoting a culture of transformation and modernization	0.0676
Management	D2	Management's interest to deliver new models	0.0575
Management	D3	Management's interest in full automation	0.0591
Production	E1	Quality standards in the production of new products	0.0601
Production	E2	Innovation in products	0.0619
Production	E3	Relationship based on trust with suppliers	0.0606
Production	E4	Flexible production capacity	0.0673

Table 7 shows the ranking of sub-factors, which are calculated based on the experts' opinions. Factors with a score of more than 7.0 were chosen. Overall, 17 factors were divided into five groups. For the first group, the workforce is the main factor, and new and existing employees' enthusiasm toward learning and training, team working, multi-skilled, and flexible staff are the sub-factors. The second main factor is market and customer, and matching customer feedback on products, flexible business, customer satisfaction rate of the new product, and respond quickly to competitors are the sub-factors. Technology is the third main factor, including diversity of equipment, technology, and operational workstations, and level of the company's information system excellence and integration of technology and information are its sub-factors. Experts chose management's interest in full automation and management's interest to deliver new models as the management's main factors while promoting a culture of transformation and modernization are the sub-factors for management. Finally, for the last main factor production, flexible production capacity, relationship based on trust with suppliers, innovation in products, and regarding quality standards in the production of new products are selected.

The weights based on the Fuzzy ANP are presented in Table 5. Zero value indicates that there is no relationship between factors or sub-factors. This table is not normalized, and it is termed as the initial supermatrix. Table 6 presents the normalized supermatrix. To obtain the weight of the factors and sub-factors, this table should be solved to get the final weights. Therefore, the final weights are provided in Table 7.

Table 7 shows the obtained weights using the VIKOR method. The data illustrated in this table is employed as the raw data for ranking the companies based on green agility with the VIKOR method. The final factors' weight is presented in Table 7. Afterward, the obtained results were applied to VIKOR in which six tables exist showing all of the steps explained with the VIKOR method.

Table 8 shows the experts' opinions after the normalization process. Then, fj$^+$ and fj$^-$ for each column are shown in Table 9. The Si and Ri of each company are demonstrated in Tables 10 and 11, respectively. In the VIKOR method, the amounts of S and R are of high importance, and their information is presented in Table 12. The Q amount is the final part of the VIKOR method, which shows the comparison between the companies as illustrated in Table 13.

Table 8. Normalized expert assessments.

Variables	Kalleh	Gella	Haraz	Saleh	Total
Multi-skilled and flexible Staff	0.329	0.111	0.231	0.329	1
Teamwork	0.314	0.151	0.22	0.314	1
New and existing employees' enthusiasm toward learning and training	0.289	0.154	0.203	0.353	1
Flexible business	0.303	0.185	0.278	0.233	1
Matching customer feedback on products	0.269	0.251	0.203	0.277	1
Customer satisfaction rate of new product	0.266	0.173	0.28	0.28	1
Respond quickly to competitors	0.327	0.192	0.231	0.25	1
Level of company's information excellence	0.303	0.284	0.117	0.297	1
Diversity of equipment, technology, and operational workstations	0.262	0.244	0.214	0.28	1
Integration of technology and information	0.297	0.251	0.205	0.247	1
Promoting a culture of transformation and modernization	0.277	0.154	0.258	0.311	1
Management's interest to deliver new models	0.263	0.16	0.277	0.3	1
Management's interest in full automation	0.286	0.157	0.272	0.286	1
Quality standards in the production of new products	0.267	0.27	0.241	0.223	1
Innovation in products	0.318	0.195	0.244	0.244	1
Relationship based on trust with suppliers	0.28	0.238	0.187	0.295	1
Flexible production capacity	0.291	0.232	0.201	0.277	1

Table 9. $F_j{}^+$ and $F_j{}^-$ for each column.

Criterion	$F_j{}^+$	$F_j{}^-$
Multi-skilled and flexible staff	0.329	0.111
Teamwork	0.314	0.151
New and existing employees' enthusiasm toward learning and training	0.353	0.154
Flexible business	0.303	0.185
Matching customer feedback on products	0.277	0.203
Customer satisfaction rate of new product	0.28	0.173
Respond quickly to competitors	0.327	0.192
Level of company's information excellence	0.303	0.303
Diversity of equipment, technology, and operational workstations	0.28	0.214
Integration of technology and information	0.297	0.205
Promoting a culture of transformation and modernization	0.311	0.154
Management's interest to deliver new models	0.3	0.16
Management's interest in full automation	0.286	0.157
Quality standards in the production of new products	0.27	0.223
Innovation in products	0.318	0.195
Relationship based on trust with suppliers	0.295	0.187
Flexible production capacity	0.291	0.201

Table 10. The values of S_i.

Company	S^i
Kalleh	0.087
Gella	0.75
Haraz	0.621
Saleh	0.204

Table 11. The values of R_i.

Company	R_i
Kalleh	0.02
Gella	0.067
Haraz	0.067
Saleh	0.067

Table 12. The ideal points S and R.

$S+$	0.0878
$S-$	0.75
$R+$	0.02
$R-$	0.067

Table 13. The final utility score Q.

Company	Q	Rank
Kalleh	0	1
Saleh	0.5	2
Haraz	0.895	3
Gella	0.9839	4

Further, the TOPSIS [78], ARAS [79], EDAS [80], and MABAC [81] methods were applied in order to test the robustness of the results. Table 14 presents the results of the comparative analysis. As one can note, there are no differences in regards to the best performing company. However, the ranking of the other companies differs across the six approaches (Tables 13 and 14).

Different Multiple Criteria Decision Making (MCDM) methods follow different principles of data aggregation and calculation of the final utility scores. The results of the ranking procedures based on different aggregation rules basically confirm the effectiveness of the approach proposed in this study. Thus, it can be applied for the multi-dimensional analysis of the performance of a dairy company.

Table 14. Comparative analysis based on different methods.

Method	Company	Utility Score	Ranking
ARAS [79]	Kalleh	0.9624	1
	Saleh	0.9352	2
	Doshe	0.7500	3
	Gella	0.6576	4
EDAS [80]	Kalleh	1	1
	Saleh	0.911	2
	Doshe	0.308	3
	Gella	0.039	4
MABAC [81]	Kalleh	0.3823	1
	Saleh	0.2660	2
	Doshe	−0.1514	3
	Gella	−0.2807	4
TOPSIS [78]	Kalleh	0.853966	1
	Saleh	0.781376	2
	Doshe	0.442795	3
	Gella	0.270683	4
VIKOR [82]	Kalleh	0	1
	Saleh	0.50	2
	Doshe	0.895	3
	Gella	0.9839	4

5. Discussion

In the present era, flexibility in the business market is an important element of any organization in order to be well survived and preserved. If an organization fail to preserve itself with the outside environment, it will lose the competition with rivals and lose their place in the markets. Green agility in any organization is the most important issue and should be taken seriously. Many elements play an important role in each organization's green agility, which we attempted to evaluate in this study. This paper investigated the factors affecting green agility in organizations and studied these factors in dairy companies, such as Kalleh, Doushe, Haraz, and Gela. We also reviewed the green agility of the four companies, compared them with each other, and ranked them based on their green agility. The model presented in this survey was formed by gathering data from several valid sources. The model was evaluated by the view of experts of the companies. After selecting a sub-factor, weighting was done using the Fuzzy ANP after the distribution of the pairwise comparison questionnaires among the experts.

Note that some previous studies, e.g., Papadopoulos [83], suggested that organizations have to adapt to business needs and leap from a traditional system to a green agile. They require improvements based on quality, control, customer perception of the final product, increased communication among team members, and better standards of employee satisfaction. Also, Tanner and Wheeling [84] reported that the effective factors for becoming an agile organization involve culture, customer involvement and commitment, stakeholders participation and sales, team structure and team logistics, project type and planning, and skill level and attitude of the team members.

The results of this study (Table 7) showed that the factors, multi-skilled and flexible employees, promoting the transformation culture and modernization, flexible production capacity, interest of new and existing employees in teaching and learning, product innovation, and relationship with suppliers based on trust and respect, are important to a movement towards green agility. Experts believe that when making a new product, considering the quality standards has the highest impact on the agility of organizations and companies. Multi-skilled and flexible employees are the most important assets of an organization. Employees who are more capable and multi-skilled have the ability to help the organization when needed [85].

An organization may require an employee to work in another part, and the multi-skilled employee will help the organization to meet its customers' needs and compete with the rivals. Promoting the culture of change and modernity: culture transformation and modernization are important factors in an organization to be well adapted as quickly as possible with their outside environment. As organizations promote the culture of change and renewal, employees will know the meaning of premature change more easily and cooperate better in order to make necessary changes to their organization. Flexible production capacity: in the present era, customers' needs are not completely consistent with the development of technology and up-to-date products. Customers are attracted to new products; thus, companies need to make their capacity flexible in order to meet the needs of previous clients, keep them, meet the needs of new customers, and to compete with their counterparts in the field of business without falling behind. The threat of falling behind and losing the business entirely to a competitor actually makes the company continue working hard [86].

The interest of existing and new employees in teaching and learning: motivation and satisfaction among employees of an organization will enhance the growth and excellence of creativity in organizations and can provide solutions to respond to competitors and the market. Discussion about employee training is one of the most important responsibilities of organizations to make their employees ready toward the threads and new opportunities in order to take the greatest advantage of their expertise and ability. Nowadays, products change and progress faster than we think. The longevity of the products has come down to such an extent that the organization has to think about what the next product will be at the beginning of launching a new product. The more a product helps the customers, the more it will be accepted by them.

Trust-based relationship with the suppliers: it is one of the most important factors in the green agility of an organization [87]. If an organization can establish a better relationship with its suppliers and rely on them in terms of preparing raw materials or making any change to the materials, it is able to respond to the market needs and be well adapted to changes that may occur in the market. The faster such adaptation occurs, the greener is the organization. Kalleh is one of the examples of such an organization. Concerning the quality standards to produce a new product: the quality is always the choice of the customer and is the sign of superiority of one company over the other. Paying attention to quality standards will enhance the quality of the products [88]. It will be easier for customers to choose the products, and with an improvement in quality, customers will choose the products more easily. Every organization needs to attempt to continuously improve the quality to have an advantage in competition with their peers.

6. Conclusions

The paper proposed a methodology for assessment of the green agility of the dairy companies. Taking the example of Iran, the results indicated that Kalleh Company was greener than the other companies because of product innovation and enthusiasm of employees to learn, which has made this company the best, as shown in Table 13. In addition to the competition in the domestic market, this company considers exporting its products outside the country borders. This shows the competency of the company in all areas of production, management, and staff. Each factor involves important sub-factors for making the selected dairy companies agile. The important sub-factors of agility are: agility in the workforce, multi-skilled and flexible employees, market dimensions and customers, the implementation of customer feedback on products in terms of the variety of machines in operation and workstation in technology dimension, agility management, promoting the culture of innovation and transformation, and trust-based relationship with the suppliers in production dimension.

This study has some limitations. First, the considered factors and expert assessments regarding their importance may vary across companies as experts might have ignored some factors. Second, expert ratings may be biased due to the goal of the study. Anyway, the results obtained are plausible given the trends prevailing in the market. Finally, data collection is highly cumbersome as the companies do not want to reveal their information due to competition. In this paper, a new method was proposed to measure the Iranian dairy company's green agility. The factors were identified related to the company's green agility. Some suggestions for further research in this field involve using Fuzzy MCDM to find the important factors in dairy companies' supply chain based on green agility. The holistic paradigm and associated methodological tools [89–92] can decrease the environmental pressures and allow responding market demand quickly. Thus, this system can decrease environmental damage and increase performance with a high level of corporate social responsibility.

Author Contributions: Investigation, A.B.; Methodology, A.M.; Writing–original draft, S.R.A. and N.Z.; Writing–review and editing, T.B., G.C.F., and D.S.

Funding: This research is funded by the European Social Fund according to the activity "Improvement of researchers' qualification by implementing world-class R&D projects" of Measure No. 09.3.3-LMT-K-712.

Acknowledgments: The authors would like to thank the Research Management Center (RMC) at Universiti Teknologi Malaysia (UTM) and Ministry of Higher Education, Malaysia for supporting and funding this research under the Fundamental Research Grant Scheme (FRGS) (Vote no. FRGS/1/2018/SS03/UTM/02/3).

Conflicts of Interest: The authors declare no conflict of interest.

References

1. Zanetti, V.; Cavalieri, S.; Pezzotta, G. Additive manufacturing and pss: A solution life-cycle perspective. *IFAC-PapersOnLine* **2016**, *49*, 1573–1578. [CrossRef]
2. Golpîra, H.; Najafi, E.; Zandieh, M.; Sadi-Nezhad, S. Robust bi-level optimization for green opportunistic supply chain network design problem against uncertainty and environmental risk. *Comput. Ind. Eng.* **2017**, *107*, 301–312. [CrossRef]
3. Lee, J.W.; Kim, Y.M.; Kim, Y.E. Antecedents of adopting corporate environmental responsibility and green practices. *J. Bus. Ethics* **2018**, *148*, 397–409. [CrossRef]
4. Rosário Cabrita, M.D.; Duarte, S.; Carvalho, H.; Cruz-Machado, V. Integration of lean, agile, resilient and green paradigms in a business model perspective: Theoretical foundations. *IFAC-PapersOnLine* **2016**, *49*, 1306–1311.
5. Singh, P.; Pandey, K.K. Green marketing: Policies and practices for sustainable development. *Integral Rev.* **2012**, *5*, 22–30.
6. Simão, L.; Lisboa, A. Green marketing and green brand—The toyota case. *Procedia Manuf.* **2017**, *12*, 183–194. [CrossRef]
7. Han, H.; Lee, M.J.; Kim, W. Antecedents of green loyalty in the cruise industry: Sustainable development and environmental management. *Bus. Strategy Environ.* **2018**, *27*, 323–335. [CrossRef]
8. Jovanović, M.; Mas, A.; Mesquida, A.-L.; Lalić, B. Transition of organizational roles in agile transformation process: A grounded theory approach. *J. Syst. Softw.* **2017**, *133*, 174–194. [CrossRef]
9. Da Silva, G.L.; Rondina, G.; Figueiredo, P.C.; Prates, G.; Savi, A.F. How quality influences in agility, flexibility, responsiveness and resilience in supply chain management. *Indep. J. Manag. Prod.* **2018**, *9*, 340–353. [CrossRef]
10. Ye, Y.; Lau, K.H. Designing a demand chain management framework under dynamic uncertainty: An exploratory study of the chinese fashion apparel industry. *Asia Pac. J. Mark. Logist.* **2018**, *30*, 198–234. [CrossRef]
11. Rodrigues, N.; Oliveira, E.; Leitão, P. Decentralized and on-the-fly agent-based service reconfiguration in manufacturing systems. *Comput. Ind.* **2018**, *101*, 81–90. [CrossRef]
12. Yusuf, Y.Y.; Adeleye, E.O.; Papadopoulos, T. Agile manufacturing practices: The role of big data and business analytics with multiple case studies au-gunasekaran, angappa. *Int. J. Prod. Res.* **2018**, *56*, 385–397.
13. Sanders, J. Process management, innovation, and efficiency performance: The moderating effect of competitive intensity. *Bus. Process Manag. J.* **2014**, *20*, 35–358.
14. Buiten, G.; Snijkers, G.; Saraiva, P.; Erikson, J.; Erikson, A.-G.; Born, A. Business data collection: Toward electronic data interchange. Experiences in portugal, canada, sweden, and the netherlands with edi. *J. Off. Stat.* **2018**, *34*, 419–443. [CrossRef]
15. Ip, W.H.; Yung, K.L.; Wang, D. A branch and bound algorithm for sub-contractor selection in agile manufacturing environment. *Int. J. Prod. Econ.* **2004**, *87*, 195–205. [CrossRef]

16. Rauch, E.; Spena, P.R.; Matt, D.T. Axiomatic design guidelines for the design of flexible and agile manufacturing and assembly systems for smes. *Int. J. Interact. Des. Manuf. (Ijidem)* **2018**, 1–22. [CrossRef]

17. Cooper, R.G.; Sommer, A.F. Agile–stage-gate for manufacturers: Changing the way new products are developed integrating agile project management methods into a stage-gate system offers both opportunities and challenges. *Res. Technol. Manag.* **2018**, *61*, 17–26. [CrossRef]

18. Jayatilleke, S.; Lai, R. A systematic review of requirements change management. *Inf. Softw. Technol.* **2018**, *93*, 163–185. [CrossRef]

19. Bouwman, H.; Heikkilä, J.; Heikkilä, M.; Leopold, C.; Haaker, T. Achieving agility using business model stress testing. *Electron. Mark.* **2018**, *28*, 149–162. [CrossRef]

20. Agostinho, O.L. Proposal of organization framework model, using business processes and hierarchical patterns to provide agility and flexibility in competitiveness environments. *Procedia Eng.* **2015**, *131*, 401–409. [CrossRef]

21. Rahnama, H.; Rajabpour, S. Factors for consumer choice of dairy products in iran. *Appetite* **2017**, *111*, 46–55. [CrossRef]

22. Singh, A.; Philip, D.; Ramkumar, J.; Das, M. A simulation based approach to realize green factory from unit green manufacturing processes. *J. Clean. Prod.* **2018**, *182*, 67–81. [CrossRef]

23. Li, T.; Liang, L.; Han, D. Research on the efficiency of green technology innovation in china's provincial high-end manufacturing industry based on the raga-pp-sfa model. *Math. Probl. Eng.* **2018**, *2018*, 13. [CrossRef]

24. Mittal, V.K.; Sindhwani, R.; Kalsariya, V.; Salroo, F.; Sangwan, K.S.; Singh, P.L. Adoption of integrated lean-green-agile strategies for modern manufacturing systems. *Procedia Cirp* **2017**, *61*, 463–468. [CrossRef]

25. Aravindraj, S.; Vinodh, S. Forty criteria based agility assessment using scoring approach in an indian relays manufacturing organization. *J. Eng. Des. Technol.* **2014**, *12*, 507–518. [CrossRef]

26. Heinonen, K.; Strandvik, T. Reflections on customers' primary role in markets. *Eur. Manag. J.* **2018**, *36*, 1–11. [CrossRef]

27. Peng, Z.; Sun, Y.; Guo, X. Antecedents of employees' extended use of enterprise systems: An integrative view of person, environment, and technology. *Int. J. Inf. Manag.* **2018**, *39*, 104–120. [CrossRef]

28. Sun, J.; Marinova, D.; Zhao, D. Evaluation of the green technology innovation efficiency of china's manufacturing industries: Dea window analysis with ideal window width au-lin, shoufu. *Technol. Anal. Strateg. Manag.* **2018**, *30*, 1166–1181.

29. Matikiti, R.; Mpinganjira, M.; Roberts-Lombard, M. Application of the technology acceptance model and the technology–organisation–environment model to examine social media marketing use in the south african tourism industry. *S. Afr. J. Inf. Manag.* **2018**, *20*, 1–12. [CrossRef]

30. Mergel, I.; Gong, Y.; Bertot, J. *Agile Government: Systematic Literature Review and Future Research*; Elsevier: Amsterdam, The Netherlands, 2018.

31. Hausman, A.; Johnston, W.J. The role of innovation in driving the economy: Lessons from the global financial crisis. *J. Bus. Res.* **2014**, *67*, 2720–2726. [CrossRef]

32. Dai, J.; Chan, H.K.; Yee, R.W.Y. Examining moderating effect of organizational culture on the relationship between market pressure and corporate environmental strategy. *Ind. Mark. Manag.* **2018**, *74*, 227–236. [CrossRef]

33. Zouaghi, F.; Sánchez, M.; Martínez, M.G. Did the global financial crisis impact firms' innovation performance? The role of internal and external knowledge capabilities in high and low tech industries. *Technol. Forecast. Soc. Chang.* **2018**, *132*, 92–104. [CrossRef]

34. Aqlan, F.; Lam, S.S.; Ramakrishnan, S. An integrated simulation–optimization study for consolidating production lines in a configure-to-order production environment. *Int. J. Prod. Econ.* **2014**, *148*, 51–61. [CrossRef]

35. Hasan, M.A.; Sarkis, J.; Shankar, R. Agility and production flow layouts: An analytical decision analysis. *Comput. Ind. Eng.* **2012**, *62*, 898–907. [CrossRef]

36. Thilak, V.; Devadasan, S.; Sivaram, N. A literature review on the progression of agile manufacturing paradigm and its scope of application in pump industry. *Sci. World J.* **2015**, *2015*. [CrossRef]

37. Bernat, S.; Karabag, S.F. Strategic alignment of technology: Organising for technology upgrading in emerging economy firms. *Technol. Forecast. Soc. Chang.* **2018**. [CrossRef]

38. Famiyeh, S.; Adaku, E.; Amoako-Gyampah, K.; Asante-Darko, D.; Amoatey, C.T. Environmental management practices, operational competitiveness and environmental performance: Empirical evidence from a developing country. *J. Manuf. Technol. Manag.* **2018**, *29*, 588–607. [CrossRef]

39. Rathi, A. Development of environmental management program in environmental impact assessment reports and evaluation of its robustness: An indian case study. *Impact Assess. Proj. Apprais.* **2018**, 1–16. [CrossRef]

40. Ibrahim, I.B.; Jaafar, H.S.B. Factors of environment management practices adoptions. *Procedia Soc. Behav. Sci.* **2016**, *224*, 353–359. [CrossRef]

41. Fernández, E.; Iglesias-Antelo, S.; López-López, V.; Rodríguez-Rey, M.; Fernandez-Jardon, C.M. Firm and industry effects on small, medium-sized and large firms' performance. *BRQ Bus. Res. Q.* **2018**. [CrossRef]

42. Ding, R.; Li, J.; Wu, Z. Government affiliation, real earnings management, and firm performance: The case of privately held firms. *J. Bus. Res.* **2018**, *83*, 138–150. [CrossRef]

43. Qin, R.; Nembhard, D.A. Workforce agility in operations management. *Surv. Oper. Res. Manag. Sci.* **2015**, *20*, 55–69. [CrossRef]

44. Pitafi, A.H.; Liu, H.; Cai, Z. Investigating the relationship between workplace conflict and employee agility: The role of enterprise social media. *Telemat. Inform.* **2018**, *35*, 2157–2172. [CrossRef]

45. Muduli, A. Workforce agility: Examining the role of organizational practices and psychological empowerment. *Glob. Bus. Organ. Excell.* **2017**, *36*, 46–56. [CrossRef]

46. Elgammal, A.; Papazoglou, M.; Krämer, B.; Constantinescu, C. Design for customization: A new paradigm for product-service system development. *Procedia Cirp* **2017**, *64*, 345–350. [CrossRef]

47. Ravichandran, T. Exploring the relationships between it competence, innovation capacity and organizational agility. *J. Strateg. Inf. Syst.* **2018**, *27*, 22–42. [CrossRef]

48. Mourtzis, D.; Vlachou, E.; Zogopoulos, V.; Gupta, R.K.; Belkadi, F.; Debbache, A.; Bernard, A. Customer feedback gathering and management tools for product-service system design. *Procedia Cirp* **2018**, *67*, 577–582. [CrossRef]

49. Pinna, C.; Galati, F.; Rossi, M.; Saidy, C.; Harik, R.; Terzi, S. Effect of product lifecycle management on new product development performances: Evidence from the food industry. *Comput. Ind.* **2018**, *100*, 184–195. [CrossRef]

50. Morgan, T.; Obal, M.; Anokhin, S. Customer participation and new product performance: Towards the understanding of the mechanisms and key contingencies. *Res. Policy* **2018**, *47*, 498–510. [CrossRef]

51. Lo, S.M.; Zhang, S.; Wang, Z.; Zhao, X. The impact of relationship quality and supplier development on green supply chain integration: A mediation and moderation analysis. *J. Clean. Prod.* **2018**, *202*, 524–535. [CrossRef]

52. Dikert, K.; Paasivaara, M.; Lassenius, C. Challenges and success factors for large-scale agile transformations: A systematic literature review. *J. Syst. Softw.* **2016**, *119*, 87–108. [CrossRef]

53. Meneses, Y.E.; Stratton, J.; Flores, R.A. Water reconditioning and reuse in the food processing industry: Current situation and challenges. *Trends Food Sci. Technol.* **2017**, *61*, 72–79. [CrossRef]

54. Zraková, D.; Kubina, M.; Koman, G. Influence of information-communication system to reputation management of a company. *Procedia Eng.* **2017**, *192*, 1000–1005. [CrossRef]

55. Carr, J.; Decreton, L.; Qin, W.; Rojas, B.; Rossochacki, T.; Yang, Y.W. Social media in product development. *Food Qual. Prefer.* **2015**, *40*, 354–364. [CrossRef]

56. Chan, A.T.L.; Ngai, E.W.T.; Moon, K.K.L. The effects of strategic and manufacturing flexibilities and supply chain agility on firm performance in the fashion industry. *Eur. J. Oper. Res.* **2017**, *259*, 486–499. [CrossRef]

57. Queiroz, M.; Tallon, P.P.; Sharma, R.; Coltman, T. The role of IT application orchestration capability in improving agility and performance. *J. Strateg. Inf. Syst.* **2018**, *27*, 4–21. [CrossRef]

58. Rungi, A.; Del Prete, D. The smile curve at the firm level: Where value is added along supply chains. *Econ. Lett.* **2018**, *164*, 38–42. [CrossRef]

59. Rasnacis, A.; Berzisa, S. Method for adaptation and implementation of agile project management methodology. *Procedia Comput. Sci.* **2017**, *104*, 43–50. [CrossRef]

60. França, C.L.; Broman, G.; Robèrt, K.-H.; Basile, G.; Trygg, L. An approach to business model innovation and design for strategic sustainable development. *J. Clean. Prod.* **2017**, *140*, 155–166. [CrossRef]

61. Liu, Y.; Tyagi, R.K. Outsourcing to convert fixed costs into variable costs: A competitive analysis. *Int. J. Res. Mark.* **2017**, *34*, 252–264. [CrossRef]

62. Paschek, D.; Rennung, F.; Trusculescu, A.; Draghici, A. Corporate development with agile business process modeling as a key success factor. *Procedia Comput. Sci.* **2016**, *100*, 1168–1175. [CrossRef]

63. Bondar, S.; Hsu, J.C.; Pfouga, A.; Stjepandić, J. Agile digitale transformation of enterprise architecture models in engineering collaboration. *Procedia Manuf.* **2017**, *11*, 1343–1350. [CrossRef]

64. Klein, T.P.; Reinhart, G. Towards agile engineering of mechatronic systems in machinery and plant construction. *Procedia Cirp* **2016**, *52*, 68–73. [CrossRef]

65. He, H.; Yu, Z. Product quality, incomplete contract and the product cycle. *Int. Rev. Econ. Financ.* **2018**, *53*, 160–167. [CrossRef]

66. Garwood, D.A.; Poole, A.H. Project management as information management in interdisciplinary research:"Lots of different pieces working together". *Int. J. Inf. Manag.* **2018**, *41*, 14–22. [CrossRef]

67. Ball, O.; Robinson, S.; Bure, K.; Brindley, D.A.; Mccall, D. Bioprocessing automation in cell therapy manufacturing: Outcomes of special interest group automation workshop. *Cytotherapy* **2018**, *20*, 592–599. [CrossRef]

68. Karpinsky, N.D.; Chancey, E.T.; Palmer, D.B.; Yamani, Y. Automation trust and attention allocation in multitasking workspace. *Appl. Ergon.* **2018**, *70*, 194–201. [CrossRef]

69. Mossalam, A.; Arafa, M. Governance model for integrating organizational project management (opm) with corporate practices. *HBRC J.* **2017**, *13*, 302–314. [CrossRef]

70. Khoshlahn, M.; Ardabili, F.S. The role of organizational agility and transformational leadership in service recovery prediction. *Procedia-Soc. Behav. Sci.* **2016**, *230*, 142–149. [CrossRef]

71. Hilt, M.J.; Wagner, D.; Osterlehner, V.; Kampker, A. Agile predevelopment of production technologies for electric energy storage systems—A case study in the automotive industry. *Procedia Cirp* **2016**, *50*, 88–93. [CrossRef]

72. Zan, J.; Hasenbein, J.J.; Morton, D.P.; Mehrotra, V. Staffing call centers under arrival-rate uncertainty with bayesian updates. *Oper. Res. Lett.* **2018**, *46*, 379–384. [CrossRef]

73. Valipour Khatir, M.; Mohammadpour Omran, M.; Akbarzadeh, Z. Evaluating indicators of organizational agility by fuzzy multi criteria decision making: (iran power development organization). *Q. Int. Innov. Enterp.* **2015**, *3*, 1–18.

74. Saaty, T.L. *The Anp for Decision Making with Dependence and Feedback*; RWS Publications: Pittsburgh, PA, USA, 1996.

75. Saaty, T.L. Making and validating complex decisions with the ahp/anp. *J. Syst. Sci. Syst. Eng.* **2005**, *14*, 1–36. [CrossRef]

76. Shemshadi, A.; Shirazi, H.; Toreihi, M.; Tarokh, M.J. A fuzzy vikor method for supplier selection based on entropy measure for objective weighting. *Expert Syst. Appl.* **2011**, *38*, 12160–12167. [CrossRef]

77. Cheng, E.W.L.; Li, H.; Yu, L. The analytic network process (anp) approach to location selection: A shopping mall illustration. *Constr. Innov.* **2005**, *5*, 83–97. [CrossRef]

78. García-Cascales, M.S.; Lamata, M.T. On rank reversal and topsis method. *Math. Comput. Model.* **2012**, *56*, 123–132. [CrossRef]

79. Reza, S.; Majid, A. Ranking financial institutions based on of trust in online banking using aras and anp method. *Int. Res. J. Appl. Basic Sci.* **2013**, *6*, 415–423.

80. Keshavarz Ghorabaee, M.; Amiri, M.; Zavadskas, E.K.; Turskis, Z.; Antucheviciene, J. A new multi-criteria model based on interval type-2 fuzzy sets and edas method for supplier evaluation and order allocation with environmental considerations. *Comput. Ind. Eng.* **2017**, *112*, 156–174. [CrossRef]

81. Sun, R.; Hu, J.; Zhou, J.; Chen, X. A hesitant fuzzy linguistic projection-based mabac method for patients' prioritization. *Int. J. Fuzzy Syst.* **2018**, *20*, 2144–2160. [CrossRef]

82. Opricovic, S.; Tzeng, G.-H. Compromise solution by mcdm methods: A comparative analysis of vikor and topsis. *Eur. J. Oper. Res.* **2004**, *156*, 445–455. [CrossRef]

83. Papadopoulos, G. Moving from traditional to agile software development methodologies also on large, distributed projects. *Procedia-Soc. Behav. Sci.* **2015**, *175*, 455–463. [CrossRef]

84. Tanner, M.; von Willingh, U. Factors Leading to the Success and Failure of Agile Projects Implemented in Traditionally Waterfall Environments. Available online: http://www.toknowpress.net/ISBN/978-961-6914-09-3/papers/ML14-618.pdf (accessed on 14 February 2019).

85. Ghobakhloo, M.; Azar, A. Business excellence via advanced manufacturing technology and lean-agile manufacturing. *J. Manuf. Technol. Manag.* **2018**, *29*, 2–24. [CrossRef]

86. Rehman, A.U.; Alkhatani, M.; Umer, U. Multi Criteria Approach to Measure Leanness of a Manufacturing Organization. *IEEE Access* **2018**, *6*, 20987–20994. [CrossRef]

87. Tilburgs, B.; Vernooij-Dassen, M.; Koopmans, R.; Weidema, M.; Perry, M.; Engels, Y. The importance of trust-based relations and a holistic approach in advance care planning with people with dementia in primary care: a qualitative study. *BMC Geriatr.* **2018**, *18*, 184. [CrossRef]

88. Kusi-Sarpong, S.; Leonilde Varela, M.; Putnik, G.; Ávila, P.; Agyemang, J. Supplier evaluation and selection: A fuzzy novel multicriteria group decision-making approach. *Int. J. Qual. Res.* **2018**, *12*, 459–486.

89. Song, M.L.; Peng, J.; Wang, J.L.; Dong, L. Better resource management: An improved resource and environmental efficiency evaluation approach that considers undesirable outputs. *Resour. Conserv. Recycl.* **2018**, *128*, 197–205. [CrossRef]

90. Chen, J.D.; Cheng, S.L.; Nikic, V.; Song, M.L. Quo Vadis? Major Players in Global Coal Consumption and Emissions Reduction. *Transform. Bus. Econ.* **2018**, *17*, 112–132.

91. Fang, S.T.; Ji, X.; Ji, X.H.; Wu, J. Sustainable urbanization performance evaluation and benchmarking: An efficiency perspective. *Manag. Environ. Qual.* **2018**, *29*, 240–254. [CrossRef]

92. Wang, W.; Lu, N.; Zhang, C. Low-carbon technology innovation responding to climate change from the perspective of spatial spillover effects. *Chin. J. Popul. Resour. Environ.* **2018**, *16*, 120–130. [CrossRef]

 © 2019 by the authors. Licensee MDPI, Basel, Switzerland. This article is an open access article distributed under the terms and conditions of the Creative Commons Attribution (CC BY) license (http://creativecommons.org/licenses/by/4.0/).

Article

Normalized Weighted Bonferroni Harmonic Mean-Based Intuitionistic Fuzzy Operators and Their Application to the Sustainable Selection of Search and Rescue Robots

Jinming Zhou [1], Tomas Baležentis [2,*] and Dalia Streimikiene [2]

[1]   School of Mathematics and Physics, Anhui Polytechnic University, Wuhu 241000, China; zjm@ahpu.edu.cn
[2]   Lithuanian Institute of Agrarian Economics, V. Kudirkos Str. 18-2, LT-03105 Vilnius, Lithuania; dalia@mail.lei.lt
*   Correspondence: tomas@laei.lt; Tel.: +370-52-262-2085

Received: 11 January 2019; Accepted: 3 February 2019; Published: 13 February 2019

Abstract: In this paper, Normalized Weighted Bonferroni Mean (NWBM) and Normalized Weighted Bonferroni Harmonic Mean (NWBHM) aggregation operators are proposed. Besides, we check the properties thereof, which include idempotency, monotonicity, commutativity, and boundedness. As the intuitionistic fuzzy numbers are used as a basis for the decision making to effectively handle the real-life uncertainty, we extend the NWBM and NWBHM operators into the intuitionistic fuzzy environment. By further modifying the NWBHM, we propose additional aggregation operators, namely the Intuitionistic Fuzzy Normalized Weighted Bonferroni Harmonic Mean (IFNWBHM) and the Intuitionistic Fuzzy Ordered Normalized Weighted Bonferroni Harmonic Mean (IFNONWBHM). The paper winds up with an empirical example of multi-attribute group decision making (MAGDM) based on triangular intuitionistic fuzzy numbers. To serve this end, we apply the IFNWBHM aggregation operator.

Keywords: Bonferroni harmonic mean; aggregation operator; intuitionistic fuzzy set; multiple attribute group decision making; search and rescue robots

---

1. Introduction

Decision making seeks to pick the best-performing option (alternative) among the feasible ones in order to satisfy a certain objective represented by an attribute. In practice, many decisions require considering more than one objective and, hence, more than one attribute. This being the case, one faces a multi-attribute decision making (MADM) problem. Basically, MADM is defined as the identification of the best-performing alternative among the feasible ones, taking multiple attributes into consideration. As multiple attributes are involved in the problem, the issue of aggregation of the decision information arises. The aggregation operators may be employed in order to summarize the decision information in MADM and, thus, consider multiple objectives simultaneously. What is more, the aggregation operators can be adjusted to account for interrelations among the decision variables.

The theory and applications of aggregation operators have been developing due to an increasing prevalence of the MADM problem in different domains [1–4]. There have been some aggregation operators available for handling MADM problems involving intuitionistic fuzzy (IF) sets [5–8]. In order to exploit multiple desirable properties of the IF sets, different types of intuitionistic fuzzy numbers (IFNs) have been established and employed for various empirical applications [9–12]. The theory of the aggregation operators has also been extended in regards to different types of IFNs. For instance, the triangular intuitionistic fuzzy numbers (TIFNs) were introduced [13,14] and applied for information aggregation by offering the corresponding extension of averaging operators, namely the intuitionistic fuzzy weighted arithmetic aggregation operator.

Yet another example regarding the aggregation operators for the IFNs was proposed by Wan and Dong [15], who developed the ordered weighted aggregation operator along with the hybrid weighted aggregation operator. The latter approach was based on the use of the measures of the expectation and expectant score determined by the position of the center of gravity of IFNs considered in the analysis. Wu and Cao [16] proposed a family of intuitionistic trapezoidal fuzzy operators weighted geometric operators (including the ordered, induced ordered, and hybrid ones).

The earlier literature has mostly opted for treating the IF information used for aggregation as showing no interdependency relations. As a result, the possible existing intercorrelation among the arguments has not been accounted for. One of the possible means for accounting for interdependence existing among the arguments of the MADM problems is the Bonferroni mean (BM) operator [17]. Yager showed that the BM may be obtained as a sum the products of arguments to be aggregated and the average value of all the arguments save the one under consideration. What is more, the arithmetic average may be replaced with the other types of means [18] including, for instance, the Choquet integral [19] or ordered weighted average operator.

Further modifications of the BM methodology were offered by Beliakov et al. [20], who developed the generalized BM. The concept of the BM has been extended for the intuitionistic fuzzy information by Xu and Yager [21] to handle the intercorrelation among the arguments throughout the aggregation. Dutta and Guha [22] proposed substituting the aggregation operators for the inner and outer means in the calculations.

While seeking to aggregate the uncertain information, the uncertain BM operator along with its ordered and Choquet integral versions were developed [23]. The generalized weighted BM operator and its intuitionistic fuzzy counterpart were introduced by Xia et al. [24]. The latter operators included expert assessments in order to improve the robustness of the aggregation. An additional technique for aggregating the IFNs—the intuitionistic fuzzy weighted power harmonic mean (IFWPHM) operator—was proposed by Das and Guha [25]. The harmonic aggregation operators for the MADM problems based upon the fuzzy information were proposed by Xu [26]. The latter group of fuzzy weighted harmonic operators includes mean, ordered mean, and hybrid mean operators. Wei [27] suggested using the order-inducing variables in the process of aggregation of the fuzzy information and devised the fuzzy induced ordered weighted harmonic mean operator. The use of the BM in the fuzzy MADM was furthered in [28] by developing the fuzzy Bonferroni harmonic mean operator and the ordered counterpart.

In the existing literature, applications of the BM operators have mostly been limited to cases where information was represented by the intuitionistic fuzzy sets established with respect to a finite universe of discourse [29–31]. However, the methods available for handling the intuitionistic fuzzy numbers, e.g., triangular intuitionistic fuzzy numbers (TIFNs), as arguments of the aggregation operators, are rather scarce in the literature. In order to extend the domain for application of the intuitionistic fuzzy information in MADM, we propose the normalized weighted triangular intuitionistic fuzzy Bonferroni harmonic mean (NWTIFBHM) operator, which is capable of aggregating the triangular intuitionistic fuzzy information. The proposed approach relies on the Bonferroni mean (BM). More specifically, we exploit the normalized weighted Bonferroni mean (NWBM) and establish the intuitionistic fuzzy normalized weighted Bonferroni harmonic mean (IFNWBHM). The proposed approach is then tested by solving a multi-attribute group decision making (MAGDM) problem involving the IFNWBHM.

The remainder of this paper unfolds as follows. Section 2 discusses the preliminary concepts and operations. Section 3 proposes the normalized weighted triangular intuitionistic fuzzy Bonferroni harmonic mean along with several important results. Section 4 presents application to MAGDM with triangular intuitionistic fuzzy information. Finally, an illustrative example is implemented with a comparative analysis of several prevalent aggregation operators with the proposed approach.

2. Preliminaries

In this section, we discuss the information carriers used for MADM, namely TIFNs. We further discuss the means for aggregations of TIFNs, which allow the utilities for the alternatives comprising the MADM problem to be derived. As the outcomes of such aggregations are also TIFNs, the ranking procedure is outlined.

2.1. TIFNs and the Associated Arithmetic Operations

Oftentimes, decision making cannot rely on precise information delivered in the form of exact (real) numbers. However, uncertain estimates can be provided regarding a certain phenomenon. Such being the case, one can embark by using the fuzzy numbers rather than crisp ones. Among different types of representation of the fuzzy information, the intuitionistic fuzzy numbers can be perceived as a generalization of the fuzzy numbers. Further on, a TIFN can be defined as an intuitionistic fuzzy set (defined in terms of a fuzzy membership function and a fuzzy non-membership function) attached to a certain real value. Mathematically, the membership and non-membership functions for a certain TIFN A are defined as [32]:

$$
\mu_A = \begin{cases} \frac{x-a}{b-a}\omega_A, & a \leq x \leq b \\ \omega_A, & x = b \\ \frac{c-x}{c-b}\omega_A, & b \leq x \leq c \\ 0, & else \end{cases}
\tag{1}
$$

and

$$
\nu_A = \begin{cases} \frac{b-x+u_A(x-a)}{b-a}, & a \leq x \leq b \\ u_A, & x = b \\ \frac{x-b+u_A(c-x)}{c-b}, & b \leq x \leq c \\ 1, & else \end{cases}
\tag{2}
$$

where parameters ω_A and u_A represent the upper limit of the value of the membership function and the minimum level of the non-membership function, respectively, with restrictions on their individual value and sum thereof given by $0 \leq \omega_A \leq 1, 0 \leq u_A \leq 1$ and $0 \leq \omega_A + u_A \leq 1$. The values of the membership and non-membership functions comprise the "core" of the degree of dependency of x to A, whereas the "uncertain" part is given by the hesitancy function $\pi_A(x) = 1 - \mu_A(x) - \nu_A(x)$, which is related to the constrains on the two functions discussed above. This definition is different from that of triangular fuzzy numbers as the latter does not involve the "uncertain part ".

In order to successfully apply the TIFNs for the MADM, the operational laws for TIFNs need to be established [32]. Let us consider the two TIFNs defined as $A_1 = ([a_1, b_1, c_1]; \omega_{A_1}, u_{A_1})$ and $A_2 = ([a_2, b_2, c_2]; \omega_{A_2}, u_{A_2})$, and assume that there exists a real number $\lambda > 0$,. Given the aforementioned variables, the following calculations serve as the operational laws for the TIFNs:

- $A_1 \oplus A_2 = ([a_1 + a_2, b_1 + b_2, c_1 + c_2]; \omega_{A_1} \wedge \omega_{A_2}, u_{A_1} \vee u_{A_2})$, where "$\wedge$" and "$\vee$" stand for the min and max operators, respectively;
- $A_1 \otimes A_2 = ([a_1 a_2, b_1 b_2, c_1 c_2]; \omega_{A_1} \wedge \omega_{A_2}, u_{A_1} \vee u_{A_2})$;
- $\lambda A_1 = ([\lambda a_1, \lambda b_1, \lambda c_1]; \omega_{A_1}, u_{A_1})$;
- $A_1^\lambda = ([a_1^\lambda, b_1^\lambda, c_1^\lambda]; \omega_{A_1}, u_{A_1})$.

The operational laws feature the following properties [32]:

- Commutativity: $A_1 \oplus A_2 = A_2 \oplus A_1, A_1 \otimes A_2 = A_2 \otimes A_1$;
- Distributivity: $\lambda(A_1 \oplus A_2) = \lambda A_1 \oplus \lambda A_2, \lambda(A_1 \otimes A_2) = \lambda A_1 \otimes A_2 = A_1 \otimes \lambda A_2$;
- Associativity: $\lambda_1 A + \lambda_2 A = (\lambda_1 + \lambda_2)A, A^{\lambda_1} \otimes A^{\lambda_2} = A^{\lambda_1 + \lambda_2}, \lambda_1 > 0, \lambda_2 > 0$.

Proof. The commutativity, distributivity, and associativity are implied by the definition of operational laws as follows:

$$A_2 \oplus A_1 = ([a_2 + a_1, b_2 + b_1, c_2 + c_1]; \omega_{A_2} \wedge \omega_{A_1}, u_{A_2} \vee u_{A_1})$$
$$= ([a_1 + a_2, b_1 + b_2, c_1 + c_2]; \omega_{A_1} \wedge \omega_{A_2}, u_{A_1} \vee u_{A_2}) = A_1 \oplus A_2,$$
$$\therefore A_1 \oplus A_2 = A_2 \oplus A_1.$$

$$A_2 \otimes A_1 = ([a_2 a_1, b_2 b_1, c_2 c_1]; \omega_{A_2} \wedge \omega_{A_1}, u_{A_2} \vee u_{A_1})$$
$$= ([a_1 a_2, b_1 b_2, c_1 c_2]; \omega_{A_1} \wedge \omega_{A_2}, u_{A_1} \vee u_{A_2}) = A_1 \otimes A_2;$$
$$\therefore A_1 \otimes A_2 = A_2 \otimes A_1.$$

$$\lambda(A_1 \oplus A_2) = ([\lambda(a_1 + a_2), \lambda(b_1 + b_2), \lambda(c_1 + c_2)]; \omega_{A_1} \wedge \omega_{A_2}, u_{A_1} \vee u_{A_2})$$
$$= ([\lambda a_1 + \lambda a_2, \lambda b_1 + \lambda b_2, \lambda c_1 + \lambda c_2]; \omega_{A_1} \wedge \omega_{A_2}, u_{A_1} \vee u_{A_2})$$
$$= ([\lambda a_1, \lambda b_1, \lambda c_1]; \omega_{A_1}, u_{A_1}) + ([\lambda a_2, \lambda b_2, \lambda c_2]; \omega_{A_2}, u_{A_2})$$
$$= \lambda([a_1, b_1, c_1]; \omega_{A_1}, u_{A_1}) + \lambda([a_2, b_2, c_2]; \omega_{A_2}, u_{A_2}) = \lambda A_1 \oplus \lambda A_2$$
$$\therefore \lambda(A_1 \oplus A_2) = \lambda A_1 \oplus \lambda A_2$$

$$\lambda(A_1 \otimes A_2) = \lambda([(a_1 a_2), (b_1 b_2), (c_1 c_2)]; \omega_{A_1} \wedge \omega_{A_2}, u_{A_1} \vee u_{A_2})$$
$$= ([(\lambda a_1) a_2, (\lambda b_1) b_2, (\lambda c_1) c_2]; \omega_{A_1} \wedge \omega_{A_2}, u_{A_1} \vee u_{A_2})$$
$$= ([a_1(\lambda a_2), b_1(\lambda b_2), c_1(\lambda c_2)]; \omega_{A_1} \wedge \omega_{A_2}, u_{A_1} \vee u_{A_2})$$
$$([(\lambda a_1) a_2, (\lambda b_1) b_2, (\lambda c_1) c_2]; \omega_{A_1} \wedge \omega_{A_2}, u_{A_1} \vee u_{A_2})$$
$$= ([\lambda a_1, \lambda b_1, \lambda c_1]; \omega_{A_1}, u_{A_1}) \otimes ([a_2, b_2, c_2]; \omega_{A_2}, u_{A_2})$$
$$= \lambda([a_1, b_1, c_1]; \omega_{A_1}, u_{A_1}) \otimes ([a_2, b_2, c_2]; \omega_{A_2}, u_{A_2}) = (\lambda A_1) \otimes A_2$$
$$([a_1(\lambda a_2), b_1(\lambda b_2), c_1(\lambda c_2)]; \omega_{A_1} \wedge \omega_{A_2}, u_{A_1} \vee u_{A_2})$$
$$= ([a_1, b_1, c_1]; \omega_{A_1}, u_{A_1}) \otimes ([\lambda a_2, \lambda b_2, \lambda c_2]; \omega_{A_2}, u_{A_2})$$
$$= ([a_1, b_1, c_1]; \omega_{A_1}, u_{A_1}) \otimes \lambda([a_2, b_2, c_2]; \omega_{A_2}, u_{A_2}) = A_1 \otimes (\lambda A_2)$$
$$\therefore \lambda(A_1 \otimes A_2) = (\lambda A_1) \otimes A_2 = A_1 \otimes (\lambda A_2)$$

$$\lambda_1 A = ([\lambda_1 a, \lambda_1 b, \lambda_1 c]; \omega_A, u_A), \quad \lambda_2 A = ([\lambda_2 a, \lambda_2 b, \lambda_2 c]; \omega_A, u_A)$$
$$\therefore \lambda_1 A + \lambda_2 A = ([(\lambda_1 + \lambda_2)a, (\lambda_1 + \lambda_2)b, (\lambda_1 + \lambda_2)c]; \omega_A, u_A) = (\lambda_1 + \lambda_2)A$$

$$A^{\lambda_1} = ([a^{\lambda_1}, b^{\lambda_1}, c^{\lambda_1}]; \omega_A, u_A), \quad A^{\lambda_2} = ([a^{\lambda_2}, b^{\lambda_2}, c^{\lambda_2}]; \omega_A, u_A).$$
$$\therefore A^{\lambda_1 + \lambda_2} = ([a^{\lambda_1 + \lambda_2}, b^{\lambda_1 + \lambda_2}, c^{\lambda_1 + \lambda_2}]; \omega_A, u_A) = ([a^{\lambda_1} a^{\lambda_2}, b^{\lambda_1} a^{\lambda_2}, c^{\lambda_1} a^{\lambda_2}]; \omega_A, u_A) = A^{\lambda_1} \otimes A^{\lambda_2}$$

The TIFNs (and fuzzy numbers in general) are rather complex structures associated with elements of the real line. Therefore, it is often useful to approximate the fuzzy numbers by assuming a certain level of the (non-)membership function and projecting the fuzzy numbers on a real line. The elements of the real set satisfying the requirements associated with the values of the (non-)membership functions are then treated as those belonging to the set approximating a certain fuzzy number (including a TIFN). The latter approach is referred to as cutting of the fuzzy numbers. An α-cut of a TIFN is a subset of crisp values which satisfy $A(\alpha) = \{x|\mu_A(x) \geq \alpha\}$ [32], where the chosen lower level of the membership function is $0 \leq \alpha \leq \omega_A$. Given Equation (1), every α-cut is a closed interval, which is obtained as

$$[A^L(\alpha), A^U(\alpha)] = \left[a + \frac{\alpha(b - a)}{\omega_A}, c - \frac{\alpha(c - b)}{\omega_A}\right] \tag{3}$$

Similarly, a β-cut of TIFN A is defined as a subset of crisp values for which the non-membership function does not exceed the upper limit, i.e., $A(\beta) = \{x|\nu_A(x) \leq \beta\}$, where the upper limit of the non-membership function is given by $0 \leq u_A \leq \beta \leq 1$. Given the properties stipulated by Equation (2), each β-cut of TIFN is a projection of a certain TIFN on the real line represented by a closed interval, as follows:

$$[A^L(\beta), A^U(\beta)] = \left[\frac{(1 - \beta)b + (\beta - u_A)a}{1 - u_A}, \frac{(1 - \beta)b + (\beta - u_A)c}{1 - u_A}\right] \tag{4}$$

Thus, one can obtain the projections of a TIFN on a real line with respect to the shape of membership and non-membership functions and the desirable level of these functions. The obtained α-cut and β-cut of a certain TIFN can be further used in, e.g., comparing the underlying TIFNs.

2.2. Bonferroni Mean

This subsection discusses the properties of the Bonferroni mean and its relevance to decision making problems. There have been different aggregation operators established in the literature, serving a number of objectives with respect to the nature of the data aggregated, preferences of the decision makers, and the interaction among the arguments. One of the topical issues the users of the aggregation operators needs to consider is the possible interrelationships among the data. This is particularly important in such cases where some deviating inputs may distort the result of aggregation and thus render a less meaningful outcome of the MADM. The deviating inputs may occur either due to measurement errors or due to biased expert ratings (whether intentionally or unintentionally). In order to avoid such situations, there have been some aggregations operators controlling for the degree of interrelationships among the data.

The BM can be applied in order to ensure that the interlinkages existing among the data are taken into account during the analysis. The BM was introduced by [17]. Later on, the BM-based aggregation operator was presented in order to allow for effective decision making based on possible interrelated data by Yager [18]. Thus, the BM aggregation operator can be employed for MADM. Indeed, the BM generalizes a family of well-known means.

Let there be two non-negative parameters $p, q \geq 0$ along with a set of n non-negative arguments $a_i, i = 1, 2, \cdots, n$. Then, if

$$
BM^{p,q}(a_1, a_2, \cdots, a_n) = \left(\frac{1}{n(n-1)} \sum_{\substack{i,j=1 \\ i \neq j}}^{n} a_i^p a_j^q \right)^{\frac{1}{p+q}}, \tag{5}
$$

$BM^{p,q}$ is termed the Bonferroni Mean (BM). Indeed, the following characteristics can be attributed to the BM:

- $BM^{p,q}(0, 0, \cdots, 0) = 0$, i.e., aggregation of the null values renders the null value too;
- (Idempotency) $BM^{p,q}(a, a, \cdots, a) = a$, i.e., aggregating a constant returns the same constant as an outcome;
- (Monotonicity) $BM^{p,q}(a_1, a_2, \cdots, a_n) \geq BM^{p,q}(b_1, b_2, \cdots, b_n)$, i.e., $BM^{p,q}$ is monotonic in its arguments for $a_i \geq b_i$, $i = 1, 2, \cdots, n$;
- (Boundedness) $\min_i \{a_i\} \leq BM^{p,q}(a_1, a_2, \cdots, a_n) \leq \max_i \{a_i\}$, i.e., the result of aggregation is bounded from below and above by the extreme values of the arguments.

The different combinations of the parameters p and q result in special cases of the BM representing various types of means. Especially setting either of the parameters to zero results in the family of mean operators involving no interactions among the arguments. Thus, setting $q = 0$ and considering Equation (1), one arrives at the following kind of aggregation:

$$
BM^{p,0}(a_1, a_2, \cdots, a_n) = \left(\frac{1}{n(n-1)} \sum_{\substack{i,j=1 \\ i \neq j}}^{n} a_i^p a_j^0 \right)^{\frac{1}{p+0}} = \left(\frac{1}{n} \sum_{i=1}^{n} a_i^p \right)^{\frac{1}{p}}, \tag{6}
$$

which represents a generalized mean operator outlined in [19]. In general, higher values of p for fixed q imply greater importance of the larger values. By further modifying the parameters governing the aggregation, one can obtain the special cases of the BM as follows:

- If one sets $p = 2, q = 0$, then the interactions are ignored and higher values of the arguments are additionally rewarded and Equation (6) becomes the square mean:

$$BM^{2,0}(a_1, a_2, \cdots, a_n) = \left(\frac{1}{n} \sum_{i=1}^{n} a_i^2 \right)^{\frac{1}{2}}. \tag{7}$$

- If one assumes $p = 1, q = 0$, then interactions remain ignored and arguments do not benefit from showing higher values, with Equation (6) becoming the arithmetic average:

$$BM^{1,0}(a_1, a_2, \cdots, a_n) = \frac{1}{n} \sum_{i=1}^{n} a_i. \tag{8}$$

- If one picks the boundary condition $p \to \infty, q = 0$, then the interactions remain ignored, with the greatest importance put on the largest argument, i.e., Equation (6) boils down to the maximum operator:

$$\lim_{p \to \infty} BM^{p,0}(a_1, a_2, \cdots, a_n) = \max_{i} \{a_i\}. \tag{9}$$

- If the boundary condition is set with $p \to 0, q = 0$, then the interactions among the arguments are ignored and the lowest values become the most important ones, with Equation (6) being reduced to the geometric mean operator:

$$\lim_{p \to 0} BM^{p,0}(a_1, a_2, \cdots, a_n) = \left(\prod_{i=1}^{n} a_i \right)^{\frac{1}{n}}. \tag{10}$$

In the case where one assumes positive values for both of the parameters, similar operators merge. However, they account for the interactions among the arguments in the latter case. Let $p = 1, q = 1$, then Equation (6) takes the following form:

$$BM^{1,1}(a_1, a_2, \cdots, a_n) = \left(\frac{1}{n} \sum_{i=1}^{n} a_i \left(\frac{1}{n-1} \sum_{\substack{j=1 \\ i \neq j}}^{n} a_j \right) \right)^{\frac{1}{2}} \tag{11}$$

Up to now, we have not included the preferences of decision makers in the analysis. In order to reflect their taste, the weights can be introduced in the decision making. In order to handle this kind of information, we can further introduce an additional instance of the BM. Let there be two parameters $p, q \geq 0$ and a vector of the arguments to be aggregated a_i (the elements of the vector are non-negative and indexed over $i = 1, 2, \cdots, n$). Furthermore, let there be vector weights $w = (w_1, w_2, \cdots, w_n)^T$, such that the weights are non-negative $w_i \geq 0$, $i = 1, 2, \cdots, n$, and normalized $\sum_{i=1}^{n} w_i = 1$. If the aggregation of the argument vector is carried out in the following manner

$$NWBM^{p,q}(a_1, a_2, \cdots, a_n) = \left(\sum_{\substack{i,j=1 \\ i \neq j}}^{n} w_i a_i^p \frac{w_j}{1 - w_i} a_j^q \right)^{\frac{1}{p+q}}$$

then $NWBM^{p,q}$ is referred to as the normalized weighted Bonferroni mean (*NWBM*) [33]. Some particular cases of the NWBM can be obtained by imposing certain conditions on the weight vector. Indeed, assuming equal weighting, i.e., $w_i = \frac{1}{n}, i = 1, 2, \cdots, n$, leads to the BM.

2.3. Normalized Weighted Bonferroni Harmonic Mean

The harmonic means are often used in the decision making due to their desirable properties. Thus, we can consider the harmonic mean in the context of the NWBM in order to improve the decision making process. Let there be two values of parameters $p, q \geq 0$ and a vector of arguments (non-negative numbers) for the aggregation $a_i, i = 1, 2, \cdots, n$, and let there be the underlying vector of the argument weights $w = (w_1, w_2, \cdots, w_n)^T$, satisfying the non-negativity condition $w_i \geq 0$, $i = 1, 2, \cdots, n$, and the normalization condition $\sum_{i=1}^{n} w_i = 1$. Given these premises, the following aggregation operator

$$NWBHM^{p,q}(a_1, a_2, \cdots, a_n) = \cfrac{1}{\left(\sum_{\substack{i,j=1 \\ i \neq j}}^{n} \frac{w_i}{a_i^p} \frac{w_j}{(1-w_i)a_j^q} \right)^{\frac{1}{p+q}}},$$

can be established and $NWBHM^{p,q}$ is referred to as the normalized weighted Bonferroni Harmonic Mean (NWBHM). The $NWBHM^{p,q}$ features similar properties to the BM; however, there are certain superiorities. In general, the NWBHM features idempotency, monotonicity, commutativity, and boundedness.

2.4. A Ranking Approach for TIFNs

As the prioritization of the alternatives remains the focus of the MADM, the ranking of fuzzy ratings is important in order to identify the most desirable decision. This can be achieved by applying certain ranking procedures for TIFNs in our case. Thus, this section presents a relatively new approach towards ranking the TIFNs. The ranking is based on the concept of the (α, β)-cut of the TIFNs. The TIFNs are represented by the interval numbers due to the applications of the (α, β)-cut, whereas the resulting interval numbers are ranked by applying the concept of the probability of dominance [34]. The ranking of the intervals representing the TIFNs allows one to draw conclusions on the ranking of the underlying TIFNs.

Let $a = [a^L, a^U]$ and $b = [b^L, b^U]$ be the two interval numbers, where the endpoints are represented by the ordered values so that $a^L \leq a^U$ and $b^L \leq b^U$. Note that if $a^L = a^U$, then the interval number degenerates to a real number a'.

Let a and b be any two real numbers, and then the probability of a > b is defined as follows:

$$p(a > b) = \begin{cases} 1, a > b; \\ 0.5, a = b; \\ 0, a < b. \end{cases}$$

Let there be the two arbitrarily chosen interval numbers, $a = [a^L, a^U]$ and $b = [b^L, b^U]$. For these two numbers, the probability of dominance of a over b, i.e., $a \geq b$, can be calculated as follows:

$$p(a \geq b) = \frac{\max\{0, L(a) + L(b) - \max\{b^U - a^L, 0\}\}}{L(a) + L(b)} \tag{12}$$

where the width of the intervals is defined as $L(a) = a^U - a^L$ and $L(b) = b^U - b^L$. The resulting probability $p(a \geq b)$ features a number of properties [34]:

(1) $0 \leq p(a \geq b) \leq 1$.
(2) $p(a \geq b) + p(a \leq b) = 1$.
(3) $p(a \geq b) = p(a \leq b) = 0.5$, if $p(a \geq b) = p(a \leq b)$.
(4) $p(a \geq b) = 0$, if $a^U \leq b^L$.
(5) Assuming there exist interval numbers a, b, and c, $p(a \geq c) \geq p(b \geq c)$ if $a \geq b$.

Up to now, we have focused on the case of two interval numbers. However, decision making often requires considering more than two interval numbers (e.g., comparison of more than two alternatives). We can, thus, extend the case of the two interval numbers to the general case of multiple interval numbers following [34]. Let there be m TIFNs defined in terms of the parameters of the membership and non-membership functions $A_i = ([a_i, b_i, c_i]; \omega_{A_i}, u_{A_i}), i = 1, 2, \cdots, m$. The ranking of the TIFNs based on the probability of dominance can be carried out in the following manner:

Step 1. For each TIFN, compute the (α, β)-cut by using Equations (3) and (4), where parameters α and β are chosen with respect to the extreme values of the membership and non-membership functions for a given set of TIFNs so that $0 \leq \alpha \leq \wedge_{i=1}^{m} \omega_{A_i}$, $\vee_{i=1}^{m} u_{A_i} \leq \beta \leq 1$ and $0 \leq \alpha + \beta \leq 1$. The resulting interval numbers representing the TIFNs are given by:

$$A_i(\alpha) = [A_i^L(\alpha), A_i^U(\alpha)], A_i(\beta) = [A_i^L(\beta), A_i^U(\beta)]$$

where the decision-maker sets the values of α, β.

Step 2. Calculate the composite interval capturing both the membership and non-membership functions for a certain TIFN:

$$\begin{aligned} A_i(\lambda) &= [A_i^L(\lambda), A_i^U(\lambda)] = \lambda A_i(\alpha) + (1 - \lambda)A_i(\beta) \\ &= [\lambda A_i^L(\alpha) + (1 - \lambda)A_i^L(\beta), \lambda A_i^U(\alpha) + (1 - \lambda)A_i^U(\beta)], (i = 1, 2, \cdots, m) \end{aligned}$$

where $\lambda \in [0, 1]$ represents the risk aversion of the decision maker as represented by the lower and upper values of the intervals covered by the membership and non-membership functions for the given levels of α and β (lower values of λ imply higher risk aversion of the decision maker).

Step 3. Establish the preference relations matrix representing pairwise comparisons among all the alternatives:

$$P = (p_{ij})_{m \times m} \tag{13}$$

where the elements of P are given as $p_{ij} = p(A_i \geq A_j) = p(A_i(\lambda) \geq A_j(\lambda))$ based on Equation (12) for $1 \leq i \leq m, 1 \leq j \leq m$.

Step 4. Aggregate results of the pairwise comparisons for each alternative by calculating the ranking indicator $RI(A_i)$ as follows [34]:

$$RI(A_i) = \frac{1}{m(m-1)}\left(\frac{m}{2} - 1 + \sum_{j=1}^{m} p_{ij}\right) \tag{14}$$

Step 5. The TIFNs are ranked with respect to the associated values of the ranking indicator $RI(A_i)$, $i = 1,2, ..., m$, so that higher values of the indicator imply higher ranking of the alternatives.

2.5. Normalized Weighted Triangular Intuitionistic Fuzzy Bonferroni Harmonic Mean

In Section 2.3, we presented the NWBHM operator for the real numbers. In order to process the TIFNs, we extend the NWBHM operator. Specifically, the NWTIFBHM operator is proposed. The proposed aggregation operator can be applied for decision making based upon the TIFNs.

For $p, q \geq 0$, let there be a collection of the TIFNs $A_i = ([a_i, b_i, c_i], \omega_{A_i}, u_{A_i})$, $i=1,2,...,n$, defined on the positive part of the real line along with the associated weight vector $w = (w_1, w_2, \cdots, w_n)^T$, such that $w_i \geq 0$, for $i = 1,2,...,n$, and $\sum_{i=1}^{n} w_i = 1$. If

$$NWTIFBHM^{p,q}(A_1, A_2, \cdots, A_n) = \cfrac{1}{\left(\underset{\substack{i,j=1 \\ i \neq j}}{\overset{n}{\oplus}} \left(\left(\frac{w_i}{(1-w_i)A_i^p}\right) \otimes \left(\frac{w_j}{A_j^q}\right)\right)\right)^{\frac{1}{p+q}}} \tag{15}$$

then $NWTIFBHM^{p,q}$ is termed the normalized weighted triangular intuitionistic fuzzy Bonferroni Harmonic mean ($NWTIFBHM$). We can derive the following results given the operational laws for the TIFNs stipulated in Equations (1)–(4).

Let there be $p, q \geq 0$ and a collection of positive TIFNs to be aggregated, $A_i = ([a_i, b_i, c_i], \omega_{A_i}, u_{A_i})$, $i = 1, 2, ... , n$, TIFNs, with weight vector $w = (w_1, w_2, \cdots, w_n)^T$, such that $w_i \geq 0$, $(i = 1,2,...,n)$ and $\sum_{i=1}^{n} w_i = 1$. The given set of TIFNs can be aggregated by the NWTIFBHM operator and the result of aggregation is also a TIFN. Specifically, the result of the aggregation is defined as follows (Proof see Appendix A):

$$NWTIFBHM^{p,q}(A_1, A_2, \cdots, A_n) = \left(\left[\frac{1}{\left(\underset{\substack{i,j=1 \\ i \neq j}}{\overset{n}{\sum}} \frac{w_i w_j}{(1-w_i)a_i^p a_j^q}\right)^{\frac{1}{p+q}}}, \frac{1}{\left(\underset{\substack{i,j=1 \\ i \neq j}}{\overset{n}{\sum}} \frac{w_i w_j}{(1-w_i)b_i^p b_j^q}\right)^{\frac{1}{p+q}}}, \frac{1}{\left(\underset{\substack{i,j=1 \\ i \neq j}}{\overset{n}{\sum}} \frac{w_i w_j}{(1-w_i)c_i^p c_j^q}\right)^{\frac{1}{p+q}}}\right], \bigwedge_{i=1}^{n}\omega_{A_i}, \bigvee_{i=1}^{n}u_{A_i}\right) \tag{16}$$

The desirable properties of the $NWTIFBHM$ operator can be proved by exploiting the relevant theorems. The main results are presented below.

Idempotency. If there exists a collection of TIFNs A_i, $i = 1,2,...,n$, where all the elements are equal to a certain value, i.e., $A_i = A = ([a, b, c], \omega_A, u_A)$, then the application of the NWTIFBHM operator results in that value:

$$NWTIFBHM^{p,q}(A_1, A_2, \cdots, A_n) = NWTIFBHM^{p,q}(A, A, \cdots, A) = A.$$

Commutativity. Let there be a set of positive TIFNs, $A_i = ([a_i, b_i, c_i], \omega_{A_i}, u_{A_i})$, $i = 1,2,...,n$, and let there be a permutation of $(A_1, A_2, \cdots, A_n)$ denoted by $(\tilde{A}_1, \tilde{A}_2, \cdots, \tilde{A}_n)$. Then, the following relationship holds:

$$NWTIFBHM^{p,q}(\tilde{A}_1, \tilde{A}_2, \cdots, \tilde{A}_n) = NWTIFBHM^{p,q}(A_1, A_2, \cdots, A_n).$$

Monotonicity. Let there be the two sets of TIFNs, $A_i = ([a_i, b_i, c_i], \omega_{A_i}, u_{A_i})$ and $\overline{A}_i = ([\overline{a}_i, \overline{b}_i, \overline{c}_i], \omega_{\overline{A}_i}, u_{\overline{A}_i})$, with $i = 1,2,...,n$. If $a_i \geq \overline{a}_i, b_i \geq \overline{b}_i, c_i \geq \overline{c}_i, \omega_{A_i} \geq \omega_{\overline{A}_i}$ and $u_{A_i} \geq u_{\overline{A}_i}$ for all i. Then, the results of aggregation are also related in the same manner. Formally,

$$NWTIFBHM^{p,q}(A_1, A_2, \cdots, A_n) \geq NWTIFBHM^{p,q}(\overline{A}_1, \overline{A}_2, \cdots, \overline{A}_n)$$

Boundedness. Let there be a collection of TIFNs denoted by $A_i = ([a_i, b_i, c_i], \omega_{A_i}, u_{A_i})$, $i = 1, 2, \cdots, n$. Furthermore, let there be negative and positive ideal solutions associated with the set defined by $A^- = ([\wedge_i a_i, \wedge_i b_i, \wedge_i c_i], \wedge_i \omega_{A_i}, \vee_i u_{A_i})$ and $A^+ = ([\vee_i a_i, \vee_i b_i, \vee_i c_i], \vee_i \omega_{A_i}, \wedge_i u_{A_i})$, respectively. Then, the result of aggregation by the NWTIFBHM is bounded by those two ideal solutions as follows:

$$A^- \leq NWTIFBHM^{p,q}(A_1, A_2, \cdots, A_n) \leq A^+$$

The ordered aggregation operators consider the position of the ordered arguments. Thus, the ordered NWTIFBHM (NWTIFOBHM) operator can be defined. Let there be $p, q \geq 0$ and let there be a set of TIFNs denoted by $A_i = ([a_i, b_i, c_i], \omega_{A_i}, u_{A_i})$, $i = 1,2,...,n$. Assume there are weights associated with the i-th largest value such that $w_i \geq 0, i = 1,2,...,n$, and $\sum_{i=1}^{n} w_i = 1$. Then, the application of the NWTIFOBHM results in a TIFN as defined below:

$$NWTIFOBHM^{p,q}(A_1, A_2, \cdots, A_n) = \cfrac{1}{\left(\displaystyle\bigoplus_{\substack{i,j=1 \\ i \neq j}}^{n} \left(\left(\frac{w_i}{(1-w_i)A_{\sigma(i)}^p} \right) \otimes \left(\frac{w_j}{A_{\sigma(j)}^q} \right) \right) \right)^{\frac{1}{p+q}}}$$

$$= \left(\left[\cfrac{1}{\left(\displaystyle\sum_{\substack{i,j=1 \\ i \neq j}}^{n} \frac{w_i w_j}{(1-w_i)a_{\sigma(i)}^p a_{\sigma(j)}^q} \right)^{\frac{1}{p+q}}}, \cfrac{1}{\left(\displaystyle\sum_{\substack{i,j=1 \\ i \neq j}}^{n} \frac{w_i w_j}{(1-w_i)b_{\sigma(i)}^p b_{\sigma(j)}^q} \right)^{\frac{1}{p+q}}}, \cfrac{1}{\left(\displaystyle\sum_{\substack{i,j=1 \\ i \neq j}}^{n} \frac{w_i w_j}{(1-w_i)c_{\sigma(i)}^p c_{\sigma(j)}^q} \right)^{\frac{1}{p+q}}} \right]; \tag{17}$$

$$\wedge_{i=1}^{n} \omega_{A_i}, \vee_{i=1}^{n} u_{A_i} \bigg)$$

where the ordered arguments are denoted by $A_{\sigma(i)} = ([a_{\sigma(i)}, b_{\sigma(i)}, c_{\sigma(i)}], \omega_{A_{\sigma(i)}}, u_{A_{\sigma(i)}})$, $i = 1, 2, \cdots, n$, and $(\sigma(1), \sigma(2), \cdots, \sigma(n))$ is a permutation of $\{1,2,...,n\}$, ensuring the ordering of the arguments, i.e., $A_{\sigma(i-1)} \geq A_{\sigma(i)}$ for $i = 2, 3, ..., n$.

3. MAGDM Based on the Triangular Intuitionistic Fuzzy Information and the NWTIFBHM Operator

This section presents the MAGDM approach based on the proposed aggregation indicators. An empirical example is provided. Finally, the comparative analysis is carried out in order to compare the proposed framework against the existing ones.

3.1. MAGDM Framework

The MAGDM problem can be solved by applying the NWTIFBHM operator to aggregate the decision information for the alternatives under consideration. This sub-section outlines the main stages of the MAGDM based upon the NWTIFBHM operator.

Let there be a finite set of n alternatives, $X = \{X_1, X_2, \cdots, X_n\}$, and a finite set of m criteria, $C = \{C_1, C_2, \cdots, C_m\}$. The MAGDM problem involves decision makers D_t, $t = 1, 2, \ldots, T$, with associated decision matrices $A_t = \left(A_{t_{ij}}\right)_{n \times m}$, where elements thereof represent the ratings of each alternative against each criterion. The ratings provided by the experts are aggregated and the organized in the aggregate decision matrix $A = \left(A_{t_{ij}}\right)_{n \times m}$.

Step 1. Establish the individual decision matrices A_t. The weights of criteria are arranged into vector w. Note that the weights can be established based on objective methods (e.g., entropy) or subjective ones (e.g., pair-wise comparisons).

Step 2. Aggregate the ratings provided by the decision makers for each alternative and criterion. The NWTIFBHM operator given by Equation (16) can be applied (assuming $p = q = 1$) for the aggregation. The resulting elements of the aggregate matrix are thus defined as:

$$A_{t_i} = NWTIFBHM^{p,q}\left(A_{t_{ij}}\right) j = 1, 2, \cdots, m; t = 1, 2, \cdots, T.$$

Step 3 Calculate the final fuzzy utility scores for each alternative considering all the criteria and experts respectively by exploiting Equation (16).

Calculate the ranking indicator defined by Equation (14) for each fuzzy utility score A_t representing the overall performance of alternative X_i, $i = 1, 2, \ldots, n$.

Step 4. Rank the alternatives based on the values of the ranking indicator $RI(A_t)$ by assigning the highest ranks to the alternatives featuring the highest values of $RI(A_t)$.

3.2. Application for the Case of Search and Rescue Robot Selection

In order to illustrate the possibilities for application of the proposed framework for the MAGDM problem, this sub-section presents its application to the case of the selection of search and rescue robots. This particular illustration is important in the sense that the performance of search and rescue robots is rather crucial for handling emergencies [35]. Accordingly, the performance of search and rescue robots should be assessed in a comprehensive manner.

Given the suggestions provided by the earlier literature [35], we consider four criteria when evaluating the performance of search and rescue robots, including: (1) viability—C_1, (2) athletic ability—C_2, (3) working ability—C_3, and (4) communication control capability—C_4. Assume there are four search and rescue robots X_i ($i = 1,2,3,4$) to be evaluated. Furthermore, the evaluation relies on expert opinions (i.e., one needs to solve an MAGDM problem). The experts provide their ratings for each alternative against the four criteria. The resulting individual decision matrices are outlined in Tables 1–4. The group of experts is assumed not to be a completely homogenous one. Accordingly, the experts are assigned with different weights arranged into vector $\eta = (0.20, 0.30, 0.35, 0.15)^{\mathrm{T}}$, where each element is associated with a corresponding expert D_t ($t = 1,2,3,4$).

Table 1. Decision matrix A_1 given by expert D_1.

Alternative	C_1	C_2	C_3	C_4
X_1	([0.05,0.1,0.15];0.7,0.2)	([0.1,0.15,0.2];0.5,0.4)	([0.1,0.2,0.25];0.6,0.4)	([0.75,0.8,0.9];0.8,0.1)
X_2	([0.2,0.25,0.3];0.6,0.3)	([0.8,0.85,0.95];0.8,0.2)	([0.15,0.2,0.25];0.7,0.2)	([0.2,0.25,0.3];0.6,0.3)
X_3	([0.1,0.2,0.3];0.5,0.4)	([0.1,0.2,0.3];0.7,0.2)	([0.85,0.9,0.95];0.6,0.3)	([0.15,0.2,0.3];0.7,0.1)
X_4	([0.85,0.9,0.95];0.5,0.3)	([0.2,0.3,0.35];0.6,0.3)	([0.15,0.3,0.4];0.5,0.2)	([0.1,0.25,0.35];0.8,0.1)

Table 2. Decision matrix A_2 given by expert D_2.

Alternative	C_1	C_2	C_3	C_4
X_1	([0.05,0.15,0.25];0.6,0.4)	([0.1,0.15,0.2];0.6,0.3)	([0.1,0.15,0.2];0.6,0.4)	([0.85,0.9,0.95];0.6,0.3)
X_2	([0.15,0.25,0.3];0.6,0.3)	([0.75,0.85,0.95];0.7,0.2)	([0.15,0.2,0.25];0.7,0.2)	([0.2,0.25,0.3];0.6,0.4)
X_3	([0.75,0.8,0.85];0.9,0.1)	([0.1,0.2,0.25];0.5,0.3)	([0.1,0.25,0.3];0.7,0.2)	([0.15,0.25,0.3];0.8,0.1)
X_4	([0.1,0.3,0.4];0.6,0.2)	([0.2,0.25,0.3];0.8,0.1)	([0.8,0.85,0.95];0.7,0.3)	([0.1,0.25,0.35];0.5,0.4)

Table 3. Decision matrix A_3 given by expert D_3.

Alternative	C_1	C_2	C_3	C_4
X_1	([0.8,0.85,0.9];0.9,0.1)	([0.2,0.25,0.3];0.5,0.4)	([0.1,0.2,0.25];0.6,0.4)	([0.15,0.2,0.3];0.8,0.1)
X_2	([0.15,0.25,0.3];0.6,0.2)	([0.1,0.15,0.2];0.6,0.2)	([0.15,0.2,0.25];0.7,0.2)	([0.8,0.85,0.95];0.8,0.2)
X_3	([0.2,0.25,0.3];0.5,0.4)	([0.05,0.1,0.15];0.7,0.2)	([0.85,0.9,0.95];0.6,0.25)	([0.15,0.2,0.25];0.7,0.1)
X_4	([0.1,0.2,0.25];0.7,0.2)	([0.75,0.8,0.9];0.6,0.2)	([0.2,0.25,0.3];0.5,0.4)	([0.1,0.25,0.3];0.6,0.3)

Table 4. Decision matrix A_4 given by expert D_4.

Alternative	C_1	C_2	C_3	C_4
X_1	([0.15,0.2,0.3];0.5,0.5)	([0.25,0.3,0.35];0.4,0.4)	([0.75,0.85,0.9];0.5,0.4)	([0.2,0.35,0.4];0.7,0.2)
X_2	([0.85,0.9,0.95];0.8,0.1)	([0.05,0.1,0.15];0.6,0.3)	([0.2,0.25,0.3];0.7,0.2)	([0.1,0.15,0.2];0.9,0.1)
X_3	([0.2,0.25,0.3];0.5,0.4)	([0.8,0.85,0.9];0.8,0.1)	([0.05,0.1,0.15];0.7,0.2)	([0.25,0.3,0.35];0.5,0.4)
X_4	([0.1,0.2,0.3];0.7,0.2)	([0.15,0.25,0.35];0.5,0.3)	([0.25,0.3,0.35];0.6,0.3)	([0.8,0.9,0.95];0.6,0.2)

The decision matrices A_t are constructed and the decision making proceeds as follows:

Step 1. Provide decision matrices A_t, $t = 1, 2, 3, 4$, and the weight vector of criteria $w = (0.22, 0.20, 0.28, 0.30)^T$.

Utilize the NWTIFBHM operator as defined by Equation (A1) with $p = q = 1$ on individual decision matrices to obtain the group ratings associated with each alternative under consideration given the assessments provided by the four experts. Table 5 presents the aggregate decision matrix.

Table 5. The overall performance value A_{t_i}, $(i, t = 1, 2, 3, 4)$ by decision makers.

Alternative	D_1	D_2	D_3	D_4
X_1	([0.1196,0.2204,0.2640]; 0.5,0.4)	([0.1196,0.2304,0.2827]; 0.6,0.4)	([0.3140,0.4742,0.5667]; 0.5,0.4)	([0.4376,0.5420,0.6837]; 0.5,0.4)
X_2	([0.3673,0.4620,0.5584]; 0.6,0.3)	([0.3225,0.4620,0.5584]; 0.6,0.4)	([0.2017,0.2990,0.3778]; 0.6,0.2)	([0.2333,0.3562,0.4641]; 0.6,0.3)
X_3	([0.2598,0.4703,0.6546]; 0.5,0.4)	([0.2328,0.4727,0.5643]; 0.5,0.3)	([0.2533,0.3826,0.4945]; 0.5,0.4)	([0.2190,0.3363,0.4401]; 0.5,0.4)
X_4	([0.3815,0.6127,0.7405]; 0.5,0.3)	([0.3420,0.5948,0.7293]; 0.5,0.4)	([0.3058,0.4600,0.5559]; 0.5,0.4)	([0.2360,0.3796,0.5107]; 0.5,0.3)

Step 2. The overall utilities are obtained for the alternatives under consideration. Decision makers' rankings of all the alternatives are calculated and the weight vector $\eta = (0.20, 0.30, 0.35, 0.15)^T$ of decision makers and the aggregated value are given as follows:

$$A_1 = (\, [0.2353\ 0.3605\ 0.4385];0.5000, 0.4000), A_2 = ([0.2758\ 0.3891\ 0.4810];0.6000, 0.4000 \,),$$
$$A_3 = (\, [0.2433\ 0.4203\ 0.5393];0.5000, 0.4000), A_4 = (\, [0.3213\ 0.5189\ 0.6381];0.5000, 0.4000 \,).$$

Step 3. The overall utility scores are expressed in the TIFNs. Therefore, we further utilize the probabilistic ranking approach outlined in Section 2.4 The ranking indicators are obtained by assuming $\alpha = \beta = \lambda = 0.5$. The following values of the ranking indicator are obtained for each alternative X_i:

$$RI(A_1) = 0.1154, \ RI(A_2) = 0.1923, \ RI(A_3) = 0.2692, \ RI(A_4) = 0.3462.$$

Step 4. Given the values of the ranking indicator, the following ranking is obtained: $RI(A_4) > RI(A_3) > RI(A_2) > RI(A_1)$. X_4 is identified as the most preferable (in the sense of the underlying fuzzy utility) search and rescue robot, as evidenced by the associated ranking indicator $RI(A_4)$ showing the largest value among the alternatives.

3.3. Comparative Analysis

In order to test the performance of the proposed operator, we solve the problem of the selection of the search and rescue robots by applying various aggregation operators, i.e., the weighted power average (TIFWPA) operator [31], weighted power geometric (TIFWPG) operator [36], weighted geometric mean (TIFWGM) operator [16], weighted power harmonic mean (TIFWPHM) operator [25], and weighted arithmetic mean (TIFWAM) operator [37] extended for the TIFNs. The comparative analysis is proceeded by implementing the procedure outlined in Section 3.1 and replacing the NWTIFBHM operator with the abovementioned aggregation operators. This results in the rankings of the alternatives associated with different aggregation operators. The results are summarized in Table 6.

Table 6. The ranking order rendered by the different methods.

Method	Ranking Order	Best Alternative
TIFWPA	$X_4 \succ X_2 \succ X_1 \succ X_3$	X_4
TIFWPG	$X_4 \succ X_2 \succ X_1 \succ X_3$	X_4
TIFWGM	$X_1 \succ X_4 \succ X_2 \succ X_3$	X_1
TIFWAM	$X_3 \succ X_4 \succ X_1 \succ X_2$	X_3
TIFWPHM	$X_4 \succ X_2 \succ X_1 \succ X_3$	X_4
NWTIFBHM	$X_4 \succ X_3 \succ X_2 \succ X_1$	X_4

The results in Table 6 clearly indicate that the use of the aggregation indicators which are not capable of handling extreme deviations in the data (i.e., the TIFWGM [16] and TIFWAM [37] operators) render rather different results from the rest of the operators. At the other end of the spectrum, the operators capable of accounting for possibly biased ratings (i.e., the proposed TIFWPHM operator, the weighted power average operator [31], and the weighted power geometric operator [36]) rendered similar results. It can be noted that all the operators belonging to the latter group can address the issue of the outlying data, yet the approach is different. Specifically, both the TIFWPA operator [31] and TIFWPG operator [36] allow low weights to be assigned for the outlying data and, thus, minimize their influence indirectly. On the other hand, the TIFWPHM operator [25] (here, it is the degenerate form of TrIFWPHM in [25]) focuses directly (due to its harmonic nature) on the outlying data to reduce the influence thereof on the final results of the aggregation. The NWTIFBHM showed the same best alternative, yet the ranking X3 appeared to be better in this case (the NWTIFBHM showed the same best alternative, yet the ranking X4 appeared to be better in this case).

Therefore, the proposed NWTIFBHM operator is suitable for dealing with situations where different importance of the arguments should be established given possibly biased rankings and the resulting inter-relationship patterns.

We further analyze the performance of the proposed NWTIFBHM operator by adjusting the underlying parameters. Specifically, parameters α and β determine the degree of uncertainty when constructing the (α, β)-cuts representing the underlying TIFNs, whereas parameter λ reflects the risk version when comparing the TIFNs. We will test the impact of changes in the values of these parameters on the results of the aggregation and ranking of the alternatives.

First, we fix the values of the parameter $\alpha = \beta = 0.5$ and allow λ to vary, i.e., $\lambda \in [0,1]$. The ranking is repeated for several values of λ and the results are summarized in Table 7. As one can note, the resulting ranking order is stable based on NWTIFBHM with fixed (α, β). Figure 1 presents the results graphically and depicts the resulting ranking indicators for each alternative under different parameter values. As it can be seen from Figure 1, as $\alpha = \beta = 0.5$, given the changes of λ (within interval defined by $\lambda \in [0, 1]$), the stability of the ranking remains rather high.

Table 7. The ordering of different λ based on NWTIFBHM operator ($\alpha = \beta = 0.5$).

λ	Ranking Index	Ranking Order
$\lambda = 0.1$	$RI(A_1) = 0.1178, RI(A_2) = 0.2004,$ $RI(A_3) = 0.2588, RI(A_4) = 0.3462$	$X_4 \succ X_3 \succ X_2 \succ X_1$
$\lambda = 0.4$	$RI(A_1) = 0.1154, RI(A_2) = 0.1945,$ $RI(A_3) = 0.2671, RI(A_4) = 0.3462$	$X_4 \succ X_3 \succ X_2 \succ X_1$
$\lambda = 0.6$	$RI(A_1) = 0.1154, RI(A_2) = 0.1923,$ $RI(A_3) = 0.2692, RI(A_4) = 0.3462$	$X_4 \succ X_3 \succ X_2 \succ X_1$
$\lambda = 0.9$	$RI(A_1) = 0.1154, RI(A_2) = 0.1923,$ $RI(A_3) = 0.2692, RI(A_4) = 0.3462$	$X_4 \succ X_3 \succ X_2 \succ X_1$

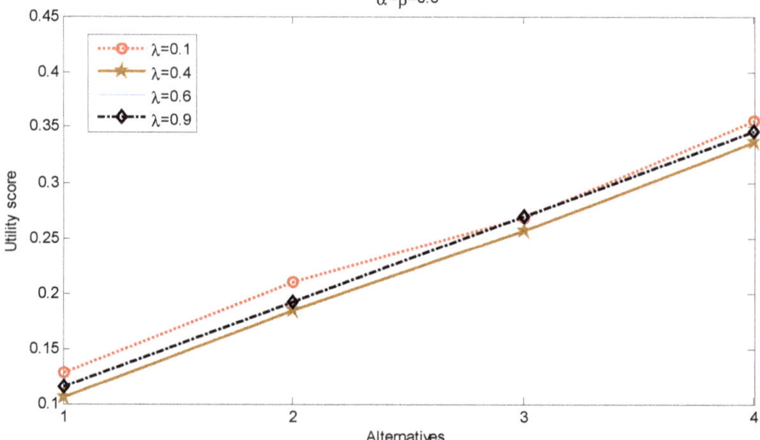

Figure 1. Sensitivity analysis of NWTIFBHM evaluation results ($\alpha = \beta = 0.5$).

Note: For the convenience of observation, the curves for $\lambda = 0.1$ and $\lambda = 0.4$ are shifted up and down by 0.01 units, respectively, and the curves for $\lambda = 0.6$ and $\lambda = 0.9$ are coincident; x-axis represents the alternatives under consideration.

Second, we allow parameters α or β to change with λ remaining fixed at 0.5 (either β or α remains fixed at 0.5 too). Since $0 \leq \alpha \leq w_A, 0 \leq u_A \leq \beta \leq 1$, we consider $\alpha \in [0, 0.5]$ and $\beta \in [0.4, 1]$ in the numerical example. The results are given in Tables 8 and 9. It is easy to see that the proposed approach is specific, with a rather high stability of the results.

Table 8. The ordering of different α based on NWTIFBHM operator ($\lambda = \beta = 0.5$).

α	Ranking Index	Ranking Order
$\alpha = 0.1$	$RI(A_1) = 0.1608, RI(A_2) = 0.2020,$ $RI(A_3) = 0.2317, RI(A_4) = 0.3285$	$X_4 \succ X_3 \succ X_2 \succ X_1$
$\alpha = 0.2$	$RI(A_1) = 0.1535, RI(A_2) = 0.1984,$ $RI(A_3) = 0.2325, RI(A_4) = 0.3387$	$X_4 \succ X_3 \succ X_2 \succ X_1$
$\alpha = 0.3$	$RI(A_1) = 0.1412, RI(A_2) = 0.1983,$ $RI(A_3) = 0.2374, RI(A_4) = 0.3462$	$X_4 \succ X_3 \succ X_2 \succ X_1$
$\alpha = 0.4$	$RI(A_1) = 0.1276, RI(A_2) = 0.1978,$ $RI(A_3) = 0.2516, RI(A_4) = 0.3462$	$X_4 \succ X_3 \succ X_2 \succ X_1$

Table 9. The ordering of different β based on NWTIFBHM operator ($\lambda = \alpha = 0.5$).

β	Ranking Index	Ranking Order
$\beta = 0.6$	$RI(A_1) = 0.1269, RI(A_2) = 0.1971,$ $RI(A_3) = 0.2530, RI(A_4) = 0.3462$	$X_4 \succ X_3 \succ X_2 \succ X_1$
$\beta = 0.7$	$RI(A_1) = 0.1363, RI(A_2) = 0.1974,$ $RI(A_3) = 0.2432, RI(A_4) = 0.3462$	$X_4 \succ X_3 \succ X_2 \succ X_1$
$\beta = 0.8$	$RI(A_1) = 0.1494, RI(A_2) = 0.1975,$ $RI(A_3) = 0.2326, RI(A_4) = 0.3435$	$X_4 \succ X_3 \succ X_2 \succ X_1$
$\beta = 0.9$	$RI(A_1) = 0.1574, RI(A_2) = 0.1976,$ $RI(A_3) = 0.2320, RI(A_4) = 0.3361$	$X_4 \succ X_3 \succ X_2 \succ X_1$

Table 8 and Figure 2 present the results when parameter α varies for the fixed values of β and λ. As shown in Figure 2, as $\lambda = \beta = 0.5$, the changes in α within $\alpha \in (0, 0.5)$ that induce greater changes in the ranking indicator for robots X_1, X_3, X_4 are affected to a higher degree, but the overall stability, sorting results remain unchanged. Thus, the changes can be considered to be more quantitative than qualitative.

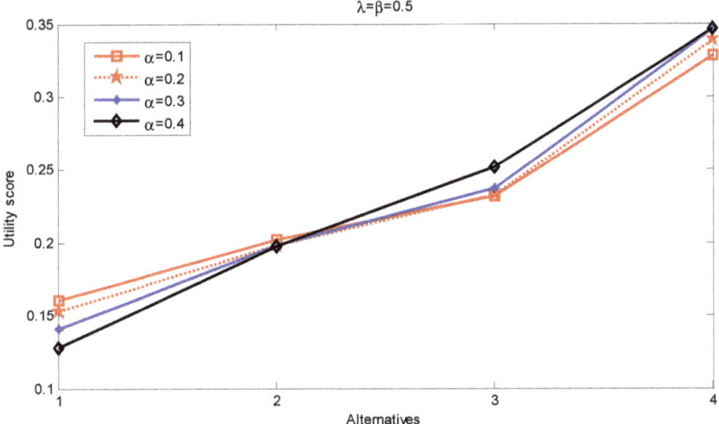

Figure 2. Sensitivity analysis of NWTIFBHM evaluation results ($\lambda = \beta = 0.5$).

Table 9 and Figure 3 deal with the case where β varies for fixed α and λ. As shown in Figure 3, as $\lambda = \alpha = 0.5$, the values of the ranking indicator for robots S_1, S_3 are more sensitive to changes in β, if opposed to the other alternatives. However, the overall ranking remains stable.

The analysis suggests that the proposed aggregation operator performs similarly to the other aggregation operators capable of accounting for the inter-relationships among the data. The changes in the parameters of the operator did not render significant changes in the rankings. Thus, the proposed model can be considered to be effective and stable.

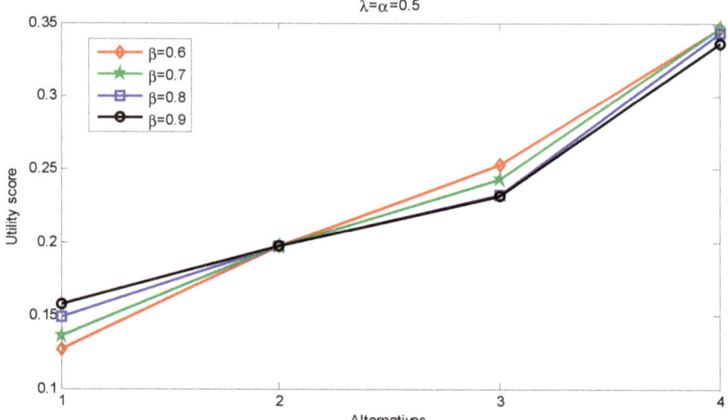

Figure 3. Sensitivity analysis of NWTIFBHM evaluation results ($\lambda = \alpha = 0.5$).

4. Conclusions

Based on the Bonferroni mean, we developed the Bonferroni harmonic mean, which addresses the inter-relationships among the data to be aggregated to a higher extent. Specifically, the outlying observations receive much lower significance without any additional processing. The normalized harmonic Bonferroni mean allows for incorporating the preferences of the decision makers regarding the importance of the arguments to be aggregated. These concepts were integrated with the triangular fuzzy numbers, allowing uncertain information in the decision making problems to be represented. As a result, we have proposed the NWTIFBHM operator.

The new operator was applied in an illustrative example on a MAGDM problem. The comparative analysis comprised two directions: comparison with the existing approaches and sensitivity to changes in the underlying parameters. The analysis showed that the proposed aggregation operator is effective and is not heavily impacted by the changes in the underlying parameters.

Future research can be directed towards extension of the proposed aggregation operator by applying the generalized normalized weighted Bonferroni mean [33], probabilistic averages [38–40], Pythagorean fuzzy sets [12], and Choquet integrals [41], along with combinations thereof [42,43]. Simulation studies can be carried out to check the performance of the proposed approach in different settings [44] and to relate it to databases for real-life situations [45,46]. From the empirical viewpoint, applications of the NWTIFNBH operator for decisions in real-life problems can be considered across different sectors.

Author Contributions: Writing—original draft, J.Z.; Writing—review & editing, T.B. and D.S.

Funding: The work was supported in part by the National Social Science Fund Project (No. 2017YYRW07), Anhui Province Natural Science in Universities of General Project (No. TSKJ2017B22), and the general project of teaching research project of Anhui Polytechnic University (No. 2018JYXM43).

Conflicts of Interest: The authors declare no conflict of interest.

Appendix A

In this appendix, we provide the proof of Equation (16).

Proof. Utilizing the principles of the operational laws for the TIFNs, one can obtain

$$\left(\frac{w_i}{(1-w_i)A_i^p}\right) \otimes \left(\frac{w_j}{A_j^q}\right) = \left(\left[\frac{w_i w_j}{(1-w_i)a_i^p a_j^q}, \frac{w_i w_j}{(1-w_i)b_i^p b_j^q}, \frac{w_i w_j}{(1-w_i)c_i^p c_j^q}\right]; \omega_{A_i} \wedge \omega_{A_j}, u_{A_i} \vee u_{A_j}\right)$$

Initially, one can derive that

$$\oplus^n_{\substack{i,j=1 \\ i \neq j}} \left(\left(\frac{w_i}{(1-w_i)A_i^p} \right) \otimes \left(\frac{w_j}{A_j^q} \right) \right)$$

$$= ([\sum_{\substack{i,j=1 \\ i \neq j}}^{n} \frac{w_i w_j}{(1-w_i)a_i^p a_j^q}, \sum_{\substack{i,j=1 \\ i \neq j}}^{n} \frac{w_i w_j}{(1-w_i)b_i^p b_j^q}, \sum_{\substack{i,j=1 \\ i \neq j}}^{n} \frac{w_i w_j}{(1-w_i)c_i^p c_j^q}]; \wedge_{i=1}^{n} \omega_{A_i}, \vee_{i=1}^{n} u_{A_i}) \tag{A1}$$

By exploiting the principle of mathematical induction upon n in the following manner:

(1)   when $n = 2$, given (15), we can show:

$$\oplus^2_{\substack{i,j=1 \\ i \neq j}} \left(\left(\frac{w_i}{(1-w_i)A_i^p} \right) \otimes \left(\frac{w_j}{A_j^q} \right) \right) = \left(\left(\frac{w_1}{(1-w_1)A_1^p} \right) \otimes \left(\frac{w_2}{A_2^q} \right) \right) \oplus \left(\left(\frac{w_2}{(1-w_2)A_2^p} \right) \otimes \left(\frac{w_1}{A_1^q} \right) \right)$$

$$= ([\frac{w_1 w_2}{(1-w_1)a_1^p a_2^q} + \frac{w_2 w_1}{(1-w_2)a_2^p a_1^q}, \frac{w_1 w_2}{(1-w_1)b_1^p b_2^q} + \frac{w_2 w_1}{(1-w_2)b_2^p b_1^q}, \frac{w_1 w_2}{(1-w_1)c_1^p c_2^q} + \frac{w_2 w_1}{(1-w_2)c_2^p c_1^q}];$$

$$\omega_{A_1} \wedge \omega_{A_2}, u_{A_1} \vee u_{A_2})$$

$$= ([\sum_{\substack{i,j=1 \\ i \neq j}}^{2} \frac{w_i w_j}{(1-w_i)a_i^p a_j^q}, \sum_{\substack{i,j=1 \\ i \neq j}}^{2} \frac{w_i w_j}{(1-w_i)b_i^p b_j^q}, \sum_{\substack{i,j=1 \\ i \neq j}}^{2} \frac{w_i w_j}{(1-w_i)c_i^p c_j^q}]; \wedge_{i=1}^{2} \omega_{A_i}, \vee_{i=1}^{2} u_{A_i})$$

(2)   assume that $n = k$ and Equation (15) holds so that

$$\oplus^k_{\substack{i,j=1 \\ i \neq j}} \left(\left(\frac{w_i}{(1-w_i)A_i^p} \right) \otimes \left(\frac{w_j}{A_j^q} \right) \right)$$

$$= ([\sum_{\substack{i,j=1 \\ i \neq j}}^{k} \frac{w_i w_j}{(1-w_i)a_i^p a_j^q}, \sum_{\substack{i,j=1 \\ i \neq j}}^{k} \frac{w_i w_j}{(1-w_i)b_i^p b_j^q}, \sum_{\substack{i,j=1 \\ i \neq j}}^{k} \frac{w_i w_j}{(1-w_i)c_i^p c_j^q}]; \wedge_{i=1}^{k} \omega_{A_i}, \vee_{i=1}^{k} u_{A_i}) \tag{A2}$$

(3)   subsequently, assume $n = k + 1$ and by the virtue of (15), get

$$\oplus^{k+1}_{\substack{i,j=1 \\ i \neq j}} \left(\left(\frac{w_i}{(1-w_i)A_i^p} \right) \otimes \left(\frac{w_j}{A_j^q} \right) \right) = \left(\oplus^k_{\substack{i,j=1 \\ i \neq j}} \left(\left(\frac{w_i}{(1-w_i)A_i^p} \right) \otimes \left(\frac{w_j}{A_j^q} \right) \right) \right) \oplus$$

$$\left(\oplus_{i=1}^{k} \left(\left(\frac{w_i}{(1-w_i)A_i^p} \right) \otimes \left(\frac{w_{k+1}}{A_{k+1}^q} \right) \right) \right) \oplus \left(\oplus_{j=1}^{k} \left(\left(\frac{w_{k+1}}{(1-w_{k+1})A_{k+1}^p} \right) \otimes \left(\frac{w_j}{A_j^q} \right) \right) \right) \tag{A3}$$

We now prove that

$$\oplus_{i=1}^{k} \left(\left(\frac{w_i}{(1-w_i)A_i^p} \right) \otimes \left(\frac{w_{k+1}}{A_{k+1}^q} \right) \right) = ([\sum_{i=1}^{k} \frac{w_i w_j}{(1-w_i)a_i^p a_j^q}, \sum_{i=1}^{k} \frac{w_i w_j}{(1-w_i)b_i^p b_j^q}, \sum_{i=1}^{k} \frac{w_i w_j}{(1-w_i)c_i^p c_j^q}];$$

$$\wedge_{i=1}^{k} (\omega_{A_i} \wedge \omega_{A_{k+1}}), \vee_{i=1}^{k} (u_{A_i} \vee u_{A_{k+1}})) \tag{A4}$$

By applying the principle of the mathematical induction upon k.

(a) Let $k = 2$, and by the virtue of Equation (A4), one can show

$$\oplus_{i=1}^{2}\left(\left(\frac{w_i}{(1-w_i)A_i^p}\right)\otimes\left(\frac{w_{k+1}}{A_{k+1}^q}\right)\right)=\left(\left(\frac{w_1}{(1-w_1)A_1^p}\right)\otimes\left(\frac{w_{k+1}}{A_{k+1}^q}\right)\right)\oplus\left(\left(\frac{w_2}{(1-w_2)A_2^p}\right)\otimes\left(\frac{w_{k+1}}{A_{k+1}^q}\right)\right)$$

$$=\left(\left[\frac{w_1 w_{k+1}}{(1-w_1)a_1^p a_{k+1}^q}+\frac{w_2 w_{k+1}}{(1-w_2)a_2^p a_{k+1}^q},\frac{w_1 w_{k+1}}{(1-w_1)b_1^p b_{k+1}^q}+\frac{w_2 w_{k+1}}{(1-w_2)b_2^p b_{k+1}^q},\right.\right. \tag{A5}$$

$$\left.\left.\frac{w_1 w_{k+1}}{(1-w_1)c_1^p c_{k+1}^q}+\frac{w_2 w_{k+1}}{(1-w_2)c_2^p c_{k+1}^q}\right];(\omega_{A_1}\wedge\omega_{A_{k+1}})\wedge(\omega_{A_2}\wedge\omega_{A_{k+1}}),(u_{A_1}\vee u_{A_{k+1}})\vee(u_{A_1}\vee u_{A_{k+1}})\right)$$

$$=\left(\left[\sum_{i=1}^{2}\frac{w_i w_j}{(1-w_i)a_i^p a_j^q},\sum_{i=1}^{2}\frac{w_i w_j}{(1-w_i)b_i^p b_j^q},\sum_{i=1}^{2}\frac{w_i w_j}{(1-w_i)c_i^p c_j^q}\right];\wedge_{i=1}^{2}(\omega_{A_i}\wedge\omega_{A_{k+1}}),\vee_{i=1}^{2}(u_{A_i}\vee u_{A_{k+1}})\right) \tag{A6}$$

(b) Assume Equation (A4) is valid for any given $k = k_0$

$$\oplus_{i=1}^{k_0}\left(\left(\frac{w_i}{(1-w_i)A_i^p}\right)\otimes\left(\frac{w_{k+1}}{A_{k+1}^q}\right)\right)$$

$$=\left(\left[\sum_{i=1}^{k_0}\frac{w_i w_j}{(1-w_i)a_i^p a_j^q},\sum_{i=1}^{k_0}\frac{w_i w_j}{(1-w_i)b_i^p b_j^q},\sum_{i=1}^{k_0}\frac{w_i w_j}{(1-w_i)c_i^p c_j^q}\right];\wedge_{i=1}^{k_0}(\omega_{A_i}\wedge\omega_{A_{k+1}}),\vee_{i=1}^{k_0}(u_{A_i}\vee u_{A_{k+1}})\right) \tag{A7}$$

(c) Subsequently, we demonstrate that the following holds for any $k = k_0 + 1$:

$$\oplus_{i=1}^{k_0+1}\left(\left(\frac{w_i}{(1-w_i)A_i^p}\right)\otimes\left(\frac{w_{k+1}}{A_{k+1}^q}\right)\right)=\oplus_{i=1}^{k_0}\left(\left(\frac{w_i}{(1-w_i)A_i^p}\right)\otimes\left(\frac{w_{k+1}}{A_{k+1}^q}\right)\right)\oplus\left(\left(\frac{w_{k_0+1}}{(1-w_{k_0+1})A_{k_0+1}^p}\right)\otimes\left(\frac{w_{k+1}}{A_{k+1}^q}\right)\right)$$

$$=\left(\left[\sum_{i=1}^{k_0}\frac{w_i w_j}{(1-w_i)a_i^p a_j^q}+\frac{w_{k_0+1}w_{k+1}}{(1-w_{k_0+1})a_{k_0+1}^p a_{k+1}^q},\sum_{i=1}^{k_0}\frac{w_i w_j}{(1-w_i)b_i^p b_j^q}+\frac{w_{k_0+1}w_{k+1}}{(1-w_{k_0+1})b_{k_0+1}^p b_{k+1}^q},\right.\right.$$

$$\left.\sum_{i=1}^{k_0}\frac{w_i w_j}{(1-w_i)c_i^p c_j^q}+\frac{w_{k_0+1}w_{k+1}}{(1-w_{k_0+1})c_{k_0+1}^p c_{k+1}^q}\right];$$

$$\left.\wedge_{i=1}^{k_0}(\omega_{A_i}\wedge\omega_{A_{k+1}})\wedge(\omega_{A_{k_0+1}}\wedge\omega_{A_{k+1}}),\vee_{i=1}^{k_0}(u_{A_i}\vee u_{A_{k+1}})\vee(u_{A_{k_0+1}}\vee u_{A_{k+1}})\right)$$

Clearly,

$$\wedge_{i=1}^{k}(\omega_{A_i}\wedge\omega_{A_{k+1}})=\wedge_{i=1}^{k+1}\omega_{A_i},\vee_{i=1}^{k}(u_{A_i}\vee u_{A_{k+1}})=\vee_{i=1}^{k+1}u_{A_i}$$

Hence,

$$\oplus_{i=1}^{k}\left(\left(\frac{w_i}{(1-w_i)A_i^p}\right)\otimes\left(\frac{w_{k+1}}{A_{k+1}^q}\right)\right)$$

$$=\left(\left[\sum_{i=1}^{k}\frac{w_i w_{k+1}}{(1-w_i)a_i^p a_{k+1}^q},\sum_{i=1}^{k}\frac{w_i w_{k+1}}{(1-w_i)b_i^p b_{k+1}^q},\sum_{i=1}^{k}\frac{w_i w_{k+1}}{(1-w_i)c_i^p c_{k+1}^q}\right];\wedge_{i=1}^{k+1}\omega_{A_i},\vee_{i=1}^{k+1}u_{A_i}\right) \tag{A8}$$

Similarly,

$$\oplus_{j=1}^{k}\left(\left(\frac{w_{k+1}}{(1-w_{k+1})A_{k+1}^p}\right)\otimes\left(\frac{w_j}{A_j^q}\right)\right)$$

$$=\left(\left[\sum_{j=1}^{k}\frac{w_{k+1}w_j}{(1-w_{k+1})a_{k+1}^p a_j^q},\sum_{j=1}^{k}\frac{w_{k+1}w_j}{(1-w_{k+1})b_{k+1}^p b_j^q},\sum_{j=1}^{k}\frac{w_{k+1}w_j}{(1-w_{k+1})c_{k+1}^p c_j^q}\right];\wedge_{j=1}^{k+1}\omega_{A_j},\vee_{j=1}^{k+1}u_{A_j}\right) \tag{A9}$$

From Equations (A3), (A8) and (A9), we get

$$\oplus_{\substack{i,j=1\\i\neq j}}^{k+1}\left(\left(\frac{w_i}{(1-w_i)A_i^p}\right)\otimes\left(\frac{w_j}{A_j^q}\right)\right)$$

$$=\left(\left[\sum_{\substack{i,j=1\\i\neq j}}^{k}\frac{w_i w_j}{(1-w_i)a_i^p a_j^q},\sum_{\substack{i,j=1\\i\neq j}}^{k}\frac{w_i w_j}{(1-w_i)b_i^p b_j^q},\sum_{\substack{i,j=1\\i\neq j}}^{k}\frac{w_i w_j}{(1-w_i)c_i^p c_j^q}\right];\wedge_{i=1}^{k}\omega_{A_i},\vee_{i=1}^{k}u_{A_i}\right)$$

$$\oplus\left(\left[\sum_{i=1}^{k}\frac{w_i w_{k+1}}{(1-w_i)a_i^p a_{k+1}^q},\sum_{i=1}^{k}\frac{w_i w_{k+1}}{(1-w_i)b_i^p b_{k+1}^q},\sum_{i=1}^{k}\frac{w_i w_{k+1}}{(1-w_i)c_i^p c_{k+1}^q}\right];\wedge_{i=1}^{k+1}\omega_{A_i},\vee_{i=1}^{k+1}u_{A_i}\right)$$

$$\oplus\left(\left[\sum_{j=1}^{k}\frac{w_{k+1} w_j}{(1-w_{k+1})a_{k+1}^p a_j^q},\sum_{j=1}^{k}\frac{w_{k+1} w_j}{(1-w_{k+1})b_{k+1}^p b_j^q},\sum_{j=1}^{k}\frac{w_{k+1} w_j}{(1-w_{k+1})c_{k+1}^p c_j^q}\right];\wedge_{j=1}^{k+1}\omega_{A_j},\vee_{j=1}^{k+1}u_{A_j}\right)$$

$$=\left(\left[\sum_{\substack{i,j=1\\i\neq j}}^{k+1}\frac{w_i w_j}{(1-w_i)a_i^p a_j^q},\sum_{\substack{i,j=1\\i\neq j}}^{k+1}\frac{w_i w_j}{(1-w_i)b_i^p b_j^q},\sum_{\substack{i,j=1\\i\neq j}}^{k+1}\frac{w_i w_j}{(1-w_i)c_i^p c_j^q}\right];\wedge_{i=1}^{k+1}\omega_{A_i},\vee_{i=1}^{k+1}u_{A_i}\right)$$

As a result, Equation (A1) is valid for $n = k+1$. Therefore, Equation (A1) is valid for any n. Considering Equation (A1) alongside operational law (3)

$$\frac{1}{\left(\oplus_{\substack{i,j=1\\i\neq j}}^{k+1}\left(\left(\frac{w_i}{(1-w_i)A_i^p}\right)\otimes\left(\frac{w_j}{A_j^q}\right)\right)\right)^{\frac{1}{p+q}}}=$$

$$\left(\left[\frac{1}{\left(\sum_{\substack{i,j=1\\i\neq j}}^{k+1}\frac{w_i w_j}{(1-w_i)a_i^p a_j^q}\right)^{\frac{1}{p+q}}},\frac{1}{\left(\sum_{\substack{i,j=1\\i\neq j}}^{k+1}\frac{w_i w_j}{(1-w_i)b_i^p b_j^q}\right)^{\frac{1}{p+q}}},\frac{1}{\left(\sum_{\substack{i,j=1\\i\neq j}}^{k+1}\frac{w_i w_j}{(1-w_i)c_i^p c_j^q}\right)^{\frac{1}{p+q}}}\right];\wedge_{i=1}^{k+1}\omega_{A_i},\vee_{i=1}^{k+1}u_{A_i}\right) \tag{A10}$$

Exploiting Equation (A10) as well as operational law (4), one can show that

$$NWTIFBHM^{p,q}(A_1,A_2,\cdots,A_n)=$$

$$\left(\left[\frac{1}{\left(\sum_{\substack{i,j=1\\i\neq j}}^{n}\frac{w_i w_j}{(1-w_i)a_i^p a_j^q}\right)^{\frac{1}{p+q}}},\frac{1}{\left(\sum_{\substack{i,j=1\\i\neq j}}^{n}\frac{w_i w_j}{(1-w_i)b_i^p b_j^q}\right)^{\frac{1}{p+q}}},\frac{1}{\left(\sum_{\substack{i,j=1\\i\neq j}}^{n}\frac{w_i w_j}{(1-w_i)c_i^p c_j^q}\right)^{\frac{1}{p+q}}}\right];\wedge_{i=1}^{n}\omega_{A_i},\vee_{i=1}^{n}u_{A_i}\right) \tag{A11}$$

As long as $a_i \leq b_i \leq c_i$, for all $i = 1,2,\cdots,n$. By the virtue of the property associated with the NWBHM, one can show that

$$\frac{1}{\left(\sum_{\substack{i,j=1\\i\neq j}}^{n}\frac{w_i w_j}{(1-w_i)a_i^p a_j^q}\right)^{\frac{1}{p+q}}}\leq\frac{1}{\left(\sum_{\substack{i,j=1\\i\neq j}}^{n}\frac{w_i w_j}{(1-w_i)b_i^p b_j^q}\right)^{\frac{1}{p+q}}}\leq\frac{1}{\left(\sum_{\substack{i,j=1\\i\neq j}}^{n}\frac{w_i w_j}{(1-w_i)c_i^p c_j^q}\right)^{\frac{1}{p+q}}} \tag{A12}$$

Also,

$$0\leq\wedge_{i=1}^{n}\omega_{A_i}+\vee_{i=1}^{n}u_{A_i}\leq 1 \tag{A13}$$

From Equations (A12) and (A13), $NWTIFBHM^{p,q}$ is a TIFN.

References

1. Atanassov, K.T. Intuitionistic fuzzy sets. *Fuzzy Sets Syst.* **1986**, *20*, 87–96. [CrossRef]
2. Atanassov, K.T. More on intuitionistic fuzzy sets. *Fuzzy Sets Syst.* **1989**, *51*, 117–118. [CrossRef]
3. Liu, P.D.; Chen, S.M. Group Decision Making Based on Heronian Aggregation Operators of Intuitionistic Fuzzy Numbers. *IEEE Trans. Cybern.* **2017**, *47*, 2514–2530. [CrossRef] [PubMed]
4. Xu, Z.S.; Yager, R.R. Some geometric aggregation operators based on intuitionistic fuzzy sets. *Int. J. Gen. Syst.* **2006**, *35*, 417–433. [CrossRef]

5. Peng, B.; Zhou, J.M.; Peng, D.H. Cloud model based approach to group decision making under uncertain pure linguistic information. *J. Intell. Fuzzy Syst.* **2017**, *32*, 1959–1968. [CrossRef]

6. Atanassov, K.T.; Pasi, G.; Yager, R.R. Intuitionistic fuzzy interpretations of multi-criteria multi-person and multi-measurement tool decision making. *Int. J. Gen. Syst.* **2005**, *36*, 859–868. [CrossRef]

7. Liang, C.; Zhao, S.; Zhang, J. Multi-criteria group decision making method based on generalized intuitionistic trapezoidal fuzzy prioritized aggregation operators. *Int. J. Mach. Learn. Cybern.* **2017**, *8*, 597–610. [CrossRef]

8. Zhao, H.; Xu, Z.S.; Ni, M.F.; Liu, S.H. Generalized Aggregation Operators for Intuitionistic Fuzzy Sets. *Int. J. Intell. Syst.* **2009**, *25*, 1–30. [CrossRef]

9. Chu, Y.C.; Liu, P.D. Some two-dimensional uncertain linguistic Heronian mean operators and their application in multiple-attribute decision making. *Neural Comput. Appl.* **2015**, *26*, 1461–1480. [CrossRef]

10. Li, D.F. A ratio ranking method of triangular intuitionistic fuzzy numbers and its application to MADM problems. *Comput. Math. Appl.* **2010**, *60*, 1557–1570. [CrossRef]

11. Su Mail, W.; Zhou, J.; Zeng, S.; Zhang, C.; Yu, K. A Novel Method for Intuitionistic Fuzzy MAGDM with Bonferroni Weighted Harmonic Means. *Recent Patents Comput. Sci.* **2017**, *10*, 178–189.

12. Zeng, S.Z.; Xiao, Y. A method based on TOPSIS and distance measures for hesitant fuzzy multiple attribute decision making. *Technol. Econ. Dev. Econ.* **2018**, *24*, 969–983. [CrossRef]

13. Wang, J.Q. Overview on fuzzy multi-criteria decision-making approach. *Control Decis.* **2008**, *23*, 601–607.

14. Fan, C.X.; Ye, J.; Hu, K.L.; Fan, E. Bonferroni Mean Operators of Linguistic Neutrosophic Numbers and Their Multiple Attribute Group Decision-Making Methods. *Information* **2017**, *8*, 107. [CrossRef]

15. Wang, J.P.; Dong, J.Y. Method of intuitionistic trapezoidal fuzzy number for multi-attribute group decision. *Control Decis.* **2010**, *25*, 773–776.

16. Wu, J.; Cao, Q.W. Same families of geometric aggregation operators with intuitionistic trapezoidal fuzzy numbers. *Appl. Math. Model.* **2013**, *37*, 318–327. [CrossRef]

17. Bonferroni, C. Sulle medie multiple di potenze. *Boll. Matemat. Ital.* **1950**, *5*, 267–270.

18. Yager, R.R. On generalized Bonferroni mean operators for multicriteria aggregation. *Int. J. Approx. Reason.* **2009**, *50*, 1279–1286. [CrossRef]

19. Choquet, G. Theory of capacities. *Ann. Inst. Fourier* **1955**, *50*, 131–295. [CrossRef]

20. Beliakov, G.; James, S.; Mordelová, J.; Rückschlossová, T.; Yager, R.R. Generalized Bonferroni mean operators in multi-criteria aggregation. *Fuzzy Sets Syst.* **2010**, *161*, 2227–2242. [CrossRef]

21. Xu, Z.S.; Yager, R.R. Intuitionistic Fuzzy Bonferroni Means. *IEEE Trans. Syst. Man Cybern. Part B Cybern.* **2011**, *41*, 568–578.

22. Dutta, B.; Guha, D. Trapezoidal Intuitionistic Fuzzy Bonferroni Means and Its Application in Multi-attribute Decision Making. In Proceedings of the 2013 IEEE International Conference on Fuzzy Systems (FUZZ), Hyderabad, India, 7–10 July 2013.

23. Xu, Z.S. Uncertain Bonferroni Mean operators. *Int. J. Comput. Intell. Syst.* **2010**, *3*, 761–769. [CrossRef]

24. Xia, M.; Xu, Z.S.; Zhu, B. Generalized Intuitionistic Fuzzy Bonferroni Means. *Int. J. Intell. Syst.* **2012**, *27*, 23–47. [CrossRef]

25. Das, S.; Guha, D. Power harmonic aggregation operator with Trapezoidal intuitionistic fuzzy numbers for solving MAGDM problems. *Iranian J. Fuzzy Syst.* **2015**, *12*, 41–74.

26. Xu, Z.S. Fuzzy harmonic mean operators. *Int. J. Intell. Syst.* **2009**, *24*, 152–172. [CrossRef]

27. Wei, G.W. FIOWHM operator and its application to multiple attribute group decision making. *Expert Syst. Appl.* **2011**, *38*, 2984–2989. [CrossRef]

28. Sun, H.; Sun, M. Generalized Bonferroni harmonic mean operators and their application to multiple attribute decision making. *J. Comput. Inf. Syst.* **2012**, *8*, 5717–5724.

29. Park, J.H.; Park, E.J. Generalized Fuzzy Bonferroni Harmonic Mean Operators and Their Applications in Group Decision Making. *J. Appl. Math.* **2013**, *2013*, 1–14. [CrossRef]

30. Liu, P.D.; Li, H.G. Interval-Valued Intuitionistic Fuzzy Power Bonferroni Aggregation Operators and Their Application to Group Decision Making. *Cogn. Comput.* **2017**, *9*, 1–19. [CrossRef]

31. Wan, S.P. Power average operators of trapezoidal intuitionistic fuzzy numbers and application to multi-attribute group decision making. *Appl. Math. Model.* **2013**, *37*, 4112–4126. [CrossRef]

32. Li, D.F. *Decision and Game Theory in Management with Intuitionistic Fuzzy Sets*; Springer: Berlin/Heidelberg, Germany, 2014.

33. Zhou, W.; He, J.M. Intuitionistic Fuzzy Normalized Weighted Bonferroni Mean and Its Application in Multicriteria Decision Making. *J. Appl. Math.* **2012**, *2012*, 1–22. [CrossRef]
34. Xu, Z.S.; Da, Q.L. A likelihood based method for priorities of interval judgment matrices. *Chin. J. Manag. Sci.* **2003**, *11*, 63–65.
35. Micire, M.J. Evolution and field performance of a rescue robot. *J. Field Robot.* **2008**, *25*, 17–30. [CrossRef]
36. Wan, S.P.; Dong, J.Y. Power geometric operators of trapezoidal intuitionistic fuzzy numbers and application to multi-attribute group decision making. *Appl. Soft Comput.* **2015**, *29*, 153–168. [CrossRef]
37. Wei, G.W. Some arithmetic aggregation operators with intuitionistic trapezoidal fuzzy numbers and their application to group decision making. *J. Comput.* **2010**, *5*, 345–351. [CrossRef]
38. Merigó, J.M.; Wei, G.W. Probabilistic aggregation operators and their application in uncertain multi-person decision making. *Technol. Econ. Dev.* **2011**, *17*, 335–351. [CrossRef]
39. Merigó, J.M. The probabilistic weighted average and its application in multi-person decision making. *Int. J. Intell. Syst.* **2012**, *27*, 457–476. [CrossRef]
40. Xu, Z.S.; Da, Q.L. An overview of operators for aggregating information. *Int. J. Intell. Syst.* **2003**, *18*, 953–968. [CrossRef]
41. Bolton, J.; Gader, P.; Wilson, J.N. Discrete Choquet integral as a distance metric. *IEEE Trans. Fuzzy Syst.* **2008**, *16*, 1107–1110. [CrossRef]
42. Xu, Z.S.; Da, Q.L. The Uncertain OWA operator. *Int. J. Intell. Syst.* **2002**, *17*, 569–575. [CrossRef]
43. Zeng, S.Z.; Su, W.H.; Zhang, C.H. A Novel Method Based on Induced Aggregation Operator for Classroom Teaching Quality Evaluation with Probabilistic and Pythagorean Fuzzy Information. *Eurasia J. Math. Sci. Technol. Educ.* **2018**, *14*, 3205–3212. [CrossRef]
44. Mukhametzyanov, I.; Pamucar, D. A sensitivity analysis in MCDM problems: A statistical approach. *Decis. Mak. Appl. Manag. Eng.* **2018**, *1*, 51–80. [CrossRef]
45. Yi, P.; Li, W.; Li, L. Evaluation and Prediction of City Sustainability Using MCDM and Stochastic Simulation Methods. *Sustainability* **2018**, *10*, 3771. [CrossRef]
46. Zhu, B.; Xu, Z. Probability-hesitant fuzzy sets and the representation of preference relations. *Technol. Econ. Dev. Econ.* **2018**, *24*, 1029–1040. [CrossRef]

 © 2019 by the authors. Licensee MDPI, Basel, Switzerland. This article is an open access article distributed under the terms and conditions of the Creative Commons Attribution (CC BY) license (http://creativecommons.org/licenses/by/4.0/).

Article

Type-2 Multi-Fuzzy Sets and Their Applications in Decision Making

Mohuya B. Kar [1], Bikashkoli Roy [2], Samarjit Kar [2], Saibal Majumder [3] and Dragan Pamucar [4,*

[1] Department of Computer Science, Heritage Institute of Technology, Kolkata 700107, India; mohuya_kar@yahoo.com
[2] Department of Mathematics, National Institute of Technology, Durgapur 713209, India; bikashroy235@gmail.com (B.R.); dr.samarjitkar@gmail.com (S.K.)
[3] Department of Computer Science and Engineering, National Institute of Technology, Durgapur 713209, India; saibaltufts@gmail.com
[4] Department of Logistics, University of Defence in Belgrade Military Academy, 11000 Belgrade, Serbia
* Correspondence: dpamucar@gmail.com

Received: 30 December 2018; Accepted: 28 January 2019; Published: 1 February 2019

Abstract: In a real-life scenario, it is undoable and unmanageable to solve a decision-making problem with the single stand-alone decision-aid method, expert assessment methodology or deterministic approaches. Such problems are often based on the suggestions or feedback of several experts. Usually, the feedback of these experts are heterogeneous imperfect information collected from various more or less reliable sources. In this paper, we introduce the concept of multi-sets over type-2 fuzzy sets. We have tried to propose an extension of type-1 multi-fuzzy sets into a type-2 multi-fuzzy set (T2MFS). After defining T2MFS, we discuss the algebraic properties of these sets including set-theoretic operations such as complement, union, intersection, and others with examples. Subsequently, we define two distance measures over these sets and illustrate a decision-making problem which uses the idea of type-2 multi-fuzzy sets. Furthermore, an application of a medical diagnosis system based on multi-criteria decision making of T2MFS is illustrated with a real-life case study.

Keywords: Multi-fuzzy sets; type-2 fuzzy sets; type-2 multi-fuzzy sets; distance measures; set operations

1. Introduction

According to Cantor, a set is a well-defined collection of distinct objects, but let us see what happens if we consider a collection of objects where one or some or all objects occur more than once. The answer to this question is a different type of set known as multi-set or m-set which was first studied by Bruijn [1]. Afterwards, Yager [2] proposed a new and more generalized concept of multi-sets named as multi-fuzzy sets (MFS) which can deal with many real life problems with some degree of ease. Sebastian and Ramakrishnan [3] also studied multi-fuzzy sets and concluded that multi-fuzzy set theory is an extension of Zadeh's fuzzy set theory, Atanassov's intuitionistic fuzzy set theory and L-fuzzy set theory. Yang et al. [4] discussed applications of multi-fuzzy soft sets in decision making. Furthermore, Das et al. [5] proposed an approach of group multi-criteria decision-making using intuitionistic multi-fuzzy sets.

A type-2 fuzzy set (T2FS) is an extension of ordinary fuzzy sets, i.e., type-1 fuzzy set (T1FS). The membership value of a type-1 fuzzy set is a real number in the closed interval $[0, 1]$. On the other hand, the membership value of a T2FS is a type-1 fuzzy set. The concept of T2FS was introduced by Zadeh [6–8]. Mizumoto and Tanaka [9,10], and Dubois and Prade [11] investigated the logical operations of T2FS. Later, many researchers did a lot of theoretical work on the properties of T2FS [12–14] and figured out many applications [15–21].

Although there are various mathematical tools such as fuzzy sets, rough sets, multi-sets, multi-fuzzy sets, intuitionistic fuzzy sets, and type-2 fuzzy sets to deal with uncertainties [22–24], there might be some physical problems where the primary membership function may have more than one secondary membership grade with the same or different values. In those particular cases, the existing mathematical tools might not be adequate. That is why we introduce a new concept of type-2 multi-fuzzy sets. Type-2 multi-fuzzy sets are supposedly a new approach which will be an extension of the existing concepts and shall be helpful to deal with problems related to uncertainties. T2MFS is a type-2 fuzzy set whose primary membership function has a sequence of secondary membership values lying in the closed interval $[0, 1]$. In this paper, we first give definitions of classical multi-sets, multi-fuzzy sets and type-2 fuzzy sets and also provide examples of each.

The main contributions of this article are highlighted as follows:

(i)   A type-2 multi-fuzzy set is proposed in this article.
(ii)   Some set-theoretic operations of T2MFS, e.g., complement, inclusion, union, and intersection of T2MFS are presented in this article.
(iii) Two distance measure: (a) Hamming distance and (b) Euclidian distance are proposed in this study.
(iv)   A suitable application based on a medical diagnosis system is presented by using the distance measure of T2MFS.

The rest of the paper is organized as follows. Some relevant studies of the literature are surveyed in Section 2. The preliminary concepts of our study are discussed in Section 3. In Section 4, we define type-2 multi-fuzzy set (T2MFS) and give examples. Subsequently, in Section 5, the algebraic operations on T2MFS like complement, inclusion, union, and intersection are discussed. Consequently, in Section 6, set-theoretic properties like idempotency, commutativity, associativity, and distributivity are verified for T2MFSs. In Section 7, we define the two distance measures of T2MFS. We provide a real-life application based on a medical diagnosis system, which applies the concept of type-2 multi-fuzzy sets in Section 8. In Section 9, we conduct a case study based on the application presented in Section 8. Finally, the study is concluded in Section 10.

2. Literature Review

In this section, we present a brief overview of different variants of the multi-fuzzy set (MFS) which have been proposed in previous studies. The survey, by no means, encompasses all the related researches in the literature. However, some related studies having significant contributions are reviewed.

The concept of MFS has originated as an extension of the fuzzy set [25], *L*-fuzzy set [26] and intuitionistic fuzzy set [27]. In a MFS, the membership function is an ordered sequence of ordinary fuzzy membership functions. Here, an element of the universe can repeat itself with possibly the same or different membership values. Motivated by the study of Yager [2] on MFS, several contributions can be observed in the field of multi-fuzzy sets and its variants. Muthuraj and Balamurugan [28] proposed some algebraic structures of multi-fuzzy subgroup and investigated their properties. Sebastian and Ramakrishnan [29] proposed the multi-fuzzy subgroup and normal multi-fuzzy subgroups. Furthermore, various bridge function like lattice homomorphisms, order homomorphisms, *L*-fuzzy lattices, and strong *L*-fuzzy lattices have been developed by Sebastian and Ramakrishnan [30]. Subsequently, multi-fuzzy topology was proposed by Sebastian and Ramakrishnan [31]. In addition, a progressive development of MFS can be observed from the contributions of several researchers [32–34].

Enlightened by the development of MFS, Dey and Pal [35] proposed multi-fuzzy complex numbers and multi-fuzzy complex sets. Using the concepts of their studies, the authors introduced multi-fuzzy complex nilpotent matrices over a distributive lattice [36]. The authors also developed multi-fuzzy vector space and multi-fuzzy linear transformation over a finite-dimensional multi-fuzzy set [37].

Shinoj and John [38] introduced the intuitionistic fuzzy multi-sets (IFMS). After that, several investigations were conducted to develop various features of IFMS. Ejegwa and Awolola [39] determined the binomial probability of IFMS, where for each trial, it was assumed that the probability of the membership degree was constant and the intuitionistic fuzzy multi-set index was negligible. Rajarajeswari and Uma [40] proposed three distance measures and their corresponding similarity measures of IFMS. These measures are based on the Hausdorff distance measure, the geometric distance measure and the normalized distance measure. Subsequently, different studies of IFMS [41–44] can be observed in the literature. Besides, Das et al. [5] proposed an efficient approach for group multi-criteria decision-making (MCDM) based on IMFS.

3. Preliminaries

Before introducing the concept of type-2 multi-fuzzy sets, we first present some essential concepts of a crisp multi-set or m-set, multi-fuzzy sets, type-2 fuzzy sets with examples, and set theoretical operations of multi-fuzzy sets.

3.1. Classical Multi-Sets

A classical multi-set or m-set (in short) is a set, where any element of the set may occur more than once. The definition can be found in the work of (Girish and John [45]). An m-set M drawn from set X is represented by a function Count-M or C_M defined as $C_M : X \to N$ where N represents the set of non-negative integers.

Here, $C_M(x)$ is the number of occurrences of the element $x \in X$ in the m-set M. We present the m-set M drawn from the set $X = \{x_1, x_2, \cdots, x_n\}$ as $M = \{m_1/x_1, m_2/x_2, \cdots, m_n/x_n\}$, where m_i is the number of occurrences of the element $x_i, i = 1, 2, \cdots, n$ in the m-set M. However, those elements which are not included in the m-set M have zero count.

Example 1: *Let us consider the universal set of some object as $X = \{a, b, c, d, e\}$ and let object a appear three times, c appear five times, d appear one time and e appear two times in set M. Then, this appearance of the objects can be represented in set form $M = \{a, a, a, c, c, c, c, c, d, e, e\}$. It can also be represented in the form as*

$$M = \{3/a, 0/b, 5/c, 1/d, 2/e\}$$

which is a m-set.

3.2. Multi-Fuzzy Set

A multi-fuzzy set (Yager [2]) is a fuzzy set, where for each element of the universal set there may be more than one membership value. It can be defined mathematically as follows. Let X be a nonempty set. A multi-fuzzy set (MFS) A on X is characterized by a function, count membership of A denoted by C_{M_A} such that $C_{M_A} : X \to Q$, where Q is the set of all crisp multi-sets drawn from the unit interval $[0, 1]$. Then, for any $x \in X$, the value $C_{M_A}(x)$ is a crisp multi-set drawn from $[0, 1]$. For each $x \in X$, the membership sequence is defined as the decreasingly ordered sequence of elements in $C_{M_A}(x)$. It is denoted by $\left(\mu_A^1(x), \mu_A^2(x), \cdots, \mu_A^p(x)\right)$, where $\mu_A^1(x) \geq \mu_A^2(x) \geq \cdots \geq \mu_A^p(x)$.

Example 2: *Let us consider fuzzy set A as follows*

$$R = \{0.7/x, 0.5/x, 0.2/x, 1.0/y, 1.0/y, 0.4/y, 0.6/y, 0.3/z\}$$

of the universal set $X = \{x, y, z\}$. From this fuzzy set, we see that the element x occurs three times with membership values 0.7, 0.5 and 0.2 respectively; the element y occurs four times with membership values

1.0, 1.0, 0.4, *and* 0.6 *respectively and the element z occurs once with a membership value* 0.3. *Thus, the set can be rewritten in the form as*

$$R = \{(0.7,\ 0.5,\ 0.2)/x, (1.0, 1.0,\ 0.4, 0.6)/y, 0.3/z\}$$

which essentially is a multi-fuzzy set.

The graphical representation of the multi-fuzzy set R is shown in Figure 1, where we consider $x = 3, y = 7$ and $z = 9$.

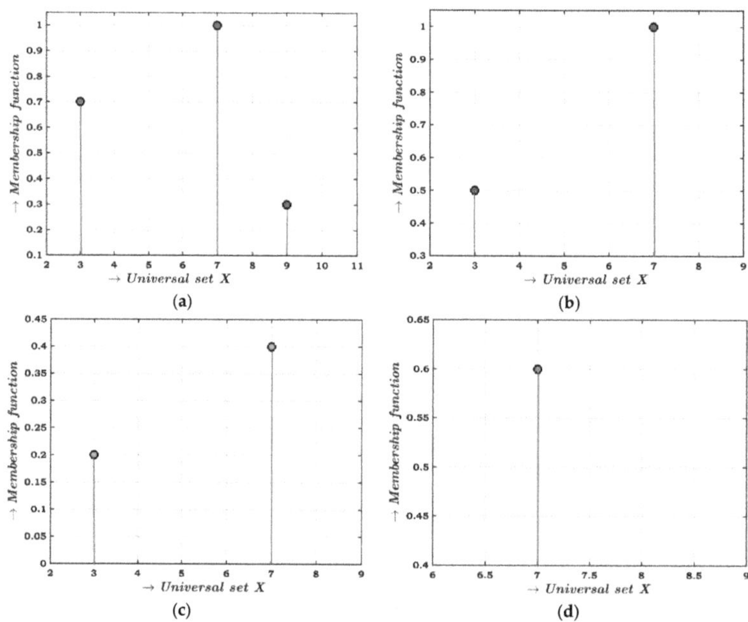

Figure 1. The membership function of the multi-fuzzy set R is represented by four different combinations of membership functions of 3, 7 and 9, where (**a**) the graphical representation of $0.7/3 + 1.0/7 + 0.3/9$, (**b**) the graphical representation of $0.5/3 + 1.0/7$, (**c**) the graphical representation of $0.2/3 + 0.4/7$ and (**d**) the graphical representation of $0.6/7$ are presented respectively.

We now discuss some basic operations such as inclusion, equality, union, and intersection of MFSs. Let A and B be two MFSs defined on X.

3.2.1. Inclusion

$$A \subseteq B \iff \mu_A^j(x) \le \mu_B^j(x),\ j = 1, 2, \cdots, L(x),\ \forall\, x \in X,$$

where $L(x) = L(x; A,\ B) = \max\{L(x; A), L(x; B)\}$ and $L(x; A) = \max\{j : \mu_A^j(x) \ne 0\}$.

Example 3: *Let us consider two multi-fuzzy sets A and B over a nonempty universe X as*

$$A = \{(0.8,\ 0.6,\ 0.5)/x, (0.6,\ 0.4,\ 0.2)/y, (0.7,\ 0.1)/z\}$$

$$B = \{(0.8,\ 0.7,\ 0.6)/x, (0.9,\ 0.8,\ 0.4)/y, (1.0, 0.8,\ 0.5)/z\}.$$

Then, from the definition it is clear that $A \subseteq B$.

3.2.2. Equality

$$A = B \iff \mu_A^j(x) = \mu_B^j(x), \; j = 1, 2, \cdots, L(x), \; \forall \, x \in X$$

3.2.3. Union

$$\mu_{A \cup B}^j(x) = \mu_A^j(x) \vee \mu_B^j(x), \; j = 1, 2, \cdots, L(x), \; \forall \, x \in X$$

3.2.4. Intersection

$$\mu_{A \cap B}^j(x) = \mu_A^j(x) \wedge \mu_B^j(x), \; j = 1, 2, \cdots, L(x), \; \forall \, x \in X.$$

Example 4: *Let us consider two multi-fuzzy sets over a nonempty universe X as*

$$A = \{(0.8, \, 0.7, \, 0.4)/x + (1.0, 1.0, 0.8, \, 0.5)/y + (0.4, \, 0.3)/z\}$$

and

$$B = \{(0.6, \, 0.5, \, 0.2)/x + (1.0, 0.9, \, 0.7, \, 0.6)/y + (0.7, 0.6, \, 0.5)/z\}.$$

Then

$$A \cup B = \left\{ \begin{array}{c} (0.8 \vee 0.6, \, 0.7 \vee 0.5, \, 0.4 \vee 0.2)/x + \\ (1.0 \vee 1.0, \, 1.0 \vee 0.9, 0.8 \vee 0.7, \, 0.5 \vee 0.6)/y + \\ (0.4 \vee 0.7, \, 0.3 \vee 0.6, 0.0 \vee 0.5)/z \end{array} \right\}$$
$$= \{(0.8, \, 0.7, \, 0.4)/x + (1.0, 1.0, \, 0.8, \, 0.6)/y + (0.7, 0.6, \, 0.5)/z\}$$

and

$$A \cap B = \left\{ \begin{array}{c} (0.8 \wedge 0.6, \, 0.7 \wedge 0.5, \, 0.4 \wedge 0.2)/x + \\ (1.0 \wedge 1.0, \, 1.0 \wedge 0.9, 0.8 \wedge 0.7, \, 0.5 \wedge 0.6)/y + \\ (0.4 \wedge 0.7, \, 0.3 \wedge 0.6, 0.0 \wedge 0.5)/z \end{array} \right\}$$
$$= \{(0.6, \, 0.5, \, 0.2)/x + (1.0, \, 0.9, 0.7, \, 0.5)/y + (0.4, 0.3)/z\}.$$

Here, $L(x) = L(x; A, B) \max\{L(x; A), L(x; B)\} = \max\{3, 3\} = 3$. Correspondingly, $L(y) = \max\{4, 4\} = 4$, $L(z) = \max\{2, 3\} = 3$.

3.3. Type-2 Fuzzy Set (T2FS)

A type-2 fuzzy set is a fuzzy set whose membership degree includes uncertainty i.e., membership degree is a type-1 fuzzy set. A T2FS introduces a third dimension to the membership function via the second membership grades. A T2FS $\tilde{A}$ is mathematically expressed as follows according to (Mendel and John [46])

$$\tilde{A} = \{((x, \, u), \mu_{\tilde{A}}(x, \, u)) : \, \forall x \in X, \, \forall \, J_x \subseteq [0, 1]\},$$

where $0 \leq \mu_{\tilde{A}}(x, \, u) \leq 1$ is the secondary membership function and J_x is the primary membership of $x \in X$, which is the domain of $\mu_{\tilde{A}}(x, \, u)$. $\tilde{A}$ can be expressed as

$$\tilde{A} = \int_{x \in X} \left(\int_{u \in J_x} \mu_{\tilde{A}}(x, \, u)/u \right)/x, \; J_x \subseteq [0, 1],$$

where $\iint$ denotes union over all admissible x and u. For a discrete universe of discourse, $\int$ is replaced by Σ.

For each value of x, the secondary membership function $\mu_{\tilde{A}}(x,\,u)$ is defined as

$$\mu_{\tilde{A}}(x,\,u) = \int_{u \in J_x} \mu_{\tilde{A}}(x,\,u)/u$$

such that for a particular $u = u' \in J_x$, the secondary membership grade of $(x,\,u)$ is called $\mu_{\tilde{A}}(x,\,u)$.

Example 5: *Let the set of infant age be represented by a type-2 fuzzy set* $\tilde{A}$. *Let youthness be the primary membership function of* $\tilde{A}$ *and the degree of youthness be the secondary membership function. Let* $E = \{8,\ 10,\ 14\}$ *be an age set with the primary membership of the members of* E *respectively being* $J_8 = \{0.8,\ 0.9,\ 1.0\}$, $J_{10} = \{0.6,\ 0.7,\ 0.8\}$, $J_{14} = \{0.4,\ 0.5,\ 0.6\}$. *The secondary membership function of 8 is* $\tilde{\mu}_{\tilde{A}}(8, u) = (0.9/0.8) + (0.7/0.9) + (0.6/1.0)$, *i.e.,* $\mu_{\tilde{A}}(8,\ 0.8) = 0.9$ *is the secondary membership grade of 8 with primary membership 0.8.*

In the same way, $\tilde{\mu}_{\tilde{A}}(10, u) = (0.8/0.6) + (0.7/0.7) + (0.6/0.8)$ and $\tilde{\mu}_{\tilde{A}}(14, u) = (0.9/0.4) + (0.8/0.5) + (0.5/0.6)$.

So the discrete type-2 fuzzy set $\tilde{A}$ can be represented by

$$\tilde{A} = (0.9/0.8) + (0.7/0.9) + (0.6/1.0)/8 + (0.8/0.6) + (0.7/0.7) + (0.6/0.8)/10$$
$$+ (0.9/0.4) + (0.8/0.5) + (0.5/0.6)/14.$$

4. Type-2 Multi-Fuzzy Sets (T2MFS)

Let X be the universe of discourse. Let A be a type-2 fuzzy set defined on X and $u \in J_x \subseteq [0,\ 1]$ be a primary membership value of an element $x \in X$. Then, A is said to be a type-2 multi-fuzzy set if it is characterised by a count function denoted by C_A and is defined as $C_A : J_x \to Q$, where Q is the set of all crisp multi-sets taken from the unit interval $[0,\ 1]$, which are the secondary membership values of $x \in X$. For each $x \in X$, the secondary membership sequence is defined in decreasing order as $\mu_A^1(x, u) \geq \mu_A^2(x, u) \geq \cdots \geq \mu_A^p(x, u)$ and is denoted by $\left(\mu_A^1(x, u), \mu_A^2(x, u), \cdots, \mu_A^p(x, u) \right)$. Then, set A can be represented as

$$A = \sum_{x \in X} \left(\sum_{u \in J_x \subseteq [0,\ 1]} \left(\mu_A^1(x, u), \mu_A^2(x, u), \ldots, \mu_A^p(x, u) \right)/u \right)/x.$$

if the universe is discrete, whereas if X is a continuous universe, then A can be written as

$$A = \int_{x \in X} \left(\int_{u \in J_x \subseteq [0,\ 1]} \left(\left(\mu_A^1(x, u), \mu_A^2(x, u), \cdots, \mu_A^p(x, u) \right) \right)/u \right)/x.$$

Let us illustrate this idea with an example.

Example 6: *Let us consider a type-2 fuzzy set T defined in the universal set* $X = \{x,\ y,\ z\}$

$$T = \frac{(0.8/0.6, 0.5/0.6, 0.2/0.6, 0.3/0.9, 0.7/0.9)/x +}{(0.7/0.3, 0.6/0.3, 0.8/0.5, 0.5/0.7, 0.3/0.7 + 0.1/0.7)/y}$$

From the structure of T, we see that three x's with primary membership values of 0.6 have secondary membership values 0.8, 0.5 and 0.2 respectively; two x's with primary membership values of 0.9 have the corresponding secondary membership values of 0.3 and 0.7; two y's with a primary membership value of 0.3 have secondary membership values of 0.7 and 0.6 respectively; one y with a primary membership value of 0.5 has a secondary membership value of 0.8; and three y's with primary membership values of 0.7 with secondary membership values 0.5, 0.3 and 0.1 respectively. Therefore, T can be represented as

$$T = \frac{((0.8, 0.5, 0.2)/0.6 + (0.3, 0.7)/0.9)/x +}{((0.7, 0.6)/0.3 + 0.8/0.5 + (0.5, 0.3, 0.1)/0.7)/y.}$$

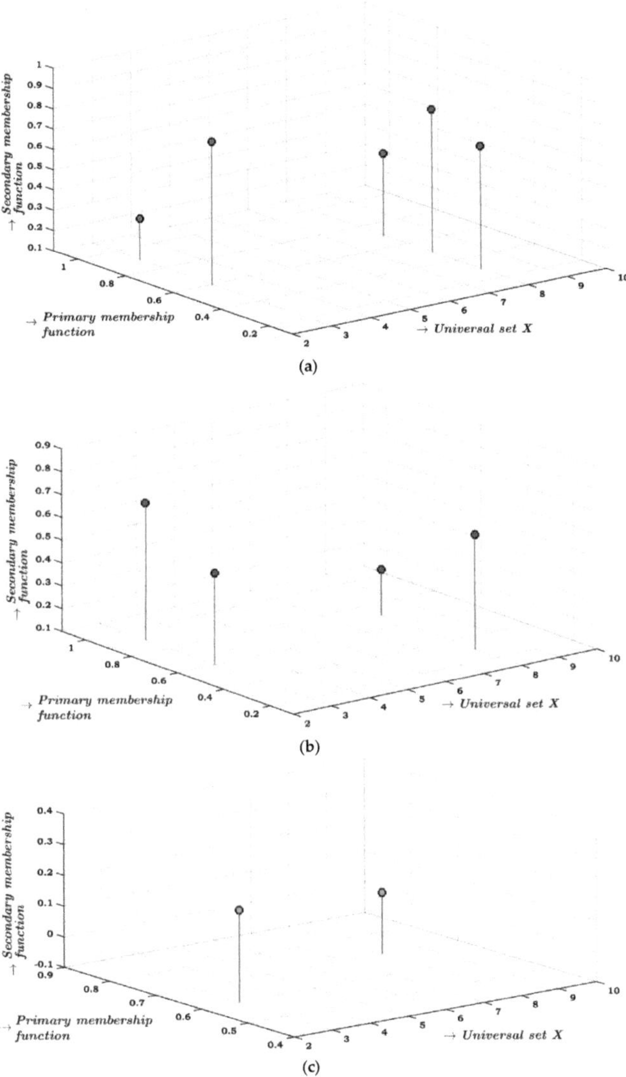

Figure 2. The membership function of T2MFS T is represented by three different combinations of primary membership and secondary membership functions of 3 and 8, where (**a**) the graphical representation of $(0.8/0.6 + 0.3/0.9)/3 + (0.7/0.3 + 0.8/0.5 + 0.5/0.7)/8$, (**b**) the graphical representation of $(0.5/0.6 + 0.7/0.9)/3 + (0.6/0.3 + 0.3/0.7)/8$ and (**c**) the graphical representation of $(0.2/0.6)/3 + (0.1/0.7)/8$ are presented respectively.

Thus, T is a type-2 multi-fuzzy set. The graphical representation of T2MFS T is shown in Figure 2, where we consider $x = 3$ and $y = 8$.

Again, let us denote the collection of all T2MFSs over the universe X by T2MF(X). When we define operations between two T2MFSs, say A and B, the lengths of their secondary membership sequences $\left(\mu_A^1(x,u), \mu_A^2(x,u), \cdots, \mu_A^p(x,u)\right)$ and $\left(\mu_B^1(x,u), \mu_B^2(x,u), \cdots, \mu_B^p(x,u)\right)$ for a particular primary membership value of A (say u) and for a particular primary membership value of B (say u') should be set to be equal. We hence define the length $L(x,u;A)$ as

$$L(x,u;A) = \max\{j : \mu_A^j(x,u) \neq 0\}$$

and

$$L(x,u,\ u';A,B) = \max\{L(x,u;A),\ L(x,u';B)\}.$$

For the sake of simplicity, we write $L(x,u,\ u';A,B)$ as $L(x,u,\ u')$. Let us describe this idea by an example.

Example 7: *Let us consider two T2MFSs, say A and B, as*

$$A = ((0.8,0.6,0.2)/0.7 + (0.9,0.5)/0.4)/x + ((0.7,0.5,0.3)/0.8 + (0.5,0.5,0.4,0.1)/0.5)/y$$

and

$$B = ((0.9,0.6)/0.8 + (1.0,1.0,0.6,0.6)/0.7)/x + ((0.6,0.2)/0.9 + 0.1/0.4)/y.$$

Then

$$L(x,0.7;A) = 3, L(x,0.4;A) = 2,\ L(y,0.8;A) = 3, L(y,0.5;A) = 4,$$

$$L(x,0.8;B) = 2, L(x,0.7;B) = 4,\ L(y,0.9;B) = 2, L(y,0.4;B) = 1.$$

Moreover,

$$L(x,0.7;0.8) = 3, L(x,0.7;0.7) = 4,\ L(x,0.4;0.8) = 2, L(x,0.4;0.7) = 4,$$

$$L(y,0.8;0.9) = 3, L(y,0.8;0.4) = 3,\ L(y,0.5;0.9) = 4, L(y,0.5;0.4) = 4.$$

5. Some Operations on T2MFS

In this section, we discuss four fundamental arithmetical operations: (i) complement, (ii) inclusion, (iii) union, and (iv) intersection of T2MFS.

5.1. Complement

Let A be a T2MFS over some universe X. Then, the complement of A denoted by A^c is defined as

$$A^c = \sum_{x \in X} \left(\sum_{v \in J_x \subseteq [0,\ 1]} \left(\mu_A^1(x,v), \mu_A^2(x,v), \cdots, \mu_A^p(x,v)\right)/v \right)/x$$

where $v = 1 - u$ and u is the primary membership function of A. Let us give an example to illustrate this idea.

Example 8: *Let us consider the T2MFS A used in Example 6.*

$$A^c = \frac{((0.8,0.5,0.2)/0.4 + (0.3,0.3)/0.1)/x+}{((0.7,0.6)/0.7 + 0.8/0.5 + (0.5,0.3,0.1)/0.3)/y.}$$

5.2. Inclusion

Let A and B be two T2MFSs over some universe X. Then we say that $A \subseteq B$ if and only if

$$u \leq u', \ \mu_A^j(x,u) \leq \mu_B^j(x,u'), j = 1,2,\cdots, L(x,u,u') \ \forall x \in X,$$

where u and u' are primary membership functions of A and B, respectively, and μ_A^j and μ_B^j are the secondary membership functions of A and B respectively. These two sets A and B are said to be equal if and only if

$$u = u', \mu_A^j(x,u) = \mu_B^j(x,u'), j = 1,2,\cdots, L(x,u,u') \ \forall x \in X.$$

Let us illustrate this idea by an example.

Example 9: *Let us consider two T2MFSs, say A and B, where*

$$A = (((0.8,0.6,0.5,0.1)/0.8 + (0.9,0.8,0.6,0.3)/0.7)/x + ((0.7,0.5,0.3)/0.9 + (0.9,0.4)/0.5)/y)$$

and

$$B = (((0.7,0.6)/0.6 + (0.8,0.5,0.4,0.1)/0.7)/x + ((0.6,0.2)/0.5 + 0.1/0.4)/y)$$

Then applying the above-mentioned definition, we can see that $B \subseteq A$.

5.3. Union

Let A and B be two T2MFSs over the universe X. Then the set $C = A \cup B$ is defined as

$$C = \sum_{x \in X} \left(\sum_{v \in J_x \subseteq [0,\ 1]} \left(\mu_A^1(x,v), \mu_A^2(x,v), \cdots, \mu_A^{L(x,u,u')}(x,v) \right) / v \right) / x,$$

where $\mu_C^j(x,v) = \mu_{A \cup B}^j(x,v) = \mu_A^j(x,u) \wedge \mu_B^j(x,u')$ and $j = 1,2,\cdots, L(x,u,u')$, $\forall x \in X$, such that $v = u \vee u'$, where u and u' are the primary membership functions of A and B respectively. $\mu_A^j(x,u)$ and $\mu_B^j(x,u')$ are the secondary membership functions of A and B respectively. Let us illustrate this idea by an example.

Example 10: *Let us consider two T2MFSs, say A and B, which are considered in Example 5. Then*

$$A = (((0.8,0.6,0.2)/0.7 + (0.9,0.5)/0.4)/x + ((0.7,0.5,0.3)/0.8 + (0.5,0.5,0.4,0.1)/0.5)/y)$$

and

$$B = (((0.9,0.6)/0.8 + (1.0,0.8,0.6,0.6)/0.7)/x + ((0.6,0.2)/0.9 + 0.1/0.4)/y).$$

Then

$A \cup B$
$= (((0.8,0.6,0.0)/0.8 + (0.8,0.6,0.2,0.0)/0.7 + (0.9,0.5)/0.8 + (0.9,0.5,0.0,0.0)/0.7)/x$
$+ ((0.6,0.2,0.0)/0.9 + (0.5,0.2,0.0,0.0)/0.9 + (0.1,0.0,0.0)/0.8 + (0.1,0.0,0.0,0.0)/0.5)/y)$
$= (((\max(0.8,0.9),\max(0.6,0.5))/0.8 + (\max(0.8,0.9),\max(0.6,0.5),\max(0.2,0.0))/0.7)/x$
$+ ((\max(0.6,0.5),\max(0.2,0.2))/0.9 + 0.1/0.8 + 0.1/0.5)/y)$
$= (((0.9,0.6)/0.8 + (0.9,0.6,0.2)/0.7)/x + ((0.6,0.2)/0.9 + 0.1/0.8 + 0.1/0.5)/y).$

5.4. Intersection

Let A and B be two T2MFSs over the universe X. Then set C is the intersection of A and B where it is denoted as $C = A \cap B$ and is defined as

$$C = \sum_{x \in X} \left(\sum_{v \in J_x \subseteq [0,\, 1]} \left(\mu_A^1(x,v), \mu_A^2(x,v), \cdots, \mu_A^{L(x,u,u')}(x,v) \right) / v \right) / x$$

Here, $\mu_C^j(x,v) = \mu_{A \cup B}^j(x,v) = \mu_A^j(x,u) \wedge \mu_B^j(x,u')$, $j = 1,2,\cdots,L(x,u,u')$, $\forall x \in X$, where $v = u \wedge u'$, and u and u' are the primary membership functions of A and B respectively. $\mu_A^j(x,u)$ and $\mu_B^j(x,u')$ are the secondary membership functions of A and B respectively. Subsequently, we illustrate this idea by an example as shown below.

Example 11: *Let us consider two T2MFSs, say A and B, where*

$$A = (((0.8,0.6,0.2)/0.7 + (0.9,0.5)/0.4)/x + ((0.7,0.5,0.3)/0.8 + (0.5,0.5,0.4,0.1)/0.5)/y)$$

$$B = (((0.9,0.6)/0.8 + (1.0,0.8,0.6,0.6)/0.7)/x + ((0.6,0.2)/0.9 + 0.1/0.4)/y)$$

Then

$A \cap B$
$= (((0.8,0.6,0.0)/0.7 + (0.8,0.6,0.2,0.0)/0.7 + (0.9,0.5)/0.4 + (0.9,0.5,0.0,0.0)/0.4)/x$
$+((0.6,0.2,0.0)/0.8 + (0.5,0.2,0.0,0.0)/0.5 + (0.1,0.0,0.0)/0.4 + (0.1,0.0,0.0,0.0)/0.4)/y)$
$= (((\max(0.8,0.8),\max(0.6,0.6),\max(0.0,0.2))/0.7 + (\max(0.9,0.9),\max(0.5,0.5))/0.4)/x$
$+((0.6,0.2)/0.8 + (0.5,0.2)/0.5 + (\max(0.1,0.1))/0.4 + 0.1/0.5)/y)$
$= (((0.8,0.6,0.2)/0.7 + (0.9,0.5)/0.4)/x + ((0.6,0.2)/0.8 + (0.5,0.2)/0.5 + 0.1/0.4)/y).$

6. Properties of T2MFS

In this section, we discuss four fundamental properties of T2MFS.

Let A, B and C be three T2MFSs over a universe X. Then the following relations hold:

(i) $A \cup A = A$, $A \cap A = A$. *(idempotency)*
(ii) $(A \cup B)^c = A^c \cap B^c$, $(A \cap B)^c = A^c \cup B^c$. *(De Morgan's law)*
(iii) $A \cup B = B \cup A$, $A \cap B = B \cap A$. *(Commutativity)*
(iv) $A \cup (B \cup C) = (A \cup B) \cup C$, $A \cap (B \cap C) = (A \cap B) \cap C$. *(Associativity)*
(v) $(A \cup B) \cap C = (A \cap C) \cup (B \cap C)$, $(A \cap B) \cup C = (A \cup C) \cap (B \cup C)$. *(Distributive law)*

The proofs of (i) to (iii) are obvious. We illustrate these results later by example. We now prove the results (iv) and (v).

Proof of (iv).

$$A \cup (B \cup C) = \sum_{x \in X} \left(\sum_{u \in J_x \subseteq [0,\,1]} \left(\mu_A^1(x,u), \mu_A^2(x,u), \ldots, \mu_A^{L(x,u;\,A)}(x,u) \right) / u \right) / x \cup$$

$$\left(\sum_{x \in X} \left(\sum_{v,w \in J_x \subseteq [0,\,1]} \left(\begin{array}{c} (\mu_B^1(x,v) \wedge \mu_C^1(x,w)), (\mu_B^2(x,v) \wedge \mu_C^2(x,w)), \ldots, \\ \left(\mu_B^{L(x,v,w)}(x,v) \wedge \mu_C^{L(x,v,w)}(x,w) \right) \end{array} \right) / (v \vee w) \right) / x \right)$$

$$= \sum_{x \in X} \left(\sum_{u,v,w \in J_x \subseteq [0,\,1]} \left(\begin{array}{c} (\mu_A^1(x,u) \wedge \mu_B^1(x,v) \wedge \mu_C^1(x,w)), \\ (\mu_A^1(x,u) \wedge \mu_B^2(x,v) \wedge \mu_C^2(x,w)), \ldots, \\ \left(\mu_A^{L(x,u,v,w)}(x,u) \wedge \mu_B^{L(x,u,v,w)}(x,v) \wedge \mu_C^{L(x,u,v,w)}(x,w) \right) \end{array} \right) / (u \vee v \vee w) \right) / x$$

$$= \left(\sum_{x \in X} \left(\sum_{u,v \in J_x \subseteq [0,\,1]} \left(\begin{array}{c} (\mu_A^1(x,u) \wedge \mu_B^1(x,v)), (\mu_A^2(x,u) \wedge \mu_B^2(x,v)), \ldots, \\ \left(\mu_A^{L(x,u,v)}(x,u) \wedge \mu_B^{L(x,u,v)}(x,v) \right) \end{array} \right) / (u \vee v) \right) / x \right) \cup$$

$$\sum_{x \in X} \left(\sum_{w \in J_x \subseteq [0,\,1]} \left(\mu_C^1(x,w), \mu_C^2(x,w), \ldots, \mu_C^{L(x,w;\,A)}(x,w) \right) / w \right) / x$$

$$= (A \cup B) \cup C.$$

Similarly, we can prove that $A \cap (B \cap C) = (A \cap B) \cap C$. $\square$

Proof of (v).

$$(A \cup B) \cap C = \left(\sum_{x \in X} \left(\sum_{u,v \in J_x \subseteq [0,\,1]} \left(\begin{array}{c} (\mu_A^1(x,u) \wedge \mu_B^1(x,v)), (\mu_A^2(x,u) \wedge \mu_B^2(x,v)), \cdots, \\ \left(\mu_A^{L(x,u,v)}(x,u) \wedge \mu_B^{L(x,u,v)}(x,v) \right) \end{array} \right) / (u \vee v) \right) / x \right) \cap$$

$$\sum_{x \in X} \left(\sum_{w \in J_x \subseteq [0,\,1]} \left(\mu_C^1(x,w), \mu_C^2(x,w), \cdots, \mu_C^{L(x,w;\,A)}(x,w) \right) / w \right) / x =$$

$$\sum_{x \in X} \left(\sum_{u,v,w \in J_x \subseteq [0,\,1]} \left(\begin{array}{c} (\mu_A^1(x,u) \wedge \mu_B^1(x,v) \wedge \mu_C^1(x,w)), \\ (\mu_A^1(x,u) \wedge \mu_B^2(x,v) \wedge \mu_C^2(x,w)), \cdots, \\ \left(\mu_A^{L(x,u,v,w)}(x,u) \wedge \mu_B^{L(x,u,v,w)}(x,v) \wedge \mu_C^{L(x,u,v,w)}(x,w) \right) \end{array} \right) / (u \vee v) \wedge w \right) / x =$$

$$\left(\sum_{x \in X} \left(\sum_{u,w \in J_x \subseteq [0,\,1]} \left(\begin{array}{c} (\mu_A^1(x,u) \wedge \mu_C^1(x,w)), (\mu_A^2(x,u) \wedge \mu_C^2(x,w)), \cdots, \\ \left(\mu_A^{L(x,u,w)}(x,u) \wedge \mu_C^{L(x,u,w)}(x,w) \right) \end{array} \right) / (u \wedge w) \right) / x \right) \cup$$

$$\left(\sum_{x \in X} \left(\sum_{v,w \in J_x \subseteq [0,\,1]} \left(\begin{array}{c} (\mu_B^1(x,v) \wedge \mu_C^1(x,w)), (\mu_B^2(x,v) \wedge \mu_C^2(x,w)), \cdots, \\ \left(\mu_B^{L(x,v,w)}(x,v) \wedge \mu_C^{L(x,v,w)}(x,w) \right) \end{array} \right) / (v \wedge w) \right) / x \right) = (A \cap C) \cup (B \cap C)$$

Similarly, we can prove that $(A \cap B) \cup C = (A \cup C) \cap (B \cup C)$. $\square$

Let us illustrate these results by an example.

Example 12: *Let us consider three T2MFSs over a non-empty universe X.*

$$A = (0.8,\ 0.7,\ 0.3)/0.7/x + (0.9,\ 0.5,\ 0.1)/0.4/y + (0.6,\ 0.5)/0.3/z$$

$$B = (1.0,\ 0.9)/0.9/x + (0.7,\ 0.6,\ 0.3)/0.6/y + (0.8,\ 0.7,\ 0.1)/0.2/z$$

$$C = (0.9,\ 0.5,\ 0.2,\ 0.0)/0.6/x + (0.8,\ 0.4)/0.9/y + (0.8,\ 0.2,\ 0.0)/0.8/z.$$

For simplicity, we have taken J_x as a singleton set for each $x \in X$.
Then

$$A \cup A = A \cap A = (0.8,\ 0.7,\ 0.3)/0.7/x + (0.9,\ 0.5,\ 0.1)/0.4/y + (0.6,\ 0.5)/0.3/z = A.$$

Therefore, the idempotent property holds.

$$(A \cup B)^c = (0.8,\ 0.7)/0.1/x + (0.7,\ 0.5,\ 0.1)/0.4/y + (0.6,\ 0.5)/0.7/z = A^c \cap B^c$$

and

$$(A \cap B)^c = (0.8,\ 0.7)/0.3/x + (0.7,\ 0.5,\ 0.1)/0.6/y + (0.6,\ 0.5)/0.8/z = A^c \cup B^c.$$

Hence, De Morgan's laws hold.

$$A \cup B = (0.8,\ 0.7)/0.9/x + (0.7,\ 0.5,\ 0.1)/0.6/y + (0.6,\ 0.5)/0.3/z = B \cup A$$

and

$$A \cap B = (0.8,\ 0.7)/0.7/x + (0.7,\ 0.5,\ 0.1)/0.4/y + (0.6,\ 0.5)/0.2/z = B \cap A.$$

Consequently, the commutative property holds.

$$A \cup (B \cup C) = (0.8,\ 0.5)/0.9/x + (0.7,\ 0.4)/0.9/y + (0.6,\ 0.2)/0.8/z = (A \cup B) \cup C$$

and

$$A \cap (B \cap C) = (0.8,\ 0.5)/0.6/x + (0.7,\ 0.4)/0.4/y + (0.6,\ 0.2)/0.2/z = (A \cap B) \cap C.$$

Hence, the associative property holds.

$$(A \cup B) \cap C = (0.8,\ 0.5)/0.6/x + (0.7,\ 0.4)/0.6/y + (0.6,\ 0.2)/0.3/z = (A \cap C) \cup (B \cap C)$$

and

$$(A \cap B) \cup C = (0.8,\ 0.5)/0.7/x + (0.7,\ 0.4)/0.9/y + (0.6,\ 0.2)/0.8/z = (A \cup C) \cap (B \cup C).$$

As a result, the distributive property holds.

7. Distance Measures of T2MFS

Let A and B be two T2MFSs. Then we can define the following distances as follows:

(1) Hamming Distance

$$d_H(A, B) = \sum_{x \in X} \sum_{u,u' \in J_x} \sum_{j=1}^{L(x,u,u';\ A,B)} \left| \mu_A^j(x,u) - \mu_B^j(x,u') \right|.$$

(2) Euclidean Distance

$$d_E(A, B) = \left[\sum_{x \in X} \sum_{u,u' \in J_x} \sum_{j=1}^{L(x,u,u';\ A,B)} \left| \mu_A^j(x,u) - \mu_B^j(x,u') \right|^2 \right]^{\frac{1}{2}},$$

where u and u' are the primary membership functions of A and B respectively, and μ_A and μ_B are the corresponding secondary membership functions of A and B. Let us now explain this idea with an example.

Example 13: *Let X be a non-empty universe. Let A and B be two T2MFSs over X which are given as*

$$A = (0.9,\ 0.6,\ 0.4)/0.8/x + (0.8,\ 0.5,\ 0.1)/0.6/y + (1.0,\ 0.7)/0.9/z$$

$$B = (0.7,\ 0.7,\ 0.2)/0.6/x + (0.9,\ 0.2,\ 0.1)/0.3/y + (0.8,\ 0.8)/0.4/z$$

$$d_H(A,B) = |0.9 - 0.7| + |0.6 - 0.7| + |0.4 - 0.2| + |0.8 - 0.9| + |0.5 - 0.2|$$
$$+ |0.1 - 0.1| + |1.0 - 0.8| + |0.7 - 0.8| = 1.2$$

and

$$d_E(A, B) = \Big[|0.9 - 0.7|^2 + |0.6 - 0.7|^2 + |0.4 - 0.2|^2 + |0.8 - 0.9|^2 + |0.5 - 0.2|^2$$
$$+ |0.1 - 0.1|^2 + |1.0 - 0.8|^2 + |0.7 - 0.8|^2 \Big]^{\frac{1}{2}} = 0.49.$$

8. Numerical Illustration

Although fuzzy logic is relatively a newer subject than classical mathematical logic, we use the former extensively in our everyday life. We primarily use fuzzy logic in decision-making problems. However, there may be some cases in decision-making where the primary membership function occurs with a sequence of same or different degrees. In that case, we need to use T2MFS to solve the problem. Let us present an example of medical diagnosis after acquiring the necessary information from Reference [38]. Let $P = \{P_1, P_2, P_3, P_4\}$ be a set of patients, $D = \{Viral\ Fever, Tuberculosis, Typhoid, Throat\ Disease\}$ be a set of diseases and $S = \{Temperature, Cough, Throat\ Pain, Headache, Body\ Pain\}$ be a set of symptoms. Let us consider the intensity of the disease symptoms to be the secondary membership functions.

Now, if we look at one set of data, there is a chance that some errors may occur since we know that at different times during a day the disease symptoms have different intensities. To minimize these errors, we study three different samples at three different times of the day. The details of one such example are provided below.

In Table 1, each symptom is described by their primary and secondary membership function.

Table 1. Symptoms vs. Diseases.

Disease	Viral Fever	Tuberculosis	Typhoid	Throat Disease
Temperature	0.8/0.7	0.3/0.1	0.4/0.6	0.3/0.2
Cough	0.4/0.3	0.8/0.9	0.2/0.4	0.4/0.4
Throat Pain	0.3/0.4	0.9/0.8	0.5/0.3	0.8/0.9
Headache	0.7/0.6	0.6/0.7	0.6/0.1	0.2/0.3
Body Pain	0.9/0.5	0.7/0.8	0.5/0.5	0.1/0.2

For proper diagnosis of each patient, we take samples at three different times in a day say at 8 AM, 2 PM and 9 PM. We use the Euclidean distance measure (cf. Section 6) to calculate the distance between the patients and the diseases. Here, the Euclidean distance is determined between each patient P_i from every symptoms S_j for each diagnosis d_k, $i, k = 1, 2, 3, 4, j = 1, 2, 3, 4, 5$.

The data reported in Table 2 is of a T2MFS, where the intensity of the symptoms form a secondary membership sequence. Moreover, in Tables 3 and 4 the Hamming distance and the Euclidean distance are determined respectively for each patient from the set of diseases.

Table 2. Patients vs. Symptoms.

Patient	Temperature	Cough	Throat Pain	Headache	Body Pain
P_1	(0.8,0.4,0.3)/0.4	(0.6,0.5,0.3)/0.5	(0.7,0.4,0.2)/0.7	(0.6,0.4,0.2)/0.3	(0.4,0.2,0.2)/0.5
P_2	(0.3,0.2,0.0)/0.2	(0.9,0.8,0.7)/0.8	(0.7,0.6,0.5)/0.6	(0.2,0.1,0.1)/0.4	(0.5,0.4,0.2)/0.7
P_3	(0.1,0.0,0.0)/0.1	(0.7,0.5,0.4)/0.6	(0.9,0.8,0.8)/0.9	(0.3,0.2,0.2)/0.3	(0.3,0.3,0.1)/0.4
P_4	(0.9,0.9,0.8)/0.7	(0.5,0.4,0.1)/0.4	(0.4,0.2,0.2)/0.3	(0.8,0.6,0.5)/0.6	(1.0,0.9,0.9)/0.8

Table 3. Hamming distance between Patients and Diseases.

Patient	Viral Fever	Tuberculosis	Typhoid	Throat Disease
P_1	2.17	2.21	1.78	1.79
P_2	2.70	1.97	2.14	1.79
P_3	2.77	2.07	2.24	1.34
P_4	1.18	2.44	2.07	2.74

Table 4. Euclidean distance between Patients and Diseases.

Patient	Viral Fever	Tuberculosis	Typhoid	Throat Disease
P_1	1.49	1.49	0.98	1.08
P_2	1.98	1.20	1.46	1.01
P_3	2.18	1.26	1.33	0.65
P_4	0.47	1.78	1.27	2.16

According to the principle of minimum distance point, a lower distance point indicates a proper diagnosis of a particular disease. While comparing the data reported in Tables 3 and 4, a similar interpretation is observed. Here, considering both these Tables 3 and 4, the lowest distance point gives the proper diagnosis, and therefore it can be inferred that patient P_1 suffers from *Typhoid* and patients P_2 and P_3 suffer from *Throat Disease*, whereas, patient P_4 suffers from *Viral Fever*. Hence, out of four patients, we observe that two patients are affected with *Throat Disease*. In addition, one patient is affected with each of the two diseases, *Viral Fever* and *Typhoid*, whereas, none of the patients are diagnosed with *Tuberculosis*.

9. Case Study

For the purpose of simulation, we consider a real case study for 500 patients of the Healthcare Hospital situated in Kolkata, India. Subsequently, we consulted with the specialist doctors of the Healthcare Hospital and received their feedback on the 500 patients at three different times in a day.

A patient suffering from a disease, when visiting a hospital, expresses his/her symptoms, e.g., *Temperature, Cough, Throat Pain, Headache* and *Body Pain* to a doctor. Based on these feedbacks, the doctor analyzes the disease (e.g., *Viral Fever, Tuberculosis, Typhoid* and *Throat Disease*) the patient is suffering from. These symptoms and the disease of the patients are represented in linguistic terms which involves uncertainty. As an example, in a day, the *Temperature* of a patient can vary in the morning, afternoon and night. Based on these recorded values of the *Temperature*, the evaluation of different doctors might change as well. For example, if the *Temperature* of a patient is recorded 100.5 Fahrenheit (F), 100.2 F and 100.7 F respectively in the morning, afternoon and night, then one doctor might analyze that the patient is having *Viral Fever*, while the other doctor might not agree that the patient is suffering from *Viral Fever* and often agree to observe the *Temperature* for some subsequent days. It can be noted here, that a particular symptom of a patient can fluctuate in a day, and based on these fluctuations, the opinion (analysis) of the experts (doctors) also varies for a particular patient. Therefore, in order to incorporate the uncertainties of the symptoms and the diseases rationally, the parameters of the symptom and the disease of a patient are represented as T2MFS. In our study, the necessary information are received from the doctors of the Healthcare Hospital. It should be mentioned here that while receiving specialist doctors feedback, we consider the set of five symptoms and the four diseases as the same, compared to the one considered in the application mentioned in the numerical illustration section. These data are presented in the tables of the supplementary file. Here, Tables 5 and 6 report the data of the Healthcare Hospital which are similar to the corresponding data presented in Tables 1 and 2. Similar to Table 4, in this case study, we determine the Euclidean distances between patients and diseases. However, instead of the corresponding data of four patients as shown in Tables 2 and 4, Table 6 and Table S1 represent the data of 500 patients. From Table 7, it is observed that 66 patients are affected with *Viral Fever*. Two-hundred-and-forty-two patients are suspected to have *Tuberculosis*; 62 patients are suffering from *Typhoid* and 130 patients are diagnosed with *Throat Disease*.

Table 5. Symptoms vs. Diseases in Healthcare Hospital.

Symptom	Viral Fever	Tuberculosis	Typhoid	Throat Disease
Temperature	0.90/0.89	0.38/0.76	0.38/0.66	0.38/0.87
Cough	0.85/0.58	0.79/0.38	0.73/0.29	0.61/0.40
Throat Pain	0.98/0.35	0.39/0.56	0.32/0.73	0.38/0.77
Headache	0.65/0.52	0.49/0.23	0.97/0.38	0.72/0.45
Body Pain	0.60/0.27	0.77/0.79	0.37/0.53	0.87/0.15

Table 6. Patients vs. Symptoms in Healthcare Hospital.

Patient	Temperature	Cough	Throat Pain	Headache	Body Pain
P_1	(0.20,0.89,0.87)/0.73	(0.13,0.40,0.22)/0.93	(0.24,0.44,0.87)/0.34	(0.99,0.85,1.00)/0.20	(0.49,0.97,0.34)/0.67
P_2	(0.28,0.80,0.23)/0.24	(0.65,0.27,0.53)/0.76	(0.64,0.11,0.49)/0.95	(0.62,0.28,0.28)/0.31	(0.61,0.40,0.58)/0.32
P_3	(0.80,0.57,0.20)/0.83	(0.47,0.66,0.27)/0.81	(0.67,0.48,0.71)/0.45	(0.47,0.74,0.11)/0.89	(0.76,0.90,0.70)/0.36
P_4	(0.22,0.60,0.31)/0.47	(0.35,0.54,0.79)/0.41	(0.49,0.88,0.14)/0.31	(0.98,0.56,0.73)/0.25	(0.99,0.42,0.48)/0.25
P_5	(0.27,0.48,0.14)/0.92	(0.36,0.21,0.62)/0.44	(0.33,0.69,0.43)/0.27	(1.00,0.51,0.75)/0.81	(1.00,0.83,0.89)/0.34
P_6	(0.47,0.41,0.83)/0.81	(0.67,0.63,0.34)/0.60	(0.14,0.47,1.00)/0.55	(0.31,0.87,0.48)/0.22	(0.19,0.96,0.88)/0.94
P_7	(0.88,0.98,0.47)/0.31	(0.65,0.84,0.48)/0.91	(0.37,0.31,0.78)/0.55	(0.62,0.45,0.41)/0.75	(0.57,0.60,0.71)/0.24
P_8	(0.69,0.49,0.27)/0.51	(0.76,0.84,0.46)/0.69	(0.56,0.16,0.63)/0.81	(0.41,0.69,0.19)/0.37	(0.69,0.88,0.39)/0.54
P_9	(0.17,0.55,0.92)/0.40	(0.90,0.22,0.27)/0.86	(0.95,0.89,0.49)/0.43	(0.67,0.71,0.21)/0.72	(0.41,0.62,0.90)/0.74
P_{10}	(0.59,0.14,0.50)/0.41	(0.82,0.47,0.65)/0.47	(0.40,0.98,0.69)/0.40	(0.61,0.66,0.31)/0.75	(0.37,0.67,0.43)/0.20
P_{11}	(0.88,0.61,0.27)/0.22	(0.96,0.51,0.84)/0.21	(0.55,0.11,0.91)/0.77	(0.69,0.94,0.97)/0.91	(0.72,0.66,0.29)/0.49
P_{12}	(0.95,0.47,0.19)/0.58	(0.38,0.39,0.25)/0.75	(0.27,0.97,0.89)/0.28	(0.45,0.38,0.44)/0.23	(0.88,0.10,0.57)/0.52
P_{13}	(0.25,0.50,0.91)/0.60	(0.83,0.25,0.33)/0.84	(0.79,0.31,0.55)/0.59	(0.97,0.52,0.73)/0.47	(0.72,0.66,0.61)/0.37
P_{14}	(0.18,0.18,0.91)/0.50	(0.98,1.00,0.65)/0.79	(0.34,0.34,0.93)/0.84	(0.26,0.78,0.68)/0.25	(0.49,0.11,0.66)/0.94
P_{15}	(0.94,0.86,0.23)/0.43	(0.89,0.11,0.34)/0.36	(0.96,0.58,0.21)/0.93	(1.00,0.89,0.16)/0.83	(0.26,0.26,0.98)/0.63
P_{16}	(0.65,0.86,0.11)/0.68	(0.66,0.44,0.31)/0.59	(0.29,0.71,0.22)/0.95	(0.34,0.17,0.77)/0.83	(0.63,0.90,0.55)/0.61
P_{17}	(0.20,0.98,0.16)/0.38	(0.40,0.74,0.85)/0.89	(0.59,0.54,0.92)/0.76	(0.63,0.18,0.91)/0.34	(0.32,0.51,0.42)/0.41
P_{18}	(0.23,0.96,0.20)/0.37	(0.55,0.74,0.18)/0.95	(0.77,0.91,0.29)/0.29	(0.21,0.13,0.29)/0.87	(0.67,0.75,0.18)/0.77
P_{19}	(0.49,0.20,0.19)/0.44	(0.48,0.41,0.94)/0.38	(0.86,0.39,0.54)/0.85	(0.74,0.14,0.57)/0.36	(0.22,0.71,0.76)/0.21
P_{20}	(0.66,0.83,0.48)/0.89	(0.86,0.68,0.80)/0.94	(0.80,0.68,0.41)/0.50	(0.42,0.33,0.53)/0.62	(0.79,0.74,0.50)/0.88
P_{21}	(0.67,0.63,0.78)/0.61	(0.11,0.42,0.45)/0.52	(0.43,0.36,0.42)/0.62	(0.34,0.45,0.83)/0.73	(0.92,0.19,0.11)/0.45
P_{22}	(0.41,0.38,0.73)/0.60	(0.78,0.99,0.61)/0.71	(0.96,0.87,0.34)/0.83	(0.24,0.87,0.61)/0.28	(0.93,0.43,0.92)/0.56
P_{23}	(0.88,0.36,0.36)/0.46	(0.17,1.00,0.49)/0.38	(0.37,0.22,0.12)/0.92	(0.31,0.10,0.28)/0.57	(0.83,0.50,0.95)/0.23
P_{24}	(0.79,0.46,0.13)/0.75	(0.79,0.56,0.50)/0.51	(0.61,0.91,0.58)/0.76	(0.99,0.24,0.85)/0.25	(0.28,0.26,0.10)/0.95
P_{25}	(0.44,0.13,0.93)/0.86	(0.37,0.14,0.73)/0.87	(0.58,0.18,0.98)/0.74	(0.94,0.32,0.79)/0.41	(0.95,0.35,0.40)/0.41
P_{26}	(0.18,0.42,0.49)/0.54	(0.49,0.74,0.71)/0.21	(0.58,0.91,0.43)/0.51	(0.57,0.40,0.51)/0.88	(0.82,0.73,0.52)/0.44
P_{27}	(0.85,0.54,0.87)/0.45	(0.54,0.61,0.33)/0.32	(0.35,0.57,0.33)/0.24	(0.49,0.11,0.35)/0.31	(0.17,0.17,0.29)/0.56
P_{28}	(0.31,0.94,0.68)/0.76	(0.89,0.52,0.71)/0.73	(0.51,0.28,0.78)/0.48	(0.66,0.12,0.12)/0.43	(0.52,0.93,0.67)/0.57
P_{29}	(0.82,0.42,0.49)/0.84	(0.17,0.49,0.90)/0.56	(0.46,0.80,0.64)/0.75	(0.17,0.74,0.36)/0.59	(0.15,0.90,0.99)/0.73
P_{30}	(0.52,0.43,0.10)/0.77	(0.13,0.32,0.23)/0.73	(0.89,0.33,0.18)/0.91	(0.20,0.63,0.58)/0.71	(0.95,0.63,0.47)/0.93
P_{31}	(0.22,0.99,0.46)/0.67	(0.61,0.19,1.00)/0.65	(0.38,0.82,0.12)/0.28	(0.92,0.19,0.57)/0.53	(0.41,0.95,0.78)/0.41
P_{32}	(0.12,0.74,0.88)/0.40	(0.92,0.21,0.21)/0.86	(0.53,0.59,0.25)/0.83	(0.71,0.89,0.52)/0.75	(0.41,0.98,0.58)/0.22
P_{33}	(0.27,0.91,0.40)/0.71	(0.81,0.97,0.98)/0.33	(0.56,0.67,0.16)/0.85	(0.33,0.92,0.41)/0.38	(0.79,0.33,0.90)/0.28
P_{34}	(0.11,0.45,0.16)/0.63	(0.26,0.49,0.50)/0.95	(0.64,0.98,0.93)/0.59	(0.35,0.66,0.23)/0.86	(0.37,0.55,0.39)/0.38
P_{35}	(0.52,0.91,0.77)/0.34	(0.84,0.48,0.20)/0.59	(0.94,0.76,0.96)/0.80	(0.45,0.81,0.30)/0.72	(0.70,0.81,0.53)/0.50
P_{36}	(0.36,0.44,0.27)/0.26	(0.76,0.72,0.32)/0.62	(0.67,0.12,0.20)/0.86	(0.68,1.00,0.42)/0.51	(0.39,0.90,0.76)/0.76
P_{37}	(0.73,0.34,0.59)/0.60	(0.73,0.30,0.21)/0.76	(0.22,0.17,0.73)/0.47	(0.43,0.55,0.40)/0.24	(0.74,0.84,0.41)/0.76
P_{38}	(0.33,0.92,0.32)/0.80	(0.69,0.65,0.14)/0.74	(0.68,0.51,0.62)/0.84	(0.31,0.91,0.26)/0.41	(0.65,0.23,0.53)/0.79
P_{39}	(0.63,0.87,0.10)/0.56	(0.76,0.81,0.53)/0.37	(0.22,0.39,0.74)/0.30	(0.23,0.27,0.48)/0.28	(0.89,0.90,0.45)/0.62
P_{40}	(0.48,0.19,0.71)/0.85	(0.33,0.45,0.73)/0.49	(0.51,0.22,0.68)/0.62	(0.43,0.80,0.99)/0.31	(0.68,0.37,0.38)/0.32
P_{41}	(0.58,0.46,0.52)/0.33	(0.24,0.28,1.00)/0.40	(0.17,0.43,0.15)/0.87	(0.86,0.11,0.48)/0.52	(0.25,0.20,0.48)/0.52
P_{42}	(0.27,0.89,0.32)/0.91	(0.96,0.58,0.54)/0.60	(0.13,0.15,0.88)/0.21	(0.83,0.72,0.45)/0.89	(0.89,0.18,0.59)/0.31
P_{43}	(0.89,0.68,0.59)/0.26	(0.59,0.25,0.50)/0.91	(0.18,0.56,0.93)/0.52	(0.41,0.44,0.48)/0.59	(0.89,0.61,0.48)/0.66
P_{44}	(0.20,0.27,0.63)/0.78	(0.62,0.92,0.48)/0.94	(0.61,0.61,0.61)/0.54	(0.38,0.10,0.20)/0.58	(0.65,0.81,0.57)/0.31
P_{45}	(0.47,0.45,0.66)/0.41	(0.75,0.79,0.32)/0.84	(0.36,0.16,0.34)/0.65	(0.33,0.27,0.59)/0.92	(0.91,0.92,0.25)/0.64
P_{46}	(0.33,0.39,0.75)/0.83	(0.34,0.54,0.54)/0.94	(0.19,0.96,0.79)/0.50	(0.85,0.16,0.99)/0.46	(0.89,0.87,0.18)/0.41
P_{47}	(0.37,0.57,0.22)/0.60	(0.83,0.39,0.35)/0.86	(0.59,0.72,0.26)/0.91	(0.98,0.94,0.90)/0.31	(1.00,1.00,0.31)/0.25
P_{48}	(0.10,0.76,0.96)/0.89	(0.83,0.97,0.63)/0.31	(0.82,0.27,0.40)/0.50	(0.60,0.81,0.48)/0.28	(0.20,0.15,0.52)/0.33
P_{49}	(0.34,0.66,0.66)/0.83	(0.37,0.90,0.47)/0.78	(0.28,0.33,0.27)/0.85	(0.99,0.22,0.52)/0.92	(0.60,0.75,0.81)/0.60
P_{50}	(0.24,0.25,0.74)/0.37	(0.69,0.62,0.76)/0.60	(0.67,1.00,0.84)/0.82	(0.20,0.88,0.20)/0.74	(0.84,0.16,0.48)/0.94

Table 6. *Cont.*

Patient	Temperature	Cough	Throat Pain	Headache	Body Pain
P_{51}	(0.17,0.44,0.47)/0.68	(0.10,0.67,0.14)/0.62	(0.84,0.80,0.21)/0.42	(0.13,0.62,0.32)/0.20	(0.30,0.55,0.59)/0.57
P_{52}	(0.45,0.43,0.16)/0.56	(0.70,0.34,0.58)/0.21	(0.78,0.54,0.45)/0.54	(0.34,0.39,0.70)/0.72	(0.93,0.21,0.87)/0.69
P_{53}	(0.99,0.96,0.29)/0.53	(0.67,0.37,0.43)/0.26	(0.59,0.43,0.35)/0.80	(0.26,0.79,0.12)/0.75	(0.84,0.49,0.97)/0.71
P_{54}	(0.26,0.50,0.36)/0.61	(0.69,0.68,0.93)/0.26	(0.99,0.31,0.27)/0.41	(0.15,0.60,0.37)/0.23	(0.28,0.60,0.91)/0.23
P_{55}	(0.93,0.26,0.61)/0.59	(0.87,0.51,0.76)/0.71	(0.44,0.92,0.73)/0.54	(0.18,0.96,0.35)/0.85	(0.95,0.54,0.18)/0.69
P_{56}	(0.69,0.56,0.30)/0.47	(0.67,0.44,0.21)/0.85	(0.79,0.97,0.88)/0.77	(0.55,0.93,0.36)/0.54	(0.69,0.21,0.40)/0.94
P_{57}	(0.12,0.49,0.55)/0.29	(0.19,0.90,0.35)/0.76	(0.31,0.13,0.12)/0.55	(1.00,0.95,0.95)/0.59	(0.86,0.30,0.79)/0.33
P_{58}	(0.97,0.58,0.20)/0.55	(0.82,0.16,0.78)/0.93	(0.43,0.74,0.34)/0.42	(0.24,0.17,0.40)/0.44	(0.42,0.63,0.95)/0.67
P_{59}	(0.92,0.76,0.17)/0.56	(0.36,0.21,0.81)/0.32	(0.34,0.45,0.60)/0.68	(0.13,0.42,0.25)/0.43	(0.22,0.80,0.55)/0.87
P_{60}	(0.48,0.31,0.55)/0.94	(0.53,0.74,0.65)/0.59	(0.61,0.92,0.15)/0.27	(0.25,0.58,0.21)/0.46	(0.91,0.62,0.22)/0.28
P_{61}	(0.25,0.85,0.70)/0.93	(0.57,0.69,0.24)/0.51	(0.61,0.16,0.99)/0.22	(0.96,0.51,0.76)/0.38	(0.98,0.24,0.71)/0.36
P_{62}	(0.63,0.50,0.87)/0.68	(0.44,0.33,0.39)/0.27	(0.71,0.17,0.90)/0.82	(0.38,0.32,0.99)/0.95	(0.99,0.61,0.99)/0.37
P_{63}	(0.81,0.14,0.41)/0.71	(0.38,1.00,0.25)/0.54	(0.21,0.98,0.60)/0.92	(0.82,0.79,0.59)/0.94	(0.97,0.29,0.74)/0.52
P_{64}	(0.87,0.30,0.92)/0.65	(0.50,0.45,1.00)/0.58	(0.75,0.57,0.62)/0.85	(0.69,0.41,0.89)/0.46	(0.57,0.43,0.16)/0.92
P_{65}	(0.61,0.17,0.88)/0.85	(0.52,0.73,0.84)/0.65	(0.19,0.36,0.23)/0.77	(0.86,0.69,0.55)/0.85	(0.48,0.76,0.16)/0.93
P_{66}	(0.25,0.79,0.28)/0.59	(0.64,0.19,0.23)/0.74	(0.89,0.63,0.95)/0.87	(0.22,0.76,0.58)/0.20	(0.56,0.64,0.73)/0.35
P_{67}	(0.34,0.76,0.89)/0.47	(0.27,0.48,0.50)/0.90	(0.22,0.52,0.91)/0.72	(0.34,0.83,0.25)/0.25	(0.77,0.84,0.63)/0.35
P_{68}	(0.85,0.79,0.17)/0.75	(0.76,0.37,0.31)/0.54	(0.87,0.46,0.25)/0.67	(0.14,0.83,0.67)/0.63	(0.15,0.50,0.82)/0.20
P_{69}	(0.64,0.18,0.62)/0.90	(0.30,0.79,0.96)/0.33	(0.31,0.51,0.43)/0.78	(0.25,0.64,0.55)/0.48	(0.78,0.83,0.36)/0.86
P_{70}	(0.48,0.80,0.31)/0.83	(0.78,0.66,0.27)/0.70	(0.81,0.78,0.83)/0.61	(0.90,0.59,0.35)/0.47	(0.44,0.87,0.66)/0.72
P_{71}	(0.47,0.92,0.42)/0.95	(0.55,0.87,0.52)/0.60	(0.37,0.35,0.95)/0.92	(0.54,0.10,0.88)/0.82	(0.28,0.17,0.62)/0.34
P_{72}	(0.27,0.88,0.66)/0.38	(0.53,0.41,0.40)/0.24	(0.76,0.92,0.99)/0.27	(0.30,0.58,0.86)/0.73	(0.29,0.70,0.97)/0.90
P_{73}	(0.49,0.82,0.39)/0.58	(0.64,0.96,0.23)/0.24	(0.23,0.66,0.45)/0.26	(0.82,0.61,0.32)/0.52	(0.55,0.72,0.53)/0.56
P_{74}	(0.99,0.84,0.11)/0.33	(0.47,0.14,0.28)/0.75	(0.68,0.52,0.67)/0.20	(0.33,0.44,0.16)/0.58	(0.53,0.56,0.75)/0.39
P_{75}	(0.58,0.84,0.71)/0.57	(0.12,0.26,0.82)/0.20	(0.33,0.85,0.30)/0.80	(0.19,0.58,0.26)/0.55	(0.81,0.50,0.79)/0.41
P_{76}	(0.89,0.65,0.99)/0.21	(0.78,0.99,0.60)/0.69	(0.18,0.37,0.42)/0.89	(0.58,0.80,0.19)/0.52	(0.36,0.79,0.88)/0.58
P_{77}	(0.22,0.57,0.71)/0.86	(0.68,0.59,0.95)/0.41	(0.92,0.55,0.87)/0.51	(0.31,0.79,0.10)/0.34	(0.85,0.86,0.83)/0.39
P_{78}	(0.13,0.26,0.51)/0.69	(0.96,0.86,0.48)/0.28	(0.53,0.33,0.89)/0.39	(0.12,0.85,0.97)/0.62	(0.75,0.46,0.27)/0.86
P_{79}	(0.84,0.84,0.86)/0.82	(0.78,0.58,0.54)/0.58	(0.38,0.95,0.74)/0.26	(0.36,0.35,0.35)/0.59	(0.35,0.51,0.11)/0.47
P_{80}	(0.70,0.19,0.82)/0.89	(0.48,0.73,0.86)/0.35	(0.68,0.81,0.68)/0.85	(0.90,0.68,1.00)/0.69	(0.55,0.28,0.95)/0.38
P_{81}	(0.85,0.20,0.17)/0.82	(0.96,0.12,0.90)/0.57	(0.38,0.31,0.24)/0.52	(0.97,0.14,0.71)/0.40	(0.26,0.59,0.25)/0.85
P_{82}	(0.16,0.44,0.35)/0.51	(0.72,0.47,0.24)/0.34	(0.88,0.54,0.11)/0.53	(0.85,0.91,0.51)/0.34	(0.31,0.82,0.56)/0.56
P_{83}	(1.00,0.19,0.57)/0.26	(0.80,0.38,0.69)/0.63	(0.92,0.92,0.35)/0.81	(0.11,0.58,0.41)/0.28	(0.88,0.32,0.40)/0.55
P_{84}	(0.81,0.22,0.42)/0.80	(0.59,0.11,0.50)/0.41	(0.29,0.33,0.10)/0.39	(0.31,0.93,0.15)/0.58	(0.92,0.75,0.29)/0.21
P_{85}	(0.39,0.46,0.72)/0.56	(0.53,0.82,0.48)/0.44	(0.93,0.85,0.21)/0.33	(0.63,0.61,0.52)/0.24	(0.26,0.56,0.73)/0.46
P_{86}	(0.58,0.78,0.99)/0.56	(0.44,0.33,0.35)/0.23	(0.43,0.59,0.40)/0.44	(0.71,0.33,0.74)/0.27	(0.92,0.73,0.55)/0.75
P_{87}	(0.26,0.30,0.98)/0.30	(0.90,0.39,0.69)/0.92	(0.54,0.55,0.45)/0.46	(0.50,0.74,0.26)/0.30	(0.69,0.90,0.45)/0.46
P_{88}	(0.56,0.22,0.34)/0.86	(0.35,0.30,0.76)/0.45	(0.42,0.21,0.20)/0.40	(0.69,0.35,0.10)/0.94	(0.75,0.81,0.56)/0.91
P_{89}	(0.40,0.81,0.39)/0.62	(0.61,0.66,0.43)/0.91	(0.18,0.87,0.88)/0.94	(0.74,0.74,0.26)/0.77	(0.34,0.26,0.61)/0.88
P_{90}	(0.97,0.37,0.22)/0.93	(0.59,0.24,0.67)/0.38	(0.51,0.11,0.22)/0.33	(0.29,0.24,0.88)/0.64	(0.70,0.56,0.77)/0.32
P_{91}	(0.53,0.42,0.12)/0.73	(0.25,0.14,0.92)/0.95	(0.38,0.20,0.51)/0.63	(0.29,0.91,0.31)/0.80	(0.92,0.83,0.40)/0.37
P_{92}	(0.83,0.95,0.55)/0.33	(0.17,0.51,0.99)/0.44	(0.32,0.66,0.21)/0.66	(0.69,1.00,0.34)/0.70	(0.66,0.41,0.59)/0.52
P_{93}	(0.56,0.62,0.89)/0.40	(0.45,0.25,0.12)/0.92	(0.44,0.53,0.61)/0.58	(0.95,0.92,0.54)/0.22	(0.12,0.30,0.53)/0.71
P_{94}	(0.79,0.86,0.75)/0.61	(0.38,0.86,0.78)/0.37	(0.26,0.94,0.10)/0.23	(0.50,0.83,0.52)/0.62	(0.34,0.32,0.56)/0.44
P_{95}	(0.79,0.18,0.70)/0.89	(0.29,0.30,0.23)/0.53	(0.63,0.11,0.86)/0.45	(0.31,0.96,0.20)/0.85	(0.94,0.44,0.90)/0.48
P_{96}	(0.83,0.85,0.24)/0.91	(0.81,0.56,0.54)/0.39	(0.61,0.45,0.52)/0.56	(0.54,0.99,0.21)/0.88	(0.48,0.91,0.73)/0.69
P_{97}	(0.92,0.49,0.91)/0.41	(0.32,0.27,0.48)/0.90	(0.13,0.37,0.29)/0.25	(0.60,0.19,0.11)/0.67	(0.37,0.33,0.11)/0.95
P_{98}	(0.72,0.43,0.75)/0.75	(0.21,0.44,0.98)/0.41	(0.59,0.51,0.69)/0.49	(0.51,0.52,0.50)/0.70	(0.62,0.84,0.87)/0.77
P_{99}	(0.98,0.33,0.81)/0.65	(0.86,0.43,0.41)/0.95	(0.14,0.67,0.46)/0.63	(0.76,0.92,0.30)/0.46	(0.46,0.21,0.53)/0.95
P_{100}	(0.64,0.20,0.81)/0.80	(0.79,0.73,0.74)/0.27	(0.16,0.59,0.29)/0.39	(0.31,0.51,0.55)/0.34	(0.53,0.71,0.15)/0.75
P_{101}	(0.21,0.59,0.17)/0.79	(0.35,0.31,0.63)/0.92	(0.32,0.27,0.14)/0.57	(0.76,0.96,0.25)/0.74	(0.55,0.87,0.70)/0.36
P_{102}	(0.82,0.92,1.00)/0.67	(0.75,0.83,0.58)/0.51	(0.67,0.33,0.59)/0.83	(0.84,0.89,0.92)/0.26	(0.54,0.49,0.30)/0.95
P_{103}	(0.94,0.61,0.15)/0.76	(0.61,0.94,0.23)/0.57	(0.94,0.62,0.86)/0.42	(0.15,0.36,0.59)/0.83	(0.55,0.53,0.16)/0.42
P_{104}	(0.61,0.88,0.33)/0.64	(0.29,0.46,0.89)/0.34	(0.43,0.46,1.00)/0.53	(0.13,0.42,0.74)/0.50	(0.82,0.37,0.39)/0.95
P_{105}	(0.83,0.36,0.18)/0.38	(1.00,0.11,0.47)/0.57	(0.79,0.22,0.93)/0.50	(0.55,0.38,0.18)/0.85	(0.94,0.11,0.38)/0.79
P_{106}	(0.58,0.26,0.73)/0.86	(0.46,0.43,0.14)/0.35	(0.67,0.91,0.82)/0.72	(0.34,0.43,0.40)/0.26	(0.19,0.21,0.81)/0.57
P_{107}	(0.85,0.85,0.56)/0.37	(0.49,0.40,0.87)/0.41	(0.53,0.58,0.51)/0.58	(0.38,0.17,0.67)/0.41	(0.24,0.48,0.36)/0.43
P_{108}	(0.82,0.78,0.40)/0.46	(0.32,0.82,0.20)/0.29	(0.12,0.51,0.31)/0.47	(0.74,0.34,0.67)/0.37	(0.38,0.78,0.61)/0.78
P_{109}	(0.29,0.45,0.71)/0.65	(0.41,0.56,0.61)/0.68	(0.72,0.98,0.70)/0.24	(0.90,0.36,0.33)/0.93	(0.50,0.38,0.32)/0.68
P_{110}	(0.31,0.24,0.26)/0.23	(0.15,0.81,0.79)/0.94	(0.79,0.98,0.34)/0.36	(0.18,0.23,0.90)/0.83	(0.60,0.46,0.11)/0.93
P_{111}	(0.62,0.77,0.46)/0.53	(1.00,0.86,0.12)/0.42	(0.10,0.59,0.58)/0.92	(0.37,0.25,0.59)/0.58	(0.76,0.82,0.28)/0.60
P_{112}	(0.33,0.89,0.87)/0.30	(0.10,0.83,0.28)/0.25	(0.90,0.33,0.11)/0.54	(0.71,0.32,0.11)/0.35	(0.41,0.56,0.87)/0.76
P_{113}	(0.68,0.55,0.16)/0.77	(0.69,0.58,0.57)/0.35	(0.84,0.34,0.87)/0.64	(0.11,0.32,0.75)/0.41	(0.45,0.12,0.17)/0.80
P_{114}	(0.72,0.44,0.83)/0.84	(0.36,0.17,0.50)/0.80	(0.60,0.11,0.18)/0.77	(0.82,0.27,0.83)/0.49	(0.16,0.52,0.31)/0.25
P_{115}	(0.45,0.87,0.93)/0.56	(0.75,0.39,0.29)/0.43	(0.53,0.95,0.93)/0.69	(0.97,0.32,0.65)/0.53	(0.72,0.68,0.42)/0.53
P_{116}	(0.83,0.91,0.61)/0.38	(0.34,0.27,0.67)/0.51	(0.81,0.33,0.44)/0.79	(0.23,0.94,0.25)/0.69	(0.30,0.38,0.49)/0.39
P_{117}	(0.55,0.13,0.18)/0.79	(0.77,0.97,0.61)/0.95	(0.27,0.69,0.63)/0.53	(0.59,0.76,0.21)/0.63	(0.51,0.52,0.62)/0.39
P_{118}	(0.51,0.63,0.35)/0.35	(0.12,0.91,1.00)/0.35	(0.60,0.50,0.20)/0.28	(0.59,0.91,0.61)/0.62	(0.51,0.62,0.74)/0.62
P_{119}	(0.79,0.76,0.81)/0.57	(0.93,0.32,0.48)/0.56	(0.43,0.11,0.33)/0.67	(0.17,0.85,0.33)/0.90	(0.62,0.89,0.77)/0.82
P_{120}	(0.33,0.64,0.59)/0.85	(0.75,0.91,0.51)/0.57	(0.75,0.31,0.28)/0.45	(0.11,0.35,0.43)/0.51	(0.59,0.95,0.43)/0.71

Table 6. Cont.

Patient	Temperature	Cough	Throat Pain	Headache	Body Pain
P_{121}	(0.28,0.81,0.28)/0.52	(0.92,0.71,0.75)/0.68	(0.33,0.43,0.84)/0.75	(0.99,0.53,0.25)/0.73	(1.00,0.10,0.30)/0.29
P_{122}	(0.13,1.00,0.81)/0.72	(0.15,0.89,0.47)/0.74	(0.35,0.27,0.64)/0.21	(0.88,0.84,0.93)/0.35	(0.53,0.90,0.22)/0.27
P_{123}	(1.00,0.88,0.78)/0.34	(0.47,0.30,0.27)/0.36	(0.90,0.83,0.82)/0.42	(0.30,0.24,0.87)/0.33	(0.65,0.13,0.75)/0.68
P_{124}	(0.63,0.96,0.55)/0.37	(0.21,0.49,0.71)/0.95	(1.00,0.33,0.58)/0.80	(0.27,0.48,0.51)/0.28	(0.41,0.66,0.49)/0.35
P_{125}	(0.83,0.44,0.57)/0.66	(0.39,0.71,0.61)/0.92	(0.31,0.35,0.63)/0.84	(0.42,0.34,0.78)/0.49	(0.25,0.65,0.54)/0.38
P_{126}	(0.39,0.22,0.30)/0.69	(0.17,0.56,0.47)/0.38	(0.33,0.96,0.77)/0.81	(0.47,0.22,0.37)/0.71	(0.13,0.75,0.35)/0.42
P_{127}	(0.37,0.56,0.50)/0.40	(0.48,0.81,0.35)/0.73	(0.64,0.68,0.49)/0.88	(0.79,0.34,0.54)/0.50	(0.17,0.45,0.40)/0.53
P_{128}	(0.60,0.94,0.38)/0.63	(0.39,0.91,0.10)/0.57	(0.25,0.36,0.73)/0.76	(0.41,0.73,0.32)/0.27	(0.96,0.11,0.73)/0.22
P_{129}	(0.80,0.88,0.60)/0.28	(0.64,0.16,0.96)/0.84	(0.97,0.88,0.49)/0.93	(0.25,0.25,0.68)/0.41	(0.81,0.51,0.81)/0.69
P_{130}	(0.95,0.59,0.91)/0.33	(0.99,0.60,0.47)/0.68	(0.11,0.99,0.45)/0.39	(0.77,0.56,0.72)/0.25	(0.32,0.65,0.90)/0.91
P_{131}	(0.49,0.29,0.63)/0.75	(0.77,0.64,0.21)/0.91	(0.36,0.89,1.00)/0.39	(0.66,0.30,0.84)/0.40	(0.92,0.53,0.31)/0.21
P_{132}	(0.67,0.32,0.45)/0.46	(0.75,0.12,0.43)/0.33	(0.27,0.31,0.31)/0.44	(0.90,0.19,0.47)/0.94	(0.82,0.99,0.73)/0.92
P_{133}	(0.68,0.66,0.92)/0.36	(0.53,0.98,0.84)/0.56	(0.22,0.87,0.99)/0.51	(0.43,0.99,0.89)/0.44	(0.28,0.93,0.22)/0.23
P_{134}	(0.25,0.43,0.70)/0.53	(0.79,0.54,0.95)/0.95	(0.49,0.97,0.52)/0.28	(0.53,0.36,0.33)/0.44	(0.50,0.50,0.37)/0.21
P_{135}	(0.96,0.80,0.17)/0.54	(0.28,0.97,0.74)/0.25	(0.74,0.31,0.25)/0.46	(0.56,0.31,0.79)/0.52	(0.42,0.24,0.42)/0.57
P_{136}	(0.10,0.50,0.75)/0.22	(0.79,0.68,0.26)/0.34	(0.52,0.94,0.22)/0.89	(0.95,0.59,0.48)/0.73	(0.66,0.68,0.37)/0.78
P_{137}	(0.32,0.82,0.73)/0.25	(0.55,0.50,0.67)/0.78	(0.28,0.54,0.91)/0.91	(0.45,0.21,0.63)/0.86	(0.52,0.90,0.32)/0.87
P_{138}	(0.15,0.52,0.22)/0.65	(0.29,0.15,0.20)/0.40	(0.61,0.58,0.13)/0.47	(0.41,0.48,0.34)/0.46	(0.76,0.94,0.88)/0.49
P_{139}	(0.21,0.99,0.98)/0.31	(0.58,0.64,0.22)/0.94	(0.55,0.37,0.21)/0.87	(0.20,0.85,0.99)/0.59	(0.61,0.25,1.00)/0.80
P_{140}	(0.78,0.94,0.20)/0.22	(0.66,0.90,0.22)/0.67	(0.57,0.65,0.18)/0.81	(0.51,0.70,0.51)/0.29	(0.89,0.52,0.14)/0.62
P_{141}	(0.13,0.61,0.83)/0.31	(0.53,0.60,0.82)/0.94	(0.13,0.94,0.27)/0.61	(0.49,0.53,0.51)/0.83	(0.94,0.81,0.60)/0.41
P_{142}	(0.64,0.27,0.49)/0.80	(0.74,0.66,0.33)/0.34	(0.87,0.11,0.18)/0.26	(0.65,0.82,0.22)/0.41	(0.68,0.22,0.47)/0.74
P_{143}	(0.65,0.50,0.79)/0.38	(0.74,0.48,0.23)/0.39	(0.59,0.99,0.38)/0.59	(0.42,0.19,0.45)/0.22	(0.46,0.11,0.83)/0.75
P_{144}	(0.95,0.18,0.51)/0.23	(0.11,0.52,0.21)/0.43	(0.93,0.95,0.76)/0.84	(0.96,0.70,0.10)/0.49	(0.98,0.99,0.89)/0.83
P_{145}	(0.63,0.11,0.45)/0.81	(0.74,0.21,0.38)/0.34	(0.47,0.69,0.64)/0.70	(0.54,0.14,0.65)/0.44	(0.73,0.82,0.18)/0.52
P_{146}	(0.27,0.54,0.48)/0.77	(0.15,0.48,1.00)/0.54	(0.22,0.67,0.50)/0.21	(0.36,0.65,0.99)/0.21	(0.11,0.71,0.90)/0.45
P_{147}	(0.23,0.33,0.93)/0.57	(0.92,0.29,0.85)/0.77	(0.75,0.35,0.75)/0.51	(1.00,0.86,0.25)/0.59	(0.61,0.80,0.78)/0.91
P_{148}	(0.76,0.65,0.98)/0.57	(0.30,0.72,0.43)/0.73	(0.49,0.48,0.83)/0.88	(0.53,0.22,0.93)/0.36	(0.17,0.39,0.64)/0.83
P_{149}	(0.95,0.85,0.10)/0.65	(0.80,0.31,0.84)/0.51	(0.39,0.86,0.32)/0.53	(0.24,0.18,0.29)/0.81	(0.56,0.18,0.96)/0.22
P_{150}	(0.84,0.56,0.56)/0.61	(0.31,0.31,0.66)/0.51	(0.88,0.83,0.37)/0.27	(0.94,0.43,0.17)/0.47	(0.85,0.12,0.71)/0.44
P_{151}	(0.22,0.28,0.64)/0.82	(0.73,0.99,0.63)/0.39	(0.30,0.63,0.18)/0.74	(0.66,0.69,0.88)/0.43	(0.67,0.14,0.91)/0.22
P_{152}	(0.14,0.14,0.69)/0.37	(0.35,0.52,0.73)/0.43	(0.80,0.86,0.88)/0.40	(0.43,0.30,0.26)/0.67	(0.23,0.65,0.96)/0.72
P_{153}	(0.11,0.89,0.23)/0.61	(0.78,0.79,0.45)/0.89	(0.65,0.53,0.83)/0.29	(0.43,0.39,0.71)/0.27	(0.45,0.54,0.56)/0.27
P_{154}	(0.28,0.33,0.53)/0.91	(0.96,0.18,0.22)/0.91	(0.59,0.86,0.91)/0.47	(0.35,0.40,0.72)/0.67	(0.90,0.69,0.40)/0.57
P_{155}	(0.46,0.56,0.84)/0.72	(0.19,0.61,0.88)/0.35	(0.89,0.40,0.27)/0.55	(0.92,0.24,0.12)/0.77	(0.51,0.65,0.53)/0.80
P_{156}	(0.22,0.19,0.77)/0.61	(0.45,0.20,0.81)/0.39	(0.35,0.62,0.65)/0.22	(0.25,0.78,0.99)/0.21	(0.77,0.70,0.25)/0.28
P_{157}	(0.49,0.67,0.90)/0.63	(0.76,0.39,0.87)/0.50	(0.45,0.11,1.00)/0.26	(0.45,0.93,0.42)/0.68	(0.11,0.67,0.84)/0.75
P_{158}	(0.63,0.42,0.22)/0.91	(0.84,0.92,0.19)/0.85	(0.28,0.55,0.38)/0.60	(0.76,0.72,0.22)/0.42	(0.94,0.16,0.56)/0.43
P_{159}	(0.46,0.74,0.60)/0.21	(0.92,0.23,0.65)/0.60	(0.16,0.13,0.63)/0.93	(0.73,0.92,0.60)/0.37	(0.29,0.16,0.89)/0.83
P_{160}	(0.54,0.82,0.97)/0.80	(0.48,0.35,0.39)/0.53	(0.36,0.70,0.83)/0.80	(0.28,0.30,0.97)/0.27	(0.74,0.68,0.66)/0.37
P_{161}	(0.52,0.85,0.83)/0.50	(0.69,0.34,0.54)/0.86	(0.18,0.94,0.80)/0.90	(0.90,0.95,0.85)/0.29	(0.90,0.11,0.14)/0.66
P_{162}	(0.11,0.66,0.22)/0.60	(0.29,0.58,0.36)/0.72	(0.83,0.33,0.91)/0.36	(0.86,0.60,0.81)/0.42	(0.74,0.32,0.54)/0.84
P_{163}	(0.19,0.53,0.72)/0.43	(0.58,0.94,0.30)/0.94	(0.27,0.98,0.22)/0.80	(0.91,0.83,0.92)/0.20	(0.87,0.15,0.96)/0.92
P_{164}	(0.14,0.80,0.54)/0.59	(0.70,0.38,0.37)/0.88	(0.36,0.78,0.75)/0.25	(0.55,0.98,0.94)/0.58	(0.49,0.68,0.43)/0.77
P_{165}	(0.73,0.42,0.13)/0.72	(0.22,0.98,0.35)/0.78	(0.59,0.46,0.12)/0.68	(0.79,0.47,0.21)/0.75	(0.46,0.19,0.23)/0.94
P_{166}	(0.66,0.97,0.54)/0.50	(0.94,0.19,0.65)/0.66	(0.99,0.59,0.45)/0.21	(0.73,0.72,0.58)/0.62	(0.57,0.59,0.34)/0.37
P_{167}	(0.26,0.45,0.29)/0.42	(0.75,0.44,0.89)/0.80	(0.52,0.46,0.13)/0.22	(0.91,0.64,0.41)/0.39	(0.59,0.29,0.15)/0.60
P_{168}	(0.19,0.68,0.47)/0.54	(0.91,0.16,0.11)/0.50	(0.52,0.70,0.71)/0.58	(0.59,0.96,0.95)/0.68	(0.12,0.65,0.10)/0.26
P_{169}	(0.72,0.11,0.97)/0.66	(0.13,0.79,0.62)/0.38	(0.77,0.89,0.58)/0.72	(0.53,0.95,0.56)/0.64	(1.00,0.26,0.82)/0.44
P_{170}	(0.41,0.69,0.69)/0.88	(0.24,0.71,0.16)/0.53	(0.57,0.41,0.53)/0.94	(0.89,0.42,0.59)/0.49	(0.72,0.65,0.44)/0.62
P_{171}	(0.21,0.24,0.63)/0.26	(0.93,0.11,0.38)/0.78	(0.41,0.60,0.60)/0.46	(0.39,0.58,0.92)/0.53	(0.60,0.52,0.65)/0.47
P_{172}	(0.37,0.86,0.49)/0.61	(0.26,0.28,0.54)/0.75	(0.59,0.37,0.82)/0.60	(0.52,0.18,1.00)/0.24	(0.59,0.34,0.89)/0.43
P_{173}	(0.20,0.73,0.63)/0.75	(0.80,0.49,0.79)/0.78	(0.46,0.83,0.71)/0.75	(0.23,0.48,0.16)/0.40	(0.36,0.44,0.79)/0.81
P_{174}	(0.91,0.74,0.84)/0.58	(0.79,0.96,0.80)/0.35	(0.38,0.12,0.10)/0.93	(0.82,0.88,0.81)/0.51	(0.57,0.30,0.94)/0.62
P_{175}	(0.18,0.24,0.33)/0.46	(0.38,0.72,0.52)/0.48	(0.69,0.62,0.51)/0.71	(0.36,0.12,0.57)/0.95	(0.82,0.65,0.71)/0.37
P_{176}	(0.52,0.12,0.77)/0.32	(0.90,0.72,0.51)/0.24	(0.52,0.39,0.90)/0.72	(0.66,0.99,0.31)/0.94	(0.20,0.54,0.10)/0.90
P_{177}	(0.64,0.16,0.66)/0.51	(0.61,0.38,0.58)/0.41	(0.15,0.53,0.90)/0.21	(0.72,0.10,0.41)/0.42	(0.72,0.75,0.33)/0.41
P_{178}	(0.39,0.63,0.12)/0.86	(0.42,0.80,0.91)/0.64	(0.41,0.57,0.64)/0.65	(0.50,0.34,0.46)/0.61	(0.22,0.56,0.76)/0.42
P_{179}	(0.71,0.84,0.71)/0.46	(0.47,0.39,0.56)/0.29	(0.31,0.46,0.47)/0.70	(0.86,0.52,0.11)/0.90	(0.19,0.91,0.63)/0.65
P_{180}	(0.47,0.63,0.37)/0.63	(0.60,0.26,0.24)/0.33	(0.55,0.49,0.58)/0.26	(0.32,0.11,0.19)/0.22	(0.96,0.82,0.72)/0.45
P_{181}	(0.21,0.51,0.32)/0.40	(0.20,0.33,0.37)/0.87	(0.11,0.63,0.87)/0.49	(0.24,0.54,0.86)/0.53	(0.45,0.47,0.62)/0.49
P_{182}	(0.75,0.50,0.62)/0.57	(0.70,0.40,0.17)/0.44	(0.75,0.85,0.75)/0.68	(0.60,0.98,0.20)/0.75	(0.86,0.43,0.40)/0.54
P_{183}	(0.11,0.59,0.12)/0.28	(0.69,0.31,0.13)/0.60	(0.76,0.36,0.71)/0.88	(0.59,0.96,0.75)/0.36	(0.67,0.48,0.48)/0.95
P_{184}	(0.22,0.73,0.90)/0.21	(0.16,0.52,0.72)/0.62	(0.89,0.63,0.52)/0.95	(0.11,0.11,0.31)/0.61	(0.29,0.65,0.42)/0.79
P_{185}	(0.20,0.17,0.65)/0.43	(0.48,0.79,0.70)/0.32	(0.89,0.51,0.30)/0.40	(0.27,0.41,0.35)/0.49	(0.85,0.93,0.80)/0.23
P_{186}	(0.34,0.86,0.58)/0.34	(0.99,0.25,0.27)/0.59	(0.12,0.35,0.56)/0.87	(0.52,0.10,0.91)/0.87	(0.35,0.52,0.68)/0.25
P_{187}	(0.93,0.95,0.52)/0.92	(0.59,0.27,0.17)/0.62	(0.35,0.37,0.49)/0.25	(0.53,0.80,0.63)/0.72	(0.18,0.44,0.27)/0.87
P_{188}	(0.78,0.84,0.18)/0.34	(0.82,0.42,0.61)/0.69	(0.16,0.26,0.48)/0.90	(0.10,0.25,0.76)/0.75	(0.89,0.90,0.94)/0.25
P_{189}	(0.50,0.46,0.33)/0.82	(0.50,0.36,0.24)/0.60	(0.42,0.53,0.83)/0.51	(0.95,0.22,0.43)/0.49	(0.98,0.82,0.40)/0.68
P_{190}	(0.43,0.99,0.85)/0.21	(0.47,1.00,0.43)/0.79	(0.57,0.88,0.83)/0.88	(0.98,0.91,0.71)/0.50	(0.52,0.30,0.80)/0.78

Table 6. *Cont.*

Patient	Temperature	Cough	Throat Pain	Headache	Body Pain
P_{191}	(0.45,0.41,0.53)/0.21	(0.15,0.82,0.56)/0.88	(0.56,1.00,0.12)/0.87	(0.28,0.24,0.53)/0.75	(0.48,0.87,0.93)/0.52
P_{192}	(0.34,0.89,0.30)/0.48	(0.83,0.36,0.47)/0.92	(0.24,0.48,1.00)/0.36	(0.30,0.78,0.55)/0.92	(0.48,0.93,0.89)/0.54
P_{193}	(0.96,0.76,0.73)/0.27	(0.56,0.13,0.44)/0.42	(0.84,0.42,0.62)/0.87	(0.47,0.41,0.13)/0.50	(0.98,0.90,0.38)/0.69
P_{194}	(0.89,1.00,0.14)/0.60	(0.22,0.53,0.33)/0.69	(0.65,0.27,0.62)/0.95	(0.75,0.88,0.60)/0.75	(0.70,0.62,0.70)/0.67
P_{195}	(0.25,0.74,0.56)/0.89	(0.98,0.11,0.85)/0.45	(0.33,0.17,0.84)/0.33	(0.32,0.43,0.61)/0.40	(0.54,0.57,0.97)/0.70
P_{196}	(0.77,0.72,0.21)/0.78	(0.29,0.12,0.99)/0.55	(0.71,0.27,0.90)/0.57	(0.18,0.48,0.37)/0.36	(0.22,0.54,0.89)/0.66
P_{197}	(0.87,0.92,0.41)/0.27	(0.47,0.36,0.59)/0.76	(0.37,0.28,0.90)/0.93	(0.99,0.84,0.44)/0.70	(0.17,0.73,0.31)/0.54
P_{198}	(0.30,0.61,1.00)/0.79	(0.81,0.81,0.42)/0.47	(0.59,0.51,0.55)/0.68	(0.26,0.98,0.50)/0.54	(0.40,1.00,0.82)/0.40
P_{199}	(0.25,0.29,0.67)/0.90	(0.26,0.94,0.48)/0.80	(0.15,0.83,0.60)/0.27	(0.37,0.40,0.18)/0.73	(0.89,0.11,0.64)/0.31
P_{200}	(0.66,0.41,0.94)/0.37	(0.24,0.51,0.95)/0.83	(0.44,0.39,0.98)/0.75	(0.47,0.96,0.83)/0.93	(0.45,0.67,0.44)/0.24
P_{201}	(0.27,0.34,0.46)/0.64	(0.85,0.96,0.37)/0.62	(0.51,0.96,0.78)/0.29	(0.63,0.39,0.18)/0.24	(0.11,0.77,0.92)/0.49
P_{202}	(0.23,0.33,0.68)/0.25	(0.69,0.62,0.42)/0.47	(0.78,0.89,0.74)/0.32	(0.67,0.52,0.80)/0.65	(0.12,0.95,0.94)/0.46
P_{203}	(0.61,0.33,0.81)/0.54	(0.90,0.30,0.91)/0.27	(0.81,0.62,0.60)/0.20	(0.17,0.38,0.67)/0.75	(0.19,0.80,0.60)/0.23
P_{204}	(0.52,0.44,0.51)/0.63	(0.67,0.91,0.43)/0.65	(0.48,0.58,0.81)/0.91	(0.50,0.35,0.76)/0.48	(0.76,0.43,0.12)/0.37
P_{205}	(0.46,0.75,0.94)/0.85	(0.82,0.25,0.96)/0.29	(0.87,0.18,0.75)/0.49	(0.17,0.86,0.55)/0.64	(0.92,0.25,0.11)/0.22
P_{206}	(0.64,0.63,0.71)/0.20	(0.17,0.51,0.20)/0.56	(0.87,0.39,0.70)/0.32	(0.73,0.70,0.27)/0.53	(0.24,0.85,0.33)/0.43
P_{207}	(0.63,0.98,0.39)/0.28	(0.74,0.78,0.12)/0.27	(0.19,0.35,0.60)/0.78	(0.64,0.53,0.11)/0.43	(0.72,0.65,0.92)/0.20
P_{208}	(0.21,0.57,0.67)/0.66	(0.78,0.32,0.38)/0.69	(0.77,0.98,0.85)/0.39	(0.49,0.78,0.43)/0.44	(0.37,0.89,0.38)/0.39
P_{209}	(0.52,0.47,0.12)/0.84	(0.91,0.27,0.94)/0.89	(0.56,0.72,0.36)/0.59	(0.16,0.93,0.43)/0.44	(0.44,0.62,0.74)/0.85
P_{210}	(0.88,0.31,0.57)/0.53	(0.97,0.45,0.19)/0.56	(0.75,0.18,0.70)/0.34	(0.18,0.30,0.50)/0.62	(0.34,0.16,0.42)/0.48
P_{211}	(0.90,0.32,0.15)/0.88	(0.98,0.53,0.88)/0.90	(0.85,0.19,0.98)/0.23	(0.25,0.54,0.76)/0.41	(0.58,0.30,0.82)/0.68
P_{212}	(0.75,0.11,0.79)/0.27	(0.28,0.73,0.16)/0.40	(0.22,0.39,0.58)/0.50	(0.96,0.76,0.10)/0.61	(0.88,0.96,0.55)/0.65
P_{213}	(0.62,0.45,0.29)/0.95	(0.88,0.43,0.95)/0.52	(0.27,0.63,0.36)/0.28	(0.27,0.17,0.90)/0.61	(0.45,0.26,1.00)/0.32
P_{214}	(0.93,0.59,0.26)/0.41	(0.75,0.99,0.20)/0.93	(0.35,0.43,0.89)/0.46	(0.24,0.82,0.89)/0.35	(0.96,0.83,0.31)/0.55
P_{215}	(0.26,0.32,0.76)/0.87	(0.74,0.74,0.98)/0.93	(0.55,0.26,0.98)/0.27	(0.15,0.14,0.99)/0.56	(0.21,0.39,0.46)/0.78
P_{216}	(0.44,0.43,0.48)/0.22	(0.58,0.62,0.11)/0.44	(0.16,0.26,0.14)/0.84	(0.81,0.55,0.28)/0.32	(0.26,0.29,0.82)/0.95
P_{217}	(0.67,0.75,0.37)/0.49	(0.54,0.23,0.69)/0.64	(0.24,0.13,0.76)/0.50	(1.00,0.69,0.78)/0.72	(0.25,0.43,0.52)/0.52
P_{218}	(0.40,0.90,0.10)/0.83	(0.69,0.73,0.32)/0.39	(0.53,0.86,0.88)/0.72	(0.58,0.78,0.38)/0.50	(0.71,0.83,0.95)/0.77
P_{219}	(0.70,0.60,0.66)/0.84	(0.58,0.52,0.82)/0.70	(0.12,0.73,0.61)/0.58	(0.33,0.79,0.74)/0.76	(0.21,0.47,0.96)/0.85
P_{220}	(0.72,0.43,0.79)/0.72	(0.95,0.83,0.49)/0.73	(0.52,0.28,0.75)/0.43	(0.21,0.32,0.98)/0.49	(0.75,0.88,0.20)/0.86
P_{221}	(0.85,0.91,0.41)/0.91	(0.78,0.27,0.45)/0.58	(0.76,0.97,0.38)/0.33	(0.51,0.40,0.80)/0.37	(0.12,0.80,0.42)/0.33
P_{222}	(0.18,0.70,0.30)/0.95	(0.42,0.69,0.67)/0.72	(0.54,0.13,0.35)/0.54	(0.80,0.71,0.29)/0.52	(0.66,0.63,0.33)/0.35
P_{223}	(0.42,0.31,0.87)/0.22	(0.71,0.27,0.44)/0.84	(0.16,0.21,0.58)/0.47	(0.41,0.93,0.37)/0.25	(0.33,0.57,0.11)/0.42
P_{224}	(0.86,0.91,0.54)/0.32	(0.36,0.31,0.47)/0.78	(0.53,0.68,0.54)/0.36	(0.85,0.12,0.88)/0.49	(0.26,1.00,0.54)/0.76
P_{225}	(0.15,0.52,0.50)/0.50	(0.52,0.88,0.75)/0.51	(0.96,0.75,0.30)/0.51	(0.12,0.35,0.78)/0.70	(0.61,0.88,0.74)/0.38
P_{226}	(0.43,0.78,0.67)/0.52	(0.26,0.11,0.25)/0.34	(0.46,0.96,0.56)/0.68	(0.85,0.43,0.60)/0.54	(0.25,0.53,0.35)/0.68
P_{227}	(0.91,0.40,0.69)/0.32	(0.35,0.88,0.91)/0.53	(0.94,0.98,0.72)/0.76	(0.45,0.41,0.84)/0.84	(0.62,0.62,0.97)/0.46
P_{228}	(0.78,0.28,0.56)/0.29	(0.19,0.99,0.80)/0.73	(0.56,0.80,0.44)/0.29	(0.14,0.28,0.87)/0.75	(0.45,0.25,0.81)/0.69
P_{229}	(0.88,0.82,0.12)/0.25	(0.56,0.37,0.97)/0.27	(0.55,0.68,0.37)/0.28	(0.93,0.12,0.22)/0.45	(0.16,0.14,0.40)/0.69
P_{230}	(0.79,0.11,0.96)/0.84	(0.19,0.56,0.57)/0.25	(0.53,0.85,0.24)/0.56	(0.57,0.29,0.40)/0.41	(0.62,0.43,0.84)/0.69
P_{231}	(0.55,0.76,0.65)/0.69	(0.12,0.97,0.43)/0.31	(0.93,0.36,0.51)/0.47	(0.98,0.58,0.79)/0.28	(0.99,0.98,0.23)/0.65
P_{232}	(0.67,0.55,0.94)/0.37	(0.33,0.98,0.49)/0.87	(0.35,0.57,0.43)/0.91	(0.82,0.20,0.30)/0.21	(0.63,0.52,0.79)/0.83
P_{233}	(0.32,0.81,0.47)/0.40	(0.11,0.19,0.14)/0.95	(0.64,0.99,0.54)/0.51	(0.36,0.19,0.76)/0.48	(0.18,0.43,0.61)/0.76
P_{234}	(0.11,0.71,0.52)/0.93	(0.46,0.96,0.76)/0.21	(0.54,0.48,0.96)/0.45	(0.80,0.22,0.83)/0.50	(0.97,0.25,0.78)/0.47
P_{235}	(0.84,0.31,0.82)/0.95	(0.75,0.39,0.40)/0.84	(0.78,1.00,0.75)/0.38	(0.65,0.48,0.32)/0.64	(0.35,0.26,0.13)/0.22
P_{236}	(0.29,0.76,0.82)/0.87	(0.81,0.84,0.21)/0.22	(0.20,0.72,0.86)/0.24	(0.27,0.42,0.97)/0.37	(0.25,0.10,0.15)/0.72
P_{237}	(0.35,0.44,0.29)/0.82	(0.18,0.25,0.77)/0.84	(0.80,0.78,0.28)/0.21	(0.32,0.44,0.27)/0.29	(0.98,0.79,0.70)/0.81
P_{238}	(0.12,0.12,0.20)/0.28	(0.68,0.62,0.76)/0.59	(0.67,0.64,0.92)/0.22	(0.64,0.22,0.76)/0.90	(0.64,0.93,0.61)/0.63
P_{239}	(0.60,0.52,0.21)/0.33	(0.90,0.71,0.33)/0.77	(0.73,0.79,0.55)/0.52	(0.47,0.24,0.38)/0.40	(0.79,0.90,0.16)/0.70
P_{240}	(0.54,0.11,0.41)/0.52	(0.58,0.12,0.68)/0.84	(0.90,1.00,0.54)/0.54	(0.45,0.85,0.98)/0.95	(0.87,0.10,0.20)/0.62
P_{241}	(0.90,0.85,0.75)/0.33	(0.15,1.00,0.96)/0.70	(0.99,0.60,0.21)/0.59	(0.81,0.23,0.68)/0.83	(0.10,0.67,0.77)/0.63
P_{242}	(0.59,0.44,0.38)/0.50	(0.27,0.12,0.50)/0.53	(0.36,0.48,0.50)/0.60	(0.86,0.82,0.39)/0.72	(0.60,0.64,0.48)/0.64
P_{243}	(0.35,0.18,0.50)/0.57	(0.69,0.30,0.88)/0.52	(0.26,0.76,0.70)/0.82	(0.54,0.30,0.54)/0.45	(0.69,0.86,0.38)/0.37
P_{244}	(0.92,0.87,0.18)/0.48	(0.13,0.10,0.12)/0.26	(0.52,0.81,0.68)/0.69	(0.79,0.53,0.27)/0.81	(0.73,0.92,0.80)/0.21
P_{245}	(0.42,0.14,0.85)/0.29	(0.84,0.20,0.57)/0.25	(0.53,0.33,0.16)/0.90	(0.66,0.91,1.00)/0.41	(0.10,0.99,0.37)/0.44
P_{246}	(0.47,0.32,0.90)/0.90	(0.43,0.52,0.23)/0.91	(0.66,0.63,0.86)/0.66	(0.93,0.83,0.56)/0.69	(0.71,0.60,0.73)/0.25
P_{247}	(0.34,0.21,0.83)/0.28	(0.53,0.67,0.32)/0.22	(0.89,0.35,0.89)/0.27	(0.43,0.94,0.77)/0.85	(0.52,0.89,0.16)/0.93
P_{248}	(0.25,0.16,0.56)/0.70	(0.99,0.43,0.27)/0.28	(0.64,0.97,0.36)/0.72	(0.45,0.44,0.62)/0.22	(0.30,0.34,0.14)/0.51
P_{249}	(0.27,0.37,0.82)/0.56	(0.31,0.64,0.20)/0.94	(0.95,0.43,0.23)/0.47	(0.54,0.28,0.46)/0.79	(0.24,0.17,0.24)/0.33
P_{250}	(0.10,0.66,0.12)/0.83	(0.80,0.88,0.65)/0.53	(0.61,0.68,0.79)/0.60	(0.83,0.73,0.27)/0.45	(0.79,0.56,0.79)/0.34
P_{251}	(0.97,0.50,0.22)/0.40	(0.10,0.45,0.24)/0.45	(0.25,0.71,0.25)/0.81	(0.62,0.54,0.78)/0.55	(0.18,0.10,0.89)/0.94
P_{252}	(0.42,0.46,0.94)/0.80	(0.48,0.87,0.32)/0.92	(0.17,0.17,0.15)/0.71	(0.96,0.98,0.42)/0.67	(0.28,0.99,0.71)/0.33
P_{253}	(0.87,0.76,0.41)/0.48	(0.52,0.64,0.83)/0.85	(0.95,0.84,0.10)/0.93	(0.31,0.18,0.96)/0.27	(0.24,0.41,0.27)/0.88
P_{254}	(0.40,0.20,0.13)/0.82	(0.79,0.83,1.00)/0.37	(0.88,0.30,0.70)/0.62	(0.42,0.36,0.62)/0.50	(0.83,0.93,0.56)/0.70
P_{255}	(0.24,0.69,0.21)/0.59	(0.80,0.37,0.57)/0.35	(0.43,0.27,0.59)/0.34	(0.32,0.72,0.48)/0.52	(0.53,0.87,0.24)/0.30
P_{256}	(0.48,0.15,0.57)/0.59	(0.41,0.35,0.98)/0.77	(0.46,0.80,0.57)/0.45	(0.18,0.14,0.77)/0.51	(0.13,0.37,0.30)/0.20
P_{257}	(0.52,0.56,0.21)/0.24	(0.10,0.85,0.87)/0.24	(0.60,0.21,0.42)/0.74	(0.92,0.83,0.49)/0.30	(0.62,0.98,0.70)/0.53
P_{258}	(0.58,0.22,0.83)/0.49	(0.82,0.70,0.37)/0.39	(0.48,0.82,1.00)/0.79	(0.17,0.25,0.65)/0.20	(0.27,0.65,0.48)/0.63
P_{259}	(0.33,0.17,0.90)/0.60	(0.62,0.13,0.11)/0.63	(0.23,0.78,0.87)/0.29	(0.82,0.15,0.61)/0.29	(0.36,0.84,0.33)/0.82
P_{260}	(0.17,0.12,0.26)/0.58	(0.89,0.36,0.92)/0.33	(0.11,0.61,0.89)/0.76	(0.71,0.83,0.67)/0.75	(0.77,0.24,0.13)/0.26

Table 6. *Cont.*

Patient	Temperature	Cough	Throat Pain	Headache	Body Pain
P_{261}	(0.36,0.84,0.41)/0.60	(0.33,0.68,0.45)/0.29	(0.15,0.70,0.22)/0.57	(0.43,0.54,0.29)/0.52	(0.59,0.50,0.46)/0.88
P_{262}	(0.33,0.60,0.51)/0.32	(0.86,0.22,0.38)/0.29	(0.42,0.21,0.75)/0.26	(0.91,0.55,0.74)/0.94	(0.94,0.16,0.31)/0.74
P_{263}	(0.95,0.74,0.65)/0.44	(0.87,0.32,0.27)/0.68	(0.89,0.82,0.57)/0.24	(0.79,0.35,0.37)/0.73	(0.11,0.12,0.65)/0.29
P_{264}	(0.44,0.33,0.92)/0.38	(0.86,0.48,0.38)/0.43	(0.10,0.73,0.67)/0.41	(0.65,0.83,0.97)/0.86	(0.96,0.42,0.49)/0.53
P_{265}	(1.00,0.96,0.72)/0.28	(0.31,0.97,0.90)/0.59	(0.52,0.80,0.90)/0.63	(0.13,0.80,0.98)/0.50	(0.83,0.21,0.91)/0.32
P_{266}	(0.44,0.81,0.81)/0.35	(0.64,0.73,0.64)/0.24	(0.36,0.55,0.17)/0.86	(0.25,0.46,0.25)/0.70	(0.23,0.99,0.95)/0.48
P_{267}	(0.69,0.22,0.79)/0.62	(0.93,0.52,0.93)/0.65	(0.67,0.51,0.93)/0.71	(0.91,0.17,0.46)/0.37	(0.80,0.52,0.23)/0.59
P_{268}	(0.74,0.61,0.80)/0.30	(0.12,0.63,0.69)/0.21	(0.56,0.47,0.95)/0.80	(0.16,0.42,0.43)/0.58	(0.12,0.44,0.41)/0.69
P_{269}	(0.42,0.98,0.86)/0.74	(0.46,0.93,0.24)/0.84	(0.45,0.65,0.59)/0.50	(0.79,0.73,0.49)/0.88	(0.38,0.96,0.99)/0.87
P_{270}	(0.46,0.52,0.62)/0.63	(0.73,0.51,0.40)/0.21	(0.30,1.00,0.41)/0.65	(0.68,0.16,0.29)/0.46	(0.30,0.62,0.60)/0.44
P_{271}	(0.77,0.58,0.55)/0.69	(0.56,0.71,0.91)/0.48	(0.15,0.68,0.38)/0.43	(0.58,0.41,0.34)/0.73	(0.34,0.25,0.27)/0.58
P_{272}	(0.93,0.36,0.45)/0.72	(0.50,0.46,0.65)/0.74	(0.94,0.12,0.74)/0.20	(0.42,0.78,0.85)/0.60	(0.43,1.00,0.83)/0.85
P_{273}	(0.82,0.16,0.99)/0.84	(0.97,0.65,0.65)/0.50	(0.71,0.80,0.47)/0.28	(0.92,0.67,0.21)/0.53	(0.36,0.39,0.65)/0.86
P_{274}	(0.25,0.53,0.28)/0.23	(0.67,1.00,0.49)/0.48	(0.41,0.78,0.63)/0.46	(0.76,0.45,0.48)/0.28	(0.90,0.86,0.52)/0.72
P_{275}	(0.22,0.85,0.54)/0.82	(0.73,0.85,0.39)/0.23	(0.17,0.23,0.62)/0.74	(0.96,0.51,0.98)/0.74	(0.59,0.37,0.21)/0.77
P_{276}	(0.49,0.43,0.70)/0.25	(0.30,0.13,0.87)/0.44	(0.58,0.42,0.58)/0.32	(0.95,0.52,0.23)/0.72	(0.10,0.31,0.41)/0.41
P_{277}	(0.95,0.73,0.82)/0.32	(0.47,0.93,0.16)/0.54	(0.91,0.77,0.41)/0.48	(0.93,0.24,0.60)/0.60	(0.38,0.80,0.59)/0.85
P_{278}	(0.36,0.29,0.57)/0.50	(0.59,0.46,0.81)/0.54	(0.62,0.72,0.25)/0.22	(0.30,0.81,0.14)/0.87	(0.33,0.18,0.30)/0.48
P_{279}	(0.71,0.59,0.19)/0.59	(0.51,0.71,0.43)/0.85	(0.34,0.30,0.94)/0.23	(0.36,0.66,0.80)/0.47	(0.82,0.15,0.36)/0.21
P_{280}	(0.89,0.49,0.29)/0.66	(0.25,0.74,0.87)/0.65	(0.30,0.67,0.66)/0.35	(0.43,0.76,0.66)/0.79	(0.68,0.73,0.53)/0.83
P_{281}	(0.14,0.13,0.81)/0.78	(0.94,0.94,0.86)/0.24	(0.36,0.15,0.97)/0.83	(0.52,0.31,0.93)/0.93	(0.98,0.31,0.95)/0.34
P_{282}	(0.92,0.98,0.34)/0.74	(0.52,0.81,0.43)/0.46	(0.28,0.33,0.16)/0.30	(0.39,0.85,0.14)/0.22	(0.71,0.32,0.47)/0.93
P_{283}	(0.92,0.17,0.97)/0.52	(0.44,0.34,0.87)/0.46	(0.47,0.66,0.25)/0.34	(0.42,0.50,0.98)/0.71	(0.11,0.56,0.20)/0.37
P_{284}	(0.75,0.78,0.98)/0.51	(0.53,0.48,0.51)/0.61	(0.49,0.93,0.22)/0.91	(0.91,0.86,0.49)/0.84	(0.98,0.29,0.94)/0.70
P_{285}	(0.59,0.89,0.84)/0.29	(0.99,0.34,0.18)/0.75	(0.18,0.44,0.89)/0.93	(0.74,0.59,0.20)/0.81	(0.13,0.79,0.31)/0.47
P_{286}	(0.63,0.61,0.89)/0.71	(0.62,0.24,0.83)/0.68	(0.96,0.67,0.73)/0.48	(0.74,0.72,0.68)/0.42	(0.47,0.90,0.60)/0.56
P_{287}	(0.83,0.15,0.22)/0.71	(0.65,0.15,0.78)/0.47	(0.62,0.97,0.49)/0.20	(0.98,0.63,0.57)/0.85	(0.15,0.84,0.99)/0.52
P_{288}	(0.59,0.95,0.19)/0.45	(0.29,0.69,0.97)/0.40	(0.83,0.65,0.79)/0.61	(0.47,0.41,0.74)/0.35	(0.62,0.94,0.66)/0.58
P_{289}	(0.16,0.26,0.92)/0.86	(0.16,0.46,0.46)/0.29	(0.53,0.17,0.89)/0.54	(0.31,0.16,0.80)/0.33	(0.51,0.87,0.46)/0.40
P_{290}	(0.63,0.57,0.52)/0.88	(0.31,0.77,1.00)/0.33	(0.78,0.83,0.59)/0.54	(0.37,0.31,0.94)/0.78	(0.30,0.12,0.87)/0.68
P_{291}	(0.44,0.41,0.56)/0.89	(0.65,0.63,0.33)/0.77	(0.91,0.36,0.61)/0.61	(0.70,0.27,0.34)/0.39	(0.27,0.51,0.93)/0.61
P_{292}	(0.81,0.76,0.15)/0.86	(0.13,0.12,0.91)/0.71	(0.37,0.52,0.85)/0.35	(0.42,0.91,0.75)/0.95	(0.66,0.49,0.38)/0.21
P_{293}	(0.22,0.25,0.81)/0.44	(0.69,0.26,0.84)/0.88	(0.15,0.10,0.35)/0.91	(0.74,0.50,0.30)/0.33	(0.64,0.61,0.33)/0.77
P_{294}	(0.19,0.99,0.93)/0.22	(0.34,0.82,0.65)/0.79	(0.92,0.85,0.30)/0.70	(0.59,0.80,0.40)/0.57	(0.96,0.20,0.55)/0.41
P_{295}	(0.11,0.46,0.41)/0.49	(0.76,0.89,0.92)/0.78	(0.55,0.93,0.91)/0.95	(0.12,0.88,0.30)/0.93	(0.38,0.71,0.86)/0.78
P_{296}	(0.55,0.23,0.29)/0.71	(0.15,0.37,0.50)/0.81	(0.84,0.70,0.15)/0.86	(0.93,0.76,0.24)/0.92	(0.49,0.85,0.69)/0.62
P_{297}	(0.62,0.40,0.50)/0.23	(0.49,0.32,0.67)/0.69	(0.98,0.16,0.14)/0.40	(0.82,0.25,0.16)/0.73	(0.22,0.56,0.97)/0.81
P_{298}	(0.78,0.19,0.33)/0.28	(0.39,0.98,0.46)/0.84	(0.46,0.29,0.65)/0.43	(0.94,0.22,0.65)/0.40	(0.67,0.96,0.80)/0.32
P_{299}	(0.75,0.73,0.61)/0.86	(0.67,0.19,0.14)/0.84	(0.42,0.35,0.43)/0.31	(0.34,0.88,0.17)/0.29	(0.39,0.56,0.62)/0.61
P_{300}	(0.94,0.92,0.47)/0.22	(0.11,0.77,0.61)/0.33	(0.51,0.51,0.79)/0.30	(0.98,0.59,0.90)/0.81	(0.80,0.50,0.53)/0.93
P_{301}	(0.50,0.77,0.82)/0.32	(0.77,0.30,0.27)/0.73	(0.27,0.52,0.32)/0.55	(0.17,0.79,0.75)/0.45	(0.70,0.26,0.60)/0.31
P_{302}	(0.93,0.82,0.90)/0.77	(0.23,0.66,0.36)/0.44	(0.92,0.96,0.31)/0.34	(0.67,0.91,0.45)/0.57	(0.84,0.67,0.22)/0.28
P_{303}	(0.81,0.93,0.46)/0.54	(0.42,0.78,0.43)/0.45	(0.16,0.90,0.78)/0.53	(0.18,0.63,0.13)/0.53	(0.74,0.55,0.10)/0.24
P_{304}	(0.89,0.18,0.22)/0.75	(0.48,0.87,0.59)/0.94	(0.58,0.70,0.78)/0.50	(0.22,0.82,0.50)/0.50	(0.37,0.78,0.56)/0.64
P_{305}	(0.13,0.72,0.86)/0.65	(0.49,0.11,0.35)/0.22	(0.66,0.35,0.88)/0.89	(0.71,0.11,0.75)/0.93	(0.44,0.14,0.28)/0.70
P_{306}	(0.68,0.54,0.79)/0.40	(0.81,0.19,0.83)/0.83	(0.59,0.17,0.27)/0.35	(0.94,0.40,0.16)/0.56	(0.87,0.10,0.25)/0.46
P_{307}	(0.14,0.74,0.79)/0.59	(0.19,0.86,0.98)/0.24	(0.71,0.42,0.78)/0.46	(0.58,0.35,0.51)/0.74	(0.71,0.72,0.43)/0.24
P_{308}	(0.10,0.65,0.46)/0.35	(0.97,0.69,0.37)/0.87	(0.64,0.45,0.78)/0.77	(0.42,0.12,0.87)/0.50	(0.24,0.26,0.22)/0.56
P_{309}	(0.66,1.00,0.62)/0.39	(0.98,0.52,0.43)/0.85	(0.20,0.45,0.89)/0.54	(0.90,0.98,0.25)/0.42	(0.58,0.94,0.60)/0.44
P_{310}	(0.29,0.14,0.20)/0.95	(0.26,0.82,0.93)/0.65	(0.98,0.96,0.10)/0.36	(0.72,0.25,0.59)/0.20	(0.12,0.37,0.25)/0.41
P_{311}	(0.11,0.19,0.75)/0.33	(0.68,0.76,0.25)/0.25	(0.88,0.86,0.98)/0.71	(0.94,0.61,0.88)/0.84	(0.70,0.79,0.42)/0.24
P_{312}	(0.71,0.88,0.95)/0.41	(0.94,0.34,0.24)/0.93	(0.59,0.73,0.35)/0.68	(0.88,0.30,0.40)/0.59	(0.66,0.28,0.90)/0.26
P_{313}	(0.83,0.35,0.45)/0.78	(0.11,0.52,0.84)/0.87	(0.41,0.96,0.60)/0.74	(0.44,0.40,0.28)/0.23	(0.90,0.10,0.25)/0.33
P_{314}	(0.96,0.81,0.82)/0.53	(0.58,0.95,0.99)/0.93	(0.33,0.71,0.63)/0.66	(0.34,0.81,0.94)/0.49	(0.94,0.88,0.89)/0.52
P_{315}	(0.41,0.11,0.10)/0.35	(0.21,0.45,0.42)/0.81	(0.29,0.13,0.92)/0.94	(0.81,0.52,0.18)/0.83	(0.26,0.32,0.50)/0.70
P_{316}	(0.84,0.99,0.18)/0.49	(0.61,0.51,0.46)/0.39	(0.48,0.76,0.19)/0.91	(0.34,0.80,0.31)/0.23	(0.31,0.61,0.12)/0.49
P_{317}	(0.56,0.45,0.69)/0.85	(0.49,0.73,0.60)/0.84	(0.25,0.58,0.53)/0.30	(0.77,0.15,0.39)/0.75	(0.34,0.72,0.61)/0.41
P_{318}	(0.34,0.54,0.49)/0.23	(0.17,0.81,0.46)/0.77	(0.64,0.43,0.48)/0.59	(0.48,0.19,0.21)/0.58	(0.94,0.99,0.43)/0.74
P_{319}	(0.98,0.19,0.26)/0.57	(0.96,0.99,0.21)/0.81	(0.94,0.48,0.16)/0.23	(0.17,0.43,0.41)/0.73	(0.14,0.63,0.95)/0.58
P_{320}	(0.23,0.26,0.79)/0.91	(0.38,0.30,0.19)/0.26	(0.78,0.19,0.63)/0.46	(0.13,0.66,0.71)/0.26	(0.23,0.95,0.74)/0.89
P_{321}	(0.20,0.21,0.16)/0.47	(0.89,0.40,1.00)/0.77	(0.29,0.14,0.56)/0.35	(0.58,0.36,0.50)/0.85	(0.79,0.11,0.96)/0.51
P_{322}	(0.70,0.70,0.24)/0.31	(0.44,0.91,0.67)/0.90	(0.28,0.19,0.12)/0.35	(0.67,0.75,0.48)/0.71	(0.30,0.36,0.17)/0.57
P_{323}	(0.46,0.23,0.61)/0.68	(0.57,0.76,0.92)/0.77	(0.99,0.80,0.16)/0.46	(0.94,0.32,0.22)/0.46	(0.77,0.25,0.51)/0.30
P_{324}	(0.87,0.35,0.49)/0.93	(0.75,0.20,0.28)/0.36	(0.44,0.35,0.70)/0.61	(0.81,0.26,0.73)/0.49	(0.60,0.56,0.14)/0.51
P_{325}	(0.92,0.28,0.27)/0.71	(0.39,0.76,0.72)/0.65	(0.56,0.18,0.74)/0.34	(0.48,0.77,0.55)/0.26	(0.37,0.68,0.26)/0.62
P_{326}	(0.48,0.21,0.35)/0.42	(0.36,0.83,0.70)/0.93	(0.84,0.10,0.43)/0.72	(0.66,0.48,0.33)/0.31	(0.37,0.60,0.29)/0.56
P_{327}	(0.66,0.45,0.69)/0.53	(0.82,0.33,0.63)/0.76	(0.94,0.62,0.53)/0.63	(0.53,0.27,0.70)/0.64	(0.26,0.49,0.25)/0.48
P_{328}	(0.37,0.85,0.59)/0.30	(0.17,0.99,0.83)/0.24	(0.89,0.28,0.77)/0.33	(0.84,0.80,0.27)/0.62	(0.10,0.50,0.92)/0.90
P_{329}	(0.23,0.24,0.67)/0.82	(0.76,0.19,0.43)/0.64	(0.88,0.93,0.51)/0.60	(0.78,0.16,0.86)/0.65	(0.94,0.54,0.29)/0.87
P_{330}	(0.45,0.88,0.42)/0.54	(0.48,0.86,0.78)/0.84	(0.11,0.57,0.78)/0.61	(0.40,0.58,0.31)/0.28	(0.95,0.38,0.88)/0.73

Table 6. *Cont.*

Patient	Temperature	Cough	Throat Pain	Headache	Body Pain
P_{331}	(0.58,0.75,0.38)/0.71	(0.55,0.83,0.71)/0.79	(0.62,0.42,0.55)/0.79	(0.40,0.51,0.21)/0.69	(0.94,0.27,0.69)/0.80
P_{332}	(0.87,0.48,0.51)/0.65	(0.78,0.30,0.70)/0.35	(0.37,0.46,0.32)/0.38	(0.78,0.80,0.71)/0.20	(0.23,0.91,0.88)/0.94
P_{333}	(0.42,0.68,0.94)/0.27	(0.64,0.47,0.96)/0.35	(0.94,1.00,0.22)/0.53	(0.35,0.31,0.31)/0.49	(0.99,0.78,0.19)/0.84
P_{334}	(0.27,0.22,0.65)/0.21	(0.10,0.88,0.56)/0.20	(0.97,0.21,0.68)/0.67	(0.81,0.43,0.27)/0.48	(0.66,0.62,0.87)/0.37
P_{335}	(0.63,0.10,0.89)/0.29	(0.18,0.53,0.47)/0.88	(0.39,0.20,0.17)/0.87	(0.50,0.82,0.90)/0.33	(0.82,0.42,0.12)/0.90
P_{336}	(0.99,0.25,0.27)/0.63	(0.64,0.28,0.76)/0.26	(0.33,0.73,0.63)/0.59	(0.63,0.54,0.37)/0.58	(0.73,0.77,0.73)/0.37
P_{337}	(1.00,0.72,0.46)/0.76	(0.82,0.48,0.55)/0.47	(0.53,0.69,0.64)/0.71	(0.53,0.95,0.32)/0.48	(0.67,1.00,0.17)/0.79
P_{338}	(0.61,0.77,0.95)/0.24	(0.64,0.96,0.72)/0.46	(0.33,0.51,0.56)/0.28	(0.82,0.23,0.38)/0.34	(0.57,0.51,0.95)/0.63
P_{339}	(0.32,0.75,0.10)/0.61	(0.56,0.42,0.87)/0.78	(0.83,0.80,0.17)/0.76	(0.57,0.82,0.61)/0.31	(0.74,0.26,0.62)/0.39
P_{340}	(0.84,0.81,0.12)/0.88	(0.11,0.60,0.98)/0.82	(0.39,0.97,0.86)/0.46	(0.60,0.49,0.86)/0.61	(0.12,0.20,0.21)/0.20
P_{341}	(0.80,0.15,0.41)/0.81	(0.13,0.37,0.73)/0.36	(0.84,0.44,0.25)/0.43	(0.43,0.97,0.24)/0.24	(0.48,0.10,0.17)/0.39
P_{342}	(0.54,0.92,0.90)/0.66	(0.27,0.47,0.87)/0.93	(0.90,0.52,0.95)/0.25	(0.82,0.92,0.51)/0.36	(0.49,0.69,0.85)/0.95
P_{343}	(0.64,0.91,0.58)/0.67	(0.30,0.21,0.61)/0.41	(0.27,0.68,0.47)/0.46	(0.29,0.53,0.86)/0.38	(0.30,0.85,0.73)/0.53
P_{344}	(0.98,0.13,0.17)/0.31	(0.61,0.87,0.15)/0.95	(0.93,0.41,0.90)/0.47	(0.63,0.26,0.35)/0.80	(0.52,0.65,0.82)/0.61
P_{345}	(0.10,0.10,0.31)/0.38	(0.91,0.57,0.79)/0.23	(0.85,0.57,0.90)/0.94	(0.25,0.74,0.78)/0.94	(0.56,0.54,0.95)/0.86
P_{346}	(0.73,0.21,0.78)/0.87	(0.18,0.41,0.56)/0.45	(0.74,0.48,0.16)/0.60	(0.75,0.98,0.70)/0.93	(0.54,0.75,0.64)/0.31
P_{347}	(0.56,0.73,0.46)/0.30	(0.24,0.73,0.10)/0.43	(0.73,0.63,0.43)/0.54	(0.34,0.95,0.24)/0.75	(0.67,0.70,0.10)/0.48
P_{348}	(0.86,0.55,0.52)/0.61	(0.55,0.51,0.32)/0.31	(0.94,0.77,0.87)/0.70	(0.33,0.24,0.61)/0.77	(0.31,0.42,0.40)/0.41
P_{349}	(0.60,0.90,0.23)/0.40	(0.79,0.39,0.70)/0.84	(0.84,0.25,0.89)/0.53	(0.47,0.51,0.44)/0.91	(0.42,0.99,0.44)/0.41
P_{350}	(0.12,0.25,0.18)/0.73	(0.14,0.30,0.49)/0.30	(0.77,0.70,0.81)/0.70	(0.26,0.43,0.10)/0.72	(0.29,0.97,0.37)/0.48
P_{351}	(0.99,0.94,0.25)/0.39	(0.35,0.30,0.30)/0.25	(0.23,0.90,0.67)/0.22	(0.99,0.95,0.75)/0.79	(0.50,0.10,0.19)/0.32
P_{352}	(0.16,0.47,0.73)/0.68	(0.35,0.59,0.73)/0.30	(0.61,0.12,0.28)/0.57	(0.30,0.47,0.43)/0.48	(0.76,0.25,0.46)/0.59
P_{353}	(0.12,0.18,0.91)/0.37	(0.61,0.48,0.53)/0.82	(0.97,0.59,0.99)/0.51	(0.13,0.57,0.14)/0.47	(0.66,0.12,0.72)/0.85
P_{354}	(0.62,0.31,0.25)/0.92	(0.94,0.85,0.73)/0.24	(0.70,0.66,0.78)/0.37	(0.19,0.83,0.64)/0.28	(0.32,0.66,0.84)/0.43
P_{355}	(0.33,0.93,0.11)/0.40	(0.56,0.41,0.53)/0.50	(0.66,0.92,0.87)/0.21	(0.85,0.21,0.84)/0.59	(0.34,0.21,0.62)/0.77
P_{356}	(0.26,0.10,0.64)/0.76	(0.80,0.86,0.70)/0.72	(0.71,0.11,0.71)/0.88	(0.90,0.86,0.66)/0.30	(0.99,0.69,0.48)/0.20
P_{357}	(0.72,0.12,0.26)/0.40	(0.82,0.40,0.25)/0.55	(0.97,0.33,0.43)/0.65	(0.11,0.97,0.22)/0.85	(0.16,0.60,0.14)/0.91
P_{358}	(0.82,0.58,0.26)/0.70	(0.52,0.43,0.91)/0.76	(0.44,0.26,0.81)/0.53	(0.64,0.97,0.37)/0.92	(0.38,0.77,0.77)/0.90
P_{359}	(0.64,0.90,0.24)/0.74	(0.65,0.57,0.87)/0.85	(0.88,0.80,1.00)/0.59	(0.57,0.57,0.80)/0.87	(0.37,0.94,0.96)/0.21
P_{360}	(0.34,0.76,0.32)/0.33	(0.45,0.58,0.14)/0.57	(0.59,0.43,0.34)/0.84	(0.75,0.19,0.50)/0.69	(0.68,0.98,0.19)/0.74
P_{361}	(0.78,0.20,0.77)/0.70	(0.27,0.93,0.81)/0.76	(0.66,0.72,0.50)/0.53	(0.18,0.32,0.33)/0.73	(0.35,0.83,0.81)/0.86
P_{362}	(0.27,0.20,0.24)/0.71	(0.67,0.65,0.84)/0.88	(0.43,0.33,0.98)/0.73	(0.22,0.66,0.83)/0.91	(0.35,0.56,0.45)/0.72
P_{363}	(0.16,0.11,0.88)/0.50	(0.21,0.97,0.92)/0.76	(0.92,0.51,0.78)/0.71	(0.54,0.51,0.69)/0.81	(0.23,0.85,0.13)/0.60
P_{364}	(0.96,0.59,0.71)/0.26	(0.52,0.66,0.47)/0.95	(0.77,0.44,0.64)/0.49	(0.16,0.97,0.49)/0.25	(0.96,0.86,0.40)/0.84
P_{365}	(0.45,0.35,0.19)/0.61	(0.55,0.29,0.44)/0.91	(0.34,0.27,0.19)/0.73	(0.21,0.18,0.90)/0.58	(0.54,0.85,0.97)/0.30
P_{366}	(0.98,0.39,0.41)/0.91	(0.96,0.22,0.92)/0.42	(0.74,0.23,0.72)/0.82	(0.56,0.35,0.89)/0.34	(0.25,1.00,0.24)/0.74
P_{367}	(0.37,0.66,0.85)/0.45	(0.85,0.24,0.21)/0.81	(0.64,0.41,0.81)/0.31	(0.66,0.57,0.66)/0.80	(0.69,0.55,0.55)/0.54
P_{368}	(0.60,0.27,0.22)/0.79	(0.85,0.26,0.58)/0.41	(0.87,0.46,0.61)/0.34	(0.77,0.26,0.79)/0.28	(0.48,0.75,0.19)/0.80
P_{369}	(0.71,0.90,1.00)/0.55	(0.24,0.39,0.70)/0.95	(0.32,0.12,0.67)/0.37	(0.94,0.19,0.59)/0.76	(0.85,1.00,0.51)/0.42
P_{370}	(0.60,0.58,0.18)/0.44	(0.92,0.12,0.42)/0.83	(0.46,0.70,0.81)/0.37	(0.81,0.95,0.12)/0.27	(0.74,0.27,0.87)/0.58
P_{371}	(0.22,0.50,0.23)/0.68	(0.79,0.46,0.55)/0.31	(0.98,0.52,0.89)/0.65	(0.61,0.93,0.16)/0.87	(0.96,0.71,0.20)/0.67
P_{372}	(0.88,0.54,0.91)/0.73	(0.22,0.18,0.26)/0.67	(0.66,0.58,0.99)/0.54	(0.70,0.36,0.52)/0.54	(0.56,0.88,0.11)/0.36
P_{373}	(0.17,0.43,0.46)/0.53	(0.48,0.84,0.65)/0.36	(0.59,0.13,0.18)/0.31	(0.85,0.81,0.65)/0.35	(0.39,0.66,0.46)/0.41
P_{374}	(0.18,0.53,0.61)/0.64	(0.52,0.96,0.84)/0.27	(0.45,0.69,0.22)/0.68	(0.34,0.30,0.79)/0.86	(0.29,0.23,0.76)/0.66
P_{375}	(0.14,0.98,0.41)/0.80	(0.49,0.35,0.45)/0.92	(0.76,0.10,0.79)/0.75	(0.72,0.81,0.15)/0.67	(0.81,0.56,0.51)/0.60
P_{376}	(0.73,0.93,0.54)/0.30	(0.21,0.39,0.53)/0.82	(0.50,0.66,0.83)/0.64	(0.72,0.19,0.38)/0.39	(0.89,0.95,0.48)/0.91
P_{377}	(0.13,0.30,0.49)/0.33	(0.33,0.22,0.38)/0.70	(0.70,0.93,0.43)/0.95	(0.27,0.23,0.34)/0.94	(0.88,0.97,0.47)/0.23
P_{378}	(0.46,0.38,0.62)/0.28	(0.79,0.94,0.19)/0.69	(0.64,0.30,0.91)/0.69	(0.81,0.54,0.20)/0.22	(0.38,0.81,0.26)/0.48
P_{379}	(0.31,0.41,0.81)/0.32	(0.53,0.32,0.53)/0.81	(0.58,0.29,0.19)/0.35	(0.32,0.70,0.38)/0.88	(0.56,0.88,0.97)/0.57
P_{380}	(0.56,0.13,0.50)/0.24	(0.71,0.95,0.15)/0.86	(0.34,0.67,0.26)/0.47	(0.96,0.69,0.13)/0.91	(0.86,0.14,0.47)/0.75
P_{381}	(0.79,0.25,0.21)/0.56	(0.98,0.95,0.45)/0.80	(0.18,0.63,0.79)/0.67	(0.89,0.50,0.28)/0.59	(0.91,0.96,0.15)/0.77
P_{382}	(0.89,0.48,0.91)/0.74	(0.98,0.75,0.48)/0.44	(1.00,0.85,0.23)/0.56	(0.99,0.92,0.87)/0.67	(1.00,0.67,0.66)/0.72
P_{383}	(0.13,0.31,0.27)/0.67	(0.15,0.60,0.63)/0.70	(0.52,0.76,0.41)/0.68	(0.85,0.34,0.95)/0.91	(0.88,0.54,0.54)/0.86
P_{384}	(0.90,0.15,0.97)/0.27	(0.37,0.86,0.33)/0.32	(0.60,0.11,0.15)/0.28	(0.81,0.87,0.92)/0.52	(0.36,0.94,0.83)/0.89
P_{385}	(0.73,0.39,0.13)/0.46	(0.95,0.94,0.97)/0.34	(0.28,0.21,0.45)/0.85	(0.88,0.79,0.91)/0.55	(0.96,0.80,0.35)/0.70
P_{386}	(0.12,1.00,0.21)/0.88	(0.52,0.83,0.21)/0.72	(0.80,0.14,0.63)/0.30	(0.44,0.53,0.53)/0.56	(0.27,0.75,0.19)/0.44
P_{387}	(0.59,0.33,0.15)/0.54	(0.31,0.87,0.66)/0.81	(0.34,0.23,0.60)/0.66	(0.97,0.39,0.49)/0.27	(0.79,0.95,0.31)/0.77
P_{388}	(0.65,0.91,0.13)/0.79	(0.98,0.17,0.86)/0.67	(0.79,0.61,0.43)/0.25	(0.49,0.31,0.62)/0.51	(0.69,0.92,0.89)/0.28
P_{389}	(0.37,0.77,0.73)/0.82	(0.40,0.23,0.82)/0.63	(0.11,0.21,0.25)/0.56	(0.18,0.78,0.25)/0.41	(0.46,0.59,0.73)/0.63
P_{390}	(0.73,0.96,0.76)/0.91	(0.13,0.73,0.73)/0.64	(0.23,0.93,0.19)/0.77	(0.68,0.11,0.71)/0.47	(0.58,0.36,0.29)/0.65
P_{391}	(0.62,0.99,0.51)/0.58	(0.18,0.81,0.68)/0.37	(0.37,0.37,0.28)/0.44	(0.70,0.82,0.15)/0.46	(0.83,0.82,0.64)/0.62
P_{392}	(0.75,0.82,0.97)/0.60	(0.57,0.81,0.70)/0.67	(0.37,0.74,0.36)/0.68	(0.94,0.56,0.47)/0.29	(0.69,0.42,0.82)/0.22
P_{393}	(0.53,0.10,0.93)/0.26	(0.16,0.82,0.71)/0.63	(0.41,0.71,0.46)/0.40	(0.50,0.76,0.37)/0.63	(0.62,0.91,0.61)/0.54
P_{394}	(0.27,0.52,0.19)/0.38	(0.42,0.31,0.31)/0.46	(0.24,0.73,0.58)/0.47	(0.99,0.95,1.00)/0.87	(0.99,0.65,0.52)/0.58
P_{395}	(0.39,0.16,0.86)/0.73	(0.56,0.17,0.50)/0.27	(0.32,0.99,0.85)/0.59	(0.54,0.12,0.68)/0.81	(0.83,0.30,0.51)/0.75
P_{396}	(0.80,0.97,0.76)/0.79	(0.40,0.10,0.74)/0.92	(0.78,0.15,0.50)/0.68	(0.78,0.60,0.26)/0.40	(0.93,0.17,0.45)/0.68
P_{397}	(0.68,0.10,0.37)/0.64	(0.91,0.34,0.35)/0.46	(0.64,0.77,0.77)/0.90	(0.32,0.44,0.81)/0.64	(0.13,0.19,0.81)/0.20
P_{398}	(0.34,0.40,0.79)/0.81	(0.35,0.71,0.99)/0.63	(0.52,0.58,0.73)/0.25	(0.79,0.81,0.55)/0.68	(0.62,0.97,0.12)/0.46
P_{399}	(0.90,0.97,0.11)/0.52	(0.68,0.92,0.49)/0.82	(0.13,0.97,0.51)/0.22	(0.75,0.38,0.47)/0.74	(1.00,0.24,0.77)/0.31
P_{400}	(0.17,0.48,0.58)/0.58	(0.69,0.68,0.57)/0.43	(0.91,0.20,0.37)/0.73	(0.62,0.23,0.14)/0.85	(0.73,0.42,0.95)/0.45

Table 6. *Cont.*

Patient	Temperature	Cough	Throat Pain	Headache	Body Pain
P_{401}	(0.14,0.15,0.77)/0.27	(0.23,0.16,0.13)/0.83	(0.44,0.68,0.55)/0.83	(0.39,0.46,0.23)/0.27	(0.48,0.55,0.49)/0.23
P_{402}	(0.52,0.69,0.95)/0.36	(0.23,0.72,0.75)/0.30	(0.25,0.90,0.67)/0.56	(0.50,0.79,0.45)/0.82	(0.71,0.15,0.71)/0.70
P_{403}	(0.21,0.24,0.60)/0.43	(0.72,0.14,0.94)/0.88	(0.56,0.33,0.77)/0.70	(0.66,0.39,0.42)/0.55	(0.48,0.67,0.42)/0.91
P_{404}	(0.59,0.76,0.42)/0.49	(0.14,0.32,0.89)/0.49	(0.37,0.61,0.97)/0.32	(0.38,0.17,0.84)/0.65	(0.72,0.92,0.52)/0.95
P_{405}	(0.60,0.58,0.16)/0.84	(0.59,0.88,0.88)/0.81	(0.42,0.47,0.75)/0.47	(0.87,0.64,0.74)/0.50	(0.62,0.61,0.55)/0.44
P_{406}	(0.43,0.31,0.89)/0.93	(0.15,0.38,0.21)/0.52	(0.55,1.00,0.45)/0.42	(0.77,0.42,0.38)/0.53	(0.32,0.15,0.40)/0.59
P_{407}	(0.42,0.11,0.72)/0.88	(0.52,0.18,0.58)/0.30	(0.10,0.22,0.92)/0.61	(0.58,0.94,0.73)/0.93	(0.98,0.67,0.43)/0.61
P_{408}	(0.10,0.88,0.46)/0.36	(0.58,0.37,0.74)/0.38	(0.83,0.77,0.83)/0.53	(0.72,0.70,0.26)/0.34	(0.34,0.80,0.49)/0.79
P_{409}	(0.85,0.88,0.89)/0.74	(0.32,0.48,0.71)/0.36	(0.69,0.25,0.34)/0.60	(0.63,0.77,0.21)/0.68	(0.11,0.20,0.66)/0.42
P_{410}	(0.26,0.62,0.70)/0.86	(0.93,0.32,0.74)/0.78	(0.29,0.36,0.89)/0.34	(0.49,0.33,0.32)/0.40	(0.92,0.43,0.97)/0.74
P_{411}	(0.55,0.35,0.37)/0.56	(0.60,0.54,0.64)/0.91	(0.30,0.25,0.78)/0.54	(0.86,0.87,0.81)/0.22	(0.78,0.55,0.67)/0.24
P_{412}	(0.42,0.91,0.16)/0.31	(0.87,0.24,0.62)/0.57	(0.30,0.69,0.86)/0.61	(0.50,0.93,0.22)/0.79	(0.62,0.66,0.27)/0.50
P_{413}	(0.61,0.62,0.38)/0.77	(0.88,0.70,0.47)/0.85	(0.76,0.13,0.34)/0.48	(0.99,0.57,0.57)/0.55	(0.32,0.33,0.97)/0.53
P_{414}	(0.23,0.89,0.15)/0.63	(0.53,0.97,0.18)/0.59	(0.10,0.36,0.44)/0.31	(0.26,0.88,0.48)/0.55	(0.47,0.37,0.66)/0.73
P_{415}	(0.49,0.80,0.65)/0.43	(0.45,0.41,0.49)/0.71	(0.28,0.27,0.50)/0.28	(0.86,0.64,0.65)/0.45	(0.85,0.74,0.18)/0.44
P_{416}	(0.75,0.90,0.84)/0.95	(0.62,0.50,0.77)/0.89	(0.93,0.77,0.64)/0.24	(0.18,0.77,0.59)/0.80	(0.92,0.93,0.33)/0.76
P_{417}	(1.00,0.37,1.00)/0.71	(0.35,0.39,0.88)/0.94	(0.84,0.73,0.30)/0.33	(0.16,0.24,0.31)/0.35	(0.71,0.57,0.85)/0.39
P_{418}	(0.57,0.22,0.11)/0.87	(0.42,0.41,0.34)/0.52	(0.63,0.56,0.45)/0.38	(0.20,0.65,0.43)/0.38	(0.29,0.38,0.93)/0.68
P_{419}	(0.92,0.66,0.67)/0.43	(0.10,0.40,0.18)/0.23	(0.64,0.60,0.74)/0.87	(0.70,0.14,0.62)/0.30	(0.65,0.22,0.91)/0.21
P_{420}	(0.92,0.82,0.78)/0.24	(0.54,0.14,0.71)/0.27	(0.98,0.38,0.39)/0.54	(0.76,0.50,0.49)/0.36	(0.39,0.69,0.19)/0.65
P_{421}	(0.26,0.12,0.46)/0.44	(0.89,0.23,0.69)/0.52	(0.69,0.51,0.23)/0.24	(0.65,0.22,0.37)/0.50	(0.83,0.45,0.84)/0.48
P_{422}	(0.18,0.96,0.63)/0.81	(0.64,0.52,0.51)/0.55	(0.94,0.66,0.26)/0.51	(0.42,0.32,0.22)/0.82	(0.85,0.58,0.48)/0.50
P_{423}	(0.67,0.50,0.85)/0.51	(0.13,0.60,0.35)/0.87	(0.82,0.38,0.88)/0.80	(0.51,0.23,0.65)/0.37	(0.26,0.21,0.75)/0.69
P_{424}	(0.76,0.96,0.25)/0.82	(0.32,0.23,0.26)/0.79	(0.23,0.56,0.60)/0.94	(0.73,0.23,0.97)/0.77	(0.37,0.28,0.67)/0.54
P_{425}	(0.88,0.77,0.48)/0.24	(0.17,0.73,0.36)/0.60	(0.71,0.64,0.49)/0.76	(0.61,0.77,0.51)/0.39	(0.17,0.54,0.67)/0.57
P_{426}	(0.97,0.76,0.43)/0.73	(0.80,0.91,0.40)/0.43	(0.39,0.58,0.20)/0.71	(0.59,0.25,0.76)/0.52	(0.61,0.25,0.58)/0.33
P_{427}	(0.63,0.86,0.70)/0.31	(0.60,0.32,0.79)/0.36	(0.42,0.25,0.22)/0.53	(0.80,0.15,0.17)/0.55	(0.51,0.28,0.41)/0.28
P_{428}	(0.69,0.82,0.52)/0.51	(0.37,0.49,1.00)/0.58	(0.49,0.19,0.29)/0.32	(0.30,0.51,1.00)/0.38	(0.71,0.56,0.16)/0.61
P_{429}	(0.12,0.75,0.12)/0.57	(0.36,0.37,0.95)/0.84	(0.79,0.61,0.48)/0.69	(0.48,0.86,0.93)/0.83	(0.99,0.29,0.66)/0.43
P_{430}	(0.73,0.74,0.81)/0.76	(0.77,0.47,0.58)/0.56	(0.88,0.79,0.41)/0.50	(0.39,0.19,0.80)/0.29	(0.41,0.14,0.63)/0.63
P_{431}	(0.37,0.64,0.37)/0.74	(0.89,0.91,1.00)/0.81	(0.83,0.98,0.61)/0.78	(0.82,0.19,1.00)/0.88	(0.51,0.14,0.70)/0.79
P_{432}	(0.43,0.19,0.16)/0.30	(0.36,0.34,0.52)/0.52	(0.30,0.98,0.75)/0.26	(0.93,0.78,0.66)/0.38	(0.81,0.44,0.27)/0.91
P_{433}	(0.54,0.28,0.75)/0.22	(0.63,0.58,0.28)/0.73	(0.59,0.74,0.72)/0.81	(0.82,0.15,0.10)/0.81	(0.80,0.18,0.24)/0.30
P_{434}	(0.58,0.39,0.63)/0.39	(0.20,0.81,0.86)/0.30	(0.88,0.26,0.51)/0.52	(0.19,0.26,0.22)/0.54	(0.79,0.40,0.55)/0.75
P_{435}	(0.22,0.49,0.61)/0.28	(0.36,0.77,0.74)/0.70	(0.28,0.14,0.58)/0.93	(0.97,0.72,0.38)/0.89	(0.38,0.48,0.73)/0.68
P_{436}	(0.23,0.91,0.44)/0.86	(0.38,0.54,0.94)/0.84	(0.38,0.68,0.49)/0.35	(0.41,0.95,0.28)/0.87	(0.50,0.89,0.30)/0.56
P_{437}	(0.70,0.50,0.72)/0.31	(0.37,0.87,0.77)/0.83	(0.17,0.55,0.83)/0.76	(0.23,0.80,0.93)/0.78	(0.45,0.21,0.47)/0.86
P_{438}	(0.89,0.49,0.26)/0.57	(0.81,0.38,0.47)/0.66	(0.85,0.17,0.14)/0.64	(0.37,0.34,0.78)/0.35	(0.54,0.69,0.84)/0.63
P_{439}	(0.41,0.25,0.54)/0.23	(0.87,0.95,0.66)/0.88	(0.55,0.25,0.95)/0.39	(0.41,0.31,0.83)/0.55	(0.55,0.97,0.11)/0.48
P_{440}	(0.95,0.90,0.18)/0.82	(0.34,0.51,0.74)/0.37	(0.72,0.92,0.38)/0.52	(0.81,0.82,0.80)/0.39	(0.57,0.12,0.72)/0.56
P_{441}	(0.20,0.54,0.85)/0.31	(0.41,0.14,0.71)/0.21	(0.10,0.92,0.54)/0.36	(0.42,0.64,0.88)/0.82	(0.50,0.58,0.52)/0.82
P_{442}	(0.16,0.47,0.73)/0.57	(0.48,0.50,0.61)/0.85	(0.17,0.13,1.00)/0.37	(0.20,0.24,0.76)/0.45	(0.19,0.98,0.99)/0.72
P_{443}	(0.37,0.56,0.92)/0.85	(0.85,0.41,0.70)/0.57	(0.66,0.65,0.43)/0.90	(0.62,0.90,0.54)/0.26	(0.36,0.10,0.85)/0.82
P_{444}	(0.69,0.69,0.86)/0.22	(0.19,0.64,0.91)/0.82	(0.84,0.54,0.24)/0.79	(0.35,0.81,0.11)/0.61	(0.54,0.81,0.41)/0.27
P_{445}	(0.67,0.67,0.95)/0.90	(0.33,0.78,0.29)/0.35	(0.71,0.13,0.96)/0.41	(0.58,0.14,0.95)/0.57	(0.77,0.79,0.28)/0.33
P_{446}	(0.23,0.18,0.78)/0.23	(0.99,0.62,0.47)/0.23	(0.18,0.16,0.72)/0.42	(0.98,0.62,0.38)/0.43	(0.78,0.15,0.26)/0.36
P_{447}	(0.64,0.77,0.24)/0.77	(0.12,0.22,0.64)/0.38	(0.13,0.94,0.20)/0.66	(0.48,0.57,0.60)/0.33	(0.88,0.70,0.40)/0.35
P_{448}	(0.14,0.24,0.89)/0.26	(0.18,1.00,0.57)/0.64	(0.96,0.67,0.26)/0.65	(0.95,0.55,0.79)/0.45	(0.39,0.57,0.31)/0.75
P_{449}	(0.58,0.47,0.49)/0.72	(0.83,0.79,0.73)/0.52	(0.82,0.32,0.18)/0.74	(0.86,0.57,0.50)/0.21	(0.75,0.89,0.97)/0.22
P_{450}	(0.61,0.85,0.32)/0.26	(0.74,0.55,0.83)/0.48	(0.79,0.93,0.83)/0.34	(0.60,0.86,0.89)/0.67	(0.84,0.47,0.50)/0.94
P_{451}	(0.65,0.85,0.94)/0.34	(0.77,0.40,0.29)/0.55	(0.92,0.84,0.19)/0.24	(0.57,0.52,0.99)/0.35	(0.53,0.54,0.52)/0.73
P_{452}	(0.82,0.34,0.87)/0.58	(0.20,0.82,0.84)/0.84	(0.51,0.97,0.97)/0.87	(0.22,0.30,0.22)/0.79	(0.71,0.14,0.76)/0.63
P_{453}	(0.99,0.85,0.73)/0.44	(0.56,0.68,0.90)/0.85	(0.28,0.92,0.24)/0.37	(0.28,0.25,0.61)/0.32	(0.62,0.78,0.88)/0.51
P_{454}	(0.65,0.42,0.14)/0.21	(0.24,0.28,0.33)/0.54	(0.67,0.87,0.24)/0.72	(0.46,0.84,0.25)/0.68	(0.10,0.28,0.12)/0.86
P_{455}	(0.71,0.73,0.71)/0.66	(0.94,0.18,0.10)/0.74	(0.46,0.79,0.78)/0.92	(0.51,0.81,0.71)/0.76	(0.15,0.50,0.71)/0.55
P_{456}	(0.15,0.64,0.66)/0.73	(0.79,0.74,0.81)/0.72	(0.53,0.91,0.27)/0.26	(0.30,0.95,0.12)/0.32	(0.50,0.14,0.12)/0.71
P_{457}	(0.99,0.34,0.41)/0.92	(0.66,0.47,0.97)/0.41	(0.36,0.27,0.50)/0.43	(0.61,0.28,0.36)/0.80	(0.18,0.56,0.38)/0.88
P_{458}	(0.57,0.21,0.72)/0.66	(0.36,0.96,0.77)/0.61	(0.25,0.22,0.43)/0.91	(0.46,0.26,0.15)/0.42	(0.96,0.33,0.33)/0.89
P_{459}	(0.16,0.11,0.53)/0.90	(0.40,0.33,0.76)/0.93	(0.51,0.67,0.80)/0.62	(1.00,0.32,0.63)/0.61	(0.13,0.49,0.20)/0.69
P_{460}	(0.15,0.15,0.47)/0.37	(0.13,0.49,0.32)/0.30	(0.47,0.73,0.60)/0.71	(0.64,0.46,0.87)/0.37	(0.12,0.85,0.44)/0.27
P_{461}	(0.15,0.63,0.43)/0.59	(0.62,0.98,0.60)/0.44	(0.90,0.44,0.45)/0.30	(0.22,0.34,0.85)/0.90	(0.89,0.65,0.89)/0.52
P_{462}	(0.66,0.85,0.33)/0.61	(0.18,0.38,0.10)/0.66	(0.88,0.89,0.42)/0.61	(0.81,0.26,0.40)/0.23	(0.69,0.39,0.50)/0.53
P_{463}	(0.22,0.12,0.29)/0.32	(0.82,0.48,0.56)/0.29	(0.22,0.56,0.52)/0.23	(0.96,0.82,0.64)/0.52	(0.42,0.62,0.58)/0.62
P_{464}	(0.39,0.38,0.77)/0.33	(0.20,0.78,0.57)/0.36	(0.96,0.79,0.88)/0.64	(0.37,0.78,0.46)/0.68	(0.74,0.21,0.43)/0.91
P_{465}	(0.74,0.29,0.98)/0.90	(0.12,0.74,0.33)/0.66	(0.75,0.98,0.75)/0.34	(0.96,0.85,0.82)/0.88	(0.94,0.58,0.51)/0.70
P_{466}	(0.60,0.58,0.41)/0.83	(0.42,0.94,0.33)/0.39	(0.86,0.54,0.11)/0.57	(0.30,0.99,0.58)/0.24	(0.58,0.93,0.98)/0.35
P_{467}	(0.68,0.55,0.65)/0.53	(0.68,0.56,0.99)/0.56	(0.74,0.61,0.90)/0.91	(0.35,0.22,0.90)/0.79	(0.57,0.18,0.92)/0.52
P_{468}	(0.90,0.60,0.69)/0.52	(0.54,0.41,0.51)/0.76	(0.24,0.26,0.82)/0.82	(0.60,0.96,0.85)/0.46	(0.15,0.54,0.13)/0.66
P_{469}	(0.78,0.41,0.45)/0.61	(0.26,0.45,0.61)/0.73	(0.55,0.59,0.71)/0.40	(0.21,0.26,0.87)/0.51	(0.50,0.68,0.50)/0.94
P_{470}	(0.69,0.50,0.61)/0.33	(0.25,0.95,0.37)/0.64	(0.89,0.52,0.50)/0.52	(0.56,0.33,0.72)/0.48	(0.78,0.23,0.36)/0.44

Table 6. *Cont.*

Patient	Temperature	Cough	Throat Pain	Headache	Body Pain
P_{471}	(0.72,0.43,0.97)/0.89	(0.83,0.34,0.23)/0.48	(0.55,0.91,0.36)/0.75	(0.44,0.76,0.47)/0.49	(0.63,0.92,0.83)/0.60
P_{472}	(0.79,0.44,0.76)/0.59	(0.30,0.78,0.56)/0.26	(0.58,0.53,0.57)/0.54	(0.93,0.70,0.10)/0.36	(0.85,0.14,0.48)/0.85
P_{473}	(0.86,0.33,0.35)/0.40	(0.34,0.47,0.74)/0.43	(0.23,0.11,0.76)/0.22	(0.47,0.94,0.57)/0.61	(0.92,0.66,0.82)/0.93
P_{474}	(0.20,0.76,0.93)/0.65	(0.18,0.64,0.26)/0.52	(0.62,0.38,0.66)/0.29	(0.53,0.68,0.49)/0.81	(0.27,0.18,0.77)/0.27
P_{475}	(0.43,0.95,0.91)/0.73	(0.65,0.65,0.32)/0.74	(0.87,0.88,0.28)/0.37	(0.74,0.30,0.85)/0.87	(0.82,0.12,0.70)/0.53
P_{476}	(0.38,0.12,0.69)/0.41	(0.94,0.20,0.22)/0.39	(0.30,0.90,0.39)/0.32	(0.12,0.62,0.23)/0.42	(0.72,0.43,0.35)/0.46
P_{477}	(0.69,0.83,0.89)/0.84	(0.35,0.70,0.21)/0.90	(0.44,0.75,0.47)/0.70	(0.36,0.80,0.44)/0.59	(0.11,0.86,0.97)/0.73
P_{478}	(0.14,0.27,0.97)/0.71	(0.31,0.81,0.69)/0.85	(0.42,1.00,0.61)/0.24	(0.26,0.63,0.56)/0.64	(0.42,0.70,0.67)/0.47
P_{479}	(0.92,0.93,0.40)/0.95	(0.13,0.30,0.36)/0.56	(0.66,0.23,0.94)/0.35	(0.85,0.82,0.97)/0.82	(0.33,0.38,0.26)/0.95
P_{480}	(0.49,0.50,0.68)/0.54	(0.19,0.45,0.40)/0.75	(0.18,0.33,0.74)/0.71	(0.31,0.42,0.31)/0.86	(0.65,0.29,0.81)/0.53
P_{481}	(0.83,0.74,0.58)/0.66	(0.48,0.40,0.76)/0.45	(0.72,0.66,0.16)/0.24	(0.64,0.77,0.31)/0.72	(0.19,0.41,0.29)/0.74
P_{482}	(0.92,0.82,0.84)/0.73	(0.31,0.31,0.45)/0.44	(0.31,0.96,0.40)/0.43	(0.56,0.91,0.67)/0.86	(0.61,0.54,0.26)/0.70
P_{483}	(0.39,0.22,0.42)/0.26	(0.97,0.78,0.77)/0.28	(0.46,0.53,0.71)/0.89	(0.17,0.48,0.75)/0.66	(0.76,0.77,0.87)/0.58
P_{484}	(0.63,0.99,0.54)/0.49	(0.48,0.97,0.40)/0.88	(0.84,0.39,0.33)/0.49	(0.44,0.22,0.64)/0.40	(0.56,0.79,0.92)/0.29
P_{485}	(0.16,0.73,0.39)/0.20	(0.50,0.29,0.39)/0.33	(0.94,0.53,0.65)/0.91	(0.59,0.89,0.19)/0.80	(0.98,0.61,0.43)/0.30
P_{486}	(0.81,0.75,0.75)/0.55	(0.87,0.22,0.93)/0.33	(0.68,0.34,0.59)/0.74	(0.10,0.85,0.87)/0.36	(0.52,0.22,0.78)/0.36
P_{487}	(0.96,0.10,0.91)/0.88	(0.66,0.78,0.80)/0.69	(0.44,0.89,0.81)/0.62	(0.95,0.57,0.38)/0.62	(0.80,0.28,0.69)/0.93
P_{488}	(0.53,0.92,0.79)/0.74	(1.00,0.22,0.85)/0.45	(0.24,0.91,0.87)/0.58	(0.93,0.31,0.53)/0.76	(0.59,0.35,0.11)/0.50
P_{489}	(0.11,0.67,0.97)/0.68	(0.78,0.75,0.66)/0.48	(0.43,0.83,0.58)/0.78	(0.48,0.26,0.80)/0.24	(1.00,0.16,0.31)/0.53
P_{490}	(0.93,0.88,0.54)/0.56	(0.19,0.71,0.37)/0.82	(0.78,0.32,0.44)/0.81	(0.97,0.49,0.60)/0.29	(0.66,0.14,0.76)/0.78
P_{491}	(0.15,0.56,0.30)/0.89	(0.93,0.82,0.54)/0.43	(0.86,0.74,0.56)/0.73	(1.00,0.96,0.96)/0.27	(0.67,0.72,0.66)/0.89
P_{492}	(0.82,1.00,0.77)/0.32	(0.80,0.72,0.68)/0.68	(0.15,0.30,0.33)/0.22	(0.26,0.97,0.96)/0.56	(0.23,0.10,0.61)/0.82
P_{493}	(0.46,0.45,0.58)/0.30	(0.73,0.68,0.53)/0.60	(0.28,0.16,0.56)/0.47	(0.53,0.51,0.92)/0.94	(0.59,0.66,0.69)/0.28
P_{494}	(0.13,0.66,0.19)/0.85	(0.23,0.44,0.83)/0.92	(0.68,0.75,0.98)/0.40	(0.12,0.54,0.93)/0.59	(0.77,0.27,0.24)/0.75
P_{495}	(0.36,0.28,0.77)/0.44	(0.34,0.69,0.76)/0.86	(0.47,0.37,0.68)/0.48	(0.43,0.62,0.78)/0.84	(0.66,0.36,0.86)/0.71
P_{496}	(0.84,0.41,0.78)/0.29	(0.85,0.22,0.22)/0.36	(0.34,0.27,0.38)/0.65	(0.93,0.61,0.56)/0.84	(0.84,0.56,0.44)/0.94
P_{497}	(0.42,0.38,0.94)/0.49	(0.91,0.78,0.54)/0.28	(0.87,0.58,0.21)/0.82	(0.40,0.44,0.40)/0.76	(0.42,0.12,0.34)/0.38
P_{498}	(0.38,0.50,0.32)/0.72	(0.25,0.60,0.53)/0.38	(0.13,0.42,0.30)/0.25	(0.87,0.55,0.75)/0.49	(0.54,0.27,0.40)/0.59
P_{499}	(0.50,0.11,0.34)/0.52	(0.76,0.96,0.27)/0.81	(0.37,0.84,0.42)/0.47	(0.43,0.31,0.76)/0.70	(0.29,0.22,0.95)/0.55
P_{500}	(0.41,0.67,0.47)/0.83	(0.46,0.75,0.14)/0.67	(0.55,0.85,0.93)/0.47	(0.61,0.27,0.48)/0.56	(0.41,0.64,0.96)/0.43

Table 7. Patients diagnosed with a particular disease in Healthcare Hospital.

Disease Diagonized	Patients Affected
Patients affected with *Viral Fever*	66
Patients affected with *Tuberculosis*	242
Patients affected with *Typhoid*	62
Patients affected with *Throat Disease*	130

10. Conclusions

In this paper, we have tried to extend the concept of multi-fuzzy set theory to type-2 multi-fuzzy sets. The T2MFS may be applied to various applications in daily life. The algebraic properties of these sets have been verified and two types of distance metrics including Hamming distance and Euclidean distance have been discussed. Moreover, a few illustrative examples and a real-life case study of the medical diagnosis system are presented in this article. In the numerical illustration, we measure the Hamming distance and Euclidean distance of each patient for the set of diseases by considering the symptoms of the disease where both types of distance measurements yield a similar diagnostic result. The lowest distance shows proper diagnosis for both the distance measurements. In addition, as an application of T2MFS, the case-study is also conducted on 500 patients undergoing treatment in a hospital.

As far as the limitation of T2MFS is concerned, it is conceptually difficult to define the T2MFS and its necessary algebraic operations in the continuous domain, since the membership functions of such continuous T2MFS will be difficult to represent.

In the future, researchers may attempt to generalize this concept further by studying higher order multi-fuzzy sets in an abstract setting. Also, this research work might be enhancing the study of T2MFS for uncertain group decision-making (GDM) problems by introducing some aggregated operators where GDM is vital due to the lack of information, the expertise of the experts, risk amendment, etc. Besides, the possible extension of T2MFS in various other domains of research including image processing and data mining can be considered as the possible future area of research.

Supplementary Materials: The following are available online at http://www.mdpi.com/2073-8994/11/2/170/s1, Table S1: Euclidean distance between Patients and Diseases in Healthcare Hospital.

Author Contributions: The individual contribution and responsibilities of the authors were as follows: M.B.K., B.R. and S.M. performed the research study, collected, pre-processed, and analyzed the data and the obtained results, and worked on the development of the paper. S.K. and D.P. provided good advice throughout the research by giving suggestions on, methodology, modelling uncertainty of the patient data, and refinement of the manuscript. All the authors have read and approved the final manuscript.

Funding: This research was funded by Department of Science & Technology (DST), Government of India (No. DST/INSPIRE Fellowship/2015/IF150410).

Conflicts of Interest: The authors declare no conflict of interest.

References

1. De Bruijn, N.G. Denumerations of Rooted trees and Multi-sets. *Discret. Appl. Math.* **1983**, *6*, 25–33. [CrossRef]
2. Yager, R.R. On the theory of bags. *Int. J. Gen. Syst.* **1986**, *13*, 23–37. [CrossRef]
3. Sebastian, S.; Ramakrishnan, T.V. Multi-fuzzy sets: An extension of fuzzy sets. *Fuzzy Inf. Eng.* **2011**, *3*, 35–43. [CrossRef]
4. Yang, Y.; Tan, X.; Meng, C. The multi-fuzzy soft set and its application in decision making. *Appl. Math. Model.* **2013**, *37*, 4915–4923. [CrossRef]
5. Das, S.; Kar, M.B.; Kar, S. Group multi-criteria decision making using intuitionistic multi-fuzzy sets. *J. Uncertain. Anal. Appl.* **2013**, *10*, 1–16. [CrossRef]
6. Zadeh, L.A. Fuzzy sets as a basis for a theory of possibility. *Fuzzy Sets Syst.* **1978**, *1*, 3–28. [CrossRef]
7. Zadeh, L.A. The concept of a linguistic variable and its application to approximate reasoning—I. *Inf. Sci.* **1975**, *8*, 199–249. [CrossRef]
8. Zadeh, L.A. The concept of a linguistic variable and its application to approximate reasoning—II. *Inf. Sci.* **1975**, *8*, 301–357. [CrossRef]
9. Mizumoto, M.; Tanaka, K. Some properties of fuzzy sets of type-2. *Inf. Control* **1976**, *31*, 312–340. [CrossRef]
10. Mizumoto, M.; Tanaka, K. Fuzzy sets of type-2 under algebraic product and algebraic sum. *Fuzzy Sets Syst.* **1981**, *5*, 277–280. [CrossRef]
11. Dubois, D.; Prade, H. *Fuzzy Sets and Systems: Theory and Application*; Academic Press: New York, NY, USA, 1980.
12. Coupland, S.; John, R. A fast geometric method for defuzzification of type-2 fuzzy sets. *IEEE Trans. Fuzzy Syst.* **2008**, *16*, 929–941. [CrossRef]
13. Greenfield, S.; John, R.I.; Coupland, S. A novel sampling method for type-2 defuzzification. In Proceedings of the UKCI, London, UK, 5–7 September 2005; pp. 120–127.
14. Karnik, N.N.; Mendel, J.M. Centroid of a type-2 fuzzy set. *Inf. Sci.* **2001**, *132*, 195–220. [CrossRef]
15. Garcia, J.C.F. Solving fuzzy linear programming problems with interval type-2 RHS. In Proceedings of the 2009 IEEE International Conference on System, Man and Cybernetics, San Antonio, TX, USA, 11–14 October 2009; pp. 262–267.
16. Stanujkic, D.; Karabasevic, D. An extension of the WASPAS method for decision-making problems with intuitionistic fuzzy numbers: a case of website evaluation. *Oper. Res. Eng. Sci. Theor. Appl.* **2018**, *1*, 29–39. [CrossRef]
17. Abdulshahed, A.M.; Badi, I. Prediction and control of surface roughness for end milling process using ANFIS. *Oper. Res. Eng. Sci. Theor. Appl.* **2018**, *1*, 1–12. [CrossRef]
18. Kundu, P.; Kar, S.; Maiti, M. Fixed charge transportation problem with type-2 fuzzy variable. *Inf. Sci.* **2014**, *255*, 170–180. [CrossRef]
19. Kundu, P.; Kar, S.; Maiti, M. Multi-item solid transportation problem with type-2 fuzzy parameters. *Appl. Soft Comput.* **2015**, *31*, 61–80. [CrossRef]
20. Kundu, P.; Majumder, S.; Kar, S.; Maiti, M. A method to solve linear programming problem with interval type-2 fuzzy parameters. *Fuzzy Optim. Decis. Mak.* **2018**. [CrossRef]
21. Pal, S.S.; Kar, S. A Hybridized Forecasting Method Based on Weight Adjustment of Neural Network Using Generalized Type-2 Fuzzy Set. *Int. J. Fuzzy Syst.* **2018**. [CrossRef]
22. Vasiljevic, M.; Fazlollahtabar, H.; Stevic, Z.; Veskovic, S. A rough multicriteria approach for evaluation of the supplier criteria in automotive industry. *Decis. Mak. Appl. Manag. Eng.* **2018**, *1*, 82–96. [CrossRef]

23. Lukovac, V.; Popovic, M. Fuzzy Delphi approach to defining a cycle for assessing the performance of military drivers. *Decis. Mak. Appl. Manag. Eng.* **2018**, *1*, 67–81. [CrossRef]

24. Pamucar, D.; Cirovic, G. Vehicle route selection with an adaptive neuro fuzzy inference system in uncertainty conditions. *Decis. Mak. Appl. Manag. Eng.* **2018**, *1*, 13–37. [CrossRef]

25. Zadeh, L.A. Fuzzy sets. *Inf. Control* **1965**, *8*, 338–353. [CrossRef]

26. Goguen, J.A. L-fuzzy sets. *J. Math. Anal. Appl.* **1967**, *18*, 145–174. [CrossRef]

27. Atanassov, K.T. Intuitionistic fuzzy sets. *Fuzzy Sets Syst.* **1986**, *20*, 87–96. [CrossRef]

28. Muthuraj, R.; Balamurugan, S. Multi-fuzzy group and its level subgroups. *Gen. Math. Notes* **2013**, *17*, 74–81.

29. Sebastian, S.; Ramakrishnan, T.V. Multi-fuzzy Subgroups. *Int. J. Contemp. Math. Sci.* **2011**, *6*, 365–372.

30. Sebastian, S.; Ramakrishnan, T.V. Multi-fuzzy extension of crisp functions using bridge functions. *Ann. Fuzzy Math. Inform.* **2011**, *2*, 1–8.

31. Sebastian, S.; Ramakrishnan, T.V. Multi-fuzzy topology. *Int. J. Appl. Math.* **2011**, *24*, 117–129.

32. Sebastian, S.; John, R. Multi-fuzzy sets and their correspondence to other sets. *Ann. Fuzzy Math. Inform.* **2016**, *11*, 341–348.

33. Sebastian, S.; Ramakrishnan, T.V. Multi-fuzzy sets. *Int. Math. Forum* **2010**, *5*, 2471–2476.

34. Sebastian, S.; Ramakrishnan, T.V. Multi-fuzzy extensions of functions. *Adv. Adapt. Data Anal.* **2011**, *3*, 339–350. [CrossRef]

35. Dey, A.; Pal, M. Multi-fuzzy complex numbers and multi-fuzzy complex sets. *Int. J. Fuzzy Syst. Appl.* **2014**, *4*, 15–27. [CrossRef]

36. Dey, A.; Pal, M. Multi-fuzzy complex nilpotent matrices. *Int. J. Fuzzy Syst. Appl.* **2016**, *5*, 52–76. [CrossRef]

37. Dey, A.; Pal, M. Multi-fuzzy vector space and multi-fuzzy linear transformation over a finite dimensional multi-fuzzy set. *J. Fuzzy Math.* **2016**, *24*, 103–116.

38. Shinoj, T.K.; John, S.J. Intuitionistic fuzzy multi-sets and its application in medical diagnosis. *Int. J. Math. Comput. Sci.* **2012**, *6*, 121–124.

39. Ejegwa, P.A.; Awolola, J.A. Intuitionistic fuzzy multiset (IFMS) in binomial distributions. *Int. J. Sci. Technol. Res.* **2014**, *3*, 335–337.

40. Rajarajeswari, P.; Uma, N. On distance and similarity measures of intuitionistic fuzzy multi-set. *IOSR J. Math.* **2013**, *5*, 19–23. [CrossRef]

41. Rajarajeswari, P.; Uma, N. Intuitionistic fuzzy multi relations. *Int. J. Math. Arch.* **2013**, *4*, 244–249.

42. Rajarajeswari, P.; Uma, N. A study of normalized geometric and normalized hamming distance measures in intuitionistic fuzzy multi sets. *Int. J. Sci. Res. Eng. Technol.* **2013**, *2*, 76–80.

43. Rajarajeswari, P.; Uma, N. Zhang and Fu's similarity measure on intuitionistic fuzzy multi sets. *Int. J. Innov. Res. Sci. Eng. Technol.* **2014**, *3*, 12309–12317.

44. Rajarajeswari, P.; Uma, N. Correlation measure for intuitionistic fuzzy multi sets. *Int. J. Res. Eng. Technol.* **2014**, *3*, 611–617.

45. Girish, K.P.; John, S.J. Multi-set topologies induced by multi-set relations. *Inf. Sci.* **2012**, *188*, 298–313. [CrossRef]

46. Mendel, J.M.; John, R.I. Type-2 fuzzy sets made simple. *IEEE Trans. Fuzzy Syst.* **2002**, *10*, 307–315. [CrossRef]

© 2019 by the authors. Licensee MDPI, Basel, Switzerland. This article is an open access article distributed under the terms and conditions of the Creative Commons Attribution (CC BY) license (http://creativecommons.org/licenses/by/4.0/).

Article

Comparative Evaluation of Sustainable Design Based on Step-Wise Weight Assessment Ratio Analysis (SWARA) and Best Worst Method (BWM) Methods: A Perspective on Household Furnishing Materials

Sarfaraz Hashemkhani Zolfani [1] **and Prasenjit Chatterjee** [2,*]

[1] School of Engineering, Catholic University of the North, Larrondo 1281, Coquimbo, Chile; sa.hashemkhani@gmail.com

[2] Department of Mechanical Engineering, MCKV Institute of Engineering, Howrah 711204, West Bengal, India

* Correspondence: prasenjit2007@gmail.com

Received: 17 December 2018; Accepted: 5 January 2019; Published: 10 January 2019

Abstract: For a few years, there has been an increasing consciousness to design structures that are concurrently economic and environmentally responsive. Eco-friendly inferences of building designs include lower energy consumption, reduction in CO_2 emissions, assimilated energy in buildings and enhancement of indoor air quality. With the aim of fulfilling design objectives, designers normally encounter a situation in which the selection of the most appropriate material from a set of various material alternatives is essential. Sustainability has been developing as a new concept in all human activities to create a better balance between social, environmental and economic issues. Designing materials based on the sustainability concept is a key step to enable a better balance because there is no need to re-structure phases and procedures to make the system more efficient in comparison to previous models. Some of the most commonly used materials are household furnishing materials, which can be electrical devices, kitchen gears or general furnishing materials. The volume of production and consumption of these materials is considerable, therefore a newer sustainable plan for a better designed system is justifiable. In the literature, the application of multi-attribute decision-making (MADM) methods has been found to be very suitable for evaluating materials and developing general plans for them. This study contributes by applying two approaches based on MADM methods for weighting the criteria related to the sustainable design of household furnishing materials. Step-Wise Weight Assessment Ratio Analysis (SWARA) and Best Worst Method (BWM) are two specialized and new methods for weighting criteria with different approaches. This paper has not only investigated the weighting of important and related criteria for sustainable design but has also evaluated the similarities and differences between the considered weighting methods. A comparative study of SWARA and BWM methods has never been conducted to date. The results show that, except pairwise comparisons, SWARA and BWM are certainly similar and in some cases SWARA can be more accurate and effective.

Keywords: sustainable design; household furnishing materials; multi-attribute decision-making (MADM); step-wise weight assessment ratio analysis (SWARA); Best Worst Method (BWM)

---

1. Introduction

Rapid movements of a large number of populations from rural to urban areas and socio-economic changes that transform the agricultural human society into an industrial society, involving the extensive economic re-organization intensify the rate of resource deficiency and ecological contamination globally. Due to uncontrolled urbanization, environmental degradation has been occurring very quickly, causing land insecurity, water quality degradation and air and noise pollution, along with severe waste disposal issues. However, due to urbanization and economic globalization, the manufacturing and construction industries are becoming the fastest developing sectors everywhere. Superior financial affordability of consumers led to an increased requirement for improved life, better aesthetics and comfort level, which ultimately asserts high demands for the interior features in terms of artifact materials [1]. Interior designers have incredible influence by deciding the types of materials and products to be used to make a viable, endurable and ecologically-clean living environment. A home designed for energy efficiency will have the advantage of the site, sunlight and natural breezes. Conventionally, the interior designing job has been confined to conformist practice, focusing only on style and extravagant design while ignoring energy savings, toxic emissions, the harmful effects on customers' psychophysical health, and environmental pollution. However, recent trends in interior design have seen a spectacular move towards design policies that not only focus on the creation of art, beauty and taste, but also a healthy and sustainable environment for consumers to live and work in [2]. Thus, sustainability is a big trend in today's construction industry for energy use, resource efficiency, material selection, safety and life-cycle management, for developing an environmentally responsible environment.

Green or sustainable household furnishing material selection plays a considerable role in the entire interior design process to ensure better product performance and diminished life-cycle impact to the surroundings and individuals' health. Green or sustainable household furnishing materials generally refer to all the physical substances that are accumulated to craft the building interior. The primary advantages of using sustainable household furnishing materials in indoor environments are the reduction of health costs, increased employee productivity in work areas and greater shrinkage in operation and maintenance costs. In addition, these materials also help to reduce the adverse impacts associated with their processing, fabrication, installation, recycling and removal. Today, most buildings are constructed from a variety of materials that have specific functionalities and multipart assembly requirements. However, selecting the most appropriate decoration material for a particular industrial application, considering multi-perspective criteria, is not an easy task and is a great challenge for the interior decorators. Architects need to consider a range of selection criteria including some basic requirements like visuals, acoustic, tactile along with environmental requirements including energy consumption, low carbon emission features, recyclability and regional reachability, reduction of cost, meeting legal requirements, accounting for operating conditions and making the new product more competitive. There are a number of reasons for the evaluation and selection of the best suited material for a given case study. These include making improvements in service performance, reduction of cost, meeting the new legal requirements, accounting for the operating conditions, making the new product more competitive, to name a few [3]. On the other hand, inappropriate material selection frequently leads to early product failure with reduced efficiency, recurrence of processes, substantial financial loss and pitiable product performance, there by negatively affecting the output, effectiveness, environment and status of the organization.

Under these circumstances, a mechanism is required to explore the alternative candidates and identify the best one. Although sustainable design has become a leading concern in the interior design business, evaluation and selection of sustainable materials in real practice is limited due to the complexities involved. The optimal solution must satisfy the decision maker's (DM's) objectives, which are often conflicting in nature. For this purpose, the DMs need to deal with a large amount of data to arrive at the decision with the most consensus. The complexity of an engineering difficulty can be eased with the development of a well-structured decision-support tool that can deliberate multiple conflicting criteria [4]. The existing literature is flooded with numerous applications of classical and trial and error-based methods while many engineering industries are reforming their evaluation and measurement systems by adopting advanced decision-making methodologies. A multi-attribute decision-making (MADM) method can thus be a useful approach to the process of material assessment and selection [5]. The MADM methods have the potential to determine a ranking pre-order of the considered material alternatives from the first ranked to the last ranked in the presence of several mutually confounding properties. As all MADM problems include multiple criteria that have different importance levels due to the preferences of the DMs, the determination of criteria weights is one of the significant issues. The problem of choosing an appropriate technique for estimating criteria weights is very important as this directly affects the outcome of the entire evaluation and selection process. That is why numerous weight elicitation methods have been developed in order to confront this cognitive issue. The purpose of this paper is to compare the results of variability between the criteria priorities for Stepwise Weight Assessment Ratio Analysis (SWARA) and Best-Worst Method (BWM) for weight elicitation and to make suggestions about the conditions of using these two methods for sustainable material selection problems. It is the first time that SWARA is compared with BWM, and this makes the study different and unique.

The paper is organized as follows: Section 2 presents the literature review on sustainable material selection; Section 3 presents the research gap; in Section 4, detailed mathematical formulations of the considered methods have been presented; a comparative study between the considered methods for a sustainable household furnishing material selection problem has been discussed in Sections 5 and 6 concludes the paper.

2. Literature Review

Numerous studies have already been directed in addressing material selection problems using different ranking and optimization techniques. However, the literature shows many fewer works that deal with the problems on green materials' selection, particularly, there is no convincing ground of scientific research on comprehensive and holistic approaches used by design engineers for the assessment of household furnishing materials [6]. Thus, the aim of this section is to study past research on sustainable material selection to understand their weaknesses and enable the DMs to reduce subjectivity and uncertainty to ensure clearer support for a strong framework. Some basic models like the Leadership in Energy and Environmental Design (LEED), an extensively deployed building assessment system and Building Research Establishment Environmental Assessment Methods (BREEAM), the foremost sustainability evaluation strategy are vastly used for materials assessment and performance prediction. The LEED system boosts the utilization of materials that have a good amount of recycled content, fast renewable periods, accountable garnering management, low poisonous substance and proper solar reflectance and emissivity indices. Whereas the BREEAM system evaluates the environmental impacts of various construction materials and considers environmental issues. User-defined weighting is also incorporated to derive the values of different environmental impacts. However, due to the lack of flexibility and requirement of superior technical expertise, the use of LEED and BREEAM systems are limited and most designers and architects select such materials based on their past knowledge, experience and perception, thereby limiting the practical applications and result in considerable disappointment. Also these tools are only environment conscious and neglect economic and social concerns when selecting materials. Castro-Lacouture et al. [7] projected a mixed integer optimization (MIO) model incorporating different constraints to maximize the values achieved in the LEED rating system. Zhou et al. [8] developed a multi-objective optimization (MOO) model for sustainable material selection through an integrated artificial neural network (ANN) and genetic algorithm (GA) approach while considering mechanical properties, process, cost, performance and environment related factors. Rahman et al. [9] employed the technique of ranking preferences by similarity to the ideal solution (TOPSIS) method to develop an integrated model for optimizing the roof material selection process. Maniya and Bhatt [10] advocated the use of a preference selection index (PSI) method to search for an appropriate material that fits with the engineering requirements. Akadiri et al. [11] presented a fuzzy extended analytical hierarchy process (FEAHP) for building material selection. Florez and Lacouture [12] explored the applicability of a MIO framework while considering the potential impact of cost constraints, design considerations; environmental requirements along with some subjective factors to help the DMs in selecting the best green material for construction projects. Baharetha et al. [13] highlighted that durability, reuse, efficient energy usage, maintainability and lower negative environment impacts are the predominant factors for a sustainable material selection process. Ribeiro et al. [14] suggested a comprehensive life cycle assessment (LCA) model for four commercial biodegradable polymers. Hosseinijou et al. [15] proposed social-LCA process for building material selection. The proposed methodology was also capable for comparative products assessment. Van der Velden et al. [16] employed a material selection model based on LCA for a wearable smart textile device to promote sustainability while considering resource diminution cost, carbon footprint and human health as the major indicators. However, LCA is very intricate to implement in real life applications, attributable to the complexities involved in data acquisition and data quality control. Zhao et al. [17] used an integrated MADM approach encompassing AHP and Grey relational analysis (GRA) methods for sustainable design of a plastic pipe. Ma et al. [18] used a combined Entropy, TOPSIS and LCA approach for an automotive material selection application. Bhowmik et al. [19] also adopted Entropy-TOPSIS model to evaluate energy-efficient materials considering some predefined material properties. Criteria weights were calculated using the Entropy method, whereas TOPSIS was applied to rank the alternative materials.

The material selection process frequently considers some archetypal factors like light weight, economic manufacturing, product function, quality, performance and aesthetics and customer satisfaction. Less attention has been paid to the environmental and social impacts of the building materials. Nowadays, the advancement of the material selection process has moved towards social and sustainable criteria. Even though the principal material selection is the same for both sustainable and regular materials, the presence of a diverse range of criteria makes the selection process quite complex and time consuming. Table 1 shows that criteria for selecting sustainable materials can be grouped into three major dimensions or criteria levels. The first dimension (economic sustainability) indicates planned designs, which can avoid the requirements of major future restorations and thus reduce costs associated with energy use, water use and maintenance. The second dimension (social sustainability) mainly helps to prevent injuries through incorporating built-in safety attributes to provide adjustability and consolation for people of alternative capabilities in distinct life phases. The third and last dimension (environmental sustainability) is intended to reduce greenhouse gas emissions, proper utilization of water and energy along with reduced waste. However, the relative significance of these attributes is very challenging to estimate.

Table 1. Comprehensive list of sustainable material selection criteria.

	Dimension	Definitions	References
C_1	Economic		
C_{1-1}	Initial costs	All primary costs related to the new production	Sahamir et al. [20]; Moghtadernejad et al. [21];
C_{1-2}	Material cost	Cost of selected materials	Lewandowska et al. [22]; He et al. [23];
C_{1-3}	Energy consumption	The rate of energy consumption in production and in case if it needs to work based on energy	Ping [24]; Sahamir et al. [20]
C_{1-4}	Maintenance cost	It is related to materials and quality of design and manufacturing quality	Halstenberg et al. [25]; Go et al. [26]
C_{1-5}	Operation cost	It depends to the level of technology and related things	Rosen & Kishawy [27]
C_{1-6}	Variety of suppliers	It is also related to the type of selected materials because resources are totally dependent	Zhang et al. [28]; Chiu & Chu [29]; Sonego et al. [30]
C_2	Social		
C_{2-1}	Safety and security	Safety and security for both workers and consumer	Jilcha & Kitaw [31]; He et al. [23]; Sahamir et al. [20]
C_{2-2}	Structure parameters	It is related to the topics such as: suitable size for consumers, ergonomic aspects	He et al. [23]
C_{2-3}	Aesthetics	The quality of appearance of final products based on manufacturing design	Bachman [32]; Cimatti et al. [33]; Moghtadernejad et al. [21]
C_{2-4}	Functionality	Possibility of do it by consumers	Sonego et al. [30]

Table 1. *Cont.*

	Dimension	Definitions	References
C_3	Environment		
$C_{3\text{-}1}$	Recyclable	Rate of using recyclable materials	He et al. [23]; Sonego et al. [30]
$C_{3\text{-}2}$	Reuse	Easy disassembly for reusing	Rosen & Kishawy [27]; Beck [34]; Sonego et al. [30]
$C_{3\text{-}3}$	Sustainable suppliers	Access to the sustainable suppliers and decreasing carbon footprint	Raoufi et al. [35]
$C_{3\text{-}4}$	Reparability	Easy to be repaired which can have social and economic advantages	Zhang et al. [28]; Yan & Feng [36]
$C_{3\text{-}5}$	Lifespan	Life cycle of the product based on manufacturing design	Qian & Zhang [37]; He et al. [23]
$C_{3\text{-}6}$	Decomposition	Rate of being environmentally friendly	Foley & Cochran [38]; Zhang et al. [39]
$C_{3\text{-}7}$	Upgrade possibility	Upgrade possibility in the future	Lumsakul et al. [40]; Sonego et al. [30]

3. Research Gap

By summarizing and exploring the existing literature, two major conclusions can be drawn. First of all, a comprehensive hierarchical structure for sustainable material criteria has never been developed by the earlier researchers and also several social and environmental dimensions have been overlooked. Secondly, for estimating criteria weights for sustainable material selection issues, previous studies have mainly adopted the Entropy and AHP methods for different assessment rationales [11,17–19]. However, a weighting method like AHP uses pairwise comparisons for preference input and it is based on the DM's knowledge about the problem. DMs usually have different persuasion, background, and knowledge levels and can hardly arrive at conformity on the relative importance of criteria which ultimately lead to deviations of criteria weights due to the involvement of subjective factors. While the objective weighting method, Entropy, is based on inherent criteria information, which has the ability to reduce man-made errors and make results in more accord with facts. The smaller the entropy value is, the smaller the disorder degree of the system is small. The Entropy method is particularly suitable for those situations in which either decision matrix is purely cardinal, or the ordinal values are converted to the comparable numbers or appropriate scales.

Since criteria weight has an enormous influence on the outcome of any MADM method, the sole pivotal dilemma is to evaluate the weights of different material characteristics. Additionally, the rationality of the weight estimation has an evidential effect on the consistency and precision of the computational results. Therefore, there is a need for an unambiguously acceptable methodology to guide the DMs to determine the criteria weights. Thus, this research aims to evaluate the sustainable material selection design and planning more objectively and realistically by introducing a hierarchical structure of the sustainable material selection criteria, as shown in Figure 1.

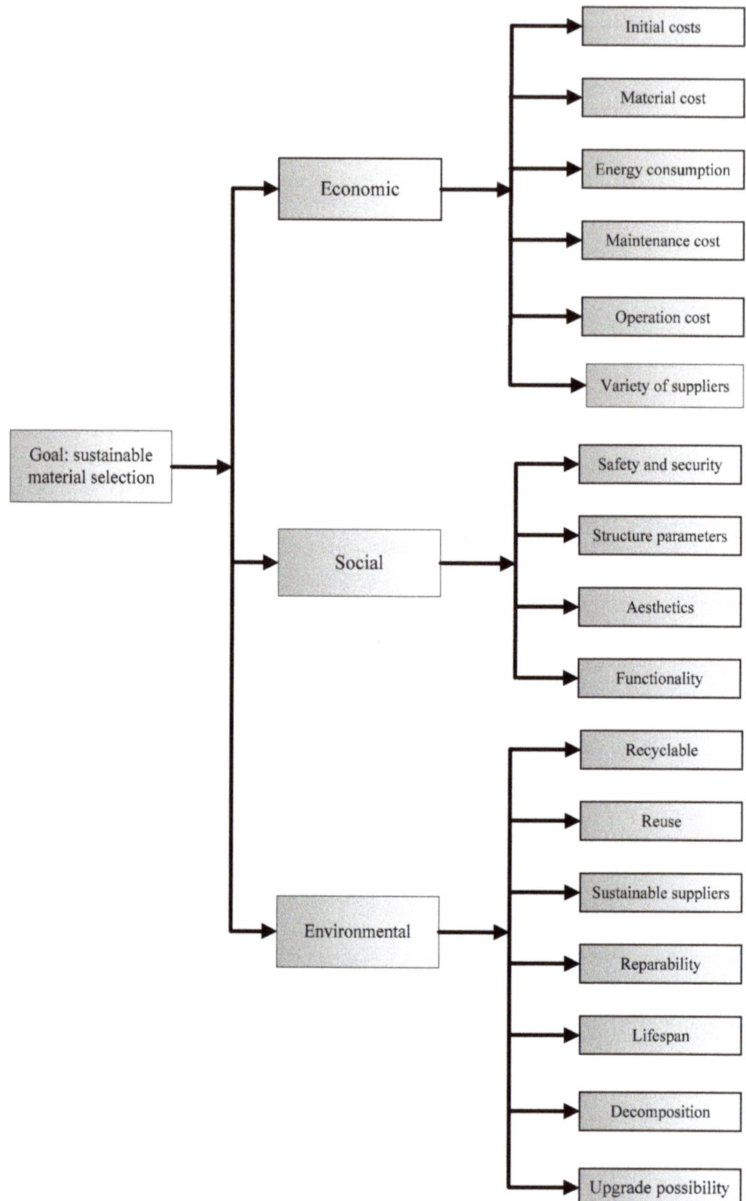

Figure 1. Proposed hierarchical criteria structure for evaluation of sustainable materials.

4. Comparative Methodologies

MADM has been developing new perspectives and methods. Mostly, new contributions on methods can be categorized in two sections. The first section mainly focuses on developing criteria weighting techniques, while the other section is centering on the ranking of alternatives. The most commonly used methods for estimating criteria weights are AHP (Saaty [41]) and the Analytic Network Process (ANP) (Saaty [42,43]), whereas, among the new criteria weight determination approaches, SWARA (Kersuliene et al. [44]), Factor Relationship (FARE) (Ginevicius, [45]), BWM (Rezaei, [46]), Extended SWARA (Hashemkhani Zolfani et al. [47]), and Full Consistency Method (FUCOM) (Pamucar et al. [48]) are worth mentioning. There are also some studies that focused on the applications of different MADM methods in sustainability challenges and decisions [49–53]. This section highlights some of the key factors for product comparison or selection and provides a tangible basis for fixing priorities with a view to tarnish the environmental footprint of any building. Generically, product-to-product demarcations are likely to be ambiguous when it deals with construction and casing or envelope materials. Product-to-product comparisons are more pragmatic for interior finishes, outfit products and interior decorations. For instance, springy or pliable floor coverings can freely be analyzed and compared with each other as far as installation materials, product cleaning, expected service life and end of a product's life are pondered. Window systems are usually mobilized to a construction site as pre-assembled elements that can be mutually compared in terms of thermal performance and other criteria, without too much regard for eclectic system speculations. However, this strategy will fail when scores are assigned to evaluate building materials in different rating systems. This is due to the fact that material scores germinate from a concurrent apprehension of environmental issues that does not necessarily go with the objective analysis. For example, according scores for recycled content in products presumes reduced environmental hazards. However, it may not always stand practical, and recycling for any specific application may have congenial impacts. It is well established that reuse or recycling can conserve hazardous waste dump, but at the same time, there is a likelihood of consuming more energy and detrimental upsets on air and water quality. The center of interest on recycling defines these unfavorable footprints and utterly gives additional weightage to solid waste and resource depletion issues as compared to global warming traits. Early conceptual design decisions are mostly concentrated to the initial material related environmental consequences of a building due to the considerations of fundamental structure and envelope elements. These are customarily the towering mass elements with remarkable manufacturing and transportation domination. Thus, it makes acceptable perception to allocate higher weightage to the environmental implications of materials at the initiation of the design procedure.

Except SWARA, other methods are based on pairwise comparisons, although there are big differences in the way of calculating the criteria weights. For example, the original BWM method, based on a non-linear model, is used only for the estimation of criteria weights [46,54]. Also, there is a simplified BWM method based on a linear model. SWARA is a policy-based method which works on weighting criteria depending up on their priority. This priority can be arranged by policy makers on the basis of descriptive future scenarios and current regulations and strategic plans [55–58].

BWM and SWARA have similar ideas with different perspectives and have not been reported yet. The notion of identifying the best and worst criterion in the BWM method is very similar to the first step of the SWARA method. When criteria are prioritized on a policy basis, a preference order emerges, which helps in identifying the best and worst criteria. When these two criteria are identified, the same pairwise comparisons as used in the SWARA method are accomplished due to similar expert opinions. BWM somehow is also a link to connect these two perspectives in weighting criteria and this research study attempted to work on this idea to check the results of SWARA and BWM methods in a single framework. During the past few years, there have been a significant number of studies associated with the applications of SWARA and BWM implementations which can be found in References [59,60].

4.1. Best Worst Method (BWM)

As one of the latest MADM methods, BWM can efficiently tackle the inconsistency derived from pairwise comparisons. This method is more consistent in comparison to the AHP, ANP, FARE and Simple Multi-Attribute Rating Technique (SMART) methods [61,62]. The BWM method has been cited 249 times-based on Google scholar information (until 19 November 2018). BWM has been applied in different studies and fields including supplier selection and development [63]; water management [64]; complex bundling configurations [65]; urban sewage sludge application [66]; social sustainability of supply chains [54]; measuring logistics performance [67]; identifying success factors [68]; cloud service selection [69]; evaluating university-industry doctoral projects [70].

The structure and basic steps of BWM method is as follows [46,71]:

Step 1: Selecting and identifying criteria in a common way; literature review, expert ideas and other probable ways.

Step 2. Identifying and selecting the best and worst criteria based on experts' ideas and opinions.

Step 3. Designing the preferences matrix based of comparing best criterion over all others by applying numbers ranging between 1 and 9.

$$A_b = (a_{1B}, a_{2B}, a_{3B}, \ldots a_{nB}) \tag{1}$$

Step 4. Designing the preferences matrix based of comparing worst criterion over all others by applying numbers between 1 and 9.

$$A_b = (a_{1W}, a_{2W}, a_{3W}, \ldots a_{nW}) \tag{2}$$

Step 5. Finding the relative importance of criteria through calculation of final and optimal weights $(w_1^*, w_2^*, w_3^*, \ldots .w_n^*)$ by solving the following optimization model.

$$Minmax_j\{|(w_B/w_j) - a_{Bj}|, |(w_j/w_w) - a_{jw}|\} \tag{3}$$

Subject to $\sum_j w_j = 1$

Model (3) can easily be converted to Model (4) to find out the optimal weights $(w_1^*, w_2^*, w_3^*, \ldots .w_n^*)$ and the optimal value of reliability level (ξ^*):

$$Min\xi$$
$$\left|\frac{w_B}{w_j} - a_{Bj}\right| \leq \xi \text{ for all } j$$
$$\left|\frac{w_j}{w_w} - a_{jw}\right| \leq \xi \text{ for all } j$$
$$\sum_j w_j = 1$$
$$w_j \geq 0 \text{ for all } j$$

where w_B and w_w indicate the weights of the best and the worst criteria respectively. a_{Bj} is the preference of the most important (best) criterion over criterion j and a_{jw} is the preference of criterion j over the least important (worst) criterion.

Step 6. Estimating the consistency ratio (K_{si}) to verify the reliability level of the pairwise comparisons using Equation (4).

Similar to the AHP method, a consistency index (CI), as shown in Table 2, helps to determine the K_{si} value. A smaller K_{si} value (close to zero) indicates superior consistency, whereas, a higher K_{si} value (close to one) indicates inferior consistency made during pairwise comparisons [46].

$$K_{si} = \frac{\xi^*}{CI} \tag{4}$$

Table 2. *CI* values for BWM method.

a_{BW}	1	2	3	4	5	6	7	8	9
CI (max ξ^*)	0	0.44	1.00	1.63	2.30	3.00	3.73	4.47	5.23

a_{BW} in Table 2 indicates the preference of the best criterion over the worst criterion. It is important to mention that CR in the AHP method is basically used to substantiate the validity of comparisons, but in BWM, its main function is to find the degree of reliability of the pairwise comparisons, thus provides more conformable results. Also, BWM employs many fewer comparisons ($2n - 3$) by forming comparison-vectors. This phenomena assures more reliability of the weights obtained by BWM as compared to the weights of AHP method. These advantages of BWM method have led the foundation of selecting it for sustainable material selection assessment. In addition to this, in BWM, no fractional numbers are used which makes the computation easier for the DMs. Rezai [46,71] statistical validated that BWM computes criteria weights appreciably better than AHP in terms of CR, total divergence and agreement.

4.2. Step-Wise Weight Assessment Ratio Analysis (SWARA)

In this method, DMs have an essential portrait of evaluations and calculation of criteria weights. The basic characteristic of this method is the assessment of expert opinion on the importance of the considered criteria for estimating their weights. Experts select the importance of each criterion and rank them in order of preference by employing their inherent experience, understanding and information. Based on this method, criterion having the highest importance is given rank 1, while the criterion with the least importance attains the last rank. The overall ranks to the group of experts are determined according to the average values. Based on Google scholar information, this method (Kersuliene et al. [44]) has been cited 210 times (until 13 December 2018). The application of the SWARA method has been developing and some of latest studies based on the SWARA method application include the evaluation of chemical wastewater purification [72]; investigating supply chain management competitive strategies [73]; evaluating construction projects [74]; flood susceptibility assessment [75]; sustainable third-party reverse logistics provider evaluation [76]; pharmacological therapy selection [77]; assessment of the railway management [78]; competency-based IT personnel selection [79].

SWARA steps are summarized as follows [80–82]:

Step 1. Sorting of criteria based on policy and expert opinion or some standards.

Step 2. Providing relative importance between criteria:

Initiating from the second criterion, experts exhibit the corresponding importance of jth criterion in congruence with the previous $(j - 1)$ criterion through comparative importance of average value (s_j) ratio.

Step 3. Computation of coefficient k_j:

$$k_j = \begin{cases} 1 & j = 1 \\ s_j + 1 & j > 1 \end{cases} \tag{5}$$

Step 4. Determination of recalculated weight w_j:

$$w_j = \begin{cases} 1 & j = 1 \\ \frac{x_{j-1}}{k_j} & j > 1 \end{cases} \tag{6}$$

Step 5. Calculation of final criteria weights:

$$q_j = \frac{w_j}{\sum_{k=1}^{n} w_j} \tag{7}$$

5. Comparative Results

This research focuses on a comparative study on criteria weight estimation using SWARA and BWM methods for sustainable household furnishing material selection problem. This part is designed according to the available literature and adopted methodologies to assess the outputs of SWARA and

BWM methods. Accordingly, five experts with at least five years of experience have been selected. To be very specific, as the main intention of this comparative analysis is to check the model of thinking of the DMs for the two different weight elicitation methods; therefore, questionnaires of SWARA and BWM methods were delivered to the experts at the same time. The most critical points of their ideas have been presented in the Appendix A. In brief, all five experts had to answer the questionnaires at the same time, but they could select one to answer first. Also, they had the opportunity to change their opinions after finishing the first questionnaire if required. Sustainable development needs a great balance between its economic, social and environmental dimensions, therefore, the proper integration between these three dimensions is an imperative need for policy-making. The main idea is to meet the boundaries and the best scenario which is aligning with 2030 agenda for sustainable development and 17 related goals [83].

Eventually, economic dimension received a weight of 0.333, social dimension received a weight of 0.333 and environment dimension received a weight of 0.334 and final weights of the sub-criteria under the three dimensions have been calculated and tabulated. The final results based on each sustainable material selection dimension are presented below. Detail calculations of SWARA and BWM methods have been added in the Appendix B.

5.1. Economic Dimension

In this section, final weights for main economic dimension as calculated using the SWARA and BWM methods-based on the opinions of the five experts are first presented in Tables 3–7 respectively. Computations of final weights based on each expert's idea are presented separately as this study intents to analyze the similarities and differences of the two above mentioned methods. From Tables 3–7, it is observed that there are no differences in sub-criteria priorities in the SWARA and BWM methods for economic dimension. In both the cases, C_{1-1} (initial costs) and C_{1-4} (maintenance cost) emerge out as the most and least important criteria under economic dimension.

Table 3. Final weights of economic dimension based on comparative study (Expert 1).

Weight	C_{1-1}	C_{1-2}	C_{1-3}	C_{1-4}	C_{1-5}	C_{1-6}
SWARA	0.075	0.064	0.045	0.041	0.058	0.05
BWM	0.121	0.077	0.031	0.014	0.051	0.039
Priority based on SWARA	1	2	5	6	3	4
Priority based on BWM	1	2	5	6	3	4
K_{si} (BWM)			0.101			

Table 4. Final weights of economic dimension based on comparative study (Expert 2).

Weight	C_{1-1}	C_{1-2}	C_{1-3}	C_{1-4}	C_{1-5}	C_{1-6}
SWARA	0.078	0.065	0.045	0.039	0.057	0.049
BWM	0.144	0.060	0.030	0.018	0.045	0.036
Priority based on SWARA	1	2	5	6	3	4
Priority based on BWM	1	2	5	6	3	4
K_{si} (BWM)			0.108			

Table 5. Final weights of economic dimension based on comparative study (Expert 3).

Weight	$C_{1\text{-}1}$	$C_{1\text{-}2}$	$C_{1\text{-}3}$	$C_{1\text{-}4}$	$C_{1\text{-}5}$	$C_{1\text{-}6}$
SWARA	0.075	0.068	0.043	0.037	0.059	0.051
BWM	0.125	0.078	0.026	0.014	0.052	0.039
Priority based on SWARA	1	2	5	6	3	4
Priority based on BWM	1	2	5	6	3	4
K_{si} (BWM)			0.089			

Table 6. Final weights of economic dimension based on comparative study (Expert 4).

Weight	$C_{1\text{-}1}$	$C_{1\text{-}2}$	$C_{1\text{-}3}$	$C_{1\text{-}4}$	$C_{1\text{-}5}$	$C_{1\text{-}6}$
SWARA	0.072	0.066	0.043	0.039	0.060	0.052
BWM	0.134	0.081	0.027	0.017	0.041	0.033
Priority based on SWARA	1	2	5	6	3	4
Priority based on BWM	1	2	5	6	3	4
K_{si} (BWM)			0.087			

Table 7. Final weights of economic dimension based on comparative study (Expert 5).

Weights	$C_{1\text{-}1}$	$C_{1\text{-}2}$	$C_{1\text{-}3}$	$C_{1\text{-}4}$	$C_{1\text{-}5}$	$C_{1\text{-}6}$
SWARA	0.078	0.068	0.043	0.039	0.057	0.047
BWM	0.143	0.061	0.031	0.015	0.046	0.037
Priority based on SWARA	1	2	5	6	3	4
Priority based on BWM	1	2	5	6	3	4
K_{si} (BWM)			0.124			

5.2. Social Dimension

Similar to the economic dimension, calculations and final weights of all sub-criteria in social dimension are exhibited in Tables 8–12 which reflect no alterations in the priorities of the different sub-criteria under this dimension, as opined by the five experts. These tables also indicate that among the four sub-criteria, $C_{2\text{-}1}$ (safety and security) becomes the most prominent criteria, whereas $C_{2\text{-}4}$ (functionality) is the least important criteria.

Table 8. Final weights of social dimension based on comparative study (Expert 1).

Weight	$C_{2\text{-}1}$	$C_{2\text{-}2}$	$C_{2\text{-}3}$	$C_{2\text{-}4}$
SWARA	0.105	0.088	0.076	0.064
BWM	0.151	0.091	0.061	0.030
Priority based on SWARA	1	2	3	4
Priority based on BWM	1	2	3	4
K_{si} (BWM)		0.091		

Table 9. Final weights of social dimension based on comparative study (Expert 2).

Weight	C_{1-1}	C_{1-2}	C_{1-3}	C_{1-4}
SWARA	0.100	0.091	0.076	0.066
BWM	0.186	0.077	0.046	0.023
Priority based on SWARA	1	2	3	4
Priority based on BWM	1	2	3	4
K_{si} (BWM)		0.140		

Table 10. Final weights of social dimension based on comparative study (Expert 3).

Weight	C_{1-1}	C_{1-2}	C_{1-3}	C_{1-4}
SWARA	0.103	0.086	0.078	0.065
BWM	0.183	0.071	0.053	0.025
Priority based on SWARA	1	2	3	4
Priority based on BWM	1	2	3	4
K_{si} (BWM)		0.092		

Table 11. Final weights of social dimension based on comparative study (Expert 4).

Weight	C_{1-1}	C_{1-2}	C_{1-3}	C_{1-4}
SWARA	0.106	0.092	0.074	0.061
BWM	0.178	0.072	0.054	0.028
Priority based on SWARA	1	2	3	4
Priority based on BWM	1	2	3	4
K_{si} (BWM)		0.117		

Table 12. Final weights of social dimension based on comparative study (Expert 5).

Weight	C_{1-1}	C_{1-2}	C_{1-3}	C_{1-4}
SWARA	0.101	0.092	0.076	0.064
BWM	0.162	0.096	0.048	0.026
Priority based on SWARA	1	2	3	4
Priority based on BWM	1	2	3	4
K_{si} (BWM)		0.092		

5.3. Environmental Dimension

Finally, the necessary information and final weights based on both the BWM and SWARA methods for the considered sub-criteria of environmental dimension are presented in Tables 13–17. It is observed that that there are no considerable differences in the sub-criteria priorities as provided by the group of experts through the SWARA and BWM methods. It is also perceived that C_{3-6} (decomposition) and C_{3-7} (upgrades possibility) criteria received the maximum and minimum weights respectively.

Table 13. Final weights of environment dimension based on comparative study (Expert 1).

Weight	$C_{3\text{-}1}$	$C_{3\text{-}2}$	$C_{3\text{-}3}$	$C_{3\text{-}4}$	$C_{3\text{-}5}$	$C_{3\text{-}6}$	$C_{3\text{-}7}$
SWARA	0.061	0.045	0.032	0.040	0.053	0.074	0.028
BWM	0.070	0.035	0.023	0.028	0.047	0.118	0.014
Priority based on SWARA	2	4	6	5	3	1	7
Priority based on BWM	2	4	6	5	3	1	7
K_{si} (BWM)				0.066			

Table 14. Final weights of environment dimension based on comparative study (Expert 2).

Weight	$C_{3\text{-}1}$	$C_{3\text{-}2}$	$C_{3\text{-}3}$	$C_{3\text{-}4}$	$C_{3\text{-}5}$	$C_{3\text{-}6}$	$C_{3\text{-}7}$
SWARA	0.062	0.045	0.033	0.039	0.050	0.075	0.030
BWM	0.070	0.035	0.028	0.028	0.047	0.112	0.014
Priority based on SWARA	2	4	6	5	3	1	7
Priority based on BWM	2	4	5	5	3	1	7
K_{si} (BWM)				0.084			

Table 15. Final weights of environment dimension based on comparative study (Expert 3).

Weight	$C_{3\text{-}1}$	$C_{3\text{-}2}$	$C_{3\text{-}3}$	$C_{3\text{-}4}$	$C_{3\text{-}5}$	$C_{3\text{-}6}$	$C_{3\text{-}7}$
SWARA	0.062	0.045	0.033	0.039	0.050	0.075	0.030
BWM	0.070	0.035	0.028	0.028	0.047	0.112	0.014
Priority based on SWARA	2	4	6	5	3	1	7
Priority based on BWM	2	4	5	5	3	1	7
K_{si} (BWM)				0.084			

Table 16. Final weights of environment dimension based on comparative study (Expert 4).

Weight	$C_{3\text{-}1}$	$C_{3\text{-}2}$	$C_{3\text{-}3}$	$C_{3\text{-}4}$	$C_{3\text{-}5}$	$C_{3\text{-}6}$	$C_{3\text{-}7}$
SWARA	0.064	0.046	0.034	0.039	0.051	0.073	0.028
BWM	0.072	0.036	0.021	0.024	0.048	0.121	0.012
Priority based on SWARA	2	4	6	5	3	1	7
Priority based on BWM	2	4	6	5	3	1	7
K_{si} (BWM)				0.072			

Table 17. Final weights of environment dimension based on comparative study (Expert 5).

Weight	$C_{3\text{-}1}$	$C_{3\text{-}2}$	$C_{3\text{-}3}$	$C_{3\text{-}4}$	$C_{3\text{-}5}$	$C_{3\text{-}6}$	$C_{3\text{-}7}$
SWARA	0.059	0.047	0.034	0.041	0.054	0.071	0.029
BWM	0.058	0.035	0.022	0.025	0.044	0.136	0.014
Priority based on SWARA	2	4	6	5	3	1	7
Priority based on BWM	2	4	6	5	3	1	7
K_{si} (BWM)				0.116			

6. Discussion and Conclusions

Conventional engineering optimization and statistical approaches are applied often on the basis of a well-developed and structured problem. Solution of engineering problems with only one objective or criterion is very easy to acquire, however, most of the real life problems consist of conflicting objectives and multiple criteria, making the process more perplexed and time consuming. It is well accepted that the weight calculation methods used for solving different MADM problems have a vital contribution to defining the criteria importance and obtaining the best and satisfying results for the DMs. In this paper, two different approaches—namely SWARA and BWM—with similar methodological structures have been adopted for computation of criteria weights for sustainable housing material selection design. The main idea is to see the differences in results. In this regard, five separated ideas based on expert opinion have been compared directly. Accordingly, after finishing the questionnaires by the experts, authors examined the general inklings of them about the two different questionnaires. In SWARA method, experts have more options to show the weightage of each criterion in comparison to the other more important criteria. The DMs can probably have a clearer idea about what they want to demonstrate in terms of criteria weights. BWM first identifies the most preferable and the least favorable criteria to make pairwise comparisons between each of them and the other considered criteria. Finally, it solves a linear optimization model to deduce the criteria weights. In BWM, the DMs probably follow the same structure as SWARA specially when there are not so many criteria for evaluation. This research is carried out with specific goals, descriptions and surveys since it endeavors to reckon the key elements in sustainable housing material evaluation process development. Moreover, in real time situations, DMs or the experts have limited domain of knowledge and expertise to present and express their ideas precisely. In case of having so many criteria, it will be certainly complicated to express the differences and priorities based on some linguistic variables or qualitative numbers. In brief, it can be said that although SWARA and BWM have different mathematical approaches, there are some similarities between them and there are some advantageous in SWARA method when the general approach (pairwise comparison or policy based) is not a big challenge and deal. SWARA and BWM methods more preferable than the AHP method which requires $n(n-1)/2$ pairwise criteria comparisons, thus complicating the application of this method. Especially, when the number of criteria is large, it becomes practicality unfeasible to perform such huge consistent pairwise comparisons in the AHP method. Also, as mathematical transitivity in the pairwise criteria comparisons is extremely important to consider the deviation from transitivity results in an increase in inconsistency in case of AHP method. However, the major issue in any decision-making process is not only finding the best alternative or criteria priorities, rather more emphasis should be given on appropriately guiding the DMs toward identifying the critical components, and proper structuring of the problem considering relevant criteria and decision alternatives. Therefore, the establishment of a comprehensive appraisal system and defining crucial decision-making points is an important and necessary step. As in this paper, equal weights for the three major three sustainable dimensions have been assumed which may also be changed as per the requirements of the DMs and the effects may also be observed in further research. Furthermore, comparative studies with other weight elicitation approaches like decision making trial and evaluation laboratory (DEMATEL), the resistance to change method and full consistency method (FUCOM) may be carried out for exploration of knowledge-base.

Author Contributions: Main idea, conceptualization, methodology, identifying criteria, interviews, appendix, calculations, discussion and editing and writing–original draft by S.H.Z.; Introduction, literature review, research gap, conclusion, editing and writing-original draft by P.C.

Funding: This research received no external funding.

Conflicts of Interest: The authors declare no conflict of interest.

Appendix A

Experts' opinions:

Expert 1

It feels like this research is doing something similar. The idea is tricky and hard to say whether the idea is completely same or not, but it seems for comparing limited criteria, one has to follow the same route. It is hard to say someone will do it for 100% but definitely for limited criteria, it should be common.

Expert 2

It seems you have two different perspectives which can lead you to a wider area. It is harder to make a proper decision in complex and larger problems based on pairwise comparisons. I think normal decisions aren't challenging and both ideas are really practical and helpful. When you have more chance to show that the exact differences of criteria, I guess you can manage a better decision while pairwise comparison is also interesting.

Expert 3

In the case of creating a ranking for the criteria, I could definitely have enough concentration of the topic and lastly, I knew what probably will happen as results.

Expert 4

When you have so many criteria, it will be really hard to do a pairwise comparison. You don't have so many differences based on a scale like 1–9.

Expert 5

Feel more concentrated when you are making decision based on a priority. How can I be sure about my assessment while I just have limited numbers to compare all criteria based on a pairwise comparison.

Appendix B

Detail calculations:

Economic dimension:

Table A1. Final weights of economic dimension based on SWARA (Expert number 1).

Criteria	The Comparative Importance of Average Value S_j	Coefficient $K_j = S_j + 1$	Recalculated Weight $w_j = \frac{x_{j-1}}{k_j}$	Weight $q_j = \frac{w_j}{\sum w_j}$	Final Weight
$C_{1\text{-}1}$	-	1	1	0.223	0.075
$C_{1\text{-}2}$	0.15	1.15	0.87	0.194	0.064
$C_{1\text{-}5}$	0.1	1.1	0.791	0.177	0.058
$C_{1\text{-}6}$	0.2	1.2	0.66	0.148	0.05
$C_{1\text{-}3}$	0.1	1.1	0.6	0.135	0.045
$C_{1\text{-}4}$	0.1	1.1	0.546	0.123	0.041

Table A2. Best criterion to other criteria for economic dimension based on BWM method (Expert number 1).

Best to Others	$C_{1\text{-}1}$	$C_{1\text{-}2}$	$C_{1\text{-}3}$	$C_{1\text{-}4}$	$C_{1\text{-}5}$	$C_{1\text{-}6}$
$C_{1\text{-}1}$	1	2	5	6	3	4

Table A3. Other criteria to the worst criterion for economic dimension based on BWM method (Expert number 1).

Others to the Worst	$C_{1\text{-}4}$
$C_{1\text{-}1}$	7
$C_{1\text{-}2}$	6
$C_{1\text{-}3}$	3
$C_{1\text{-}4}$	1
$C_{1\text{-}5}$	4
$C_{1\text{-}6}$	5

Table A4. Final results and weights of main criteria for economic dimension based on BWM method (Expert number 1).

Weight	$C_{1\text{-}1}$	$C_{1\text{-}2}$	$C_{1\text{-}3}$	$C_{1\text{-}4}$	$C_{1\text{-}5}$	$C_{1\text{-}6}$
	0.362	0.232	0.093	0.043	0.154	0.116
Final weight	0.121	0.077	0.031	0.014	0.051	0.039
K_{si}			0.101			

Table A5. Comparative results (Expert number 1).

Weight	$C_{1\text{-}1}$	$C_{1\text{-}2}$	$C_{1\text{-}3}$	$C_{1\text{-}4}$	$C_{1\text{-}5}$	$C_{1\text{-}6}$
SWARA	0.075	0.064	0.045	0.041	0.058	0.05
BWM	0.121	0.077	0.031	0.014	0.051	0.039
Priority based on SWARA	1	2	5	6	3	4
Priority based on BWM	1	2	5	6	3	4

Table A6. Final weights of economic dimension based on SWARA (Expert number 2).

Criteria	The Comparative Importance of Average Value S_j	Coefficient $K_j = S_j + 1$	Recalculated Weight $w_j = \frac{x_{j-1}}{k_j}$	Weight $q_j = \frac{w_j}{\sum w_j}$	Final Weight
$C_{1\text{-}1}$	-	1	1	0.235	0.078
$C_{1\text{-}2}$	0.2	1.2	0.833	0.196	0.065
$C_{1\text{-}5}$	0.15	1.15	0.725	0.170	0.057
$C_{1\text{-}6}$	0.15	1.15	0.630	0.148	0.049
$C_{1\text{-}3}$	0.1	1.1	0.573	0.134	0.045
$C_{1\text{-}4}$	0.15	1.15	0.498	0.117	0.039

Table A7. Best criterion to other criteria for economic dimension based on BWM method (Expert number 2).

Best to Others	$C_{1\text{-}1}$	$C_{1\text{-}2}$	$C_{1\text{-}3}$	$C_{1\text{-}4}$	$C_{1\text{-}5}$	$C_{1\text{-}6}$
$C_{1\text{-}1}$	1	3	6	7	4	5

Table A8. Other criteria to the worst criterion for economic dimension based on BWM method (Expert number 2).

Others to the Worst	$C_{1\text{-}4}$
$C_{1\text{-}1}$	6
$C_{1\text{-}2}$	5
$C_{1\text{-}3}$	2
$C_{1\text{-}4}$	1
$C_{1\text{-}5}$	3
$C_{1\text{-}6}$	4

Table A9. Final results and weights of main criteria for economic dimension based on BWM method (Expert number 2).

Weight	$C_{1\text{-}1}$	$C_{1\text{-}2}$	$C_{1\text{-}3}$	$C_{1\text{-}4}$	$C_{1\text{-}5}$	$C_{1\text{-}6}$
	0.432	0.180	0.090	0.054	0.135	0.108
Final weight	0.144	0.060	0.030	0.018	0.045	0.036
K_{si}			0.108			

Table A10. Comparative results (Expert number 2).

Weight	$C_{1\text{-}1}$	$C_{1\text{-}2}$	$C_{1\text{-}3}$	$C_{1\text{-}4}$	$C_{1\text{-}5}$	$C_{1\text{-}6}$
SWARA	0.078	0.065	0.045	0.039	0.057	0.049
BWM	0.144	0.060	0.030	0.018	0.045	0.036
Priority based on SWARA	1	2	5	6	3	4
Priority based on BWM	1	2	5	6	3	4

Table A11. Final weights of economic dimension based on SWARA (Expert number 3).

Criteria	The Comparative Importance of Average Value S_j	Coefficient $K_j = S_j + 1$	Recalculated Weight $w_j = \frac{x_{j-1}}{k_j}$	Weight $q_j = \frac{w_j}{\sum w_j}$	Final Weight
$C_{1\text{-}1}$	-	1	1	0.224	0.075
$C_{1\text{-}2}$	0.1	1.1	0.909	0.204	0.068
$C_{1\text{-}5}$	0.15	1.15	0.791	0.177	0.059
$C_{1\text{-}6}$	0.15	1.15	0.687	0.154	0.051
$C_{1\text{-}3}$	0.2	1.2	0.573	0.128	0.043
$C_{1\text{-}4}$	0.15	1.15	0.498	0.112	0.037

Table A12. Best criterion to other criteria for economic dimension based on BWM method (Expert number 3).

Best to Others	$C_{1\text{-}1}$	$C_{1\text{-}2}$	$C_{1\text{-}3}$	$C_{1\text{-}4}$	$C_{1\text{-}5}$	$C_{1\text{-}6}$
$C_{1\text{-}1}$	1	2	6	7	3	4

Table A13. Other criteria to the worst criterion for economic dimension based on BWM method (Expert number 3).

Others to the Worst	C_{1-4}
C_{1-1}	7
C_{1-2}	6
C_{1-3}	2
C_{1-4}	1
C_{1-5}	3
C_{1-6}	5

Table A14. Final results and weights of main criteria for economic dimension based on BWM method (Expert number 3).

Weight	C_{1-1}	C_{1-2}	C_{1-3}	C_{1-4}	C_{1-5}	C_{1-6}
	0.377	0.233	0.078	0.041	0.155	0.116
Final weight	0.125	0.078	0.026	0.014	0.052	0.039
K_{si} (BWM)	0.089					

Table A15. Comparative results (Expert number 3).

Weight	C_{1-1}	C_{1-2}	C_{1-3}	C_{1-4}	C_{1-5}	C_{1-6}
SWARA	0.075	0.068	0.043	0.037	0.059	0.051
BWM	0.125	0.078	0.026	0.014	0.052	0.039
Priority based on SWARA	1	2	5	6	3	4
Priority based on BWM	1	2	5	6	3	4

Table A16. Final weights of economic dimension based on SWARA (Expert number 4).

Criteria	The Comparative Importance of Average Value S_j	Coefficient $K_j = S_j + 1$	Recalculated Weight $w_j = \frac{x_{j-1}}{k_j}$	Weight $q_j = \frac{w_j}{\sum w_j}$	Final Weight
C_{1-1}	-	1	1	0.218	0.072
C_{1-2}	0.1	1.1	0.909	0.198	0.066
C_{1-5}	0.1	1.1	0.826	0.180	0.060
C_{1-6}	0.15	1.15	0.719	0.156	0.052
C_{1-3}	0.2	1.2	0.599	0.130	0.043
C_{1-4}	0.1	1.1	0.544	0.118	0.039

Table A17. Best criterion to other criteria for economic dimension based on BWM method (Expert number 4).

Best to Others	C_{1-1}	C_{1-2}	C_{1-3}	C_{1-4}	C_{1-5}	C_{1-6}
C_{1-1}	1	2	6	6	4	5

Table A18. Other criteria to the worst criterion for economic dimension based on BWM method (Expert number 4).

Others to the Worst	$C_{1\text{-}4}$
$C_{1\text{-}1}$	6
$C_{1\text{-}2}$	5
$C_{1\text{-}3}$	3
$C_{1\text{-}4}$	1
$C_{1\text{-}5}$	4
$C_{1\text{-}6}$	3

Table A19. Final results and weights of main criteria for economic dimension based on BWM method (Expert number 4).

Weight	$C_{1\text{-}1}$	$C_{1\text{-}2}$	$C_{1\text{-}3}$	$C_{1\text{-}4}$	$C_{1\text{-}5}$	$C_{1\text{-}6}$
	0.402	0.244	0.081	0.052	0.122	0.098
Final weight	0.134	0.081	0.027	0.017	0.041	0.033
K_{si} (BWM)	0.087					

Table A20. Comparative results (Expert number 4).

Weight	$C_{1\text{-}1}$	$C_{1\text{-}2}$	$C_{1\text{-}3}$	$C_{1\text{-}4}$	$C_{1\text{-}5}$	$C_{1\text{-}6}$
SWARA	0.072	0.066	0.043	0.039	0.060	0.052
BWM	0.134	0.081	0.027	0.017	0.041	0.033
Priority based on SWARA	1	2	5	6	3	4
Priority based on BWM	1	2	5	6	3	4

Table A21. Final weights of economic dimension based on SWARA (Expert number 5).

Criteria	The Comparative Importance of Average Value S_j	Coefficient $K_j = S_j + 1$	Recalculated Weight $w_j = \frac{x_{j-1}}{k_j}$	Weight $q_j = \frac{w_j}{\sum w_j}$	Final Weight
$C_{1\text{-}1}$	-	1	1	0.236	0.078
$C_{1\text{-}2}$	0.15	1.15	0.870	0.205	0.068
$C_{1\text{-}5}$	0.2	1.2	0.725	0.171	0.057
$C_{1\text{-}6}$	0.2	1.2	0.604	0.142	0.047
$C_{1\text{-}3}$	0.1	1.1	0.549	0.129	0.043
$C_{1\text{-}4}$	0.1	1.1	0.499	0.118	0.039

Table A22. Best criterion to other criteria for economic dimension based on BWM method (Expert number 5).

Best to Others	$C_{1\text{-}1}$	$C_{1\text{-}2}$	$C_{1\text{-}3}$	$C_{1\text{-}4}$	$C_{1\text{-}5}$	$C_{1\text{-}6}$
$C_{1\text{-}1}$	1	3	6	7	4	5

Table A23. Other criteria to the worst criterion for economic dimension based on BWM method (Expert number 5).

Others to the Worst	C_{1-4}
C_{1-1}	7
C_{1-2}	6
C_{1-3}	3
C_{1-4}	1
C_{1-5}	6
C_{1-6}	4

Table A24. Final results and weights of main criteria for economic dimension based on BWM method (Expert number 5).

Weight	C_{1-1}	C_{1-2}	C_{1-3}	C_{1-4}	C_{1-5}	C_{1-6}
	0.430	0.185	0.092	0.044	0.138	0.111
Final weights	0.143	0.061	0.031	0.015	0.046	0.037
K_{si} (BWM)			0.124			

Table A25. Comparative results (Expert number 5).

Weight	C_{1-1}	C_{1-2}	C_{1-3}	C_{1-4}	C_{1-5}	C_{1-6}
SWARA	0.078	0.068	0.043	0.039	0.057	0.047
BWM	0.143	0.061	0.031	0.015	0.046	0.037
Priority based on SWARA	1	2	5	6	3	4
Priority based on BWM	1	2	5	6	3	4

Social dimension:

Table A26. Final weights of social dimension based on SWARA (Expert number 1).

Criteria	The Comparative Importance of Average Value S_j	Coefficient $K_j = S_j + 1$	Recalculated Weight $w_j = \frac{x_{j-1}}{k_j}$	Weight $q_j = \frac{w_j}{\Sigma w_j}$	Final Weight
C_{2-1}	-	1	1	0.316	0.105
C_{2-2}	0.2	1.2	0.833	0.264	0.088
C_{2-3}	0.15	1.15	0.725	0.229	0.076
C_{2-4}	0.2	1.2	0.604	0.191	0.064

Table A27. Best criterion to other criteria for social dimension based on BWM method (Expert number 1).

Best to Others	C_{2-1}	C_{2-2}	C_{2-3}	C_{2-4}
C_{2-1}	1	2	3	4

Table A28. Other criteria to the worst criterion for social dimension based on BWM method (Expert number 1).

Others to the Worst	$C_{2\text{-}4}$
$C_{2\text{-}1}$	5
$C_{2\text{-}2}$	4
$C_{2\text{-}3}$	3
$C_{2\text{-}4}$	1

Table A29. Final results and weights of main criteria for social dimension based on BWM method (Expert number 1).

Weight	C_1	C_2	C_3	C_4
	0.455	0.273	0.182	0.091
Final weight	0.151	0.091	0.061	0.030
K_{si} (BWM)		0.091		

Table A30. Comparative results (Expert number 1).

Weight	$C_{1\text{-}1}$	$C_{1\text{-}2}$	$C_{1\text{-}3}$	$C_{1\text{-}4}$
SWARA	0.105	0.088	0.076	0.064
BWM	0.151	0.091	0.061	0.030
Priority based on SWARA	1	2	3	4
Priority based on BWM	1	2	3	4

Table A31. Final weights of social dimension based on SWARA (Expert number 2).

Criteria	The Comparative Importance of Average Value S_j	Coefficient $K_j = S_j + 1$	Recalculated Weight $w_j = \frac{x_{j-1}}{k_j}$	Weight $q_j = \frac{w_j}{\sum w_j}$	Final Weight
$C_{2\text{-}1}$	-	1	1	0.301	0.100
$C_{2\text{-}2}$	0.1	1.1	0.909	0.273	0.091
$C_{2\text{-}3}$	0.2	1.2	0.758	0.228	0.076
$C_{2\text{-}4}$	0.15	1.15	0.659	0.198	0.066

Table A32. Best criterion to other criteria for social dimension based on BWM method (Expert number 2).

Best to Others	$C_{2\text{-}1}$	$C_{2\text{-}2}$	$C_{2\text{-}3}$	$C_{2\text{-}4}$
$C_{2\text{-}1}$	1	3	5	7

Table A33. Other criteria to the worst criterion for social dimension based on BWM method (Expert number 2).

Others to the Worst	$C_{2\text{-}4}$
$C_{2\text{-}1}$	6
$C_{2\text{-}2}$	5
$C_{2\text{-}3}$	4
$C_{2\text{-}4}$	1

Table A34. Final results and weights of main criteria for social dimension based on BWM method (Expert number 2).

Weight	C_1	C_2	C_3	C_4
	0.558	0.223	0.140	0.070
Final weight	0.186	0.077	0.046	0.023
K_{si} (BWM)	0.140			

Table A35. Comparative results (Expert number 2).

Weight	$C_{1\text{-}1}$	$C_{1\text{-}2}$	$C_{1\text{-}3}$	$C_{1\text{-}4}$
SWARA	0.100	0.091	0.076	0.066
BWM	0.186	0.077	0.046	0.023
Priority based on SWARA	1	2	3	4
Priority based on BWM	1	2	3	4

Table A36. Final weights of social dimension based on SWARA (Expert number 3).

Criteria	The Comparative Importance of Average Value S_j	Coefficient $K_j = S_j + 1$	Recalculated Weight $w_j = \frac{x_{j-1}}{k_j}$	Weight $q_j = \frac{w_j}{\sum w_j}$
$C_{2\text{-}1}$	-	1	1	0.103
$C_{2\text{-}2}$	0.2	1.2	0.833	0.086
$C_{2\text{-}3}$	0.1	1.1	0.758	0.078
$C_{2\text{-}4}$	0.2	1.2	0.631	0.065

Table A37. Best criterion to other criteria for social dimension based on BWM method (Expert number 3).

Best to Others	$C_{2\text{-}1}$	$C_{2\text{-}2}$	$C_{2\text{-}3}$	$C_{2\text{-}4}$
$C_{2\text{-}1}$	1	3	4	6

Table A38. Other criteria to the worst criterion for social dimension based on BWM method (Expert number 3).

Others to the Worst	$C_{2\text{-}4}$
$C_{2\text{-}1}$	6
$C_{2\text{-}2}$	4
$C_{2\text{-}3}$	3
$C_{2\text{-}4}$	1

Table A39. Final results and weights of main criteria for social dimension based on BWM method (Expert number 3).

Weight	C_1	C_2	C_3	C_4
	0.550	0.214	0.160	0.076
Final weight	0.183	0.071	0.053	0.025
K_{si} (BWM)		0.092		

Table A40. Comparative results (Expert number 3).

Weight	$C_{1\text{-}1}$	$C_{1\text{-}2}$	$C_{1\text{-}3}$	$C_{1\text{-}4}$
SWARA	0.103	0.086	0.078	0.065
BWM	0.183	0.071	0.053	0.025
Priority based on SWARA	1	2	3	4
Priority based on BWM	1	2	3	4

Table A41. Final weights of social dimension based on SWARA (Expert number 4).

Criteria	The Comparative Importance of Average Value S_j	Coefficient $K_j = S_j + 1$	Recalculated Weight $w_j = \frac{x_{j-1}}{k_j}$	Weight $q_j = \frac{w_j}{\sum w_j}$	Final Weight
$C_{2\text{-}1}$	-	1	1	0.318	0.106
$C_{2\text{-}2}$	0.15	1.15	0.870	0.276	0.092
$C_{2\text{-}3}$	0.25	1.25	0.696	0.221	0.074
$C_{2\text{-}4}$	0.2	1.2	0.580	0.184	0.061

Table A42. Best criterion to other criteria for social dimension based on BWM method (Expert number 4).

Best to Others	$C_{2\text{-}1}$	$C_{2\text{-}2}$	$C_{2\text{-}3}$	$C_{2\text{-}4}$
$C_{2\text{-}1}$	1	3	4	5

Table A43. Other criteria to the worst criterion for social dimension based on BWM method (Expert number 4).

Others to the Worst	$C_{2\text{-}4}$
$C_{2\text{-}1}$	5
$C_{2\text{-}2}$	4
$C_{2\text{-}3}$	3
$C_{2\text{-}4}$	1

Table A44. Final results and weights of main criteria for social dimension based on BWM method (Expert number 4).

Weight	C_1	C_2	C_3	C_4
	0.536	0.218	0.163	0.084
Final weight	0.178	0.072	0.054	0.028
K_{si} (BWM)	0.117			

Table A45. Comparative results (Expert number 4).

Weight	$C_{1\text{-}1}$	$C_{1\text{-}2}$	$C_{1\text{-}3}$	$C_{1\text{-}4}$
SWARA	0.106	0.092	0.074	0.061
BWM	0.178	0.072	0.054	0.028
Priority based on SWARA	1	2	3	4
Priority based on BWM	1	2	3	4

Table A46. Final weights of social dimension based on SWARA (Expert number 5).

Criteria	The Comparative Importance of Average Value S_j	Coefficient $K_j = S_j + 1$	Recalculated Weight $w_j = \frac{x_{j-1}}{k_j}$	Weight $q_j = \frac{w_j}{\sum w_j}$	Final Weight
$C_{2\text{-}1}$	-	1	1	0.303	0.101
$C_{2\text{-}2}$	0.1	1.1	0.909	0.276	0.092
$C_{2\text{-}3}$	0.2	1.2	0.758	0.230	0.076
$C_{2\text{-}4}$	0.2	1.2	0.631	0.191	0.064

Table A47. Best criterion to other criteria for social dimension based on BWM method (Expert number 5).

Best to Others	$C_{2\text{-}1}$	$C_{2\text{-}2}$	$C_{2\text{-}3}$	$C_{2\text{-}4}$
$C_{2\text{-}1}$	1	2	4	5

Table A48. Other criteria to the worst criterion for social dimension based on BWM method (Expert number 5).

Others to the Worst	$C_{2\text{-}4}$
$C_{2\text{-}1}$	5
$C_{2\text{-}2}$	4
$C_{2\text{-}3}$	3
$C_{2\text{-}4}$	1

Table A49. Final results and weights of main criteria for social dimension based on BWM method (Expert number 5).

Weight	C_1	C_2	C_3	C_4
	0.487	0.289	0.145	0.079
Final weight	0.162	0.096	0.048	0.026
K_{si} (BWM)	0.092			

Table A50. Comparative results (Expert number 5).

Weight	$C_{1\text{-}1}$	$C_{1\text{-}2}$	$C_{1\text{-}3}$	$C_{1\text{-}4}$
SWARA	0.101	0.092	0.076	0.064
BWM	0.162	0.096	0.048	0.026
Priority based on SWARA	1	2	3	4
Priority based on BWM	1	2	3	4

Environmental dimension:

Table A51. Final weights of environment dimension based on SWARA (Expert number 1).

Criteria	The Comparative Importance of Average Value S_j	Coefficient $K_j = S_j + 1$	Recalculated Weight $w_j = \frac{x_{j-1}}{k_j}$	Weight $q_j = \frac{w_j}{\sum w_j}$	Final Weight
$C_{3\text{-}6}$	-	1	1	0.221	0.074
$C_{3\text{-}1}$	0.2	1.2	0.833	0.184	0.061
$C_{3\text{-}5}$	0.15	1.15	0.725	0.160	0.053
$C_{3\text{-}2}$	0.2	1.2	0.604	0.133	0.045
$C_{3\text{-}4}$	0.1	1.1	0.549	0.121	0.040
$C_{3\text{-}3}$	0.25	1.25	0.439	0.097	0.032
$C_{3\text{-}7}$	0.15	1.15	0.382	0.084	0.028

Table A52. Best criterion to other criteria for environment dimension based on BWM method (Expert number 1).

Best to Others	$C_{3\text{-}1}$	$C_{3\text{-}2}$	$C_{3\text{-}3}$	$C_{3\text{-}4}$	$C_{3\text{-}5}$	$C_{3\text{-}6}$	$C_{3\text{-}7}$
$C_{3\text{-}6}$	2	4	6	5	3	1	7

Table A53. Other criteria to the worst criterion for environment dimension based on BWM method (Expert number 1).

Others to the Worst	$C_{3\text{-}7}$
$C_{3\text{-}1}$	6
$C_{3\text{-}2}$	4
$C_{3\text{-}3}$	2
$C_{3\text{-}4}$	3
$C_{3\text{-}5}$	5
$C_{3\text{-}6}$	7
$C_{3\text{-}7}$	1

Table A54. Final results and weights of main criteria for environment dimension based on BWM method (Expert number 1).

Weight	$C_{3\text{-}1}$	$C_{3\text{-}2}$	$C_{3\text{-}3}$	$C_{3\text{-}4}$	$C_{3\text{-}5}$	$C_{3\text{-}6}$	$C_{3\text{-}7}$
	0.209	0.105	0.070	0.084	0.139	0.353	0.041
Final weight	0.070	0.035	0.023	0.028	0.047	0.118	0.014
K_{si} (BWM)				0.066			

Table A55. Comparative results (Expert number 1).

Weight	$C_{3\text{-}1}$	$C_{3\text{-}2}$	$C_{3\text{-}3}$	$C_{3\text{-}4}$	$C_{3\text{-}5}$	$C_{3\text{-}6}$	$C_{3\text{-}7}$
SWARA	0.061	0.045	0.032	0.040	0.053	0.074	0.028
BWM	0.070	0.035	0.023	0.028	0.047	0.118	0.014
Priority based on SWARA	2	4	6	5	3	1	7
Priority based on BWM	2	4	6	5	3	1	7

Table A56. Final weights of environment dimension based on SWARA (Expert number 2).

Criteria	The Comparative Importance of Average Value S_j	Coefficient $K_j = S_j + 1$	Recalculated Weight $w_j = \frac{x_{j-1}}{k_j}$	Weight $q_j = \frac{w_j}{\sum w_j}$	Final Weight
C_{3-6}	-	1	1	0.224	0.075
C_{3-1}	0.2	1.2	0.833	0.186	0.062
C_{3-5}	0.25	1.25	0.667	0.149	0.050
C_{3-2}	0.1	1.1	0.606	0.136	0.045
C_{3-4}	0.15	1.15	0.527	0.118	0.039
C_{3-3}	0.2	1.2	0.439	0.098	0.033
C_{3-7}	0.1	1.1	0.399	0.089	0.030

Table A57. Best criterion to other criteria for environment dimension based on BWM method (Expert number 2).

Best to Others	C_{3-1}	C_{3-2}	C_{3-3}	C_{3-4}	C_{3-5}	C_{3-6}	C_{3-7}
C_{3-6}	2	4	5	5	3	1	6

Table A58. Other criteria to the worst criterion for environment dimension based on BWM method (Expert number 2).

Others to the Worst	C_{3-7}
C_{3-1}	6
C_{3-2}	4
C_{3-3}	2
C_{3-4}	4
C_{3-5}	5
C_{3-6}	7
C_{3-7}	1

Table A59. Final results and weights of main criteria for environment dimension based on BWM method (Expert number 2).

Weight	C_{3-1}	C_{3-2}	C_{3-3}	C_{3-4}	C_{3-5}	C_{3-6}	C_{3-7}
	0.210	0.105	0.084	0.084	0.140	0.336	0.042
Final weight	0.070	0.035	0.028	0.028	0.047	0.112	0.014
K_{si} (BWM)				0.084			

Table A60. Comparative results (Expert number 2).

Weight	C_{3-1}	C_{3-2}	C_{3-3}	C_{3-4}	C_{3-5}	C_{3-6}	C_{3-7}
SWARA	0.062	0.045	0.033	0.039	0.050	0.075	0.030
BWM	0.070	0.035	0.028	0.028	0.047	0.112	0.014
Priority based on SWARA	2	4	6	5	3	1	7
Priority based on BWM	2	4	5	5	3	1	7

Table A61. Final weights of environment dimension based on SWARA (Expert number 3).

Criteria	The Comparative Importance of Average Value S_j	Coefficient $K_j = S_j + 1$	Recalculated Weight $w_j = \frac{x_{j-1}}{k_j}$	Weight $q_j = \frac{w_j}{\sum w_j}$	Final Weight
$C_{3\text{-}6}$	-	1	1	0.208	0.070
$C_{3\text{-}1}$	0.1	1.1	0.909	0.189	0.063
$C_{3\text{-}5}$	0.15	1.15	0.791	0.165	0.055
$C_{3\text{-}2}$	0.2	1.2	0.659	0.137	0.046
$C_{3\text{-}4}$	0.2	1.2	0.549	0.114	0.038
$C_{3\text{-}3}$	0.15	1.15	0.477	0.099	0.033
$C_{3\text{-}7}$	0.15	1.15	0.415	0.086	0.029

Table A62. Best criterion to other criteria for environment dimension based on BWM method (Expert number 3).

Best to Others	$C_{3\text{-}1}$	$C_{3\text{-}2}$	$C_{3\text{-}3}$	$C_{3\text{-}4}$	$C_{3\text{-}5}$	$C_{3\text{-}6}$	$C_{3\text{-}7}$
$C_{3\text{-}6}$	2	4	5	5	3	1	6

Table A63. Other criteria to the worst criterion for environment dimension based on BWM method (Expert number 3).

Others to the Worst	$C_{3\text{-}7}$
$C_{3\text{-}1}$	6
$C_{3\text{-}2}$	4
$C_{3\text{-}3}$	2
$C_{3\text{-}4}$	4
$C_{3\text{-}5}$	5
$C_{3\text{-}6}$	7
$C_{3\text{-}7}$	1

Table A64. Final results and weights of main criteria for environment dimension based on BWM method (Expert number 3).

Weight	$C_{3\text{-}1}$	$C_{3\text{-}2}$	$C_{3\text{-}3}$	$C_{3\text{-}4}$	$C_{3\text{-}5}$	$C_{3\text{-}6}$	$C_{3\text{-}7}$
	0.210	0.105	0.084	0.084	0.140	0.336	0.042
Final weights	0.070	0.035	0.028	0.028	0.047	0.112	0.014
K_{si} (BWM)				0.084			

Table A65. Comparative results (Expert number 3).

Weight	$C_{3\text{-}1}$	$C_{3\text{-}2}$	$C_{3\text{-}3}$	$C_{3\text{-}4}$	$C_{3\text{-}5}$	$C_{3\text{-}6}$	$C_{3\text{-}7}$
SWARA	0.062	0.045	0.033	0.039	0.050	0.075	0.030
BWM	0.070	0.035	0.028	0.028	0.047	0.112	0.014
Priority based on SWARA	2	4	6	5	3	1	7
Priority based on BWM	2	4	5	5	3	1	7

Table A66. Final weights of environment dimension based on SWARA (Expert number 4).

Criteria	The Comparative Importance of Average Value S_j	Coefficient $K_j = S_j + 1$	Recalculated Weight $w_j = \frac{x_{j-1}}{k_j}$	Weight $q_j = \frac{w_j}{\sum w_j}$	Final Weight
C_{3-6}	-	1	1	0.219	0.073
C_{3-1}	0.15	1.15	0.870	0.190	0.064
C_{3-5}	0.25	1.25	0.696	0.152	0.051
C_{3-2}	0.1	1.1	0.632	0.139	0.046
C_{3-4}	0.2	1.2	0.527	0.115	0.039
C_{3-3}	0.15	1.15	0.458	0.100	0.034
C_{3-7}	0.2	1.2	0.382	0.084	0.028

Table A67. Best criterion to other criteria for environment dimension based on BWM method (Expert number 4).

Best to Others	C_{3-1}	C_{3-2}	C_{3-3}	C_{3-4}	C_{3-5}	C_{3-6}	C_{3-7}
C_{3-6}	2	4	7	6	3	1	8

Table A68. Other criteria to the worst criterion for environment dimension based on BWM method (Expert number 4).

Others to the Worst	C_{3-7}
C_{3-1}	7
C_{3-2}	3
C_{3-3}	2
C_{3-4}	4
C_{3-5}	5
C_{3-6}	8
C_{3-7}	1

Table A69. Final results and weights of main criteria for environment dimension based on BWM method (Expert number 4).

Weight	C_{3-1}	C_{3-2}	C_{3-3}	C_{3-4}	C_{3-5}	C_{3-6}	C_{3-7}
	0.216	0.108	0.062	0.072	0.144	0.361	0.036
Final weight	0.072	0.036	0.021	0.024	0.048	0.121	0.012
K_{si} (BWM)	0.072						

Table A70. Comparative results (Expert number 4).

Weight	C_{3-1}	C_{3-2}	C_{3-3}	C_{3-4}	C_{3-5}	C_{3-6}	C_{3-7}
SWARA	0.064	0.046	0.034	0.039	0.051	0.073	0.028
BWM	0.072	0.036	0.021	0.024	0.048	0.121	0.012
Priority based on SWARA	2	4	6	5	3	1	7
Priority based on BWM	2	4	6	5	3	1	7

Table A71. Final weights of environment dimension based on SWARA (Expert number 5).

Criteria	The Comparative Importance of Average Value S_j	Coefficient $K_j = S_j + 1$	Recalculated Weight $w_j = \frac{x_{j-1}}{k_j}$	Weight $q_j = \frac{w_j}{\sum w_j}$	Final Weight
$C_{3\text{-}6}$	-	1	1	0.212	0.071
$C_{3\text{-}1}$	0.2	1.2	0.833	0.177	0.059
$C_{3\text{-}5}$	0.1	1.1	0.758	0.161	0.054
$C_{3\text{-}2}$	0.15	1.15	0.659	0.140	0.047
$C_{3\text{-}4}$	0.15	1.15	0.573	0.121	0.041
$C_{3\text{-}3}$	0.2	1.2	0.477	0.101	0.034
$C_{3\text{-}7}$	0.15	1.15	0.415	0.088	0.029

Table A72. Best criterion to other criteria for environment dimension based on BWM method (Expert number 5).

Best to Others	$C_{3\text{-}1}$	$C_{3\text{-}2}$	$C_{3\text{-}3}$	$C_{3\text{-}4}$	$C_{3\text{-}5}$	$C_{3\text{-}6}$	$C_{3\text{-}7}$
$C_{3\text{-}6}$	3	5	8	7	4	1	8

Table A73. Other criteria to the worst criterion for environment dimension based on BWM method (Expert number 5).

Others to the Worst	$C_{3\text{-}7}$
$C_{3\text{-}1}$	7
$C_{3\text{-}2}$	4
$C_{3\text{-}3}$	3
$C_{3\text{-}4}$	3
$C_{3\text{-}5}$	5
$C_{3\text{-}6}$	7
$C_{3\text{-}7}$	1

Table A74. Final results and weights of main criteria for environment dimension based on BWM method (Expert number 5).

Weight	$C_{3\text{-}1}$	$C_{3\text{-}2}$	$C_{3\text{-}3}$	$C_{3\text{-}4}$	$C_{3\text{-}5}$	$C_{3\text{-}6}$	$C_{3\text{-}7}$
	0.175	0.105	0.066	0.075	0.131	0.408	0.042
Final weight	0.058	0.035	0.022	0.025	0.044	0.136	0.014
K_{si} (BWM)				0.116			

Table A75. Comparative results (Expert number 5).

Weight	$C_{3\text{-}1}$	$C_{3\text{-}2}$	$C_{3\text{-}3}$	$C_{3\text{-}4}$	$C_{3\text{-}5}$	$C_{3\text{-}6}$	$C_{3\text{-}7}$
SWARA	0.059	0.047	0.034	0.041	0.054	0.071	0.029
BWM	0.058	0.035	0.022	0.025	0.044	0.136	0.014
Priority based on SWARA	2	4	6	5	3	1	7
Priority based on BWM	2	4	6	5	3	1	7

References

1.	Tian, G.; Zhang, H.H.; Feng, Y.X.; Wang, D.Q.; Peng, Y.; Jia, H.F. Green decoration materials selection under interior environment characteristics: A grey-correlation based hybrid MCDM method. *Renew. Sustain. Energy Rev.* **2018**, *81*, 682–692. [CrossRef]
2.	Hayles, C.S. Environmentally sustainable interior design: A snapshot of current supply of and demand for green, sustainable or Fair Trade products for interior design practice. *Int. J. Sustain. Built Environ.* **2015**, *4*, 100–108. [CrossRef]
3.	Farag, M.M. Quantitative methods of materials selection. In *Handbook of Materials Selection*; Kutz, M., Ed.; John Wiley & Sons: Hoboken, NJ, USA, 2002; pp. 3–24.
4.	Cables, E.; Lamata, M.T.; Verdegay, J.L. RIM-reference ideal method in multi criteria decision making. *Inf. Sci.* **2016**, *337–338*, 1–10. [CrossRef]
5.	Yazdani, M.; Jahan, A.; Zavadskas, E.K. Analysis in Material Selection: Influence of Normalization Tools on COPRAS-G. *Econ. Comput. Econ. Cybern. Stud. Res.* **2017**, *51*, 59–74.
6.	Ogunkah, I.; Yang, J. Investigating Factors Affecting Material Selection: The Impacts on Green Vernacular Building Materials in the Design-Decision Making Process. *Buildings* **2012**, *2*, 1–32. [CrossRef]
7.	Castro-Lacouture, D.; Sefair, J.; Flórez, L.; Medaglia, A.L. Optimization model for the selection of materials using a LEED-based green building rating system in Colombia. *Build. Environ.* **2009**, *44*, 1162–1170. [CrossRef]
8.	Zhou, C.C.; Yin, G.F.; Hu, X.B. Multi-objective optimization of material selection for sustainable products: Artificial neural networks and genetic algorithm approach. *Mater. Des.* **2009**, *30*, 1209–1215. [CrossRef]
9.	Rahman, S.; Srinath, P.; Henry, O.; Yaxin, B. A knowledge-based decision support system for roofing materials selection and cost estimating: A conceptual framework and data modelling. In Proceedings of the 25th Annual ARCOM Conference, Nottingham, UK, 7–9 September 2009.
10.	Maniya, K.; Bhatt, M.G. A selection of material using a novel type decision-making method: Preference selection index method. *Mater. Des.* **2010**, *31*, 1785–1789. [CrossRef]
11.	Akadiri, P.O.; Olomolaiye, P.O.; Chinyio, E.A. Multi-criteria evaluation model for the selection of sustainable materials for building projects. *Autom. Constr.* **2013**, *30*, 113–125. [CrossRef]
12.	Florez, L.; Castro-Lacouture, D. Optimization model for sustainable materials selection using objective and subjective factors. *Mater. Des.* **2013**, *46*, 310–321. [CrossRef]
13.	Baharetha, S.M.; Al-Hammad, A.A.; Alshuwaikhat, H.M. Towards a unified set of sustainable building materials criteria. In Proceedings of the International Conference on Sustainable Design, Engineering, and Construction, Fort Worth, TX, USA, 7–9 November 2013; pp. 732–740.
14.	Ribeiro, I.; Peças, P.; Henriques, E. A life cycle framework to support material selection for Ecodesign: A case study on biodegradable polymers. *Mater. Des.* **2013**, *51*, 300–308. [CrossRef]
15.	Hosseinijou, S.A.; Mansour, S.; Shirazi, M.A. Social life cycle assessment for material selection: A case study of building materials. *Int. J. Life Cycle Assess.* **2014**, *19*, 620–645. [CrossRef]
16.	Van der Velden, N.M.; Kuusk, K.; Köhler, A.R. Life cycle assessment and eco-design of smart textiles: The importance of material selection demonstrated through e-textile product redesign. *Mater. Des.* **2015**, *84*, 313–324. [CrossRef]
17.	Zhao, R.; Su, H.; Chen, X.; Yu, Y. Commercially available materials selection in sustainable design: An integrated multi-attribute decision making approach. *Sustainability* **2016**, *8*, 79. [CrossRef]
18.	Ma, F.; Zhao, Y.; Pu, Y.; Li, J. A comprehensive multi-criteria decision making model for sustainable material selection considering life cycle assessment method. *IEEE Access* **2018**, *6*, 58338–58354. [CrossRef]
19.	Bhowmik, C.; Gangwar, S.; Bhowmik, S.; Ray, A. Optimum Selection of Energy-Efficient Material: A MCDM-Based Distance Approach. In *Soft Computing Applications. Studies in Computational Intelligence*; Ray, K., Pant, M., Bandyopadhyay, A., Eds.; Springer: Singapore, 2018; p. 761.
20.	Sahamir, S.R.; Zakaria, R.; Rooshdi, R.R.; Adenan, M.A.; Shamsuddin, S.M.; Abidin, N.I.; Ismail, N.A.; Aminudin, E. Multi-Criteria Decision Analysis for Evaluating Sustainable Lifts Design of Public Hospital Buildings. *Chem. Eng. Trans.* **2018**, *63*, 199–204.
21.	Moghtadernejad, S.; Chouinard, L.E.; Mirza, M.S. Multi-criteria decision-making methods for preliminary design of sustainable facades. *J. Build. Eng.* **2018**, *19*, 181–190. [CrossRef]

22. Lewandowska, A.; Branowsk, B.; Joachimiak-Lechman, K.; Kurczewski, P.; Selech, J.; Zablocki, M. Sustainable Design: A Case of Environmental and Cost Life Cycle Assessment of a Kitchen Designed for Seniors and Disabled People. *Sustainability* **2017**, *9*, 1329. [CrossRef]

23. He, B.; Niu, Y.; Hou, S.; Li, F. Sustainable design from functional domain to physical domain. *J. Clean. Prod.* **2018**, *197*, 1296–1306. [CrossRef]

24. Ping, L.X. Application of green concept in mechanical design and manufacture. In *IOP Conference Series: Earth and Environmental Science*; IOP Publishing: Bristol, UK, 2017; Volume 94, p. 012128.

25. Halstenberg, F.A.; Buchert, T.; Bonvoisin, J.; Lindow, K.; Stark, R. Target-oriented Modularization-Addressing Sustainability Design Goals in Product Modularization. *Procedia CIRP* **2015**, *29*, 603–608. [CrossRef]

26. Go, T.F.; Wahab, D.A.; Hishamuddin, H. Multiple generation life-cycles for product sustainability: The way forward. *J. Clean. Prod.* **2015**, *95*, 16–29. [CrossRef]

27. Rosen, M.A.; Kishawy, H.A. Sustainable Manufacturing and Design: Concepts, Practices and Needs. *Sustainability* **2012**, *4*, 154–174. [CrossRef]

28. Zhang, C.; Xie, G.; Liu, C.; Lu, C. Assessment of soil erosion under woodlands using USLE in China. *Front. Earth Sci.* **2011**, *5*, 150–161. [CrossRef]

29. Chiu, M.C.; Chu, C.H. Review of sustainable product design from life cycle perspectives. *Int. J. Precis. Eng. Manuf.* **2012**, *13*, 1259–1272. [CrossRef]

30. Sonego, M.; Soares Echeveste, M.E.; Debarba, H.G. The role of modularity in sustainable design: A systematic review. *J. Clean. Prod.* **2018**, *176*, 196–209. [CrossRef]

31. Jilcha, K.; Kitaw, D. Industrial occupational safety and health innovation for sustainable development. *Eng. Sci. Technol.* **2016**, *20*, 372–380. [CrossRef]

32. Bachman, L.R. Sustainable Design and Postindustrial Society: Our Ethical and Aesthetic Crossroads. *Enquiry* **2016**, *13*, 30–38. [CrossRef]

33. Cimatti, B.; Campana, G.; Carluccio, L. Eco Design and Sustainable Manufacturing in Fashion: A Case Study in the Luxury Personal Accessories Industry. *Procedia Manuf.* **2017**, *8*, 393–400. [CrossRef]

34. Beck, L.A. New as Renewal: A Framework for Adaptive Reuse in the Sustainable Paradigm. Master's Thesis, University of Massachusetts Amherst, Amherst, MA, USA, 2014.

35. Raoufi, K.; Park, K.; Hasan Khan, M.T.; Haapala, K.R.; Psenka, C.E.; Jackson, K.L.; Okudan Kremer, G.E. A Cyber Learning Platform for Enhancing Undergraduate Engineering Education in Sustainable Product Design. *J. Clean. Prod.* **2018**, *211*, 730–741. [CrossRef]

36. Yan, J.; Feng, C. Sustainable design-oriented product modularity combined with 6R concept: A case study of rotor laboratory bench. *Clean Technol. Environ. Policy* **2014**, *16*, 95–109. [CrossRef]

37. Qian, X.; Zhang, H.C. Design for Environment: An Environmentally Conscious Analysis Model for Modular Design. *IEEE Trans. Electron. Packag. Manuf.* **2009**, *32*, 164–175. [CrossRef]

38. Foley, J.T.; Cochran, D.S. Manufacturing System Design Decomposition: An Ontology for Data Analytics and System Design Evaluation. *Procedia CIRP* **2017**, *60*, 175–180. [CrossRef]

39. Zhang, X.; Zhao, Y.; Sun, Q.; Wang, C. Decomposition and Attribution Analysis of Industrial Carbon Intensity Changes in Xinjiang. China. *Sustainability* **2017**, *9*, 459. [CrossRef]

40. Lumsakul, P.; Sheldrick, L.; Rahimifard, S. The Sustainable Co-Design of Products and Production Systems. *Procedia Manuf.* **2018**, *21*, 854–861. [CrossRef]

41. Saaty, T.L. *The Analytical Hierarchy Process*; McGraw-Hill: New York, NY, USA, 1980.

42. Saaty, T.J. *Decision Making in Complex Environments, The Analytical Hierarchy Process for Decision Making with Dependence and Dependence and Feedback*; RWS Publications: Pittsburgh, PA, USA, 1996.

43. Saaty, T.L. Fundamentals of analytic network process. In Proceedings of the International Symposium on the Analytic Hierarchy Process, Japan, Kobe, 12–14 August 1999; pp. 348–379.

44. Keršulienė, V.; Zavadskas, E.K.; Turskis, Z. Selection of rational dispute resolution method by applying new step-wise weight assessment ratio analysis (SWARA). *J. Bus. Econ. Manag.* **2010**, *11*, 243–258. [CrossRef]

45. Ginevicius, R. A New Determining Method for the Criteria Weights in Multi-Criteria Evaluation. *Int. J. Inf. Technol. Decis. Mak.* **2011**, *10*, 1067–1095. [CrossRef]

46. Rezaei, J. Best-worst multi-criteria decision-making method. *Omega* **2015**, *53*, 49–57. [CrossRef]

47. Hashemkhani Zolfani, S.; Yazdani, M.; Zavadskas, E.K. An extended stepwise weight assessment ratio analysis (SWARA) method for improving criteria prioritization process. *Soft Comput.* **2018**, *22*, 7399–7405. [CrossRef]

48. Pamucar, D.; Stevic, Z.; Sremac, S. A New Model for Determining Weight Coefficients of Criteria in MCDM Models: Full Consistency Method (FUCOM). *Symmetry* **2018**, *10*, 393. [CrossRef]

49. Govindan, K. Application of multi-criteria decision making/operations research techniques for sustainable management in mining and minerals. *Resour. Policy* **2015**, *46*, 1–5. [CrossRef]

50. Zavadskas, E.K.; Govindan, K.; Antucheviciene, J.; Turskis, Z. Hybrid multiple criteria decision-making methods: A review of applications for sustainability issues. *Econ. Res.* **2016**, *29*, 857–887. [CrossRef]

51. Kumar, A.; Sah, B.; Singh, A.R.; Deng, Y.; He, X.; Kumar, P.; Bancal, R.C. A review of multi criteria decision making (MCDM) towards sustainable renewable energy development. *Renew. Sustain. Energy Rev.* **2017**, *69*, 596–609. [CrossRef]

52. Siksnelyte, I.; Zavadskas, E.K.; Streimikiene, D.; Sharma, D. An Overview of Multi-Criteria Decision-Making Methods in Dealing with Sustainable Energy Development Issues. *Energies* **2018**, *11*, 2754. [CrossRef]

53. Shen, K.Y.; Tzeng, G.H. Advances in Multiple Criteria Decision Making for Sustainability: Modeling and Applications. *Sustainability* **2018**, *10*, 1600. [CrossRef]

54. Badri Ahmadi, H.; Kusi-Sarpong, S.; Rezaei, J. Assessing the social sustainability of supply chains using Best Worst Method. *Resour. Conserv. Recycl.* **2017**, *126*, 99–106. [CrossRef]

55. Zolfani, S.H.; Saparauskas, J. New Application of SWARA Method in Prioritizing Sustainability Assessment Indicators of Energy System. *Eng. Econ.* **2013**, *24*, 408–414. [CrossRef]

56. Hashemkjani Zolfani, S.; Bahrami, M. Investment prioritizing in high tech industries based on SWARA-COPRAS approach. *Technol. Econ. Dev. Econ.* **2014**, *20*, 534–553. [CrossRef]

57. Ghorshi Nezhad, M.R.; Hashemkhani Zolfani, S.; Moztarzadeh, F.; Zavadskas, E.K.; Bahrami, M. Planning the priority of high tech industries based on SWARA-WASPAS methodology: The case of the nanotechnology industry in Iran. *Econ. Res.* **2015**, *28*, 1111–1137. [CrossRef]

58. Hashemkhani Zolfani, S.; Salimi, J.; Maknoon, R.; Kildiene, S. Technology foresight about R&D projects selection; application of SWARA method at the policy making level. *Eng. Econ.* **2015**, *26*, 571–580.

59. Mardani, A.; Nilashi, M.; Zakuan, N.; Loganathan, N.; Soheilirad, S.; Saman, M.Z.M.; Ibrahim, O. A systematic review and meta-Analysis of SWARA and WASPAS methods: Theory and applications with recent fuzzy developments. *Appl. Soft Comput.* **2017**, *57*, 265–292. [CrossRef]

60. Sadjadia, S.J.; Karimi, M. Best-worst multi-criteria decision-making method: A robust approach. *Decis. Sci. Lett.* **2018**, *7*, 323–340. [CrossRef]

61. Gupta, H.; Barua, M.K. Identifying enablers of technological innovation for Indian MSMEs using best–worst multi criteria decision making method. *Technol. Forecast. Soc. Chang.* **2016**, *107*, 69–79. [CrossRef]

62. Yadav, G.; Mangla, S.K.; Luthra, S.; Jakhar, S. Hybrid BWM-ELECTRE-based decision framework for effective offshore outsourcing adoption: A case study. *Int. J. Prod. Res.* **2018**, *56*, 6259–6278. [CrossRef]

63. Rezaei, J.; Wang, J.; Tavasszy, L. Linking supplier development to supplier segmentation using Best Worst Method. *Expert Syst. Appl.* **2015**, *42*, 9152–9164. [CrossRef]

64. Chitsaz, N.; Azarivand, A. Water Scarcity Management in Arid Regions Based on an Extended Multiple Criteria Technique. *Water Resour. Manag.* **2016**, *31*, 233–250. [CrossRef]

65. Rezaei, J.; Hemmes, A.; Tavasszy, L.A. Multi-criteria decision-making for complex bundling configurations in surface transportation of air freight. *J. Air Transp. Manag.* **2017**, *61*, 95–105. [CrossRef]

66. Ren, J.; Liang, H.; Chan, F.T.S. Urban sewage sludge, sustainability, and transition for Eco-City: Multi-criteria sustainability assessment of technologies based on best-worst method. *Technol. Forecast. Soc. Chang.* **2017**, *116*, 29–39. [CrossRef]

67. Rezaei, J.; van Roekel, W.S.; Tavasszy, L.A. Measuring the relative importance of the logistics performance index indicators using Best Worst Method. *Transp. Policy* **2018**, *68*, 158–169. [CrossRef]

68. Van de Kaan, G.; Janssen, M.; Rezaei, J. Standards battles for business-to-government data exchange: Identifying success factors for standard dominance using the Best Worst Method. *Technol. Forecast. Soc. Chang.* **2018**, *137*, 182–189. [CrossRef]

69. Nawaz, F.; Asadabadi, M.R.; Janjua, N.K.; Hussain, O.K.; Chang, E.; Saberi, M. An MCDM method for cloud service selection using a Markov chain and the best-worst method. *Knowl.-Based Syst.* **2018**, *159*, 120–131. [CrossRef]

70. Salimi, N.; Rezaei, J. Measuring efficiency of university-industry Ph.D. projects using best worst method. *Scientometrics* **2016**, *109*, 1911–1938. [CrossRef] [PubMed]

71. Rezaei, J. Best-worst Multi-criteria Decision-Making Method: Some Properties and a Linear Model. *Omega* **2016**, *62*, 126–130. [CrossRef]
72. Khodadadi, M.R.; Hashemkhani Zolfani, S.; Yazdani, M.; Zavadskas, E.K. A hybrid MADM analysis in evaluating process of chemical wastewater purification regarding to advance oxidation processes. *J. Environ. Eng. Landsc. Manag.* **2017**, *25*, 277–288. [CrossRef]
73. Jamali, G.; Karimi Asl, E.; Hashemkhani Zolfani, S.; Saparauskas, J. Analysing large supply chain management competitive strategies in Iranian cement industries. *E+M Ekon. Manag.* **2017**, *20*, 70–84. [CrossRef]
74. Hashemkhani Zolfani, S.; Pourhossein, M.; Yazdani, M.; Zavadskas, E.K. Evaluating construction projects of hotels based on environmental sustainability with MCDM framework. *Alex. Eng. J.* **2018**, *57*, 357–365. [CrossRef]
75. Hong, H.; Panahi, M.; Shirzadi, A.; Ma, T.; Liu, J.; Zhu, A.; Chen, W.; Kougias, I.; Kazakis, N. Flood susceptibility assessment in Hengfeng area coupling adaptive neuro-fuzzy inference system with genetic algorithm and differential evolution. *Sci. Total Environ.* **2018**, *621*, 1124–1141. [CrossRef]
76. Zarbakhshnia, N.; Soleimani, H.; Ghaderi, H. Sustainable third-party reverse logistics provider evaluation and selection using fuzzy SWARA and developed fuzzy COPRAS in the presence of risk criteria. *Appl. Soft Comput.* **2018**, *65*, 307–319. [CrossRef]
77. Eghbali-Zarch, M.; Tavakkoli-Moghaddam, R.; Esfahanian, F.; Sepehri, M.M.; Azaron, A. Pharmacological therapy selection of type 2 diabetes based on the SWARA and modified MULTIMOORA methods under a fuzzy environment. *Artif. Intell. Med.* **2018**, *87*, 20–33. [CrossRef]
78. Vesković, S.; Stević, Z.; Stojić, G.; Vasiljević, M.; Milinković, S. Evaluation of the railway management model by using a new integrated model DELPHI-SWARA-MABAC. *Decis. Mak.* **2018**, *1*, 34–50. [CrossRef]
79. Heidary Dahooie, J.; Beheshti Jazan Abadi, E.; Salar Vanaki, A. Competency-based IT personnel selection using a hybrid SWARA and ARAS-G methodology. *Hum. Factors Ergon. Manuf. Serv. Ind.* **2018**, *28*, 5–16.
80. Keršulienė, V.; Turskis, Z. Integrated fuzzy multiple criteria decision making model for architect selection. *Technol. Econ. Dev. Econ.* **2011**, *17*, 645–666. [CrossRef]
81. Vafaeipour, M.; Hashemkhani Zolfani, S.; Varzandeh, M.H.; Derakhti, A.; Eshkalag, M.K. Assessment of regions priority for implementation of solar projects in Iran: New application of a hybrid multi-criteria decision making approach. *Energy Convers. Manag.* **2014**, *86*, 653–663. [CrossRef]
82. Stanujkic, D.; Karabasevic, D.; Zavadskas, E.K. A Framework for the Selection of a Packaging Design Based on the SWARA Method. *Eng. Econ.* **2015**, *26*, 181–187. [CrossRef]
83. ESCAP. *Integrating the Three Dimensions of Sustainable Development: A Framework and Tools*; ST/ESCAP/2737; United Nations Publication: New York, NY, USA, 2015.

© 2019 by the authors. Licensee MDPI, Basel, Switzerland. This article is an open access article distributed under the terms and conditions of the Creative Commons Attribution (CC BY) license (http://creativecommons.org/licenses/by/4.0/).

 *symmetry*

Article

Methods for Multiple-Attribute Group Decision Making with q-Rung Interval-Valued Orthopair Fuzzy Information and Their Applications to the Selection of Green Suppliers

Jie Wang [1], Hui Gao [1], Guiwu Wei [1,*] and Yu Wei [2,*]

[1] School of Business, Sichuan Normal University, Chengdu 610101, China; JW970326@163.com (J.W.); gaohuisxy@sicnu.edu.cn (H.G.)

[2] School of Finance, Yunnan University of Finance and Economics, Kunming 650221, China

* Correspondence: weiguiwu1973@sicnu.edu.cn (G.W.); ywei@home.swjtu.edu.cn (Y.W.)

Received: 15 November 2018; Accepted: 26 December 2018; Published: 6 January 2019

Abstract: In the practical world, there commonly exist different types of multiple-attribute group decision making (MAGDM) problems with uncertain information. Symmetry among some attributes' information that is already known and unknown, and symmetry between the pure attribute sets and fuzzy attribute membership sets, can be an effective way to solve this type of MAGDM problem. In this paper, we investigate four forms of information aggregation operators, including the Hamy mean (HM) operator, weighted HM (WHM) operator, dual HM (DHM) operator, and the dual-weighted HM (WDHM) operator with the q-rung interval-valued orthopair fuzzy numbers (q-RIVOFNs). Then, some extended aggregation operators, such as the q-rung interval-valued orthopair fuzzy Hamy mean (q-RIVOFHM) operator; q-rung interval-valued orthopairfuzzy weighted Hamy mean (q-RIVOFWHM) operator; q-rung interval-valued orthopair fuzzy dual Hamy mean (q-RIVOFDHM) operator; and q-rung interval-valued orthopair fuzzy weighted dual Hamy mean (q-RIVOFWDHM) operator are presented, and some of their precious properties are studied in detail. Finally, a real example for green supplier selection in green supply chain management is provided, to demonstrate the proposed approach and to verify its rationality and scientific nature.

Keywords: multiple attribute group decision making (MAGDM); Pythagorean fuzzy set (PFSs); q-rung orthopair fuzzy sets (q-RIVOFSs); q-RIVOFWHM operator; q-RIVOFWDHM operator; green suppliers selection

1. Introduction

For the indeterminacy of decision makers and decision-making issues, we cannot always give accurate evaluation values for alternatives to select the best project in real multiple-attribute decision making (MADM) problems. To overcome this disadvantage, fuzzy set theory, as defined by Zadeh [1] in 1965, originally used the membership function to describe the estimation results, rather than an exact real number. Atanassov [2,3] presents the intuitionistic fuzzy set (IFS) by considering another measurement index which names a non-membership function. Hereafter, the IFS and its extension has aroused the attention of a large number of scholars since its appearance [4–25]. More recently, the Pythagorean fuzzy set (PFS) [26,27] has emerged as a useful tool for describing the indeterminacy of the MADM problems. Zhang and Xu [28] proposed the detailed mathematical expression for PFS and presented the definition of Pythagorean fuzzy numbers (PFNs). Wei and Lu [29] proposed some Maclaurin Symmetric Mean Operators with PFNs. Peng and Yang [30] studied the division and subtraction operations of PFNs. Wei and Lu [31] defined some power aggregation operators with PFNs based on the traditional power aggregation operators [32–37]. Beliakov and James [38] presented the average aggregation functions of PFNs. Reformat and Yager [39] studied the collaborative-based recommender system under the Pythagorean fuzzy environment. Gou et al. [40] proposed some desirable properties of the continuous Pythagorean fuzzy number. Wei and Wei [41] defined some similar measures of Pythagorean fuzzy sets, based on cosine functions with traditional similarity measures [42–45]. Ren et al. [46] applied the Pythagorean fuzzy TODIM model in MADM. Garg [47] combines the Einstein Operations and Pythagorean fuzzy information to propose a new aggregation operator. Zeng et al. [48] provided a Pythagorean fuzzy hybrid method to study MADM. Garg [49] presents a novel accuracy function based on interval-valued Pythagorean fuzzy information for solving MADM problems. Wei et al. [50] propose the Pythagorean hesitant fuzzy Hamacher operators in MADM. Wei and Lu [51] develop the dual hesitant Pythagorean fuzzy Hamacher operators in MADM. Lu et al. [52] develop the hesitant Pythagorean fuzzy Hamacher aggregation operators in MADM.

In addition to this, based on the fundamental theories of IFS and PFS, Yager [53] further defined the q-rung orthopair fuzzy sets (q-ROFSs), in which the sum of the qth power of the degrees of membership and the qth power of the degrees of non-membership is satisfied the condition $\mu^q + v^q \leq 1$. It is clear that the q-ROFSs are more general for IFSs and PFSs, as they are all special cases. Therefore, we can express a wider range of fuzzy information by using q-ROFSs. Liu and Wang [54] develop the q-rung orthopair, fuzzy weighted averaging (q-ROFWA) operator and the q-rung orthopair, fuzzy weighted geometric (q-ROFWG) operator to fuse the evaluation information. Liu and Liu [55] proposes a q-rung orthopair, fuzzy Bonferroni mean (q-ROFBM) aggregation operator, by considering the q-rung orthopair fuzzy information and the Bonferroni mean (BM) operator. Wei et al. [56] combine the q-rung orthopair fuzzy numbers (q-ROFNs) with a generalized Heronian mean (GHM) operator to present some aggregation operators, and applied them into MADM problems. Wei et al. [57] define some q-rung orthopair, fuzzy Maclaurin symmetric mean operators for the potential evaluation of emerging technology commercialization.

Nevertheless, in many practical decision-making problems, for the uncertainty of the decision-making environment and the subjectivity of decision makers (DMs), it is always difficult for DMs to exactly describe their views with a precise number; however, they can be expressed by an interval number within [0, 1]. This denotes that it is necessary to introduce the definition of q-rung interval-valued orthopair fuzzy sets (q-RIVOFSs), of which the degrees of positive membership and negative membership are given by an interval value. This kind of situation is more or less like that encountered in interval-valued, intuitionistic fuzzy environments [58,59]. It should be noted that when the upper and lower limits of the interval values are same, q-RIVOFSs reduce to q-ROFSs, meaning that the latter is a special case of the former.

This research has four main purposes. The first is to develop a comprehensive MAGDM method for selecting the best green supplier with q-RIVOFNs. The second purpose lies in exploring several aggregation operators based on traditional Hamy mean (HM) operators with q-RIVOFNs. The third is to establish an integrated outranking decision-making method by the q-RIVOFWHM (q-RIVOFWDHM) operators. The final purpose is to demonstrate the application, practicality, and effectiveness of the proposed MADM method for selecting the best green supplier.

To further study the q-RIVOFSs, our paper combines the Hamy mean (HM) operator, which considers the relationship between the attribute's estimation values with q-rung interval-valued orthopair fuzzy numbers to investigate MAGDM problems. For the sake of clarity, the rest of this research is organized as follows. Firstly, we briefly introduce the fundamental theories, such as definition, score, and accuracy functions, and operational laws of the q-ROFSs and q-RIVOFSs in Section 2. Then, based on q-RIVOFSs and Hamy mean (HM) operators, we propose four aggregation operators, including the q-rung interval-valued orthopair, fuzzy Hamy mean (q-RIVOFHM) operator; the q-rung interval-valued orthopair, fuzzy weighted Hamy mean (q-RIVOFWHM) operator; the q-rung interval-valued orthopair, fuzzy dual Hamy mean (q-RIVOFDHM) operator; and the q-rung interval-valued orthopair, fuzzy weighted dual Hamy mean (q-RIVOFWDHM) operator in Section 3. Meanwhile, some important properties of these operators are also studied. Thereafter, the models which apply the proposed aggregation operators to solve MAGDM problems are presented in Section 4, and an illustrative example to select the best green supplier is developed. Some comments are provided to summarize this article in Section 5.

2. Preliminaries

2.1. q-Rung Interval-Valued Orthopair Fuzzy Sets (q-RIVOFSs)

According to the q-rung orthopair fuzzy sets (q-ROFSs) [53] and interval-valued Pythagorean fuzzy sets (IVPFSs) [49], we develop the definition of the q-rung interval-valued orthopair fuzzy sets (q-RIVOFSs).

Definition 1. *Let X be a fixed set. A q-RIVOFS is an object having the form*

$$\tilde{Q} = \left\{ \left\langle x, \left(\tilde{\mu}_{\tilde{Q}}(x), \tilde{v}_{\tilde{Q}}(x) \right) \right\rangle | x \in X \right\} \tag{1}$$

where $\tilde{\mu}_{\tilde{Q}}(x) \subset [0,1]$ and $\tilde{v}_{\tilde{Q}}(x) \subset [0,1]$ are interval numbers, and $\tilde{\mu}_{\tilde{Q}}(x) = \left[\mu^L_{\tilde{Q}}(x), \mu^R_{\tilde{Q}}(x) \right]$, $\tilde{v}_{\tilde{Q}}(x) = \left[v^L_{\tilde{Q}}(x), v^R_{\tilde{Q}}(x) \right]$ with the condition $0 \leq \left(\mu^R_{\tilde{Q}}(x) \right)^q + \left(v^R_{\tilde{Q}}(x) \right)^q \leq 1$, $\forall x \in X, q \geq 1$. The numbers $\tilde{\mu}_{\tilde{Q}}(x), \tilde{v}_{\tilde{Q}}(x)$ represent, respectively, the function of positive membership degree (PMD) and negative membership degree (NMD) of the element x to $\tilde{Q}$. Then, for $x \in X$, $\tilde{\pi}_{\tilde{Q}}(x) = \left[\pi^L_{\tilde{Q}}(x), \pi^R_{\tilde{Q}}(x) \right] = \left[\sqrt[q]{1 - \left(\left(\mu^R_{\tilde{Q}}(x) \right)^q + \left(v^R_{\tilde{Q}}(x) \right)^q \right)}, \sqrt[q]{1 - \left(\left(\mu^L_{\tilde{Q}}(x) \right)^q + \left(v^L_{\tilde{Q}}(x) \right)^q \right)} \right]$ denotes the function of the refusal membership degree (RMD) of the element x to $\tilde{Q}$.

As a matter of convenience, we called $\tilde{q} = \left(\left[u^L_{\tilde{q}}, u^R_{\tilde{q}} \right], \left[v^L_{\tilde{q}}, v^R_{\tilde{q}} \right] \right)$ a q-rung interval-valued orthopair fuzzy number (q-RIVOFN). Let $\tilde{q} = \left(\left[u^L_{\tilde{q}}, u^R_{\tilde{q}} \right], \left[v^L_{\tilde{q}}, v^R_{\tilde{q}} \right] \right)$ be a q-RIVOFN, then $S(\tilde{q}) = \frac{1}{4} \left[\left(1 + \left(u^L_{\tilde{q}} \right)^q - \left(v^L_{\tilde{q}} \right)^q \right) + \left(1 + \left(u^R_{\tilde{q}} \right)^q - \left(v^R_{\tilde{q}} \right)^q \right) \right]$ and $H(\tilde{q}) = \frac{\left(u^L_{\tilde{q}} \right)^q + \left(u^R_{\tilde{q}} \right)^q + \left(v^L_{\tilde{q}} \right)^q + \left(v^R_{\tilde{q}} \right)^q}{2}$ are the score and accuracy function of a q-RIVOFN $\tilde{q}$.

Definition 2. *Let $\tilde{q}_1 = \left(\left[u^L_{\tilde{q}_1}, u^R_{\tilde{q}_1} \right], \left[v^L_{\tilde{q}_1}, v^R_{\tilde{q}_1} \right] \right)$ and $\tilde{q}_2 = \left(\left[u^L_{\tilde{q}_2}, u^R_{\tilde{q}_2} \right], \left[v^L_{\tilde{q}_2}, v^R_{\tilde{q}_2} \right] \right)$ be two q-RIVOFNs; $S(\tilde{q}_1)$ and $S(\tilde{q}_2)$ be the scores of $\tilde{q}_1$ and $\tilde{q}_2$, respectively; and let $H(\tilde{q}_1)$ and $H(\tilde{q}_2)$ be the accuracy degrees of $\tilde{q}_1$ and $\tilde{q}_2$, respectively. Then, if $S(\tilde{q}_1) < S(\tilde{q}_2)$, then $\tilde{q}_1 < \tilde{q}_2$; if $S(\tilde{q}_1) = S(\tilde{q}_2)$, then (1) if $H(\tilde{q}_1) = H(\tilde{q}_2)$, then $\tilde{q}_1 = \tilde{q}_2$; (2) if $H(\tilde{q}_1) < H(\tilde{q}_2)$, then $\tilde{q}_1 < \tilde{q}_2$.*

Definition 3. *Let $\tilde{q}_1 = \left(\left[u^L_{\tilde{q}_1}, u^R_{\tilde{q}_1} \right], \left[v^L_{\tilde{q}_1}, v^R_{\tilde{q}_1} \right] \right)$, $\tilde{q}_2 = \left(\left[u^L_{\tilde{q}_2}, u^R_{\tilde{q}_2} \right], \left[v^L_{\tilde{q}_2}, v^R_{\tilde{q}_2} \right] \right)$, and $\tilde{q} = \left(\left[u^L_{\tilde{q}}, u^R_{\tilde{q}} \right], \left[v^L_{\tilde{q}}, v^R_{\tilde{q}} \right] \right)$ be three q-RIVOFNs, and some basic operation rules for them are shown as follows:*

$$(1)\ \widetilde{q}_1 \oplus \widetilde{q}_2 = \left(\left[\sqrt[q]{\left(u_{\widetilde{q}_1}^L\right)^q + \left(u_{\widetilde{q}_2}^L\right)^q - \left(u_{\widetilde{q}_1}^L\right)^q\left(u_{\widetilde{q}_2}^L\right)^q},\ \sqrt[q]{\left(u_{\widetilde{q}_1}^R\right)^q + \left(u_{\widetilde{q}_2}^R\right)^q - \left(u_{\widetilde{q}_1}^R\right)^q\left(u_{\widetilde{q}_2}^R\right)^q} \right],\ \left[v_{\widetilde{q}_1}^L v_{\widetilde{q}_2}^L, v_{\widetilde{q}_1}^R v_{\widetilde{q}_2}^R \right] \right);$$

$$(2)\ \widetilde{q}_1 \otimes \widetilde{q}_2 = \left(\left[\mu_{\widetilde{q}_1}^L v_{\widetilde{q}_2}^L, \mu_{\widetilde{q}_1}^R v_{\widetilde{q}_2}^R \right],\ \left[\sqrt[q]{\left(v_{\widetilde{q}_1}^L\right)^q + \left(v_{\widetilde{q}_2}^L\right)^q - \left(v_{\widetilde{q}_1}^L\right)^q\left(v_{\widetilde{q}_2}^L\right)^q},\ \sqrt[q]{\left(v_{\widetilde{q}_1}^R\right)^q + \left(v_{\widetilde{q}_2}^R\right)^q - \left(v_{\widetilde{q}_1}^R\right)^q\left(v_{\widetilde{q}_2}^R\right)^q} \right] \right);$$

$$(3)\ \lambda \widetilde{q} = \left(\left[\sqrt[q]{1 - \left(1 - \left(u_{\widetilde{q}}^L\right)^q\right)^\lambda},\ \sqrt[q]{1 - \left(1 - \left(u_{\widetilde{q}}^R\right)^q\right)^\lambda} \right],\ \left[\left(v_{\widetilde{q}}^L\right)^\lambda, \left(v_{\widetilde{q}}^R\right)^\lambda \right] \right), \lambda > 0;$$

$$(4)\ \left(\widetilde{q}\right)^\lambda = \left(\left[\left(\mu_{\widetilde{q}}^L\right)^\lambda, \left(\mu_{\widetilde{q}}^R\right)^\lambda \right],\ \left[\sqrt[q]{1 - \left(1 - \left(v_{\widetilde{q}}^L\right)^q\right)^\lambda},\ \sqrt[q]{1 - \left(1 - \left(v_{\widetilde{q}}^R\right)^q\right)^\lambda} \right] \right), \lambda > 0;$$

$$(5)\ \widetilde{q}^c = \left(\left[v_{\widetilde{q}}^L, v_{\widetilde{q}}^R \right],\ \left[\mu_{\widetilde{q}}^L, \mu_{\widetilde{q}}^R \right] \right).$$

2.2. Hamy Mean Operator

Definition 4 [60]. *The HM operator is defined as follows:*

$$\mathrm{HM}^{(x)}(\widetilde{q}_1, \widetilde{q}_2, \cdots, \widetilde{q}_n) = \frac{\displaystyle\sum_{1 \le i_1 < \ldots < i_x \le n} \left(\prod_{j=1}^x \widetilde{q}_{i_j} \right)^{\frac{1}{x}}}{C_n^x} \tag{2}$$

where x is a parameter and $x = 1, 2, \ldots, n$, $i_1, i_2, \ldots, i_x$ are x integer values taken from the set $\{1, 2, \ldots, n\}$ of k integer values; C_n^x denotes the binomial coefficient and $C_n^x = \frac{n!}{x!(n-x)!}$.

3. Some Hamy Mean Operators with *q*-RIVOFNs

3.1. *q*-RIVOFHM Operator

In this chapter, consider both HM operator and *q*-RIVOFNs, we propose the *q*-rung interval-valued orthopair fuzzy Hamy mean (*q*-RIVOFHM) operator.

Definition 5. *Let $\widetilde{q}_j = \left(\left[u_{\widetilde{q}_j}^L, u_{\widetilde{q}_j}^R \right], \left[v_{\widetilde{q}_j}^L, v_{\widetilde{q}_j}^R \right] \right) (j = 1, 2, \ldots, n)$ be a set of q-RIVOFNs. The q-RIVOFHM operator is*

$$q\text{-RIVOFHM}^{(x)}(\widetilde{q}_1, \widetilde{q}_2, \cdots, \widetilde{q}_n) = \frac{\displaystyle\bigoplus_{1 \le i_1 < \ldots < i_x \le n} \left(\bigotimes_{j=1}^x \widetilde{q}_{i_j} \right)^{\frac{1}{x}}}{C_n^x} \tag{3}$$

Theorem 1. *Let $\widetilde{q}_j = \left(\left[u_{\widetilde{q}_j}^L, u_{\widetilde{q}_j}^R \right], \left[v_{\widetilde{q}_j}^L, v_{\widetilde{q}_j}^R \right] \right) (j = 1, 2, \ldots, n)$ be a set of q-RIVOFNs. The fused value by using q-RIVOFHM operator is also a q-RIVOFN, where*

$$q\text{-RIVOFHM}^{(x)}(\widetilde{q}_1,\widetilde{q}_2,\cdots,\widetilde{q}_n) = \frac{\underset{1\leq i_1<\ldots<i_x\leq n}{\oplus}\left(\overset{x}{\underset{j=1}{\otimes}}\widetilde{q}_{i_j}\right)^{\frac{1}{x}}}{C_n^x}$$

$$= \left\{ \begin{array}{c} \left[\sqrt[q]{1-\left(\underset{1\leq i_1<\ldots<i_x\leq n}{\prod}\left(1-\left(\overset{x}{\underset{j=1}{\prod}}u_{\widetilde{q}_j}^{L}\right)^{\frac{q}{x}}\right)\right)^{\frac{1}{C_n^x}}},\ \sqrt[q]{1-\left(\underset{1\leq i_1<\ldots<i_x\leq n}{\prod}\left(1-\left(\overset{x}{\underset{j=1}{\prod}}u_{\widetilde{q}_j}^{R}\right)^{\frac{q}{x}}\right)\right)^{\frac{1}{C_n^x}}}\right], \\ \left[\left(\underset{1\leq i_1<\ldots<i_x\leq n}{\prod}\sqrt[q]{1-\left(\overset{x}{\underset{j=1}{\prod}}\left(1-\left(v_{\widetilde{q}_j}^{L}\right)^q\right)\right)^{\frac{1}{x}}}\right)^{\frac{1}{C_n^x}},\ \left(\underset{1\leq i_1<\ldots<i_x\leq n}{\prod}\sqrt[q]{1-\left(\overset{x}{\underset{j=1}{\prod}}\left(1-\left(v_{\widetilde{q}_j}^{L}\right)^q\right)\right)^{\frac{1}{x}}}\right)^{\frac{1}{C_n^x}}\right] \end{array} \right\} \tag{4}$$

Proof.

$$\overset{x}{\underset{j=1}{\otimes}}\widetilde{q}_{i_j} = \left\{\left[\overset{x}{\underset{j=1}{\prod}}u_{\widetilde{q}_j}^{L},\overset{x}{\underset{j=1}{\prod}}u_{\widetilde{q}_j}^{R}\right],\left[\sqrt[q]{1-\overset{x}{\underset{j=1}{\prod}}\left(1-\left(v_{\widetilde{q}_j}^{L}\right)^q\right)},\ \sqrt[q]{1-\overset{x}{\underset{j=1}{\prod}}\left(1-\left(v_{\widetilde{q}_j}^{R}\right)^q\right)}\right]\right\} \tag{5}$$

Thus,

$$\left(\overset{x}{\underset{j=1}{\otimes}}\widetilde{q}_{i_j}\right)^{\frac{1}{x}} = \left\{ \left[\left(\overset{x}{\underset{j=1}{\prod}}u_{\widetilde{q}_j}^{L}\right)^{\frac{1}{x}},\left(\overset{x}{\underset{j=1}{\prod}}u_{\widetilde{q}_j}^{R}\right)^{\frac{1}{x}}\right],\left[\begin{array}{c}\sqrt[q]{1-\left(\overset{x}{\underset{j=1}{\prod}}\left(1-\left(v_{\widetilde{q}_j}^{L}\right)^q\right)\right)^{\frac{1}{x}}},\\ \sqrt[q]{1-\left(\overset{x}{\underset{j=1}{\prod}}\left(1-\left(v_{\widetilde{q}_j}^{R}\right)^q\right)\right)^{\frac{1}{x}}}\end{array}\right]\right\} \tag{6}$$

Thereafter,

$$\underset{1\leq i_1<\ldots<i_x\leq n}{\oplus}\left(\overset{x}{\underset{j=1}{\otimes}}\widetilde{q}_{i_j}\right)^{\frac{1}{x}}$$

$$= \left\{ \begin{array}{c} \left[\sqrt[q]{1-\underset{1\leq i_1<\ldots<i_x\leq n}{\prod}\left(1-\left(\overset{x}{\underset{j=1}{\prod}}u_{\widetilde{q}_j}^{L}\right)^{\frac{q}{x}}\right)},\ \sqrt[q]{1-\underset{1\leq i_1<\ldots<i_x\leq n}{\prod}\left(1-\left(\overset{x}{\underset{j=1}{\prod}}u_{\widetilde{q}_j}^{R}\right)^{\frac{q}{x}}\right)}\right], \\ \left[\underset{1\leq i_1<\ldots<i_x\leq n}{\prod}\sqrt[q]{1-\left(\overset{x}{\underset{j=1}{\prod}}\left(1-\left(v_{\widetilde{q}_j}^{L}\right)^q\right)\right)^{\frac{1}{x}}},\ \underset{1\leq i_1<\ldots<i_x\leq n}{\prod}\sqrt[q]{1-\left(\overset{x}{\underset{j=1}{\prod}}\left(1-\left(v_{\widetilde{q}_j}^{R}\right)^q\right)\right)^{\frac{1}{x}}}\right] \end{array} \right\} \tag{7}$$

Therefore,

$$q\text{-RIVOFHM}^{(x)}(\widetilde{q}_1,\widetilde{q}_2,\cdots,\widetilde{q}_n) = \frac{\underset{1\leq i_1<\ldots<i_x\leq n}{\oplus}\left(\overset{x}{\underset{j=1}{\otimes}}\widetilde{q}_{i_j}\right)^{\frac{1}{x}}}{C_n^x}$$

$$= \left\{ \begin{array}{c} \left[\sqrt[q]{1-\left(\underset{1\leq i_1<\ldots<i_x\leq n}{\prod}\left(1-\left(\overset{x}{\underset{j=1}{\prod}}u_{\widetilde{q}_j}^{L}\right)^{\frac{q}{x}}\right)\right)^{\frac{1}{C_n^x}}},\ \sqrt[q]{1-\left(\underset{1\leq i_1<\ldots<i_x\leq n}{\prod}\left(1-\left(\overset{x}{\underset{j=1}{\prod}}u_{\widetilde{q}_j}^{R}\right)^{\frac{q}{x}}\right)\right)^{\frac{1}{C_n^x}}}\right], \\ \left[\left(\underset{1\leq i_1<\ldots<i_x\leq n}{\prod}\sqrt[q]{1-\left(\overset{x}{\underset{j=1}{\prod}}\left(1-\left(v_{\widetilde{q}_j}^{L}\right)^q\right)\right)^{\frac{1}{x}}}\right)^{\frac{1}{C_n^x}},\ \left(\underset{1\leq i_1<\ldots<i_x\leq n}{\prod}\sqrt[q]{1-\left(\overset{x}{\underset{j=1}{\prod}}\left(1-\left(v_{\widetilde{q}_j}^{R}\right)^q\right)\right)^{\frac{1}{x}}}\right)^{\frac{1}{C_n^x}}\right] \end{array} \right\} \tag{8}$$

Hence, Equation (4) is kept.

Then, we need to prove that Equation (4) is a q-RIVOFN. We need to prove $0 \leq \left(\mu_{\widetilde{Q}}^{R}(x)\right)^q + \left(v_{\widetilde{Q}}^{R}(x)\right)^q \leq 1$.

Let

$$\mu_{\widetilde{Q}}^{R}(x) = \sqrt[q]{1 - \left(\prod_{1 \le i_1 < \ldots < i_x \le n}\left(1 - \left(\prod_{j=1}^{x} u_{\widetilde{q}_j}^{R}\right)^{\frac{q}{x}}\right)\right)^{\frac{1}{C_n^x}}}$$

$$\nu_{\widetilde{Q}}^{R}(x) = \left(\prod_{1 \le i_1 < \ldots < i_x \le n} \sqrt[q]{1 - \left(\prod_{j=1}^{x}\left(1 - \left(v_{\widetilde{q}_j}^{R}\right)^{q}\right)\right)^{\frac{1}{x}}}\right)^{\frac{1}{C_n^x}}$$

$\square$

Proof.

$$0 \le \left(\mu_{\widetilde{Q}}^{R}(x)\right)^{q} + \left(\nu_{\widetilde{Q}}^{R}(x)\right)^{q}$$

$$= 1 - \left(\prod_{1 \le i_1 < \ldots < i_x \le n}\left(1 - \left(\prod_{j=1}^{x} u_{\widetilde{q}_j}^{R}\right)^{\frac{q}{x}}\right)\right)^{\frac{1}{C_n^x}} + \left(\prod_{1 \le i_1 < \ldots < i_x \le n}\left(1 - \left(\prod_{j=1}^{x}\left(1 - \left(v_{\widetilde{q}_j}^{R}\right)^{q}\right)\right)^{\frac{1}{x}}\right)\right)^{\frac{1}{C_n^x}}$$

$$\le 1 - \left(\prod_{1 \le i_1 < \ldots < i_x \le n}\left(1 - \left(\prod_{j=1}^{x}\left(1 - \left(v_{\widetilde{q}_j}^{R}\right)^{q}\right)\right)^{\frac{1}{x}}\right)\right)^{\frac{1}{C_n^x}} + \left(\prod_{1 \le i_1 < \ldots < i_x \le n}\left(1 - \left(\prod_{j=1}^{x}\left(1 - \left(v_{\widetilde{q}_j}^{R}\right)^{q}\right)\right)^{\frac{1}{x}}\right)\right)^{\frac{1}{C_n^x}}$$

$$= 1$$

So $0 \le \left(\mu_{\widetilde{Q}}^{R}(x)\right)^{q} + \left(\nu_{\widetilde{Q}}^{R}(x)\right)^{q} \le 1$ is maintained. $\square$

Example 1. *Let* $([0.5, 0.8], [0.4, 0.5])$, $([0.3, 0.5], [0.6, 0.7])$, $([0.5, 0.7], [0.2, 0.3])$ *and* $([0.4, 0.8], [0.1, 0.2])$ *be four q-RIVOFNs, and suppose* $x = 2$, $q = 3$—*then, according to Equation (4), we have*

$$q\text{-RIVOFHM}^{(x)}(\widetilde{q}_1, \widetilde{q}_2, \cdots, \widetilde{q}_n) = \frac{\underset{1 \le i_1 < \ldots < i_x \le n}{\oplus}\left(\underset{j=1}{\overset{x}{\otimes}}\widetilde{q}_{ij}\right)^{\frac{1}{x}}}{C_n^x}$$

$$= \left(\begin{bmatrix}\left[\sqrt[3]{1 - \left(\begin{array}{l}\left(1 - (0.5 \times 0.3)^{\frac{3}{2}}\right) \times \left(1 - (0.5 \times 0.5)^{\frac{3}{2}}\right) \times \left(1 - (0.5 \times 0.4)^{\frac{3}{2}}\right) \\ \times \left(1 - (0.3 \times 0.4)^{\frac{3}{2}}\right) \times \left(1 - (0.3 \times 04)^{\frac{3}{2}}\right) \times \left(1 - (0.5 \times 0.4)^{\frac{3}{2}}\right)\end{array}\right)^{\frac{1}{C_4^2}}},\right. \\ \left.\sqrt[3]{1 - \left(\begin{array}{l}\left(1 - (0.8 \times 0.5)^{\frac{3}{2}}\right) \times \left(1 - (0.8 \times 0.7)^{\frac{3}{2}}\right) \times \left(1 - (0.8 \times 0.8)^{\frac{3}{2}}\right) \\ \times \left(1 - (0.5 \times 0.7)^{\frac{3}{2}}\right) \times \left(1 - (0.5 \times 0.8)^{\frac{3}{2}}\right) \times \left(1 - (0.7 \times 0.8)^{\frac{3}{2}}\right)\end{array}\right)^{\frac{1}{C_4^2}}}\right], \\ \left[\left(\sqrt[3]{\begin{array}{l}\left(1 - ((1 - 0.4^3) \times (1 - 0.6^3))^{\frac{1}{2}}\right) \times \left(1 - ((1 - 0.4^3) \times (1 - 0.2^3))^{\frac{1}{2}}\right) \\ \times \left(1 - ((1 - 0.4^3) \times (1 - 0.1^3))^{\frac{1}{2}}\right) \times \left(1 - ((1 - 0.6^3) \times (1 - 0.2^3))^{\frac{1}{2}}\right) \\ \times \left(1 - ((1 - 0.6^3) \times (1 - 0.1^3))^{\frac{1}{2}}\right) \times \left(1 - ((1 - 0.2^3) \times (1 - 0.1^3))^{\frac{1}{2}}\right)\end{array}}\right)^{\frac{1}{C_4^2}},\right. \\ \left.\left(\sqrt[3]{\begin{array}{l}\left(1 - ((1 - 0.5^3) \times (1 - 0.7^3))^{\frac{1}{2}}\right) \times \left(1 - ((1 - 0.5^3) \times (1 - 0.3^3))^{\frac{1}{2}}\right) \\ \times \left(1 - ((1 - 0.5^3) \times (1 - 0.2^3))^{\frac{1}{2}}\right) \times \left(1 - ((1 - 0.7^3) \times (1 - 0.3^3))^{\frac{1}{2}}\right) \\ \times \left(1 - ((1 - 0.7^3) \times (1 - 0.2^3))^{\frac{1}{2}}\right) \times \left(1 - ((1 - 0.3^3) \times (1 - 0.2^3))^{\frac{1}{2}}\right)\end{array}}\right)^{\frac{1}{C_4^2}}\right]\end{bmatrix}\right)$$

$$= ([0.4261, 0.7072], [0.3604, 0.4605])$$

The q-RIVOFHM satisfies the following three properties.

Property 1. *Idempotency: if* $\widetilde{q}_j = \left(\left[u_{\widetilde{q}_j}^{L}, u_{\widetilde{q}_j}^{R}\right], \left[v_{\widetilde{q}_j}^{L}, v_{\widetilde{q}_j}^{R}\right]\right)(j = 1, 2, \ldots, n)$ *are equal, then*

$$q\text{-RIVOFHM}^{(x)}(\widetilde{q}_1, \widetilde{q}_2, \cdots, \widetilde{q}_n) = \widetilde{q} \tag{9}$$

Proof. Since $\widetilde{q}_j = \widetilde{q} = \left(\left[u_{\widetilde{q}}^L, u_{\widetilde{q}}^R \right], \left[v_{\widetilde{q}}^L, v_{\widetilde{q}}^R \right] \right)$, then

$$q\text{-RIVOFHM}^{(x)}(\widetilde{q}, \widetilde{q}, \cdots, \widetilde{q}) = \frac{\underset{1 \le i_1 < \dots < i_x \le n}{\oplus} \left(\overset{x}{\underset{j=1}{\otimes}} \widetilde{q} \right)^{\frac{1}{x}}}{C_n^x}$$

$$= \left\{ \left[\sqrt[q]{1 - \left(\prod_{1 \le i_1 < \dots < i_x \le n} \left(1 - \left(\prod_{j=1}^{x} u_{\widetilde{q}}^L \right)^{\frac{x}{x}} \right) \right)^{\frac{q}{x}}}^{\frac{1}{C_n^x}}, \sqrt[q]{1 - \left(\prod_{1 \le i_1 < \dots < i_x \le n} \left(1 - \left(\prod_{j=1}^{x} u_{\widetilde{q}}^R \right)^{\frac{x}{x}} \right) \right)^{\frac{q}{x}}}^{\frac{1}{C_n^x}} \right], \right.$$

$$\left. \left[\left(\prod_{1 \le i_1 < \dots < i_x \le n} \sqrt[q]{1 - \left(\prod_{j=1}^{x} \left(1 - \left(v_{\widetilde{q}}^L \right)^q \right) \right)^{\frac{1}{x}}} \right)^{\frac{1}{C_n^x}}, \left(\prod_{1 \le i_1 < \dots < i_x \le n} \sqrt[q]{1 - \left(\prod_{j=1}^{x} \left(1 - \left(v_{\widetilde{q}}^R \right)^q \right) \right)^{\frac{1}{x}}} \right)^{\frac{1}{C_n^x}} \right] \right\}$$

$$= \left\{ \left[\sqrt[q]{1 - \left(\left(1 - \left(\left(u_{\widetilde{q}}^L \right)^x \right)^{\frac{q}{x}} \right)^{C_n^x} \right)^{\frac{1}{C_n^x}}}, \sqrt[q]{1 - \left(\left(1 - \left(\left(u_{\widetilde{q}}^R \right)^x \right)^{\frac{q}{x}} \right)^{C_n^x} \right)^{\frac{1}{C_n^x}}} \right], \right.$$

$$\left. \left[\left(\left(\sqrt[q]{1 - \left(\left(1 - \left(v_{\widetilde{q}}^L \right)^q \right)^x \right)^{\frac{1}{x}}} \right)^{C_n^x} \right)^{\frac{1}{C_n^x}}, \left(\left(\sqrt[q]{1 - \left(\left(1 - \left(v_{\widetilde{q}}^R \right)^q \right)^x \right)^{\frac{1}{x}}} \right)^{C_n^x} \right)^{\frac{1}{C_n^x}} \right] \right\}$$

$$= \left\{ \left[u_{\widetilde{q}}^L, u_{\widetilde{q}}^R \right], \left[v_{\widetilde{q}}^L, v_{\widetilde{q}}^R \right] \right\} = \widetilde{q}$$

$\square$

Property 2. *Monotonicity: let* $\widetilde{q}_j = \left(\left[u_{\widetilde{q}_j}^L, u_{\widetilde{q}_j}^R \right], \left[v_{\widetilde{q}_j}^L, v_{\widetilde{q}_j}^R \right] \right) (j = 1, 2, \dots, n)$ *and* $\widetilde{q}_j' = \left(\left[\left(u_{\widetilde{q}_j}^L \right)', \left(u_{\widetilde{q}_j}^R \right)' \right], \left[\left(v_{\widetilde{q}_j}^L \right)', \left(v_{\widetilde{q}_j}^R \right)' \right] \right) (j = 1, 2, \dots, n)$ *be two sets of q-RIVOFNs. If* $u_{\widetilde{q}_j}^L \le \left(u_{\widetilde{q}_j}^L \right)', u_{\widetilde{q}_j}^R \le \left(u_{\widetilde{q}_j}^R \right)', v_{\widetilde{q}_j}^L \ge \left(v_{\widetilde{q}_j}^L \right)'$ *and* $v_{\widetilde{q}_j}^L \ge \left(v_{\widetilde{q}_j}^R \right)'$ *hold for all j, then*

$$q\text{-RIVOFHM}^{(x)}(\widetilde{q}_1, \widetilde{q}_2, \cdots, \widetilde{q}_n) \le q\text{-RIVOFHM}^{(x)}(\widetilde{q}_1', \widetilde{q}_2', \cdots, \widetilde{q}_n') \tag{10}$$

Proof. Given that $u_{\widetilde{q}_j}^L \le \left(u_{\widetilde{q}_j}^L \right)'$, we can obtain

$$\left(\prod_{j=1}^{x} u_{\widetilde{q}}^L \right)^{\frac{q}{x}} \le \left(\prod_{j=1}^{x} \left(u_{\widetilde{q}}^L \right)' \right)^{\frac{q}{x}} \tag{11}$$

$$\left(\prod_{1 \le i_1 < \dots < i_x \le n} \left(1 - \left(\prod_{j=1}^{x} u_{\widetilde{q}}^L \right)^{\frac{q}{x}} \right) \right)^{\frac{1}{C_n^x}} \ge \left(\prod_{1 \le i_1 < \dots < i_x \le n} \left(1 - \left(\prod_{j=1}^{x} \left(u_{\widetilde{q}}^L \right)' \right)^{\frac{q}{x}} \right) \right)^{\frac{1}{C_n^x}} \tag{12}$$

Thereafter,

$$\sqrt[q]{1 - \left(\prod_{1 \le i_1 < \dots < i_x \le n} \left(1 - \left(\prod_{j=1}^{x} u_{\widetilde{q}}^L \right)^{\frac{q}{x}} \right) \right)^{\frac{1}{C_n^x}}} \le \sqrt[q]{1 - \left(\prod_{1 \le i_1 < \dots < i_x \le n} \left(1 - \left(\prod_{j=1}^{x} \left(u_{\widetilde{q}}^L \right)' \right)^{\frac{q}{x}} \right) \right)^{\frac{1}{C_n^x}}} \tag{13}$$

That means $u_{\widetilde{q}}^L \le \left(u_{\widetilde{q}}^L \right)'$. Similarly, we can obtain $u_{\widetilde{q}}^R \le \left(u_{\widetilde{q}}^R \right)', v_{\widetilde{q}}^L \ge \left(v_{\widetilde{q}}^L \right)'$ and $v_{\widetilde{q}}^L \ge \left(v_{\widetilde{q}}^R \right)'$. Thus, the proof is complete. $\square$

Property 3. *Boundedness: let* $\widetilde{q}_j = \left(\left[u^L_{\widetilde{q}_j}, u^R_{\widetilde{q}_j} \right], \left[v^L_{\widetilde{q}_j}, v^R_{\widetilde{q}_j} \right] \right) (j = 1, 2, \ldots, n)$ *be a set of q-RIVOFNs. If* $\widetilde{q}^+ = \left(\left[max_i\left(u^L_{\widetilde{q}_j} \right), max_i\left(u^R_{\widetilde{q}_j} \right) \right], \left[min_i\left(v^L_{\widetilde{q}_j} \right), min_i\left(v^R_{\widetilde{q}_j} \right) \right] \right)$ *and* $\widetilde{q}^- = \left(\left[min_i\left(u^L_{\widetilde{q}_j} \right), min_i\left(u^R_{\widetilde{q}_j} \right) \right], \left[max_i\left(v^L_{\widetilde{q}_j} \right), max_i\left(v^R_{\widetilde{q}_j} \right) \right] \right)$ *then*

$$\widetilde{q}^- \leq q\text{-RIVOFHM}^{(x)}(\widetilde{q}_1, \widetilde{q}_2, \cdots, \widetilde{q}_n) \leq \widetilde{q}^+ \tag{14}$$

From Property 1,

$$q\text{-RIVOFHM}^{(x)}\left(\widetilde{q}_1^-, \widetilde{q}_2^-, \cdots, \widetilde{q}_n^-\right) = \widetilde{q}^-$$
$$q\text{-RIVOFHM}^{(x)}\left(\widetilde{q}_1^+, \widetilde{q}_2^+, \cdots, \widetilde{q}_n^+\right) = \widetilde{q}^+$$

From Property 2,

$$\widetilde{q}^- \leq q\text{-RIVOFHM}^{(x)}(\widetilde{q}_1, \widetilde{q}_2, \cdots, \widetilde{q}_n) \leq \widetilde{q}^+$$

3.2. The q-RIVOFWHM Operator

In practical MADM problems, it is important to take the attribute weights into account. This section will develop the q-rung interval-valued orthopair, fuzzy weighted Hamy mean (q-RIVOFWHM) operator.

Definition 6. *Let* $\widetilde{q}_j = \left(\left[u^L_{\widetilde{q}_j}, u^R_{\widetilde{q}_j} \right], \left[v^L_{\widetilde{q}_j}, v^R_{\widetilde{q}_j} \right] \right) (j = 1, 2, \ldots, n)$ *be a set of q-RIVOFNs, with their weight vector as* $w_i = (w_1, w_2, \ldots, w_n)^T$, *thereby satisfying* $w_i \in [0, 1]$ *and* $\sum_{i=1}^n w_i = 1$. *Then we can define the q-RIVOFWHM operator as follows:*

$$q\text{-RIVOFWHM}_w^{(x)}(\widetilde{q}_1, \widetilde{q}_2, \ldots, \widetilde{q}_n) = \frac{\underset{1 \leq i_1 < \ldots < i_x \leq n}{\oplus} \left(\underset{j=1}{\overset{x}{\otimes}} \left(\widetilde{q}_{i_j} \right)^{w_{i_j}} \right)^{\frac{1}{x}}}{C_n^x} \tag{15}$$

Theorem 2. *Let* $\widetilde{q}_j = \left(\left[u^L_{\widetilde{q}_j}, u^R_{\widetilde{q}_j} \right], \left[v^L_{\widetilde{q}_j}, v^R_{\widetilde{q}_j} \right] \right) (j = 1, 2, \ldots, n)$ *be a set of q-RIVOFNs. The fused value obtained by using q-RIVOFWHM operator is also a q-RIVOFN, where*

$$q\text{-RIVOFWHM}_w^{(x)}(\widetilde{q}_1, \widetilde{q}_2, \ldots, \widetilde{q}_n) = \frac{\underset{1 \leq i_1 < \ldots < i_x \leq n}{\oplus} \left(\underset{j=1}{\overset{x}{\otimes}} \left(\widetilde{q}_{i_j} \right)^{w_{i_j}} \right)^{\frac{1}{x}}}{C_n^x}$$
$$= \left\{ \left[\sqrt[q]{1 - \left(\underset{1 \leq i_1 < \ldots < i_x \leq n}{\prod} \left(1 - \left(\overset{x}{\underset{j=1}{\prod}} \left(u^L_{\widetilde{q}_j} \right)^{w_{i_j}} \right)^{\frac{q}{x}} \right) \right)^{\frac{1}{C_n^x}}}, \sqrt[q]{1 - \left(\underset{1 \leq i_1 < \ldots < i_x \leq n}{\prod} \left(1 - \left(\overset{x}{\underset{j=1}{\prod}} \left(u^R_{\widetilde{q}_j} \right)^{w_{i_j}} \right)^{\frac{q}{x}} \right) \right)^{\frac{1}{C_n^x}}} \right], \right. $$
$$\left. \left[\left(\underset{1 \leq i_1 < \ldots < i_x \leq n}{\prod} \left(\sqrt[q]{1 - \left(\overset{x}{\underset{j=1}{\prod}} \left(1 - \left(v^L_{\widetilde{q}_j} \right)^q \right)^{w_{i_j}} \right)^{\frac{1}{x}}} \right) \right)^{\frac{1}{C_n^x}}, \left(\underset{1 \leq i_1 < \ldots < i_x \leq n}{\prod} \left(\sqrt[q]{1 - \left(\overset{x}{\underset{j=1}{\prod}} \left(1 - \left(v^R_{\widetilde{q}_j} \right)^q \right)^{w_{i_j}} \right)^{\frac{1}{x}}} \right) \right)^{\frac{1}{C_n^x}} \right] \right\} \tag{16}$$

Proof. From Definition 3, we can obtain

$$\left(\widetilde{q}_{i_j} \right)^{w_{i_j}} = \left\{ \left[\left(u^L_{\widetilde{q}_j} \right)^{w_{i_j}}, \left(u^R_{\widetilde{q}_j} \right)^{w_{i_j}} \right], \left[\sqrt[q]{1 - \left(1 - \left(v^L_{\widetilde{q}_j} \right)^q \right)^{w_{i_j}}}, \sqrt[q]{1 - \left(1 - \left(v^R_{\widetilde{q}_j} \right)^q \right)^{w_{i_j}}} \right] \right\} \tag{17}$$

Thus,

$$\overset{x}{\underset{j=1}{\otimes}}\left(\widetilde{q}_{i_j}\right)^{w_{i_j}} = \left\{ \begin{array}{c} \left[\prod_{j=1}^{x}\left(u_{\widetilde{q}_j}^{L}\right)^{w_{i_j}}, \prod_{j=1}^{x}\left(u_{\widetilde{q}_j}^{R}\right)^{w_{i_j}}\right], \\ \left[\sqrt[q]{1-\prod_{j=1}^{x}\left(1-\left(v_{\widetilde{q}_j}^{L}\right)^{q}\right)^{w_{i_j}}}, \sqrt[q]{1-\prod_{j=1}^{x}\left(1-\left(v_{\widetilde{q}_j}^{R}\right)^{q}\right)^{w_{i_j}}}\right] \end{array} \right\} \tag{18}$$

Therefore,

$$\left(\overset{x}{\underset{j=1}{\otimes}}\left(\widetilde{q}_{i_j}\right)^{w_{i_j}}\right)^{\frac{1}{x}} = \left\{ \begin{array}{c} \left[\left(\prod_{j=1}^{x}\left(u_{\widetilde{q}_j}^{L}\right)^{w_{i_j}}\right)^{\frac{1}{x}}, \left(\prod_{j=1}^{x}\left(u_{\widetilde{q}_j}^{R}\right)^{w_{i_j}}\right)^{\frac{1}{x}}\right], \\ \left[\sqrt[q]{1-\left(\prod_{j=1}^{x}\left(1-\left(v_{\widetilde{q}_j}^{L}\right)^{q}\right)^{w_{i_j}}\right)^{\frac{1}{x}}}, \sqrt[q]{1-\left(\prod_{j=1}^{x}\left(1-\left(v_{\widetilde{q}_j}^{R}\right)^{q}\right)^{w_{i_j}}\right)^{\frac{1}{x}}}\right] \end{array} \right\} \tag{19}$$

Thereafter,

$$\underset{1\leq i_1<...<i_x\leq n}{\oplus}\left(\overset{x}{\underset{j=1}{\otimes}}\left(\widetilde{q}_{i_j}\right)^{w_{i_j}}\right)^{\frac{1}{x}}$$
$$= \left\{ \begin{array}{c} \left[\sqrt[q]{1-\underset{1\leq i_1<...<i_x\leq n}{\prod}\left(1-\left(\prod_{j=1}^{x}\left(u_{\widetilde{q}_j}^{L}\right)^{w_{i_j}}\right)^{\frac{q}{x}}\right)}, \sqrt[q]{1-\underset{1\leq i_1<...<i_x\leq n}{\prod}\left(1-\left(\prod_{j=1}^{x}\left(u_{\widetilde{q}_j}^{R}\right)^{w_{i_j}}\right)^{\frac{q}{x}}\right)}\right], \\ \left[\underset{1\leq i_1<...<i_x\leq n}{\prod}\left(\sqrt[q]{1-\left(\prod_{j=1}^{x}\left(1-\left(v_{\widetilde{q}_j}^{L}\right)^{q}\right)^{w_{i_j}}\right)^{\frac{1}{x}}}\right), \underset{1\leq i_1<...<i_x\leq n}{\prod}\left(\sqrt[q]{1-\left(\prod_{j=1}^{x}\left(1-\left(v_{\widetilde{q}_j}^{R}\right)^{q}\right)^{w_{i_j}}\right)^{\frac{1}{x}}}\right)\right] \end{array} \right\} \tag{20}$$

Furthermore,

$$q\text{-RIVOFWHM}_{w}^{(x)}\left(\widetilde{q}_1,\widetilde{q}_2,\ldots,\widetilde{q}_n\right) = \frac{\underset{1\leq i_1<...<i_x\leq n}{\oplus}\left(\overset{x}{\underset{j=1}{\otimes}}\left(\widetilde{q}_{i_j}\right)^{w_{i_j}}\right)^{\frac{1}{x}}}{C_n^x}$$
$$= \left\{ \begin{array}{c} \left[\sqrt[q]{1-\left(\underset{1\leq i_1<...<i_x\leq n}{\prod}\left(1-\left(\prod_{j=1}^{x}\left(u_{\widetilde{q}_j}^{L}\right)^{w_{i_j}}\right)^{\frac{q}{x}}\right)\right)^{\frac{1}{C_n^x}}}, \sqrt[q]{1-\left(\underset{1\leq i_1<...<i_x\leq n}{\prod}\left(1-\left(\prod_{j=1}^{x}\left(u_{\widetilde{q}_j}^{R}\right)^{w_{i_j}}\right)^{\frac{q}{x}}\right)\right)^{\frac{1}{C_n^x}}}\right], \\ \left[\left(\underset{1\leq i_1<...<i_x\leq n}{\prod}\left(\sqrt[q]{1-\left(\prod_{j=1}^{x}\left(1-\left(v_{\widetilde{q}_j}^{L}\right)^{q}\right)^{w_{i_j}}\right)^{\frac{1}{x}}}\right)\right)^{\frac{1}{C_n^x}}, \left(\underset{1\leq i_1<...<i_x\leq n}{\prod}\left(\sqrt[q]{1-\left(\prod_{j=1}^{x}\left(1-\left(v_{\widetilde{q}_j}^{R}\right)^{q}\right)^{w_{i_j}}\right)^{\frac{1}{x}}}\right)\right)^{\frac{1}{C_n^x}}\right] \end{array} \right\} \tag{21}$$

Hence, Equation (16) is kept.

Then we need to prove that Equation (16) is a q-RIVOFN. We need to prove that $0 \leq \left(\mu_{\widetilde{Q}}^{R}(x)\right)^{q} + \left(v_{\widetilde{Q}}^{R}(x)\right)^{q} \leq 1$.

Let

$$\mu_{\widetilde{Q}}^{R}(x) = \sqrt[q]{1-\left(\underset{1\leq i_1<...<i_x\leq n}{\prod}\left(1-\left(\prod_{j=1}^{x}\left(u_{\widetilde{q}_j}^{R}\right)^{w_{i_j}}\right)^{\frac{q}{x}}\right)\right)^{\frac{1}{C_n^x}}}$$

$$v_{\widetilde{Q}}^{R}(x) = \left(\underset{1\leq i_1<...<i_x\leq n}{\prod}\left(\sqrt[q]{1-\left(\prod_{j=1}^{x}\left(1-\left(v_{\widetilde{q}_j}^{R}\right)^{q}\right)^{w_{i_j}}\right)^{\frac{1}{x}}}\right)\right)^{\frac{1}{C_n^x}}$$

$\square$

Proof.

$$0 \leq \left(\mu_{\tilde{Q}}^{R}(x)\right)^{q} + \left(\nu_{\tilde{Q}}^{R}(x)\right)^{q}$$

$$= 1 - \left(\prod_{1 \leq i_1 < \ldots < i_x \leq n} \left(1 - \left(\prod_{j=1}^{x}\left(u_{\tilde{q}_j}^{R}\right)^{w_{ij}}\right)^{\frac{q}{x}}\right)\right)^{\frac{1}{C_n^x}} + \left(\prod_{1 \leq i_1 < \ldots < i_x \leq n} \left(1 - \left(\prod_{j=1}^{x}\left(1 - \left(v_{\tilde{q}_j}^{R}\right)^{q}\right)^{w_{ij}}\right)^{\frac{1}{x}}\right)\right)^{\frac{1}{C_n^x}}$$

$$\leq 1 - \left(\prod_{1 \leq i_1 < \ldots < i_x \leq n} \left(1 - \left(\prod_{j=1}^{x}\left(1 - \left(v_{\tilde{q}_j}^{R}\right)^{q}\right)^{w_{ij}}\right)^{\frac{1}{x}}\right)\right)^{\frac{1}{C_n^x}} + \left(\prod_{1 \leq i_1 < \ldots < i_x \leq n} \left(1 - \left(\prod_{j=1}^{x}\left(1 - \left(v_{\tilde{q}_j}^{R}\right)^{q}\right)^{w_{ij}}\right)^{\frac{1}{x}}\right)\right)^{\frac{1}{C_n^x}}$$

$$= 1$$

Therefore, $0 \leq \left(\mu_{\tilde{Q}}^{R}(x)\right)^{q} + \left(\nu_{\tilde{Q}}^{R}(x)\right)^{q} \leq 1$ is maintained. $\square$

Example 2. *Let* $([0.5, 0.8], [0.4, 0.5])$, $([0.3, 0.5], [0.6, 0.7])$, $([0.5, 0.7], [0.2, 0.3])$ *and* $([0.4, 0.8], [0.1, 0.2])$ *be four q-RIVOFNs, and* $w = (0.2, 0.1, 0.3, 0.4)$; *in addition, suppose* $x = 2$, $q = 3$. *Then, according to Equation (16), we have*

$$q\text{-RIVOFWHM}_{w}^{(x)}(\tilde{q}_1, \tilde{q}_2, \ldots, \tilde{q}_n) = \frac{\displaystyle\mathop{\oplus}_{1 \leq i_1 < \ldots < i_x \leq n} \left(\mathop{\otimes}_{j=1}^{x}\left(\tilde{q}_{ij}\right)^{w_{ij}}\right)^{\frac{1}{x}}}{C_n^x}$$

$$= \left(\left[\begin{array}{c} \sqrt[3]{1 - \left(\begin{array}{c}\left(1 - (0.5^{0.2} \times 0.3^{0.1})^{\frac{3}{2}}\right) \times \left(1 - (0.5^{0.2} \times 0.5^{0.3})^{\frac{3}{2}}\right) \times \left(1 - (0.5^{0.2} \times 0.4^{0.4})^{\frac{3}{2}}\right) \\ \times \left(1 - (0.3^{0.1} \times 0.5^{0.3})^{\frac{3}{2}}\right) \times \left(1 - (0.3^{0.1} \times 0.4^{0.4})^{\frac{3}{2}}\right) \times \left(1 - (0.5^{0.3} \times 0.4^{0.4})^{\frac{3}{2}}\right)\end{array}\right)^{\frac{1}{C_4^2}}}, \\ \sqrt[3]{1 - \left(\begin{array}{c}\left(1 - (0.8^{0.2} \times 0.5^{0.1})^{\frac{3}{2}}\right) \times \left(1 - (0.8^{0.2} \times 0.7^{0.3})^{\frac{3}{2}}\right) \times \left(1 - (0.8^{0.2} \times 0.8^{0.4})^{\frac{3}{2}}\right) \\ \times \left(1 - (0.5^{0.1} \times 0.7^{0.3})^{\frac{3}{2}}\right) \times \left(1 - (0.5^{0.1} \times 0.8^{0.4})^{\frac{3}{2}}\right) \times \left(1 - (0.7^{0.3} \times 0.8^{0.4})^{\frac{3}{2}}\right)\end{array}\right)^{\frac{1}{C_4^2}}} \end{array}\right], \right.$$

$$\left.\left[\begin{array}{c} \sqrt[3]{\left(\left(\begin{array}{c}\left(1 - \left((1 - 0.4^3)^{0.2} \times (1 - 0.6^3)^{0.1}\right)^{\frac{1}{2}}\right) \times \left(1 - \left((1 - 0.4^3)^{0.2} \times (1 - 0.2^3)^{0.3}\right)^{\frac{1}{2}}\right) \\ \times \left(1 - \left((1 - 0.4^3)^{0.2} \times (1 - 0.1^3)^{0.4}\right)^{\frac{1}{2}}\right) \times \left(1 - \left((1 - 0.6^3)^{0.1} \times (1 - 0.2^3)^{0.3}\right)^{\frac{1}{2}}\right) \\ \times \left(1 - \left((1 - 0.6^3)^{0.1} \times (1 - 0.1^3)^{0.4}\right)^{\frac{1}{2}}\right) \times \left(1 - \left((1 - 0.2^3)^{0.3} \times (1 - 0.1^3)^{0.4}\right)^{\frac{1}{2}}\right)\end{array}\right)\right)^{\frac{1}{C_4^2}}}, \\ \sqrt[3]{\left(\left(\begin{array}{c}\left(1 - \left((1 - 0.5^3)^{0.2} \times (1 - 0.7^3)^{0.1}\right)^{\frac{1}{2}}\right) \times \left(1 - \left((1 - 0.5^3)^{0.2} \times (1 - 0.3^3)^{0.3}\right)^{\frac{1}{2}}\right) \\ \times \left(1 - \left((1 - 0.5^3)^{0.2} \times (1 - 0.2^3)^{0.4}\right)^{\frac{1}{2}}\right) \times \left(1 - \left((1 - 0.7^3)^{0.1} \times (1 - 0.3^3)^{0.3}\right)^{\frac{1}{2}}\right) \\ \times \left(1 - \left((1 - 07^3)^{0.1} \times (1 - 0.2^3)^{0.4}\right)^{\frac{1}{2}}\right) \times \left(1 - \left((1 - 0.3^3)^{0.3} \times (1 - 0.2^3)^{0.4}\right)^{\frac{1}{2}}\right)\end{array}\right)\right)^{\frac{1}{C_4^2}}} \end{array}\right]\right)$$

$$= ([0.8204, 0.9266], [0.1983, 0.2589])$$

The q-RIVOFWHM operator satisfies the following properties.

Property 4. *Monotonicity: let* $\tilde{q}_j = \left(\left[u_{\tilde{q}_j}^{L}, u_{\tilde{q}_j}^{R}\right], \left[v_{\tilde{q}_j}^{L}, v_{\tilde{q}_j}^{R}\right]\right)(j = 1, 2, \ldots, n)$ *and* $\tilde{q}_j' = \left(\left[\left(u_{\tilde{q}_j}^{L}\right)', \left(u_{\tilde{q}_j}^{R}\right)'\right], \left[\left(v_{\tilde{q}_j}^{L}\right)', \left(v_{\tilde{q}_j}^{R}\right)'\right]\right)(j = 1, 2, \ldots, n)$ *be two sets of q-RIVOFNs. If* $u_{\tilde{q}_j}^{L} \leq \left(u_{\tilde{q}_j}^{L}\right)', u_{\tilde{q}_j}^{R} \leq \left(u_{\tilde{q}_j}^{R}\right)', v_{\tilde{q}_j}^{L} \geq \left(v_{\tilde{q}_j}^{L}\right)'$ *and* $v_{\tilde{q}_j}^{L} \geq \left(v_{\tilde{q}_j}^{R}\right)'$ *hold for all* j, *then*

$$q\text{-RIVOFWHM}^{(x)}(\tilde{q}_1, \tilde{q}_2, \cdots, \tilde{q}_n) \leq q\text{-RIVOFWHM}^{(x)}(\tilde{q}_1', \tilde{q}_2', \cdots, \tilde{q}_n') \tag{22}$$

The proof is similar to q-RIVOFHM, so it is omitted here.

Property 5. *Boundedness: let* $\widetilde{q}_j = \left(\left[u^L_{\widetilde{q}_j}, u^R_{\widetilde{q}_j}\right], \left[v^L_{\widetilde{q}_j}, v^R_{\widetilde{q}_j}\right]\right)(j = 1, 2, \ldots, n)$ *be a set of q-RIVOFNs. If* $\widetilde{q}^+ = \left(\left[max_i\left(u^L_{\widetilde{q}_j}\right), max_i\left(u^R_{\widetilde{q}_j}\right)\right], \left[min_i\left(v^L_{\widetilde{q}_j}\right), min_i\left(v^R_{\widetilde{q}_j}\right)\right]\right)$ *and* $\widetilde{q}^- = \left(\left[min_i\left(u^L_{\widetilde{q}_j}\right), min_i\left(u^R_{\widetilde{q}_j}\right)\right], \left[max_i\left(v^L_{\widetilde{q}_j}\right), max_i\left(v^R_{\widetilde{q}_j}\right)\right]\right)$ *then*

$$\widetilde{q}^- \leq q\text{-RIVOFWHM}^{(x)}(\widetilde{q}_1, \widetilde{q}_2, \cdots, \widetilde{q}_n) \leq \widetilde{q}^+ \tag{23}$$

From Theorem 2, we get

$$q\text{-RIVOFWHM}^{(x)}\left(\widetilde{q}_1^-, \widetilde{q}_2^-, \cdots, \widetilde{q}_n^-\right)$$

$$= \left\{\left[\begin{array}{c} \sqrt[q]{1 - \left(\prod\limits_{1\leq i_1 <\ldots<i_x\leq n}\left(1 - \left(\prod\limits_{j=1}^{x}\left(\min\left(u^L_{\widetilde{q}_j}\right)\right)^{w_{i_j}}\right)^{\frac{q}{x}}\right)\right)^{\frac{1}{C_n^x}}}, \\[4mm] \sqrt[q]{1 - \left(\prod\limits_{1\leq i_1 <\ldots<i_x\leq n}\left(1 - \left(\prod\limits_{j=1}^{x}\left(\min\left(u^R_{\widetilde{q}_j}\right)\right)^{w_{i_j}}\right)^{\frac{q}{x}}\right)\right)^{\frac{1}{C_n^x}}} \end{array}\right], \left[\begin{array}{c} \left(\prod\limits_{1\leq i_1 <\ldots<i_x\leq n}\left(\sqrt[q]{1 - \left(\prod\limits_{j=1}^{x}\left(1 - \left(\max\left(v^L_{\widetilde{q}_j}\right)\right)^q\right)^{w_{i_j}}\right)^{\frac{1}{x}}}\right)\right)^{\frac{1}{C_n^x}}, \\[4mm] \left(\prod\limits_{1\leq i_1 <\ldots<i_x\leq n}\left(\sqrt[q]{1 - \left(\prod\limits_{j=1}^{x}\left(1 - \left(\max\left(v^R_{\widetilde{q}_j}\right)\right)^q\right)^{w_{i_j}}\right)^{\frac{1}{x}}}\right)\right)^{\frac{1}{C_n^x}} \end{array}\right]\right\} \tag{24}$$

$$q\text{-RIVOFWHM}^{(x)}\left(\widetilde{q}_1^+, \widetilde{q}_2^+, \cdots, \widetilde{q}_n^+\right)$$

$$= \left\{\left[\begin{array}{c} \sqrt[q]{1 - \left(\prod\limits_{1\leq i_1 <\ldots<i_x\leq n}\left(1 - \left(\prod\limits_{j=1}^{x}\left(\max\left(u^L_{\widetilde{q}_j}\right)\right)^{w_{i_j}}\right)^{\frac{q}{x}}\right)\right)^{\frac{1}{C_n^x}}}, \\[4mm] \sqrt[q]{1 - \left(\prod\limits_{1\leq i_1 <\ldots<i_x\leq n}\left(1 - \left(\prod\limits_{j=1}^{x}\left(\max\left(u^R_{\widetilde{q}_j}\right)\right)^{w_{i_j}}\right)^{\frac{q}{x}}\right)\right)^{\frac{1}{C_n^x}}} \end{array}\right], \left[\begin{array}{c} \left(\prod\limits_{1\leq i_1 <\ldots<i_x\leq n}\left(\sqrt[q]{1 - \left(\prod\limits_{j=1}^{x}\left(1 - \left(\min\left(v^L_{\widetilde{q}_j}\right)\right)^q\right)^{w_{i_j}}\right)^{\frac{1}{x}}}\right)\right)^{\frac{1}{C_n^x}}, \\[4mm] \left(\prod\limits_{1\leq i_1 <\ldots<i_x\leq n}\left(\sqrt[q]{1 - \left(\prod\limits_{j=1}^{x}\left(1 - \left(\min\left(v^R_{\widetilde{q}_j}\right)\right)^q\right)^{w_{i_j}}\right)^{\frac{1}{x}}}\right)\right)^{\frac{1}{C_n^x}} \end{array}\right]\right\} \tag{25}$$

From Property 4, we get

$$\widetilde{q}^- \leq q\text{-RIVOFWHM}^{(x)}(\widetilde{q}_1, \widetilde{q}_2, \cdots, \widetilde{q}_n) \leq \widetilde{q}^+ \tag{26}$$

It is obvious that the q-RIVOFWHM operator lacks the property of idempotency.

3.3. The q-RIVOFDHM Operator

Wu et al. [61] define the dual Hamy mean (DHM) operator.

Definition 7 [61]. *The DHM operator can be defined as:*

$$\mathrm{DHM}^{(x)}(\tilde{q}_1, \tilde{q}_2, \cdots, \tilde{q}_n) = \left(\prod_{1 \le i_1 < \ldots < i_x \le n} \left(\frac{\sum\limits_{j=1}^{x} \tilde{q}_{i_j}}{x} \right) \right)^{\frac{1}{C_n^x}} \tag{27}$$

where x is a parameter, and $x = 1, 2, \ldots, n$, $i_1, i_2, \ldots, i_x$ *are x integer values taken from the set* $\{1, 2, \ldots, n\}$ *of k integer values;* C_n^x *denotes the binomial coefficient and* $C_n^x = \frac{n!}{x!(n-x)!}$.

In this section, we will propose the *q*-rung interval-valued orthopair, fuzzy DHM (*q*-RIVOFDHM) operator.

Definition 8. *Let* $\tilde{q}_j = \left(\left[u_{\tilde{q}_j}^L, u_{\tilde{q}_j}^R \right], \left[v_{\tilde{q}_j}^L, v_{\tilde{q}_j}^R \right] \right) (j = 1, 2, \ldots, n)$ *be a set of q-RIVOFNs. The q-RIVOFDHM operator is*

$$q\text{-RIVOFDHM}^{(x)}(\tilde{q}_1, \tilde{q}_2, \cdots, \tilde{q}_n) = \left(\bigotimes_{1 \le i_1 < \ldots < i_x \le n} \left(\frac{\bigoplus\limits_{j=1}^{x} \tilde{q}_{i_j}}{x} \right) \right)^{\frac{1}{C_n^x}} \tag{28}$$

Theorem 3. *Let* $\tilde{q}_j = \left(\left[u_{\tilde{q}_j}^L, u_{\tilde{q}_j}^R \right], \left[v_{\tilde{q}_j}^L, v_{\tilde{q}_j}^R \right] \right) (j = 1, 2, \ldots, n)$ *be a set of q-RIVOFNs. The fused value by using q-RIVOFDHM operators is also a q-RIVOFN, where*

$$q\text{-RIVOFDHM}^{(x)}(\tilde{q}_1, \tilde{q}_2, \cdots, \tilde{q}_n) = \left(\bigotimes_{1 \le i_1 < \ldots < i_x \le n} \left(\frac{\bigoplus\limits_{j=1}^{x} \tilde{q}_{i_j}}{x} \right) \right)^{\frac{1}{C_n^x}}$$

$$= \left\{ \begin{array}{l} \left[\left(\prod_{1 \le i_1 < \ldots < i_x \le n} \sqrt[q]{1 - \left(\prod\limits_{j=1}^{x} \left(1 - \left(u_{\tilde{q}_j}^L \right)^q \right) \right)^{\frac{1}{x}}} \right)^{\frac{1}{C_n^x}}, \left(\prod_{1 \le i_1 < \ldots < i_x \le n} \sqrt[q]{1 - \left(\prod\limits_{j=1}^{x} \left(1 - \left(u_{\tilde{q}_j}^R \right)^q \right) \right)^{\frac{1}{x}}} \right)^{\frac{1}{C_n^x}} \right], \\[24pt] \left[\sqrt[q]{1 - \left(\prod\limits_{1 \le i_1 < \ldots < i_x \le n} \left(1 - \left(\prod\limits_{j=1}^{x} v_{\tilde{q}_j}^L \right)^{\frac{q}{x}} \right) \right)^{\frac{1}{C_n^x}}}, \sqrt[q]{1 - \left(\prod\limits_{1 \le i_1 < \ldots < i_x \le n} \left(1 - \left(\prod\limits_{j=1}^{x} v_{\tilde{q}_j}^R \right)^{\frac{q}{x}} \right) \right)^{\frac{1}{C_n^x}}} \right] \end{array} \right\} \tag{29}$$

Proof.

$$\bigoplus_{j=1}^{x} \tilde{q}_{i_j} = \left\{ \left[\sqrt[q]{1 - \prod_{j=1}^{x} \left(1 - \left(u_{\tilde{q}_j}^L \right)^q \right)}, \sqrt[q]{1 - \prod_{j=1}^{x} \left(1 - \left(u_{\tilde{q}_j}^R \right)^q \right)} \right], \left[\prod_{j=1}^{x} v_{\tilde{q}_j}^L, \prod_{j=1}^{x} v_{\tilde{q}_j}^R \right] \right\} \tag{30}$$

Thus,

$$\frac{\bigoplus\limits_{j=1}^{x} \tilde{q}_{i_j}}{x} = \left\{ \left[\sqrt[q]{1 - \left(\prod\limits_{j=1}^{x} \left(1 - \left(u_{\tilde{q}_j}^L \right)^q \right) \right)^{\frac{1}{x}}}, \sqrt[q]{1 - \left(\prod\limits_{j=1}^{x} \left(1 - \left(u_{\tilde{q}_j}^R \right)^q \right) \right)^{\frac{1}{x}}} \right], \left[\left(\prod\limits_{j=1}^{x} v_{\tilde{q}_j}^L \right)^{\frac{1}{x}}, \left(\prod\limits_{j=1}^{x} v_{\tilde{q}_j}^R \right)^{\frac{1}{x}} \right] \right\} \tag{31}$$

Thereafter,

$$
\bigotimes_{1\leq i_1<...<i_x\leq n}\left(\frac{\overset{x}{\underset{j=1}{\oplus}}\tilde{q}_{i_j}}{x}\right)
$$

$$
=\left\{\left[\prod_{1\leq i_1<...<i_x\leq n}\sqrt[q]{1-\left(\prod_{j=1}^{x}\left(1-\left(u^{L}_{\tilde{q}_j}\right)^q\right)\right)^{\frac{1}{x}}},\ \prod_{1\leq i_1<...<i_x\leq n}\sqrt[q]{1-\left(\prod_{j=1}^{x}\left(1-\left(u^{R}_{\tilde{q}_j}\right)^q\right)\right)^{\frac{1}{x}}}\right]\right. \tag{32}
$$
$$
\left.\left[\sqrt[q]{1-\prod_{1\leq i_1<...<i_x\leq n}\left(1-\left(\prod_{j=1}^{x}v^{L}_{\tilde{q}_j}\right)^{\frac{q}{x}}\right)},\ \sqrt[q]{1-\prod_{1\leq i_1<...<i_x\leq n}\left(1-\left(\prod_{j=1}^{x}v^{R}_{\tilde{q}_j}\right)^{\frac{q}{x}}\right)}\right]\right\}
$$

Therefore,

$$
q\text{-RIVOFDHM}^{(x)}(\tilde{q}_1,\tilde{q}_2,\cdots,\tilde{q}_n)=\left(\bigotimes_{1\leq i_1<...<i_x\leq n}\left(\frac{\overset{x}{\underset{j=1}{\oplus}}\tilde{q}_{i_j}}{x}\right)\right)^{\frac{1}{C_n^x}}
$$

$$
=\left\{\left[\left(\prod_{1\leq i_1<...<i_x\leq n}\sqrt[q]{1-\left(\prod_{j=1}^{x}\left(1-\left(u^{L}_{\tilde{q}_j}\right)^q\right)\right)^{\frac{1}{x}}}\right)^{\frac{1}{C_n^x}},\ \left(\prod_{1\leq i_1<...<i_x\leq n}\sqrt[q]{1-\left(\prod_{j=1}^{x}\left(1-\left(u^{R}_{\tilde{q}_j}\right)^q\right)\right)^{\frac{1}{x}}}\right)^{\frac{1}{C_n^x}}\right],\right. \tag{33}
$$
$$
\left.\left[\left(\sqrt[q]{1-\left(\prod_{1\leq i_1<...<i_x\leq n}\left(1-\left(\prod_{j=1}^{x}v^{L}_{\tilde{q}_j}\right)^{\frac{q}{x}}\right)\right)}\right)^{\frac{1}{C_n^x}},\ \left(\sqrt[q]{1-\left(\prod_{1\leq i_1<...<i_x\leq n}\left(1-\left(\prod_{j=1}^{x}v^{R}_{\tilde{q}_j}\right)^{\frac{q}{x}}\right)\right)}\right)^{\frac{1}{C_n^x}}\right]\right\}
$$

Hence, Equation (29) is kept.

Then, we need to prove that Equation (29) is a q-RIVOFN. We need to prove that $0\leq\left(\mu^{R}_{\tilde{Q}}(x)\right)^q+\left(v^{R}_{\tilde{Q}}(x)\right)^q\leq 1$.

Let

$$
\mu^{R}_{\tilde{Q}}(x)=\left(\prod_{1\leq i_1<...<i_x\leq n}\sqrt[q]{1-\left(\prod_{j=1}^{x}\left(1-\left(u^{R}_{\tilde{q}_j}\right)^q\right)\right)^{\frac{1}{x}}}\right)^{\frac{1}{C_n^x}}
$$

$$
v^{R}_{\tilde{Q}}(x)=\sqrt[q]{1-\left(\prod_{1\leq i_1<...<i_x\leq n}\left(1-\left(\prod_{j=1}^{x}v^{R}_{\tilde{q}_j}\right)^{\frac{q}{x}}\right)\right)^{\frac{1}{C_n^x}}}
$$

$\square$

Proof.

$$
0\leq\left(\mu^{R}_{\tilde{Q}}(x)\right)^q+\left(v^{R}_{\tilde{Q}}(x)\right)^q
$$
$$
=1-\left(\prod_{1\leq i_1<...<i_x\leq n}\left(1-\left(\prod_{j=1}^{x}v^{R}_{\tilde{q}_j}\right)^{\frac{q}{x}}\right)\right)^{\frac{1}{C_n^x}}+\left(\prod_{1\leq i_1<...<i_x\leq n}\left(1-\left(\prod_{j=1}^{x}\left(1-\left(u^{R}_{\tilde{q}_j}\right)^q\right)\right)^{\frac{1}{x}}\right)\right)^{\frac{1}{C_n^x}}
$$
$$
\leq 1-\left(\prod_{1\leq i_1<...<i_x\leq n}\left(1-\left(\prod_{j=1}^{x}\left(1-\left(u^{R}_{\tilde{q}_j}\right)^q\right)\right)^{\frac{1}{x}}\right)\right)^{\frac{1}{C_n^x}}+\left(\prod_{1\leq i_1<...<i_x\leq n}\left(1-\left(\prod_{j=1}^{x}\left(1-\left(u^{R}_{\tilde{q}_j}\right)^q\right)\right)^{\frac{1}{x}}\right)\right)^{\frac{1}{C_n^x}}
$$
$$
=1
$$

Therefore, $0\leq\left(\mu^{R}_{\tilde{Q}}(x)\right)^q+\left(v^{R}_{\tilde{Q}}(x)\right)^q\leq 1$ is maintained. $\square$

Example 3. *Let* $([0.5, 0.8], [0.4, 0.5])$, $([0.3, 0.5], [0.6, 0.7])$, $([0.5, 0.7], [0.2, 0.3])$ *and* $([0.4, 0.8], [0.1, 0.2])$ *be four q-RIVOFNs, and suppose* $x = 2$, $q = 3$; *then according to Equation (29), we have*

$$q\text{-RIVOFDHM}^{(x)}(\widetilde{q}_1, \widetilde{q}_2, \cdots, \widetilde{q}_n) = \left(\bigotimes_{1 \le i_1 < \ldots < i_x \le n} \left(\frac{\overset{x}{\underset{j=1}{\oplus}} \widetilde{q}_{i_j}}{x} \right) \right)^{\frac{1}{C_n^x}}$$

$$= \left(\left[\left(\sqrt[3]{\left(\begin{array}{c} \left(1 - ((1 - 0.5^3) \times (1 - 0.3^3))^{\frac{1}{2}}\right) \times \left(1 - ((1 - 0.5^3) \times (1 - 0.5^3))^{\frac{1}{2}}\right) \\ \times \left(1 - ((1 - 0.5^3) \times (1 - 0.4^3))^{\frac{1}{2}}\right) \times \left(1 - ((1 - 0.3^3) \times (1 - 0.5^3))^{\frac{1}{2}}\right) \\ \times \left(1 - ((1 - 0.3^3) \times (1 - 0.4^3))^{\frac{1}{2}}\right) \times \left(1 - ((1 - 0.5^3) \times (1 - 0.4^3))^{\frac{1}{2}}\right) \end{array} \right)} \right)^{\frac{1}{C_4^2}} \right], \right.$$

$$\left. \left(\sqrt[3]{\left(\begin{array}{c} \left(1 - ((1 - 0.8^3) \times (1 - 0.5^3))^{\frac{1}{2}}\right) \times \left(1 - ((1 - 0.8^3) \times (1 - 0.7^3))^{\frac{1}{2}}\right) \\ \times \left(1 - ((1 - 0.8^3) \times (1 - 0.8^3))^{\frac{1}{2}}\right) \times \left(1 - ((1 - 0.5^3) \times (1 - 0.7^3))^{\frac{1}{2}}\right) \\ \times \left(1 - ((1 - 0.5^3) \times (1 - 0.8^3))^{\frac{1}{2}}\right) \times \left(1 - ((1 - 0.7^3) \times (1 - 0.8^3))^{\frac{1}{2}}\right) \end{array} \right)} \right)^{\frac{1}{C_4^2}} \right],$$

$$\left[\left(\sqrt[3]{1 - \left(\begin{array}{c} \left(1 - (0.4 \times 0.6)^{\frac{3}{2}}\right) \times \left(1 - (0.4 \times 0.2)^{\frac{3}{2}}\right) \times \left(1 - (0.4 \times 0.1)^{\frac{3}{2}}\right) \\ \times \left(1 - (0.6 \times 0.2)^{\frac{3}{2}}\right) \times \left(1 - (0.6 \times 0.1)^{\frac{3}{2}}\right) \times \left(1 - (0.2 \times 0.1)^{\frac{3}{2}}\right) \end{array} \right)} \right)^{\frac{1}{C_4^2}}, \right.$$

$$\left. \left(\sqrt[3]{1 - \left(\begin{array}{c} \left(1 - (0.5 \times 0.7)^{\frac{3}{2}}\right) \times \left(1 - (0.5 \times 0.3)^{\frac{3}{2}}\right) \times \left(1 - (0.5 \times 0.2)^{\frac{3}{2}}\right) \\ \times \left(1 - (0.7 \times 0.3)^{\frac{3}{2}}\right) \times \left(1 - (0.7 \times 0.2)^{\frac{3}{2}}\right) \times \left(1 - (0.3 \times 0.2)^{\frac{3}{2}}\right) \end{array} \right)} \right)^{\frac{1}{C_4^2}} \right] \right)$$

$$= ([0.4348, 0.7214], [0.3283, 0.4291])$$

The q-RIVOFDHM has the following three operators.

Property 6. *Idempotency: if* $\widetilde{q}_j = \left(\left[u_{\widetilde{q}_j}^L, u_{\widetilde{q}_j}^R \right], \left[v_{\widetilde{q}_j}^L, v_{\widetilde{q}_j}^R \right] \right) (j = 1, 2, \ldots, n)$ *are equal, then*

$$q\text{-RIVOFDHM}^{(x)}(\widetilde{q}_1, \widetilde{q}_2, \cdots, \widetilde{q}_n) = \widetilde{q} \tag{34}$$

Proof. Since $\widetilde{q}_j = \widetilde{q} = \left(\left[u_{\widetilde{q}}^L, u_{\widetilde{q}}^R \right], \left[v_{\widetilde{q}}^L, v_{\widetilde{q}}^R \right] \right)$, then

$$q\text{-RIVOFDHM}^{(x)}(\tilde{q}_1, \tilde{q}_2, \cdots, \tilde{q}_n) = \left(\underset{1 \le i_1 < \ldots < i_x \le n}{\otimes} \left(\frac{\overset{x}{\underset{j=1}{\oplus}} \tilde{q}_{i_j}}{x} \right) \right)^{\frac{1}{C_n^x}}$$

$$= \left\{ \begin{array}{l} \left[\left(\underset{1 \le i_1 < \ldots < i_x \le n}{\prod} \sqrt[q]{1 - \left(\prod_{j=1}^{x} \left(1 - \left(u_{\tilde{q}_j}^L \right)^q \right) \right)^{\frac{1}{x}}} \right)^{\frac{1}{C_n^x}}, \left(\underset{1 \le i_1 < \ldots < i_x \le n}{\prod} \sqrt[q]{1 - \left(\prod_{j=1}^{x} \left(1 - \left(u_{\tilde{q}_j}^R \right)^q \right) \right)^{\frac{1}{x}}} \right)^{\frac{1}{C_n^x}} \right], \\[30pt] \left[\sqrt[q]{1 - \left(\underset{1 \le i_1 < \ldots < i_x \le n}{\prod} \left(1 - \left(\prod_{j=1}^{x} v_{\tilde{q}_j}^L \right)^{\frac{q}{x}} \right) \right)^{\frac{1}{C_n^x}}}, \sqrt[q]{1 - \left(\underset{1 \le i_1 < \ldots < i_x \le n}{\prod} \left(1 - \left(\prod_{j=1}^{x} v_{\tilde{q}_j}^R \right)^{\frac{q}{x}} \right) \right)^{\frac{1}{C_n^x}}} \right] \end{array} \right\}$$

$$= \left\{ \begin{array}{l} \left[\left(\left(\sqrt[q]{1 - \left(\left(1 - \left(u_{\tilde{q}}^L \right)^q \right)^x \right)^{\frac{1}{x}}} \right)^{C_n^x} \right)^{\frac{1}{C_n^x}}, \left(\left(\sqrt[q]{1 - \left(\left(1 - \left(u_{\tilde{q}}^R \right)^q \right)^x \right)^{\frac{1}{x}}} \right)^{C_n^x} \right)^{\frac{1}{C_n^x}} \right], \\[30pt] \left[\sqrt[q]{1 - \left(\left(1 - \left(\left(v_{\tilde{q}}^L \right)^x \right)^{\frac{q}{x}} \right)^{C_n^x} \right)^{\frac{1}{C_n^x}}}, \sqrt[q]{1 - \left(\left(1 - \left(\left(v_{\tilde{q}}^R \right)^x \right)^{\frac{q}{x}} \right)^{C_n^x} \right)^{\frac{1}{C_n^x}}} \right] \end{array} \right\}$$

$$= \left\{ \left[u_{\tilde{q}}^L, u_{\tilde{q}}^R \right], \left[v_{\tilde{q}}^L, v_{\tilde{q}}^R \right] \right\} = \tilde{q}$$

□

Property 7. *Monotonicity: let* $\tilde{q}_j = \left(\left[u_{\tilde{q}_j}^L, u_{\tilde{q}_j}^R \right], \left[v_{\tilde{q}_j}^L, v_{\tilde{q}_j}^R \right] \right) (j = 1, 2, \ldots, n)$ *and* $\tilde{q}_j' = \left(\left[\left(u_{\tilde{q}_j}^L \right)', \left(u_{\tilde{q}_j}^R \right)' \right], \left[\left(v_{\tilde{q}_j}^L \right)', \left(v_{\tilde{q}_j}^R \right)' \right] \right) (j = 1, 2, \ldots, n)$ *be two sets of q-RIVOFNs. If* $u_{\tilde{q}_j}^L \le \left(u_{\tilde{q}_j}^L \right)', u_{\tilde{q}_j}^R \le \left(u_{\tilde{q}_j}^R \right)', v_{\tilde{q}_j}^L \ge \left(v_{\tilde{q}_j}^L \right)'$ *and* $v_{\tilde{q}_j}^L \ge \left(v_{\tilde{q}_j}^R \right)'$ *hold for all j, then*

$$q\text{-RIVOFDHM}^{(x)}(\tilde{q}_1, \tilde{q}_2, \cdots, \tilde{q}_n) \le q\text{-RIVOFDHM}^{(x)}(\tilde{q}_1', \tilde{q}_2', \cdots, \tilde{q}_n') \tag{35}$$

Proof. Given that $u_{\tilde{q}_j}^L \le \left(u_{\tilde{q}_j}^L \right)'$, we can obtain

$$\prod_{j=1}^{x} \left(1 - \left(u_{\tilde{q}_j}^L \right)^q \right) \ge \prod_{j=1}^{x} \left(1 - \left(\left(u_{\tilde{q}_j}^L \right)' \right)^q \right) \tag{36}$$

$$1 - \left(\prod_{j=1}^{x} \left(1 - \left(u_{\tilde{q}_j}^L \right)^q \right) \right)^{\frac{1}{x}} \le 1 - \left(\prod_{j=1}^{x} \left(1 - \left(\left(u_{\tilde{q}_j}^L \right)' \right)^q \right) \right)^{\frac{1}{x}} \tag{37}$$

Thereafter,

$$\left(\underset{1 \le i_1 < \ldots < i_x \le n}{\prod} \sqrt[q]{1 - \left(\prod_{j=1}^{x} \left(1 - \left(u_{\tilde{q}_j}^L \right)^q \right) \right)^{\frac{1}{x}}} \right)^{\frac{1}{C_n^x}} \le \left(\underset{1 \le i_1 < \ldots < i_x \le n}{\prod} \sqrt[q]{1 - \left(\prod_{j=1}^{x} \left(1 - \left(\left(u_{\tilde{q}_j}^L \right)' \right)^q \right) \right)^{\frac{1}{x}}} \right)^{\frac{1}{C_n^x}} \tag{38}$$

That means that $u_{\tilde{q}}^L \le \left(u_{\tilde{q}}^L \right)'$. Similarly, we can obtain $u_{\tilde{q}}^R \le \left(u_{\tilde{q}}^R \right)', v_{\tilde{q}}^L \ge \left(v_{\tilde{q}}^L \right)'$ and $v_{\tilde{q}}^L \ge \left(v_{\tilde{q}}^R \right)'$. Thus, the proof is complete. □

Property 8. *Boundedness: let* $\widetilde{q}_j = \left(\left[u_{\widetilde{q}_j}^L, u_{\widetilde{q}_j}^R \right], \left[v_{\widetilde{q}_j}^L, v_{\widetilde{q}_j}^R \right] \right) (j = 1, 2, \ldots, n)$ *be a set of q-RIVOFNs.* *If* $\widetilde{q}^+ = \left(\left[max_i \left(u_{\widetilde{q}_j}^L \right), max_i \left(u_{\widetilde{q}_j}^R \right) \right], \left[min_i \left(v_{\widetilde{q}_j}^L \right), min_i \left(v_{\widetilde{q}_j}^R \right) \right] \right)$ *and* $\widetilde{q}^- = \left(\left[min_i \left(u_{\widetilde{q}_j}^L \right), min_i \left(u_{\widetilde{q}_j}^R \right) \right], \left[max_i \left(v_{\widetilde{q}_j}^L \right), max_i \left(v_{\widetilde{q}_j}^R \right) \right] \right)$ *then*

$$\widetilde{q}^- \leq q\text{-RIVOFDHM}^{(x)} (\widetilde{q}_1, \widetilde{q}_2, \cdots, \widetilde{q}_n) \leq \widetilde{q}^+ \tag{39}$$

From Property 6,

$$q\text{-RIVOFDHM}^{(x)} \left(\widetilde{q}_1^-, \widetilde{q}_2^-, \cdots, \widetilde{q}_n^- \right) = \widetilde{q}^-$$
$$q\text{-RIVOFDHM}^{(x)} \left(\widetilde{q}_1^+, \widetilde{q}_2^+, \cdots, \widetilde{q}_n^+ \right) = \widetilde{q}^+$$

From Property 7,

$$\widetilde{q}^- \leq q\text{-RIVOFDHM}^{(x)} (\widetilde{q}_1, \widetilde{q}_2, \cdots, \widetilde{q}_n) \leq \widetilde{q}^+$$

3.4. The q-RIVOFWDHM Operator

In real MADM problems, it's of necessity to take attribute weights into account; we will propose the q-rung interval-valued orthopair fuzzy weighted DHM (q-RIVOFWDHM) operator in this chapter.

Definition 9. *Let* $\widetilde{q}_j = \left(\left[u_{\widetilde{q}_j}^L, u_{\widetilde{q}_j}^R \right], \left[v_{\widetilde{q}_j}^L, v_{\widetilde{q}_j}^R \right] \right) (j = 1, 2, \ldots, n)$ *be a set of q-RIVOFNs, with their weight vector as* $w_i = (w_1, w_2, \ldots, w_n)^T$, *thereby satisfying* $w_i \in [0, 1]$ *and* $\sum_{i=1}^n w_i = 1$. *If*

$$q\text{-RIVOFWDHM}^{(x)} (\widetilde{q}_1, \widetilde{q}_2, \cdots, \widetilde{q}_n) = \left(\underset{1 \leq i_1 < \ldots < i_x \leq n}{\otimes} \left(\frac{\overset{x}{\underset{j=1}{\oplus}} w_{i_j} \widetilde{q}_{i_j}}{x} \right) \right)^{\frac{1}{C_n^x}} \tag{40}$$

Theorem 4. *Let* $\widetilde{q}_j = \left(\left[u_{\widetilde{q}_j}^L, u_{\widetilde{q}_j}^R \right], \left[v_{\widetilde{q}_j}^L, v_{\widetilde{q}_j}^R \right] \right) (j = 1, 2, \ldots, n)$ *be a set of q-RIVOFNs. The fused value by using q-RIVOFWDHM operators is also a q-RIVOFN, where*

$$q\text{-RIVOFWDHM}^{(x)} (\widetilde{q}_1, \widetilde{q}_2, \cdots, \widetilde{q}_n) = \left(\underset{1 \leq i_1 < \ldots < i_x \leq n}{\otimes} \left(\frac{\overset{x}{\underset{j=1}{\oplus}} w_{i_j} \widetilde{q}_{i_j}}{x} \right) \right)^{\frac{1}{C_n^x}}$$
$$= \left\{ \begin{array}{c} \left[\left(\underset{1 \leq i_1 < \ldots < i_x \leq n}{\prod} \left(\sqrt[q]{1 - \prod_{j=1}^x \left(1 - \left(u_{\widetilde{q}_j}^L \right)^q \right)^{w_{i_j}}} \right)^{\frac{1}{x}} \right)^{\frac{1}{C_n^x}}, \left(\underset{1 \leq i_1 < \ldots < i_x \leq n}{\prod} \left(\sqrt[q]{1 - \prod_{j=1}^x \left(1 - \left(u_{\widetilde{q}_j}^R \right)^q \right)^{w_{i_j}}} \right)^{\frac{1}{x}} \right)^{\frac{1}{C_n^x}} \right] \\ \left[\sqrt[q]{1 - \left(\underset{1 \leq i_1 < \ldots < i_x \leq n}{\prod} \left(1 - \left(\prod_{j=1}^x \left(v_{\widetilde{q}_j}^L \right)^{w_{i_j}} \right)^{\frac{1}{x}} \right) \right)^{\frac{1}{C_n^x}}}, \sqrt[q]{1 - \left(\underset{1 \leq i_1 < \ldots < i_x \leq n}{\prod} \left(1 - \left(\prod_{j=1}^x \left(v_{\widetilde{q}_j}^R \right)^{w_{i_j}} \right)^{\frac{q}{x}} \right) \right)^{\frac{1}{C_n^x}}} \right] \end{array} \right\} \tag{41}$$

Proof.

$$w_{i_j} \widetilde{q}_{i_j} = \left\{ \left[\sqrt[q]{1 - \left(1 - \left(u_{\widetilde{q}_j}^L \right)^q \right)^{w_{i_j}}}, \sqrt[q]{1 - \left(1 - \left(u_{\widetilde{q}_j}^R \right)^q \right)^{w_{i_j}}} \right], \left[\left(v_{\widetilde{q}_j}^L \right)^{w_{i_j}}, \left(v_{\widetilde{q}_j}^R \right)^{w_{i_j}} \right] \right\} \tag{42}$$

Thus,

$$\mathop{\oplus}_{j=1}^{x}\left(w_{i_j}\widetilde{q}_{i_j}\right) = \left\{ \begin{array}{l} \left[\sqrt[q]{1-\prod_{j=1}^{x}\left(1-\left(u_{\widetilde{q}_j}^L\right)^q\right)^{w_{i_j}}},\sqrt[q]{1-\prod_{j=1}^{x}\left(1-\left(u_{\widetilde{q}_j}^R\right)^q\right)^{w_{i_j}}}\right], \\ \left[\prod_{j=1}^{x}\left(v_{\widetilde{q}_j}^L\right)^{w_{i_j}},\prod_{j=1}^{x}\left(v_{\widetilde{q}_j}^R\right)^{w_{i_j}}\right] \end{array} \right\} \tag{43}$$

Therefore,

$$\frac{\mathop{\oplus}_{j=1}^{x}\left(w_{i_j}\widetilde{q}_{i_j}\right)}{x} = \left\{ \begin{array}{l} \left[\sqrt[q]{1-\left(\prod_{j=1}^{x}\left(1-\left(u_{\widetilde{q}_j}^L\right)^q\right)^{w_{i_j}}\right)^{\frac{1}{x}}},\sqrt[q]{1-\left(\prod_{j=1}^{x}\left(1-\left(u_{\widetilde{q}_j}^R\right)^q\right)^{w_{i_j}}\right)^{\frac{1}{x}}}\right], \\ \left[\left(\prod_{j=1}^{x}\left(v_{\widetilde{q}_j}^L\right)^{w_{i_j}}\right)^{\frac{1}{x}},\left(\prod_{j=1}^{x}\left(v_{\widetilde{q}_j}^R\right)^{w_{i_j}}\right)^{\frac{1}{x}}\right] \end{array} \right\} \tag{44}$$

Thereafter,

$$\mathop{\otimes}_{1\le i_1<...<i_x\le n}\left(\frac{\mathop{\oplus}_{j=1}^{x}w_{i_j}\widetilde{q}_{i_j}}{x}\right)$$
$$= \left\{ \begin{array}{l} \left[\prod_{1\le i_1<...<i_x\le n}\left(\sqrt[q]{1-\left(\prod_{j=1}^{x}\left(1-\left(u_{\widetilde{q}_j}^L\right)^q\right)^{w_{i_j}}\right)^{\frac{1}{x}}}\right),\prod_{1\le i_1<...<i_x\le n}\left(\sqrt[q]{1-\left(\prod_{j=1}^{x}\left(1-\left(u_{\widetilde{q}_j}^R\right)^q\right)^{w_{i_j}}\right)^{\frac{1}{x}}}\right)\right], \\ \left[\sqrt[q]{1-\prod_{1\le i_1<...<i_x\le n}\left(1-\left(\prod_{j=1}^{x}\left(v_{\widetilde{q}_j}^L\right)^{w_{i_j}}\right)^{\frac{q}{x}}\right)},\sqrt[q]{1-\prod_{1\le i_1<...<i_x\le n}\left(1-\left(\prod_{j=1}^{x}\left(v_{\widetilde{q}_j}^R\right)^{w_{i_j}}\right)^{\frac{q}{x}}\right)}\right] \end{array} \right\} \tag{45}$$

Furthermore,

$$q\text{-RIVOFWDHM}^{(x)}(\widetilde{q}_1,\widetilde{q}_2,\cdots,\widetilde{q}_n) = \left(\mathop{\otimes}_{1\le i_1<...<i_x\le n}\left(\frac{\mathop{\oplus}_{j=1}^{x}w_{i_j}\widetilde{q}_{i_j}}{x}\right)\right)^{\frac{1}{C_n^x}}$$
$$= \left\{ \begin{array}{l} \left[\left(\prod_{1\le i_1<...<i_x\le n}\left(\sqrt[q]{1-\left(\prod_{j=1}^{x}\left(1-\left(u_{\widetilde{q}_j}^L\right)^q\right)^{w_{i_j}}\right)^{\frac{1}{x}}}\right)\right)^{\frac{1}{C_n^x}},\left(\prod_{1\le i_1<...<i_x\le n}\left(\sqrt[q]{1-\left(\prod_{j=1}^{x}\left(1-\left(u_{\widetilde{q}_j}^R\right)^q\right)^{w_{i_j}}\right)^{\frac{1}{x}}}\right)\right)^{\frac{1}{C_n^x}}\right], \\ \left[\sqrt[q]{1-\left(\prod_{1\le i_1<...<i_x\le n}\left(1-\left(\prod_{j=1}^{x}\left(v_{\widetilde{q}_j}^L\right)^{w_{i_j}}\right)^{\frac{q}{x}}\right)\right)^{\frac{1}{C_n^x}}},\sqrt[q]{1-\left(\prod_{1\le i_1<...<i_x\le n}\left(1-\left(\prod_{j=1}^{x}\left(v_{\widetilde{q}_j}^R\right)^{w_{i_j}}\right)^{\frac{q}{x}}\right)\right)^{\frac{1}{C_n^x}}}\right] \end{array} \right\} \tag{46}$$

Hence, Equation (41) is kept.

Then, we need to prove that Equation (41) is a q-RIVOFN. We need to prove that $0\le\left(\mu_{\widetilde{Q}}^R(x)\right)^q+\left(v_{\widetilde{Q}}^R(x)\right)^q\le 1$.

Let

$$\mu_{\widetilde{Q}}^R(x) = \left(\prod_{1\le i_1<...<i_x\le n}\left(\sqrt[q]{1-\left(\prod_{j=1}^{x}\left(1-\left(u_{\widetilde{q}_j}^R\right)^q\right)^{w_{i_j}}\right)^{\frac{1}{x}}}\right)\right)^{\frac{1}{C_n^x}}$$

$$v_{\widetilde{Q}}^R(x) = \sqrt[q]{1-\left(\prod_{1\le i_1<...<i_x\le n}\left(1-\left(\prod_{j=1}^{x}\left(v_{\widetilde{q}_j}^R\right)^{w_{i_j}}\right)^{\frac{q}{x}}\right)\right)^{\frac{1}{C_n^x}}}$$

$\square$

Proof.

$$0 \leq \left(\mu_{\tilde{Q}}^R(x)\right)^q + \left(\nu_{\tilde{Q}}^R(x)\right)^q$$

$$= \left(\prod_{1 \leq i_1 < \ldots < i_x \leq n}\left(1 - \left(\prod_{j=1}^x\left(1 - \left(u_{\tilde{q}_j}^R\right)^q\right)^{w_{ij}}\right)^{\frac{1}{x}}\right)\right)^{\frac{1}{C_n^x}} + 1 - \left(\prod_{1 \leq i_1 < \ldots < i_x \leq n}\left(1 - \left(\prod_{j=1}^x\left(v_{\tilde{q}_j}^R\right)^{w_{ij}}\right)^{\frac{q}{x}}\right)\right)^{\frac{1}{C_n^x}}$$

$$\leq \left(\prod_{1 \leq i_1 < \ldots < i_x \leq n}\left(1 - \left(\prod_{j=1}^x\left(1 - \left(u_{\tilde{q}_j}^R\right)^q\right)^{w_{ij}}\right)^{\frac{1}{x}}\right)\right)^{\frac{1}{C_n^x}} + 1 - \left(\prod_{1 \leq i_1 < \ldots < i_x \leq n}\left(1 - \left(\prod_{j=1}^x\left(1 - \left(u_{\tilde{q}_j}^R\right)^q\right)^{w_{ij}}\right)^{\frac{1}{x}}\right)\right)^{\frac{1}{C_n^x}}$$

$$= 1$$

Therefore, $0 \leq \left(\mu_{\tilde{Q}}^R(x)\right)^q + \left(\nu_{\tilde{Q}}^R(x)\right)^q \leq 1$ is maintained. $\square$

Example 4. *Let* $([0.5, 0.8], [0.4, 0.5])$, $([0.3, 0.5], [0.6, 0.7])$, $([0.5, 0.7], [0.2, 0.3])$ *and* $([0.4, 0.8], [0.1, 0.2])$ *be four q-RIVOFNs; suppose* $x = 2$, $q = 3$, *and* $\omega = (0.2, 0.1, 0.3, 0.4)$. *Then, based on Equation (41), we can get*

$$q\text{-RIVOFWDHM}^{(x)}(\tilde{q}_1, \tilde{q}_2, \cdots, \tilde{q}_n) = \left(\bigotimes_{1 \leq i_1 < \ldots < i_x \leq n}\left(\frac{\overset{x}{\underset{j=1}{\oplus}} w_{ij}\tilde{q}_{ij}}{x}\right)\right)^{\frac{1}{C_n^x}}$$

$$= \left(\left[\left[\left(\sqrt{\begin{array}{l}\left(1 - \left((1-0.5^3)^{0.2} \times (1-0.3^3)^{0.1}\right)^{\frac{1}{2}}\right) \times \left(1 - \left((1-0.5^3)^{0.2} \times (1-0.5^3)^{0.3}\right)^{\frac{1}{2}}\right) \\ \times \left(1 - \left((1-0.5^3)^{0.2} \times (1-0.4^3)^{0.4}\right)^{\frac{1}{2}}\right) \times \left(1 - \left((1-0.3^3)^{0.1} \times (1-0.5^3)^{0.3}\right)^{\frac{1}{2}}\right) \\ \times \left(1 - \left((1-0.3^3)^{0.1} \times (1-0.4^3)^{0.4}\right)^{\frac{1}{2}}\right) \times \left(1 - \left((1-0.5^3)^{0.3} \times (1-0.4^3)^{0.4}\right)^{\frac{1}{2}}\right)\end{array}}\right)^{\frac{1}{C_4^2}}\right],$$

$$\left[\left(\sqrt{\begin{array}{l}\left(1 - \left((1-0.8^3)^{0.2} \times (1-0.5^3)^{0.1}\right)^{\frac{1}{2}}\right) \times \left(1 - \left((1-0.8^3)^{0.2} \times (1-0.7^3)^{0.3}\right)^{\frac{1}{2}}\right) \\ \times \left(1 - \left((1-0.8^3)^{0.2} \times (1-0.8^3)^{0.4}\right)^{\frac{1}{2}}\right) \times \left(1 - \left((1-0.5^3)^{0.1} \times (1-0.7^3)^{0.3}\right)^{\frac{1}{2}}\right) \\ \times \left(1 - \left((1-0.5^3)^{0.1} \times (1-0.8^3)^{0.4}\right)^{\frac{1}{2}}\right) \times \left(1 - \left((1-0.7^3)^{0.3} \times (1-0.8^3)^{0.4}\right)^{\frac{1}{2}}\right)\end{array}}\right)^{\frac{1}{C_4^2}}\right],$$

$$\left[1 - \left(\sqrt{\begin{array}{l}\left(1 - (0.4^{0.2} \times 0.6^{0.1})^{\frac{3}{2}}\right) \times \left(1 - (0.4^{0.2} \times 0.2^{0.3})^{\frac{3}{2}}\right) \times \left(1 - (0.4^{0.2} \times 0.1^{0.4})^{\frac{3}{2}}\right) \\ \times \left(1 - (0.6^{0.1} \times 0.2^{0.3})^{\frac{3}{2}}\right) \times \left(1 - (0.6^{0.1} \times 0.1^{0.4})^{\frac{3}{2}}\right) \times \left(1 - (0.2^{0.3} \times 0.1^{0.4})^{\frac{3}{2}}\right)\end{array}}\right)^{\frac{1}{C_4^2}}\right],$$

$$\left[1 - \left(\sqrt{\begin{array}{l}\left(1 - (0.5^{0.2} \times 0.7^{0.1})^{\frac{3}{2}}\right) \times \left(1 - (0.5^{0.2} \times 0.3^{0.3})^{\frac{3}{2}}\right) \times \left(1 - (0.5^{0.2} \times 0.2^{0.4})^{\frac{3}{2}}\right) \\ \times \left(1 - (0.7^{0.1} \times 0.3^{0.3})^{\frac{3}{2}}\right) \times \left(1 - (0.7^{0.1} \times 0.2^{0.4})^{\frac{3}{2}}\right) \times \left(1 - (0.3^{0.3} \times 0.2^{0.4})^{\frac{3}{2}}\right)\end{array}}\right)^{\frac{1}{C_4^2}}\right]\right)$$

$$= ([0.2819, 0.4954], [0.7249, 0.7855])$$

We will then study some precious properties of q-RIVOFWDHM operator.

Property 9. *Monotonicity: let* $\tilde{q}_j = \left(\left[u_{\tilde{q}_j}^L, u_{\tilde{q}_j}^R\right], \left[v_{\tilde{q}_j}^L, v_{\tilde{q}_j}^R\right]\right)(j = 1, 2, \ldots, n)$ *and* $\tilde{q}_j' = \left(\left[\left(u_{\tilde{q}_j}^L\right)', \left(u_{\tilde{q}_j}^R\right)'\right], \left[\left(v_{\tilde{q}_j}^L\right)', \left(v_{\tilde{q}_j}^R\right)'\right]\right)(j = 1, 2, \ldots, n)$ *be two sets of q-RIVOFNs. If* $u_{\tilde{q}_j}^L \leq \left(u_{\tilde{q}_j}^L\right)', u_{\tilde{q}_j}^R \leq \left(u_{\tilde{q}_j}^R\right)', v_{\tilde{q}_j}^L \geq \left(v_{\tilde{q}_j}^L\right)'$ *and* $v_{\tilde{q}_j}^L \geq \left(v_{\tilde{q}_j}^R\right)'$ *hold for all j, then*

$$q\text{-RIVOFWDHM}^{(x)}(\tilde{q}_1, \tilde{q}_2, \cdots, \tilde{q}_n) \leq q\text{-RIVOFWDHM}^{(x)}(\tilde{q}_1', \tilde{q}_2', \cdots, \tilde{q}_n') \tag{47}$$

This proof is similar to q-RIVOFDHM, so it is omitted here.

Property 10. *(Boundedness) Let* $\widetilde{q}_j = \left(\left[u^L_{\widetilde{q}_j}, u^R_{\widetilde{q}_j}\right], \left[v^L_{\widetilde{q}_j}, v^R_{\widetilde{q}_j}\right]\right)(j = 1, 2, \ldots, n)$ *be a set of* q-RIVOFNs. *If* $\widetilde{q}^+ = \left(\left[max_i\left(u^L_{\widetilde{q}_j}\right), max_i\left(u^R_{\widetilde{q}_j}\right)\right], \left[min_i\left(v^L_{\widetilde{q}_j}\right), min_i\left(v^R_{\widetilde{q}_j}\right)\right]\right)$ *and* $\widetilde{q}^- = \left(\left[min_i\left(u^L_{\widetilde{q}_j}\right), min_i\left(u^R_{\widetilde{q}_j}\right)\right], \left[max_i\left(v^L_{\widetilde{q}_j}\right), max_i\left(v^R_{\widetilde{q}_j}\right)\right]\right)$ *then*

$$\widetilde{q}^- \leq q\text{-RIVOFWDHM}^{(x)}(\widetilde{q}_1, \widetilde{q}_2, \cdots, \widetilde{q}_n) \leq \widetilde{q}^+ \tag{48}$$

From Theorem 4, we get

$$q\text{-RIVOFWDHM}^{(x)}\left(\widetilde{q}_1^-, \widetilde{q}_2^-, \cdots, \widetilde{q}_n^-\right)$$

$$= \left\{ \left[\left(\prod_{1 \leq i_1 < \ldots < i_x \leq n} \left(\sqrt[q]{1 - \left(\prod_{j=1}^{x} \left(1 - \left(\min\left(u^L_{\widetilde{q}_j}\right)\right)^q\right)^{w_{i_j}}\right)^{\frac{1}{x}}} \right) \right)^{\frac{1}{C_n^x}}, \right.\right.$$
$$\left. \left(\prod_{1 \leq i_1 < \ldots < i_x \leq n} \left(\sqrt[q]{1 - \left(\prod_{j=1}^{x} \left(1 - \left(\min\left(u^R_{\widetilde{q}_j}\right)\right)^q\right)^{w_{i_j}}\right)^{\frac{1}{x}}} \right) \right)^{\frac{1}{C_n^x}} \right],$$
$$\left[\sqrt[q]{1 - \left(\prod_{1 \leq i_1 < \ldots < i_x \leq n} \left(1 - \left(\prod_{j=1}^{x} \left(\max\left(v^L_{\widetilde{q}_j}\right)\right)^{w_{i_j}}\right)^{\frac{q}{x}}\right) \right)^{\frac{1}{C_n^x}}}, \right.$$
$$\left.\left. \sqrt[q]{1 - \left(\prod_{1 \leq i_1 < \ldots < i_x \leq n} \left(1 - \left(\prod_{j=1}^{x} \left(\max\left(v^R_{\widetilde{q}_j}\right)\right)^{w_{i_j}}\right)^{\frac{q}{x}}\right) \right)^{\frac{1}{C_n^x}}} \right] \right\} \tag{49}$$

$$q\text{-RIVOFWDHM}^{(x)}\left(\widetilde{q}_1^+, \widetilde{q}_2^+, \cdots, \widetilde{q}_n^+\right)$$

$$= \left\{ \left[\left(\prod_{1 \leq i_1 < \ldots < i_x \leq n} \left(\sqrt[q]{1 - \left(\prod_{j=1}^{x} \left(1 - \left(\max\left(u^L_{\widetilde{q}_j}\right)\right)^q\right)^{w_{i_j}}\right)^{\frac{1}{x}}} \right) \right)^{\frac{1}{C_n^x}}, \right.\right.$$
$$\left. \left(\prod_{1 \leq i_1 < \ldots < i_x \leq n} \left(\sqrt[q]{1 - \left(\prod_{j=1}^{x} \left(1 - \left(\max\left(u^R_{\widetilde{q}_j}\right)\right)^q\right)^{w_{i_j}}\right)^{\frac{1}{x}}} \right) \right)^{\frac{1}{C_n^x}} \right],$$
$$\left[\sqrt[q]{1 - \left(\prod_{1 \leq i_1 < \ldots < i_x \leq n} \left(1 - \left(\prod_{j=1}^{x} \left(\min\left(v^L_{\widetilde{q}_j}\right)\right)^{w_{i_j}}\right)^{\frac{q}{x}}\right) \right)^{\frac{1}{C_n^x}}}, \right.$$
$$\left.\left. \sqrt[q]{1 - \left(\prod_{1 \leq i_1 < \ldots < i_x \leq n} \left(1 - \left(\prod_{j=1}^{x} \left(\min\left(v^R_{\widetilde{q}_j}\right)\right)^{w_{i_j}}\right)^{\frac{q}{x}}\right) \right)^{\frac{1}{C_n^x}}} \right] \right\} \tag{50}$$

From Property 9, we get

$$\widetilde{q}^- \leq q\text{-RIVOFWDHM}^{(x)}(\widetilde{q}_1, \widetilde{q}_2, \cdots, \widetilde{q}_n) \leq \widetilde{q}^+ \tag{51}$$

It is obvious that the q-RIVOFWDHM operator is short of the property of idempotency.

4. Application of Green Supplier Selection

4.1. Numerical Example

With the rapid development of economic globalization, and the growing enterprise competition environment, the competition between modern enterprises has become the competition between supply chains. The diversity of the people consuming is increasing, and the new product life cycles are getting shorter. The volatility of the demand market and from external factors drives enterprises for effective supply chain integration and management, as well as strategic alliances with other enterprises to enhance core competitiveness and resist external risk. The key measure to achieving this goal is supplier selection. Therefore, the supplier selection problem has gained a lot of attention, whether in regard to supply chain management theory or in actual production management problems [62–70]. In order to illustrate our proposed method in this article, we provide a numerical example for selecting green suppliers in green supply chain management using q-RIVOFNs. There is a panel with five possible green suppliers in green supply chain management to select: $\widetilde{Q}_i(i = 1, 2, 3, 4, 5)$. The experts select four attributes to evaluate the five possible green suppliers: (1) C_1 is the product quality factor; (2) C_2 is the environmental factors; (3) C_3 is the delivery factor; and (4) C_4 is the price factor. The five possible green suppliers $\widetilde{Q}_i(i = 1, 2, 3, 4, 5)$ are to be evaluated by the decision maker using the q-RIVOFNs, under the above four attributes (whose weighting vector $\omega = (0.3, 0.2, 0.3, 0.2)$, and expert weighting vector $\omega = (0.2, 0.2, 0.6)$) which are listed in Tables 1–3.

Table 1. The q-RIVOFN decision matrix 1 (R_1) by expert one.

Alternatives	C_1	C_2	C_3	C_4
$\widetilde{Q}_1$	([0.4,0.5],[0.5,0.7])	([0.6,0.7],[0.2,0.3])	([0.3,0.5],[0.4,0.6])	([0.7,0.8],[0.2,0.4])
$\widetilde{Q}_2$	([0.2,0.3],[0.4,0.5])	([0.1,0.2],[0.6,0.7])	([0.6,0.8],[0.2,0.3])	([0.5,0.6],[0.5,0.7])
$\widetilde{Q}_3$	([0.7,0.9],[0.1,0.2])	([0.4,0.5],[0.2,0.3])	([0.5,0.7],[0.3,0.4])	([0.6,0.7],[0.1,0.2])
$\widetilde{Q}_4$	([0.3,0.5],[0.4,0.6])	([0.2,0.3],[0.1,0.2])	([0.5,0.6],[0.1,0.5])	([0.3,0.4],[0.2,0.3])
$\widetilde{Q}_5$	([0.3,0.6],[0.2,0.4])	([0.4,0.6],[0.2,0.3])	([0.1,0.2],[0.4,0.5])	([0.2,0.4],[0.1,0.3])

Table 2. The q-RIVOFN decision matrix 1 (R_2) by expert two.

Alternatives	C_1	C_2	C_3	C_4
$\widetilde{Q}_1$	([0.3,0.4],[0.4,0.6])	([0.7,0.8],[0.3,0.4])	([0.2,0.4],[0.3,0.5])	([0.8,0.9],[0.3,0.5])
$\widetilde{Q}_2$	([0.1,0.2],[0.3,0.4])	([0.2,0.3],[0.7,0.8])	([0.5,0.7],[0.1,0.2])	([0.6,0.7],[0.6,0.8])
$\widetilde{Q}_3$	([0.6,0.8],[0.1,0.2])	([0.5,0.6],[0.3,0.4])	([0.4,0.6],[0.2,0.3])	([0.7,0.8],[0.2,0.3])
$\widetilde{Q}_4$	([0.2,0.4],[0.3,0.5])	([0.3,0.4],[0.2,0.3])	([0.4,0.5],[0.1,0.4])	([0.4,0.5],[0.3,0.4])
$\widetilde{Q}_5$	([0.2,0.5],[0.1,0.3])	([0.5,0.7],[0.3,0.4])	([0.1,0.2],[0.3,0.4])	([0.3,0.5],[0.2,0.4])

Table 3. The q-RIVOFN decision matrix 1 (R_3) by expert three.

Alternatives	C_1	C_2	C_3	C_4
$\widetilde{Q}_1$	([0.5,0.6],[0.6,0.8])	([0.5,0.6],[0.1,0.2])	([0.4,0.6],[0.5,0.7])	([0.6,0.7],[0.1,0.3])
$\widetilde{Q}_2$	([0.3,0.4],[0.5,0.6])	([0.1,0.2],[0.5,0.6])	([0.7,0.9],[0.3,0.4])	([0.4,0.5],[0.4,0.6])
$\widetilde{Q}_3$	([0.8,0.9],[0.2,0.3])	([0.3,0.4],[0.1,0.2])	([0.6,0.8],[0.4,0.5])	([0.5,0.6],[0.1,0.2])
$\widetilde{Q}_4$	([0.4,0.6],[0.5,0.7])	([0.1,0.2],[0.1,0.2])	([0.6,0.7],[0.2,0.6])	([0.2,0.3],[0.1,0.2])
$\widetilde{Q}_5$	([0.4,0.7],[0.3,0.5])	([0.3,0.5],[0.1,0.2])	([0.2,0.3],[0.5,0.6])	([0.1,0.3],[0.1,0.2])

In the following, we utilize the approach developed to select green suppliers in green supply chain management.

Step 1. According to q-RIVOFNs $\widetilde{q}_{ij}(i = 1, 2, 3, 4, 5, j = 1, 2, 3, 4)$, we can aggregate all q-RIVOFNs $\widetilde{q}_{ij}$ by using the q-RIVOFWA (q-RIVOFWG) operator, to get the overall q-RIVOFNs $\widetilde{Q}_i(i = 1, 2, 3, 4, 5)$ of the green suppliers $\widetilde{Q}_i$. Then, the fused values are given in Table 4. (Let $q = 3$).

Definition 10. *Let $\widetilde{q}_j = \left(\left[u_{\widetilde{q}_j}^L, u_{\widetilde{q}_j}^R \right], \left[v_{\widetilde{q}_j}^L, v_{\widetilde{q}_j}^R \right] \right) (j = 1, 2, \ldots, n)$ be a set of q-RIVOFNs, with their weight vector as $w_i = (w_1, w_2, \ldots, w_n)^T$, thereby satisfying $w_i \in [0,1]$ and $\sum_{i=1}^n w_i = 1$. Then we can obtain*

$$q\text{-RIVOFWA}(\widetilde{q}_1, \widetilde{q}_2, \ldots, \widetilde{q}_n) = \sum_{j=1}^n w_j \widetilde{q}_j$$
$$= \left\langle \left[\sqrt[q]{1 - \prod_{j=1}^n \left(1 - u_{\widetilde{q}_j}^L\right)^{w_j}}, \sqrt[q]{1 - \prod_{j=1}^n \left(1 - u_{\widetilde{q}_j}^R\right)^{w_j}} \right], \left[\prod_{j=1}^n \left(v_{\widetilde{q}_j}^L\right)^{w_j}, \prod_{j=1}^n \left(v_{\widetilde{q}_j}^R\right)^{w_j} \right] \right\rangle \tag{52}$$

$$q\text{-RIVOFWG}(\widetilde{q}_1, \widetilde{q}_2, \ldots, \widetilde{q}_n) = \prod_{j=1}^n \left(\widetilde{q}_j\right)^{w_j}$$
$$= \left\langle \left[\prod_{j=1}^n \left(u_{\widetilde{q}_j}^L\right)^{w_j}, \prod_{j=1}^n \left(u_{\widetilde{q}_j}^R\right)^{w_j} \right], \left[\sqrt[q]{1 - \prod_{j=1}^n \left(1 - v_{\widetilde{q}_j}^L\right)^{w_j}}, \sqrt[q]{1 - \prod_{j=1}^n \left(1 - v_{\widetilde{q}_j}^R\right)^{w_j}} \right] \right\rangle \tag{53}$$

Table 4. The fused results from the q-RIVOFWA operator.

Alternatives	C_1	C_2
$\widetilde{Q}_1$	([0.7637,0.8175],[0.5335,0.7354])	([0.8283,0.8756],[0.1431,0.2491])
$\widetilde{Q}_2$	([0.6249,0.7011],[0.4317,0.5335])	([0.4945,0.6047],[0.5547,0.6554])
$\widetilde{Q}_3$	([0.9089,0.9601],[0.1516,0.2551])	([0.7149,0.7756],[0.1431,0.2491])
$\widetilde{Q}_4$	([0.7011,0.8175],[0.4317,0.6346])	([0.5474,0.6420],[0.1149,0.2169])
$\widetilde{Q}_5$	([0.7011,0.8654],[0.2221,0.4317])	([0.7149,0.8283],[0.1431,0.2491])

Alternatives	C_3	C_4
$\widetilde{Q}_1$	([0.7011,0.8175],[0.4317,0.6346])	([0.8756,0.9197],[0.1431,0.3519])
$\widetilde{Q}_2$	([0.8654,0.9498],[0.2221,0.3288])	([0.7756,0.8283],[0.4536,0.6554])
$\widetilde{Q}_3$	([0.8175,0.9089],[0.3288,0.4317])	([0.8283,0.8756],[0.1149,0.2169])
$\widetilde{Q}_4$	([0.8175,0.8654],[0.1516,0.5335])	([0.6420,0.7149],[0.1431,0.2491])
$\widetilde{Q}_5$	([0.5445,0.6396],[0.4317,0.5335])	([0.5474,0.7149],[0.1149,0.2491])

Step 2. Based on Table 4, we can fuse all q-RIVOFNs $\widetilde{q}_{ij}$ by the q-RIVOFWHM (q-RIVOFWDHM) operator to get the results of q-RIVOFNs. Let $x = 2$, then the fused values are given in Table 5.

Table 5. The fused values of the q-rung interval-valued orthopair, fuzzy weighted Hamy mean (q-RIVOFWHM) and the q-rung interval-valued orthopair, fuzzy weighted dual Hamy mean (q-RIVOFWDHM)) operators.

Alternatives	q-**RIVOFWHM**	q-**RIVOFWDHM**
$\widetilde{Q}_1$	([0.9422,0.9616],[0.2248,0.3558])	([0.5409,0.6039],[0.7423,0.8415])
$\widetilde{Q}_2$	([0.9148,0.9418],[0.2710,0.3562])	([0.4842,0.5611],[0.7959,0.8530])
$\widetilde{Q}_3$	([0.9536,0.9720],[0.1237,0.1901])	([0.5790,0.6546],[0.6536,0.7346])
$\widetilde{Q}_4$	([0.9112,0.9379],[0.1415,0.2910])	([0.4637,0.5318],[0.6697,0.8006])
$\widetilde{Q}_5$	([0.8903,0.9356],[0.1575,0.2505])	([0.4140,0.5250],[0.6861,0.7805])

Step 3. Based on the fused values given in Table 5, and the score functions of q-RIVOFNs, the green suppliers' scores are shown in Table 6.

Table 6. The score values $s\left(\widetilde{Q}_i\right)$ of the green suppliers.

Alternatives	q-RIVOFWHM	q-RIVOFWDHM
$\widetilde{Q}_1$	0.9172	0.3434
$\widetilde{Q}_2$	0.8840	0.2914
$\widetilde{Q}_3$	0.9442	0.4497
$\widetilde{Q}_4$	0.8885	0.3591
$\widetilde{Q}_5$	0.8762	0.3543

Step 4. Rank all the alternatives by the values of Table 6, and the ordering results are shown in Table 7. Obviously, the best selection is $\widetilde{Q}_3$.

Table 7. Ordering of the green suppliers.

Methods	Ordering
q-RIVOFWHM	$\widetilde{Q}_3 > \widetilde{Q}_1 > \widetilde{Q}_4 > \widetilde{Q}_2 > \widetilde{Q}_5$
q-RIVOFWDHM	$\widetilde{Q}_3 > \widetilde{Q}_4 > \widetilde{Q}_5 > \widetilde{Q}_1 > \widetilde{Q}_2$

4.2. Influence of the Parameter x

In order to show the effects on the ranking results, by changing parameters of x in the q-RIVOFWHM (q-RIVOFWDHM) operators, all of the results are shown in Tables 8 and 9. (Let $q = 3$).

Table 8. Ordering results for different x values by the q-RIVOFWHM operator.

Parameters	$S(\widetilde{Q}_1)$	$S(\widetilde{Q}_2)$	$S(\widetilde{Q}_3)$	$S(\widetilde{Q}_4)$	$S(\widetilde{Q}_5)$	Ordering
$x = 1$	0.9306	0.8993	0.9476	0.8941	0.8844	$\widetilde{Q}_3 > \widetilde{Q}_1 > \widetilde{Q}_2 > \widetilde{Q}_4 > \widetilde{Q}_5$
$x = 2$	0.9172	0.8840	0.9442	0.8885	0.8762	$\widetilde{Q}_3 > \widetilde{Q}_1 > \widetilde{Q}_4 > \widetilde{Q}_2 > \widetilde{Q}_5$
$x = 3$	0.9290	0.8959	0.9454	0.8947	0.8786	$\widetilde{Q}_3 > \widetilde{Q}_1 > \widetilde{Q}_2 > \widetilde{Q}_4 > \widetilde{Q}_5$
$x = 4$	0.9080	0.8772	0.9419	0.8839	0.8703	$\widetilde{Q}_3 > \widetilde{Q}_1 > \widetilde{Q}_4 > \widetilde{Q}_2 > \widetilde{Q}_5$

Table 9. Ordering results for different x values by the q-RIVOFWDHM operator.

Parameters	$S(\widetilde{Q}_1)$	$S(\widetilde{Q}_2)$	$S(\widetilde{Q}_3)$	$S(\widetilde{Q}_4)$	$S(\widetilde{Q}_5)$	Ordering
$x = 1$	0.3330	0.2579	0.4340	0.3424	0.3464	$\widetilde{Q}_3 > \widetilde{Q}_5 > \widetilde{Q}_4 > \widetilde{Q}_1 > \widetilde{Q}_2$
$x = 2$	0.3434	0.2914	0.4497	0.3591	0.3543	$\widetilde{Q}_3 > \widetilde{Q}_4 > \widetilde{Q}_5 > \widetilde{Q}_1 > \widetilde{Q}_2$
$x = 3$	0.2557	0.2292	0.3406	0.3005	0.3024	$\widetilde{Q}_3 > \widetilde{Q}_5 > \widetilde{Q}_4 > \widetilde{Q}_1 > \widetilde{Q}_2$
$x = 4$	0.3486	0.3150	0.4585	0.3679	0.3586	$\widetilde{Q}_3 > \widetilde{Q}_4 > \widetilde{Q}_5 > \widetilde{Q}_1 > \widetilde{Q}_2$

4.3. Influence of the Parameter q

In order to show the effects on the ranking results by changing the parameters of q in the q-RIVOFWHM (q-RIVOFWDHM) operators, all of the results are shown in Tables 10 and 11. (Let $x = 2$).

Table 10. Ordering results for different q by the q-RIVOFWHM operator.

Parameters	$S(\widetilde{Q}_1)$	$S(\widetilde{Q}_2)$	$S(\widetilde{Q}_3)$	$S(\widetilde{Q}_4)$	$S(\widetilde{Q}_5)$	Ordering
$q = 1$	0.9090	0.8899	0.9481	0.9147	0.9121	$\widetilde{Q}_3 > \widetilde{Q}_4 > \widetilde{Q}_5 > \widetilde{Q}_1 > \widetilde{Q}_2$
$q = 2$	0.9244	0.8982	0.9555	0.9109	0.9031	$\widetilde{Q}_3 > \widetilde{Q}_1 > \widetilde{Q}_4 > \widetilde{Q}_5 > \widetilde{Q}_2$
$q = 3$	0.9172	0.8840	0.9442	0.8885	0.8762	$\widetilde{Q}_3 > \widetilde{Q}_1 > \widetilde{Q}_4 > \widetilde{Q}_2 > \widetilde{Q}_5$
$q = 4$	0.9033	0.8634	0.9293	0.8627	0.8469	$\widetilde{Q}_3 > \widetilde{Q}_1 > \widetilde{Q}_2 > \widetilde{Q}_4 > \widetilde{Q}_5$
$q = 5$	0.8872	0.8412	0.9139	0.8371	0.8187	$\widetilde{Q}_3 > \widetilde{Q}_1 > \widetilde{Q}_2 > \widetilde{Q}_4 > \widetilde{Q}_5$
$q = 6$	0.8705	0.8193	0.8989	0.8127	0.7926	$\widetilde{Q}_3 > \widetilde{Q}_1 > \widetilde{Q}_2 > \widetilde{Q}_4 > \widetilde{Q}_5$
$q = 7$	0.8540	0.7983	0.8844	0.7899	0.7687	$\widetilde{Q}_3 > \widetilde{Q}_1 > \widetilde{Q}_2 > \widetilde{Q}_4 > \widetilde{Q}_5$
$q = 8$	0.8380	0.7785	0.8704	0.7687	0.7468	$\widetilde{Q}_3 > \widetilde{Q}_1 > \widetilde{Q}_2 > \widetilde{Q}_4 > \widetilde{Q}_5$
$q = 9$	0.8225	0.7600	0.8570	0.7490	0.7270	$\widetilde{Q}_3 > \widetilde{Q}_1 > \widetilde{Q}_2 > \widetilde{Q}_4 > \widetilde{Q}_5$
$q = 10$	0.8078	0.7427	0.8441	0.7308	0.7089	$\widetilde{Q}_3 > \widetilde{Q}_1 > \widetilde{Q}_2 > \widetilde{Q}_4 > \widetilde{Q}_5$

Table 11. Ordering results for different q by the q-RIVOFWDHM operator.

Parameters	$S(\widetilde{Q}_1)$	$S(\widetilde{Q}_2)$	$S(\widetilde{Q}_3)$	$S(\widetilde{Q}_4)$	$S(\widetilde{Q}_5)$	Ordering
$q = 1$	0.2814	0.2415	0.3520	0.2756	0.2655	$\widetilde{Q}_3 > \widetilde{Q}_1 > \widetilde{Q}_4 > \widetilde{Q}_5 > \widetilde{Q}_2$
$q = 2$	0.3107	0.2617	0.4074	0.3188	0.3110	$\widetilde{Q}_3 > \widetilde{Q}_4 > \widetilde{Q}_5 > \widetilde{Q}_1 > \widetilde{Q}_2$
$q = 3$	0.3434	0.2914	0.4497	0.3591	0.3543	$\widetilde{Q}_3 > \widetilde{Q}_4 > \widetilde{Q}_5 > \widetilde{Q}_1 > \widetilde{Q}_2$
$q = 4$	0.3722	0.3204	0.4788	0.3913	0.3893	$\widetilde{Q}_3 > \widetilde{Q}_5 > \widetilde{Q}_4 > \widetilde{Q}_1 > \widetilde{Q}_2$
$q = 5$	0.3962	0.3464	0.4978	0.4161	0.4163	$\widetilde{Q}_3 > \widetilde{Q}_5 > \widetilde{Q}_4 > \widetilde{Q}_1 > \widetilde{Q}_2$
$q = 6$	0.4157	0.3689	0.5098	0.4350	0.4369	$\widetilde{Q}_3 > \widetilde{Q}_5 > \widetilde{Q}_4 > \widetilde{Q}_1 > \widetilde{Q}_2$
$q = 7$	0.4314	0.3881	0.5171	0.4494	0.4524	$\widetilde{Q}_3 > \widetilde{Q}_5 > \widetilde{Q}_4 > \widetilde{Q}_1 > \widetilde{Q}_2$
$q = 8$	0.4441	0.4044	0.5211	0.4604	0.4641	$\widetilde{Q}_3 > \widetilde{Q}_5 > \widetilde{Q}_4 > \widetilde{Q}_1 > \widetilde{Q}_2$
$q = 9$	0.4543	0.4181	0.5230	0.4688	0.4729	$\widetilde{Q}_3 > \widetilde{Q}_5 > \widetilde{Q}_4 > \widetilde{Q}_1 > \widetilde{Q}_2$
$q = 10$	0.4625	0.4297	0.5235	0.4754	0.4795	$\widetilde{Q}_3 > \widetilde{Q}_5 > \widetilde{Q}_4 > \widetilde{Q}_1 > \widetilde{Q}_2$

4.4. Comparative Analysis

In this chapter, we compare the q-RIVOFWHM and q-RIVOFWDHM operators with the q-RIVOFWA and q-RIVOFWG operators. The comparative results are shown in Table 12.

Table 12. Comparative results.

Methods	Ordering
q-RIVOFWA	$\widetilde{Q}_3 > \widetilde{Q}_4 > \widetilde{Q}_1 > \widetilde{Q}_2 > \widetilde{Q}_5$
q-RIVOFWG	$\widetilde{Q}_3 > \widetilde{Q}_2 > \widetilde{Q}_5 > \widetilde{Q}_1 > \widetilde{Q}_4$

From above, we can see that we get the same optimal green suppliers, which shows the practicality and effectiveness of the proposed approaches. However, the q-RIVOFWA operator and q-RIVOFWG operator do not consider the information about the relationship between arguments being aggregated, and thus cannot eliminate the influence of unfair arguments on decision results. Our proposed q-RIVOFWHM and q-RIVOFWDHM operators consider the information about the relationship among arguments being aggregated.

At the same time, Liu and Wang [54] develop the q-rung orthopair, fuzzy weighted averaging (q-ROFWA) operator, as well as the q-rung orthopair, fuzzy weighted geometric (q-ROFWG) operator. Liu and Liu [55] propose some q-rung orthopair, fuzzy Bonferroni mean (q-ROFBM) aggregation operators. Wei et al. [56] define the generalized Heronian mean (GHM) operator to present some aggregation operators, and apply them into MADM problems. Wei et al. [57] define some q-rung orthopair, fuzzy Maclaurin symmetric mean operators. However, all of these operators can only deal with q-rung orthopair fuzzy sets (q-ROFSs), and cannot deal with q-rung interval-valued orthopair fuzzy sets (q-RIVOFSs). The main contribution of this paper is to study the MAGDM problems based on the q-rung interval-valued orthopair fuzzy sets (q-RIVOFSs), and to utilize the Hamy mean

(HM) operator, weighted Hamy mean (WHM) operator, dual Hamy mean (DHM) operator, and weighted dual Hamy mean (WDHM) operator, to develop some Hamy mean aggregation operators with q-RIVOFNs.

5. Conclusions

In this paper, we study the MAGDM problems with q-RIVOFNs. Then, we utilize the Hamy mean (HM) operator, weighted Hamy mean (WHM) operator, dual Hamy mean (DHM) operator, and weighted dual Hamy mean (WDHM) operator, in order to develop some Hamy mean aggregation operators with q-RIVOFNs. The prominent characteristic of each of these proposed operators is studied. Then, we have utilized these operators to develop some approaches to solve the MAGDM problems with q-RIVOFNs. Finally, a practical example for green supplier selection is given to show the developed approach. Using the illustrated example, we have roughly shown the effects on the ranking results by changing parameters in the q-RIVOFWHM (q-RIVOFWDHM) operators. In the future, the application of the proposed fused operators of q-RIVOFNs needs to be explored in decision making [71–74], risk analysis [75,76], and many other fields under uncertain environments [77–81].

Author Contributions: J.W., H.G., G.W. and Y.W. conceived and worked together to achieve this work, J.W. compiled the computing program by Excel and analyzed the data, J.W. and G.W. wrote the paper. Finally, all the authors have read and approved the final manuscript.

Funding: The work was supported by the National Natural Science Foundation of China under Grant No. 71571128 and the Humanities and Social Sciences Foundation of Ministry of Education of the People's Republic of China (17XJA630003) and the Construction Plan of Scientific Research Innovation Team for Colleges and Universities in Sichuan Province (15TD0004).

Conflicts of Interest: The authors declare no conflicts of interest.

References

1. Zadeh, L.A. Fuzzy sets. *Inf. Control* **1965**, *8*, 338–356. [CrossRef]
2. Atanassov, K. Intuitionistic fuzzy sets. *Fuzzy Sets Syst.* **1986**, *20*, 87–96. [CrossRef]
3. Atanassov, K. Two theorems for intuitionistic fuzzy sets. *Fuzzy Sets Syst.* **2000**, *110*, 267–269. [CrossRef]
4. Wu, L.; Wei, G.; Gao, H.; Wei, Y. Some interval-valued intuitionistic fuzzy dombi hamy mean operators and their application for evaluating the elderly tourism service quality in tourism destination. *Mathematics* **2018**, *6*, 294. [CrossRef]
5. Wang, R.; Wang, J.; Gao, H.; Wei, G. Methods for MADM with picture fuzzy muirhead mean operators and their application for evaluating the financial investment risk. *Symmetry* **2019**, *11*, 6. [CrossRef]
6. Li, Z.; Gao, H.; Wei, G. Methods for multiple attribute group decision making based on intuitionistic fuzzy dombi hamy mean operators. *Symmetry* **2018**, *10*, 574. [CrossRef]
7. Deng, X.M.; Wei, G.W.; Gao, H.; Wang, J. Models for safety assessment of construction project with some 2-tuple linguistic Pythagorean fuzzy Bonferroni mean operators. *IEEE Access* **2018**, *6*, 52105–52137. [CrossRef]
8. Gao, H. Pythagorean fuzzy hamacher prioritized aggregation operators in multiple attribute decision making. *J. Intell. Fuzzy Syst.* **2018**, *35*, 2229–2245. [CrossRef]
9. Wei, G.; Wei, Y. Some single-valued neutrosophic dombi prioritized weighted aggregation operators in multiple attribute decision making. *J. Intell. Fuzzy Syst.* **2018**, *35*, 2001–2013. [CrossRef]
10. Wang, J.; Wei, G.; Gao, H. Approaches to multiple attribute decision making with interval-valued 2-tuple linguistic pythagorean fuzzy information. *Mathematics* **2018**, *6*, 201. [CrossRef]
11. Wei, G.W. TODIM method for picture fuzzy multiple attribute decision making. *Informatica* **2018**, *29*, 555–566. [CrossRef]
12. Xu, Z.S. intuitionistic fuzzy aggregation operators. *IEEE Trans. Fuzzy Syst.* **2007**, *15*, 1179–1187.
13. Wei, G.W.; Garg, H.; Gao, H.; Wei, C. Interval-Valued pythagorean fuzzy maclaurin symmetric mean operators in multiple attribute decision making. *IEEE Access* **2018**, *6*, 67866–67884. [CrossRef]
14. Wei, G.W.; Wei, C.; Gao, H. multiple attribute decision making with interval-valued bipolar fuzzy information and their application to emerging technology commercialization evaluation. *IEEE Access* **2018**, *6*, 60930–60955. [CrossRef]

15. Wang, J.; Wei, G.; Lu, M. An Extended VIKOR Method for Multiple Criteria Group Decision Making with Triangular Fuzzy Neutrosophic Numbers. *Symmetry* **2018**, *10*, 497. [CrossRef]
16. Wang, J.; Wei, G.; Lu, M. TODIM Method for multiple attribute group decision making under 2-Tuple Linguistic Neutrosophic Environment. *Symmetry* **2018**, *10*, 486. [CrossRef]
17. Li, Z.; Wei, G.; Lu, M. Pythagorean fuzzy hamy mean operators in multiple attribute group decision making and their application to supplier selection. *Symmetry* **2018**, *10*, 505. [CrossRef]
18. Wei, G.W. Some geometric aggregation functions and their application to dynamic multiple attribute decision making in intuitionistic fuzzy setting. *Int. J. Uncertain. Fuzz. Knowl.-Based Syst.* **2009**, *17*, 179–196. [CrossRef]
19. Li, Z.; Wei, G.; Gao, H. Methods for multiple attribute decision making with interval-valued pythagorean fuzzy information. *Mathematics* **2018**, *6*, 228. [CrossRef]
20. Deng, X.; Wang, J.; Wei, G.; Lu, M. Models for multiple attribute decision making with some 2-tuple linguistic pythagorean fuzzy hamy mean operators. *Mathematics* **2018**, *6*, 236. [CrossRef]
21. Tang, X.Y.; Wei, G.W. Models for green supplier selection in green supply chain management with Pythagorean 2-tuple linguistic information. *IEEE Access* **2018**, *6*, 18042–18060. [CrossRef]
22. Ye, J. Fuzzy decision-making method based on the weighted correlation coefficient under intuitionistic fuzzy environment. *Eur. J. Oper. Res.* **2010**, *205*, 202–204. [CrossRef]
23. Huang, Y.H.; Wei, G.W. TODIM Method for Pythagorean 2-tuple linguistic multiple attribute decision making. *J. Intell. Fuzzy Syst.* **2018**, *35*, 901–915. [CrossRef]
24. Wei, G.W.; Gao, H.; Wang, J.; Huang, Y.H. Research on risk evaluation of enterprise human capital investment with Interval-valued bipolar 2-tuple linguistic Information. *IEEE Access* **2018**, *6*, 35697–35712. [CrossRef]
25. Liang, X.; Wei, C. An Atanassov's intuitionistic fuzzy multi-attribute group decision making method based on entropy and similarity measure. *Int. J. Mach. Learn. Cybern.* **2014**, *5*, 435–444. [CrossRef]
26. Yager, R.R. Pythagorean fuzzy subsets. In Proceedings of the Joint IFSA World Congress and NAFIPS Annual Meeting, Emonton, AB, Canada, 21 June 2013; pp. 57–61.
27. Yager, R.R. Pythagorean membership grades in multicriteria decision making. *IEEE Trans. Fuzzy Syst.* **2014**, *22*, 958–965. [CrossRef]
28. Zhang, X.L.; Xu, Z.S. Extension of TOPSIS to multiple criteria decision making with Pythagorean fuzzy sets. *Int. J. Intell. Syst.* **2014**, *29*, 1061–1078. [CrossRef]
29. Wei, G.W.; Lu, M. Pythagorean Fuzzy Maclaurin Symmetric Mean Operators in multiple attribute decision making. *Int. J. Intell. Syst.* **2018**, *33*, 1043–1070. [CrossRef]
30. Peng, X.; Yang, Y. Some results for Pythagorean Fuzzy Sets. *Int. J. Intell. Syst.* **2015**, *30*, 1133–1160. [CrossRef]
31. Wei, G.W.; Lu, M. Pythagorean fuzzy power aggregation operators in multiple attribute decision making. *Int. J. Intell. Syst.* **2018**, *33*, 169–186. [CrossRef]
32. Wei, G.W. Interval valued hesitant fuzzy uncertain linguistic aggregation operators in multiple attribute decision making. *Int. J. Mach. Learn. Cybern.* **2016**, *7*, 1093–1114. [CrossRef]
33. Wei, G.W. Picture fuzzy Hamacher aggregation operators and their application to multiple attribute decision making. *Fundam. Inf.* **2018**, *157*, 271–320. [CrossRef]
34. Gao, H.; Lu, M.; Wei, G.W.; Wei, Y. Some Novel Pythagorean Fuzzy Interaction Aggregation Operators in Multiple Attribute Decision Making. *Fundam. Inf.* **2018**, *159*, 385–428. [CrossRef]
35. Wei, G.W. Interval valued hesitant fuzzy uncertain linguistic aggregation operators in multiple attribute decision making. *Int. J. Mach. Learn. Cybern.* **2016**, *7*, 1093–1114. [CrossRef]
36. Wei, G.W.; Zhao, X.F.; Wang, H.J.; Lin, R. Fuzzy power aggregating operators and their application to multiple attribute group decision making. *Technol. Econ. Dev. Econ.* **2013**, *19*, 377–396. [CrossRef]
37. Wei, G.W. Some linguistic power aggregating operators and their application to multiple attribute group decision making. *J. Intell. Fuzzy Syst.* **2013**, *25*, 695–707.
38. Beliakov, G.; James, S. Averaging aggregation functions for preferences expressed as Pythagorean membership grades and fuzzy orthopairs. In Proceedings of the 2014 IEEE International Conference on Fuzzy Systems (FUZZ-IEEE), Beijing, China, 6–11 July 2014; pp. 298–305.
39. Reformat, M.; Yager, R.R. Suggesting Recommendations Using Pythagorean Fuzzy Sets illustrated Using Netflix Movie Data. In Proceedings of the International Conference on Information Processing and Management of Uncertainty in Knowledge-Based Systems, Montpellier, France, 15–19 July 2014; pp. 546–556.
40. Gou, X.; Xu, Z.; Ren, P. The Properties of Continuous Pythagorean Fuzzy Information. *Int. J. Intell. Syst.* **2016**, *31*, 401–424. [CrossRef]

41. Wei, G.W.; Wei, Y. Similarity measures of Pythagorean fuzzy sets based on cosine function and their applications. *Int. J. Intell. Syst.* **2018**, *33*, 634–652. [CrossRef]
42. Wei, G.W.; Gao, H. The Generalized Dice Similarity Measures for Picture Fuzzy Sets and Their Applications. *Informatica* **2018**, *29*, 107–124. [CrossRef]
43. Wei, G.W. Some similarity measures for picture fuzzy sets and their applications. *Iran. J. Fuzzy Syst.* **2018**, *15*, 77–89.
44. Wei, G.W. Some Cosine Similarity Measures for Picture Fuzzy Sets and Their Applications to Strategic Decision Making. *Informatica* **2017**, *28*, 547–564. [CrossRef]
45. Wei, G.W.; Lin, R.; Wang, H.J. Distance and similarity measures for hesitant interval-valued fuzzy sets. *J. Intell. Fuzzy Syst.* **2014**, *27*, 19–36.
46. Ren, P.; Xu, Z.; Gou, X. Pythagorean fuzzy TODIM approach to multi-criteria decision making. *Appl. Soft Comput.* **2016**, *42*, 246–259. [CrossRef]
47. Garg, H. A New Generalized Pythagorean Fuzzy Information Aggregation Using Einstein Operations and Its Application to Decision Making. *Int. J. Intell. Syst.* **2016**, *31*, 886–920. [CrossRef]
48. Zeng, S.; Chen, J.; Li, X. A Hybrid Method for Pythagorean Fuzzy Multiple-Criteria Decision Making. *Int. J. Inf. Technol. Decis. Mak.* **2016**, *15*, 403–422. [CrossRef]
49. Garg, H. A novel accuracy function under interval-valued Pythagorean fuzzy environment for solving multicriteria decision making problem. *J. Intell. Fuzzy Syst.* **2016**, *31*, 529–540. [CrossRef]
50. Wei, G.W.; Lu, M.; Tang, X.Y.; Wei, Y. Pythagorean Hesitant Fuzzy Hamacher Aggregation Operators and Their Application to Multiple Attribute Decision Making. *Int. J. Intell. Syst.* **2018**, *33*, 1197–1233. [CrossRef]
51. Wei, G.W.; Lu, M. Dual hesitant pythagorean fuzzy Hamacher aggregation operators in multiple attribute decision making. *Arch. Control Sci.* **2017**, *27*, 365–395. [CrossRef]
52. Lu, M.; Wei, G.W.; Alsaadi, F.E.; Hayat, T.; Alsaedi, A. Hesitant Pythagorean fuzzy Hamacher aggregation operators and their application to multiple attribute decision making. *J. Intell. Fuzzy Syst.* **2017**, *33*, 1105–1117. [CrossRef]
53. Yager, R.R. Generalized orthopair fuzzy sets. *IEEE Trans. Fuzzy Syst.* **2017**, *25*, 1222–1230. [CrossRef]
54. Liu, P.D.; Wang, P. Some q-Rung Orthopair Fuzzy Aggregation Operators and their Applications to Multiple-Attribute Decision Making. *Int. J. Intell. Syst.* **2018**, *32*, 259–280. [CrossRef]
55. Liu, P.D.; Liu, J.L. Some q-Rung Orthopai Fuzzy Bonferroni Mean Operators and Their Application to Multi-Attribute Group Decision Making. *Int. J. Intell. Syst.* **2018**, *33*, 315–347. [CrossRef]
56. Wei, G.W.; Gao, H.; Wei, Y. Some q-Rung Orthopair Fuzzy Heronian Mean Operators in Multiple Attribute Decision Making. *Int. J. Intell. Syst.* **2018**, *33*, 1426–1458. [CrossRef]
57. Wei, G.W.; Wei, C.; Wang, J.; Gao, H.; Wei, Y. Some q-rung orthopair fuzzy maclaurin symmetric mean operators and their applications to potential evaluation of emerging technology commercialization. *Int. J. Intell. Syst.* **2019**, *34*, 50–81. [CrossRef]
58. Atanassov, K.; Gargov, G. Interval-valued intuitionistic fuzzy sets. *Fuzzy Sets Syst.* **1989**, *31*, 343–349. [CrossRef]
59. Atanassov, K. Operators over interval-valued intuitionistic fuzzy sets. *Fuzzy Sets Syst.* **1994**, *64*, 159–174. [CrossRef]
60. Hara, T.; Uchiyama, M.; Takahasi, S.E. A refinement of various mean inequalities. *J. Inequal. Appl.* **1998**, *2*, 387–395. [CrossRef]
61. Wu, S.; Wang, J.; Wei, G.; Wei, Y. Research on Construction Engineering Project Risk Assessment with Some 2-Tuple Linguistic Neutrosophic Hamy Mean Operators. *Sustainability* **2018**, *10*, 1536. [CrossRef]
62. Wang, J.; Wei, G.W.; Wei, Y. Models for Green Supplier Selection with Some 2-Tuple Linguistic Neutrosophic Number Bonferroni Mean Operators. *Symmetry* **2018**, *10*, 131. [CrossRef]
63. Gao, H.; Wei, G.W.; Huang, Y.H. Dual hesitant bipolar fuzzy Hamacher prioritized aggregation operators in multiple attribute decision making. *IEEE Access* **2018**, *6*, 11508–11522. [CrossRef]
64. Wei, G.W.; Alsaadi, F.E.; Hayat, T.; Alsaedi, A. Projection models for multiple attribute decision making with picture fuzzy information. *Int. J. Mach. Learn. Cybern.* **2018**, *9*, 713–719. [CrossRef]
65. Wang, S.Y. Applying 2-tuple multigranularity linguistic variables to determine the supply performance in dynamic environment based on product-oriented strategy. *IEEE Trans. Fuzzy Syst.* **2008**, *16*, 29–39. [CrossRef]
66. Wei, G.W.; Alsaadi, F.E.; Hayat, T.; Alsaedi, A. Bipolar fuzzy Hamacher aggregation operators in multiple attribute decision making. *Int. J. Fuzzy Syst.* **2018**, *20*, 1–12. [CrossRef]

67. You, X.Y.; You, J.X.; Liu, H.C.; Zhen, L. Group multi-criteria supplier selection using an extended VIKOR method with interval 2-tuple linguistic information. *Expert Syst. Appl.* **2015**, *42*, 1906–1916. [CrossRef]

68. Wei, G.W.; Alsaadi, F.E.; Hayat, T.; Alsaedi, A. Picture 2-tuple linguistic aggregation operators in multiple attribute decision making. *Soft Comput.* **2018**, *22*, 989–1002. [CrossRef]

69. Santos, L.; Osiro, L.; Lima, R.H.P. A model based on 2-tuple fuzzy linguistic representation and Analytic Hierarchy Process for supplier segmentation using qualitative and quantitative criteria. *Expert Syst. Appl.* **2017**, *79*, 53–64. [CrossRef]

70. Wei, G.W.; Wang, J.M. A comparative study of robust efficiency analysis and Data Envelopment Analysis with imprecise data. *Expert Syst. Appl.* **2017**, *81*, 28–38. [CrossRef]

71. Wei, G.W. Picture uncertain linguistic Bonferroni mean operators and their application to multiple attribute decision making. *Kbernetes* **2017**, *46*, 1777–1800. [CrossRef]

72. Wei, G.W. Picture 2-tuple linguistic Bonferroni mean operators and their application to multiple attribute decision making. *Int. J. Fuzzy Syst.* **2017**, *19*, 997–1010. [CrossRef]

73. Wei, G.W.; Alsaadi, F.E.; Hayat, T.; Alsaedi, A. Hesitant bipolar fuzzy aggregation operators in multiple attribute decision making. *J. Intell. Fuzzy Syst.* **2017**, *33*, 1119–1128. [CrossRef]

74. Li, L.; Zhang, R.; Wang, J.; Shang, X.; Bai, K. A Novel Approach to Multi-Attribute Group Decision-Making with q-Rung Picture Linguistic Information. *Symmetry* **2018**, *10*, 172. [CrossRef]

75. Wei, Y.; Liu, J.; Lai, X.; Hu, Y. Which determinant is the most informative in forecasting crude oil market volatility: Fundamental, speculation, or uncertainty? *Energy Econ.* **2017**, *68*, 141–150. [CrossRef]

76. Wei, Y.; Yu, Q.; Liu, J.; Cao, Y. Hot money and China's stock market volatility: Further evidence using the GARCH-MIDAS model. *Physica A* **2018**, *492*, 923–930. [CrossRef]

77. Alcantud, J.C.R.; de Andrés Calle, R.; Cascón, J.M. A unifying model to measure consensus solutions in a society. *Math. Comput. Model.* **2013**, *57*, 1876–1883. [CrossRef]

78. Alcantud, J.C.R.; de Andrés Calle, R.; Cascón, J.M. On measures of cohesiveness under dichotomous opinions: Some characterizations of approval consensus measures. *Inf. Sci.* **2013**, *240*, 45–55. [CrossRef]

79. Ullah, K.; Mahmood, T.; Jan, N. Similarity Measures for T-Spherical Fuzzy Sets with Applications in Pattern Recognition. *Symmetry* **2018**, *10*, 193. [CrossRef]

80. Zhu, H.; Zhao, J.B.; Xu, Y. 2-dimension linguistic computational model with 2-tuples for multi-attribute group decision making. *Knowl.-Based Syst.* **2016**, *103*, 132–142. [CrossRef]

81. Wei, C.P.; Liao, H.C. A Multigranularity Linguistic Group Decision-Making Method Based on Hesitant 2-Tuple Sets. *Int. J. Intell. Syst.* **2016**, *31*, 612–634. [CrossRef]

 © 2019 by the authors. Licensee MDPI, Basel, Switzerland. This article is an open access article distributed under the terms and conditions of the Creative Commons Attribution (CC BY) license (http://creativecommons.org/licenses/by/4.0/).

 *symmetry*

Article

New Analytic Solutions of Queueing System for Shared–Short Lanes at Unsignalized Intersections

Ilija Tanackov [1,*], Darko Dragić [1], Siniša Sremac [1], Vuk Bogdanović [1], Bojan Matić [1] and Milica Milojević [2]

[1]   Faculty of technical sciences, University of Novi Sad, Trg Dositeja Obradovića 6, 21 000 Novi Sad, Serbia; darkoenator@gmail.com (D.D.); sremacs@uns.ac.rs (S.S.); vuk@uns.ac.rs (V.B.); bojanm@uns.ac.rs (B.M.)
[2]   Faculty of Architecture, University of Belgrade, Bulevar kralja Aleksandra 73, 11 000 Belgrade, Serbia; m.milojevic@arh.bg.ac.rs
*   Correspondence: ilijat@uns.ac.rs; Tel.: +381-60-7696-0000

Received: 24 November 2018; Accepted: 25 December 2018; Published: 6 January 2019

Abstract: Designing the crossroads capacity is a prerequisite for achieving a high level of service with the same sustainability in stochastic traffic flow. Also, modeling of crossroad capacity can influence on balancing (symmetry) of traffic flow. Loss of priority in a left turn and optimal dimensioning of shared-short line is one of the permanent problems at intersections. A shared–short lane for taking a left turn from a priority direction at unsignalized intersections with a homogenous traffic flow and heterogeneous demands is a two-phase queueing system requiring a first in–first out (FIFO) service discipline and single-server service facility. The first phase (short lane) of the system is the queueing system $M(p\lambda)/M(\mu)/1/\infty$, whereas the second phase (shared lane) is a system with a binomial distribution service. In this research, we explicitly derive the probability of the state of a queueing system with a short lane of a finite capacity for taking a left turn and shared lane of infinite capacity. The presented formulas are under the presumption that the system is Markovian, i.e., the vehicle arrivals in both the minor and major streams are distributed according to the Poisson law, and that the service of the vehicles is exponentially distributed. Complex recursive operations in the two-phase queueing system are explained and solved in manuscript.

Keywords: sustainability; left turn; intersections; lane capacity

---

1. Introduction

The initial considerations of the queuing systems for shared–short lanes at unsignalized intersections were based on the proven procedure by Harders [1], where the lengths of the short lanes are considered either as infinite or zero. In his paper, Harders had presented a limiting analytic frame of the queuing system.

Wu [2] used a pure queuing system, considering in detail the Markovian and non-Markovian systems depending on the distribution of the service, in both the steady and unsteady states of the working regimes of an unsignalized intersection. Wu noticed that the relation between the vehicles in the minor stream of a shared–short lane introduces very complex recursive operations in the queuing system, especially for left turns.

A basic problem while dimensioning shared–short Lanes is the occurrence of the "short lane domino effect" phenomena, queue overflow/short lane saturation [2] and inevitable consequences, time delay. This phenomenon is inevitable when demand exceeds its capacity. Increasing the capacity of road engineering according to Li et al. [3] has become an important way of solving traffic problems. Obligatory consequence of saturation is [4] increases in fuel consumption and [5] increased air pollution at intersections [6].

For now research on signalized intersectionsare dominant. Applying queuing theory in solutions for signalized intersections has a classic theme status and tradition longer than 60 years [7,8], with developed analytical models car-following models, macro traffic flow models, complex networks approaches, cellular automata models, traffic sensing technologies-based approaches, etc. [9]. Apart from Markovian queuing systems non-Markovian queuing systems can be found on signalized intersections [10]. Key spot in these solutions is presented in the form of shared–short Lanes for the left turn [11,12].

The solution of Shared–Short Lanes optimization for unsignalized intersections [13] is more difficult than it is for signalized intersections. Detailed explanation for different analytical approach has been given by Nielsen, Frederiksen and Simonsen [14]. Due to sequential distribution of priority, on signalized intersections deterministic solutions can be applied as well [15]. A probabilistic solution in deterministic time sequences of signalized intersection [16] and right turn solutions [17] cannot be applied for left turn. Unlike unsignalized intersections, on signalized intersections solutions can be found even under conditions of great variations of capacity [18]. One of the basic reasons is the relation between priorities [19].

A homogenous vehicle flow entering an unsignalized intersection system is characterized by simultaneous heterogeneous demands (left turn, though, and right turn). Owing to the traffic rules, the flow is separated into vehicles that are prioritized (through and right turn) and those that lose their priority (left turn). This differentiation of the flow is associated with the significant role of the binomial distribution of the arrival process. A queuing system with such specificity has been observed and explained by Yajima and Phung-Duc [20]. The previously mentioned complex recursive operations noticed by Wu [2] are based on such a binomial distribution, which determines the values and approaches to the transition between the different states of the system. Binomial distribution laws are determined on intersections as well [21] but are signalized. However binomial distribution analytically "favours" Markovian process. Poisson expression of binomial probabilities directly introduces Poisson flows into analytical tools of the queuing system, and with it exponential distribution according to the Palm theory.

The capacity of a short lane performs the spatial selection of vehicles based on the demands alters their prioritization. The basic objective is to ensure that the flow of vehicles that lose priority (left turn) according to the traffic rules do not slow down or entirely block the flow of the vehicles that retain their priority (through and right turn) owing to the first in-first out (FIFO) discipline of the service. Therefore, the proper dimensioning of a short lane has a significant effect on the capacity of the intersection and losses in time.

New analytic solutions of a Queuing system for shared–short lanes at unsignalized intersections have been presented through the following chapters after introduction:

2. Queueing system phenomenon

3. Unsignalised intersections and queueing

4. Limiting analytic framework of a queuing system

5. Calculation of the maximal number of vehicles in a system with only a shared lane

    5.1. Binomial distribution of the vehicle service in a system with only a finite-capacity shared lane

    5.2. Solving queuing systems with only an infinite-capacity shared lane

6. Solving a two-phase queuing system with a finite-capacity short lane *i=constant* and an infinite-capacity shared lane $j \in [1, \infty)$

    6.1. Probabilities of the states of the two-phase queuing system

    6.2. Validation of probability $P_{0,0}$ of a state of the two-phase queuing system

    6.3. Average number of vehicle in short and share lane

7. Resultsfor maximal lane capacity of unsignalized intersection

8. Discussion

9. Conclusions

2. Queueing System Phenomenon

Queuing theory is generally considered a branch of operations research as sub-field of applied mathematics. This was founded just over 100 years ago, by the publication of works and by successful practical application by the Danish mathematician, statistician and engineer Agner Krarup Erlang (1878–1929). However, after initial success in its application, this avant-garde probabilistic methods for making decisions about the resources needed to provide a service, has provided numerous analytical limitations.

Queuing theory is based on elementary system theory, on entity structure and relations. A dominant part in queuing systems is occupied by relations—randomly distributed continuous-time. Entities are system states. They are always whole numbers and represent number of clients in the system. Based on relations between system's intersections and systems entities, the probability of each system state is calculated.

Primary classification of queuing system depends on probabilistic distribution of time. If distribution density is exponential $f(t) = \lambda(t)e^{-\lambda(t)}$, queuing system is Markovian. In case of any other time distribution, the system is non-Markovian. Markovian systems are by rule analytically available. Otherwise, if distribution density is not exponential, analytical calculation is extremely difficult and in some cases even today unsolvable. This classification has been established as an honor to Andrei Andreyevich Markov (1856–1922). David George Kendall (1918–2007) adjusted basic systematization and notation of queuing systems to primary classification.

Secondary classification has also been based on a system's relations. If the average value of probabilistic distribution of time is constant, a queuing system is stationary. Stationary Markovian queuing system has exponential distribution density $f(t) = \lambda e^{-\lambda t}$, $\lambda(t) = \lambda = $ const. The method for the analytical solution of unstationary Markovian queuing system was presented in 1931 by Nikolaevich Kolmogorov (1903–1987) [22]. Solution determined by Kolmogorov for Markovian queuing systems is principally same as for non-Markovian systems. It is based on a system of differential equations. The number of the equation is always equal to number of states, which can be infinite as well! Application of Laplace transformation for solving system of differential equation is much easier in case of Markovian queuing systems. Also, it is understood that queuing system is ergodic.

Tertiary classification is based on the use of system entity, for service and waiting. During this the queuing system can have different service disciplines: FIFO, LIFO (last in–first out), stochastic choice of service, group service, priorities in service etc. For waiting as a rule the FQFS (first in queue–first on service) discipline is used. The basic structure of the queuing system is dominantly based on tertiary classification.

Quartic classification is based on client flow. This structure can be homogenous or inhomogeneous or in other words heterogeneous. Classification has a dual nature. The simplest queuing system concept is when homogenous clients demand homogenous service. In any case of inhomogeneousness of clients and service, queuing system structure becomes delicate to solve.

Many great mathematicians and engineers had contributed to development of queuing theory: Félix Pollaczek (1892–1981) [23],Aleksandr YakovlevichKhinchin (1894–1959), our contemporary, Sir John Frank Charles Kingman (born 1939) [24], David George Kendall (1918–2007), and our other contemporary Jonh Dutton Conant Little (born 1928), etc. Their research has been dominantly pointed towards solving non-Markovian queuing systems. However, an approach towards solving nonstationary non-Markovian queuing systems has been lacking. The development of personal computers of the 1980s and 1990s made the prognosis that each queuing system could be solved by the use of simulations. This attitude has somewhat discouraged further efforts in the analytical approach of the queuing theory and was consistently described by Koenigsberg in the set and reasoned antithesis [25]. His absolutely correct assessment of the necessity of analytical approach and positive development prognosis, confirmed Schwartz, Selinka and Stoletz, especially for non-stationary time-dependent non-Markovian queuing systems [26]. The analytical approach to solving the queuing system remains an imperative. This imperative does not exist in itself, it is encouraged by the practical application of the queuing system and the lifeblood of queuing theory lies in its applications [27].

3. Unsignalised Intersections and Queueing

During the first research in the 1930s, probabilistic nature of traffic had been determined. Determined Poisson distribution in research of road infrastructure capacity in papers by Kinzer [28] and Adams [29] had for the first time proven Markovian structure of traffic flow through Conrad Palma's (1907–1951) theorem. Whole number clients (vehicles in traffic) and exponential distribution of time between consecutive cars in free traffic flow, presented an ideal basis for the application of queuing theory. Traffic flow intensity is by rule time-dependent, or unstationary. However, dimensioning traffic infrastructure capacity of most frequent intensity or maximal intensity can be chosen and declared as stationary. This depends on the solving strategy of queuing system. This effectively expands the first two classifications.

An intersection is a queuing system. However, circumstances on intersections get extremely complicated in the parts of the third and the fourth classification. Thw parallel approach of numerous exponential flows, priority distributions, client heterogeneousness (pedestrians, cyclists, different vehicles: cars, busses, trucks, etc.), different demands (driving straight, left turn, right turn) results in a large number of interactions and complex probabilistic conditioning. Apart from this permanent imperative traffic safety, always presents additional conditions into complex probabilistic conditionality.

This conditionality is greater on unsignalized intersections. On signalized intersections in calculated time sequences, priorities are strictly distributed, which in great measure reduces probabilistic conditionality of antagonistic flows.

This paper treats intersection as Markovian stationary queuing system with FIFO service discipline, one service channel, the final capacity of short lane, endless number of places in queue/shared lane, homogenous vehicle flow with heterogeneous demands: driving straight and left turn. Demand distribution is a stationary discrete random variable of binominal distribution. Even though only one intersection segment had been considered, very complex recursive operations assumed by Wu [2], had been solved within this manuscript after 25 years.

4. Limiting Analytic Framework of a Queueing System

The probability "p" with which vehicles from the priority direction decide to make a left turn can be statistically determined based on the classic Laplacian definition of probability. It is equal to the quotient of number of vehicles turning left and total number of vehicles arriving at the intersection.

If the flow of vehicles is independent, then the Poisson flow with arbitrarily assumed average arrival rate λ can be described as two independent Poisson flows (1):

$$\lambda = p\lambda + (1 - p)\lambda \tag{1}$$

Distribution of the service time for taking a left turn has been the subject of various analyses [30,31] starting from the first concrete application of the queueing systems to the latest research results. The approaches for the utilization of these intervals are in the domain of the time differences from minimally accepted to maximally rejected intervals of priority Poisson flow. The approximation for the service rate of a left turn as an exponential distribution of intensity μ is not very effective for a queuing system at unsignalized intersections owing to the dispersed data points. However, for the first complete analytical solution of the complex recursive operations, it is necessary to remain in the Markovian domain [12,32–34]. Thus, the service rate exponential distribution has been adopted here, and its derived solutions have a complete theoretical and practical relevance.

According to Harders [1], depending on the number of locations on an individual lane for taking a left turn, there are two limiting cases: minimal and maximal average number of vehicles in the system.

A minimal average number of vehicles in the system is achieved if the intersection is designed with a separate lane of unlimited capacity for taking a left turn. The vehicles that plan to move forward in the intersection have separate reserved server, and based on the priority achieve the maximal level of service. A queue is formed only by those vehicles planning to take a left turn at the intersection. In this system, average arrival rate is "$p\lambda$" and average service rate is "μ". The discipline of the service is FIFO. There is a single server and an unlimited number of positions in the queue. This system is a classic Erlang system in which no vehicles are rejected with Kendal markings $M(p\lambda)/M(\mu)/1/\infty$ (Figure 1), and it has already been considered by Wu [2]. The states are determined by the number of vehicles in a separate lane for making a left turn. Accordingly, there are two indices in $X_{i,j}$, where index "i" denotes the number of vehicles in the lane for making a left turn (short lane), whereas index "j" denotes the number of locations in the lane for all the vehicles in the traffic (shared lane). In the queueing system (Figure 1), the indices of the states have values $i\in[0,\infty)$ and $j = 0$.

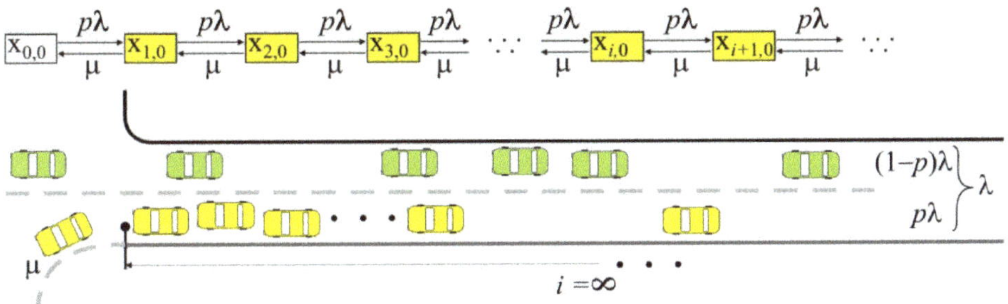

Figure 1. Queueing system with a separate lane for left–turn maneuvering.

According to the formula from for queueing system with Kendal denotation $M(p\lambda)/M(\mu)/1/\infty$, the minimal average number of vehicles in the system is defined in (2).

$$k_{\min} = \left(\frac{\mu - p\lambda}{\mu}\right) \sum_{i=0}^{\infty} i \left(\frac{p\lambda}{\mu}\right)^{i} = \frac{p\lambda}{\mu - p\lambda} \qquad (2)$$

The maximal average number of vehicles in the system is achieved for all the vehicles that are present only in a single-shared lane. In the queueing system (Figure 2), the indices of the states have values $i = 0$ and $j\in[0,\infty)$. In this system, the arrival rate is not the same for all the states. The system crosses from state $X_{0,0}$ into state $X_{0,1}$ only on arrival of a vehicle that plans to make a left turn at the intersection with probability "p." In this case, the arrival rate equals "$p\lambda$". When waiting for a service owing to the FIFO discipline of the service, the server is occupied for all the other vehicles arriving at the intersection with λ intensity, and a queue is formed in the system by all the vehicles. The intensity of the service is not equal for all the states. Only the intensity of service μ from state $X_{0,1}$ is known for the vehicles performing the left-turn maneuver.

If a consecutive vehicle in state $X_{0,2}$ at the intersection plans a left–turn maneuver with probability "p", then after servicing the vehicles from state $X_{0,1}$ in the queuing system, it switches from state $X_{0,2}$ into state $X_{0,1}$ with intensity "$p\mu$."

However, if a consecutive vehicle in state $X_{0,2}$ at the intersection plans to go forward with probability $(1 - p)$, then after servicing the vehicles from state $X_{0,1}$ of the queuing system, it directly transfers from state $X_{0,2}$ to $X_{0,0}$ state with "$(1-p)\mu$" intensity.

Depending on the binomial distribution of the vehicles, the system from state $X_{0,3}$ can transite to states $X_{0,2}$, $X_{0,1}$, or $X_{0,0}$ with different service levels. In general, the system can transite from state $X_{0,j}$ into any of the previous states with different intensities, with a final summation of "μ" (Figure 2). It should be noticed that each state can be achieved from any of the following states.

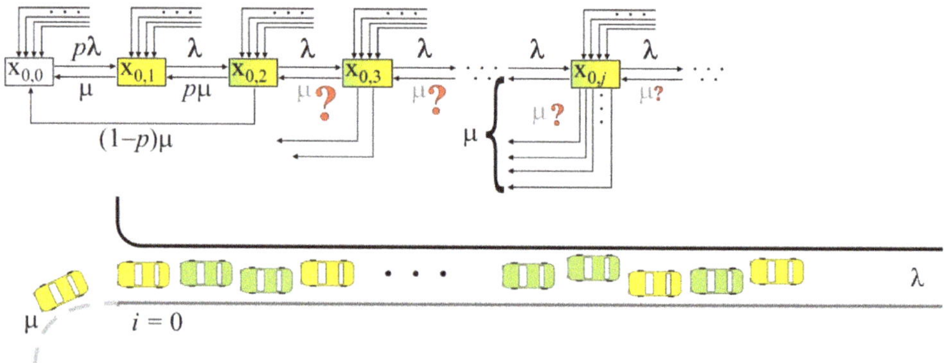

Figure 2. Queueing system with a shared lane.

The maximal average number of vehicles in this system is obtained by (3), which will be explained later in Section 5.2.

$$k_{\max} = \frac{\mu - p\lambda}{p((1-p)\lambda + \mu)} \sum_{j=0}^{\infty} j \left(\frac{\lambda}{(1-p)\lambda + \mu} \right)^j = \frac{\lambda}{\mu - p\lambda} \tag{3}$$

For an intersection that has a short lane designed for taking left turns with final capacity "i" and has a shared lane with unlimited number of shared places $j \in [1,\infty)$ for cars, the queueing system is presented in Figure 3.

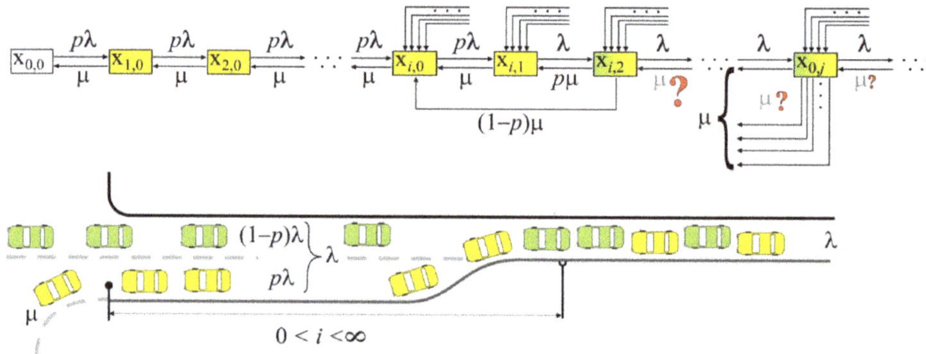

Figure 3. Queueing system with a shared–short lane.

The average number of vehicles in such a queuing system is within the limits of the minimal (2) and maximal (3). The relation between the limiting values can be obtained from expression (4).

$$k_{\min} = \frac{p\lambda}{\mu - p\lambda} < k_i < \frac{\lambda}{\mu - p\lambda} = \frac{k_{\min}}{p} = k_{\max} \tag{4}$$

The procedure for the calculation of expression (4) will be presented in detail in Section 6.3.

The average time that a vehicle spends in this system can be calculated based on the Little formula. From Figure 3, it is obvious that this is a two-phase queueing system. The first phase corresponds to the filling of the short lane for taking left turns, from state $X_{0,0}$ to state $X_{i,0}$. State $X_{i,0}$ is the state connecting Phases I and II. The first phase finishes and the Phase II starts in the same state ($X_{i,0}$).

5. Calculation of the Maximal Number of Vehicles in a System with Only a Shared Lane

5.1. Binomial Distribution of the Vehicle Service in a System with Only a Finite-Capacity Shared Lane

Until now we have explained the transition from state $X_{0,2}$ into $X_{0,0}$. Therefore, we next discuss the intensity of the vehicle service with capacity $j = 3$.

A queue in a joint lane is formed by vehicles that plan to take a left turn at the intersection with probability "p" and those vehicles that plan to drive with priority with probability "$(1 - p)$."

Therefore, the arrival rate is $\lambda = p\lambda + (1 - p)\lambda$ (1). If the system is in state $X_{0,0}$, on arrival of a vehicle with arrival rate priority "$(1 - p)\lambda$" it will not change its state (Figure 4).

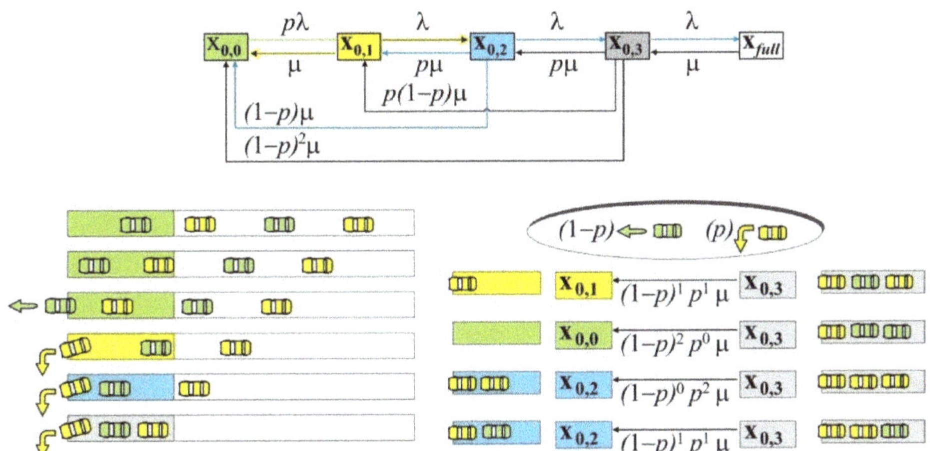

Figure 4. Probabilities and intensities of the transitions from x_{k+3} state to the remaining states.

The system can only transit to state $X_{0,1}$ with the arrival of a vehicle that plans to make a left turn with arrival rate "$p\lambda$" and service intensity "μ." This vehicle shuts down the server. Therefore, all the vehicles form a queue with arrival rate "λ." If the system is in state $X_{0,1}$, i.e., $j = 1$, it can only transition into state $X_{0,0}$ with intensity of left turn μ.

If the system is in state $X_{0,2}$, it can undergo two transitions:

- into state $X_{0,0}$ with intensity "$(1 - p)\mu$" if the second vehicle plans to drive with priority;
- into state $X_{0,1}$ with intensity "$p\mu$" if the second vehicle in the queue plans a left-turn maneuver.
  If the system is in state $X_{0,3}$, it can make four possible transitions (Figure 4):
- into state $X_{0,0}$ with intensity "$(1 - p)^2\mu$" if the second and third vehicles plan to drive with priority;
- into state $X_{0,1}$ with intensity "$(1 - p)p\mu$" if the second vehicle in the queue plans to driveride with priority;
- into state $X_{0,2}$ with intensity "$p^2\mu$" if both the second and third vehicles plan a left-turn maneuver;
- into the $X_{0,2}$ with intensity "$(1 - p)p\mu$" if the second vehicle plans a left-turn maneuver whereas the third vehicle plans to drive with priority.

It should be noted that the total intensity of the transitions into state $X_{0,2}$ is $p^2\mu + (1 - p)p\mu = p\mu$.

It can be generalized that from each state $X_{0,j}$, there are 2^{j-1} possible transitions to each of the previous states. The balance (differential) equations of the steady-states are expressed in (5).

$$P'_{0,0}(t) = 0 = -\lambda p P_{0,0} + \mu P_{0,1} + (1-p)\mu P_{0,2} + (1-p)^2 \mu P_{0,3}$$
$$P'_{0,1}(t) = 0 = +\lambda p P_{0,0} - \lambda P_{0,1} - \mu P_{0,1} + p\mu P_{0,2} + p(1-p)\mu P_{0,3}$$
$$P'_{0,2}(t) = 0 = +\lambda P_{0,1} - \lambda P_{0,2} - p\mu P_{0,2} - (1-p)\mu P_{0,2} + p\mu P_{0,3}$$
$$P'_{0,3}(t) = 0 = +\lambda P_{0,2} - p\mu P_{0,3} - (1-p)p\mu P_{0,3} - (1-p)^2 \mu P_{0,3} + \mu P_{full}$$
$$P'_{full}(t) = 0 = +\lambda P_{0,3} - \mu P_{full}$$

$$(5)$$

To observe the binomial laws, a system with capacity $j = 4$ is also considered (Figure 5).

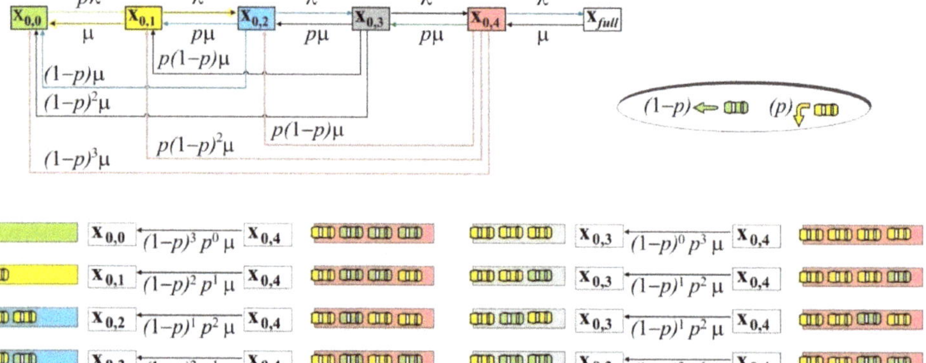

Figure 5. Probabilities and intensities of the transitions from x_{k+4} state to the remaining states.

The balance (differential) equations of the steady-state of the queueing system with four places in the queue for vehicles in a joint traffic lane are given in (6).

$$P'_{0,0}(t) = 0 = -\lambda p P_{0,0} + \mu P_{0,1} + (1-p)^1 \mu P_{0,2} + (1-p)^2 \mu P_{0,3} + (1-p)^3 \mu P_{0,4}$$
$$P'_{0,1}(t) = 0 = +\lambda p P_{0,0} - \lambda P_{0,1} - \mu P_{0,1} + p\mu P_{0,2} + p(1-p)\mu P_{0,3} + p(1-p)^2 \mu P_{0,4}$$
$$P'_{0,2}(t) = 0 = +\lambda P_{0,1} - \lambda P_{0,2} - p\mu P_{0,2} + (1-p)\mu P_{0,2} + p\mu P_{0,3} + p(1-p)\mu P_{0,4}$$
$$P'_{0,3}(t) = 0 = +\lambda P_{0,2} - P_{0,3}(\lambda + p\mu + p\mu(1-p) + \mu(1-p)^2) + p\mu P_{0,4}$$
$$P'_{0,4}(t) = 0 = +\lambda P_{0,3} - P_{0,4}(\lambda + p\mu + p\mu(1-p) + p\mu(1-p)^2 + \mu(1-p)^3) + \mu P_{full}$$
$$P'_{full}(t) = 0 = +\lambda P_{0,4} - \mu P_{full}$$

$$(6)$$

The system can switch from state $X_{0,4}$ into state $X_{0,3}$ in four ways, which are included in the binomial expression in (7).

$$\mu\left(p^3(1-p)^0 + 2p^2(1-p)^1 + p^1(1-p)^2\right) = p\mu \sum_{k=0}^{2} \binom{2}{k} p^k (1-p)^{2-k} \tag{7}$$

From each $X_{0,j}$ state there are $(j-2)$ ways to switch to $X_{0,j-1}$ state, which are included in the binomial expression for complementary probabilities, as expressed in (8):

$$p\mu \sum_{k=0}^{j-2} \binom{j-2}{k} p^k (1-p)^{j-2-k} = p\mu(p + (1-p))^{j-2} = p\mu(1)^{j-2} = p\mu \tag{8}$$

and there are $(n-1)$ ways to switch from each states $X_{0,j \in [2,\,\infty]}$ to state $X_{0,n \in [1,\,j-1]}$, which are included in the binomial expression of complementary probabilities given in (9), for $k \in N$.

$$p(1-p)^{j-n-1} \mu \sum_{k=0}^{n-1} \binom{n-1}{k} p^k (1-p)^{n-1-k} = p\mu(p + (1-p))^{n-2} = p(1-p)^{j-n-1}\mu \tag{9}$$

The exception is state $X_{0,0}$ whose intensities are expressed as the product of the probabilities of each state in a geometric series (10).

$$\mu \sum_{j=0}^{\infty} (1-p)^j P_{0,j} \tag{10}$$

From each state $j \in [1, k]$, $k \in N$, intensity of the vehicle service "μ" "spills" according to the partial geometric dependence defined in (11).

$$\mu(1-p)^{j-1} + p\mu \sum_{k=0}^{j-2} (1-p)^k + = \mu(1-p)^{j-1} + p\mu \frac{1-(1-p)^{j-1}}{1-(1-p)} = \mu \tag{11}$$

5.2. Solving Queueing Systems with Only an Infinite-Capacity Shared Lane

A queueing system in which no vehicles are rejected, because of the absence of a short lane ($i = 0$), in a heterogeneous vehicle flow with an infinite-capacity shared lane has states as presented in Figure 6.

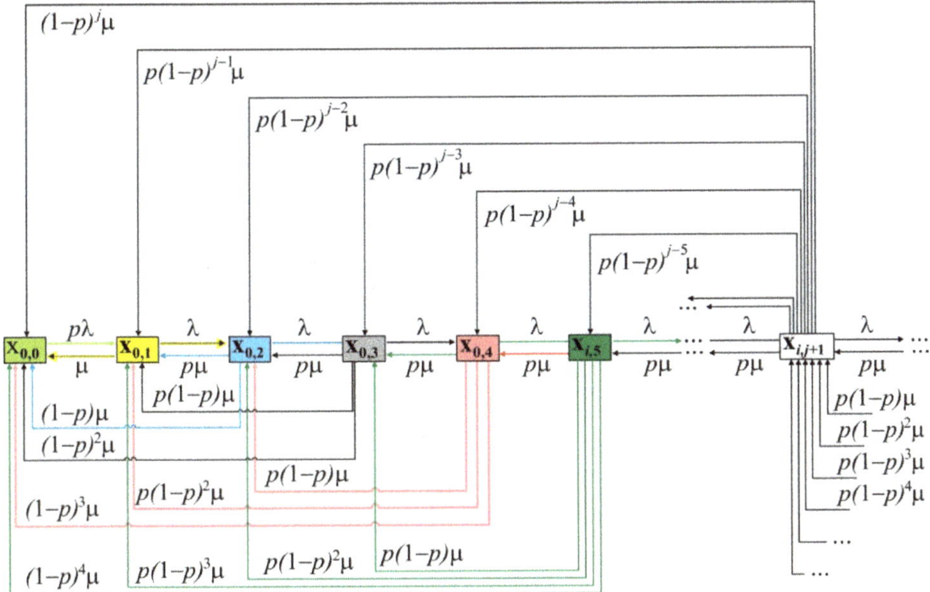

Figure 6. Queueing system for a shared lane only–connection between the state of the system and intensity.

The system of balance equations of the corresponding steady-state is given in (12), $k \in N$.

$$\begin{aligned}
P'_{0,0}(t) &= 0 = -p\lambda P_{0,0} + \frac{1}{p}\sum_{j=1}^{\infty} p(1-p)^{j-1}\mu P_{0,j} \\
P'_{0,1}(t) &= 0 = p\lambda P_{0,0} - (\lambda+\mu)P_{0,1} + \sum_{j=2}^{\infty} p(1-p)^{j-2}\mu P_{0,j} \\
P'_{0,2}(t) &= 0 = \lambda P_{0,1} - (\lambda+\mu)P_{0,2} + \sum_{j=3}^{\infty} p(1-p)^{j-3}\mu P_{0,j} \\
&\quad \ldots \\
P'_{0,k+1}(t) &= 0 = \lambda P_{0,k} - (\lambda+\mu)P_{0,k+1} + \sum_{j=k+2}^{\infty} p(1-p)^{j-(k+2)}\mu P_{0,j} \\
&\quad \ldots
\end{aligned} \tag{12}$$

From the balance equations of the steady-state of the system in (12), the relations between the probabilities and sums of the series of the geometric products of the probabilities are obtained, as given in (13).

$$\left. \begin{aligned}
p\lambda P_{0,0} &= \tfrac{1}{p} \sum_{j=1}^{\infty} p(1-p)^{j-1}\mu P_{0,j} \\
(\lambda+\mu)P_{0,1} - \lambda P_{0,0} &= \sum_{j=2}^{\infty} p(1-p)^{j-2}\mu P_{0,j} \\
(\lambda+\mu)P_{0,2} - \lambda P_{0,1} &= \sum_{j=3}^{\infty} p(1-p)^{j-3}\mu P_{0,j} \\
&\cdots \\
(\lambda+\mu)P_{0,k} - \lambda P_{0,k-1} &= \sum_{j=k+1}^{\infty} p(1-p)^{j-k-1}\mu P_{0,j} \\
&\cdots
\end{aligned} \right\}
\tag{13}$$

The relations between the sums should be noticed (14).

$$\left. \begin{aligned}
\sum_{j=1}^{\infty} p(1-p)^{j-1}\mu P_j &= p\mu P_{0,1} + (1-p)\sum_{j=2}^{\infty} p(1-p)^{j-2}\mu P_{0,j} \\
\sum_{j=2}^{\infty} p(1-p)^{n-2}\mu P_{0,j} &= p\mu P_{0,2} + (1-p)\sum_{j=3}^{\infty} p(1-p)^{j-3}\mu P_{0,j} \\
\sum_{j=3}^{\infty} p(1-p)^{n-3}\mu P_{0,j} &= p\mu P_{0,3} + (1-p)\sum_{j=4}^{\infty} p(1-p)^{j-4}\mu P_{0,j} \\
&\cdots \\
\sum_{j=k}^{\infty} p(1-p)^{j-k}\mu P_{0,j} &= p\mu P_{0,k} + (1-p)\sum_{j=k+1}^{\infty} p(1-p)^{j-(k+1)}\mu P_{0,j} \\
&\cdots
\end{aligned} \right\}
\tag{14}$$

From the first balance equation of the steady-state in (12), expression (15) can be obtained.

$$p\lambda P_{0,0} = \frac{1}{p}\sum_{j=1}^{\infty} p(1-p)^{j-1}\mu P_{0,j} \Leftrightarrow p^2\lambda P_{0,0} = p\mu P_{0,1} + (1-p)\sum_{j=2}^{\infty} p(1-p)^{j-2}\mu P_{0,j} \tag{15}$$

From the second relation of the sums in (14), we can obtain expression (16).

$$\begin{aligned}
p^2\lambda P_{0,0} &= p\mu P_{0,1} + (1-p)[(\lambda+\mu)P_{0,1} - p\lambda P_{0,0}] \\
p\lambda P_{0,0} &= \lambda P_{0,1} + \mu P_{0,1} - p\lambda P_{0,1} \Leftrightarrow P_{0,1} = \tfrac{p\lambda}{(1-p)\lambda+\mu}P_{0,0}
\end{aligned} \tag{16}$$

From the second balance equation in (12) of the steady-state, expression (17) can be obtained.

$$(\lambda+\mu)P_{0,1} - p\lambda P_{0,0} = p\mu P_{0,2} + (1-p)\sum_{j=3}^{\infty} p(1-p)^{j-3}\mu P_{0,j} \tag{17}$$

From the third relation of the sums in (14), expression (18) can be derived.

$$\begin{aligned}
(\lambda+\mu)P_{0,1} - p\lambda P_{0,0} &= p\mu P_{0,2} + (1-p)[(\lambda+\mu)P_{0,2} - \lambda P_{0,1}] \\
[(1-p)\lambda + \mu]P_{0,2} &= \tfrac{[2\lambda-p\lambda+\mu]-[\lambda-p\lambda+\mu]}{(1-p)\lambda+\mu}p\lambda P_{0,0} \Leftrightarrow P_{0,2} = \tfrac{p\lambda^2 P_{0,0}}{[(1-p)\lambda+\mu]^2}
\end{aligned} \tag{18}$$

Furthermore, from the third balance equation of the steady-state of the system, expression (19) can be obtained.

$$(\lambda+\mu)P_{0,2} - \lambda P_1 = \sum_{j=3}^{\infty} p(1-p)^{j-3}\mu P_{0,j} = p\mu P_{0,3} + (1-p)\sum_{j=4}^{\infty} p(1-p)^{j-4}\mu P_{0,j} \tag{19}$$

From the fourth relation of the sums, expression (20) can be obtained.

$$(\lambda + \mu)P_{0,2} - \lambda P_1 = p\mu P_{0,3} + (1-p)[(\lambda+\mu)P_{0,3} - \lambda P_{0,2}]$$

$$[(1-p)\lambda + \mu]P_{0,3} = [(2-p)\lambda + \mu]P_{0,2} - \lambda P_{0,1} \Leftrightarrow P_{0,3} = \frac{p\lambda^3 P_{0,0}}{[(1-p)\lambda+\mu]^3} \tag{20}$$

From (16), (18) and (20) the recurrent relation for the probabilities of the states, $P_{0,j}$ is obtained analogously (21), and can be proved by mathematical induction.

$$P_{0,j} = \frac{p\lambda^j P_{0,0}}{[(1-p)\lambda + \mu]^j} \tag{21}$$

From the condition in (22):

$$P_{0,0} + P_{0,1} + P_{0,2} + \ldots = \sum_{j=0}^{\infty} P_{0,j} \tag{22}$$

The probability of the initial state in (23) is obtained. Under the stability condition $\mu \geq p\lambda$ and $0 \leq p \leq 1$, probability of the state without a vehicle is always $0 \leq P_{0,0} \leq 1$.

$$P_{0,0} = \frac{1}{p \sum\limits_{j=0}^{\infty} \left(\frac{\lambda}{(1-p)\lambda+\mu}\right)^j} = \frac{1}{p}\left(1 - \frac{\lambda}{(1-p)\lambda+\mu}\right) = \frac{\mu - p\lambda}{p((1-p)\lambda+\mu)} \tag{23}$$

From the recurrent relation of the probabilities of the states for $i = 0$ and $j \in [0, \infty)$, (24) is obtained.

$$P_{0,j} = \frac{\mu - p\lambda}{p((1-p)\lambda + \mu)} \frac{p\lambda^j}{[(1-p)\lambda + \mu]^j} = \frac{\mu - p\lambda}{((1-p)\lambda + \mu)} \frac{\lambda^j}{[(1-p)\lambda + \mu]^j} \tag{24}$$

The average number of vehicles in the system is defined in (25)

$$k_{\max} = \sum_{j=0}^{\infty} k P_{0,j} = \frac{\frac{\mu-p\lambda}{(1-p)\lambda+\mu}}{1 - \frac{\mu-p\lambda}{(1-p)\lambda+\mu}} = \frac{\lambda}{\mu - p\lambda} \tag{25}$$

This average number of vehicles is the maximal average number of vehicles $k_{\max}$ in (2) achieved in the queueing system for the given values of λ, μ, and p. It is obvious that when $p\lambda = \mu$, the average number of vehicles diverges, and under the stability condition $\mu \geq p\lambda$, the system does not fulfill its objective (26).

$$\lim_{p\lambda \to \mu} k_{\max} = \lim_{p\lambda \to \mu} \frac{\lambda}{\mu - p\lambda} = \infty \tag{26}$$

6. Solving a Two-Phase Queueing System with a Finite-Capacity Short Lane i = Const and an Infinite-Capacity Shared Lane $j \in [1, \infty)$

This queueing system is the usual state in practical conditions. The system has one server for taking a left turn. A separate lane for the left turn is designed, and it has finite capacity "i". A separate lane fills with a homogenous vehicle flow with a homogenous demand that becomes a heterogeneous demandon at unsignalized intersections when vehicles plan a left-turn maneuver. If product "$p\lambda$" converges to service rate "μ" or if there is high participation of "p" in the incoming flow, vehicles fill all the places "i" in the short lane, and then form a queue of heterogeneous vehicles in the shared lane with intensity "λ." The graph of the states of the queueing system is presented in Figure 7.

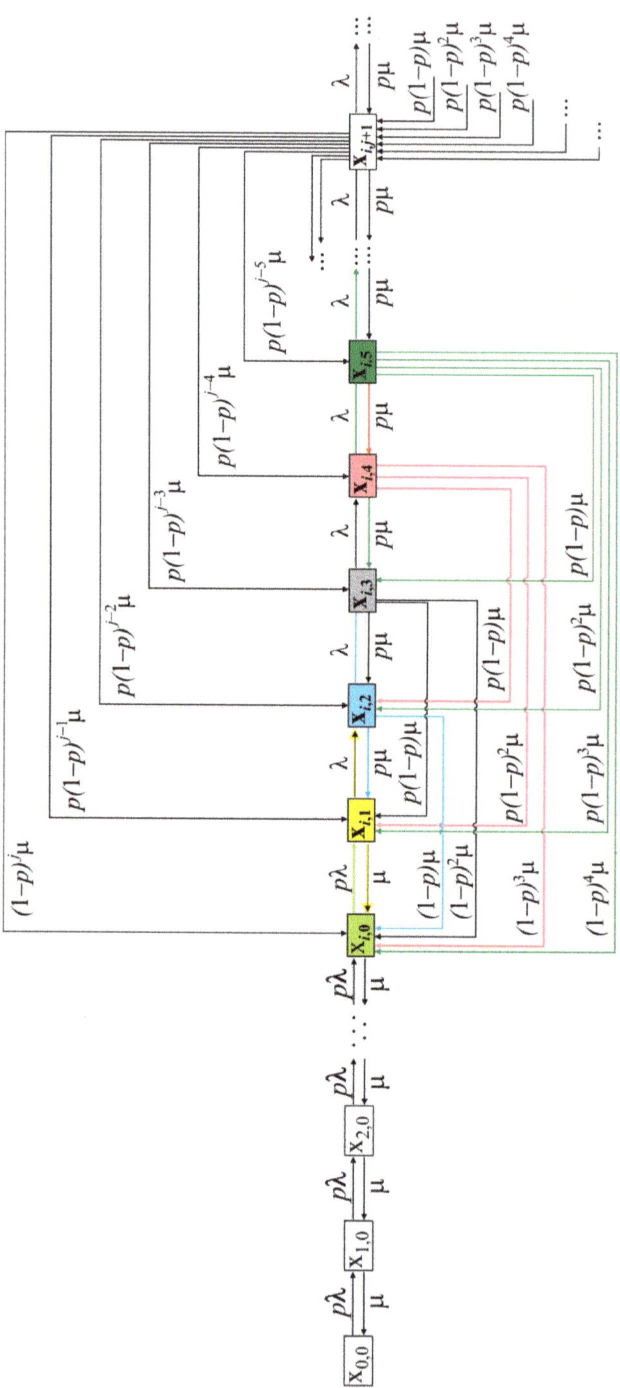

Figure 7. Two-phase queueing system of shared–short lane.

6.1. Probabilities of the States of the Two-Phase Queueing System

The connecting phases I and II is state $P_{i,0}$. The balance equations of the steady-state of the two-phase system are given in (27).

$$
\left.
\begin{aligned}
&P'_{0,0}(t) = 0 = -p\lambda P_{0,0} + \mu P_{1,0} \\
&P'_{1,0}(t) = 0 = +p\lambda P_{0,0} - p\lambda P_{1,0} - \mu P_{1,0} + \mu P_{2,0} \\[8pt]
&\cdots \\
&P'_{i-1,0}(t) = 0 = +p\lambda P_{i-2,0} - p\lambda P_{i-1,0} - \mu P_{i-1,0} + \mu P_{i,0} \\
&P'_{i,0}(t) = 0 = +p\lambda P_{i-1,0} - p\lambda P_{i,0} - \mu P_{i,0} + \tfrac{1}{p}\sum_{j=1}^{\infty} p(1-p)^{j-1}\mu P_{i,j} \\
&P'_{i,1}(t) = 0 = \lambda P_{i,0} - (\lambda+\mu)P_{i,1} + \sum_{j=2}^{\infty} p(1-p)^{j-2}\mu P_{i,j} \\
&P'_{i,2}(t) = 0 = \lambda P_{i,1} - (\lambda+\mu)P_{i,2} + \sum_{j=3}^{\infty} p(1-p)^{j-3}\mu P_{i,j} \\
&P'_{i,3}(t) = 0 = \lambda P_{i,2} - (\lambda+\mu)P_{i,3} + \sum_{j=4}^{\infty} p(1-p)^{j-4}\mu P_{i,j} \\[6pt]
&\cdots
\end{aligned}
\right\}
\tag{27}
$$

From state $X_{0,0}$ to state $X_{i,0}$, there are known relations based on the system $M(p\lambda)/M(\mu)/i/\infty$ (28).

$$
\left.
\begin{aligned}
&p\lambda P_{0,0} = \mu P_{1,0} \Leftrightarrow P_{1,0} = \tfrac{p\lambda}{\mu} P_{0,0} \\
&\lambda \tfrac{p\lambda}{\mu} P_{0,0} + \mu \tfrac{p\lambda}{\mu} P_{0,0} - p\lambda P_{0,0} = \mu P_{2,0} \Leftrightarrow P_{2,0} = \left(\tfrac{p\lambda}{\mu}\right)^2 P_{0,0} \\
&P_{3,0} = \left(\tfrac{p\lambda}{\mu}\right)^3 P_{0,0} \\
&\cdots \\
&P_{i-1,0} = \left(\tfrac{p\lambda}{\mu}\right)^{i-1} P_{0,0} \\
&P_{i,0} = \tfrac{p\lambda}{\mu} P_{i-1,0} = \left(\tfrac{p\lambda}{\mu}\right)^{i} P_{0,0}
\end{aligned}
\right\}
\tag{28}
$$

For further solving the new relations between the sums, the expressions in (29) need to be noted ($k \in N$).

$$
\left.
\begin{aligned}
&\sum_{j=1}^{\infty} p(1-p)^{j-1}\mu P_{i,j} = p\mu P_{i,1} + (1-p)\sum_{j=2}^{\infty} p(1-p)^{j-2}\mu P_{i,j} \\
&\sum_{j=2}^{\infty} p(1-p)^{j-2}\mu P_{i,j} = p\mu P_{i,2} + (1-p)\sum_{j=3}^{\infty} p(1-p)^{j-3}\mu P_{i,j} \\
&\sum_{j=3}^{\infty} p(1-p)^{j-3}\mu P_{i,j} = p\mu P_{i,3} + (1-p)\sum_{j=4}^{\infty} p(1-p)^{j-4}\mu P_{i,j} \\
&\cdots \\
&\sum_{j=k}^{\infty} p(1-p)^{j-k}\mu P_{i,k} = p\mu P_{i,k} + (1-p)\sum_{j=k+1}^{\infty} p(1-p)^{j-(k+1)}\mu P_{i,j} \\
&\cdots
\end{aligned}
\right\}
\tag{29}
$$

From the equation for $P_{i,1}$, by changing the sum for $j = 3$, (30) is obtained.

$$
\begin{aligned}
&(\lambda+\mu)P_{i,1} - \lambda P_{i,0} = \sum_{j=2}^{\infty} p(1-p)^{j-2}\mu P_{i,j} \\
&(\lambda+\mu)P_{i,1} - \lambda P_{i,0} = p\mu P_{i,2} + (1-p)\sum_{j=3}^{\infty} p(1-p)^{j-3}\mu P_{i,j} \\
&(\lambda+\mu)P_{i,1} - \lambda P_{i,0} = p\mu P_{i,2} - (1-p)(\lambda P_{i,1} - (\lambda+\mu)P_{i,2}) \\
&[(2-p)\lambda + \mu]P_{i,1} - \lambda P_{i,0} = [(1-p)\lambda + \mu]P_{i,2} \Leftrightarrow \lambda P_{i,0} = [(2-p)\lambda + \mu]P_{i,1} - [(1-p)\lambda + \mu]P_{i,2} \\
&P_{i,0} = [(2-p) + \tfrac{\mu}{\lambda}]P_{i,1} - [(1-p) + \tfrac{\mu}{\lambda}]P_{i,2}
\end{aligned}
\tag{30}
$$

From the equation for $P_{i,2}$, by changing the sum for $j = 4$, (31) is obtained.

$$(\lambda + \mu)P_{i,2} - \lambda P_{i,1} = \sum_{j=3}^{\infty} p(1-p)^{j-3}\mu P_{i,j}$$
$$(\lambda + \mu)P_{i,2} - \lambda P_{i,1} = p\mu P_{i,3} + (1-p)\sum_{j=4}^{\infty} p(1-p)^{j-4}\mu P_{i,j}$$
$$(\lambda + \mu)P_{i,2} - \lambda P_{i,1} = p\mu P_{i,3} - (1-p)(\lambda P_{i,2} - (\lambda+\mu)P_{i,3})$$
$$[(2-p)\lambda + \mu]P_{i,2} - \lambda P_{i,1} = [(1-p)\lambda + \mu]P_{i,3}$$
$$P_{i,1} = \left[(2-p) + \tfrac{\mu}{\lambda}\right]P_{i,2} - \left[(1-p) + \tfrac{\mu}{\lambda}\right]P_{i,3} \tag{31}$$

Furthermore, a recurrent relation given in (32) is obtained.

$$P_{i,k} = \left[(2-p) + \frac{\mu}{\lambda}\right]P_{i,k+1} - \left[(1-p) + \frac{\mu}{\lambda}\right]P_{i,k+2} \tag{32}$$

From the normative condition in (33):

$$\sum_{k=0}^{i-1} P_{k,0} + \sum_{j=0}^{\infty} P_{i,j} = (P_{0,0} + P_{1,0} + \ldots + P_{i-1,0}) + (P_{i,0} + P_{i,1} + P_{i,2} + P_{i,3}\ldots) = 1 \tag{33}$$

and applying the recurrent equation in (32), (34) can be derived.

$$\sum_{k=0}^{i-1} P_{k,0} + \left[(2-p) + \tfrac{\mu}{\lambda}\right]P_{i,1} \underbrace{-\left[(1-p) + \tfrac{\mu}{\lambda}\right]P_{i,2} + \left[(2-p) + \tfrac{\mu}{\lambda}\right]P_{i,2}}_{P_{i,2}}$$
$$\underbrace{-\left[(1-p) + \tfrac{\mu}{\lambda}\right]P_{i,3} + \left[(2-p) + \tfrac{\mu}{\lambda}\right]P_{i,3}}_{P_{i,3}} \underbrace{-\left[(1-p) + \tfrac{\mu}{\lambda}\right]P_{i,4} + \ldots}_{P_{i,4}} = 1 \tag{34}$$

Then the normative condition becomes (35).

$$\sum_{k=0}^{i-1} P_{k,0} + \left[(2-p) + \frac{\mu}{\lambda}\right]P_{i,1} + P_{i,2} + P_{i,3} + \ldots = 1 \tag{35}$$

Since,

$$\left[(2-p) + \frac{\mu}{\lambda}\right]P_{i,1} = 2P_{i,1} - pP_{i,1} + \frac{\mu}{\lambda}P_{i,1} = \left[(1-p) + \frac{\mu}{\lambda}\right]P_{i,1} + P_{i,1} \tag{36}$$

The normative condition can be given as (37).

$$\sum_{k=0}^{i-1} P_{k,0} + \left[(1-p) + \frac{\mu}{\lambda}\right]P_{i,1} + P_{i,1} + P_{i,2} + \ldots = 1 \tag{37}$$

By expanding (37) with $(\pm P_{i,0})$, the direct relation between probabilities $P_{i,0}$ and $P_{i,1}$ is obtained from the normative condition, and it is given in (38).

$$\sum_{k=0}^{i-1} P_{k,0} + \left[(1-p) + \frac{\mu}{\lambda}\right]P_{i,1} + \underbrace{(-P_{i,0} + P_{i,0}) + P_{i,1} + P_{i,2} + P_{i,3} + \ldots}_{\sum_{j=0}^{\infty} P_{i,j}} = 1 \tag{38}$$

As a part of the normative condition in (33) is contained in (38), we can now obtain (39).

$$\underbrace{\sum_{k=0}^{i-1} P_{k,0} + \sum_{j=0}^{\infty} P_{i,j}}_{from\ (33)=1} + \left[(1-p) + \frac{\mu}{\lambda}\right]P_{i,1} - P_{i,0} = 1 \tag{39}$$

By accepting the last balance equation given in (28), the value of first probability of the Phase II, $P_{i,1}$ is obtained through $P_{0,0}$, as expressed in (40).

$$[(1-p)\lambda + \mu]P_{i,1} = \lambda P_{i,0} \Leftrightarrow P_{i,1} = \frac{\lambda P_{i,0}}{[(1-p)\lambda+\mu]} = \left(\frac{p\lambda}{\mu}\right)^i \frac{\lambda}{(1-p)\lambda+\mu} P_{0,0}$$

$$P_{i,j} = \frac{\lambda P_{i,0}}{[(1-p)\lambda+\mu]} = \left(\frac{p\lambda}{\mu}\right)^i \left(\frac{\lambda}{(1-p)\lambda+\mu}\right)^j P_{0,0} \tag{40}$$

The relation between the consecutive probabilities is maintained in the Phase II of the system, and so, it is the same as in (24). Therefore, the final expression for the normative condition through initial probability $P_{0,0}$ becomes:

$$P_{0,0}\sum_{k=0}^{i-1}\left(\frac{p\lambda}{\mu}\right)^k + P_{0,0}\left(\frac{p\lambda}{\mu}\right)^i \sum_{j=0}^{\infty}\left(\frac{\lambda}{(1-p)\lambda+\mu}\right)^j = 1 \tag{41}$$

The sums of the finite geometric series of the first phase and infinite geometric series of the Phase II are given in (42):

$$P_{0,0}\frac{1-\left(\frac{p\lambda}{\mu}\right)^i}{1-\left(\frac{p\lambda}{\mu}\right)} + P_{0,0}\left(\frac{p\lambda}{\mu}\right)^i\frac{(1-p)\lambda+\mu}{\mu-p\lambda} = 1$$

$$P_{0,0}\frac{\mu-\mu\left(\frac{p\lambda}{\mu}\right)^i}{\mu-p\lambda} + P_{0,0}\frac{\lambda\left(\frac{p\lambda}{\mu}\right)^i - p\lambda\left(\frac{p\lambda}{\mu}\right)^i + \mu\left(\frac{p\lambda}{\mu}\right)^i}{\mu-p\lambda} = 1 \tag{42}$$

and they yield final probability of the initial state $P_{0,0}$ (43):

$$P_{0,0} = \frac{\mu-p\lambda}{(1-p)\lambda\left(\frac{p\lambda}{\mu}\right)^i + \mu} \tag{43}$$

6.2. Validation of Probability $P_{0,0}$ of a State of the Two-Phase Queueing System

The validation of probability $P_{0,0}$ of a state of a two-phase queuing system can be achieved within the limiting conditions. Apart from the stability condition $\mu \geq p\lambda$, the first limiting condition is that when the number of places in the separate lane tends toward infinity, the well-known value of $P_{0,0}$ is obtained for queueing system with Kendal denotation $M(p\lambda)/M(\mu)/1/\infty$, i.e., for a system with a distinct separate lane for left turns with infinite capacity (44).

$$\lim_{i\to\infty} P_{0,0} = \lim_{i\to\infty} \frac{\mu-p\lambda}{(1-p)\lambda\left(\frac{p\lambda}{\mu}\right)^i + \mu} = \frac{\mu-p\lambda}{\mu} = 1 - \frac{p\lambda}{\mu} \tag{44}$$

The second limiting case is when the number of places in the separate lane converges towards 0. The relation obtained in this case has already been defined in (24), with a difference in the value of "p" and in the denominator, derived from the differences in the two-phase queueing system with input intensity "$p\lambda$" for state $P_{i,0}$ when "i" converge to 0 ($i\to0$) (45).

$$\lim_{i\to0} P_{0,0} = \lim_{i\to0} \frac{\mu-p\lambda}{(1-p)\lambda\left(\frac{p\lambda}{\mu}\right)^i + \mu} = \frac{\mu-p\lambda}{(1-p)\lambda+\mu} \tag{45}$$

6.3. Average Number of Vehicle in Short and Share Lane

The general expression for the average number of vehicle switch capacity of short lane "i" is equal to the sum of the average number of vehicles per phase (46). It should be noticed that in the second phase, the short lane of capacity "i" fills up.

$$k_i = k_{i(I)} + k_{i(II)} = \underbrace{\sum_{k=0}^{i-1} kP_{k,0}}_{\text{phase I}} + \underbrace{\sum_{j=0}^{\infty} (i+j)P_{i,j}}_{\text{phase II}} \tag{46}$$

The first phase has a known value of the average number of vehicles based on the $M(p\lambda)/M(\mu)/1/i$ queueing system (47):

$$k_{i(I)} = \sum_{k=0}^{i-1} kP_{k,0} = 1 - \frac{1 - \frac{p\lambda}{\mu}}{1 - \left(\frac{p\lambda}{\mu}\right)^{i+1}} + \left(\frac{p\lambda}{\mu}\right)^2 \frac{1 - \left(\frac{p\lambda}{\mu}\right)^{i-1}\left((i-1)(1-\frac{p\lambda}{\mu})+1\right)}{\left(1 - \frac{p\lambda}{\mu}\right) \cdot \left[1 - \left(\frac{p\lambda}{\mu}\right)^{i+1}\right]} \tag{47}$$

Phase II can be separated into two sums (first phase is filled with "*i*" vehicles) (48):

$$k_{i(II)} = \sum_{j=0}^{\infty} (i+j)P_{i,j} = \underbrace{i\sum_{j=0}^{\infty} P_{i,j}}_{\text{Full short lane}} + \sum_{j=0}^{\infty} jP_{i,j} \tag{48}$$

The first sum is a pure geometric series, whereas the second is a known series for a system with infinite number of states, as given in (49).

$$k_{i(II)} = \frac{i}{1 - \frac{\lambda}{(1-p)\lambda+\mu}} + \frac{\frac{\lambda}{(1-p)\lambda+\mu}}{1 - \frac{\lambda}{(1-p)\lambda+\mu}} = \frac{i((1-p)\lambda+\mu)+\lambda}{\mu - p\lambda} \tag{49}$$

7. Results for Maximal Lane Capacity of Unsignalized Intersection

Figure 8 presents the average number of vehicles in the system for $\lambda = 500$ vehicle/h, $\mu = 300$ vehicle/h ($\lambda+\mu = 800$ vehicle/h, maximal lane capacity of unsignalized intersection established by Lakkundi [35], $i \in [5, 20]$, and $p \in [0.10, 0.50]$. The maximal number of vehicles in the system is marked in the figure at $i = 0$ and calculated according to (25) for different probabilities "p" (emphasized by red dots). The average number of vehicles has been calculated according to expressions (47)–(49). From Figure 8, it can be seen that probability "p" (with which vehicles from the priority direction decide to make a left turn) has more effect than number of places in the short lane "i".

For the given parameters, namely, $\lambda = 500$ vehicle/h and $\mu = 300$ vehicle/h, two nomographic distributions for the queue lengths can be expressed by using the cumulative probabilities - analogous to the usage of ($P_{0,0} + \ldots + P_{i,0} + P_{i,1} + P_{i,2} + \ldots$). A 3D function of the independent variable of the short lane "i" and probability of the left-turn maneuver "p" is presented in Figure 9.

In the first case, for $i = 5$, it can be noted that 99% of the vehicles will be satisfied up to $p = 0.20$. This implies that for a short lane with a capacity of five places for vehicles, 99% of the vehicles will not use the capacity of the shared lane for forming a queue if $p = 0.20$, without queue overflow/short lane saturation. This is apriori consideration of capacity. Table 1 presents the calculation of cumulative probability. It is obvious that for $p > 0.2$ at given flow shared lane saturation begins.

In the second nomograph in Figure 9, an a posteriori problem is considered. Assumed, or in concrete case, statistical determined distribution of left turn is $p = 0.27$. From $p = 0.27$ ordinate on the surface of nomograph we reach intersection with line for percentile 99 - point "a". By following the 99 percentile line we reach point "b". From it, using the surface of nomograph we descend using ordinate reaching Queue lengths value which asses the value of capacity of short lane $i \approx 10$! From the Table 1 concrete value can be calculated by interpolation. It is obvious that for $i = 10$ and $p = 0.25$ cumulative probability is $0.994 > 0.99$, but for $i = 10$ and $p = 0.30$ cumulative probability is $0.986 > 0.99$.

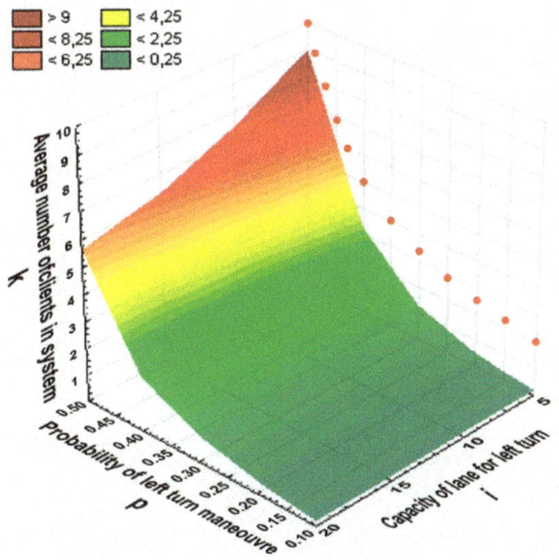

Figure 8. Average number of vehicles in the system for the chosen parameters.

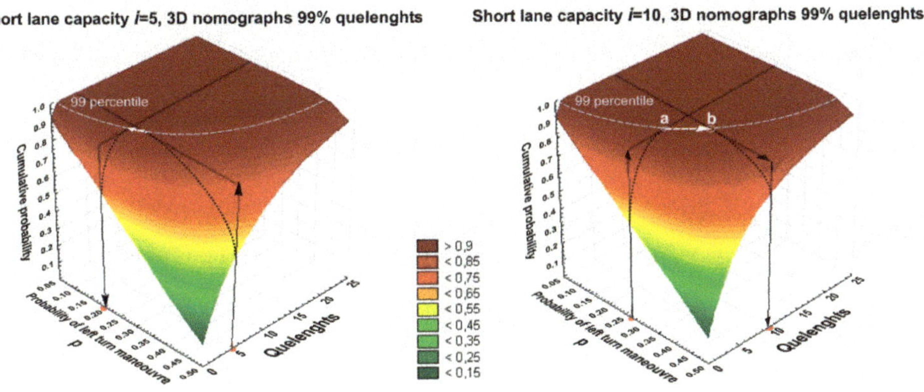

Figure 9. Nomographs for parameters $\lambda = 500$ vehicle/h, $\mu = 300$ vehicle/h, and short lane capacity $i = 5$ and $i = 10$.

The chosen examples have a simple message: to increase probability from $p = 0.2$ to $p = 0.27$ or for 35%, short lane capacity doubled respectively from $i = 5$ to $i = 10$! Further growth of probability "p" disproportionally increases needed capacity of short lane. For example, for $p = 0.35$, $\lambda = 500$ vehicle/h and $\mu = 300$ vehicle/h, short lane will satisfy 99% of traffic flow if the designed short lane capacity $i \geq 17$ (see values in Table 1, values on gray background are less than 99 percentile). For values of limiting $p_{max} = 0.375$ (50) queue overflow/short lane saturation is permanent and separate lane for left-turn maneuvering (Figure 1) needs to be designed.

$$p_{\max} = \frac{\mu}{\lambda + \mu} = \frac{300}{800} = 0.375 \tag{50}$$

To calculate the average time that a vehicle spends in a two-phase queueing system by using the Little formula, it is necessary to note that the intensity "$p\lambda$" changes to "λ" at state $X_{i,1}$ (Figure 7). Thus, the average time that a vehicle spends in the system is defined in (51), and $k_{i(I)}$ is calculated according to (47) from states $X_{i,0}$ to $X_{i-1,0}$ (see Figure 7). Intensity "$p\lambda$" is the arrival rate for states $X_{i,0}$ and $X_{i,1}$. Consequently, Little's formula has the form:

$$\bar{t} = \frac{k_{i(I)} + iP_{i,0} + (i+1)P_{i,1}}{p\lambda} + \frac{\sum\limits_{j=2}^{\infty} (j+i)P_{i,j}}{\lambda} \tag{51}$$

The average time that a vehicle spends in the system is largely, but not entirely, proportional to the average number of vehicles in the system.

Table 1. Cumulative probability calculation for 3D nomograph $i=5$, short lane capacity $i\in[1, 20]$, left turn probability $p\in[0.05, 0.40]$.

i	$p = 0.05$	$p = 0.10$	$p = 0.15$	$p = 0.20$	$p = 0.25$	$p = 0.30$	$p = 0.35$	$p = 0.40$
1	0.917	0.833	0.749	0.663	0.574	0.482	0.388	0.295
2	0.993	0.972	0.936	0.884	0.814	0.724	0.615	0.491
3	0.999	0.995	0.983	0.958	0.913	0.844	0.747	0.622
4	1.000	0.999	0.995	0.982	0.955	0.905	0.824	0.709
5	1.000	1.000	0.998	0.990	0.972	0.935	0.869	0.767
6	1.000	1.000	0.998	0.993	0.979	0.950	0.895	0.806
7	1.000	1.000	0.999	0.995	0.985	0.961	0.916	0.838
8	1.000	1.000	0.999	0.997	0.989	0.970	0.933	0.865
9	1.000	1.000	0.999	0.998	0.992	0.977	0.946	0.888
10	1.000	1.000	1.000	0.998	0.994	0.982	0.957	0.906
11	1.000	1.000	1.000	0.999	0.995	0.986	0.966	0.922
12	1.000	1.000	1.000	0.999	0.997	0.990	0.973	0.935
13	1.000	1.000	1.000	0.999	0.997	0.992	0.978	0.946
14	1.000	1.000	1.000	1.000	0.998	0.994	0.982	0.955
15	1.000	1.000	1.000	1.000	0.999	0.995	0.986	0.962
16	1.000	1.000	1.000	1.000	0.999	0.996	0.989	0.969
17	1.000	1.000	1.000	1.000	0.999	0.997	0.991	0.974
18	1.000	1.000	1.000	1.000	0.999	0.998	0.993	0.978
19	1.000	1.000	1.000	1.000	1.000	0.998	0.994	0.982
20	1.000	1.000	1.000	1.000	1.000	0.999	0.995	0.985

8. Discussion

Existing methods for the calculation of shared–short lanes capacity are based on simulation software such as Highway Capacity Manual (HCM), Sidra Intersection, VISIT, etc. These methods are predominantly designed for signalized intersections.

For unsignalized intersections there is a previous method based on solving approximately queuing system from Wu [36] which is also verified by simulation. Development of this method lead to the presented analytical solution of complex recursive operations noted by Wu [2], too.

Transportation planners and policy makers have at their disposal a new analytical method for calculation of shared/short lanes for left turn capacity. According to Koenigsberg [25] it should be more precise in calculation and more sensitive to parameter variations when making decisions on design and reconstruction of capacity of shared/short lanes for the left turn.

For this reason, we emphasize that the key role in capacity calculation for shared/short lanes has the intensity of left turn "μ". In theory, if this intensity is large or $\mu >> p\lambda$, needed capacity "i" of shared/short lanes for left turn converges to zero and all the needs are satisfied by the shared lane. However in the opposite case, capacity of shared/short lanes for left turn diverges for minimal values of "p" and "λ".This poses a question of justifiability of short lane for left turn design. Capacity with intensity "μ" is conditioned by density of priority (opposite) flow which can be calculated on standard unsignalized intersections [34] and non-standard unsignalized intersections [37]. Ultimately, intensity of left turn "μ" is an indirect product of traffic safety!

9. Conclusions

The significance of signalized intersection has proven to be more important than unsignalized intersections. Signalized intersections service a higher part of the traffic flow. This is the reason why the participation of signalized intersections in traffic safety and service quality is greater. The prevalent direction of researchers is justifiably in the direction of signalized intersections. For these reasons there are significant difference in solutions for signalized and unsignalized intersection occurs. If the fact that there are significantly more unsignalized intersections than signalized ones is taken into consideration, created difference can be declared as theoreticaldeficit.

Optimal dimensioning of short-shared line for left turn from priority direction is the predominant problem of unsignalized intersections. The proposed model is solved only for homogenous car flows, but can easily be adapted for heterogeneous flows with cyclists and pedestrians. For the generalization and calculation of heterogeneous system it is not necessary to change the structure of thequeuing system. Analytic background for adaption is in the well-known characteristic of adding up Poisson flows of different intensities which is proven in Raikov's theorem from 1937. Based on a stated theoretical base, left turns from non−priority directions of unsignalized intersections can be easily solved.

With the possibility of practical application in intersection planning, this paper has brought a new solution in queuing theory. So far, binomial distribution through its relation with Poisson distribution has been used in monophasic homogenous Markovian queuing systems. The presented solution brings a new vision for solving multiphase heterogeneous queuing systems in which a client selectively decides on the service. In a small corps of existing solutions, once again it has been confirmed that the queuing system can be solved if client selection has binomial distribution. Future research can integrate different approaches like multi-criteria decision making methods [38,39].

Author Contributions: Each author has participated and contributed sufficiently to take public responsibility for appropriate portions of the content.

Funding: This research was funded by the Ministry of Science and Technological Development of Serbia, grant number TR 36012.

Conflicts of Interest: The authors declare no conflict of interest.

References

1. Harders, J. *Die Leistungsfähigkeit Nicht Signalgeregelter Städtischer Verkehrsknoten*; Schriftenreihe "Straßenbau und Straßenverkehrstechnik": Heft, Germany, 1968; p. 76.
2. Wu, N. An approximation for the distribution of queue lengths at unsignalized intersections. In *Proceedings of the Second International Symposium on Highway Capacity*; Australian Road Research Board Ltd.: Victoria, Australia, 1994.
3. Li, Y.; Zhao, L.; Suo, J. Comprehensive assessment on sustainable development of highway transportation capacity based on entropy weight and TOPSIS. *Sustainability* **2014**, *6*, 4685–4693. [CrossRef]
4. Chand, S.; Gupta, N.J.; Velmurugan, S. Development of Saturation Flow Model at Signalized Intersection for Heterogeneous Traffic. *Transp. Res. Procedia* **2017**, *25*, 1662–1671. [CrossRef]
5. Tang, T.Q.; Yi, Z.Y.; Lin, Q.F. Effects of signal light on the fuel consumption and emissions under car-following model. *Physica A* **2017**, *469*, 200–205. [CrossRef]
6. Wang, Z.; Zhong, S.; He, H.-D.; Peng, Z.-R.; Cai, M. Fine-scale variations in PM2.5 and black carbon concentrations and corresponding influential factors at an urban road intersection. *Build. Environ.* **2018**, *141*, 215–225. [CrossRef]
7. Haight, F.A. Overflow at a traffic light. *Biometrika* **1959**, *46*, 420–424. [CrossRef]
8. Darroch, J. On the traffic-light queue. *Ann. Math. Stat.* **1964**, *35*, 380–388. [CrossRef]

9. Yang, Q.; Shi, Z. Effects of the design of waiting areas on the dynamic behavior of queues at signalized intersections. *Physica A* **2018**, *509*, 181–195. [CrossRef]
10. Hu, X.N.; Tang, L.C.; Ong, H.L. A M/Dx/1 vacation queue model for a signalized intersection. *Comput. Ind. Eng.* **1997**, *33*, 801–804. [CrossRef]
11. Ma, W.J.; Liu, Y.; Zhao, J.; Wu, N. Increasing the capacity of signalized intersections with left-turn waiting areas. *Transp. Res. Part A* **2017**, *105*, 181–196. [CrossRef]
12. Yang, Q.L.; Shi, Z.K.; Yu, S.W.; Zhou, J. Analytical evaluation of the use of left-turn phasing for single left-turn lane only. *Transp. Res. Part B* **2018**, *111*, 266–303. [CrossRef]
13. Heidemann, D.; Wegmann, H. Queueing at unsignalized intersections. *Transp. Res. Part B Methodol.* **1997**, *31*, 239–263. [CrossRef]
14. Nielsen, O.A.; Frederiksen, R.D.; Simonsen, N. Stochastic user equilibrium traffic assignment with turn-delays in intersections. *Int. Trans. Oper. Res.* **1998**, *5*, 555–568. [CrossRef]
15. Guell, D.L. Additional through lanes at signalized intersections. *J. Transp. Eng.* **1983**, *109*, 499–505. [CrossRef]
16. Wu, N. Total approach capacity at signalized intersections with shared-short lanes—A generalized model based on simulation study. *Transp. Res. Rec.* **2007**, 19–26. [CrossRef]
17. Tian, Z.Z.; Wu, N. Probabilistic model for signalized intersection capacity with a short right-turn lane. *J. Transp. Eng.* **2006**, *132*, 205–212. [CrossRef]
18. Chodur, J.; Ostrowski, K.; Tracz, M. Variability of Capacity and Traffic Performance at Urban and Rural Signalised Intersections. *Transp. Res. Procedia* **2016**, *15*, 87–99. [CrossRef]
19. Troutbeck, R.J.; Kako, S. Limited priority merge at unsignalized intersections. *Transp. Res. Part A* **1999**, *33*, 291–304. [CrossRef]
20. Yajima, M.; Phung-Duc, T. Batch arrival single-server queue with variable service speed and setup time. *Queueing Syst.* **2017**, *86*, 241–260. [CrossRef]
21. Twaddle, H.; Busch, F. Binomial and multinomial regression models for predicting the tactical choices of bicyclists at signalised intersections. *Transp. Res. Part F* **2019**, *60*, 47–57. [CrossRef]
22. Kolmogorov, A. Sur le problème d'attente. *Matematicheskii Sbornik* **1931**, *38*, 101–106. (In French)
23. Pollaczek, F. Ueber eine Aufgabe der Wahrscheinlichkeits theory. *Mathematische Zeitschrift* **1930**, *32*, 64–100. [CrossRef]
24. Kingman, J.F.C. The single server queue in heavy traffic. In *Mathematical Proceedings of the Cambridge Philosophical Society*; Cambridge University Press: Cambridge, UK, 1961; Volume 57, pp. 902–904. [CrossRef]
25. Koenigsberg, E. Is queuing theory dead? *Omega* **1991**, *19*, 69–78. [CrossRef]
26. Schwarz, J.A.; Selinka, G.; Stolletz, R. Performance analysis of time-dependent queuing systems: Survey and classification. *Omega* **2016**, *63*, 170–189. [CrossRef]
27. Whitt, W. A broad view of queuing theory through one issue. *Queueing Syst.* **2018**, *89*, 3–14. [CrossRef]
28. Kinzer, J.P. Application of the Theory of Probability to Problems of Highway Traffic. Doctoral Thesis, Polytechnic Press of the Polytechnic Institute of Brooklyn, New York, NY, USA, 1933.
29. Adams, W.F. Road Traffic Considered as A Random Series. *J. Inst. Civ. Eng.* **1936**, *4*, 121–130. [CrossRef]
30. Daganzo, C.F. Estimation of gap acceptance parameters within and across the population from direct roadside observation. *Transp. Res. Part B Methodol.* **1981**, *15*, 1–15. [CrossRef]
31. Troutbeck, R.J. *Estimating the Critical Acceptance Gap from Traffic Movements*; Physical Infrastructure Centre, Research Report; Queensland University of Technology: Brisbane, Australia, 1992; pp. 92–95.
32. Brilon, W.; Wu, N. Delays at fixed-time traffic signals under time-dependent traffic conditions. *Traffic Eng. Control* **1999**, *31*, 623–631.
33. Alfa, A.S.; Neuts, M.F. Modelling vehicular traffic using the discrete time Markovian arrival process. *Transp. Sci.* **1995**, *29*, 109–117. [CrossRef]
34. Tanackov, I.; Deretić, N.; Bogdanović, V.; Ruškić, N.; Jović, S. Safety time in critical gap of left turn manoeuvre from priority approach at TWSC unsignalized intersections. *Physica A* **2018**, *505*, 1196–1211. [CrossRef]
35. Lakkundi, V.R.; Park, B.; Garber, N.J.; Fontaine, M.D. *Development of Left-Turn Lane Guidelines for Signalized and Unsignalized Intersections Final Report*; Mid-Atlantic Universities Transportation Center: Charlottesville, VA, USA, 2004.
36. Wu, N. Capacity of shared-short lanes at unsignalized intersections. *Transp. Res. Part A* **1999**, *33*, 255–274. [CrossRef]

37. Bogdanović, V.; Ruškić, N.; Basarić, V.; Tanackov, I. Capacity analysis procedure for Four-Leg Non-Standard Unsignalised intersections. *Traffic Transp.* **2017**, *29*, 543–550. [CrossRef]
38. Liu, F.; Aiwu, G.; Lukovac, V.; Vukic, M. A multicriteria model for the selection of the transport service provider: A single valued neutrosophic DEMATEL multicriteria model. *Decis. Mak. Appl. Manag. Eng.* **2018**, *1*, 121–130. [CrossRef]
39. Pamučar, D.; Lukovac, V.; Božanić, D.; Komazec, N. Multi-criteria FUCOM-MAIRCA model for the evaluation of level crossings: Case study in the Republic of Serbia. *Oper. Res. Eng. Sci.* **2018**, *1*, 108–129. [CrossRef]

© 2019 by the authors. Licensee MDPI, Basel, Switzerland. This article is an open access article distributed under the terms and conditions of the Creative Commons Attribution (CC BY) license (http://creativecommons.org/licenses/by/4.0/).

Article

Evaluation of the Influencing Factors on Job Satisfaction Based on Combination of PLS-SEM and F-MULTIMOORA Approach

Abteen Ijadi Maghsoodi [1,*], Iman Azizi-ari [1], Zahra Barzegar-Kasani [2], Mehdi Azad [3], Edmundas Kazimieras Zavadskas [4] and Jurgita Antucheviciene [5]

[1] Department of Industrial Engineering, Science and Research Branch, Islamic Azad University, Tehran 1477893855, Iran; imanazizi65@gmail.com
[2] Department of Management, Central Tehran Branch, Islamic Azad University, Tehran 1477893855, Iran; zahrabarzegar18@yahoo.com
[3] Department of Humanities, Faculty of Humanities and Psychology, Shahid Chamran University, Ahvaz 6135743136, Iran; azadmehdi@ymail.com
[4] Institute of Sustainable Construction, Vilnius Gediminas Technical University, LT-10223 Vilnius, Lithuania; edmundas.zavadskas@vgtu.lt
[5] Department of Construction Management and Real Estate, Vilnius Gediminas Technical University, LT-10223 Vilnius, Lithuania; jurgita.antucheviciene@vgtu.lt
[*] Correspondence: aimaghsoodi@srbiau.ac.ir; Tel.: +98-912-643-3448

Received: 6 December 2018; Accepted: 24 December 2018; Published: 28 December 2018

Abstract: A primary issue that is being discussed nowadays in organizations is continuous improvement of the organization itself, because the procedure of periodic evaluation is an important tool to maintain the improvement of the organization. An essential factor in any organization is the human resources as a key asset to guide organizations to sustain their competitive advantages by employing particular knowledge and skills to form a comprehensive and sustainable human resource management. Evaluation of job satisfaction has become a part of the strategic approach toward incorporating business policies and human resource actions in modern day organizations. The current research study presents a novel hybrid validation framework to evaluate and appraise the factors influencing job satisfaction based on the fuzzy MULTIMOORA approach and partial least-squares path modeling (PLS-PM). The proposed fuzzy MCDM technique and statistical method validate each other to present an optimal assessment of influencing factors in job satisfaction. Eventually, a real-world case study in regard to influential factors in job satisfaction has been suggested in this study, to show that the proposed framework is a practical and accurate method to tackle an assessment problem in a real-world application of influencing factors in job satisfaction in a cross-industrial multi-national construction and geotechnical engineering organization in Iran.

Keywords: job satisfaction; modelling in management engineering; sustainable human resource management; structural equation model (SEM); partial least-squares path modeling (PLS-PM); fuzzy multiple criteria decision-making (FMCDM); F-MULTIMOORA

---

1. Introduction

In today's competitive and dynamic atmosphere in organizations all around the world, satisfaction is one of the important issues that have been through a huge paradigm shift in the definition. Satisfaction has been defined in many different ways. "Fulfilment of a need or want", "a source or means of enjoyment" and "the quality or state of being satisfied" are defined in Webster dictionary, however, these are only a few of the definitions of satisfaction. Locke [1] described job satisfaction as a positive feeling and emotional attitude of individuals against their current job and employment status.

Spector [2] described job satisfaction as a general feeling about the job or as a related set of attitudes and approaches about different aspects of the job and employment status.

Job satisfaction is one of the most widely overviewed concepts that has been investigated in an industrial/organizational psychology research context, and hundreds of definitions of job satisfaction have been made since this day [3]. There are various variables and elements that can influence job satisfaction such as job salary, organization environment, and type of work activity. In this study, the significance of various factors for job satisfaction has been analyzed based on a real-world investigation of the job satisfaction concept. Before analyzing the effect of each factor on job satisfaction, an assessment of proposed variables was obtained. Whenever the subject of evaluation and ranking about a subject is discussed, one of the best solutions is Multiple Criteria Decision-Making (MCDM). These techniques are one of the popular methodologies which can help decision-makers rank their evaluation and assessment problems. The implication of MCDM methods in human resource management activities has been studied widely in the past few years [4–6].

The current research study is focused on investigating the effects of multiple independent factors influencing job satisfaction. The first phase of this study is a presentation of an MCDM methodology for evaluating factors affecting job satisfaction in order to measure the importance of each criterion. To achieve this target, after the determination of independent factors from experts comments and previous research, the Multi-Objective Optimization on the basis of Ratio Analysis plus the full multiplicative form (MULTIMOORA) has been implied based on a fuzzy data structure. In the second phase, after evaluating the effect of each factor on job satisfaction, a calculation of elements influencing the job satisfaction has been obtained by utilizing the Structural Equation Model (SEM). Ultimately, based on the proposed approaches and techniques, a real-world case study has been analyzed in a cross-industrial multinational company in Iran.

The current paper is structured as the following. Section 2.1 overviews a short survey on job satisfaction developments. Section 2.2 presents a short survey on applications and developments on MULTIMOORA approach. Section 2.3 discusses research gaps and contributions of the current research. Section 3 overviews the research methodology. Section 4 presents the application of the proposed framework in a real-world case study as well as finding and managerial discussions of the proposed case study, while Section 5 offers conclusions and recommendations for the further research.

2. Research Background and Literature Overview

2.1. Survey on Job Satisfaction and Proposed Evaluations

Job satisfaction is one of the most important direct measurements of effectiveness in order to obtain the productivity of employees considering their current jobs which concerns the continuous improvement [7]. Presumed job satisfaction has an enormous amount of behavioral consequences and huge concerns of various factors. Since job satisfaction has been introduced in the human resources literature, it has been theorized and applied by psychologists, organizational scientist, and management experts. Many theoretical frameworks and real-world applications and developments have been suggested in this regard. Metle [8] proposed an exploration of the relationship between job satisfaction and educational factors considering the field of education and the level of education in a private banking sector among Kuwaiti women employees utilizing need fulfillment theory and Herzberg's two-factor theory. Moreover, there are many studies which have analyzed the relationship between the institutional factors and job satisfaction [9,10].

Hann et al. [11] performed a secondary data analysis with a view to obtaining the relationship amongst job satisfaction, intentions to leave family practice and actually leaving among family physicians in the National Health Service (NHS) in Britain. Mansor et al. [12] analyzed with a view to investigating the job satisfaction among bankers in the Islamic financial institution in the eastern region of Malaysia. Elshout et al. [13] suggested a consolidated method design with quantitative and qualitative research structure to analyze the link between leadership style, employment satisfaction, and absenteeism in a mental health care institution in the Netherlands. Ruchman et al. [14] evaluated the job satisfaction of the Accreditation Council for Graduate Medical Education (ACGME)-approved diagnostic radiology programs. Frey et al. [15] proposed an experimental study and a dyadic field study to measure the influence of customer satisfaction on employee satisfaction and retention in professional services at two western European business schools, in which students are in business administration. Naqbi et al. [16] identified the evaluation factors of employee satisfaction with services of Human Resources (HR) Departments in the Fujairah Medical District (FMD) in the UAE.

Tansel and Gazîoğlu [17] investigated the job satisfaction considering managerial attitudes concerning employees and firm size in Britain. Matthies-Baraibar et al. [18] measured the employee satisfaction considering organizational development in the implementation of the European Foundation for Quality Management (EFQM) model in 30 healthcare organizations including hospitals, primary care and mental health providers in Osakidetza. Tso et al. [19] suggested an exploratory research on identifying elements influencing employee satisfaction in Chinese resource-based state-owned enterprises. Leder et al. [20] presented a measurement to assess the effect of office environment on employee satisfaction considering open-plan and private offices. Mathieu et al. [21] proposed an evaluation of the relationship between supervisory behavior, job satisfaction and organizational commitment to employee turnover in small and medium-sized enterprises as well as large enterprises. Jacobs et al. [22] explored the influence of internal communication and employee satisfaction on supply chain integration in China.

Barakat et al. [23] presented a measurement in order to evaluate the effect of corporate social responsibility on employee satisfaction in Brazil. Boddy and Taplin [24] investigated the relationship between job satisfaction and workplace psychopathy factor. Koklic et al. [25] suggested an examination of customer satisfaction with low-cost and full-service airline companies. Tarcan et al. [26] presented an assessment of the connection between burnout and job satisfaction among emergency health professionals in emergency services in two public hospitals in Turkey. Holmberg et al. [27] applied the two-factor theory in order to evaluate job satisfaction among Swedish mental health nursing personnel. Bae and Yang [28] proposed an assessment of the influential factors of family-friendly policies in job satisfaction and organizational commitment in Korea. Hafez et al. [29] examined the effect of talent management on employee retention and job satisfaction for personnel administration at Ain-Shams University in Egypt.

Hayes et al. [30] presented a comprehensive review of literature of factors contributing to nursing job satisfaction which was conducted between 2004 and 2009 to identify elements contributing to satisfaction for nurses working in acute hospital settings. Zhu [31] overviewed the concept of job satisfaction and the influential factors for the proposed concept in a-decade period. Furthermore, Hantula [32] suggested a literature review on job satisfaction and influencing factors. Özpehlivan and Acar [3] proposed and validated a multidimensional job satisfaction scale in different cultures which has been collected from well-known Turkish and Russian people in the business.

2.2. Survey on Applications and Developments on the MULTIMOORA Method

The multi-objective optimization on the basis of ratio analysis plus the full multiplicative form (MULTIMOORA) method is an efficient and robust technique that has been proposed by Brauers and Zavadskas [33]. The MULITMOORA method is established based on four phases, including three subordinate techniques and one dominant theory that combines the three subordinate ranks. This methodology has proven to be more robust and accurate ranks than traditional MADM methods and its previous structure, MOORA method. The robustness of the MULTIMOORA approach has been investigated for the first time by Brauers and Zavadskas [34] in a project management problem. Since the introduction of the MULTIMOORA approach, many research studies have utilized this methodology in various decision-making and evaluation problems.

Balezentis and Balezentis [35] suggested a comprehensive survey on the extensions of MULTIMOORA approach based on the fuzzy environment and group decision-making.

Zavadskas et al. [36] suggested a novel extension of the MULTIMOORA method by combining interval-valued intuitionistic fuzzy terms. Stanujkic et al. [37] suggested the extended MULTIMOORA approach based on a triangular fuzzy data set in order to solve a communication circuits design selection problem. Awasthi and Baležentis [38] presented a combinative methodology using BOCR criteria integrated with F-MULTIMOORA approach based on a Monte Carlo simulation sensitivity analysis, in order to determine the robustness of the proposed method to variation in criterion and decision-maker weights in a logistics service provider selection. Stanujkic et al. [39] suggested an extended MULTIMOORA method integrated with single-valued neutrosophic terms that was resulted as an efficient methodology in terms of solving complex predictive problems. The extension was applied by Zavadskas et al. [40]. Gou et al. [41] presented an application of the MULTIMOORA approach combined with a hesitant fuzzy linguistic structure as well as a double-hierarchy hesitant fuzzy linguistic structure in a selection of the optimal city in China with evaluation of the implementation status of haze controlling measures.

Ceballos et al. [42] proposed a comparison of rankings calculated by F-MULTIMOORA, F-TOPSIS, F-VIKOR, and F-WASPAS to answer the primary question which occurs in every MCDM problem, which is "which method should be used to solve the ranking problem", and the result of this question is still open. Stević et al. [43] applied a hybrid novel MCDM methodology including DEMATEL, EDAS, COPRAS and MULTIMOORA methods in a supplier selection problem in regard to a construction organization. Maghsoodi et al. [44] suggested an application of the MULTIMOORA approach integrated with Shannon's entropy in ranking and selecting the best performance appraisal method in an organization in Iran. Maghsoodi et al. [45] proposed a combinative approach called cluster analysis for improving multiple criteria decision analysis (CLUS-MCDA) including the k-means clustering technique and MULTIMOORA method in order to tackle a big data supplier selection problem in an ICT company in Iran. Hafezalkotob et al. [46] presented a hybrid decision support system based on a combination of BWM approach with target-based MULTIMOORA and WASPAS applied in an olive harvester machines and equipment selection problem. Chen et al. [47] applied the MULTIMOORA method along with linguistic evaluations in order to form a multi-attribute group decision-making method which was applied in a wastewater treatment assessment problem.

2.3. Research Gap and Contributions of the Current Study

There are many studies that have analyzed and investigated the influencing factors on job satisfaction. In almost all of the previous studies regarding the factors influencing job satisfaction, no more than four factors have been analyzed at the same time. Additionally, to the best of the authors' knowledge, no single study has considered a framework based on a fuzzy MCDM approach combined with a statistical approach alongside. Ultimately, no single research study has conducted an MCDM approach in an assessment and evaluation problem regarding influencing factors of job satisfaction. Specifically, there is not a single study that shows an application of the F-MULTIMOORA method in an evaluation considering influencing elements of job satisfaction, which means the current study proposed a novel application of the F-MULTIMOORA method in order to assess the influential factors of job satisfaction and to validate a statistical method along with multiple other contributions. The focus of this paper is to provide a framework based on an F-MULTIMOORA approach, and PLS-SEM-based CFA, in order to evaluate the influence measurement of each factor affecting job satisfaction with respect to nine factors. In this study, in order to present a clear image of the procedure, a real-world case study has been utilized which is a cross-industrial multinational company in Iran. A questionnaire has been made by the researchers and experts in the mentioned organization to gather the related data of computation in statistical phase. To be fairly concise, the current research study presents a comprehensive proposition of a practical application considering consolidation of a fuzzy MCDM method and a statistical approach to validate both approaches in a real-world organization problem.

3. Research Methodology

3.1. F-MULTIMOORA Approach

Brauers and Zavadskas [33] developed the multi-objective optimization by ratio analysis (MOORA) which was composed from the combination of the ratio system and the reference point approach, later on, the full multiplicative form added to the MOORA technique in order to form an extended version of the MOORA approach, called MULTIMOORA method which is a more robust procedure for ranking and assessment of similar complex problems [34]. The F-MULTIMOORA approach has been suggested by Brauers et al. [48] that has been implied in the current study. A feasible solution to tackle the uncertainty in the modeling problems is the implication of the fuzzy sets and fuzzy logic. Zadeh [49] presented and developed the fuzzy set theory as an extension of the classical set theory due to the uses of natural language application of the exact data, which is impossible in real-world problems. The current study utilized a triangular fuzzy number (TFN). It should also be noted that while using such data structure can be useful for utilizing fuzzy sets, there are also other approaches that can be utilized with MADM approaches. For example, utilization of adaptive neuro-fuzzy system can be very useful in projection and prediction of decision-making problems [50–53].

Consequently, the fuzzy set theory can be employed on these specified problems considering uncertain environments. The first phase in the F-MULTIMOORA approach is forming the fuzzy decision matrix $\widetilde{X}$, where $\widetilde{x}_{ij} = (x_{ij1}, x_{ij2}, x_{ij3})$ presents the performance index of ith alternative respecting j^{th} attribute $i = 1, 2, \ldots, m$ and $j = 1, 2, \ldots, n$, and the $\widetilde{x}_{ij}$ values are aggregated by using a fuzzy weighted averaging (FWA) operator, i.e. Equation (2), where $\widetilde{\Psi}_k$ denotes the fuzzy coefficient of significance for the kth decision-maker:

$$X = [\widetilde{x}_{ij}]_{m \times n'} \tag{1}$$

$$\widetilde{x}_{ij} = \frac{\sum_{k=1}^{K} \widetilde{\Psi}_k \widetilde{x}_{ij}^k}{\sum_{k=1}^{K} \widetilde{\Psi}_k}, \tag{2}$$

The calculation parameters of the F-MULTIMOORA technique have to be without dimensions in order to compare the performance indices. Consequently, the fuzzy decision matrix is a normalization ratio based on a comparison amongst responses of an alternative to a criterion as a numerator, and a denominator that is a representative for all alternative performances on that attribute. The dimension dominator is performed by comparing the appropriate values of fuzzy numbers:

$$\widetilde{x}_{ij}^* = \left(\widetilde{x}_{ij1}^*, \widetilde{x}_{ij2}^*, \widetilde{x}_{ij3}^* \right) = \left(\frac{x_{ij1}}{\sqrt{\sum_{i=1}^{m} x_{ij1}^2}}, \frac{x_{ij2}}{\sqrt{\sum_{i=1}^{m} x_{ij2}^2}}, \frac{x_{ij3}}{\sqrt{\sum_{i=1}^{m} x_{ij3}^2}} \right), \; for \; all \; i, j, \tag{3}$$

where $\widetilde{x}_{ij}^*$ signifies the normalized performance index of ith alternative with respect to j^{th} attribute $i = 1, 2, \ldots, m$ and $j = 1, 2, \ldots, n$. The proposed normalization approach is the most robust selection among various normalization methods considering the F-MULTIMOORA technique

3.1.1. The Fuzzy Ratio System

In the current assessment and selection approach, the fuzzy normalization structure, i.e., Equation (3), justifies the foundations of F-MULTIMOORA method as the fuzzy ratio system, and the fuzzy normalized performance indices are added in case of maximization and subtracted in the event of minimization according to mathematical calculations between fuzzy values:

$$\widetilde{y}_i^* = \sum_{j=1}^{g} \widetilde{x}_{ij}^* \ominus \sum_{j=g+1}^{n} \widetilde{x}_{ij}^*, \tag{4}$$

where g represents the objectives being maximized, $(n - g)$ indicates the objectives being minimized, and $\widetilde{y}_i^*$ shows the total fuzzy assessment of alternative i which can be positive or negative based on the totals of calculations. Each ratio $\widetilde{y}_i^* = (\widetilde{y}_{i1}^*, \widetilde{y}_{i2}^*, \widetilde{y}_{i3}^*)$ is defuzzified by applying the best non-fuzzy performance value BNP_i of the ith alternative:

$$BNP_{\widetilde{y}_i^*} = (\frac{(\widetilde{y}_{i3}^* - \widetilde{y}_{i1}^*) + (\widetilde{y}_{i2}^* - \widetilde{y}_{i1}^*)}{3} + \widetilde{y}_{i1}^*), \tag{5}$$

The optimal alternative of the fuzzy ratio system A_{FRS}^* is an ordinal ranking of the $BNP_{\widetilde{y}_i^*}$ that has the maximum (or highest) assessment value:

$$A_{FRS}^* = \{A_i | max_i BNP_{\widetilde{y}_i^*}\}. \tag{6}$$

3.1.2. The Fuzzy Reference Point Approach

The second phase of the F-MULTIMOORA method is established based on the fuzzy normalization technique, as demonstrated in Equation (3). A maximal fuzzy objective reference point is also included in the method which is obtained as the following:

$$\widetilde{r}_j = \begin{cases} \widetilde{x}_j^+ = (max_i x_{ij1}^*, max_i x_{ij2}^*, max_i x_{ij3}^*) in\ case\ of\ maximization\ (j \leq g) \\ \widetilde{x}_j^+ = (min_i x_{ij1}^*, min_i x_{ij2}^*, min_i x_{ij3}^*)\ in\ case\ of\ minimization\ (j > g) \end{cases}, \tag{7}$$

where $\widetilde{r}_j$ signifies the ith co-ordinate of the fuzzy maximal objective reference point vector.

Deviation of a performance index from the reference point r_j can be obtained by calculation of the distance between the fuzzy values of $\widetilde{r}_j$ and $\widetilde{x}_{ij}^*$. Subsequently, the maximum value of the deviation of each alternative z_i^* can be obtained as follows:

$$\widetilde{z}_i^* = max_j | d(\widetilde{r}_j, \widetilde{x}_{ij}^*) |, \tag{8}$$

Moreover, the optimal alternative A_{FRP}^* in the fuzzy reference point method can be calculated by ranking the lowest assessment value $\widetilde{z}_i^*$ which is demonstrated in Equation (9):

$$A_{FRP}^* = \{A_i | min_i\ \widetilde{z}_i^*\}. \tag{9}$$

3.1.3. The Fuzzy Full Multiplicative Form

The third phase of the MULTIMOORA approach is extended by Brauers and Zavadskas [34] based on an idea in economic mathematics. Later on, the fuzzy form of the proposed technique was suggested by Brauers et al. [48]. The crisp format of the full multiplicative form can be demonstrated as showed in Equation (10), where g denotes the objectives being maximized and $(n - g)$ indicates the objectives being minimized. The numerator of Equation (10) indicates the product of performance indices of ith alternative related to beneficial attributes. The denominator of Equation (10) shows the product of performance indices of ith alternative regarding non-beneficial attributes with respect to the weights of the subjective coefficients w_j^s (if available):

$$U_i' = \frac{\prod_{j=1}^{g} (x_{ij})^{w_j^s}}{\prod_{j=g+1}^{n} (x_{ij})^{w_j^s}}, \tag{10}$$

The overall utility of the ith alternative can be presented as dimensionless values by employing the division of a fuzzy number in order to form the fuzzy full multiplicative form assessment value:

$$\widetilde{U}_i' = \widetilde{A}_i \oslash \widetilde{B}_i, \tag{11}$$

$$\widetilde{A}_i = (A_{i1}, A_{i2}, A_{i3}) = \prod_{j=1}^{g} \widetilde{x}_{ij}^*, \ i = 1, 2, \ldots m \ and \ g = 1, 2, \ldots n, \tag{12}$$

$$\widetilde{B}_i = (B_{i1}, B_{i2}, B_{i3}) = \prod_{j=g+1}^{n} \widetilde{x}_{ij}^*, \ i = 1, 2, \ldots m \ and \ g = 1, 2, \ldots n, \tag{13}$$

where $\widetilde{A}_i$ describes the product of objectives of the ith alternative to be maximized with g, being the number of objectives (structural indicators) to be maximized, and $\widetilde{A}_i$ denotes the product of objectives of the ith alternative to be minimized with $n - g$ value being the number of objectives (structural indicators) to be minimized. Each value of $\widetilde{U}_i^{*\prime} = (\widetilde{U}_{i1}^{*\prime}, \widetilde{U}_{i2}^{*\prime}, \widetilde{U}_{i3}^{*\prime})$ is defuzzified by applying the best non-fuzzy performance value BNP_i of the ith alternative:

$$BNP_{\widetilde{U}_i^{*\prime}} = (\frac{(\widetilde{U}_{i3}^{*\prime} - \widetilde{U}_{i1}^{*\prime}) + (\widetilde{U}_{i2}^{*\prime} - \widetilde{U}_{i1}^{*\prime})}{3} + \widetilde{U}_{i1}^{*\prime}), \tag{14}$$

The optimal alternative A_{FMF}^* in the fuzzy full multiplicative form is obtained based on the searching for the highest value among all assessment values of $BNP_{\widetilde{U}_i^{*\prime}}$ as demonstrated:

$$A_{FMF}^* = \left\{ A_i \middle| max_i \ BNP_{\widetilde{U}_i^{*\prime}} \right\}. \tag{15}$$

3.1.4. The Dominance Theory

The dominance theory was developed as a methodology for ranking subordinate alternatives in the MULTIMOORA approach [54,55]. After the computation of each subordinate technique rankings, they can be integrated into a final rank, which is the final result of the MULTIMOORA approach. In dominance theory, a summary of the classification of the three MULTIMOORA methods (which are the fuzzy values in the current study) is made based on cardinal and ordinal scales, where rankings rules should be applied including dominance, transitivity and equability [54]. Theory of dominance can be discussed as: (1) "the plurality rule assisted with a kind of lexicographic method", and (2) "the method of correlation of ranks", which has been described by Brauers and Zavadskas [54]. For a more detailed explanation of the dominance theory, refer to study of Brauers and Zavadskas [54].

3.2. Structural Equation Modeling (SEM) Based on Partial Least Squares (PLS)

A popular approach to statistic modeling procedures, which is used in an enormous array of complex modeling areas, is path analysis developed by Wright [56]. Path analysis is one of the special cases of the SEM which is a structure from an extensive set of mathematical and statistical algorithms and models that contains confirmatory factor analysis, path analysis, partial least squares path modeling, and latent growth modeling. In this study, components-based SEM, or partial least squares path modeling (partial least-squares path modeling (PLS-PM) or PLS-SEM) has been utilized that allows calculating complex cause–effect relationship models with latent variables. SEMs are multiple-equation regression models considering the response variable in one regression equation can present an explanatory variable in another equation, which contains two main variables that have effects on each another, reciprocally, directly, or indirectly, considering a feedback loop [57]. On the other hand, SEMs are known as a tool to analyze multivariate problems; this model goes beyond ordinary regression models to subsume multiple independent and dependent variables along with hypothetical latent constructs that clusters of observed variables might represent [58]. A review about applications of SEM to solve sustainability problems is presented [59].

Before describing the SEM procedure and related factors, there are few classes of variables that need to be explained. Starting with endogenous variables y_α, and y_β that are the response variables of the model, and exogenous variables x_α and x_β, which are determined outside of the model, it means there are only explanatory variables in the structural equations, and structural errors ε_α, and ε_β denote the aggregated omitted causes of the endogenous variables and possibly intrinsic randomness in the endogenous variables. There can be two types of structural coefficients: structural coefficients of an exogenous variable on an endogenous variable which is denoted as $\gamma_{\alpha\alpha}$, and structural coefficients of an endogenous variable on another endogenous variable $\beta_{\alpha\beta}$. The same rules is also applicable for covariance, i.e., two exogenous variables $\sigma_{\alpha\beta}$ and two error variables $\sigma_{\alpha\beta}$. The SEM of a specified model can be read directly from the path diagram but sometimes it is useful (e.g. for generality) to cast a structural-equation model in an equation form, which can be presented as follow:

$$B_{(n\times n)}y_{i_{(n\times 1)}} + \Gamma_{(n\times m)}x_{i_{(m\times 1)}} = \varepsilon_{i_{(n\times 1)}}. \tag{16}$$

where there are n endogenous variables, also n errors and m exogenous variables, and B and Γ are matrices of structural coefficients.

As mentioned earlier, PLS-PM was applied in the current study. The PLS-PM approach aims to define a system of weights to be applied to each variable, i.e., endogenous or exogenous variables. Optimization criteria behind the PLS-PM approach based on the outward directed links and inward directed links can be written as a general form, which is demonstrated as follows:

$$
\begin{aligned}
argmax_{w_n} &= \{\textstyle\sum_{n\neq m} c_{nm}g(cov(x_nw_n, x_mw_m))\} = \\
argmax_{w_n} &= \{\textstyle\sum_{n\neq m} c_{nm}g[cor(x_nw_n, x_mw_m)\sqrt{var(x_nw_n)}\sqrt{var(x_mw_m)}\}, \\
st. &\parallel x_nw_n \parallel^2, \parallel w_n \parallel^2. \\
where\ c_{nm} &= \begin{cases} 0 & if\ x_n\ and\ x_m\ are\ conected \\ 1 & otherwide \end{cases}, \\
g &= \begin{cases} Square & Factorial\ scheme \\ Abolute\ Value & centeroid\ scheme \end{cases}.
\end{aligned}
\tag{17}
$$

where w_n denotes the outer weights of the nth variable, and x_n presents the latent variable. As aforementioned in the current study, a factor analysis is utilized by this study. To be precise, confirmatory factor analysis (CFA) was used; CFA is a form of factor analysis where it is used to test to which extent a construct or a factor is consistent with the proposed understanding of the nature of the mentioned construct or factor. Before applying a CFA, a hypothesis has to be developed which is about the underlying measures of proposed hypothesis; by implying these constraints, the proposition has forced the model to be consistent with the theory [60]. Furthermore, Exploratory Factor Analysis (EFA) determines the nature and number of latent variables that account for observed variation and covariation among a set of observed elements [60]. Two types of CFA output are available in general, which are: (a) unstandardized and (b) standardized versions. The unstandardized form predicts scale-sensitive original item response which is very useful to compare solutions across groups or time, demonstrated in Equation (18). Consequently, when the solution is transformed to $var(y_i) = 1$ and $var(F) = 1$, CFA could be obtained in a standardized form which is useful when comparing items within a solution and on the same scale, measurement model per item ($i = 1, 2, \ldots n$) for subject n obtain as follow:

$$x_{in} = \mu_i + \lambda_i F_n + \varepsilon_{in}, \tag{18}$$

where x_{in} denotes the main component or variable, μ_i is the item's intercept factor, λ_i is the factor loading, $F_n = var(x_{in})_n$ is the factor variance, and ε_{in} demonstrates the error variance. To sum up, in the current study, an SEM model based on the PLS-PM (or PLS-SEM) logic has been applied considering CFA reasoning. The suggested approach is a path analysis proposition. Figure 1 illustrates a conceptual SEM considering a PLS path model based on the CFA approach.

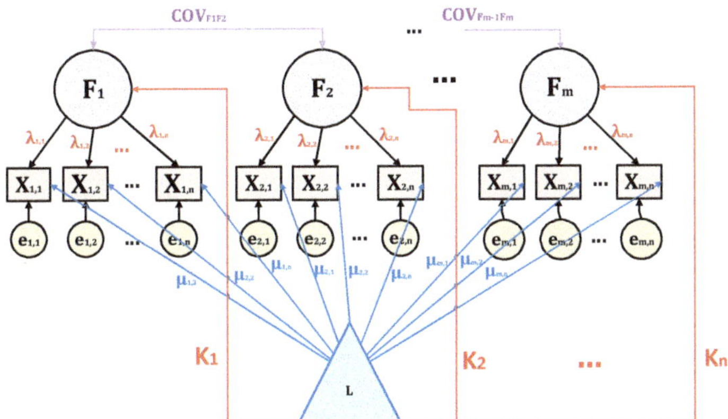

Figure 1. A conceptual CFA model based on PLS-SEM including factor means and item intercepts.

To describe how well the statistical model fits a set of observations (between observed values and the values expected in the model), the global model fit needs to be calculated. Table 1 describes the global model fit indices for SEM-based PLS-SEM considering CFA.

Table 1. Indices of global model fit.

Fit Indices	Statistical Notation	Formula	
Construct Reliability	CR	$CR = \dfrac{\sum (\lambda_{n,m})^2}{\sum (\lambda_{n,m})^2 + \sum (\mathrm{Var}(\varepsilon_{n,m}))}$	(19)
Average Variance Extracted	AVE	$Ave = \dfrac{\sum (Communalities)^2}{n}$	(20)
Blindfolding Criteria Index	F2(A/B)	$F^2\left(\dfrac{A}{B}\right) = \dfrac{R^2_{included} - R^2_{excluded}}{1 - R^2_{included}}$	(21)
Goodness of Fit	GOF	$GOF = \sqrt{Communality * R^2}$	(22)
Coefficient of Determination	R2	$R^2 = 1 - \left(\dfrac{L(0)}{L(\alpha)}\right)^{2/n}$	(23)

4. A Case Study: Evaluation of Factors Influencing Job Satisfaction in an Organization

The type of this study is a practical and validation experiment, regarding its objective as a presentation of a novel procedure in order to evaluate a hybrid framework, i.e., statistical approach and FMCDM technique. However, the proposed algorithm could be also extended based on a specific input data structure in terms of mathematical calculations. Considering the content and data collection, this study is descriptive and quantitative, and the type of review is a real-world case study of an organization in Iran. Given that the success of assessment and evaluation of the influencing factors of job satisfaction in an organization is taken from the perspective of the human resource manager, high-level managers, and strategic management, the study population included the specialists, experts, and officials, which are concerned by the evaluation of the influencing factors of job satisfaction in a multi-national cross-industrial company with the major activities in construction and transportation infrastructure, along with technical consultancy of geotechnical management in Iran.

A set of criteria was collected in order to process the first stage of the current study, which is application of the FMCDM method, i.e. F-MULTIMOORA, in assessing the influencing factors of job satisfaction, which was identified and collected from the previous research and expert comments that are classified in Table 2, in which the description of each criterion is available. Table 3 shows the candidate alternatives, i.e., influencing factors which were selected to be included in the ranking procedure to obtain the degree of importance based on the calculated final ranking.

Table 2. Criteria definitions for evaluating the influencing factors in job satisfaction.

ID	Criteria	Functional Requirement	Description
C1	Organizational Strategies	Maximum	The definition and description of job satisfaction could be different based on the organization's practical strategies, in which the core concept of job satisfaction is defined in the organization's predetermined objectives and vision. Therefore, it is crucial to include the idea of job satisfaction in the functional strategy.
C2	Importance of the Influencing Criteria	Maximum	In order to rank and present a comprehensive assessment of each influencing factor on job satisfaction in the current study, instead of using the weighting technique in the MCDM approach, specific criteria have been assigned in the ranking procedure.
C3	Cost of Improvement	Minimum	Cost improvement is an important factor in evaluating the influencing factors in job satisfaction, and is the cost of the improvement. Because it is critical for every organization to calculate how much it takes to improve the specific deficiencies.
C4	Comprehensiveness of Criteria (All Levels)	Maximum	To rank the influencing factor in job satisfaction with an MCDM procedure, it is important to choose the right factors. This criterion evaluates the measure of the accuracy of the suggested factor in every job level in the organization.
C5	The Degree of Difficulty	Minimum	The difficulty level of measuring the specific criteria is varied due to the influencing elements and sub-criteria of specific factors. Therefore, there might be few sub-criteria which could have an effect on the proposed factor and make the measurement difficult.
C6	Fitness of criteria with maturity level	Maximum	Criteria fitness with the maturity standards of the organization is one of the main reasons that a factor is selected among many influencing factors in the job satisfaction in the previous literature, which is based on the measurements of the maturity levels of each organization by itself.
C7	Compatibility with Organizational	Maximum	If an organization wants to survive or improve in the competitive environments, the periodic change should take place. To maintain the satisfaction of the employees, it is imperative to select the compatible influencing factors in job satisfaction.
C8	Comprehensiveness of Criteria (Project Level)	Maximum	The suggested case study in this research is an organization which has both the traditional and the project-based structure. This criterion evaluates the degree of accuracy of the suggested factor in the project structure job level in the organization.
C9	Physical and Mental Health Factors	Maximum	To obtain an accurate evaluation of an influencing factor in the job satisfaction, it is critical to include the physical and psychological health setting of the factor in the job satisfaction, and it is important to select the appropriate factor to maintain the physical and mental health factors.
C10	Managers Standards	Maximum	One of the important criteria for evaluating the influencing factors in job satisfaction is the opinions of the supervisors. There might be many factors which could have enormous effects on job satisfaction generally, but there are specific elements that would have the accuracy to evaluate the job satisfaction in different organizations.

Table 3. Influencing factors in job satisfaction, i.e., candidate alternatives for assessment.

ID	Influencing Factor	Description
A1	Organizational Commitment	One of the important indicators of advantages in an organization is the employees' commitment. The more employees are satisfied in an organization, the more commitment each employee shows in their behavior. Furthermore, abandonment of employees will have high costs for the organization which can be prevented by increasing the satisfaction levels of the employees.
A2	Leadership/Supervisory Method	From the beginning of the management science, leadership evolves many times. The method of supervisory is an important factor in maintaining the satisfaction of employees. By development in technology, the paradigm of the leadership has been changed and the expectations undergo many changes, which is a significant factor in improving and implying corrective actions.

Table 3. *Cont.*

ID	Influencing Factor	Description
A3	Job Security	With a view to protecting the employees from fluctuations of wage and salary, and to keeping the job positions safe, the factor of job security arises. Job security is a mental aspect which is directly connected to job satisfaction. Consequently, whenever the job security levels are high, it will have positive results on job satisfaction.
A4	Wage and Salary	To pay back the labor contribution in any type, wage and salaries have been raised. One of the significant elements in any organization for any workforce, in general, is the number of salaries and wage. Ultimately, there are many employees who are only motivated by the wage and salary factor in achieving higher job satisfaction.
A5	Job Stress	An important influencing factor in job satisfaction is the level of job stress. In order to survive in today's dynamic and competitive organizations, an employee will suffer from an enormous amount of stress. Job stress is one of the main reasons for the existence of the burnout concept, because there are many people who are willing to tolerate a high amount of stress in order to maintain a normal life quality.
A6	Individual Development Possibility	Personal growth is one of the key issues that a person attends a job position. A human being is an ideal creature which is continuously searching for self-development and possibilities to grow. Therefore, availability of the individual development possibility in an organization results in structuring a pleasing environment for the workforce.
A7	Amenities	In order to make positive enforcement in organizations, exclusive services are offered to employees. The amenities suggested to employees may vary considering different job positions. Consequently, lack of these specific services for employees is one of the reasons for abandonment in many organizations.
A8	Personnel Relationship	One of the features of a healthy person is their skills and abilities to communicate with other people. The work environment, in general, is the second home for many individuals because they spend most of their adult life in such environments. Therefore, lack of communication in the workplace results in a decrease in job satisfaction in general.
A9	Educational and Learning Opportunity	Availability of learning and educational opportunities is one of the important factors influencing job satisfaction in general. The reason is that every human being is searching for an opportunity to develop and grow. Therefore, if in a job position the educational and learning opportunity is absent, eventually the situation becomes impractical and purposeless.

Flyvbjerg [61] suggested that to employ in-depth research on any topic, "one can study only one case, and the result can be generalized." Ultimately, the mentioned case study in this research was not chosen randomly. It intended and targeted to select a specific organization to be able to obtain certain understandings that other organizations would not be able to offer which, in this case study, is collecting an input data, where validation of a statistical model and FMCDM model in a hybrid framework is meaningful. Moreover, expert judgments are valid enough to check the proposition of the mentioned framework. As aforementioned, the present study was structured from two main phases. First, the evaluation process of the influencing factors in job satisfaction was based on the F-MULTIMOORA approach. Second, the assessment procedure of the similar approach utilizd a statistical technique, i.e., PLS-SEM. Consequently, to validate the proposed techniques, a hybrid framework was implied. Ultimately, the final ranking results of the mentioned techniques were compared to the expert judgments in the real-world case study. The comparison was applied using the Spearman's correlation coefficients.

The necessary research data for the current research in this case study in the FMCDM phase were collected through interview using a Question & Answer (Q&A) approach based on a linguistic structure presented in Table 4. The primary objective of the Q&A approach was to complete the decision matrix which builds on the linguistic terms and the corresponding numbers in Table 4. The statistical approach was obtained by random sampling through a custom researcher-made questionnaire based on a population of 400 (N = 400), considering a reliable sample size of 200 (S = 200) concerning on sample size determination through the Krejcie and Morgan table. Characteristics of the respondents in both the statistical approach and FMCDM method regarding their education, occupation, and experience are high-level management employees, supervisors, and experts in human resource management. Additionally, the calculation process of the F-MULTIMOORA approach was computed by Microsoft Excel 2013, and in the statistical approach, the input data were processed with IBM SPSS 2015, and the SEM path modeling was computed with the SmartPLS software. To present a better understanding of the proposed validation framework which also evaluates the job satisfaction levels, Figure 2 illustrates a comprehensive discerption of the proposed method.

Table 4. Linguistic terms and the corresponding numbers.

Linguistic Term	Alphabetical Value of Verbal Comments	Numerical Value of Verbal Comments
Very Poor	VP	(1,1,1)
Poor	P	(1,2,3)
Moderate	M	(2,3,4)
Good	G	(3,4,5)
Very Good	VG	(4,5,6)

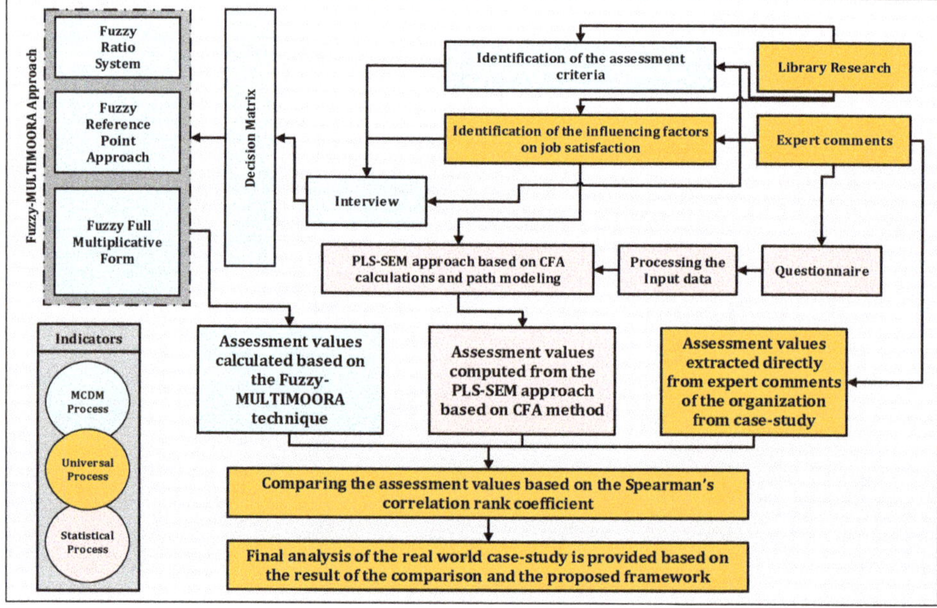

Figure 2. Flow diagram of the evaluation process regarding influencing factors of job satisfaction.

4.1. Finding and Results

The current section investigates the usability and accuracy of the proposed methodology that have been introduced in the previous section in a multinational cross-industrial organization in Iran. Accordingly, the fundamental purpose of identifying the alternatives and criteria, along with a linguistic term set that was described in Table 4, was to complete the fuzzy decision matrix based on the numerical values of the linguistic terms, which were obtained and presented in Table 5, based on the assessment criteria and candidate alternatives regarding the influencing job satisfaction factors and elements.

Table 5. The fuzzy decision matrix to evaluate the influencing factors in job satisfaction.

Influencing Factors	Criteria									
	C1 (Max)	C2 (Max)	C3 (Min)	C4 (Max)	C5 (Min)	C6 (Max)	C7 (Max)	C8 (Max)	C9 (Max)	C10 (Max)
A1	(2,3,4)	(4,5,6)	(3,4,5)	(4,5,6)	(1,1,1)	(3,4,5)	(1,2,3)	(3,4,5)	(1,2,3)	(3,4,5)
A2	(1,2,3)	(2,3,4)	(1,2,3)	(3,4,5)	(3,4,5)	(2,3,4)	(1,2,3)	(2,3,4)	(3,4,5)	(1,2,3)
A3	(4,5,6)	(3,4,5)	(1,2,3)	(4,5,6)	(2,3,4)	(2,3,4)	(3,4,5)	(3,4,5)	(4,5,6)	(1,2,3)
A4	(4,5,6)	(3,4,5)	(2,3,4)	(3,4,5)	(1,2,3)	(3,4,5)	(4,5,6)	(4,5,6)	(4,5,6)	(4,5,6)
A5	(1,2,3)	(3,4,5)	(3,4,5)	(2,3,4)	(3,4,5)	(2,3,4)	(2,3,4)	(1,2,3)	(2,3,4)	(3,4,5)
A6	(4,5,6)	(1,2,3)	(3,4,5)	(1,2,3)	(3,4,5)	(2,3,4)	(3,4,5)	(3,4,5)	(2,3,4)	(3,4,5)
A7	(1,2,3)	(1,1,1)	(3,4,5)	(4,5,6)	(1,2,3)	(3,4,5)	(2,3,4)	(3,4,5)	(4,5,6)	(3,4,5)
A8	(4,5,6)	(1,2,3)	(3,4,5)	(3,4,5)	(3,4,5)	(2,3,4)	(1,2,3)	(2,3,4)	(3,4,5)	(3,4,5)
A9	(4,5,6)	(1,2,3)	(4,5,6)	(3,4,5)	(1,2,3)	(2,3,4)	(2,3,4)	(3,4,5)	(3,4,5)	(2,3,4)

As mentioned before, to proceed the F-MULTIMOORA process, the decision matrix (which has different dimensions and measurement units) has to transform into a dimensionless matrix. The normalization procedure applied based on Equation (3) dominates the dimensions in the fuzzy decision matrix to compare numbers to each other. The normalized fuzzy decision matrix is demonstrated in Table 6.

Table 6. The normalized fuzzy decision matrix to evaluate the influencing factors in job satisfaction.

Influencing Factors	Criteria									
	C1 (Max)	C2 (Max)	C3 (Min)	C4 (Max)	C5 (Min)	C6 (Max)	C7 (Max)	C8 (Max)	C9 (Max)	C10 (Max)
A1	(0.22, 0.25, 0.27)	(0.56, 0.51, 0.48)	(0.36, 0.36, 0.35)	(0.42, 0.40, 0.39)	(0.15, 0.10, 0.08)	(0.42, 0.39, 0.38)	(0.14, 0.20, 0.23)	(0.35, 0.35, 0.35)	(0.10, 0.16, 0.20)	(0.36, 0.36, 0.35)
A2	(0.11, 0.17, 0.20)	(0.28, 0.30, 0.32)	(0.12, 0.18, 0.21)	(0.31, 0.32, 0.32)	(0.45, 0.43, 0.41)	(0.28, 0.29, 0.30)	(0.14, 0.20, 0.23)	(0.23, 0.26, 0.28)	(0.32, 0.33, 0.33)	(0.12, 0.18, 0.21)
A3	(0.44, 0.42, 0.41)	(0.42, 0.41, 0.40)	(0.12, 0.18, 0.21)	(0.42, 0.40, 0.39)	(0.30, 0.32, 0.33)	(0.28, 0.29, 0.30)	(0.42, 0.40, 0.39)	(0.35, 0.35, 0.35)	(0.43, 0.41, 0.40)	(0.12, 0.18, 0.21)
A4	(0.44, 0.42, 0.41)	(0.42, 0.41, 0.40)	(0.24, 0.27, 0.28)	(0.31, 0.32, 0.32)	(0.15, 0.10, 0.08)	(0.42, 0.39, 0.38)	(0.57, 0.51, 0.47)	(0.47, 0.44, 0.42)	(0.43, 0.41, 0.40)	(0.48, 0.45, 0.42)
A5	(0.11, 0.17, 0.20)	(0.42, 0.41, 0.40)	(0.36, 0.36, 0.35)	(0.21, 0.24, 0.26)	(0.45, 0.43, 0.41)	(0.28, 0.29, 0.30)	(0.28, 0.30, 0.31)	(0.11, 0.17, 0.21)	(0.21, 0.24, 0.26)	(0.36, 0.36, 0.35)
A6	(0.44, 0.42, 0.41)	(0.14, 0.20, 0.24)	(0.36, 0.36, 0.35)	(0.10, 0.16, 0.19)	(0.45, 0.43, 0.41)	(0.28, 0.29, 0.30)	(0.42, 0.40, 0.39)	(0.35, 0.35, 0.35)	(0.21, 0.24, 0.26)	(0.36, 0.36, 0.35)
A7	(0.11, 0.17, 0.20)	(0.41, 0.10, 0.08)	(0.36, 0.36, 0.35)	(0.42, 0.40, 0.39)	(0.15, 0.10, 0.08)	(0.42, 0.39, 0.38)	(0.28, 0.30, 0.31)	(0.35, 0.35, 0.35)	(0.43, 0.41, 0.40)	(0.36, 0.36, 0.35)
A8	(0.33, 0.34, 0.34)	(0.14, 0.20, 0.24)	(0.36, 0.36, 0.35)	(0.31, 0.32, 0.32)	(0.45, 0.43, 0.41)	(0.28, 0.29, 0.30)	(0.14, 0.20, 0.23)	(0.23, 0.26, 0.28)	(0.32, 0.33, 0.33)	(0.36, 0.36, 0.35)
A9	(0.44, 0.42, 0.41)	(0.14, 0.20, 0.24)	(0.48, 0.45, 0.42)	(0.31, 0.32, 0.32)	(0.15, 0.10, 0.08)	(0.28, 0.29, 0.30)	(0.28, 0.30, 0.31)	(0.35, 0.35, 0.35)	(0.32, 0.33, 0.33)	(0.24, 0.27, 0.28)

Consequently, the assessment values of each stage of the F-MULTIMOORA method were calculated based on the aforementioned procedures in Section 3.1. The assessment values and the final rankings of each stage as well as the dominance theory are shown in the Table 7.

Table 7. Assessment values and rankings of the F-MULTIMOORA regarding job satisfaction factors.

Influencing Factors	Assessment Values			Rankings			
	$BNP_{\tilde{y}_i^*}$	z_i^*	$BNP_{\tilde{u}_i^{*'}}$	FRS	FRP	FMF	Final Rank
A1	2.18823	0.32349	0.00235	3	8	3	3
A2	1.47133	0.27160	0.00021	8	3	8	8
A3	2.39508	0.27160	0.00392	2	2	2	2
A4	2.89269	0.30618	0.01636	1	4	1	1
A5	1.42310	0.25630	0.00015	9	1	9	9
A6	1.67263	0.30779	0.00035	6	6	6	6
A7	1.93577	0.41039	0.00058	4	9	5	5
A8	1.53960	0.30779	0.00028	7	7	7	7
A9	1.85043	0.30779	0.00084	5	5	4	4

As aforementioned, the PLS-SEM approach based on the CFA technique was utilized in the current study. Figure 3 illustrates the SEM model of the proposed case study based on nine influencing factors, which were identified considering expert comments, discusses the profiles of the corresponding employees, which are demonstrated in Table 8 and library research.

Table 8. Profile of the respondents related to the statistical procedure.

Demographic Items	Frequency	Percentile
Gender		
Male	200	100
Female	0	0
Mariel Status		
Single	112	56
Married	88	44
Age		
Less than 20	14	7
20–30	67	33.5
31–40	87	43.5
41–50	22	11
50 & Above	10	5
Job Level		
Employee	61	30.5
Expert	71	35.5
Supervisor	26	13
Manager	26	13
Top Manager	10	5
CEO	6	3

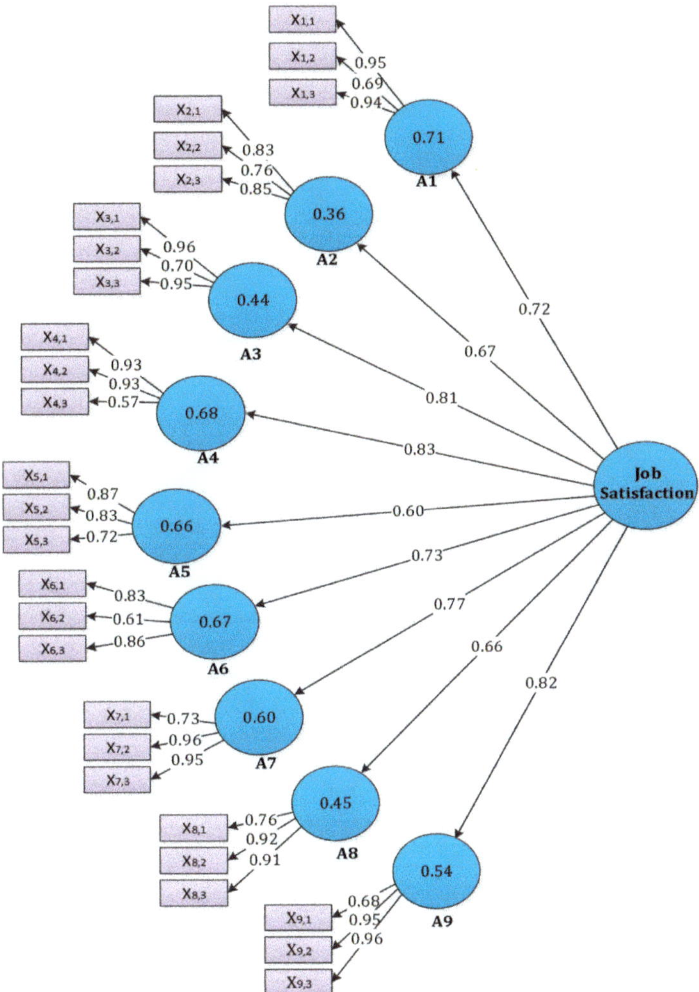

Figure 3. The influential factors in job satisfaction based on a PLS-SEM model.

Moreover, Table 9 demonstrates the reliability analysis, overview of the statistical factors and indices of the proposed model based on the PLS-SEM approach in the suggested case study.

Table 9. Construct measures and statistics for nine influencing factors in the job satisfaction.

Research Construct		Sample Mean	Standard Deviation	Standard Error	T Statistics	Cronbach's Alpha	C.R Value	AVE Value	Factor Loading
A1(F1)	$X_{1,1}$ $X_{1,2}$ $X_{1,3}$	0.71	0.05	0.05	13.85	0.84	0.91	0.77	0.95 0.69 0.94
A2(F2)	$X_{2,1}$ $X_{2,2}$ $X_{2,3}$	0.66	0.07	0.07	8.41	0.75	0.86	0.67	0.83 0.76 0.85

Table 9. *Cont.*

Research Construct			Sample Mean	Standard Deviation	Standard Error	T Statistics	Cronbach's Alpha	C.R Value	AVE Value	Factor Loading
Job Satisfaction	A3(F3)	$X_{3,1}$ $X_{3,2}$ $X_{3,3}$	0.8	0.07	0.07	9.52	0.85	0.91	0.78	0.96 0.7 0.95
	A4(F4)	$X_{4,1}$ $X_{4,2}$ $X_{4,3}$	0.82	0.03	0.03	23.66	0.74	0.86	0.69	0.93 0.93 0.57
	A5(F5)	$X_{5,1}$ $X_{5,2}$ $X_{5,3}$	0.6	0.04	0.04	22.83	0.73	0.85	0.65	0.87 0.83 0.72
	A6(F6)	$X_{6,1}$ $X_{6,2}$ $X_{6,3}$	0.72	0.03	0.03	26.63	0.71	0.84	0.63	0.83 0.61 0.86
	A7(F7)	$X_{7,1}$ $X_{7,2}$ $X_{7,3}$	0.76	0.05	0.05	15.69	0.85	0.92	0.79	0.73 0.96 0.95
	A8(F8)	$X_{8,1}$ $X_{8,2}$ $X_{8,3}$	0.65	0.08	0.08	8.23	0.84	0.90	0.76	0.76 0.92 0.91
	A9(F9)	$X_{9,1}$ $X_{9,2}$ $X_{9,3}$	0.81	0.05	0.05	15.86	0.83	0.90	0.76	0.68 0.95 0.96

Table 9 demonstrates the related factor loading values of the questions are higher than 0.40. On the other hand, the factor loading values are fluctuating between 0.57 and 0.96. Additionally, the results of the T test in every factor loading variable are meaningful and accurate lower than 0.01. Furthermore, based on Cronbach's [62], the acceptable value of alpha for model reliability is 0.70, in which in the current study the alpha value for every variable is more than 0.70. Consequently, the composite reliability (CR) values for variables are obtained as more than 0.70 in each variable. The average variance-extracted (AVE) values describe that the convergent validity of the variables is convenient. Ultimately, to calculate the discriminant validity, the cross loading (CL) test, along with Fornell–Larcker [63] test, has been applied. Table 10 demonstrates the CL factors of the current study, in which Qi is the indicator of the questions from the questionnaire used to gather the input data in the present study.

Table 10. Demonstration of the convergent validity of factors (the blue numbers are cross loading factors).

	A1	A2	A3	A4	A5	A6	A7	A8	A9
Q4	0.95	0.28	0.30	0.49	0.46	0.47	0.42	0.47	0.52
Q5	0.69	0.34	0.17	0.39	0.45	0.42	0.38	0.45	0.38
Q6	0.95	0.28	0.30	0.49	0.46	0.47	0.42	0.47	0.52
Q10	0.31	0.84	0.39	0.50	0.27	0.22	0.45	0.29	0.22
Q11	0.20	0.76	0.39	0.38	0.21	0.23	0.61	0.22	0.17
Q12	0.31	0.85	0.27	0.44	0.30	0.25	0.45	0.32	0.23
Q13	0.30	0.32	0.96	0.57	0.35	0.44	0.45	0.22	0.34
Q15	0.19	0.51	0.70	0.47	0.22	0.25	0.44	0.18	0.36
Q16	0.30	0.32	0.96	0.57	0.35	0.44	0.45	0.22	0.34
Q20	0.49	0.49	0.30	0.93	0.53	0.45	0.54	0.42	0.51
Q23	0.49	0.49	0.30	0.93	0.53	0.45	0.54	0.42	0.51
Q24	0.30	0.32	0.96	0.57	0.35	0.44	0.45	0.22	0.34
Q25	0.49	0.22	0.28	0.45	0.87	0.86	0.41	0.58	0.50
Q27	0.43	0.24	0.36	0.50	0.83	0.83	0.47	0.46	0.55
Q28	0.34	0.33	0.20	0.45	0.72	0.49	0.29	0.34	0.55
Q31	0.43	0.24	0.36	0.50	0.83	0.83	0.47	0.46	0.55
Q35	0.32	0.23	0.44	0.35	0.45	0.68	0.31	0.33	0.39

Table 10. *Cont.*

	A1	A2	A3	A4	A5	A6	A7	A8	A9
Q36	0.49	0.22	0.28	0.45	0.87	0.86	0.41	0.58	0.50
Q37	0.37	0.51	0.51	0.46	0.38	0.44	0.73	0.30	0.34
Q39	0.43	0.56	0.42	0.59	0.46	0.45	0.96	0.46	0.43
Q42	0.43	0.56	0.42	0.59	0.46	0.45	0.96	0.46	0.43
Q43	0.40	0.23	0.35	0.41	0.62	0.61	0.48	0.76	0.45
Q44	0.48	0.32	0.11	0.35	0.42	0.44	0.35	0.92	0.23
Q48	0.48	0.32	0.11	0.35	0.42	0.44	0.35	0.92	0.23
Q51	0.53	0.21	0.21	0.37	0.46	0.43	0.42	0.37	0.68
Q53	0.46	0.23	0.40	0.53	0.61	0.57	0.38	0.30	0.95
Q54	0.46	0.23	0.40	0.53	0.61	0.57	0.38	0.30	0.95

The crosswise factors and discriminant validity have been presented in Table 11, which demonstrates that the suggested SEM model achieved an acceptable validity. Table 12 describes the path coefficients which were calculated from the path analysis, in which the final rankings were calculated and ranked from the path coefficients.

Table 11. The discriminant validity of the factors.

	A1	A2	A3	A4	A5	A6	A7	A8	A9
A1	0.88								
A2	0.34	0.82							
A3	0.30	0.43	0.88						
A4	0.53	0.54	0.61	0.83					
A5	0.52	0.32	0.35	0.57	0.81				
A6	0.52	0.29	0.44	0.55	0.71	0.79			
A7	0.47	0.62	0.50	0.62	0.49	0.50	0.89		
A8	0.53	0.34	0.24	0.44	0.58	0.58	0.46	0.87	
A9	0.55	0.25	0.39	0.56	0.65	0.61	0.45	0.37	0.87

Table 12. The final results of the SEM approach regarding influencing factors in job satisfaction.

Influencing Factors	Path coefficients	T Statistics	R Square	Rejected/Supported	Final Rank
A1	0.72	13.85	0.71	Supported	6
A2	0.67	9.52	0.44	Supported	8
A3	0.81	22.83	0.66	Supported	3
A4	0.83	23.66	0.68	Supported	1
A5	0.60	8.41	0.36	Supported	9
A6	0.73	15.86	0.54	Supported	5
A7	0.77	15.69	0.60	Supported	4
A8	0.66	8.23	0.45	Supported	7
A9	0.82	26.63	0.67	Supported	2

Furthermore, based on the factor loadings that have been calculated to evaluate the influencing factors in job satisfaction, the wage and salary properties obtained the highest rank, which means it is the most important element to assess the job satisfaction in the case study. Consequently, as aforementioned, to test the proposed statistical algorithm, a goodness-of-fit (GOF) index is suggested by Tenenhaus et al. [64], considering PLS logic which has been demonstrated in Equation (23).

Moreover, the GOF index value was calculated to be 0.612, which means the GOF value obtained a high and acceptable in the current study, i.e., the present model consists of great fit index. Moreover, in order to validate the proposed framework in the current study and to compare the results of applied methods in the scenario of the real-world case study, a comparison of mentioned propositions is presented in Table 13.

Table 13. Comparison between rankings of the FMCDM approach, PLS-SEM method, and expert judgments.

Influencing Factors	Final Ranks		
	FMCDM: Fuzzy-MULTIMOORA	Statistical Approach: PLS-SEM	Expert Judgements
A1	3	6	3
A2	8	8	9
A3	2	3	8
A4	1	1	1
A5	9	9	5
A6	6	5	2
A7	5	4	6
A8	7	7	7
A9	4	2	4

The Spearman's rank correlation coefficient simplifies the evaluation process of the similarity of the rankings. A coefficient is a real number ranging between −1 and 1. The Spearman's coefficient equal to 1 denotes identical rankings and −1 indicates opposite rankings. Figure 4 illustrates the correlation between ranking lists by utilizing the Spearman's rank correlation coefficient, which is based on the rankings presented in Table 13. The correlation coefficient of the Spearman's rank based on similarity of rankings was calculated as follows:

$$r_s = 1 - \frac{\sum_{i=1}^{n} D^2}{n(n^2 - 1)} \tag{19}$$

where D is differences between the two ranks and n denotes the sample size.

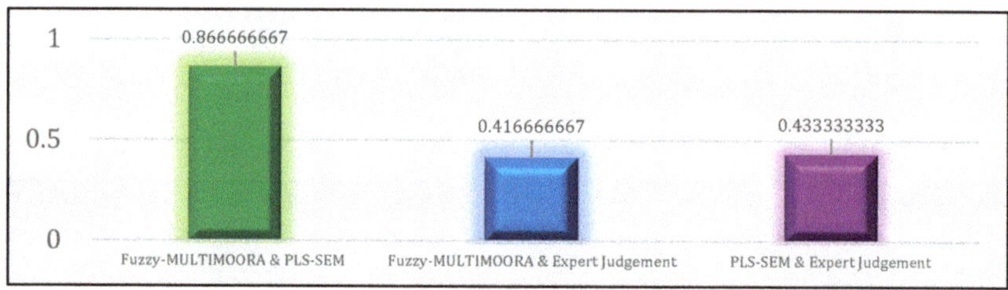

Figure 4. Correlation between the rankings based on the Spearman's correlation coefficient.

Figure 4 shows that the correlation between the FMCDM approach and the statistical technique proposed in the current study is 0.86, which is a high correlation compared to other ranks. Furthermore, the high correlation of the proposed method in this study shows the suggested framework, in order to validate the FMCDM method and the statistical approach at the same time, i.e., the proposed hybrid validation is accurate. It is also worth mentioning that a decent solution to such practical case can include preferences of every employee of the organization in the job satisfaction evaluation. In this case, while the items and factors of job satisfaction might be the same, the input data will change based on the mentioned structure. In such cases, group decision-making approaches combined with fuzzy ontologies and multi-granular linguistic modelling methods can become useful [65,66]. It is recommended to apply such methodologies and compare it to this study, in order to see if there is any correlation between high-level mangers comments and employees preferences in regard to factors influencing job satisfaction. Additionally, along with the abovementioned methodology, it is also suggested to use novel data-gathering approaches such as user experience questionnaires (UEQ-S) [67].

4.2. Discusssions and Managerial Implications

With a view to presenting a comprehensive analysis of the final result of the current study, which has been obtained based on the proposed algorithm using a combination of F-MULTIMOORA and a statistical approach in the suggested real-world case study, Table 14 demonstrates a detailed opportunity for improvement (OFI) plan based on the obtained evaluation of each influencing factor in the job satisfaction in the mentioned organization. Consequently, the proposed detailed analysis of each factor could play a vital role in improving the evaluation of influencing factors in job satisfaction, and improving the satisfaction of employees considering adjusting factors, and in planning a corrective action. To sum up, the final analyses of the case study, a sensitivity analysis based on the proposed methods, i.e., FMCDM, and the statistical approach are provided in Figures 5 and 6.

Table 14. Comprehensive analysis of the opportunity for improvement.

Influencing Factor in Job Satisfaction	Comprehensive Data Analysis	
	Root Cause Description	**Opportunity for Improvement**
Organization Commitment (A1)	Convergence and lobbying to recruit personnel, lack of clear goals in the organization, lack of performance appraisal system for mid-level managers, deficiency in reward and punishment system.	Employing talented and motivated personnel, job compliance establishment in the organization, creating a supportive atmosphere in the organization, and enhancing long-term payment systems.
Leadership (A2)	The main reason for the relative satisfaction of this factor in the organization is the performance evaluation centers evaluating the performance of the high-level managers and leaders of the organization. However, still, there is lack of satisfaction in the leadership in the project-based structure.	To keep the satisfaction levels in the current situation or enhance the satisfaction in employees considering leadership factors, the long term performance appraisal mechanisms should be planned and continued.
Job Security (A3)	In general, the high-level supervisors and managers usually have a high level of job security due to their key competencies. In lower level positions due to the lack of transparency in the organization, the job security is low.	Designing preventive mechanisms is one of the furthermost effective approaches towards preventing the fear of job loss in any organization level. Preventive mechanisms include: long-term contracts, long-term loans, converting the status of day-to-day employees to the contractor and contractors to the official.
Wage and Salary (A4)	Transparency in contracts is very unclear among high-level managers in the current organization. Also, payment quality and legal issues for mid-level managers and supervisors are very controversial. Furthermore, justice in calculating salaries is vague for the technicians and labors. Consequently, the obscurity of upstream rules in the organization results in low level of satisfaction of this factor.	This factor can be examined and improved from three different points of views: (a) justice—lack of discrimination to determine rights and to observe all laws; (b) transparency—with a view to developing transparency among all levels of organization, it is suggested to increase the knowledge level of employees considering their legal rights; (c) quality—ensuring the correct way of calculating rights.
Job Stress (A5)	The working pressures in this organization in most of the sectors are under control, except the project-based sector in which the job stress due to the working conditions is very high among project managers and experts.	In order to decrease the level of stress for every occupational level, it is suggested to plan a physio-mental analysis for every employee (which is not available in the current case study) to control the mental health of personnel.
Individual Development Possibility (A6)	Regarding management in the current case study, it appears that specialized training, personal growth and individual development do not matter, and individual development mechanisms are defective.	Designing competency systems and implying the proposed system in the organizations is an effective approach to increasing the personal growth and individual development for employees.
Amenities (A7)	In the current organization, welfare services have a low position in the employee's point of view. Furthermore, the management of these kinds of services is not important to be provided for the personnel.	Needs assessment of occupational levels and job positions is an accurate way of understanding the specified need of employees and offering the most required amenities to enhance satisfaction.
Personal Relationship (A8)	The workplace environment in the current organization is a single-sex environment, in which only male employees are working. Consequently, expectations of employees for having a productive and happy workplace decrease. Furthermore, the quality of life in such environments is deficient, due to the specific structure of the workplace.	To enhance the quality of the workplace environment, the culture and the core environment have to change. Therefore, it is suggested to plan a close-up time for employees in order to make more communications. Furthermore, designing a committee of practice to plan an effective communication for employees is also worthy.
Education and Learning Opportunity (A9)	Disagreement over ideas and comments among experts and management level is evident concerning this factor. Because a big part of the organization is project-based, the long-term planning for education and learning is considered as a waste in this organization, and in their point of view, training at the service seems enough.	It is suggested that with a long-term plan based on the job analysis of every job position, the specific needs of each job position are obtained in order to imply the empowerment through education and learning opportunities.

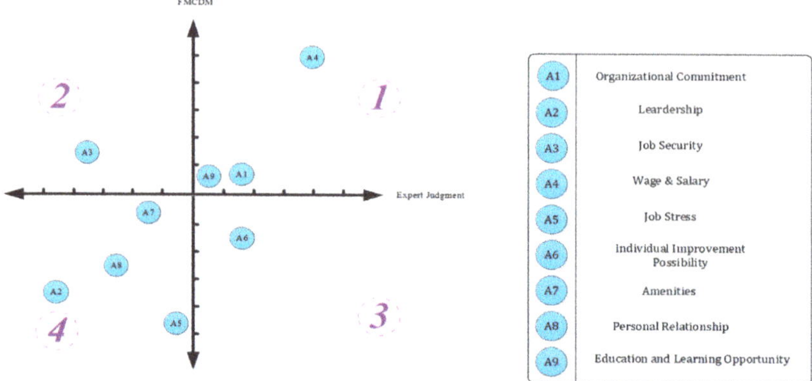

Figure 5. Sensitivity analysis of FMCDM approach and expert judgments.

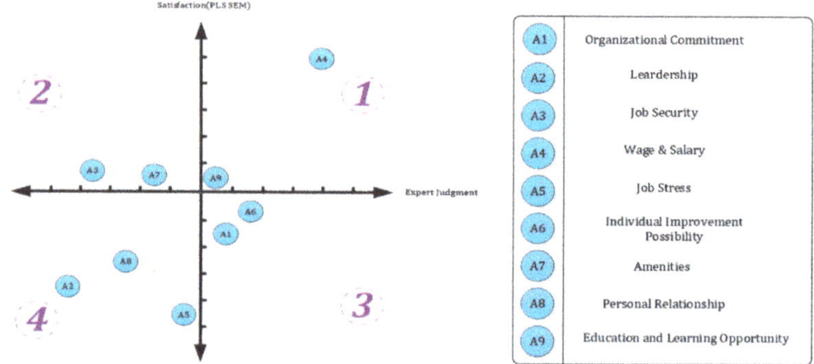

Figure 6. Sensitivity analysis of the statistical approach and expert judgments.

Based on Figures 5 and 6, four areas are identified and available in this analysis, in which area (1) demonstrates a high level of importance measures and a high level of impacts on satisfactions. The factors that are located in this area are dominant and supreme factors because any improvement in the factors in this area results in a huge amount of satisfaction from employees in the organization.

Area (2) shows that the factors located in this area have a high level of impacts on satisfactions, but based on expert comments, these factors have a lack of importance. Therefore, improvement of factors located in this area may result in ineffectual circumstances or may be harmful to the organizations.

Area (3) illustrates a high level of importance along with a low level of impacts on satisfactions for employees. The factors locating in this area are the main reason why employees abandoned the organization, especially the experts. Therefore, area (3) is a critical zone, and to improve the satisfaction, employees' need for long-term planning is fundamental.

Ultimately, in area (4), the level of importance and level of impacts on satisfactions are shallow. Consequently, improving the factors that are located in this area imposes unnecessary costs to the organization, which results in idle circumstances to make progress in the organization.

5. Conclusions

In any competitive organization, it is a crucial factor in maintaining primary improvement tools of competitiveness such as the human factor. To achieve a better understanding of the work-force in any organization, evaluation of the job satisfaction is provided. Therefore, evaluating the influential factors in the job satisfaction is significantly important. The current study presented a novel hybrid validation framework based on the F-MULTIMOORA method and PLS-SEM approach considering CFA, to analyse the influential factors in job satisfaction in a real-world case study. The criteria and the influential factors in job satisfaction as candidate alternatives in the current study were identified, and a comprehensive description of each criterion and alternative was provided. Based on the F-MULTIMOORA approach, an assessment of the influencing factors has been provided. Additionally, a CFA based on the PLS-SEM approach has been provided to evaluate the measurement of the influencing factors in job satisfaction. The correlations between the rankings of the statistical approach, FMCDM method and expert judgments have been examined applying the Spearman's rank correlation coefficients. Finally, a comprehensive analysis of the real-world case study has been provided based on the proposed framework and the expert judgments from the case study, in which a detailed opportunity for improvement plan has been presented based on the systematic analysis of the current scenario in the case study, in order to utilize the procedure with corrective action.

Future suggestions and developments of the current study may be as follows. First, the input data of the MCDM approach, which have been suggested in the current study, can be extended to the cases, in which the data of the problem have different mathematical forms such as extensions of fuzzy sets, interval data structure, and granular data structures. Second, the validation framework in this study utilized the F-MULTIMOORA approach, in which the MCDM framework could be replaced by other MCDM techniques such as Axiomatic Design (AD), the Technique for Order of Preference by Similarity to Ideal Solution (TOPSIS), and Decision-Making Trial and Evaluation Laboratory (DEMATEL). Third, the statistical process in the proposed study is based on the PLS-SEM approach; for the future studies, the covariance-based methods (CB-SEM) can also be applied. Fourth, significance coefficients of attributes could be combined with the assessment phase to achieve more expert-based decisions, which may be attained subjectively or objectively. Subjective significance coefficients may be computed by applying various methods like the ANP, AHP, and BWM. It is recommended to use hierarchical weighting methods such as the hierarchical group BWM suggested by Maghsoodi et al. [68], in order to reach the optimal weights based on a group decision-making approach. Fifth, the validation framework could also be implemented in many other applications with the input content of statistical values, such as the effect of influential factors on performance appraisal.

Author Contributions: Conceptualization, A.I.M. and I.A.; methodology, A.I.M and M.A.; validation, A.I.M., I.A., Z.B. and M.A.; investigation, A.I.M and M.A.; resources, I.A. and Z.B.; data curation, I.A.; writing of the original draft preparation, A.I.M.; writing of review and editing, E.K.Z. and J.A.; supervision, E.K.Z. and J.A.; project administration, A.I.M. and E.K.Z.; funding acquisition, A.I.M., E.K.Z. and J.A.

Funding: This research received no external funding.

Conflicts of Interest: The authors declare no conflicts of interest.

References

1. Locke, E.A. The nature and causes of job satisfaction. In *Handbook of Industrial and Organizational Psychology*; American Psychological Association: Washington, DC, USA, 1976; pp. 1297–1349.
2. Spector, P.E. *Job Satisfaction: Application, Assessment, Cause, and Consequences*; SAGE Publications, Inc.: Thousand Oaks, CA, USA, 1997.
3. Özpehlivan, M.; Acar, A.Z. Development and validation of a multidimensional job satisfaction scale in different cultures. *Cogent Soc. Sci.* **2016**, *30*, 1–20. [CrossRef]
4. Baležentis, A.; Baležentis, T.; Brauers, W.K.M. Personnel selection based on computing with words and fuzzy Multimoora. *Expert Syst. Appl.* **2012**, *39*, 7961–7967. [CrossRef]
5. Shaout, A.; Yousif, M.K. Performance Evaluation—Methods and Techniques Survey. *Int. J. Comput. Inf. Technol.* **2014**, *3*, 966–979.
6. Keshavarz-Ghorabaee, M.; Amiri, M.; Zavadskas, E.K.; Turskis, Z.; Antucheviciene, J. An extended step-wiseweight assessment ratio analysis with symmetric interval type-2 fuzzy sets for determining the subjective weights of criteria in multi-criteria decision-making problems. *Symmetry (Basel)* **2018**, *10*, 91. [CrossRef]

7. Tumen, S.; Zeydanli, T. Social Interactions in Job Satisfaction. *Int. J. Manpow.* **2016**, *37*, 426–455. [CrossRef]
8. Metle, M.K. Education, job satisfaction and gender in Kuwait. *Int. J. Hum. Resour. Manag.* **2001**, *12*, 311–332. [CrossRef]
9. Asrar-ul-Haq, M.; Kuchinke, K.P.; Iqbal, A. The relationship between corporate social responsibility, job satisfaction, and organizational commitment: Case of Pakistani higher education. *J. Clean. Prod.* **2017**, *142*, 2352–2363. [CrossRef]
10. Lee, Y.; Sabharwal, M. Education–Job Match, Salary, and Job Satisfaction Across the Public, Non-Profit, and For-Profit Sectors: Survey of recent college graduates. *Public Manag. Rev.* **2016**, *18*, 40–64. [CrossRef]
11. Hann, M.; Reeves, D.; Sibbald, B. Relationships between job satisfaction, intentions to leave family practice and actually leaving among family physicians in England. *Eur. J. Public Health* **2011**, *21*, 499–503. [CrossRef]
12. Mansor, N.; Mohd Noor, J.M.; Nik Hassan, N.F. Job satisfaction among the bankers: An investigation on Islamic financial institution in eastern region of Malaysia. *Asian Soc. Sci.* **2012**, *8*, 186–197. [CrossRef]
13. Elshout, R.; Scherp, E.; van der Feltz-Cornelis, C.M. Understanding the link between leadership style, employee satisfaction, and absenteeism: A mixed methods design study in a mental health care institution. *Neuropsychiatr. Dis. Treat.* **2013**, *9*, 823–837. [PubMed]
14. Ruchman, R.B.; Kwak, A.J.; Jaeger, J.; Sayegh, A. Job satisfaction of program directors in radiology: A survey of current program directors. *Am. J. Roentgenol.* **2013**, *200*, 238–247. [CrossRef] [PubMed]
15. Frey, R.-V.; Bayon, T.; Totzek, D. How Customer Satisfaction Affects Employee Satisfaction and Retention in a Professional Services Context. *J. Serv. Res.* **2013**, *16*, 503–517. [CrossRef]
16. Naqbi, S.A.A.; Hamidi, S.A.; Younis, M.Z.; Rivers, P.A. Determinants of Employee Satisfaction with Services of Human Resource Departments in the Fujairah Medical District (FMD) of the United Arab Emirates. *J. Health Manag.* **2014**, *16*, 271–287. [CrossRef]
17. Tansel, A.; Gazîoğlu, Ş. Management-employee relations, firm size and job satisfaction. *Int. J. Manpow.* **2014**, *35*, 1260–1275. [CrossRef]
18. Matthies-Baraibar, C.; Arcelay-Salazar, A.; Cantero-González, D.; Colina-Alonso, A.; García-Urbaneja, M.; González-Llinares, R.M.; Letona-Aranburu, J.; Martínez-Carazo, C.; Mateos-Del Pino, M.; Nuño-Solinís, R.; et al. Is organizational progress in the EFQM model related to employee satisfaction? *BMC Health Serv. Res.* **2014**, *14*, 468. [CrossRef] [PubMed]
19. Tso, G.K.F.; Liu, F.; Li, J. Identifying Factors of Employee Satisfaction: A Case Study of Chinese Resource-Based State-Owned Enterprises. *Soc. Indic. Res.* **2015**, *123*, 567–583. [CrossRef]
20. Leder, S.; Newsham, G.R.; Veitch, J.A.; Mancini, S.; Charles, K.E. Effects of office environment on employee satisfaction: A new analysis. *Build. Res. Inf.* **2016**, *44*, 34–50. [CrossRef]
21. Mathieu, C.; Fabi, B.; Lacoursière, R.; Raymond, L. The role of supervisory behavior, job satisfaction and organizational commitment on employee turnover. *J. Manag. Organ.* **2016**, *22*, 113–129. [CrossRef]
22. Jacobs, M.A.; Yu, W.; Chavez, R. The effect of internal communication and employee satisfaction on supply chain integration. *Int. J. Prod. Econ.* **2016**, *171*, 60–70. [CrossRef]
23. Barakat, S.R.; Isabella, G.; Boaventura, J.M.G.; Mazzon, J.A. The influence of corporate social responsibility on employee satisfaction. *Manag. Decis.* **2016**, *54*, 2325–2339. [CrossRef]
24. Boddy, C.R.; Taplin, R. The influence of corporate psychopaths on job satisfaction and its determinants. *Int. J. Manpow.* **2016**, *37*, 965–988. [CrossRef]
25. Koklic, M.K.; Kukar-Kinney, M.; Vegelj, S. An investigation of customer satisfaction with low-cost and full-service airline companies. *J. Bus. Res.* **2017**, *80*, 188–196. [CrossRef]
26. Tarcan, G.Y.; Tarcan, M.; Top, M. An analysis of relationship between burnout and job satisfaction among emergency health professionals. *Total Qual. Manag. Bus. Excell.* **2017**, *28*, 1339–1356. [CrossRef]
27. Holmberg, C.; Caro, J.; Sobis, I. Job satisfaction among Swedish mental health nursing personnel: Revisiting the two-factor theory. *Int. J. Ment. Health Nurs.* **2018**, *27*, 581–592. [CrossRef] [PubMed]
28. Bin Bae, K.; Yang, G. The Effects of Family-Friendly Policies on Job Satisfaction and Organizational Commitment. *Public Pers. Manag.* **2017**, *46*, 25–40.
29. Hafez, E.; AbouelNeel, R.; Elsaid, E. An Exploratory Study on How Talent Management Affects Employee Retention and Job Satisfaction for Personnel Administration in Ain Shams University Egypt. *J. Manag. Strateg.* **2017**, *8*, 1–17. [CrossRef]

30. Hayes, B.; Bonner, A.; Pryor, J. Factors contributing to nurse job satisfaction in the acute hospital setting: A review of recent literature. *J. Nurs. Manag.* **2010**, *18*, 804–814. [CrossRef]
31. Zhu, Y. A review of job satisfaction. *Asian Soc. Sci.* **2013**, *9*, 293–298. [CrossRef]
32. Hantula, D.A. Job Satisfaction: The Management Tool and Leadership Responsibility. *J. Organ. Behav. Manag.* **2015**, *35*, 81–94. [CrossRef]
33. Brauers, W.K.M.; Zavadskas, E.K. The MOORA method and its application to privatization in a transition economy by A new method: The MOORA method. *Control Cybern.* **2006**, *35*, 445–469.
34. Brauers, W.K.M.; Zavadskas, E.K. Project management by multimoora as an instrument for transition economies. *Technol. Econ. Dev. Econ.* **2010**, *16*, 5–24. [CrossRef]
35. Baležentis, T.; Baležentis, A. A Survey on Development and Applications of the Multi-criteria Decision Making Method Multimoora. *J. Multi-Crit. Decis. Anal.* **2014**, *21*, 209–222. [CrossRef]
36. Zavadskas, E.K.; Antucheviciene, J.; Razavi Hajiagha, S.H.; Hashemi, S.S. The interval-valued intuitionistic fuzzy MultimoorA method for group decision making in engineering. *Math. Probl. Eng.* **2015**, *2015*, 560690. [CrossRef]
37. Stanujkic, D.; Karabasevic, D.; Zavadskas, E.K.; Brauers, W.K.M. An extension of the MULTIMOORA method for solving complex decision-making problems based on the use of interval-valued triangular fuzzy numbers. *Transform. Bus. Econ.* **2015**, *14*, 355–375.
38. Awasthi, A.; Baležentis, T. A hybrid approach based on BOCR and fuzzy Multimoora for logistics service provider selection. *Int. J. Logist. Syst. Manag.* **2017**, *27*, 261–282. [CrossRef]
39. Stanujkic, D.; Zavadskas, E.K.; Smarandache, F.; Brauers, W.K.M.; Karabasevic, D. A Neutrosophic Extension of the Multimoora Method. *Informatica* **2017**, *28*, 181–192. [CrossRef]
40. Zavadskas, E.K.; Baušys, R.; Juodagalvienė, B.; Garnytė-Sapranavičienė, I. Model for residential house element and material selection by neutrosophic Multimoora method. *Eng. Appl. Artif. Intell.* **2017**, *64*, 315–324. [CrossRef]
41. Gou, X.; Liao, H.; Xu, Z.; Herrera, F. Double hierarchy hesitant fuzzy linguistic term set and Multimoora method: A case of study to evaluate the implementation status of haze controlling measures. *Inf. Fusion* **2017**, *38*, 22–34. [CrossRef]
42. Ceballos, B.; Lamata, M.T.; Pelta, D.A. Fuzzy Multicriteria Decision-Making Methods: A Comparative Analysis. *Int. J. Intell. Syst.* **2016**, *32*, 722–738. [CrossRef]
43. Stević, Ž.; Pamučar, D.; Vasiljević, M.; Stojić, G.; Korica, S. Novel integrated multi-criteria model for supplier selection: Case study construction company. *Symmetry (Basel)* **2017**, *9*, 279. [CrossRef]
44. Ijadi Maghsoodi, A.; Abouhamzeh, G.; Khalilzadeh, M.; Zavadskas, E.K. Ranking and selecting the best performance appraisal method using the Multimoora approach integrated Shannon's entropy. *Front. Bus. Res. China* **2018**, *12*, 1–21. [CrossRef]
45. Ijadi Maghsoodi, A.; Kavian, A.; Khalilzadeh, M.; Brauers, W.K.M. CLUS-MCDA: A Novel Framework based on Cluster Analysis and Multiple Criteria Decision Theory in a Supplier Selection Problem. *Comput. Ind. Eng.* **2018**, *118*, 409–422. [CrossRef]
46. Hafezalkotob, A.; Hami-Dindar, A.; Rabie, N.; Hafezalkotob, A. A decision support system for agricultural machines and equipment selection: A case study on olive harvester machines. *Comput. Electron. Agric.* **2018**, *148*, 207–216. [CrossRef]
47. Chen, X.; Zhao, L.; Liang, H. A novel multi-attribute group decision-making method based on the Multimoora with linguistic evaluations. *Soft Comput.* **2018**, *22*, 5347–5361. [CrossRef]
48. Brauers, W.K.M.; Baležentis, A.; Baležentis, T. *Multimoora* for the EU member states updated with fuzzy number theory. *Technol. Econ. Dev. Econ.* **2011**, *17*, 259–290. [CrossRef]
49. Zadeh, L.A. Fuzzy sets. *Inf. Control* **1965**, *8*, 338–353. [CrossRef]
50. Joelianto, E.; Widiyantoro, S.; Ichsan, M. Time series estimation on earthquake events using ANFIS with mapping function. *Int. J. Artif. Intell.* **2009**, *3*, 37–63.
51. Ćirović, G.; Pamučar, D. Decision support model for prioritizing railway level crossings for safety improvements: Application of the adaptive neuro-fuzzy system. *Expert Syst. Appl.* **2013**, *40*, 2208–2223. [CrossRef]
52. Moallem, P.; Mousavi, B.S.; Naghibzadeh, S.S. Fuzzy inference system optimized by genetic algorithm for robust face and pose detection. *Int. J. Artif. Intell.* **2015**, *13*, 73–88.

53. Nowaková, J.; Prílepok, M.; Snášel, V. Medical Image Retrieval Using Vector Quantization and Fuzzy S-tree. *J. Med. Syst.* **2017**, *41*, 18. [CrossRef] [PubMed]

54. Brauers, W.; Zavadskas, E.K. Multimoora optimization used to decide on a bank loan to buy property. *Technol. Econ. Dev. Econ.* **2011**, *17*, 174–188. [CrossRef]

55. Ijadi Maghsoodi, A.; Ijadi Maghsoodi, A.; Mosavi, A.; Rabczuk, T.; Zavadskas, E.K. Renewable Energy Technology Selection Problem Using Integrated H-SWARA-MULTIMOORA Approach. *Sustainability* **2018**, *10*, 4481. [CrossRef]

56. Wright, S. Correlation and Causation. *J. Agric. Res.* **1921**, *20*, 557–585.

57. Fox, J. An Introduction to Structural Equation Modeling With the sem Package in R. *Struct. Equ. Model.* **2006**, *13*, 465–486. [CrossRef]

58. Savalei, V.; Bentler, P.M. Structural Equation Modeling. In *Handbook of Marketing Research: Uses, Misuses, and Future Advances*; Sage Publications: Thousand Oaks, CA, USA, 2006; pp. 330–364.

59. Mardani, A.; Streimikiene, D.; Zavadskas, E.K.; Cavallaro, F.; Nilashi, M.; Jusoh, A.; Zare, H. Application of Structural Equation Modeling (SEM) to Solve Environmental Sustainability Problems: A Comprehensive Review and Meta-Analysis. *Sustainability* **2017**, *9*, 1814. [CrossRef]

60. Statistics Solutions. Confirmatory Factor Analysis. 2016. Available online: https://www.statisticssolutions.com/factor-analysis-sem-exploratory-factor-analysis/ (accessed on 5 December 2018).

61. Flyvbjerg, B. Five Misunderstandings About Case-Study Research. *Qual. Inq.* **2006**, *12*, 219–245. [CrossRef]

62. Cronbach, L.J.; Meehl, P.E. Conctruct validity in psychological tests. *Psychol. Bull.* **1955**, *52*, 281–302. [CrossRef]

63. Fornell, C.; Larcker, D.F. Structural Equation Models with Unobservable Variables and Measurement Error: Algebra and Statistics. *J. Mark. Res.* **1981**, *18*, 382–388. [CrossRef]

64. Tenenhaus, M. PLS path modeling. *Comput. Stat. Data Anal.* **2005**, *48*, 159–205. [CrossRef]

65. Morente-Molinera, J.A.; Kou, G.; González-Crespo, R.; Corchado, J.M.; Herrera-Viedma, E. Solving multi-criteria group decision making problems under environments with a high number of alternatives using fuzzy ontologies and multi-granular linguistic modelling methods. *Knowl.-Based Syst.* **2017**, *137*, 54–64. [CrossRef]

66. Morente-Molinera, J.A.; Kou, G.; González-Crespo, R.; Corchado, J.M.; Managing Multi-Criteria Group. Decision Making Environments with High Number of Alternatives Using Fuzzy Ontologies. *SoMeT* **2018**, *303*, 493–506.

67. Schrepp, M.; Hinderks, A.; Thomaschewski, J. Design and Evaluation of a Short Version of the User Experience Questionnaire (UEQ-S). *Int. J. Interact. Multimed. Artif. Intell.* **2017**, *4*, 103. [CrossRef]

68. Ijadi Maghsoodi, A.; Mosavat, M.; Hafezalkotob, A.; Hafezalkotob, A. Hybrid hierarchical fuzzy group decision-making based on information axioms and BWM: Prototype design selection. *Comput. Ind. Eng.* **2018**, in press. [CrossRef]

 © 2018 by the authors. Licensee MDPI, Basel, Switzerland. This article is an open access article distributed under the terms and conditions of the Creative Commons Attribution (CC BY) license (http://creativecommons.org/licenses/by/4.0/).

Article

Probabilistic Linguistic Preference Relation-Based Decision Framework for Multi-Attribute Group Decision Making

R. Krishankumar [1], K. S. Ravichandran [1], M. Ifjaz Ahmed [1], Samarjit Kar [2,*] and Sanjay K. Tyagi [3]

[1] School of Computing, SASTRA University, Thanjavur-613401, Tamil Nadu, India;
    krishankumar@sastra.ac.in (R.K.); raviks@sastra.edu (K.S.R.); ifjazahmed@it.sastra.edu (M.I.A.)
[2] Department of Mathematics, National Institute of Technology, Durgapur-713209, West Bengal, India
[3] Department of General Studies, Higher College of Technology, Fujairah-4114, UAE; skumar1@hct.ac.ae
* Correspondence: samarjit.kar@maths.nitdgp.ac.in; Tel.: +91-943-445-3186

Received: 26 November 2018; Accepted: 17 December 2018; Published: 20 December 2018

Abstract: With trending competition in decision-making process, linguistic decision-making is gaining attractive attention. Previous studies on linguistic decision-making have neglected the occurring probability (relative importance) of each linguistic term which causes unreasonable ranking of objects. Further, decision-makers' (DMs) often face difficulties in providing apt preference information for evaluation. Motivated by these challenges, in this paper, we set our proposal on probabilistic linguistic preference relation (PLPR)-based decision framework. The framework consists of two phases viz., (a) *missing value entry phase* and (b) *ranking phase*. In phase (a), the missing values of PLPR are filled using a newly proposed automatic procedure and consistency of PLPR is ensured using a consistency check and repair mechanism. Following this, in phase (b), objects are ranked using newly proposed analytic hierarchy process (AHP) method under PLPR context. The practicality of the proposal is validated by using two numerical examples viz., green supplier selection problem for healthcare and the automobile industry. Finally, the strength and weakness of the proposal are discussed by comparing with similar methods.

Keywords: analytic hierarchy process; consistency measure; group decision-making; probabilistic linguistic preference relation

---

1. Introduction

Decision-making is an inevitable aspect of human life which involves uncertainty and vagueness. The process of selecting a suitable object for the task brings cognitive thought processes into the picture, which is dynamic and competing in nature [1]. DMs often face difficulties in expressing their opinions in a sensible manner and to alleviate the issue to a certain extent; they adopt linguistic preference information [2]. Previous studies on linguistic decision process [2–4] have claimed that (a) linguistic preferences are simple and straightforward information which can be directly obtained from the DM, (b) also, these linguistic preferences mitigate the cost of inaccuracies to some extent. Motivated by these claims; scholars presented different decision-making framework under the linguistic context, of which, some are reviewed here. Zadeh [5] framed the genesis of a linguistic variable and applied the same for approximate reasoning. Later, Herrera et al. [2] fabricated the initial idea of the linguistic decision process and proposed a sequential decision framework in a linguistic environment. Following this, Herrera et al. [3,6] presented the consensus model for group decision-making under the linguistic context. Inspired by the power of linguistic theory, Xu [7] extended the geometric mean and ordered weighted geometric aggregation operator for a linguistic domain. Further, He et al. [8] put forward a new entropy measure for linguistic-based group decision-making.

Though linguistic decision-making is an attractive concept, DMs still face difficulties in rationally rating the objects. The main reason for this difficulty is the cognitive behavior of the human mind, which encourages pair-wise comparative analysis rather than a standalone rating [9]. Motivated by the power of pair-wise comparison and linguistic term set (LTS), Herrera et al. [10,11] proposed the linguistic preference relation (LPR) concept for group decision-making and investigated some choice functions for the same. Following this, Xu [12] put forward some deviation measures for decision-making process under LPR context. Recently, Molinera et al. [13] developed new fuzzy ontologies for linguistic preference information and applied the same for decision-making. Wang and Xu [14] put forward an interactive algorithm for filling the missing LPR values using consistency measures and repaired the consistency of the same using the repairing mechanism.

Inspired by the power of linguistic information and its substantial use in decision-making, Rodriguez [15] proposed the hesitant fuzzy linguistic term set (HFLTS) which is an extension to LTS under hesitant fuzzy environment [16]. The HFLTS allowed DMs to give different choices of preference for the same instance, which managed uncertainty to some extent. Motivated by the power of HFLTS, Zhu and Xu [17] put forward the hesitant fuzzy linguistic preference relation (HFLPR), which is an extension to preference relation under HFLTS context. They also investigated some consistency measures for the same. Following this, Wang and Xu [18] presented the concept of extended hesitant fuzzy preference relation and studied some consistency measures for the same. Wu [19] presented a consensus model based on possibility distribution for HFLPR and validated the applicability of same for decision-making process. Recently, Song and Hu [20] proposed a decision framework for handling incomplete HFLPR and applied the same for real-time group decision-making problem. Tuysuz and Simsek [21] extended the popular AHP method under HFLTS context and applied the same for assessing the performance of cargo factory.

Though the HFLPR is able to manage DMs' hesitation in preference information, the occurring probability (distribution assessment) of each linguistic term in the decision-making process is neglected. In many practical applications, all linguistic choices by the DM do not bear the same importance and hence, ignoring the occurring probability of each linguistic term is unreasonable and illogical. To circumvent this challenge, Zhang et al. [22] introduced the concept of linguistic distribution assessment (LDA) and associated symbolic proportion for each linguistic term. Later, Pang et al. [23] generalized the idea of LDA by allowing partial ignorance ($\sum_i p_i \leq 1$) in preference elicitation and termed it as probabilistic linguistic term set (PLTS) which is an extension to HFLTS with probability concept. Recently, Zhang et al. [24] put forward the concept of incomplete LDA which is similar to PLTS and used in for decision-making. Inspired by the superiority of PLTS in associating occurring probability to each linguistic term; Bai et al. [25] presented a new comparison method use area concept for PLTS. Later, Liao et al. [26] extended the programming model to PLTS for multi-attribute decision-making (MADM). Liu and Teng [27] extended the Muirhead mean aggregation operator to PLTS for group decision-making. Zhang et al. [28,29] put forward the probabilistic linguistic preference relation (PLPR) concept which is an extension to preference relation under PLTS context and some additive consistency and consensus reaching measures were also investigated. Recently, Xie et al. [30] proposed probabilistic uncertain preference relation and applied the same for virtual reality application. Recently, attracted by the power of PLPR, Wu and Liao [31] proposed gain-lost dominance score method under PLTS for consensus reaching. Xie et al. [32] extended AHP (analytic hierarchy process) method and applied the same for assessing the performance of a new area. Since the concept of PLPR just began, we gained motivation to throw some light towards this concept and set our research focus in this direction.

Based on the review conducted above, some genuine challenges/lacunas are identified which are presented in a nutshell below:

(1) Investigation of decision process using the preference relation proves to be effective than investigation using attribute driven methods [28]. The reason for this is evident from the ease of pair-wise comparison mechanism, which allows DMs to produce sensible preference information

about each object with respect to a specific criterion. Also, the process of pair-wise comparison closely resembles with the practical decision process. Thus, motivated by the power of pair-wise comparison, we set our proposal in this context.

(2) Since PLPR is a recent research topic, the challenge of automatic filling of missing values under PLPR context needs to be addressed. DMs often get confused between objects (alternatives) due to external pressure and lack of sufficient knowledge. This forces DMs to be ignorant and hesitant towards a certain pair of objects which eventually leads to missing values in the preference relation(s).

(3) Checking and repairing the consistency of PLPRs in an automated fashion by using a systematic procedure is also an interesting challenge to be addressed. The consistency of preference relation is substantial aspect for rational and reasonable decision-making. Due to various external pressures, DMs often face difficulty in providing a consistent preference relation for evaluation and manual repairing of the preference relation is an ordeal and unreasonable. Though, Xie et al. [32] presented a method for consistency check and repair, they are complex and computationally intensive as they involve logarithmic function and iterative calculation of Eigen vectors.

(4) Furthermore, extension of ranking methods under PLPR context is also an attractive challenge to be addressed for sensible prioritization of objects. DMs prefer systematic scientific procedure for selection of objects rather than random guess. Though, Xie et al. [32] extended AHP method, they converted the PLTS information into single value by using possibility degree measure which causes potential loss of information leading to unreasonable prioritization of objects.

Motivated by these challenges and with the view of alleviating these challenges, in this paper, we propose a new scientific decision framework, which consists of two phases viz., (1) *missing value entry phase* and (2) *ranking phase*. Xu [33] clearly pointed out that, (i) DMs are often unwilling to reconstruct the evaluation matrix and (ii) also the chance for the manually reconstructed matrix to be consistent is very less. Thus, motivated by these claims,

(1) In the first phase of the proposal, a new automated procedure for filling the missing values is presented.

(2) Following this, a new systematic procedure is proposed for checking the consistency of PLPRs and inconsistent PLPRs are repaired automatically in an iterative manner. Unlike method discussed in [32], the proposed procedure uses simple and straightforward operational law(s) of PLEs.

(3) Further, in the second phase of the proposal, a new extension to AHP method under PLPR context is presented for suitable selection of the object from the set of objects. Unlike method [32], the proposed extension for AHP retains the PLTS information throughout the formulation and mitigates information loss which allows reasonable prioritization of objects.

(4) Finally, the practicality, strength, and weakness of the proposal are realized by using green supplier selection problem.

The rest of the paper is constructed as Section 2 for preliminaries, Section 3 for calculation of missing values and ranking of objects. Section 4 presents a numerical example for demonstrating the practical use of the framework. Section 5 presents the comparative study and Section 6 gives the concluding remarks and future works.

2. Preliminaries

Let us review some basics of LTS and PLTS concepts.

Definition 1 ([12]). *Consider a LTS S defined by $\{s_\alpha | \alpha \in [-n, n]\}$ with n being the limits of the term set and s_{-n} and s_n are the lower and upper bounds of the term set. The s_α is a linguistic term set with the following characteristics:*

(a) s_α and s_β are two linguistic term sets with $s_\alpha > s_\beta$ only if $\alpha > \beta$.

(b)   *The negation of s_α is denoted by $neg(s_\alpha)$ and is given by $neg(s_\alpha) = s_{-\alpha}$. As a special case, $neg(s_0) = s_0$.*

Definition 2 ([23]). *Consider a LTS S defined by $\{s_\alpha | \alpha \in [-n, n]\}$, then the PLTS is defined by:*

$$L(p) = \left\{ L^t(p^t) | L^t \in S, 0 \leq p^t \leq 1, t = 1, 2, \ldots, \#L(p), \sum_{i=1}^{\#L(s)} p_i^t \leq 1 \right\} \tag{1}$$

where L^t is the tth linguistic term and p^t is the associated occurring probability of the tth linguistic term.

Note 1: The concept of PLTS [23] is a generalization to LDA [22] that allows partial ignorance ($\sum_i p_i \leq 1$) in preference elicitation and the concept of incomplete LDA [24] is similar to PLTS.

Remark 1. *For brevity of representation, we denote a probabilistic linguistic element (PLE) as $\{r^t(p^t)\}$ where r is the subscript of the linguistic term, p is the corresponding probability of the linguistic term and t is the number of instances.*

Definition 3 ([29]). *The PLPR is a square matrix of the form $R = \left(L_{ij}^t \left(p_{ij}^t \right) \right)_{n \times n}$ with $L_{ii}^t = \{s_0\}$, $p_{ji}^t = p_{ij}^t$ and $L_{ji}^t = neg\left(L_{ij}^t \right)$.*

Definition 4 ([23]). *Consider two PLEs, $L_1(p)$ and $L_2(p)$ as defined before. Then,*

$$L_1(p) \oplus L_2(p) = \{r_1^t + r_2^t; p_1^t \times p_2^t\} = \{r_3^t(p_3^t)\} = L_3^t(p) \tag{2}$$

$$\lambda L_1(p) = \{r_1^t \times \lambda; p_1^t\} = L_3^t(p) \tag{3}$$

where $t = 0, 1, \ldots, \#L(p)$.

Remark 2. *The operational laws defined in Definition 4 are valid only when the length of the PLEs is equal. If the length is unequal, we apply method from [23] to make the length of PLEs equal. Also from Equations (2) and (3), we observe that the linguistic term of PLE sometimes gets outside the boundary which can be transformed to PLTS within the boundary by using [28].*

3. Proposed Decision Framework under Probabilistic Linguistic Preference Relation (PLPR) Context

3.1. Proposed Architecture of PLPR Based Decision Framework

The architecture of the proposed scientific decision framework is presented in Figure 1 which is simple and straightforward to understand.

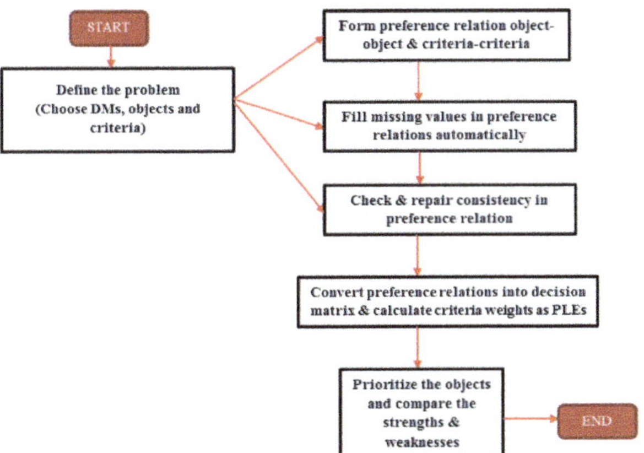

Figure 1. Architecture of proposed scientific decision framework.

3.2. Proposed Automatic Procedure for Filling Missing Values and Consistency Check and Repair for PLPRs

In this section, the procedure for finding the missing values of a PLPR is presented. Generally, DMs find pairwise comparison as an easier option for rating alternatives [28]. DMs rate the alternatives upon each criterion and sometimes they are unwilling or confused between alternatives' performance over a specific criterion and this forces them to ignore such rating. As a result, the decision matrix is now incomplete and further processing becomes difficult. To circumvent this issue, an automated procedure is proposed which automatically fits a value to the missing information. Zhang et al. [33] claimed that "(a) manual entry of missing values by some random information is unreasonable and causes potential loss of information and (b) returning of decision matrix to the DM for re-entry is also unreasonable and computationally ineffective". Motivated by such claims, in this paper, an automated procedure is presented under PLTS context for filling missing values.

The procedure for automated filling of missing value is given below:

Step 1: Consider a PLPR $R = \left\{ L_{ij}^t \left(p_{ij}^t \right) \right\}_{n \times n}$ which has PLEs. Identify the instance which is missing. If $j > i + 1$, then the missing instance can be automatically estimated (follow steps below), else follow Equation (4).

$$R_{ij} = \left\{ \frac{\sum_{i=1}^{m} r_{ij}}{m}, \frac{\sum_{i=1}^{m} p_{ij}}{m} \right\} \forall j \leq i + 1 \tag{4}$$

where r_{ij} is the subscript of the linguistic term, p_{ij} is the associated occurring probability of the linguistic term and m is the order of the matrix.

Step 2: When $j > i + 1$, apply Equation (5) to automatically estimate the missing values.

$$R_{ij} = min\left(\left(\oplus_{k=1}^{j-i-1} \left\{ r_{(i+k)(i+k+1)} \right\} \right), \left(\oplus_{k=1}^{j-i-1} \left\{ r_{(i+k)(i+k+1)} \right\} \oplus \oplus_{k=1}^{j-i-1} \left\{ \left(1 - r_{(i+k)(i+k+1)} \right) \right\} \right) \right) \text{ and}$$

$$min\left(\left(\bigoplus_{k=1}^{j-i-1} \left\{ p_{(i+k)(i+k+1)} \right\} \right), \left(\bigoplus_{k=1}^{j-i-1} \left\{ p_{(i+k)(i+k+1)} \right\} \oplus \bigoplus_{k=1}^{j-i-1} \left\{ \left(1 - p_{(i+k)(i+k+1)} \right) \right\} \right) \right) \tag{5}$$

where $\oplus$ is an operator given in Definition 4.

Note 2: The result from Equation (5) is also a PLE and the values that go out of bounds when $\oplus$ operator is applied are transformed using Remark 2.

Step 3: Check the consistency of the matrix $R = \left(\left\{ L_{ij}^t \left(p_{ij}^t \right) \right\} \right)_{n \times n}$ by using Equations (6) and (7).

$$R_{ij}^z = \left(\left\{ L_{ij}^{*t} \left(p_{ij}^{*t} \right) \right\} \right)_{n \times n} \tag{6}$$

where L_{ij}^{*t} and p_{ij}^{*t} can be calculated by using Equation (7).

$$L_{ij}^{*t}\left(p_{ij}^{*t}\right) = \begin{cases} \frac{1}{m}\left(\bigoplus_{e=1}^{m}\left(L_{ie}(p)\oplus L_{ej}(p)\right)\right) \ \forall i \neq j \\ \{s_0\} \ otherwise \end{cases} \tag{7}$$

where $\oplus$ is an operator given in Definition 4.

Here linguistic terms are added as per Definition 4 and transformation procedure is applied to those terms that exceed the limits. However, the corresponding probability terms are calculated by using weighted geometry method to avoid unreasonable probability values. The personal opinion on each alternative is given by the DM with $\sum_{i=1}^{n}\omega_i = 1$.

Step 4: Calculate the distance between R_{ij} and R_{ij}^z by using Equation (8) to determine the consistency index (CI).

$$CI(R) = d\left(R_{ij}, R_{ij}^z\right) = \sqrt{\frac{2}{m(m-1)}\sum_{i=1}^{m}\sum_{j=i+1}^{m}\left(\sum_{t=1}^{\#L(p)}\left(p_{ij}^t \times p_{ij}^{t*}\right)\left(\frac{r_{ij}^t - r_{ij}^{t*}}{T}\right)\right)^2} \tag{8}$$

where T is the cardinality of the LTS, r is the subscript of the PLTS and p is the corresponding probability of the term set.

Note 3: The distance formula described in Equation (8) obeys the desirable distance properties viz., non-negative, non-degenerate, symmetric and transitive.

Step 5: The consistency values obtained from step 4 $(CI(R))$ are compared with the standard consistency value $\left(\widetilde{CI}(R)\right)$ (suggested as 0.05 by DMs). If $CI(R) \leq \widetilde{CI}(R)$ then, R is acceptable; else R is unacceptable and automatic repairing must be done by following the steps below.

Step 6: Repair the inconsistent PLPR automatically by using Equation (9).

$$R_{ij}^{z+1} = \frac{\left(L_{ij}(p)\right)^{1-\tau\sigma}\oplus\left(L_{ij}^*(p)\right)^{\tau\sigma}}{\left\{\left(L_{ij}(p)\right)^{1-\tau\sigma}\oplus\left(L_{ij}^*(p)\right)^{\tau\sigma}\right\}\oplus\left\{\left(1-L_{ij}(p)\right)^{1-\tau\sigma}\oplus\left(1-L_{ij}^*(p)\right)^{\tau\sigma}\right\}} \tag{9}$$

where $L_{ij}(p) = \left\{r_{ij}\left(p_{ij}\right)\right\}$, $L_{ij}^*(p) = \left\{r_{ij}^*\left(p_{ij}^*\right)\right\}$, τ and σ are parameters in the range [0,1].

Note that this repairing is an iterative process and until consistent matrix is obtained, we apply the procedure.

Step 7: Repeat the steps 5 and 6 iteratively till a PLPR of acceptable consistency is obtained.

3.3. Proposed Analytic Hierarchy Process (AHP) Method under PLPR Context

Analytic hierarchy process (AHP) is a classical ranking method that is based on the pairwise comparison concept [34]. This ranking method works with preference relations and weight of each alternative is determined. Based on the weight values, alternative are ranked and the suitable object is selected for the process. Recently, Emrouznejad and Marra [35] conducted a comprehensive review on AHP method and identified its diverse applicability in MCDM and the interesting variants of AHP. Clearly from the review, extension of AHP to PLTS context is a new idea for exploration and the work of Xie et al. [32] framed the genesis for the same. Some lacunas are discussed in Section 1 which motivates the proposed extension of AHP under PLPR context.

Now, we present the procedure for ranking objects using the proposed extension to AHP under PLPR context.

Step 1: Define the problem under multi-attributes decision-making context and determine the number of objects, attributes and DMs. Use PLEs as preference information.

Step 2: Suppose, m objects and n attributes are considered, n PLPRs of order $(m \times m)$ is formed. Following this, a PLPR of order $(n \times n)$ is formed for the attributes.

Step 3: Check the consistency of all PLPRs using the procedure presented in Section 3.2 and repair the inconsistent PLPR. Apply Equation (2) to the PLPR of order $(n \times n)$. This forms a weight vector for the attributes which is probabilistic linguistic in nature.

Step 4: Following step 3, we aggregate the PLEs from $(m \times m)$ matrices using Equation (2) to form a decision matrix with PLTS information of order $(m \times n)$ where m is the number of alternatives and n is the number of attributes.

Step 5: The attribute weights and decision matrix are taken from steps 3 and 4 respectively and Equation (2) is applied to obtain a vector of order $(m \times 1)$ for each of the m alternatives.

Step 6: The vector obtained from step 6 contains PLTS information which is used for the final ranking by applying Equation (10).

$$\varphi_i = \sum_{k=1}^{\#L(p)} \left(r_i^k \times p_i^k \right) \tag{10}$$

where r_i is the subscript of the ith object and p_i is the probability of the corresponding ith object.

Thus, the object which has large φ_i value is ranked first and so on.

4. Numerical Example

4.1. Green Supplier Selection for Healthcare Center

Indian healthcare industries are gaining high interest in recent times because of its diverse spectrum of high-tech equipment, highly skilled professionals, eco-friendly infrastructure etc. On April 2015, IBEF (Indian brand equity foundation) conducted a survey and identified that Indian healthcare industries are a big asset for the nation with an outreach of USD 280 billion by 2020. The report also showed that India is ranked third in the global healthcare sector. With the motive of igniting the spirit, GoI (government of India) started many interesting and innovative initiatives (www.ibef.org) like *"signing of MoA (memorandum of agreement) with WHO (world health organization) for promoting public health in India, signing MoU (memorandum of understanding) with medical agencies of BRICS to facilitate healthy medical products"*. A study by Healthcare Design magazine showed that *"each year, expenditure on energy usage by healthcare is USD 8 billion"* which drives them to place a concrete carbon footprint. To better reduce the CO_2 emission and energy usage, healthcare must tune their thoughts towards green technologies and selection of equipment suppliers who follow green standards ISO 14000 and 14001 actively.

With this train of thought, we consider a healthcare center in Tirchy that wants to expand its service and hospitality for the betterment of the people in and around the region and also reduce its contribution in carbon footprint by adopting green technology. To do so, the management decides to renovate certain policies of the hospital which include proper and hygienic service to patients, proper and effective resource management, purchase of equipment from green suppliers and intense and sensible care at critical times. Surfing through the previous reports, the management finds an urgent need to make a reasonable decision with regards to the purchase of surgical equipment for the health center. An expert committee of three members viz., chief doctor (E_1), senior stock manager (E_2) and chief technical officer (E_3) is formed and suitable supplier is chosen using a systematic scientific approach. Initially, seven green suppliers are chosen for the process and out of these seven suppliers four green suppliers who actively follow ISO 14000 and 14001 standards are selected based on the pre-screening test. Now, the committee decides four attributes for evaluation of four green suppliers. The committee plans to do the pairwise comparison and used PLTS information for rating. The details of these four attributes are given below:

- **Hygiene and safety** (C_1): This attribute measures the amount of care given by the suppliers in adhering to the green technologies and standards.
- **Quality of equipment** (C_2): The longevity and correctness of the product is determined from this attribute.

- **On-time delivery** (C_3): Delivery of product at right time under critical scenario is determined from this attribute.
- **Cost of equipment** (C_4): This attribute determines the total cost involved during the product life cycle.

Let us now consider the following procedure for evaluation:

Step 1: Construct four PLPRs of order (4×4) with PLTS information. Each criterion is taken and the DMs form pairwise comparison matrices with each supplier over a specific criterion.

The missing values in Table 1 are determined using Equation (5) and it is shown in Table 2. Clearly, the missing values which are calculated is also a PLE.

Table 1. Probabilistic Linguistic Preference Relation (PLPR) matrices for each criterion.

Attributes	Supplier	S_1	S_2	S_3	S_4
C_1	S_1		$\{2,(0.3),\ 1,(0.2),\ -2,(0.42)\}$	$\{2,(0.33),\ 1,(0.44),\ -1,(0.2)\}$	$\{1,(0.35),\ -1,(0.25),\ -2,(0.3)\}$
	S_2			$\{-1,(0.3),\ 1,(0.25),\ -2,(0.42)\}$	$\{2,(0.22),\ 1,(0.4),\ -2,(0.35)\}$
	S_3				$\{2,(0.42),\ -2,(0.25),\ -1,(0.15)\}$
	S_4				
C_2	S_1		$\{2,(0.28),\ 1,(0.35),\ -1,(0.3)\}$	$\{0,(0.22),\ -2,(0.45),\ -1,(0.15)\}$	$\{2,(0.18),\ 1,(0.35),\ -1,(0.25)\}$
	S_2			$\{-2,(0.4),\ 0,(0.33),\ 1,(0.25)\}$	X
	S_3				$\{1,(0.25),\ -2,(0.33),\ -1,(0.25)\}$
	S_4				
C_3	S_1		$\{1,(0.45),\ 0,(0.3),\ -1,(0.11)\}$	$\{2,(0.25),\ 1,(0.37),\ -1,(0.33)\}$	X
	S_2			$\{2,(0.28),\ -1,(0.35),\ 0,(0.3)\}$	$\{1,(0.35),\ -1,(0.25),\ -2,(0.3)\}$
	S_3				$\{-2,(0.33),\ 1,(0.4),\ 2,(0.22)\}$
	S_4				
C_4	S_1		$\{-1,(0.3),\ 1,(0.42),\ -2,(0.25)\}$	$\{-2,(0.33),\ 1,(0.4),\ 2,(0.22)\}$	$\{2,(0.33),\ 0,(0.35),\ -1,(0.2)\}$
	S_2			$\{-2,(0.3),\ 2,(0.24),\ 1,(0.4)\}$	$\{2,(0.3),\ -1,(0.35),\ 1,(0.25)\}$
	S_3				$\{1,(0.4),\ 0,(0.22),\ -1,(0.15)\}$
	S_4				

Table 2. PLPR matrices after finding the missing values.

Attributes	Supplier	S_1	S_2	S_3	S_4
C_1	S_1		$\{2,(0.3),\ 1,(0.2),\ -2,(0.42)\}$	$\{2,(0.33),\ 1,(0.44),\ -1,(0.2)\}$	$\{1,(0.35),\ -1,(0.25),\ -2,(0.3)\}$
	S_2			$\{-1,(0.3),\ 1,(0.25),\ -2,(0.42)\}$	$\{2,(0.22),\ 1,(0.4),\ -2,(0.35)\}$
	S_3				$\{2,(0.42),\ -2,(0.25),\ -1,(0.15)\}$
	S_4				
C_2	S_1		$\{2,(0.28),\ 1,(0.35),\ -1,(0.3)\}$	$\{0,(0.22),\ -2,(0.45),\ -1,(0.15)\}$	$\{1,(0.19),\ -2,(0.22),\ -1,(0.19)\}$
	S_2			$\{-2,(0.4),\ 0,(0.33),\ 1,(0.25)\}$	$\{-2,(0.4),\ 0,(0.33),\ 1,(0.25)\}$
	S_3				$\{1,(0.25),\ -2,(0.33),\ -1,(0.25)\}$
	S_4				
C_3	S_1		$\{1,(0.45),\ 0,(0.3),\ -1,(0.11)\}$	$\{2,(0.25),\ 1,(0.37),\ -1,(0.33)\}$	$\{0,(0.06),\ 0,(0.08),\ 1,(0.05)\}$
	S_2			$\{2,(0.28),\ -1,(0.35),\ 0,(0.3)\}$	$\{1,(0.35),\ -1,(0.25),\ -2,(0.3)\}$
	S_3				$\{-2,(0.33),\ 1,(0.4),\ 2,(0.22)\}$
	S_4				
C_4	S_1		$\{-1,(0.3),\ 1,(0.42),\ -2,(0.25)\}$	$\{-2,(0.33),\ 1,(0.4),\ 2,(0.22)\}$	$\{2,(0.33),\ 0,(0.35),\ -1,(0.2)\}$
	S_2			$\{-2,(0.3),\ 2,(0.24),\ 1,(0.4)\}$	$\{2,(0.3),\ -1,(0.35),\ 1,(0.25)\}$
	S_3				$\{1,(0.4),\ 0,(0.22),\ -1,(0.15)\}$
	S_4				

Step 2: Construct one PLPR matrix of order (4×4) to determine the weights of the attributes. The Equation (2) is used to determine the weight of each criterion. The weight values are probabilistic linguistic in nature.

From Table 3 we obtain the weight value (relative importance) for each criterion. By applying Equation (2) we get the weight values as PLEs and it is given by $C_1 = \{2,(0.32), 1,(0.37), 2,(0.41)\}$, $C_2 = \{1,(0.49), 0,(0.37), 2,(0.38)\}$, $C_3 = \{1,(0.30), 1,(0.32), 2,(0.38)\}$ and $C_4 = \{1,(0.35), 2,(0.45), 0,(0.44)\}$.

Step 3: Check the consistency of all PLPRs and repair those PLPRs that are inconsistent in nature. Follow the procedure from Section 3.2 for automatic repairing of inconsistent PLPR.

All the above four PLPR matrices are checked for consistency by using the Equations (6) and (7). The child PLPR matrices (R_i^z) are initially formed from all four parent PLPR matrices (R_i) and the distance between each of these PLPRs is calculated. These distance values are shown in Table 4. Just for an example, let us consider the child matrix corresponding to C_1 and the distance between the parent and child matrices are calculated and it is given in Table 4.

$$R_1^z = \begin{pmatrix} \left\{ \begin{matrix} 0,(1), \\ 0,(1), \\ 0,(1) \end{matrix} \right\} & \left\{ \begin{matrix} 2,(0.15), \\ 0,(0.14), \\ -2,(0.2) \end{matrix} \right\} & \left\{ \begin{matrix} 2,(0.17), \\ 2,(0.12), \\ -2,(0.12) \end{matrix} \right\} & \left\{ \begin{matrix} 2,(0.17), \\ -1,(0.15), \\ -2,(0.15) \end{matrix} \right\} \\ \left\{ \begin{matrix} -2,(0.15), \\ 0,(0.14), \\ 2,(0.2) \end{matrix} \right\} & \left\{ \begin{matrix} 0,(1), \\ 0,(1), \\ 0,(1) \end{matrix} \right\} & \left\{ \begin{matrix} -2,(0.16), \\ 2,(0.15), \\ -2,(0.16) \end{matrix} \right\} & \left\{ \begin{matrix} 2,(0.16), \\ -1,(0.18), \\ -2,(0.2) \end{matrix} \right\} \\ \left\{ \begin{matrix} -2,(0.17), \\ -2,(0.12), \\ 2,(0.12) \end{matrix} \right\} & \left\{ \begin{matrix} 2,(0.16), \\ -2,(0.15), \\ 2,(0.16) \end{matrix} \right\} & \left\{ \begin{matrix} 0,(1), \\ 0,(1), \\ 0,(1) \end{matrix} \right\} & \left\{ \begin{matrix} 2,(0.18), \\ -2,(0.16), \\ -2,(0.12) \end{matrix} \right\} \\ \left\{ \begin{matrix} -2,(0.17), \\ 1,(0.15), \\ 2,(0.15) \end{matrix} \right\} & \left\{ \begin{matrix} -2,(0.16), \\ 1,(0.18), \\ 2,(0.2) \end{matrix} \right\} & \left\{ \begin{matrix} -2,(0.18), \\ 2,(0.16), \\ 2,(0.12) \end{matrix} \right\} & \left\{ \begin{matrix} 0,(1), \\ 0,(1), \\ 0,(1) \end{matrix} \right\} \end{pmatrix}$$

Table 3. Attributes weight estimation matrix.

Attributes	C_1	C_2	C_3	C_4
C_1		$\left\{\begin{matrix} 2,(0.4), \\ 1,(0.32), \\ -1,(0.25) \end{matrix}\right\}$	$\left\{\begin{matrix} -1,(0.25), \\ 0,(0.3), \\ 1,(0.35) \end{matrix}\right\}$	$\left\{\begin{matrix} -1,(0.15), \\ 1,(0.25), \\ 2,(0.4) \end{matrix}\right\}$
C_2			$\left\{\begin{matrix} 1,(0.35), \\ -1,(0.12), \\ 0,(0.25) \end{matrix}\right\}$	$\left\{\begin{matrix} 0,(0.35), \\ -2,(0.32), \\ 2,(0.25) \end{matrix}\right\}$
C_3				$\left\{\begin{matrix} -1,(0.18), \\ 1,(0.4), \\ 2,(0.33) \end{matrix}\right\}$
C_4				

Table 4. Calculation of distance values.

$CI(R)$	Value(s)
$d\left(R_1, R_1^z\right)$	0.0675
$d\left(R_2, R_2^z\right)$	0.1267
$d\left(R_3, R_3^z\right)$	0.0717
$d\left(R_4, R_4^z\right)$	0.0453

These values are compared against the standard value $\widetilde{CI}(R)$ (0.05). Since the distance values of the first three PLPRs are greater than 0.05, it is inconsistent and so we apply Equation (9) to repair these matrices. Just for an example, consider the child matrix of C_1 which becomes consistent in the second iteration with the distance value of 0.0136 which is less than the threshold 0.05. Further, child matrix of C_2 and C_3 are also inconsistent which are repaired using Equation (9) and the distance value is given by 0.018 (second iteration) and 0.019 (first iteration) which is less than the threshold value 0.05.

$$R_1^{z+1} = \begin{pmatrix} \left\{ \begin{array}{c} 0,(1), \\ 0,(1), \\ 0,(1) \end{array} \right\} & \left\{ \begin{array}{c} 2,(0.13), \\ 0,(0.12), \\ -2,(0.16) \end{array} \right\} & \left\{ \begin{array}{c} 2,(0.14), \\ 2,(0.12), \\ -2,(0.11) \end{array} \right\} & \left\{ \begin{array}{c} 2,(0.14), \\ -2,(0.13), \\ -2,(0.13) \end{array} \right\} \\ \left\{ \begin{array}{c} -2,(0.13), \\ 0,(0.12), \\ 2,(0.16) \end{array} \right\} & \left\{ \begin{array}{c} 0,(1), \\ 0,(1), \\ 0,(1) \end{array} \right\} & \left\{ \begin{array}{c} -2,(0.14), \\ 2,(0.13), \\ -2,(0.14) \end{array} \right\} & \left\{ \begin{array}{c} 2,(0.13), \\ -1,(0.15), \\ -2,(0.16) \end{array} \right\} \\ \left\{ \begin{array}{c} -2,(0.14), \\ -2,(0.12), \\ 2,(0.11) \end{array} \right\} & \left\{ \begin{array}{c} 2,(0.14), \\ -2,(0.13), \\ 2,(0.14) \end{array} \right\} & \left\{ \begin{array}{c} 0,(1), \\ 0,(1), \\ 0,(1) \end{array} \right\} & \left\{ \begin{array}{c} 2,(0.15), \\ -2,(0.13), \\ -2,(0.1) \end{array} \right\} \\ \left\{ \begin{array}{c} -2,(0.14), \\ 2,(0.13), \\ 2,(0.13) \end{array} \right\} & \left\{ \begin{array}{c} -2,(0.13), \\ 1,(0.15), \\ 2,(0.16) \end{array} \right\} & \left\{ \begin{array}{c} -2,(0.15), \\ 2,(0.13), \\ 2,(0.1) \end{array} \right\} & \left\{ \begin{array}{c} 0,(1), \\ 0,(1), \\ 0,(1) \end{array} \right\} \end{pmatrix}$$

Step 4: Apply the proposed ranking method from Section 3.3 over the consistent PLPRs and obtain a suitable supplier for the process. The four PLPR matrices of order $(m \times m)$ and attributes weight matrix of order $(n \times 1)$ are aggregated using the $\oplus$ operator defined in Definition 4. The resultant matrix is given in Table 5.

Table 5. Decision matrix with probabilistic linguistic term set (PLTS) information.

Supplier vs. Attributes	C_1	C_2	C_3	C_4
S_1	2, (0.55), 2, (0.55), 2, (0.57)	2, (0.56), 2, (0.57), 2, (0.53)	2, (0.53), 1, (0.54), 2, (0.52)	2, (0.58), 2, (0.62), 2, (0.55)
S_2	2, (0.49), 2, (0.51), 2, (0.53)	1, (0.54), 0, (0.51), 2, (0.49)	2, (0.51), 1, (0.51), 2, (0.51)	2, (0.53), 2, (0.57), 2, (0.53)
S_3	2, (0.55), 1, (0.55), 2, (0.56)	2, (0.57), 2, (0.57), 2, (0.53)	1, (0.56), 2, (0.58), 2, (0.58)	2, (0.59), 2, (0.60), 0, (0.56)
S_4	2, (0.49), 2, (0.51), 2, (0.52)	1, (0.53), 2, (0.51), 2, (0.49)	2, (0.47), 2, (0.48), 2, (0.49)	1, (0.53), 2, (0.57), 2, (0.51)

Table 5 is formed after applying Equation (2) over the attributes-alternative pair. When Equation (2) is applied, a vector of order $(1 \times n)$ is obtained for all m suppliers and finally, a decision matrix of order $(m \times n)$ with PLTS information is shown in Table 5. By using the procedure given in Section 3.3 on Table 5, we obtain final rank values as shown in Table 6.

From Table 6, we observe that the ranking order is given by $S_1 > S_4 > S_2 > S_3$ and S_1 is chosen as a suitable supplier for the healthcare. Further, suppliers S_4, S_2 and S_3 are for backup plans. When method from [21] is applied, the ranking order becomes $S_4 > S_1 > S_2 > S_3$ which is different from the ranking order obtained by the proposed framework. This is evident from the fact that the method discussed in [21] does not contain occurring probability values.

Step 5: Compare the strength and weakness of the proposal with state of the art methods. Readers are encouraged to refer Section 5 for the same.

Table 6. Final rank values.

Supplier(s)	PLTS Information	Ranking Value(s)
S_1	2, (0.56), 1, (0.57), 2, (0.54)	2.7899
S_2	1, (0.52), 0, (0.52), 2, (0.52)	1.5612
S_3	1, (0.57), 1, (0.58), 0, (0.55)	1.1554
S_4	1, (0.51), 2, (0.52), 2, (0.5)	2.5677

4.2. Green Supplier Selection for Automobile Industry in India

Automobile industries in India are booming at a faster pace providing economic growth and global market improvement. These industries drive avenues of employment to approximately 13 million people in India. As per the 2013–14 annual report on automobiles, a grand total of 21,500,165 vehicles were produced which eventually boomed the revenue for India. Despite the attractive advantages, the pollution caused by these industries is huge which affect the living beings and the environment as a whole. A study found that by 2020, almost half of the cars in India will use diesel and roughly 620,000 people will die due to respiratory issues (https://community.data.gov.in/automobiles-and-pollution-in-india/). This alarming analysis motivates automobile industries to choose green suppliers for purchasing their raw materials. Green suppliers actively monitor their system and practices to ensure limited emission of environmental pollutants. These suppliers strongly follow the ISO 14000 and 14001 standards pertaining to the adoption of green practices and technologies.

Motivated by this background, in this paper, we plan to provide a systematic framework for suitable selection of green supplier from the set of suppliers for leading automobile industry in India (name anonymous). Let $E = (e_1, e_2, e_3)$ be a set of three DMs who constitute the expert committee. $G = (g_1, g_2, g_3, g_4)$ and $C = (c_1, c_2, c_3, c_4)$ be the set of green suppliers and the corresponding evaluation attributes respectively. Since the attributes used in Section 4.1 adhere to the green standards, we adopt the same in this example also. Initally, eight green suppliers were chosen for the process and based on pre-screening and Delphi method, four green suppliers are finalized for evaluation. These suppliers actively obey ISO 14,000 and 14,001 standards and they are evaluated under four attributes adapted from Section 4.1. Following steps are presented for the systematic selection of green supplier:

Step 1. Form the PLPRs supplier wise for each criterion. This produces four matrices of order 4×4 that correspond to one preference relation for each criterion.

Step 2: Fill the missing values by using the proposed procedure given in Section 3.2. The missing values are represented by "X" in Table 7 and these values are filled systematically using procedure proposed in Section 3.2 and the values are PLEs (refer Table 8).

Table 7. PLPR information supplier to supplier for each attribute.

Attributes	Supplier	S_1	S_2	S_3	S_4
C_1	S_1		$\left\{\begin{array}{l} 3,(0.35), \\ 2,(0.22), \\ -2,(0.40) \end{array}\right\}$	$\left\{\begin{array}{l} 2,(0.35), \\ 1,(0.44), \\ -2,(0.25) \end{array}\right\}$	$\left\{\begin{array}{l} 3,(0.35), \\ -2,(0.25), \\ 2,(0.3) \end{array}\right\}$
	S_2			$\left\{\begin{array}{l} 1,(0.35), \\ 2,(0.3), \\ -2,(0.40) \end{array}\right\}$	$\left\{\begin{array}{l} 3,(0.25), \\ -1,(0.3), \\ 2,(0.45) \end{array}\right\}$
	S_3				$\left\{\begin{array}{l} -2,(0.45), \\ 2,(0.30), \\ -3,(0.25) \end{array}\right\}$
	S_4				
C_2	S_1		$\left\{\begin{array}{l} -2,(0.38), \\ 1,(0.44), \\ -1,(0.30) \end{array}\right\}$	$\left\{\begin{array}{l} 0,(0.25), \\ 2,(0.45), \\ -1,(0.45) \end{array}\right\}$	$\left\{\begin{array}{l} 3,(0.18), \\ -2,(0.35), \\ -1,(0.25) \end{array}\right\}$
	S_2			$\left\{\begin{array}{l} -2,(0.45), \\ 0,(0.27), \\ 1,(0.35) \end{array}\right\}$	X
	S_3				$\left\{\begin{array}{l} 1,(0.25), \\ 2,(0.35), \\ -1,(0.35) \end{array}\right\}$
	S_4				
C_3	S_1		$\left\{\begin{array}{l} 1,(0.45), \\ 0,(0.35), \\ -2,(0.18) \end{array}\right\}$	$\left\{\begin{array}{l} 2,(0.25), \\ 1,(0.44), \\ -2,(0.35) \end{array}\right\}$	X
	S_2			$\left\{\begin{array}{l} 2,(0.42), \\ 1,(0.35), \\ 0,(0.35) \end{array}\right\}$	$\left\{\begin{array}{l} 1,(0.35), \\ -1,(0.45), \\ -2,(0.50) \end{array}\right\}$
	S_3				$\left\{\begin{array}{l} -2,(0.35), \\ 1,(0.42), \\ 0,(0.32) \end{array}\right\}$
	S_4				
C_4	S_1		$\left\{\begin{array}{l} -1,(0.3), \\ 1,(0.42), \\ -2,(0.25) \end{array}\right\}$	$\left\{\begin{array}{l} -2,(0.35), \\ -1,(0.42), \\ 2,(0.25) \end{array}\right\}$	$\left\{\begin{array}{l} 3,(0.43), \\ 0,(0.35), \\ 1,(0.25) \end{array}\right\}$
	S_2			$\left\{\begin{array}{l} 2,(0.35), \\ -2,(0.25), \\ 1,(0.45) \end{array}\right\}$	$\left\{\begin{array}{l} 2,(0.4), \\ -1,(0.35), \\ 3,(0.25) \end{array}\right\}$
	S_3				$\left\{\begin{array}{l} 2,(0.45), \\ 0,(0.35), \\ -1,(0.25) \end{array}\right\}$
	S_4				

Table 8. Filling of missing values in PLPRs.

Attributes	Supplier	S_1	S_2	S_3	S_4
C_1	S_1		3, (0.35), 2, (0.22), −2, (0.40)	2, (0.35), 1, (0.44), −2, (0.25)	3, (0.35), −2, (0.25), 2, (0.3)
	S_2			1, (0.35), 2, (0.3), −2, (0.40)	3, (0.25), −1, (0.3), 2, (0.45)
	S_3				−2, (0.45), 2, (0.30), −3, (0.25)
	S_4				
C_2	S_1		−2, (0.38), 1, (0.44), −1, (0.30)	0, (0.25), 2, (0.45), −1, (0.45)	3, (0.18), −2, (0.35), −1, (0.25)
	S_2			−2, (0.45), 0, (0.27), 1, (0.35)	1, (0.18), 1, (0.23), −1, (0.23)
	S_3				1, (0.25), 2, (0.35), −1, (0.35)
	S_4				
C_3	S_1		1, (0.45), 0, (0.35), −2, (0.18)	2, (0.25), 1, (0.44), −2, (0.35)	0, (0.11), 2, (0.09), 0, (0.08)
	S_2			2, (0.42), 1, (0.35), 0, (0.35)	1, (0.35), −1, (0.45), −2, (0.50)
	S_3				−2, (0.35), 1, (0.42), 0, (0.32)
	S_4				
C_4	S_1		−1, (0.3), 1, (0.42), −2, (0.25)	−2, (0.35), −1, (0.42), 2, (0.25)	3, (0.43), 0, (0.35), 1, (0.25)
	S_2			2, (0.35), −2, (0.25), 1, (0.45)	2, (0.4), −1, (0.35), 3, (0.25)
	S_3				2, (0.45), 0, (0.35), −1, (0.25)
	S_4				

Step 3: Determine the consistency of each PLPR and repair the inconsistent PLPR iteratively using the proposed procedure given in Section 3.2. The $d\left(R_1, R_1^z\right)$ is 0.13 which is inconsistent and it is made consistent in two iterations with $d\left(R_1, R_1^z\right)$ as 0.013. Further, $d\left(R_2, R_2^z\right)$ is 0.091 which is inconsistent and it is made consistent in two iterations with $d\left(R_2, R_2^z\right)$ as 0.03. The $d\left(R_3, R_3^z\right)$ and $d\left(R_4, R_4^z\right)$ are 0.14 and 0.11 respectively which is inconsistent and it is made consistent with in a single iteration with $d\left(R_3, R_3^z\right)$ as 0.021 and 0.025 respectively.

$$
R_1^{z+1} = \begin{pmatrix}
\left\{ \begin{matrix} 0,(1), \\ 0,(1), \\ 0,(1) \end{matrix} \right\} & \left\{ \begin{matrix} 2,(0.15), \\ 2,(0.12), \\ -2,(0.17) \end{matrix} \right\} & \left\{ \begin{matrix} 2,(0.16), \\ 2,(0.13), \\ -2,(0.13) \end{matrix} \right\} & \left\{ \begin{matrix} 2,(0.15), \\ -1,(0.13), \\ 0,(0.15) \end{matrix} \right\} \\[4pt]
\left\{ \begin{matrix} -2,(0.15), \\ -2,(0.12), \\ 2,(0.17) \end{matrix} \right\} & \left\{ \begin{matrix} 0,(1), \\ 0,(1), \\ 0,(1) \end{matrix} \right\} & \left\{ \begin{matrix} 2,(0.16), \\ 0,(0.14), \\ 0,(0.16) \end{matrix} \right\} & \left\{ \begin{matrix} 2,(0.16), \\ -2,(0.14), \\ 2,(0.19) \end{matrix} \right\} \\[4pt]
\left\{ \begin{matrix} -2,(0.16), \\ -2,(0.13), \\ 2,(0.13) \end{matrix} \right\} & \left\{ \begin{matrix} -2,(0.16), \\ 0,(0.14), \\ 0,(0.16) \end{matrix} \right\} & \left\{ \begin{matrix} 0,(1), \\ 0,(1), \\ 0,(1) \end{matrix} \right\} & \left\{ \begin{matrix} 2,(0.17), \\ 0,(0.15), \\ 1,(0.15) \end{matrix} \right\} \\[4pt]
\left\{ \begin{matrix} -2,(0.15), \\ 1,(0.13), \\ 0,(0.15) \end{matrix} \right\} & \left\{ \begin{matrix} -2,(0.16), \\ 2,(0.14), \\ -2,(0.19) \end{matrix} \right\} & \left\{ \begin{matrix} -2,(0.17), \\ 0,(0.15), \\ -1,(0.15) \end{matrix} \right\} & \left\{ \begin{matrix} 0,(1), \\ 0,(1), \\ 0,(1) \end{matrix} \right\}
\end{pmatrix}
$$

Just as an example, the consistent PLPR $R_1^{(2)}$ after second iteration is shown above.

Step 4: From step 3, we obtain consistent PLPRs which are used for prioritizing green suppliers and selection of a suitable green supplier for the automobile industry. The extended AHP under PLPR context (from Section 3.3) is used for prioritization of green suppliers.

Table 9 shows the decision matrix which is formed by applying Equation (2) over Table 8. The elements of Table 9 are PLEs and the order of the matrix is 4×4.

Table 9. Decision matrix with PLTS information.

Supplier vs. Attributes	C_1	C_2	C_3	C_4
S_1	2, (0.57), 2, (0.56), 2, (0.59)	2, (0.58), 2, (0.58), 2, (0.57)	2, (0.55), 2, (0.55), 2, (0.55)	2, (0.61), 2, (0.64), 2, (0.60)
S_2	2, (0.50), 1, (0.51), 2, (0.55)	2, (0.55), 1, (0.52), 2, (0.52)	2, (0.52), 1, (0.52), 2, (0.55)	2, (0.54), 2, (0.58), 2, (0.60)
S_3	2, (0.56), 1, (0.57), 2, (0.58)	2, (0.60), 2, (0.58), 2, (0.59)	1, (0.57), 1, (0.59), 2, (0.61)	2, (0.61), 2, (0.63), 0, (0.61)
S_4	2, (0.51), 2, (0.51), 2, (0.54)	1, (0.53), 0, (0.52), 2, (0.51)	2, (0.49), 1, (0.50), 2, (0.52)	1, (0.55), 2, (0.59), 2, (0.56)

The green suppliers are prioritized by applying Equation (2) on Table 9. Table 9 depicts the PLPR values after step 5 of Section 3.3. We again apply Equation (2) on Table 9 to obtain a vector (order 4×1) of PLEs corresponding to each green supplier. The green suppliers are prioritized using the vector and it is given by: $S_1 = \{2, (0.57), 2, (0.58), 2, (0.58)\}$; $S_2 = \{2, (0.53), 2, (0.54), 2, (0.55)\}$; $S_3 = \{1, (0.59), 1, (0.59), 0, (0.60)\}$ and $S_1 = \{1, (0.52), 0, (0.53), 0, (0.54)\}$.

By applying Equation (10) on this vector, we obtain the ranking order as $S_1 \succ S_2 \succ S_4 \succcurlyeq S_3$.

Step 5: Compare the superiority and weakness of the proposed framework with other methods (refer Section 5 for details).

5. Comparative Analysis: PLPR Based Decision Framework vs. Others

In this section, we make a comparative analysis of the proposed decision framework with [32] and [21]. The method [32] presents an extension to AHP method under PLTS context and method [21] extends AHP method to HFLTS context. In order to maintain homogeneity in the process of comparison, the proposed decision framework is compared with [32] and [21]. Table 10 shows the analysis of these methods under the theoretic and numeric perspectives. The theoretic factors are chosen based on intuition and the numeric factors are chosen from [36].

The strengths of the proposed decision framework are:

(1) Unlike methods [21,27], the proposed framework can handle missing values in the PLPR in a much sensible and rational manner by automatically filling the missing values using a systematic procedure.

(2) Though, method [27] presents a procedure for consistency check and repair, it is complex and computationally intensive as it involves Eigen vector calculation and uses logarithmic function. To circumvent the issue, the proposed framework presents a systematic procedure for consistency check and repairing inconsistent PLPRs. The procedure automatically repairs inconsistency in an iterative manner with less intervention from DMs. The proposed procedure is computationally feasible as it uses operational law(s) of PLTS.

(3) Method [21] extends AHP under HFLTS context for ranking objects which loses potential probability information and hence, produces unreasonable ranking of objects. Further, method [27] extends AHP under PLTS context but, loses some information when transforming PLTS information to single values using possibility degree. To circumvent the issue, the proposed framework presents a method for ranking objects by extending the popular AHP under PLPR context. The preference information is retained throughout the formulation and hence, information loss is mitigated in an effective manner.

(4) The practicality of the proposed framework is also realized by solving green supplier selection problem for a healthcare center.

(5) Also, from the time complexity analysis, we can observe that proposed decision framework and method [27] has three crucial operations viz., (a) filling missing values, (b) check & repair of inconsistent and (c) ranking of objects with m objects and n attributes. Operation (a) takes $O\left(m^2\right)$ time complexity, operation (b) takes $O\left(m^2\right)$ time complexity and operation (c) takes $O\left(m^2(n+1)\right)$. So, the complexity of the proposed decision framework is $O\left(3m^2 + nm^2\right) \approx O\left(m^2\right)$. In contrary, the complexity of [27] (by similar analysis) is $O\left(m^3 + m^2 + nm^2\right) \approx O\left(m^3\right)$ which is evidently complex than the proposed decision framework.

Some weaknesses of the proposed framework are:

(1) It is computationally complex because of the idea of pair-wise comparison.
(2) Also, the agility for judgment is slow (refer Table 7) because of the pair-wise comparison.

Table 10. Investigation of features: Proposed vs. Others.

Context(s)	Method(s)		
	Proposed	Xie et al. [32]	Tuysuz and Simsek [21]
Input	PLTS information	PLTS information	HFLTS information
Aggregation	Ring sum operator	PLWG	no
Weight calculation	Ring sum operator	Eigen vectors	no
Fuzziness	yes	yes	yes
Occurring probability	yes	yes	no
Total preorder	yes	yes	yes
Missing value(s)	Filled automatically using a systematic procedure	no	no
Consistency	Check & repair using systematic procedure	Check & repair using expectation measure for geometric consistency index	no
Adequacy test	Causes rank reversal issue when adequate changes are made to objects and attributes.		
Scalability	Obeys Saaty's principle [37]		
Agility	$(n(n-1)/2) + n(m(m-1)/2)$ where n is the number of attributes and m is the number of objects.		
Ranking principle	Pair-wise comparison and Equation (10) are used for ranking objects.	Pair-wise comparison and possibility degree measure are used for ranking objects.	Pair-wise comparison and pessimistic & optimistic preference evaluation.
Information loss	Mitigated to a great extent by retaining the PLTS information throughout the decision process	Some information is lost when PLTS information is converted into a single value using possibility degree	Crucial occurring probability value is missing in HFLTS context

6. Conclusions

This paper presents a new scientific decision framework under the PLPR context for rational decision-making under critical situations. The missing values are sensibly filled by using a systematic approach. Also, the consistency of the PLPR is determined and inconsistent PLPRs are repaired using the proposed method. Finally, the AHP method is extended to PLPR for selecting a suitable object from the set of objects. The practicality of the proposed decision framework is demonstrated by solving equipment supplier selection problem for a healthcare center. Also, the strengths and weaknesses of the proposal are realized by comparison with other methods under both theoretic and numeric perspectives.

Some managerial implications are presented in a nutshell below:

(1) The proposed framework can be used as a "ready-to-use" framework for rational decision-making under uncertain situations.
(2) Also, the consistency of the information is ensured by using a systematic procedure without loss of substantial preference information.
(3) This framework can be used by the managers for proper planning of inventory and management of profit and risk the organization.
(4) Further, customers can use this framework as a supplementary aid for making rational decisions.

As a part of the future scope, we plan the following research directions: (i) to present new methods for ranking under pair-wise comparison ideas; (ii) to enhance the consistency of the PLPRs under both additive and multiplicative context; (iii) to develop methods for consensus reaching by gaining motivation from [38,39] and strategic weight calculation inspired by [40,41].

Author Contributions: The individual contribution and responsibilities of the authors were as follows: R.K., K.S.R. and M.I.A. designed the research model, collected, pre-processed, and analyzed the data and the obtained results, and worked on the development of the paper. S.K. and S.K.T. provided good advice throughout the research by giving suggestions on model design, methodology, and inferences, and refined the manuscript. All the authors have read and approved the final manuscript.

Funding: This research was funded by University Grants Commission (UGC), India and Department of Science & Technology (DST), India under grant number F./2015-17/RGNF-2015-17-TAM-83 and SR/FST/ETI-349/2013.

Acknowledgments: Authors thank the editors and the anonymous reviewers for their insightful comments which improved the quality of the paper.

Conflicts of Interest: The authors declare no conflict of interest.

References

1. Triantaphyllou, E.; Shu, B. Multi-attributes decision-making: an operations research approach. *Encycl. Electr. Electron. Eng.* **1998**, *15*, 175–186.
2. Herrera, F.; Herrera-Viedma, E.; Verdegay, J.L. A sequential selection process in group decision-making with a linguistic assessment approach. *Inf. Sci.* **1995**, *239*, 223–239. [CrossRef]
3. Herrera, F.; Herrera-Viedma, E.; Verdegay, J.L. Linguistic measures based on fuzzy coincidence for reaching consensus in group decision-making. *Int. J. Approx. Reason.* **1997**, *16*, 309–334. [CrossRef]
4. Gou, X.; Xu, Z. Novel basic operational laws for linguistic terms, hesitant fuzzy linguistic term sets and probabilistic linguistic term sets. *Inf. Sci.* **2016**, *372*, 407–427. [CrossRef]
5. Zadeh, L.A. The concept of a linguistic variable and its application to approximate reasoning–I. *Inf. Sci.* **1975**, *8*, 199–249. [CrossRef]
6. Herrera, F.; Herrera-Viedma, E.; Verdegay, J.L. A model of consensus in group decision-making under linguistic assessments. *Fuzzy Sets Syst.* **1996**, *78*, 73–87. [CrossRef]
7. Xu, Z. An approach based on the uncertain LOWG and induced uncertain LOWG operators to group decision-making with uncertain multiplicative linguistic preference relations. *Decis. Support. Syst.* **2006**, *41*, 488–499. [CrossRef]
8. He, Y.; Guo, H.; Jin, M.; Ren, P. A linguistic entropy weight method and its application in linguistic multi-attribute group decision-making. *Nonlinear Dyn.* **2016**, *84*, 399–404. [CrossRef]
9. Saaty, T.L. How to make a Decision: Analytical hierarchy process. *Eur. J. Oper. Res.* **1990**, *48*, 9–26. [CrossRef]
10. Herrera, F.; Herrera-Viedma, E. Choice functions and mechanisms for linguistic preference relations. *Eur. J. Oper. Res.* **2000**, *120*, 144–161. [CrossRef]
11. Herrera, F.; Herrera-Viedma, E. Linguistic decision analysis: steps for solving decision problems under linguistic information. *Fuzzy Sets Syst.* **2000**, *115*, 67–82. [CrossRef]
12. Xu, Z. Deviation measures of linguistic preference relations in group decision-making. *Omega* **2005**, *33*, 249–254. [CrossRef]
13. Morente-Molinera, J.A.; Perez, I.J.; Ureña, M.R.; Herrera-Viedma, E. Building and managing fuzzy ontologies with heterogeneous linguistic information. *Knowl.-Based Syst.* **2015**, *88*, 154–164. [CrossRef]
14. Wang, H.; Xu, Z. Interactive algorithms for improving incomplete linguistic preference relations based on consistency measures. *Appl. Soft Comput.* **2016**, *32*, 66–79. [CrossRef]
15. Rodriguez, R.M.; Martinez, L.; Herrera, F. Hesitant fuzzy linguistic term sets for decision-making. *IEEE Trans. Fuzzy Syst.* **2012**, *20*, 109–119. [CrossRef]
16. Torra, V. Hesitant fuzzy sets. *Int. J. Intell. Syst.* **2010**, *25*, 529–539. [CrossRef]
17. Zhu, B.; Xu, Z. Consistency measures for hesitant fuzzy linguistic preference relations. *IEEE Trans. Fuzzy Syst.* **2014**, *22*, 35–45. [CrossRef]
18. Wang, H.; Xu, Z. Some consistency measures of extended hesitant fuzzy linguistic preference relations. *Inf. Sci.* **2015**, *297*, 316–331. [CrossRef]
19. Wu, Z. A consensus process for hesitant fuzzy linguistic preference relations. In Proceedings of the 2015 IEEE International Conference on Fuzzy System, Istanbul, Turkey, 2–5 August 2015.
20. Song, Y.; Hu, J. A group decision-making model based on incomplete comparative expressions with hesitant linguistic terms. *Appl. Soft Comput.* **2017**, *59*, 174–181. [CrossRef]
21. Tüysüz, F.; Şimşek, B. A hesitant fuzzy linguistic term sets-based AHP approach for analyzing the performance evaluation factors: An application to cargo sector. *Complex. Intell. Syst.* **2017**, 1–9. [CrossRef]

22. Zhang, G.; Dong, Y.; Xu, Y. Consistency and consensus measures for linguistic preference relations based on distribution assessments. *Inf. Fusion* **2014**, *17*, 46–55. [CrossRef]

23. Pang, Q.; Wang, H.; Xu, Z. Probabilistic linguistic term sets in multi-attribute group decision-making. *Inf. Sci.* **2016**, *369*, 128–143. [CrossRef]

24. Zhang, B.; Liang, H.; Gao, Y.; Zhang, G. The optimization-based aggregation and consensus with minimum-cost in group decision-making under incomplete linguistic distribution context. *Knowl.-Based Syst.* **2018**, 1–36. [CrossRef]

25. Bai, C.; Zhang, R.; Qian, L.; Wu, Y. Comparisons of probabilistic linguistic term sets for multi-attributes decision-making. *Knowl.-Based Syst.* **2016**, *119*, 284–291. [CrossRef]

26. Liao, H.; Jiang, L.; Xu, Z.; Xu, J.; Herrera, F. A probabilistic linguistic linear programming method in hesitant qualitative multiple attributes decision-making. *Inf. Sci.* **2017**, *416*, 341–355. [CrossRef]

27. Liu, P.; Teng, F. Some Muirhead mean operators for probabilistic linguistic term sets and their applications to multiple attribute decision-making. *Appl. Soft Comput.* **2018**, *68*, 396–431. [CrossRef]

28. Zhang, Y.; Xu, Z.; Wang, H.; Liao, H. Consistency-based risk assessment with probabilistic linguistic preference relation. *Appl. Soft Comput.* **2016**, *49*, 817–833. [CrossRef]

29. Zhang, Y.; Xu, Z.; Liao, H. A consensus process for group decision-making with probabilistic linguistic preference relations. *Inf. Sci.* **2017**, *414*, 260–275. [CrossRef]

30. Xie, W.; Ren, Z.; Xu, Z.; Wang, H. The consensus of probabilistic uncertain linguistic preference relations and the application on the virtual reality industry. *Knowl.-Based Syst.* **2018**, *162*, 14–28. [CrossRef]

31. Wu, X.; Liao, H. A consensus-based probabilistic linguistic gained and lost dominance score method. *Eur. J. Oper. Res.* **2019**, *272*, 1017–1027. [CrossRef]

32. Xie, W.; Xu, Z.; Ren, Z.; Wang, H. Probabilistic linguistic analytic hierarchy process and its application on the performance assessment of Xiongan new area. *Int. J. Inf. Technol. Decis. Mak.* **2018**, *16*, 1–32. [CrossRef]

33. Liao, H.; Xu, Z.; Xia, M. Multiplicative consistency of interval-valued intuitionistic fuzzy preference relation. *J. Intell. Fuzzy Syst.* **2014**, *27*, 2969–2985.

34. Saaty, T.L. *The Analytic Hierarchy Process*, 1st ed.; Mc-Graw Hill: New York, NY, USA, 1980; pp. 579–606.

35. Emrouznejad, A.; Marra, M. The state of the art development of AHP (1979–2017): A literature review with a social network analysis. *Int. J. Prod. Res.* **2017**, *55*, 6653–6675. [CrossRef]

36. Lima Junior, F.R.; Osiro, L.; Carpinetti, L.C.R. A comparison between Fuzzy AHP and Fuzzy TOPSIS methods to supplier selection. *Appl. Soft Comput.* **2014**, *21*, 194–209. [CrossRef]

37. Saaty, T.L.; Ozdemir, M.S. Why the magic number seven plus or minus two. *Math. Comput. Model.* **2003**, *28*, 233–244. [CrossRef]

38. Zhang, H.; Dong, Y.; Chen, X. The 2-rank consensus reaching model in the multigranular linguistic multiple-attribute group decision-making. *IEEE Trans. Syst. Man Cybern. Syst.* **2017**, *48*, 1–15. [CrossRef]

39. Li, C.C.; Dong, Y.; Herrera, F. A consensus model for large-scale linguistic group decision-making with a feedback recommendation based on clustered personalized individual semantics and opposing consensus groups. *IEEE Trans. Fuzzy Syst.* **2018**, 1–13. [CrossRef]

40. Liu, Y.; Dong, Y.; Liang, H.; Chiclana, F.; Herrera-Viedma, E. Multiple attribute strategic weight manipulation with minimum cost in a group decision-making context with interval attribute weights information. *IEEE Trans. Syst. Man Cybern. Syst.* **2018**, 1–12. [CrossRef]

41. Dong, Y.; Liu, Y.; Liang, H.; Chiclana, F.; Herrera-Viedma, E. Strategic weight manipulation in multiple attribute decision-making. *Omega* **2018**, *75*, 1339–1351. [CrossRef]

© 2018 by the authors. Licensee MDPI, Basel, Switzerland. This article is an open access article distributed under the terms and conditions of the Creative Commons Attribution (CC BY) license (http://creativecommons.org/licenses/by/4.0/).

Article

A New Methodology for Improving Service Quality Measurement: Delphi-FUCOM-SERVQUAL Model

Olegas Prentkovskis [1,*], Živko Erceg [2], Željko Stević [2], Ilija Tanackov [3], Marko Vasiljević [2] and Mladen Gavranović [2]

[1] Department of Mobile Machinery and Railway Transport, Faculty of Transport Engineering, Vilnius Gediminas Technical University, Plytinės g. 27, LT-10105 Vilnius, Lithuania
[2] Faculty of Transport and Traffic Engineering, University of East Sarajevo, Vojvode Mišića 52, 74000 Doboj, Bosnia and Herzegovina; zivko.erceg@sf.ues.rs.ba (Ž.E.); zeljko.stevic@sf.ues.rs.ba (Ž.S.); marko.vasiljevic@sf.ues.rs.ba (M.V.); gavranovicrsv@gmail.com (M.G.)
[3] Faculty of Technical sciences, University of Novi Sad, Trg Dositeja Obradovića 6, 21000 Novi Sad, Serbia; ilijat@uns.ac.rs
* Correspondence: olegas.prentkovskis@vgtu.lt

Received: 25 November 2018; Accepted: 12 December 2018; Published: 16 December 2018

Abstract: The daily requirements and needs imposed on the executors of logistics services imply the need for a higher level of quality. In this, the proper execution of all sustainability processes and activities plays an important role. In this paper, a new methodology for improving the measurement of the quality of the service consisting of three phases has been developed. The first phase is the application of the Delphi method to determine the quality dimension ranking. After that, in the second phase, using the FUCOM (full consistency method), we determined the weight coefficients of the quality dimensions. The third phase represents determining the level of quality using the SERVQUAL (service quality) model, or the difference between the established gaps. The new methodology considers the assessment of the quality dimensions of a large number of participants (customers), on the one hand, and experts' assessments on the other hand. The methodology was verified through the research carried out in an express post company. After processing and analyzing the collected data, the Cronbach alpha coefficient for each dimension of the SERVQUAL model for determining the reliability of the response was calculated. To determine the validity of the results and the developed methodology, an extensive statistical analysis (ANOVA, Duncan, Signum, and chi square tests) was carried out. The integration of certain methods and models into the new methodology has demonstrated greater objectivity and more precise results in determining the level of quality of sustainability processes and activities.

Keywords: quality; sustainability processes; Delphi; FUCOM (full consistency method); SERVQUAL (service quality); new methodology

---

1. Introduction

According to Nowotarski [1], it can be said that quality is directly connected with meeting requirements, expectations, and needs of customers. By applying different tools and techniques, it is possible to manage a quality level in one way. Measuring the quality of all processes that make a coherent whole can greatly affect the full quality of service in all areas. Whether a service will be reused also depends on adequate quality, especially nowadays, when production processes are of approximate and high quality. In such conditions, proper and perfect execution of logistics services can have a crucial impact on its reuse. It is important to strive constantly for higher goals and their achievements. It requires also an adequate methodology that can help improve the quality measurement of logistics services.

The research domain is the logistics of express post, including all the activities and processes carried out within it, from the aspect of logistics service quality. The activities included into the domain of research are the activities of informing customers of express post services until the end that implies a logistics service provided. The survey was conducted on a sample of 70 respondents, permanent customers of services of the express post company, as well as customers who used services on a one-time basis. Introducing the types of express post services to customers leads to the creation of a certain degree of expectation, which may differ from the perception of the service provided. The subject of the research is to determine the quality of the logistics service of express post based on a new developed methodology. The motivation for execution of this research can be explained through two main reasons. The first reason represents a lack of universal methodology for service quality assessment that considers the nature of input parameters, needs, and requests of customers' ability of companies and other uncertainties. The second reason is the possibility for improving the efficiency of a company that is the object of research by developing a new methodology, which can be useful for strategic management and planning. This paper has several goals. The first one relates to the development of a new methodology that treats input and output parameters with precision and provides results that are more objective. The first goal is achieved throughout three different phases, which, when integrated, create the developed model. The advantages of the Delphi method are used first, whereby a total of 70 customers provide weighted dimension values, based on which, a ranking is made. Thereafter, the FUCOM (full consistency method) for determining the weight dimension values is applied, allowing consistent evaluations by the experts involved in this process. The second goal is to enrich the methodology for improving service quality measurement by applying the new developed model. This provides an adequate methodology for future research in this area. In addition, the third goal of the paper is to determine the difference between expectations and perceptions of the formed dimensions of the modified SERVQUAL (service quality) model and the possibility of identifying and improving critical factors of the logistics service in an express post company, which is the object of research.

After introductory considerations where the significance of research and goals are presented, the paper is structured throughout six more sections. Section 2 provides a review of the application of the SERVQUAL model in various areas for measuring the quality of different processes. Section 3 presents the new developed methodology that implies the integration of three different methods to provide the most accurate outputs. There is a flow chart of the study with an explanation of all phases and steps. Section 4 is a case study where the input parameters are defined, quality measurements are presented, the initial dimension ranking is provided, and the weighted values of all five dimensions are calculated. Section 5 presents the results of the research using the developed methodology, while Section 5.3 provides a comprehensive statistical analysis that establishes the regularities and conditions of expectation and perception processes. Section 6 is a conclusion, with an emphasis on the scientific contribution of this research and guidelines for future research.

2. Literature Review

A model that is often used to measure the quality of service is the SERVQUAL model. Motivated by the need to measure the contribution of the SERVQUAL model, Wang et al. [2] conducted a study, which proved that the SERVQUAL model was one of the major research topics for academic researchers in the period from 1998 to 2013 and that the model contributed significantly to the research on service quality.

2.1. Quality Measurement in Logistics and Transport

According to Kersten and Koch [3], in the past decades, the scope of logistics services has broadened from the provision of isolated services, such as transport and warehousing, to the management and handling of the flow of goods for entire companies. In such conditions of the market, service quality has a large influence on company efficiency. One of the most applied models for service quality is the SERVQUAL model. This model was applied in the field of passenger traffic [4], where the stated hypotheses were disproven because of a negative gap, and the SERVQUAL model pointed to critical business functions and the possibility of their improvement. In [5], the SERVQUAL model was based on 10 logistics service attributes for estimating performances in the field of refrigerated transport. The proof of how much the SERVQUAL model is used in all areas was shown by Roslan et al. [6], where the model measured the quality of service of logistics centers in Iskandar, Malaysia. For the same purpose, in research [7], authors developed a new hybrid MCDM (multi-criteria decision-making) model, consisting of an analytic hierarchy process (AHP), decision-making trial, and evaluation laboratory (DEMATEL), and analytic network process (ANP) methods. A combined approach integrating gap analysis, quality function deployment (QFD), and AHP for improving logistics service quality was applied in [8]. In research [9], an extension of the three-column format SERVQUAL instrument was extended for evaluation of passenger rail service quality. Three new transport dimensions (comfort, connection, and convenience) were added to the original five SERVQUAL dimensions.

For the evaluation of service quality in logistics and other fields, the Kano model [10–12], QFD method [13,14], six sigma [15,16] etc. can be applied, or, for example, a new developed Agro-Logistic Analysis and Design Instrument (ALADIN) model, which involves logistics, sustainability, and food quality analysis [17].

2.2. Quality Measurement in Other Fields

In their paper, Cho et al. [18] explored ways to improve services in service centers of electronics companies. They introduced and modified the SERVQUAL model to understand customers' demands for all service centers. According to Paryani et al. [19], the SERVQUAL model is also a very useful tool for identifying customers' demands. The evidence of how much the SERVQUAL model is present in studies is also shown in [20], where the authors used the model to assess patients' satisfaction by providing services at Sunyani Regional Hospital in Ghana; Behdioğlu et al. [21], who evaluated the quality of services at Yoncalı Physiotherapy and Rehabilitation Hospital in Kutahya, Turkey; Singh and Prasher [22], who measured the quality of services in hospitals from the Punjab state of India; as well as Khan et al. [23], who also measured the quality of services in hospitals. Using the SERVQUAL model, Chou et al. [24] have proved that the quality of service largely depends on the subjective assessment of service customers. To rank life insurance companies and assess the quality of services provided, Saeedpoor et al. [25] also used the SERVQUAL model. Additionally, the SERVQUAL model was used to measure the impact of technology on the quality of banking services and to measure the level of customer satisfaction [26]. Using the SERVQUAL model, Long [27] and Apornak [28] have shown that there is a significant link between technology used in providing services and the quality of services. The SERVQUAL model has also been used in a number of studies to rate the quality of banking services provided [29–33]. Wang et al. [34] also used five dimensions of the SERVQUAL model (tangibles, reliability, responsiveness, assurance, and empathy) to measure the service quality of an e-learning system. Moreover, those five dimensions were used by Yang and Zhu [35] to highlight the quality of community-based service provided by university-affiliated stadiums, as well as Luo et al. [36] while measuring satisfaction of outward-bound tourists.

2.3. Integrated MCDM-SERVQUAL Model for Quality Measurement

To measure the perception of service quality, Altuntas et al. [37] used the SERVQUAL model and two of the most known methods of MCDM method-based scales. By applying MCDM methods, it is possible to choose appropriate strategies, rationalize certain logistics and other processes, and make appropriate decisions that affect the operations of companies or their subsystems, as proved by the following research [38–51]. These methods can be easily integrated into other approaches, such as integration with SWOT (strengths, weaknesses, opportunities, and threats) analysis [42] or with the SERVQUAL model, as is the case in this paper. Rezaei et al. [52] integrated the SERVQUAL model with the best worst method, while Xuehua [53] applied a combined fuzzy AHP-SERVQUAL (analytic hierarchy process for service quality) model for evaluation of express service quality. The model was based on 14 indicators divided into five standard dimensions.

3. New Methodology: DELPHI-FUCOM-SERVQUAL Model

3.1. The Proposed Methodology

The developed methodology (Figure 1) for improving service quality measurement consists of four phases, with 18 steps in total. The first phase refers to the collection and preparation of data, which consists of six steps. First, it is necessary to form a SERVQUAL questionnaire on which the results of the research depend to a significant extent. It is necessary to consider the interdependence of certain elements of the questionnaire, which may influence the reliability of subsequent results. In this research, two important elements are taken into consideration when forming the questionnaire, the satisfaction of both the scientific and professional aspects.

Accordingly, scientists were consulted and the opinions of the management of the express post company were taken. A classic SQ (SERVQUAL) questionnaire consisting of 22 expectation questions and the same number of questions for perceptions was devised. The first contribution of this methodology is the modification of the SQ questionnaire for a specific case and the formation of a total of 25 elements for expectations and perceptions. It is recommended that this number is 20–30, depending on a specific situation. Subsequently, in the second step, the questionnaire was sent to customers to carry out their assessment in the fourth step, while the team of experts for evaluating the main dimensions of the SQ questionnaire was formed in the third step. Then, in the fifth step, hypotheses were defined, the number of which may vary depending on the area of application and a specific problem. It is possible to form hypotheses for each SQ dimension or for the overall SQ gap. In the last sixth step of the first phase, the data were processed and prepared for the next phase. The second phase implies the integration of different approaches into a new methodology consisting of nine steps. It is necessary to apply the Delphi method in the first step to allow customers to express their preferences regarding the main dimensions, i.e., their significance. After the results were obtained using the Delphi method, a ranking of all five dimensions was performed, so that a team of experts could determine their preferences. In the second step, the FUCOM for obtaining the weight values of SQ dimensions was applied. As it is group decision-making, all steps of this method should be implemented in the third step for each expert individually. In the fifth step, the averaging of the values obtained in the previous step to gain the final weight values of dimensions was performed. The sixth step determines the mean value of customers' responses for all dimensions regarding expectations, while, in the seventh step, the same was performed for perceptions. In the eighth step, the mean values obtained in the previous two steps were multiplied by the weight values obtained by the FUCOM. In the final step of this phase, the difference between perceptions and expectations was determined by taking into consideration the previously obtained values. The third phase implies the determination of the model reliability, which is defined by two steps: The calculation of the Cronbach alpha coefficient for all SQ dimensions and the performance of statistical analysis. The choice of adequate statistical tests is conditioned by the allocations of customers' responses, so it is impossible to define a universal one for application in this phase. Finally, the application of an adequate statistical test, and confirmation or rejection of previously set hypotheses was performed.

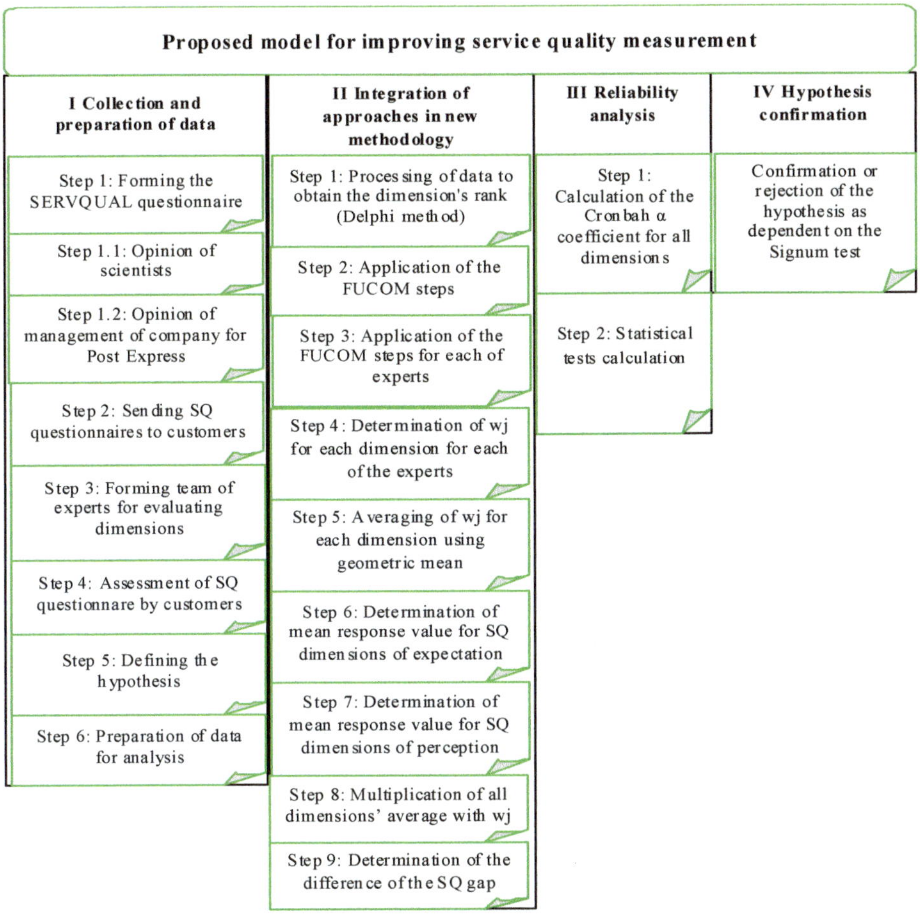

<table>
<tr>
<td colspan="4" align="center">Proposed model for improving service quality measurement</td>
</tr>
<tr>
<td align="center">I Collection and preparation of data</td>
<td align="center">II Integration of approaches in new methodology</td>
<td align="center">III Reliability analysis</td>
<td align="center">IV Hypothesis confirmation</td>
</tr>
<tr>
<td>Step 1: Forming the SERVQUAL questionnaire</td>
<td>Step 1: Processing of data to obtain the dimension's rank (Delphi method)</td>
<td>Step 1: Calculation of the Cronbah α coefficient for all dimensions</td>
<td>Confirmation or rejection of the hypothesis as dependent on the Signum test</td>
</tr>
<tr>
<td>Step 1.1: Opinion of scientists</td>
<td>Step 2: Application of the FUCOM steps</td>
<td></td>
<td></td>
</tr>
<tr>
<td>Step 1.2: Opinion of management of company for Post Express</td>
<td>Step 3: Application of the FUCOM steps for each of experts</td>
<td>Step 2: Statistical tests calculation</td>
<td></td>
</tr>
<tr>
<td>Step 2: Sending SQ questionnaires to customers</td>
<td>Step 4: Determination of wj for each dimension for each of the experts</td>
<td></td>
<td></td>
</tr>
<tr>
<td>Step 3: Forming team of experts for evaluating dimensions</td>
<td>Step 5: Averaging of wj for each dimension using geometric mean</td>
<td></td>
<td></td>
</tr>
<tr>
<td>Step 4: Assessment of SQ questionnare by customers</td>
<td>Step 6: Determination of mean response value for SQ dimensions of expectation</td>
<td></td>
<td></td>
</tr>
<tr>
<td>Step 5: Defining the hypothesis</td>
<td>Step 7: Determination of mean response value for SQ dimensions of perception</td>
<td></td>
<td></td>
</tr>
<tr>
<td>Step 6: Preparation of data for analysis</td>
<td>Step 8: Multiplication of all dimensions' average with wj</td>
<td></td>
<td></td>
</tr>
<tr>
<td></td>
<td>Step 9: Determination of the difference of the SQ gap</td>
<td></td>
<td></td>
</tr>
</table>

Figure 1. New methodology for improving service quality measurement.

In this paper, a new Delphi-FUCOM-SERVQUAL methodology has been developed to improve the process of service quality measurement. The advantages of the new methodology developed are reflected in that it provides precise treatment of input and output parameters, obtaining results that are more objective. Firstly, the advantages of Delphi method were used, whereby a total of 70 customers provided weight values of dimensions and based on which their ranking was made. Thereafter, the FUCOM for determining the weight values of dimensions was applied, allowing consistent evaluations of the experts involved in this process to determine finally the difference between perceptions and expectations of the modified SERVQUAL model. Mentioned advantages make this method better than other similar approaches because of the way data is handled. The developed methodology can be applied without any restrictions in various research fields. In addition, it is possible to determine the quality and efficiency of the companies which are the objects of research based on the satisfaction of its customers, but it also enables further application and re-application of this methodology. This methodology can be very helpful for strategic management of the company to improve their efficiency. This methodology enusres more precise treatment of input parameters and achieves better results than traditional quality measurement methods.

3.2. Delphi Method

The Delphi method does the study of and gives projections of uncertain or possible future situations for which we are unable to perform objective statistical legalities, to form a model, or apply a formal method. These phenomena are very difficult to quantify because they are mainly qualitative in their nature, i.e., there are not enough statistical data regarding them that could be used as the basis for our studies. The Delphi method is one of the basic forecasting methods, the most famous and most widely used expert judgment method. Methods of experts' assessments represent a significant improvement of the classical ways of obtaining the forecast by joint consultation of an expert group for a certain studied phenomenon. In other words, this is a methodologically organized use of experts' knowledge to predict future states and phenomena. A typical group in one Delphi session ranges from a few to 30 experts. Each interviewed expert, a participant in the method, relies on knowledge, experience, and his/her own opinion. The goal of the Delphi method is to exploit the collective, group thinking of experts about a certain field. The goal is to reach a consensus on an event by group thinking. This is a method of indirect collective testing, but with a return link. It consists of eight steps:

- Step 1: Selection of the prognostic task, defining basic questions and fields for it;
- Step 2: Selection of experts;
- Step 3: Preparation of questionnaires;
- Step 4: Delivery of questionnaires to experts;
- Step 5: Collecting responses and their evaluation;
- Step 6: Analysis and interpretation of responses;
- Step 7: Re-exams; and
- Step 8: Interpretation of responses and setting up of the final forecast.

3.3. Full Consistency Method (FUCOM)

The FUCOM was developed by Pamučar, Stević, and Sremac, [54] for the determination of weights of criteria. It represents a new method that, according to the authors, represents a better method than AHP (analytical hierarchy process) and BWM (best worst method). For now, it has been applied in research by Nunić (2018). It consists of the three following steps.

Step 1. In the first step, the criteria from the predefined set of the evaluation criteria, $C = \{C_1, C_2, \ldots, C_n\}$, are ranked. The ranking is performed according to the significance of the criteria, i.e., starting from the criterion that is expected to have the highest weight coefficient to the criterion of the least significance. Thus, the criteria ranked according to the expected values of the weight coefficients are obtained:

$$C_{j(1)} > C_{j(2)} > \ldots > C_{j(k)} \tag{1}$$

where k represents the rank of the observed criterion. If there is a judgment of the existence of two or more criteria with the same significance, the sign of equality is placed instead of ">" between these criteria in expression (1).

Step 2. In the second step, a comparison of the ranked criteria is carried out and the comparative priority ($\varphi_{k/(k+1)}$, $k = 1, 2, \ldots, n$, where k represents the rank of the criteria) of the evaluation criteria is determined. The comparative priority of the evaluation criteria ($\varphi_{k/(k+1)}$) is an advantage of the criterion of $C_{j(k)}$ rank compared to the criterion of $C_{j(k+1)}$ rank. Thus, the vectors of the comparative priorities of the evaluation criteria are obtained, as in expression (2):

$$\Phi = \left(\varphi_{1/2}, \varphi_{2/3}, \ldots, \varphi_{k/(k+1)} \right) \tag{2}$$

where $\varphi_{k/(k+1)}$ represents the significance (priority) of the criterion of $C_{j(k)}$ rank compared to the criterion of $C_{j(k)}$ rank.

Step 3. In the third step, the final values of the weight coefficients of the evaluation criteria $(w_1, w_2, \ldots, w_n)^T$ are calculated. The final values of the weight coefficients should satisfy the two conditions:

(1) That the ratio of the weight coefficients is equal to the comparative priority among the observed criteria ($\varphi_{k/(k+1)}$) defined in *Step 2*, i.e., that the following condition is met:

$$\frac{w_k}{w_{k+1}} = \varphi_{k/(k+1)} \tag{3}$$

(2) In addition to condition (3), the final values of the weight coefficients should satisfy the condition of mathematical transitivity:

$$\frac{w_k}{w_{k+2}} = \varphi_{k/(k+1)} \otimes \varphi_{(k+1)/(k+2)} \tag{4}$$

Full consistency, i.e., minimum DFC (deviation from full consistency) (χ) is satisfied only if transitivity is fully respected. Based on the defined settings, the final model for determining the final values of the weight coefficients of the evaluation criteria can be defined:

$$\begin{aligned}
& \min \chi \\
& s.t. \\
& \left| \frac{w_{j(k)}}{w_{j(k+1)}} - \varphi_{k/(k+1)} \right| \leq \chi, \ \forall j \\
& \left| \frac{w_{j(k)}}{w_{j(k+2)}} - \varphi_{k/(k+1)} \otimes \varphi_{(k+1)/(k+2)} \right| \leq \chi, \ \forall j \\
& \sum_{j=1}^{n} w_j = 1, \ \forall j \\
& w_j \geq 0, \ \forall j
\end{aligned} \tag{5}$$

By solving the model (5), the final values of the evaluation criteria $(w_1, w_2, \ldots, w_n)^T$ and the degree of DFC (χ) are generated.

3.4. SERVQUAL Model

The model was developed in 1985 [55] and purified and improved in 1988 [56] and 1994 [57]. In current practice, it has become one of the most distinguished models in the area of service quality. It is expressed by the "perception minus expectation" algorithm.

The SERVQUAL model includes five basic quality dimensions: Tangibles, reliability, responsiveness, assurance, and empathy. Each of these dimensions is described by its attributes. The SERVQUAL model quality function is expressed by Equation (6):

$$SQ_i = \Sigma W_j \left(P_{ij} - E_{ij} \right) \tag{6}$$

where: SQ_i—perceived dimension quality; W_j—attribute importance factor; P_{ij}—perception of dimension i in relation to attribute j; E_{ij}—expected level of attribute; and j, which is a normative of dimension i.

Five SERVQUAL dimensions (reliability, responsiveness, assurance, empathy, and tangibles) concisely represent an essential criterion used by customers when assessing the quality of services. The value of the dimensions in a classic SERVQUAL model is determined based on a questionnaire that contains 44 quality characteristics, 22 of which refer to expectations (E) and 22 to perceptions (P).

In this paper, as already mentioned, a modification of the SERVQUAL model has been carried out, which contains a total of 50 quality characteristics arranged equally for expectations and perceptions.

4. Case Study: Measuring the Quality of Logistics Service in a Company of Express Post

In this paper, the quality of logistics service was determined by applying the developed Delphi-FUCOM-SERVQUAL model. The aim of the research from the aspect of the company for which it was conducted was to determine the current level of logistics service quality and to improve it. For the survey of customers, a "Google forms" online application was used. The questionnaire consisted of 25 questions, including five dimensions: Reliability, assurance, tangibles, empathy, and responsiveness. Prior to filling in the questionnaire, respondents provided information, such as: Customer's status, gender, age, and employment status. The survey was conducted using the questionnaire in which a Likert scale was applied, including points from one to five. At the end of the questionnaire, the customer determined the values of weight coefficients depending on which dimension was most important to them. For every individual, each dimension that determines the quality level of service was different importance.

Regarding the status of respondents, 42 out of 70 customers were natural persons, while the remaining 28 were legal entities, i.e., 60% of respondents were natural persons, and 40% of respondents were legal entities. Division by gender shows that customers of both genders were approximately of the same percentage. Then, the highest number of respondents were aged 35–50, i.e., 25 respondents of 70. The percentage of 30% was taken by those aged 24–35, namely 21 respondents. Out of 70 respondents, the highest percentage of 68% belongs to the employed customers of the company's express post services. The target group are young entrepreneurial people with a frequent need for express post services. Table 1 shows the questions included into the questionnaire from the aspect of customer expectations.

Table 1. A questionnaire form from the aspect of customer expectations.

Order No.	Questions
1.	The company will provide a service at the expected time.
2.	Employees in the company will show interest in customers' problems.
3.	The company will provide a service as promised.
4.	Delivery of the shipment will be carried out on the first attempt.
5.	The company will reliably carry out delivery of large value shipment.
6.	The company will deliver the shipment at the expected time for long distance.
7.	Employees' conduct will create trust of customers.
8.	Customers will be safe while using services.
9.	Senders/receivers will be informed if the service is not possible.
10.	Couriers will pick up and/or deliver the shipment at the expected time.
11.	The cost of the service will be acceptable.
12.	Couriers in the company will be kind.
13.	Company's delivery vehicles will be visually appealing.
14.	Packaging of delivered shipment will be clean and neat.
15.	Employees in the company will look neat.
16.	Delivery vehicles will be modern and will have all necessary equipment.
17.	Individual attention will be given to the customer.
18.	Customers will feel comfortable in contact with employees.
19.	Employees in the company will show understanding.
20.	The company will recognize the needs of customers.
21.	The working hours of the company will be appropriate and acceptable to customers.
22.	Employees in the company will be willing and able to help.
23.	Customers will obtain right answers to their questions.
24.	Employees at the Call Center will provide all necessary information to customers.
25.	Upon request, customers will respond quickly and reliably.

Table 1 presents all the questions that were used to test the degree of customer satisfaction. The questions are related to customer expectations about the services provided by the express post company. The questions are divided into five basic dimensions, i.e., the questions from one to six relate to the dimension of reliability, from seven to 10 to the dimension of assurance. The questions from 11 to 16 relate to the dimension of tangibles, from 17 to 21, to the dimension of empathy, and from 22 to 25 to the dimension of responsiveness. In this part of the questionnaire, questions are written in the future tense as they relate to customer expectations for the quality of logistics service. The form of questions for both aspects, expectations and perceptions, is the same, but questions in terms of perceptions are set in the past tense. Perception questions define the real customer perception of the quality of the service provided.

Based on all the above, a hypothesis of the research was set: *There is no significant difference between expectations and perceptions of the SERVQUAL model in providing logistics services.* In addition to the main hypothesis in the paper, some regularity of certain questions and attitudes of the customers has been established.

The dimension of reliability is mainly related to the timely delivery of a service that directly affects the quality of express post-delivery. The questions from the order number seven to 10 relate to the dimension of assurance. Within this dimension, it can be seen the degree of quality that refers to the trust and confidence of customers regarding the services of the express post company. The dimension of tangibles includes the questions that relate exclusively to couriers, delivery vehicles of the company, and the cost of the service. The results of the tangible dimension significantly provide information about the company where the research was conducted. This dimension also carries useful information on the real degree of quality of logistics service. Particular attention should be paid to each customer. Throughout the dimension of empathy, we can see how much the company really focuses on customers, their needs, and their problems. By understanding customers and anticipating their needs, the company can strive for an extremely high quality of service. Within the dimension of responsiveness, there are questions solely related to both daily and extraordinary situations. These are questions related to all necessary information and customers' requests, which can be obtained by employees in the company.

4.1. Determining Dimension Ranks by Supplying the Delphi Method

At the end of the questionnaire, the percentage of the dimensions most important for each customer were determined. The total sum of the assessed dimensions should be 100%. While assessing, customers considered which dimension was personally the most important factor affecting the quality of the logistics service. Table 2 shows the rank of SQ dimensions, used as a basis to create prerequisites for applying the FUCOM.

Table 2 shows the ranks obtained by the customers' responses. The method used to obtain these ranks is as follows: At the end of the questionnaire, all respondents determined the percentage for each dimension. After that, the sum of all values for one dimension was divided by 7000. The coefficient values for each dimension were obtained in the same way. Table 3 shows the percentages of dimensions for each dimension stated.

From Table 3, we can see that the sum of all percentage values is 7000. The procedure to obtain the weight coefficients is as follows: The sum of the percentage values of one dimension was divided by the sum of percentage values for all dimensions. The following example shows how to calculate the value of the weight coefficient for the dimension of assurance (w_j—weight coefficient).

The weight coefficient value for the dimension of assurance is 0.2629:

$$w_j = \frac{sum\ of\ percentage\ values\ for\ the\ dimension\ of\ assurance}{sum\ of\ percentage\ values\ for\ all\ dimensions}$$
$$w_j = \frac{1840}{7000} = 0.2629$$

Table 2. The ranks of dimensions by applying the Delphi method.

Dimension	Rank
Reliability	1
Assurance	2
Tangibles	4
Empathy	5
Responsiveness	3

Table 3. Percentage values of five dimensions by 70 respondents.

Main Indicators	Reliability	Assurance	Tangibles	Empathy	Responsiveness	Σ
Respondent 1	25	20	15	15	25	100
Respondent 2	30	30	10	10	20	100
Respondent 3	25	15	15	20	25	100
Respondent 4	50	30	5	5	10	100
Respondent 5	25	25	15	15	20	100
Respondent 6	25	25	25	15	10	100
Respondent 7	40	30	5	5	20	100
Respondent 8	20	20	20	20	20	100
Respondent 9	20	20	20	20	20	100
Respondent 10	20	20	20	20	20	100
Respondent 11	25	20	20	15	20	100
			...			
Respondent 67	80	10	0	0	10	100
Respondent 68	20	50	0	0	30	100
Respondent 69	20	20	20	0	40	100
Respondent 70	25	20	5	30	20	100
SUM	1860	1840	895	775	1630	7000
w_j	0.2657	0.2629	0.1279	0.1107	0.2329	1
Rank	1	2	3	4	5	

4.2. Determining the Weight Values of Dimensions Applying the FUCOM

Step 1. In the first step, decision-makers need to rank criteria (dimensions). Compared to the original FUCOM, where the experts themselves perform the ranking, in this paper, the same was performed using the Delphi method based on the responses of 70 customers of logistics service. The dimensions ranking is as follows: $D_1 > D_2 > D_5 > D_3 > D_4$.

Step 2. In the second step (Step 2b), the decision-maker performed the pairwise comparison of the ranked dimensions from Step 1. The comparison was made with respect to the first-ranked D1 dimension. In this step, it a team of five experts was formed who assessed previously ranked dimensions. The experts carried out the assessment based on the scale $[1, 9]$. Thus, the priorities of the dimensions $(\omega_{C_{j(k)}})$ by the first decision-maker for all the criteria ranked in Step 1 were obtained (Table 4). Based on the obtained priorities of the dimensions, the comparative priorities of the dimensions were calculated: $\varphi_{C_1/C_2} = 1.2/1 = 1.200$, $\varphi_{C_2/C_5} = 1.5/1.2 = 1.250$, $\varphi_{C_5/C_3} = 2.7/1.5 = 1.800$, and $\varphi_{C_3/C_4} = 3.2/2.7 = 1.185$.

Table 4. Priorities of dimensions.

Dimension	D_1	D_2	D_5	D_3	D_4
$\omega_{C_{j(k)}}$	1	1.2	1.5	2.7	3.2

Step 3. The final values of the weight coefficients should meet the following two conditions:

(1) The final values of the weight coefficients should meet condition (3), i.e., that $\frac{w_1}{w_2} = 1.2$, $\frac{w_2}{w_5} = 1.250$, $\frac{w_5}{w_3} = 1.800$ and $\frac{w_3}{w_4} = 1.185$.

(2) In addition to condition (3), the final values of the weight coefficients should meet the condition of mathematical transitivity, i.e., that $\frac{w_1}{w_5} = 1.2 \times 1.25 = 1.500$, $\frac{w_2}{w_3} = 1.25 \times 1.8 = 2.250$, and $\frac{w_5}{w_4} = 1.8 \times 1.185 = 2.133$. By applying expression (5), the final model for determining the weight coefficients can be defined as:

$$\min \chi$$

$$s.t. \begin{cases} \left| \frac{w_1}{w_2} - 1.200 \right| \le \chi, \left| \frac{w_2}{w_5} - 1.250 \right| \le \chi, \left| \frac{w_5}{w_3} - 1.800 \right| \le \chi, \left| \frac{w_3}{w_4} - 1.185 \right| \le \chi, \\ \left| \frac{w_1}{w_5} - 1.500 \right| \le \chi, \left| \frac{w_2}{w_3} - 2.250 \right| \le \chi, \left| \frac{w_5}{w_4} - 2.133 \right| \le \chi, \\ \sum_{j=1}^{5} w_j = 1, w_j \ge 0, \forall j \end{cases}$$

By solving this model, the final values of the weight coefficients $(0.315, 0.263, 0.210, 0.113, 0.099)^T$ and DFC of the results $\chi = 0.000$ were obtained. Weight coefficient values are shown in the ranked order of dimensions from the first step. The individual values of weight coefficients for all dimensions were obtained in the same way. Table 5 shows dimension ratings according to all criteria and values of weight coefficients using the previously demonstrated steps. The final values of weight coefficients of the dimension of reliability ($D_1 = 0.291$), assurance ($D_2 = 0.259$), tangibles ($D_3 = 0.130$), empathy ($D_4 = 0.109$), and responsiveness ($D_5 = 0.207$) were calculated using the geometric mean.

Table 5. Priorities of dimensions by five experts and obtained weights of dimensions.

E_1						
Dimension	D_1	D_2	D_5	D_3	D_4	*DFC*
$\omega_{C_{j(k)}}$	1	1.2	1.5	2.7	3.2	
Weights	0.315	0.263	0.210	0.113	0.099	0.000
E_2						
Dimension	D_1	D_2	D_5	D_3	D_4	*DFC*
$\omega_{C_{j(k)}}$	1	1.3	1.5	2.7	3.2	
Weights	0.337	0.260	0.178	0.116	0.109	0.000
E_3						
Dimension	D_1	D_2	D_5	D_3	D_4	*DFC*
$\omega_{C_{j(k)}}$	1	1.05	1.15	1.8	2.2	
Weights	0.261	0.248	0.227	0.145	0.119	0.000
E_4						
Dimension	D_1	D_2	D_5	D_3	D_4	*DFC*
$\omega_{C_{j(k)}}$	1	1	1.2	1.9	2.6	
Weights	0.267	0.267	0.222	0.141	0.103	0.000
E_5						
Dimension	D_1	D_2	D_5	D_3	D_4	*DFC*
$\omega_{C_{j(k)}}$	1	1.1	1.4	2	2.4	
Weights	0.282	0.257	0.202	0.141	0.118	0.000

4.3. The Frequency of Responses

The frequency of the occurrence of a response is called the frequency of responses. As mentioned earlier, when filling out a questionnaire, customers used a Likert scale, or more precisely for each question, they assigned a point from one to five: 1—completely disagree; 2—partially disagree; 3—have no opinion; 4—partially agree; 5—completely agree. According to the frequency of responses, customers had extremely high expectations because they only responded 14 times with a rating of 1 and 769 times with the highest rating. Compared to the frequency of responses in terms of expectations, a significant difference can be noticed for rating 1, but also a difference for the highest rating. Based on the response frequency, it can be assumed that there will be significant differences between customer expectations and perceptions of the service provided. There were 33 responses with a rating of 1, which further implies that a certain number of customers are dissatisfied with the service provided. In addition, while perceiving the service provided, customers mostly gave a rating of 5, and then a rating of 4.

Figure 2 shows a graph of customers' responses in terms of expectations and perceptions. Regarding expectations, only one customer assigned the lowest rating to Q_4, while we had more responses with the lowest rating referring to perceptions. Q_3 did not record any of the lowest ratings regarding either expectations or perceptions. From the aspect of perceptions, Q_1 recorded the highest number of answers with the highest rating, namely 35, compared to the expectations where 30 customers responded with a rating of 5. For Q_2, customers expressed great satisfaction, where in terms of expectations, 26 customers responded with a rating of 5, while 34 customers responded with the same rating for the same question regarding perceptions. The lowest rating for Q_5 was given by two customers, while no response with a rating of 1 was given for expectations. The number of customers for the same question with the highest rating from the aspect of perception was 33, while 31 customers marked 5 regarding expectations for the question. Customers also expressed satisfaction with Q_6 with a rating of 4, i.e., regarding expectations, 18 customers marked 4, while in response to perceptions, 25 customers responded by that rating. Generally, it can be noticed that the quality of the service provided is very high for this dimension.

After the dimension of reliability, high satisfaction was expressed for the dimension of assurance (Figure 3). No significant difference in customers' responses regarding expectations and perceptions was noticed for Q_7. For each question of that dimension, the response with the lowest rating was recorded. Q_8 recorded 35 responses with a rating of 5 when perceived by customers, while the same rating was assigned to expectations by 30 customers. Three customers responded by rating 1 for $Q9$ regarding perception. For the same question, there is a difference in rating 5, where the highest rating was given by 33 customers regarding the perception, and when the expectation was recorded, the rating was recorded by 29 customers. $Q10$, the last question in the dimension of assurance, had the highest number of responses, with a rating of 5. Namely, customer satisfaction can be seen by the number of customers' responses, with a rating of 2 and 5. Regarding expectations, 12 customers responded with a rating of 2, while five customers less responded with the same rating for perceptions. The highest mark, rating 5, was selected by 31 customers for expectations while regarding perceptions, 37 customers responded to Q_{10} with a rating of 5.

The results of the dimension of tangibles (Figure 4) are specific because of customers' responses to Q_{11}. Generally, the Q_{11} results did not significantly affect the overall customer satisfaction. Concerning expectations, three customers selected a rating of 1, while 8 customers responded with the same rating for perceptions. Additionally, a rating of 2 was given by six customers, while 14 customers responded with a rating of 2 for perceptions. The customer dissatisfaction for Q_{11} can be noticed by the number of customers' responses with a rating of 4 where, in reference to expectations, 28 customers responded with that rating, while after the service provided, 18 customers responded with a rating of 4. The total satisfaction of the customers for assessing the tangibles was influenced by the results of Q_{12}. For question Q_{12}, after the service was provided, 38 customers responded with a rating of 5, while for the same question, when responding to expectations, 29 customers answered with a rating of 5. For question Q_{13}, it is also possible to notice the difference in customers' responses for the highest rating. With regard to expectations, 29 customers selected a rating of 5, while the same rating after the perception of the service was selected by 39 customers. Rating 4 for Q_{14} was chosen by the same number of customers, namely 26. After the service was provided, 34 customers answered with a rating of 5 for that question, while 32 customers responded with the highest rating regarding perceptions. The great satisfaction of customers concerning the dimension of tangibles was expressed for Q_{15}. Before the service was provided, a rating of 5 was selected by 26 customers, and after the service was perceived, 39 customers answered with the highest rating. For Q_{16}, there was also a significant difference expressed by rating 4 and 5. For that question, before the service was provided, 23 customers answered with a rating of 4, and 27 customers after its realization. Rating 5 was given by three customers more after the service was provided.

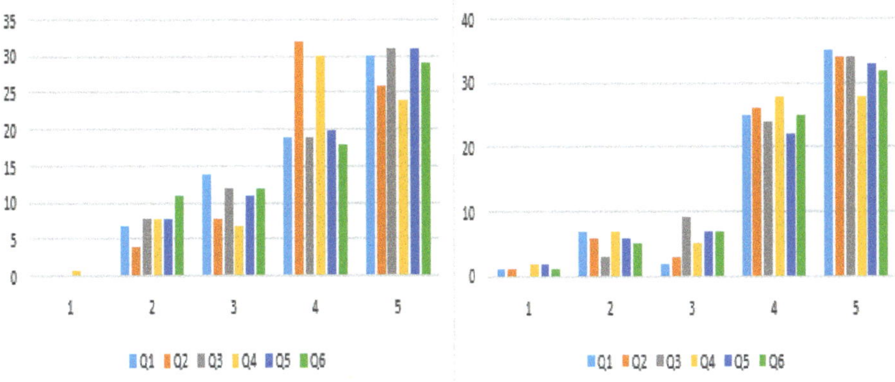

Figure 2. Graph of customers' responses regarding the dimension of reliability (**left**-expectations and **right**- perceptions).

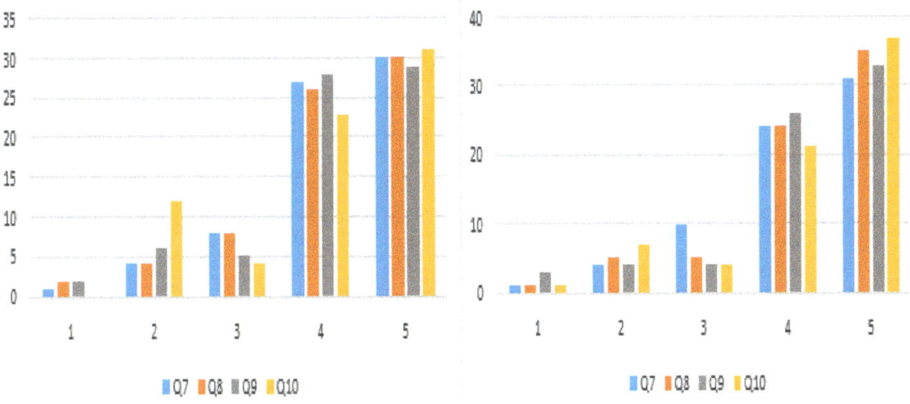

Figure 3. Graph of customers' responses regarding the dimension of assurance (**left**-expectations and **right**-perceptions).

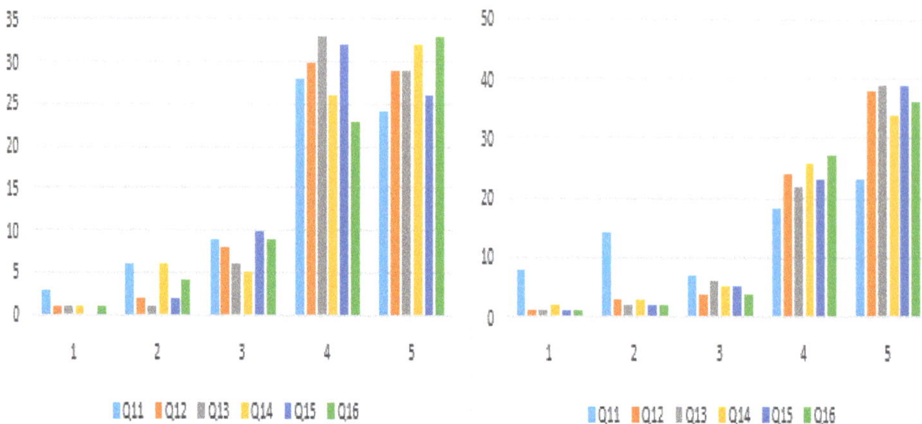

Figure 4. Graph of customers' responses regarding the dimension of tangibles (**left**-expectations and **right**-perceptions).

From Figure 5, it can be seen that there is very little positive difference in terms of customer perception. For customer expectations, one response with a rating of 1 was noted for Q_{21}. Customers expressed satisfaction for Q_{17}, where 35 customers responded with a rating of 5 after the service was provided, and 30 customers responded with the same rating before its realization. Concerning Q_{18}, a large number of customers (34) responded with a rating of 5, while 39 customers answered with the same rating after the service was provided. Question Q_{19} was the only question that customers did not answer with a rating of 1 after the service was perceived. In addition, customers had high expectations for Q_{19}, i.e., 33 customers answered with a rating of 5, while 38 customers answered with the same rating after the service was provided. In reference to Q_{20}, 32 customers responded with a rating of 4 in terms of perceptions, while 25 responses were recorded with the same rating regarding expectations. For the same question, the diagram shows a much larger number of responses with a rating of 3 from the aspect of customer expectations, where 11 customers responded with that rating, and after the realization, only four customers responded with a rating of 3. Concerning question Q_{21}, customers had very high expectations, with 44 customers responding with a rating of 5. A slight decrease in satisfaction could be noticed after the service was provided, where 44 respondents answered Q_{21} with a rating of 5.

In Figure 6, in terms of customer expectations, it can be noticed that there were no responses with a rating of 1. Final survey results indicated that there was no difference between the customer expectations and perceptions of the quality of the service provided. Concerning Q_{22}, 41 customers responded with a rating of 5 for perceptions, while regarding expectations, 33 customers answered with a rating of 5 for the same question. For Q_{23}, customers did not generally express satisfaction, where 37 customers responded with a rating of 5 regarding expectations, and 34 customers responded with the highest rating after the service was provided. The number of customers' responses to Q_{23} with a rating of 4 was the same, i.e., 24 responses for both aspects of the SERVQUAL model. Question Q_{24} showed a small positive difference in customer satisfaction, i.e., 35 respondents answered with a rating of 5 prior to the service being provided, and after its realization, 38 customers responded with the highest rating. In the diagram, the biggest positive difference can be identified for question Q_{25}. The number of customers who answered with a rating of 5 for that question regarding expectations was 28, and after the service was provided, 38 customers responded with a rating of 5. The positive difference regarding the last question, Q_{25}, had a significant impact on the ultimate result of the dimension of responsiveness.

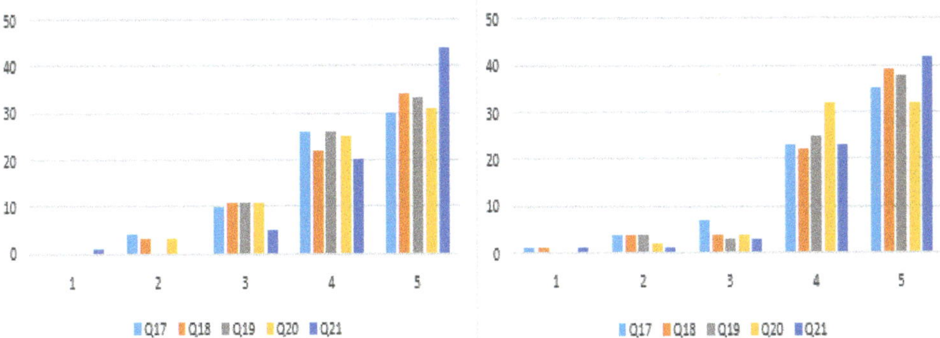

Figure 5. Graph of customers' responses regarding the dimension of empathy (**left**-expectations and **right**-perceptions).

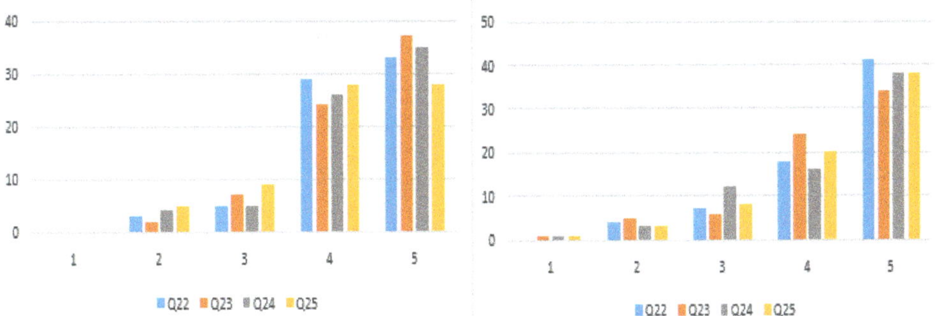

Figure 6. Graph of customers' responses regarding the dimension of responsiveness (**left**-expectations and **right**-perceptions).

5. Research Results

5.1. The Results of Dimensions in Terms of Customer Expectations

Table 6 shows the results of dimensions with expectations. Dimensions are presented in terms of expectations and their average value, standard deviation, weight coefficients, and the value of the Cronbach alpha coefficient.

Table 6. The results of dimensions with customer expectations.

Dimension	AV	SD	W_j	Cronbach Alpha Coefficient
Reliability	4.029	1.010	0.291	0.918
Assurance	4.100	1.022	0.259	0.891
Tangibles	4.150	0.924	0.130	0.845
Empathy	4.260	0.829	0.109	0.851
Responsiveness	4.282	0.831	0.207	0.875
SERVQUAL (1)	**4.164**	**0.923**	**1**	**0.876**

The Cronbach alpha test is considered positive only if coefficients above 0.7 are obtained. Certain sources state that a reliable value of the Cronbach alpha test is 0.5. From the table of percentage values, it can be seen that customers have the highest expectations regarding the reliability dimension, and the least expectations regarding the dimension of empathy. The average value for the dimension of responsiveness was 4.282, which is the highest average value. For the dimension of reliability, there is the smallest average value and it was 4.029. The standard deviation for all dimensions was 0.923 and the average value of the Cronbach alpha test for all dimensions was 0.876. The weight coefficient values for both expectations and perceptions were the same.

5.2. Results of Dimensions in Terms of Customer Perceptions

Table 7 shows the results of dimensions regarding customer perceptions.

From the previous two tables, it can be seen that the value of the Cronbach alpha test was far above 0.7, which means that the dimensions are reliable. The highest quality perceptions were for the dimension of empathy, 4.360, and then for the dimension of responsiveness, 4.282. The lowest perceptions were related to the reliability dimension and were 4.176. It can be seen that the values for the dimension of responsiveness were the same for both expectations and perceptions, which means that there were no significant changes in relation to the quality of the service provided. In addition, it can be seen that the values for the dimension of responsiveness were the least from both aspects.

Table 8 shows the results obtained by using the developed Delphi-FUCOM-SERVQUAL methodology.

Table 8 shows the difference between customer perceptions and expectations. Generally, customers are satisfied with the quality of the logistics service of the express post company. For all dimensions except for the dimension of responsiveness, the result is positive. It can be noticed that the greatest satisfaction of customers was expressed for the dimension of reliability. According to customers' percentage rating, the dimension of reliability is the most important of all the five dimensions for customers. The results of the dimension of responsiveness remains the same, as the expectations of customers are equal to their perceptions. The questions of the reliability dimension included a part of the logistics service where the company can create the biggest improvements.

Table 7. The results of dimensions regarding customer perceptions.

Dimension	AV	SD	W_j	Cronbach Alpha Coefficient
Reliability	4.176	0.995	0.291	0.947
Assurance	4.196	1.006	0.259	0.889
Tangibles	4.200	1.040	0.130	0.824
Empathy	4.360	0.844	0.109	0.891
Responsiveness	4.282	0.944	0.207	0.894
SERVQUAL (2)	**4.243**	**0.966**	**1**	**0.889**

Table 8. Research results.

Delphi-FUCOM-SERVQUAL			
Dimensions	PER	EXP	Gap
Reliability	1.172	1.215	0.043
Assurance	1.062	1.087	0.025
Tangibles	0.540	0.546	0.006
Empathy	0.464	0.475	0.011
Responsiveness	0.886	0.886	0.000
Total			0.017

Table 9 outlines the questions of responsiveness from the aspect of customer perceptions. According to the results, after the responsiveness dimension, the dimension of tangibles with +0.064 also has opportunity for improvement. From the results obtained, it can be seen that Q_{11} has a significant impact on the quality of this dimension. Namely, three customers from the aspect of expectations gave the lowest rating for this question, and after the service was provided, eight customers gave the lowest rating for that question. Concerning expectations, for the same question, six customers selected a rating of 2, or "partially disagree", while regarding perceptions, 14 customers responded with "partially disagree".

According to the analysis, customers expressed the highest satisfaction for the dimension of reliability, with +0.0392. The dimension of reliability is focused on delivery quality and delivery time. For all six questions of the reliability dimension, customers gave higher ratings than the ratings in terms of customer expectations.

Table 9. Statements for the dimension of responsiveness in terms of customer perceptions.

Responsiveness
22. Employees in the company are willing and able to help.
23. Customers obtained right answers to their questions.
24. Employees at the Call Center provided all necessary information to the customers.
25. Customer requests are responded quickly and reliably.

5.3. Statistical Analysis

For the set of expectations and perceptions, $n \in N = 70$ (Table 10), so that the parameter of binomial distribution for $n \in [1, 5]$ can fully substitute mathematical expectation.

Table 10. Verification of distribution for the set of expectations and perceptions.

	Expectations			Perceptions		Signum Test	Correlation Coefficient	ANOVA
	Binomial Distribution Parameter	Verification by χ^2 Test		Binomial Distribution Parameter	Verification by χ^2 Test			
E_{01}	0.8057	0.0184	P_{01}	08457	0.2460	0.6264	+0.0920	0.5038
E_{02}	0.8285	0.3297	P_{02}	0.8457	0.3392	0.3613	+0.1183	0.4417
E_{03}	0.8085	0.0177	P_{03}	0.8542	0.4359	0.4291	+0.1176	0.6555
E_{04}	0.7942	0.3839	P_{04}	0.8085	0.2737	0.6434	−0.1183	0.6561
E_{05}	0.8114	0.0331	P_{05}	0.8228	0.0783	1.0000	+0.2656	0.0168
E_{06}	0.7857	0.0068	P_{06}	0.8343	0.3552	0.2683	+0.1855	0.1438
E_{07}	0.8314	0.5879	P_{07}	0.8285	0.2804	0.8596	+0.1340	0.0764
E_{08}	0.7856	0.3754	P_{08}	0.8485	0.3036	0.7277	+0.3101	0.0131
E_{09}	0.8171	0.3079	P_{09}	0.8343	0.2301	0.6170	+0.3256	0.0061
E_{10}	0.8085	0.0780	P_{10}	0.8457	0.0793	0.4291	+0.0804	0.4144
E_{11}	0.7826	0.0359	P_{11}	0.6971	0.0000	0.0743	+0.5181	0.0000
E_{12}	0.8400	0.5127	P_{12}	0.8714	0.4980	0.1762	+0.4371	0.0001
E_{13}	0.8514	0.1261	P_{13}	0.8742	0.3553	0.1762	+0.4264	0.0000
E_{14}	0.8342	03614	P_{14}	0.8485	0.4148	0.5107	+0.3628	0.0003
E_{15}	0.8342	0,3381	P_{15}	0.8771	0.5093	0.0311	+0.4857	0.0001
E_{16}	0.8371	0.2123	P_{16}	0.8714	0.4989	0.3239	+0.2936	0.0105
E_{17}	0.8342	0.6026	P_{17}	0.8485	0.2523	0.7353	+0.1732	0.1517
E_{18}	0.8485	0.1715	P_{18}	0.8685	0.2670	0.1884	+0.1809	0.0311
E_{19}	0.8628	0.7082	P_{19}	0.8771	0.6105	0.2299	+0.1327	0.0129
E_{20}	0.8400	0.4780	P_{20}	0.8685	0.1061	0.2962	+0.2851	0.0055
E_{21}	0.9028	0.4991	P_{21}	0.8971	0.6446	0.8598	+0.4327	0.0000
E_{22}	0.8628	0.4420	P_{22}	0.8742	0.0515	0.8638	+0.1411	0.4380
E_{23}	0.8742	0.6696	P_{23}	0.8428	0.2887	0.4576	+0.3871	0.0005
E_{24}	0.8628	0.6829	P_{24}	0.8485	0.0045	0.7277	+0.2907	0.0172
E_{25}	0.8257	0.7607	P_{25}	0.8600	0.0942	0.0909	+0.3487	0.0053

The mean value of the binomial distribution parameter for expectations was $p_E = 0.8307$, and the mean value of the binomial distribution parameter for perceptions was $p_p = 0.8477$. Between these values, there was a high significant correlation of mean values of $p = 0.9303$.

Although the distributions of expectations and perceptions are the same and in most cases, they have nonparametric correlation (not in one case out of 25, E_{15}/P_{15}), it should be noticed that the coefficients of liner correlations are, on average, small and of a normal distribution N (0.2562; 0.0233), with the significance threshold of $p = 0.5708$. This means that there is a large fluctuation between expectations and perceptions, i.e., there is a large number of respondents who had high expectations, but were disappointed with perceptions and vice versa.

The ANOVA test significantly confirmed that there are 17 out of 25 such cases. The Duncan test of post-hoc ANOVA was used to determine the mean values of perceptions for the factor of expectation estimates for all variables, where the variance analysis showed significant differences.

In Table 11, the values of the mean estimates for perceptions for given estimates of expectation where the ANOVA analysis has identified significant differences are given.

Based on the established mean values of binomial distribution parameters for expectations, $p_E = 0.8307$ and $p_P = 0.8477$ and for $N = 70$, we can estimate the mean number of respondents who, according to binomial distribution, assigned the ratings $n \in [1, 5]$. The expected average number of respondents who selected one of the ratings for expectations and perceptions is given in Table 12.

Table 11. The values of mean estimates for perceptions for given estimates of expectations where the ANOVA analysis has identified significant differences.

Score for $P(n)$	Rating 1	Rating 2	Rating 3	Rating 4	Rating 5
Score P_{05}	2.0000	4.1667	4.0000	3.8182	4.3333
Score P_{07}	5.0000	4.5000	3.6000	3.9583	4.4194
Score P_{08}	1.0000	3.8000	4.0000	4.0000	4.3428
Score P_{09}	2.0000	4.2500	4.0000	4.0385	4.3030
Score P_{11}	2.8750	3.6428	2.8571	4.1111	4.6087
Score P_{12}	1.0000	3.6667	4.2500	3.9583	4.4737
Score P_{13}	1.0000	4.0000	4.0000	4.1364	4.4615
Score P_{14}	1.5000	4.6667	3.8000	4.0385	4.4414
Score P_{15}	2.0000	4.0000	3.6000	3.8696	4.4872
Score P_{16}	1.0000	4.0000	4.2500	4.0741	4.3611
Score P_{18}	3.0000	4.7500	3.7500	3.9091	4.4615
Score P_{19}	-	4.7500	3.3333	4.1200	4.4737
Score P_{20}	-	4.5000	3.0000	4.0625	4.4688
Score P_{21}	1.0000	5.0000	3.6667	4.5217	4.6429
Score P_{23}	4.0000	3.6000	4.5000	4.0000	4.7353
Score P_{24}	2.0000	4.3333	3.9167	4.4375	4.4474
Score P_{25}	2.0000	4.3333	3.5000	3.9500	4.3947
Mean value	2.0916	4.2329	3.7661	4.0590	4.4621
Std. deviation	1.1919	0.4256	0.4320	0.1814	0.1127

Table 12. Calculation of the average number of respondents based on the binomial distribution parameters, p_E and p_P, the number of respondents, $N = 70$, and ratings, $n \in [1, 5]$.

	Rating 1	Rating 2	Rating 3	Rating 4	Rating 5	Σ
Expectations E	0.0574	1.1269	8.2996	27.1676	33.3484	70
Perceptions P	0.0376	0.8381	6.9988	25.9748	36.1506	70

Further as follows:

The respondents with the expected rating, $E(n) = 1$, provided an average perception estimate of 2.0916, so we can conclude that the increase in the values of estimates was not significant, with $p = 0.1616$ (the minimum number of respondents adopted for one side test difference between two means was two for expectations and perceptions).

Respondents with the expected rating, $E(n) = 2$, provided an average perception estimate of 4.2329, so we can conclude that the increase in values of estimates was significant, with $p = 0.0176$ (the minimum number of respondents adopted for two side test differences between two means was two for expectations and perceptions). Respondents with the expected rating, $E(n) = 3$, provided an average perception estimate of 3.7661, so we can conclude that the increase in values of estimates was significant, with $p = 0.0002$ (the number of respondents adopted for two side test differences between two means was eight for expectations and seven for perceptions). Respondents with the excepted rating, $E(n) = 4$, provided an average perception estimate of 4.0590, so we can conclude that the increase in values of estimates was not significant with $p = 0.0970$ (the number of respondents adopted for two side test differences between two means was 27 for expectations and 26 for perceptions).

Respondents with the expected rating, $E(n) = 5$, provided an average perception estimate of 4.4621, so we can conclude that the decrease in values of estimates was significant, with $p = 0.0000$ (the number of respondents adopted for one side test difference between two means was 33 for expectations and 36 for perceptions)

To conclude:

Respondents who had low expectations of 2 or 3 significantly identified a perception increase to 4.0590 or 3.7761, respectively, but respondents with high expectations of 5, significantly reduced their perceptions to 4.4621.

Respondents who had expectations of 4 significantly maintained the same level of 4.0590. Considering the above, there is a stable rating of 4 for expectations, which can be adopted as the company's final assessment.

Regarding the system of expectation and perception assessment:

- There were no significant quantitative differences between expectations and perceptions. Most of the estimates were significantly binomially distributed with approximately the same parameter, as confirmed by the Signum test in 24 out of 25 estimates;
- there were significant qualitative differences in assessing the expectations and perceptions contained in the fluctuation according to a stable rating of "4". These differences are in favor of the objectivity of respondents and the concept of assessment, the correctness of the questions asked, etc., and realistically assess the company with ratings of 4.

The impact of expectations (E) as a factor on reliability, assurance, tangibles, empathy, and responsiveness is given in the Table 13. The variance analysis identified one significant case of the impact of expectations on reliability (E_{03}), four on empathy (E_{05}, E_{08}, E_{13}, E_{15}, and E_{18}), and two on responsiveness (E_{03} and E_{20}). The expectations of assurance and tangibles had no impact.

Table 13. The variance analysis of influencing factors of expectations on reliability, assurance, tangibles, empathy, and responsiveness.

	Reliability	Assurance	Tangibles	Empathy	Responsiveness
E_{01}	0.2333	0.3283	0.9282	0.1498	0.8148
E_{02}	0.7551	0.4521	0.2927	0.0902	0.2.855
E_{03}	0.0298	0.9813	0.7376	0.7055	0.0342
E_{04}	0.9925	0.8959	0.8593	0.4429	0.7441
E_{05}	0.2810	0.6594	0.1281	0.0254	0.4000
E_{06}	0.1904	0.5799	0.7243	0.1564	0.9151
E_{07}	0.8086	0.2367	0.4655	0.2065	0.7040
E_{08}	0.6519	0.7103	0.6425	0.0136	0.8606
E_{09}	0.4248	0.5802	0.6967	0.8821	0.3657
E_{10}	0.5789	0.9385	0.5314	0.5716	0.4891
E_{11}	0.4554	0.7911	0.2901	0.1673	0.7318
E_{12}	0.6603	0.9752	0.4280	0.4646	0.7597
E_{13}	0.5509	0.1355	0.2993	0.0195	0.8361
E_{14}	0.9289	0.4619	0.9377	0.9059	0.2006
E_{15}	0.4364	0.1113	0.1549	0.0008	0.8963
E_{16}	0.4509	0.3688	0.9011	0.5942	0.2297
E_{17}	0.2614	0.9714	0.4900	0.6001	0.4829
E_{18}	0.3586	0.4111	0.5082	0.0327	0.4736
E_{19}	0.7623	0.1867	0.6461	0.5103	0.1482
E_{20}	0.2489	0.3789	0.2294	0.1240	0.0114
E_{21}	0.4091	0.9642	0.1852	0.4320	0.7649
E_{22}	0.8918	0.3974	0.1006	0.0827	0.3012
E_{23}	0.8118	0.1397	0.5539	0.1420	0.2649
E_{24}	0.3725	0.3800	0.8176	0.9727	0.0757
E_{25}	0.8864	0.5691	0.9378	0.9155	0.6695

Duncan's test of post-hoc ANOVA revealed the values that led to the emphasis of factors as follows (Table 14):

- Expectation E_{03} with a rating of "3" was significantly the lowest mean value for reliability, "15.000";
- expectation E_{03} with a rating of "3" was significantly the highest mean value for responsiveness, "33.333";
- expectation E_{05} with a rating of "4" was significantly the lowest mean value for empathy, "6.500";
- expectation E_{08} with a rating of "3" was significantly the lowest mean value for empathy, "3.750";
- Expectation E_{13} with a rating of "2" had significantly the highest mean value for Empathy of "30.000";
- expectation E_{15} with a rating of "2" was significantly the highest mean value of empathy, "20.000";
- expectation E_{18} with a rating of "3" and "4" was significantly the lowest mean value for empathy, "9.545" and "7.954". There were no significant differences between these values; and
- expectation E_{20} with a rating of "3" was significantly the highest mean value for responsiveness, "35.445".

For a significant influence of expectations, it is necessary to have at least three significant differences ($p(2) = 0.0745 > 0.05$, no significant influence, $p(3) = 0.0370 < 0.05$ influence was significant) for one of the dimensions (reliability, assurance, tangibles, empathy, and responsiveness). With the significance threshold of $p(5) = 0.0092$, we confirm the significant influence of expectations on empathy.

The impact of perceptions (P) as a factor on reliability, assurance, tangibles, empathy, and responsiveness is given in Table 15. The variance analysis determined one significant case of the impact of expectation on assurance (P_{18}), two on tangibles (P_{09}, P_{21}), and three on empathy (P_{08}, P_{10}, and P_{23}). Perceptions had no influence on reliability and responsiveness.

Table 14. Calculation of the attribute mean values for a given rating of significant expectation influence.

		1	2	3	4	5
E_{03}	Reliability	-	30.625	15.000	26.597	30.000
E_{03}	Responsiveness	-	18.125	33.333	23.421	20.645
E_{05}	Empathy	-	11.250	13.636	6.500	13.065
E_{08}	Empathy	15.000	16.250	3.750	9.423	15.500
E_{13}	Empathy	10.000	30.000	14.167	8.333	12.931
E_{15}	Empathy	-	20.000	8.500	7.812	15.358
E_{18}	Empathy	-	20.000	9.545	7.954	12.794
E_{20}	Responsiveness	-	15.000	35.445	20.800	21.774

Table 15. The influence of perceptions (P) on dimensions using ANOVA.

	Reliability	Assurance	Tangibles	Empathy	Responsiveness
P_{01}	0.6222	0.8900	0.9662	0.1628	0.5745
P_{02}	0.7485	0.8557	0.2543	0.0965	0.3145
P_{03}	0.2446	0.6583	0.2228	0.1344	0.2166
P_{04}	0.6516	0.9724	0.8606	0.1406	0.3944
P_{05}	0.9525	0.2886	0.7884	0.1146	0.7364
P_{06}	0.1071	0.7545	0.2750	0.1366	0.5060
P_{07}	0.6714	0.2601	0.2585	0.0611	0.9407
P_{08}	0.3100	0.2353	0.3748	0.0495	0.5877
P_{09}	0.7329	0.2739	0.0449	0.0681	0.7172
P_{10}	0.1876	0.5262	0.1879	0.0037	0.3463
P_{11}	0.9835	0.5157	0.7200	0.5412	0.4980
P_{12}	0.4387	0.7817	0.2365	0.1570	0.9112
P_{13}	0.3607	0.4781	0.1240	0.1666	0.2573
P_{14}	0.1763	0.6297	0.0715	0.1430	0.8599
P_{15}	0.2135	0.3281	0.2578	0.0605	0.3694
P_{16}	0.9670	0.5075	0.8083	0.9810	0.4574
P_{17}	0.7442	0.3776	0.5458	0.2159	0.3887
P_{18}	0.7128	0.0117	0.2318	0.1094	0.4439
P_{19}	0.2658	0.6042	0.2876	0.2885	0.8120
P_{20}	0.9867	0.0873	0.4312	0.0697	0.5619
P_{21}	0.5551	0.7017	0.0404	0.4582	0.8648
P_{22}	0.5802	0.6291	0.3115	0.1567	0.5822
P_{23}	0.0832	0.4412	0.0810	0.0074	0.2661
P_{24}	0.5726	0.6054	0.7890	0.1816	0.5781
P_{25}	0.3734	0.8136	0.8965	0.0647	0.6609

Duncan's test of post-hoc ANOVA revealed the values that led to the emphasis of factors as follows (Table 16):

— Perception P_{08} with a rating of "2" was significantly the lowest mean value for empathy, "2.000";
— perception P_{09} with a rating of "2" was significantly the lowest mean value for tangibles, "2.500";
— perception P_{10} with a rating of "2" was significantly the lowest mean value for empathy, "0.714";
— perception P_{18} with a rating of "3" was significantly the highest mean value for assurance, "52.500";
— perception P_{21} with a rating of "2" was significantly the lowest mean value for tangibles, "0.000"; and
— perception P_{23} with a rating of "2" was significantly the lowest mean value for empathy, "3.000".

Table 16. Calculation of the attribute mean values for given ratings of significant perception influence.

		1	2	3	4	5
P_{08}	Empathy	10.000	2.000	6.000	12.083	12.429
P_{09}	Tangibles	13.333	2.500	18.750	14.615	11.818
P_{10}	Empathy	10.000	0.714	10.000	10.476	13.514
P_{18}	Assurance	25.000	35.000	52.500	25.000	23.426
P_{21}	Tangibles	10.000	0.000	23.333	10.217	13.810
P_{23}	Empathy	20.000	3.000	9.166	8.541	14.118

Regarding expectations, for a significant influence of perceptions, it is necessary to have at least three significant differences for one of the dimensions (reliability, assurance, tangibles, empathy, and responsiveness), which was only recorded for empathy. With the significance threshold, $p(3) = 0.0370$, we confirm the significant influence of perceptions on empathy.

A particularly specific case is the empathy function as a variable depending on expectation E_{08} and perception P_{08}, which at the same time, had a significant impact on empathy. From the graph, it is evident that respondents who had low expectations (1 or 2) and identified great perceptions (4 or 5) had excessively high empathy (15 to 20), which was likely to be generated as a reactive compensation to the determined difference between perceptions and expectations (Figure 7).

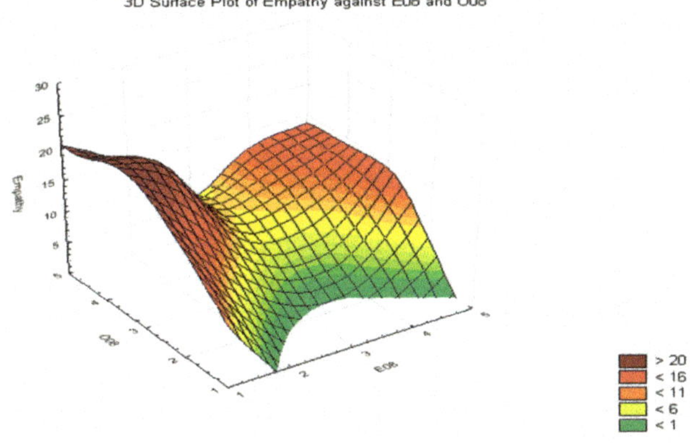

Figure 7. Empathy function as a variable depending on expectation E_{08} and perception P_{08}.

Here, it is necessary to recall that a significant difference between expectation E_{08} and perception P_{08} was determined in the distribution changes, and it is also necessary to notice that the difference of binomial parameters (which is analogous to mathematical expectation) had the largest increase (+0.0629), particularly in the difference between the binomial parameters for P_{08} and E_{08}.

Between the impact of expectations on reliability (1 of 25) and the impact of perceptions on reliability (0 of 25), there was no significant difference, $p = 0.3145 > 0.05$.

Between the impact of expectations on assurance (0 of 25) and the impact of perceptions on assurance (1 of 25), there was no significant difference, $p = 0.3145 > 0.05$.

Between the impact of expectations on tangibles (0 of 25) and the impact of perceptions on tangibles (2 of 25), there was no significant difference, $p = 0.1525 > 0.05$.

Between the impact of expectations on empathy (5 of 25) and the impact of perceptions on empathy (3 of 25), there was no significant difference, $p = 0.5755 > 0.05$ (expectations and perceptions had a significant impact on empathy, but there were no differences between their significant impacts).

Between the impact of expectations on responsiveness (2 of 25) and the impact of perceptions on responsiveness (0 of 25), there was no significant difference $p = 0.1525 > 0.05$.

The others (R, A, T, R) had no significant impact, and there was no significant difference between them, too.

6. Conclusions

By applying appropriate scientific tools and techniques, it is possible to make improvements from a professional aspect in different areas, one of which is certainly quality management. In this paper, therefore, a new Delphi-FUCOM-SERVQUAL methodology was developed to improve the process of service quality measurement. The company where the case study was conducted provides express post services, so it can be said that this paper has a twofold contribution. The first contribution relates to a scientific aspect that implies the development of an integrated methodology to improve a quality measurement process that can be applied without any restrictions in various areas. The advantages of the developed methodology are reflected in the fact that it enables precision treatment of input and output parameters and provides results that are more objective. In addition, from a professional aspect of the study, it is possible to determine the quality and efficiency of the company based on the satisfaction of its customers, but it also enables further application and re-application of this methodology. This methodology can be very helpful for strategic management of the company to improve their efficiency. Considering all the relevant factors, it is possible to conclude that this paper contributes to the overall literature, enriching it in a certain way, as it provides future researchers with a new methodology that more precisely treats input parameters and achieves better results than traditional quality measurement methods.

All contributions and conclusions were confirmed throughout a comprehensive and detailed statistical analysis in which even the regularity of interaction between certain questions was established. The Cronbach alpha coefficient showed the reliability of the formed questionnaire, while ANOVA showed that there was a large fluctuation between expectations and perceptions, i.e., there was a large number of respondents who had high expectations and were disappointed with perceptions, and vice versa. Considering it at the general level, the research conducted on the system of estimating expectations and perceptions shows that: *There were no significant quantitative differences between expectations and perceptions,* which means that the hypothesis set in the paper was confirmed. Most of the estimates were significantly binomially distributed with approximately the same parameter, as confirmed by the Signum test in 24 out of 25 estimates. From the aspect of qualitative differences, there was significance in assessing expectations and perceptions, which was contained in the fluctuation towards a stable rating of "4". These differences support the objectivity of the respondents and the concept of assessment, the correctness of the questions asked, etc., and realistically evaluate the company with a rating of 4. Future research related to this paper may imply the improvement of the proposed methodology by defining a universal linguistics scale for expressing customer satisfaction. In addition, depending on specific cases, it is possible to modify the structure of dimensions within the SQ questionnaire.

Author Contributions: Each author has participated and contributed sufficiently to take public responsibility for appropriate portions of the content.

Conflicts of Interest: The authors declare no conflict of interest.

References

1. Nowotarski, P.; Pasławski, J.; Kadler, A. Quality Management Systems as a key element for company strategy selection—Case study. *MATEC Web Conf.* **2018**, *222*, 01012. [CrossRef]
2. Wang, Y.L.; Luor, T.; Luarn, P.; Lu, H.P. Contribution and Trend to Quality Research—A literature review of SERVQUAL model from 1998 to 2013. *Inform. Econ.* **2015**, *19*, 34. [CrossRef]
3. Kersten, W.; Koch, J. The effect of quality management on the service quality and business success of logistics service providers. *Int. J. Qual. Reliabil. Manag.* **2010**, *27*, 185–200. [CrossRef]
4. Memić, Z.; Vasiljević, M.; Stević, Ž.; Tanackov, I. Measuring the quality of logistics services in the transport company using the SERVQUAL MODEL. In Proceedings of the 2nd International Conference on Management, Engineering and Environment, Belgrade, Serbia, 11–12 October 2018.
5. Gajewska, T.; Grigoroudis, E. Estimating the performance of the logistics services attributes influencing customer satisfaction in the field of refrigerated transport. *Int. J. Shipp. Transport Logist.* **2017**, *9*, 540–561. [CrossRef]
6. Roslan, N.A.A.; Wahab, E.; Abdullah, N.H. Service Quality: A case study of logistics sector in Iskandar Malaysia using SERVQUAL Model. *Procedia Soc. Behav. Sci.* **2015**, *172*, 457–462. [CrossRef]
7. Tsai, J.Y.; Ding, J.F.; Liang, G.S.; Ye, K.D. Use of a hybrid MCDM method to evaluate key solutions influencing service quality at a port logistics center in Taiwan. *Brodogradnja Teorija i Praksa Brodogradnje i Pomorske Tehnike* **2018**, *69*, 89–105. [CrossRef]
8. Awasthi, A.; Sayyadi, R.; Khabbazian, A. A combined approach integrating gap analysis, QFD and AHP for improving logistics service quality. *Int. J. Logist. Syst. Manag.* **2018**, *29*, 190–214. [CrossRef]
9. Cavana, R.Y.; Corbett, L.M.; Lo, Y.L. Developing zones of tolerance for managing passenger rail service quality. *Int. J. Qual. Reliabil. Manag.* **2007**, *24*, 7–31. [CrossRef]
10. Sohn, J.I.; Woo, S.H.; Kim, T.W. Assessment of logistics service quality using the Kano model in a logistics-triadic relationship. *Int. J. Logist. Manag.* **2017**, *28*, 680–698. [CrossRef]
11. Lin, F.H.; Tsai, S.B.; Lee, Y.C.; Hsiao, C.F.; Zhou, J.; Wang, J.; Shang, Z. Empirical research on Kano's model and customer satisfaction. *PLoS ONE* **2017**, *12*, e0183888. [CrossRef] [PubMed]
12. Hu, K.C.; Lee, P.T.W. Novel 3D model for prioritising the attributes of port service quality: Cases involving major container ports in Asia. *Int. J. Shipp. Transport Logist.* **2017**, *9*, 673–695. [CrossRef]
13. Bulut, E.; Duru, O.; Huang, S.T. A multidimensional QFD design for the service quality assessment of Kansai International Airport, Japan. *Total Qual. Manag. Bus. Excell.* **2018**, *29*, 202–224. [CrossRef]
14. Huang, S.T.; Su, I. Applying Multilayer QFD to Assess Quality of Short Sea Shipping: An Empirical Study on Maritime Express Service between Taiwan and Mainland China. In Proceedings of the International Forum on Shipping, Ports and Airports (IFSPA) 2017: Innovative Transport Logistics in Shaping the Future of Supply Chains, Hong Kong, China, 22–25 May 2017.
15. Al-Aomar, R.; Chaudhry, S. Simulation-based Six Sigma value function for system-level performance assessment and improvement. *Int. J. Product. Perform. Manag.* **2018**, *67*, 66–84. [CrossRef]
16. Raja Sreedharan, V.; Raju, R.; Rajkanth, R.; Nagaraj, M. An empirical assessment of Lean Six Sigma Awareness in manufacturing industries: Construct development and validation. *Total Qual. Manag. Bus. Excell.* **2018**, *29*, 686–703. [CrossRef]
17. Van Der Vorst, J.G.; Tromp, S.O.; Zee, D.J.V.D. Simulation modelling for food supply chain redesign; integrated decision making on product quality, sustainability and logistics. *Int. J. Prod. Res.* **2009**, *47*, 6611–6631. [CrossRef]
18. Cho, I.J.; Kim, Y.J.; Kwak, C. Application of SERVQUAL and fuzzy quality function deployment to service improvement in service centres of electronics companies. *Total Qual. Manag. Bus. Excell.* **2016**, *27*, 368–381. [CrossRef]
19. Paryani, K.; Masoudi, A.; Cudney, E.A. QFD application in the hospitality industry: A hotel case study. *Qual. Manag. J.* **2010**, *17*, 7–28. [CrossRef]

20. Peprah, A.A.; Atarah, B.A. Assessing patient's satisfaction using SERVQUAL model: A case of Sunyani Regional hospital, Ghana. *Int. J. Bus. Soc. Res.* **2014**, *4*, 133–143. [CrossRef]
21. Behdioğlu, S.; Acar, E.; Burhan, H.A. Evaluating service quality by fuzzy SERVQUAL: A case study in a physiotherapy and rehabilitation hospital. *Total Qual. Manag. Bus. Excell.* **2017**, 1–19. [CrossRef]
22. Singh, A.; Prasher, A. Measuring healthcare service quality from patients' perspective: Using Fuzzy AHP application. *Total Qual. Manag. Bus. Excell.* **2017**, 1–17. [CrossRef]
23. Khan, A.M.R.; Prasad, P.N.; Rajamanoharane, S. A decision-making framework for service quality measurements in hospitals. *Int. J. Enterp. Netw. Manag.* **2010**, *4*, 80–91. [CrossRef]
24. Chou, C.C.; Liu, L.J.; Huang, S.F.; Yih, J.M.; Han, T.C. An evaluation of airline service quality using the fuzzy weighted SERVQUAL method. *Appl. Soft Comput.* **2011**, *11*, 2117–2128. [CrossRef]
25. Saeedpoor, M.; Vafadarnikjoo, A.; Mobin, M.; Rastegari, A. A servqual model approach integrated with fuzzy AHP and fuzzy TOPSIS methodologies to rank life insurance firms. In Proceedings of the International Annual Conference of the American Society for Engineering Management, Indianapolis, IN, USA, 7–10 October 2015; American Society for Engineering Management (ASEM): Huntsville, AL, USA.
26. Ahmed, R.R.; Vveinhardt, J.; Štreimikienė, D.; Ashraf, M.; Channar, Z.A. Modified SERVQUAL model and effects of customer attitude and technology on customer satisfaction in banking industry: Mediation, moderation and conditional process analysis. *J. Bus. Econ. Manag.* **2017**, *18*, 974–1004. [CrossRef]
27. Long, S. The Correlation Research between Express Company Service Quality and Undergraduate Customer Satisfaction Degree Based on SERVQUAL Model. *Korean Rev. Corp. Manag.* **2016**, *7*, 1–23. [CrossRef]
28. Apornak, A. Customer satisfaction measurement using SERVQUAL model, integration Kano and QFD approach in an educational institution. *Int. J. Product. Perform. Manag.* **2017**, *21*, 129–141. [CrossRef]
29. Oskooii, N.; Albonaiemi, E. Measuring the customer satisfaction based on SERVQUAL model (case study: Mellat Bank in Tehran city). *Innov. Mark.* **2017**, *13*, 13–22. [CrossRef]
30. Bourne, P.A. Customer Satisfaction of Policing the Jamaican Society: Using SERVQUAL to Evaluate Customer Satisfaction. *J. Healthc. Commun.* **2016**, *1*, 25. [CrossRef]
31. Raza, S.A.; Jawaid, S.T.; Hassan, A. Internet banking and customer satisfaction in Pakistan. *Qual. Res. Financ. Mark.* **2015**, *7*, 24–36. [CrossRef]
32. Ananda, S.; Devesh, S. Service quality and customer satisfaction: A study in the perception of retail banking customers in oman. In Proceedings of the 17th International Scientific Conference on Economic and Social Development—Managerial Issues in Modern Business, Warsaw, Poland, 20–21 October 2016; p. 333.
33. Ali, M.; Raza, S.A. Service quality perception and customer satisfaction in Islamic banks of Pakistan: The modified SERVQUAL model. *Total Qual. Manag. Bus. Excell.* **2017**, *28*, 559–577. [CrossRef]
34. Wang, R.; Yan, Z.; Liu, K. An Empirical Study: Measuring the service quality of an e-learning system with the model of ZOT SERVQUAL. In Proceedings of the IEEE 2010 International Conference on E-Business and E-Government (ICEE), Guangzhou, China, 7–9 May 2010; pp. 5379–5382. [CrossRef]
35. Yang, X.S.; Zhu, Y.Y. SERVQUAL-based evaluation study on the quality of community-based service provided by university-affiliated stadiums. In Proceedings of the IEEE 2010 International Conference on Management and Service Science (MASS), Wuhan, China, 24–26 August 2010; pp. 1–4. [CrossRef]
36. Luo, F.; Zhong, Y.; Zhang, X. Perceived performance measurement of outward bound tourists: An empirical study of SERVQUAL model. In Proceedings of the IEEE 2010 International Conference on Management and Service Science (MASS), Wuhan, China, 24–26 August 2010; pp. 1–4. [CrossRef]
37. Altuntas, S.; Dereli, T.; Yilmaz, M.K. Multi-criteria decision making methods based weighted SERVQUAL scales to measure perceived service quality in hospitals: A case study from Turkey. *Total Qual. Manag. Bus. Excell.* **2012**, *23*, 1379–1395. [CrossRef]
38. Badi, I.; Ballem, M. Supplier selection using the rough BWM-MAIRCA model: A case study in pharmaceutical supplying in Libya. *Decis. Mak. Appl. Manag. Eng.* **2018**, *1*, 16–33. [CrossRef]
39. Liu, F.; Aiwu, G.; Lukovac, V.; Vukic, M. A multicriteria model for the selection of the transport service provider: A single valued neutrosophic DEMATEL multicriteria model. *Decis. Mak. Appl. Manag. Eng.* **2018**, *1*, 121–130. [CrossRef]
40. Petrović, I.; Kankaraš, M. DEMATEL-AHP multi-criteria decision making model for the selection and evaluation of criteria for selecting an aircraft for the protection of air traffic. *Decis. Mak. Appl. Manag. Eng.* **2018**, *1*, 93–110. [CrossRef]

41. Pamučar, D.; Božanić, D.; Lukovac, V.; Komazec, N. Normalized weighted geometric bonferroni mean operator of interval rough numbers—Application in interval rough DEMATEL-COPRAS. *Facta Univ. Ser. Mech. Eng.* **2018**, *16*, 171–191. [CrossRef]

42. Stević, Ž.; Stjepanović, Ž.; Božičković, Z.; Das, D.K.; Stanujkić, D. Assessment of Conditions for Implementing Information Technology in a Warehouse System: A Novel Fuzzy PIPRECIA Method. *Symmetry* **2018**, *10*, 586. [CrossRef]

43. Paddeu, D.; Fancello, G.; Fadda, P. An experimental Customer Satisfaction Index to evaluate the performance of city logistics services. *Transport* **2017**, *32*, 262–271. [CrossRef]

44. Nunić, Z. Evaluation and selection of the PVC carpentry Manufacturer using the FUCOM-MABAC model. *Oper. Res. Eng. Sci. Theory Appl.* **2018**, *1*, 13–28.

45. Sharma, H.; Roy, J.; Kar, S.; Prentkovskis, O. Multi Criteria Evaluation Framework for Prioritizing Indian Railway Stations Using Modified Rough AHP-Mabac Method. *Transport Telecommun. J.* **2018**, *19*, 113–127. [CrossRef]

46. Verseckiene, A.; Palsaitis, R.; Yatskiv, I. Evaluation of Alternatives to Integrate Special Transportation Services for People with Movement Disorders. *Transport Telecommun. J.* **2017**, *18*, 263–274. [CrossRef]

47. Csiszár, C.; Sándor, Z. Method for analysis and prediction of dwell times at stops in local bus transportation. *Transport* **2017**, *32*, 302–313. [CrossRef]

48. Stanujkić, D.; Karabašević, D. An extension of the WASPAS method for decision-making problems with intuitionistic fuzzy numbers: A case of website evaluation. *Oper. Res. Eng. Sci. Theory Appl.* **2018**, *1*, 29–39.

49. Nathanail, E.; Gogas, M.; Adamos, G. Assessing the Contribution of Urban Freight Terminals in Last Mile Operations. *Transport Telecommun. J.* **2016**, *17*, 231–241. [CrossRef]

50. Grzegorzewski, P. On Separability of Fuzzy Relations. *Int. J. Fuzzy Logic Intell. Syst.* **2017**, *17*, 137–144. [CrossRef]

51. Stojaković, M.; Twrdy, E. A decision support tool for container terminal optimization within the berth subsystem. *Transport* **2016**, *31*, 29–40. [CrossRef]

52. Rezaei, J.; Kothadiya, O.; Tavasszy, L.; Kroesen, M. Quality assessment of airline baggage handling systems using SERVQUAL and BWM. *Tour. Manag.* **2018**, *66*, 85–93. [CrossRef]

53. Ji, X. SERVQUAL-Model-Based Fuzzy Evaluation of Express Service Quality. *Int. J. Transp. Eng. Technol.* **2018**, *4*, 20. [CrossRef]

54. Pamučar, D.; Stević, Ž.; Sremac, S. A New Model for Determining Weight Coefficients of Criteria in MCDM Models: Full Consistency Method (FUCOM). *Symmetry* **2018**, *10*, 393. [CrossRef]

55. Parasuraman, A.; Zeithaml, V.A.; Berry, L.L. A conceptual model of service quality and its implications for future research. *J. Mark.* **1985**, *49*, 41–50. [CrossRef]

56. Parasuraman, A.; Zeithaml, V.A.; Berry, L.L. SERVQUAL: A multiple-item scale for measuring consumer perceptions of service quality. *J. Retail.* **1988**, *64*, 12–40.

57. Parasuraman, A.; Zeithaml, V.A.; Berry, L.L. Reassessment of expectations as a comparison standard in measuring service quality: Implications for further research. *J. Mark.* **1994**, *58*, 111–124. [CrossRef]

© 2018 by the authors. Licensee MDPI, Basel, Switzerland. This article is an open access article distributed under the terms and conditions of the Creative Commons Attribution (CC BY) license (http://creativecommons.org/licenses/by/4.0/).

Article

A Novel Approach for Green Supplier Selection under a q-Rung Orthopair Fuzzy Environment

Rui Wang [1] and Yanlai Li [1,2,*]

[1] School of Transportation and Logistics, Southwest Jiaotong University, Chengdu 610031, China; wryuedi@my.swjtu.edu.cn

[2] National Lab of Railway Transportation, Southwest Jiaotong University, Chengdu 610031, China

* Correspondence: yanlaili@home.swjtu.edu.cn; Tel.: +86-156-8007-2505

Received: 9 November 2018; Accepted: 26 November 2018; Published: 1 December 2018

Abstract: With environmental issues becoming increasingly important worldwide, plenty of enterprises have applied the green supply chain management (GSCM) mode to achieve economic benefits while ensuring environmental sustainable development. As an important part of GSCM, green supplier selection has been researched in many literatures, which is regarded as a multiple criteria group decision making (MCGDM) problem. However, these existing approaches present several shortcomings, including determining the weights of decision makers subjectively, ignoring the consensus level of decision makers, and that the complexity and uncertainty of evaluation information cannot be adequately expressed. To overcome these drawbacks, a new method for green supplier selection based on the q-rung orthopair fuzzy set is proposed, in which the evaluation information of decision makers is represented by the q-rung orthopair fuzzy numbers. Combined with an iteration-based consensus model and the q-rung orthopair fuzzy power weighted average (q-ROFPWA) operator, an evaluation matrix that is accepted by decision makers or an enterprise is obtained. Then, a comprehensive weighting method can be developed to compute the weights of criteria, which is composed of the subjective weighting method and a deviation maximization model. Finally, the TODIM (TOmada de Decisao Interativa e Multicritevio) method, based on the prospect theory, can be extended into the q-rung orthopair fuzzy environment to obtain the ranking result. A numerical example of green supplier selection in an electric automobile company was implemented to illustrate the practicability and advantages of the proposed approach.

Keywords: green supplier selection; q-rung orthopair fuzzy set; consensus-reaching process; the q-ROFPWA operator; TODIM method

---

1. Introduction

During the past decades, environmental issues have been receiving more and more attention; certain enterprises, especially in the developing countries, have made great efforts in the fields of sustainable development and pollution prevention to face the environmental pressures [1]. These environmental pressures are rooted in two aspects, namely, through government or consumer [2]. The governments have promulgated a series of environmental laws and regulations to restrict the behavior of enterprises; consumers may take the environmental impact of different enterprises into account when making their choices. Therefore, more and more enterprises apply the novel environmental management mode of green supply chain management (GSCM) to reduce the pollution during the operation processes of supply chains [3–6]. GSCM involves many aspects of a supply chain, namely, product design, supplier selection, production, packaging, transportation, marketing, and recycling [7,8]. Among the different segments, green suppliers are the initial link of a supply chain and affect the efficiency and environmental performance of the supply chain; thus, the green supplier selection plays a key role in GSCM [9–11]. To solve the complex green supplier selection problems in practice, many scholars have proposed different green supplier selection approaches [11].

Essentially, green supplier selection can be regarded as a multiple criteria group decision making (MCGDM) problem where decision makers evaluate several potential green suppliers with respect to some criteria to determine the best alternative [8,12,13]. In practice, the evaluation information may be uncertain and incomplete. The fuzzy set (FS) and its generalized forms have been widely used in the current literature to solve this problem [14–16]. Q-rung orthopair fuzzy set (q-ROFS), which was developed by Yager [17], can express the membership, non-membership, and indeterminacy membership degrees of decision makers, simultaneously. Scholars have introduced the q-ROFS to many practical fields, such as investment, enterprise resource planning system selection, and so on [18–20]. Therefore, to deal with the increasing complexity of green supplier selection, decision makers can express a wider range of evaluation information by using the q-ROFS to evaluate the potential green suppliers.

During the process of green supplier selection, decision makers may differentiate from the research fields and practical experiences; thus, the evaluation information of different decision makers will vary widely. However, under the premise of cooperation between decision makers, the ranking result with a relatively high consensus level is desirable [21,22]. In real life, unanimity is difficult or impossible to achieve; the concept of soft consensus was proposed to solve these MCGDM problems. Furthermore, the consensus model has been applied in many practical areas [23–25]. To the best of our knowledge, little attention has been paid to the green supplier selection approaches that include the consensus-reaching process. Therefore, we developed an iteration-based consensus model under the q-rung orthopair fuzzy (q-ROF) environment, which can offer suggestions for decision makers on how to revise their non-consensus evaluation information in each iteration round. Consequently, the consensus model is used during the green supplier selection process to obtain a more accurate ranking result.

The individual acceptable consensus evaluation matrix of each decision maker can be obtained by the consensus model; thus, the next issue is how to aggregate this evaluation information to determine a collective evaluation matrix. Due to the different backgrounds of decision makers in practice, the weights of them will always be difficult to determine simply. Most existing green supplier selection approaches determine the weights of decision makers using the subjective weighting methods or assume that the decision makers are equivalent important, which is inconsistent with the actual situations and may lead to an inaccurate ranking result. To address this problem, Yager [26] proposed the power average (PA) operator, in which the weights of aggregated arguments are determined by the support degrees of them objectively. Since then, the PA operator has been investigated by many scholars to propose its generalized forms under different fuzzy environments; the decision maker weights can be determined by considering the subjective and objective factors, simultaneously. In this paper, the q-rung orthopair fuzzy power weighted average (q-ROFPWA) operator, which was proposed by Liu et al. [27], is utilized during green supplier selection to complete the information fusion effectively.

Since the collective evaluation matrix of potential green suppliers was determined, we needed to obtain the ranking index of each green supplier. Because the evaluation behavior of decision makers is bounded rational, the attitude towards gain and loss of decision makers should be considered while determining the final ranking of green suppliers [8]. Nevertheless, most existing green supplier selection approaches ignored this bounded rationality behavior of decision makers. Inspired by the literature [8], we introduced the TODIM (TOmada de Decisao Interativa e Multicritevio) method to deal with these situations. Gomes and Lima [28] developed this TODIM method, in which the bounded rationality is considered according to the prospect theory [29]. The utility function is introduced to compute the dominance degree of each alternative over all the alternatives; then, the global values of alternatives can be obtained to determine the best alternative. Therefore, in this paper, the q-rung orthopair fuzzy TODIM (q-ROF-TODIM) method was put forward to determine the ranking result of green suppliers.

According to the discussion above, this paper proposes an improved green supplier selection approach based on q-ROFS and TODIM method. The main contributions of this study are presented as in the following. (1) The q-ROFS was used to express the evaluation information of decision makers, which can deal with the uncertainty and complexity of evaluation information in practice effectively. (2) The non-consensus evaluation information could be improved by an efficient iteration-based consensus model to obtain a ranking result that was accepted by decision makers or enterprise. (3) Considering the objective and subjective factors of the decision maker weights, the q-ROFPWA operator was introduced to aggregate the individual evaluation information. (4) The TODIM method under q-ROF environment was constructed to obtain the ranking that reflects the bounded rationality of decision makers. To achieve this, the rest of this paper is presented as follows. The related literature is reviewed in Section 2. The definition, operations, comparison method, distance measure, and aggregation operator of q-ROFS are introduced in Section 3. Section 4 proposes a novel approach for green supplier selection. Section 5 applies a numerical example to show the feasibility and validity of the proposed approach. Some conclusions are summarized in Section 6.

2. Literature Review

2.1. Green Supplier Selection Approaches

As the MCGDM problems become more and more complicated, many novel approaches based on MCGDM methods or soft computing were investigated [30–32]. Similarly, due to the features of green supplier selection, many scholars have researched the green supplier selection method by regarding it as a complex MCGDM problem; thus, a series of MCGDM methods under fuzzy environments have been applied into the research of green supplier selection. For example, Lee et al. [33] developed a fuzzy analytic hierarchy process (AHP) approach for green supplier selection in a high-tech industry. Both Chen et al. [34] and Yazdani [35] constructed an integrated fuzzy multiple criteria decision making approach to obtain the best green supplier, which is composed of fuzzy AHP and technique for order performance by similarity to ideal solution (TOPSIS) methods. Combined with AHP and entropy, elimination and choice expressing the reality III (ELECTRE III) methods, Tsui and Wen [36] proposed an approach for selecting a green supplier, and several improvement suggestions were presented to raise the performance of suppliers. Kannan et al. [9] determined the best green supplier for an engineering plastic material manufacturer in Singapore by using a fuzzy axiomatic design method. Dobos and Vörösmarty [37] evaluated green suppliers with respect to composite indicators based on the data envelopment analysis (DEA) method. Hashemi et al. [38] determined the ranking of green suppliers in GSCM by a comprehensive method that consisted of the analytic network process (ANP) and grey relational analysis (GRA) methods. Kuo et al. [39] integrated the artificial neural network (ANN), ANP, and DEA methods to choose suppliers by considering the environmental regulations. Kuo et al. [40] utilized the decision-making trial and evaluation laboratory (DEMATEL)-based ANP method to investigate the relationships between the criteria and compute the weights of criteria, and then selected the green suppliers combined with the VIKOR (VlseKriterijumska Optimizacija I Kompromisno Resenje) method. To discuss the applications of fuzzy green supplier selection approaches, Banaeian et al. [14] evaluated the green suppliers in the agri-food industry by using the TOPSIS, VIKOR, and GRA methods, respectively. Both Qin et al. [8] and Sang and Liu [41] expressed the uncertainty of evaluation information using an interval type-2 fuzzy set, then utilized the TODIM method to obtain the ranking of green suppliers. Govindan et al. [42] put forward a green supplier selection method based on the revised Simos procedure and preference ranking organization method for enrichment evaluation (PROMETHEE) method. Quan et al. [43] investigated the green supplier selection with a large-scale group of decision makers and developed an integrated method combined with ant colony algorithms and multi-objective optimizations by ratio analysis plus the full multiplicative form (MULTIMOORA) method.

2.2. Q-ROFS

In practice, the related qualitative and quantitative data of green suppliers are always incomplete and complex; thus, crisp numbers cannot express the uncertainty of evaluation information given by decision makers. To solve this problem, Zadeh [44] developed the FS theory to represent the evaluation information; the generalized fuzzy numbers, including triangular fuzzy numbers and type-2 fuzzy numbers, were widely used in approaches for green supplier selection [8,14,33,41]. However, the FS ignores the non-membership degree of evaluation information. For instance, a business manager evaluates an investment before investing; they might think the probability of profit is 0.6, and the probability of loss is 0.3. Obviously, the FS cannot represent the aforementioned evaluation information. Therefore, Atanassov [45] applied the non-membership degree to improve the FS, and proposed the intuitionistic fuzzy set (IFS). Consequently, the evaluation information of the business manager can be expressed by an IFS, i.e., the membership and non-membership degrees are 0.6 and 0.3, respectively. Afterwards, IFS has been applied into green supplier selection [16,46]. Yager [47] proposed a generalized form of IFS called the Pythagorean fuzzy set (PFS), in which the sum of squares of membership and non-membership degrees is less than 1. Furthermore, to provide decision makers with a more relaxed evaluation environment, Yager [17] put forward the q-ROFS theory to express more potential evaluation information of decision makers. Then, the generalized form of q-ROFS, i.e., the q-rung picture linguistic set was proposed [48]. The q-ROFS theory can be regarded as a generalized form of IFS [45] and PFS [47], and the space of acceptable orthopairs increased with the increasing rung q as shown in Figure 1. Combined with the refusal membership degree, Cuong [49] developed the picture fuzzy set; subsequently, the similarity measures of the generalized picture fuzzy sets that including spherical fuzzy sets and T-spherical fuzzy sets were investigated [50]. However, picture fuzzy set is more applicable to model phenomena like voting.

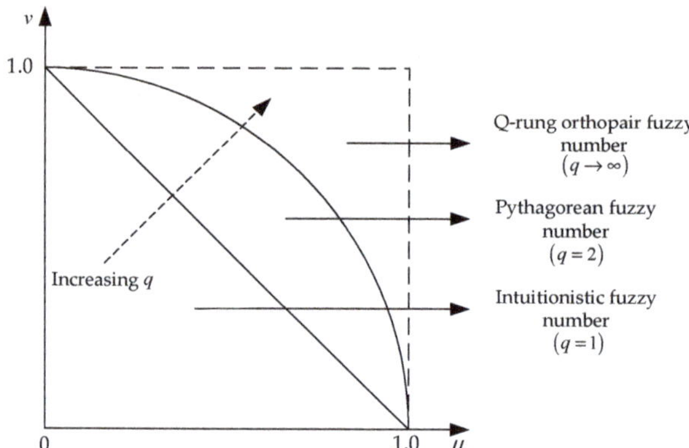

Figure 1. Geometric space range of the intuitionistic fuzzy set (IFS), Pythagorean fuzzy set (PFS), and q-rung orthopair fuzzy set (q-ROFS).

2.3. Consensus Model

During the green supplier selection process, decision makers may come from different research fields of GSCM and have varying degrees of domain experience. Therefore, the non-consensus evaluation information, which is far from group opinions, will be inevitably revealed; however, a ranking result with a low consensus level may be obtained, which is not desirable. In recent years, the consensus model of MCGDM problems has been a hot topic. The existing consensus models can be divided into two categories; one is the iteration-based consensus model. For example,

Herrera-Viedma et al. [51] defined the consensus and proximity measures of different preference structures to construct an iteration-based consensus support system. Herrera-Viedma et al. [52] investigated the consistency and consensus of incomplete fuzzy preference relations; a feedback mechanism was put forward to revise the consistency and consensus levels, simultaneously. With respect to intuitionistic fuzzy preference relations, Chu et al. [53] developed two iteration-based algorithms to improve the consistency and consensus levels, respectively. Wu et al. [54] put forward an iteration-based consensus model to revise the incomplete linguistic information, in which the trust degree was used to complete the decision matrices and adjust the weights of decision makers. Wu and Xu [55] developed a consensus measure of hesitant fuzzy linguistic information to complete the consensus-reaching process. Another kind of consensus model is the optimization-based consensus model. For instance, Dong et al. [56] constructed an optimization programming model for minimizing the number of adjusted simple terms to complete the consensus-reaching process under hesitant linguistic environment. Gong et al. [57] constructed two consensus models according to the minimum cost of consensus and maximum return to achieve a relatively high consensus level. Gong et al. [58] put forward a consensus model for optimizing the economic efficiency. Based on the multiplicative consistency, Xu et al. [59] and Zhang and Pedrycz [60] proposed goal programming models to improve the consistency and consensus levels of intuitionistic fuzzy preference and intuitionistic multiplicative preference relations, respectively. For green supplier selection issues, Zhu and Li [12] introduced a consensus model to put forward a novel green supplier selection approach; nevertheless, the consensus model can only provide suggestions to one of the decision makers for revising the non-consensus evaluation information in each iteration round, an thus, it can take a lot of time to achieve consensus in a complex environment.

2.4. The PA Operator

Based on the individual evaluation information, the collective information of each green supplier with respect to the criteria can be obtained by aggregation tools. Considering the relationships between the input information, Yager [26] developed the PA operator to aggregate the individual information. According to the PA operator and generalized weighted average operator, Zhou et al. [61] proposed the generalized power weighted average (GPWA) operator, in which the weight vectors were obtained by the subjective weight values and support degrees between different aggregated arguments. Xu [62] extended the PA operator into the intuitionistic fuzzy (IF) and interval-valued intuitionistic fuzzy environment; then, the generalized power weighted average operators were defined. Wan [63] put forward the MCGDM method by using a trapezoidal intuitionistic fuzzy power weighted average (TIFPWA) operator. Furthermore, Liu and Liu [64] investigated the generalized form of a TIFPWA operator. He et al. [65] discussed the properties of the interval-valued intuitionistic fuzzy power weighted average (IVIFPWA) operator and developed a novel MCGDM approach based on the IVIFPWA operator. According to the Frank operational laws, Zhang et al. [66] proposed a new form of PA operator. Wei and Lu [67] developed the Pythagorean fuzzy power weighted average (PFPWA) operator. To aggregate the q-rung orthopair fuzzy numbers (q-ROFNs), Liu et al. [27] extended the PA operator to the q-ROFPWA operator. Furthermore, the generalized PA operators have been applied to many practical areas to solve MCGDM problems [62,68–71].

3. Preliminaries

To make this paper as self-contained as possible, this section introduces the definition, operational laws, comparison method, Minkowski distance, and aggregation operator of q-ROFS, that will be utilized in the subsequent research.

3.1. q-ROFS

Based on the IFS and PFS, Yager [17] proposed a more general form, i.e., q-ROFS, and developed the operations of q-ROFS.

Definition 1 [17]. *Let X be a non-empty and finite set, A q-ROFS Q on X is defined by:*

$$Q = \{\langle x, (\mu_Q(x), v_Q(x))\rangle | x \in X\}, \tag{1}$$

where the functions $\mu_Q : X \to [0,1]$ and $v_Q : X \to [0,1]$ represent the membership and non-membership degrees of $x \in X$ to Q, respectively, and they satisfy the condition of $(\mu_Q(x))^q + (v_Q(x))^q \leq 1, q \geq 1$. Furthermore, function $\pi_Q(x) = \sqrt[q]{(\mu_Q(x))^q + (v_Q(x))^q - (\mu_Q(x))^q(v_Q(x))^q}$ indicates the indeterminacy membership degree. For convenience, we call $a = (\mu, v)$ a q-ROFN.

Definition 2 [17]. *Let $a = (\mu, v)$, $a_1 = (\mu_1, v_1)$, and $a_2 = (\mu_2, v_2)$ be three q-ROFNs, $\lambda > 0$, and a^c is the complementary set of a, then:*

$$a^c = (v, \mu); \tag{2}$$

$$a_1 \oplus a_2 = \left(\sqrt[q]{(\mu_1)^q + (\mu_2)^q - (\mu_1)^q(\mu_2)^q}, v_1 v_2 \right); \tag{3}$$

$$a_1 \otimes a_2 = \left(\mu_1 \mu_2, \sqrt[q]{(v_1)^q + (v_2)^q - (v_1)^q(v_2)^q} \right); \tag{4}$$

$$\lambda a = \left(\sqrt[q]{1 - (1 - \mu^q)^\lambda}, v^\lambda \right); \tag{5}$$

$$a^\lambda = \left(\mu^\lambda, \sqrt[q]{1 - (1 - v^q)^\lambda} \right). \tag{6}$$

Example 1. *Suppose that $a_1 = (0.6500, 0.8298)$ and $a_2 = (0.5000, 0.7500)$ are two q-ROFNs, $q = 3$ and $\lambda = 2$, then:*

(1) $(a_1)^c = (0.8298, 0.6500)$, $(a_2)^c = (0.7500, 0.5000)$;
(2) $a_1 \oplus a_2 = (0.7149, 0.6224)$;
(3) $a_1 \otimes a_2 = (0.3250, 0.9094)$;
(4) $\lambda a_1 = (0.7796, 0.6886)$, $\lambda a_2 = (0.6166, 0.5625)$;
(5) $(a_1)^\lambda = (0.4225, 0.9346)$, $(a_2)^\lambda = (0.2500, 0.8732)$.

Liu et al. [18] and Wei et al. [19] investigated the score and accuracy functions of q-ROFS, then, the comparison method of q-ROFNs was put forward.

Definition 3 [18,19]. *Let $a = (\mu, v)$ be a q-ROFN, the score and accuracy functions of a are respectively given by:*

$$s(a) = (1 + \mu^q - v^q)/2; \tag{7}$$

$$h(a) = \mu^q + v^q. \tag{8}$$

Definition 4 [19]. *Let $a_1 = (\mu_1, v_1)$ and $a_2 = (\mu_2, v_2)$ be two q-ROFNs, then:*

(1) *If $s(a_1) < s(a_2)$, then $a_1 < a_2$;*
(2) *If $s(a_1) = s(a_2)$, then*

    a.   *$h(a_1) < h(a_2)$, then $a_1 < a_2$;*
    b.   *$h(a_1) = h(a_2)$, then $a_1 = a_2$.*

Later, Du [72] developed the Minkowski distance measure of q-ROFNs.

Definition 5 [72]. *Let $a_1 = (\mu_1, v_1)$ and $a_2 = (\mu_2, v_2)$ be two q-ROFNs, then the Minkowski distance between them is defined by:*

$$d(a_1, a_2) = \left(\frac{1}{2}|\mu_1 - \mu_2|^p + \frac{1}{2}|v_1 - v_2|^p \right)^{1/p}. \tag{9}$$

Example 2. *Suppose that $a_1 = (0.6500, 0.8298)$ and $a_2 = (0.5000, 0.7500)$ are two q-ROFNs, $q = 3$; according to Definition 3, we have $s(a_1) = s(a_2) = 0.3516$, $h(a_1) = 0.8460$, and $h(a_2) = 0.5469$, then $a_1 > a_2$. In addition, the Minkowski distance between them can be computed as $d(a_1, a_2) = 0.1248$.*

3.2. The q-ROFPWA Operator

Considering the relationship between the aggregated values, Yager [26] proposed the PA operator to fuse the information.

Definition 6 [26]. *Let $a_i(i = 1, 2, \ldots, n)$ be a collection of evaluation values, the PA operator is a mapping $\Omega^n \to \Omega$ as:*

$$PA(a_1, a_2, \ldots, a_n) = \sum_{i=1}^{n} \frac{(1 + T(a_i))a_i}{\sum_{j=1}^{n}(1 + T(a_j))}. \tag{10}$$

where $T(a_i) = \sum_{j=1, j \neq i}^{n} Sup(a_i, a_j)$ and $Sup(a_i, a_j)$ is the support degree for a_i from a_j that satisfies the conditions as follows: (1) $Sup(a_i, a_j) \in [0, 1]$; (2) $Sup(a_i, a_j) = Sup(a_j, a_i)$; (3) If $|a_i - a_j| > |a_s - a_t|$, then $Sup(a_i, a_j) \leq Sup(a_s, a_t)$.

The PA operator can reflect the relationship between the aggregated values during the information fusion; however, it can only aggregate a crisp number. Therefore, Liu et al. [27] extended the PA operator into the q-ROF environment to propose the q-ROFPWA operator.

Definition 7 [27]. *Let $a_i = (\mu_i, v_i)(i = 1, 2, \ldots, n)$ be a collection of q-ROFNs; the q-ROFPWA operator is a mapping $\Omega^n \to \Omega$ as:*

$$q - ROFPWA(a_1, a_2, \ldots, a_n) = \bigoplus_{i=1}^{n} \frac{w_i(1 + T(a_i))a_i}{\sum_{j=1}^{n}(w_j(1 + T(a_j)))}. \tag{11}$$

where $w = (w_1, w_2, \ldots, w_n)^T$ is the weight vector of the q-ROFNs a_i, $T(a_i) = \sum_{j=1, j \neq i}^{n}(w_j Sup(a_i, a_j))$, and $Sup(a_i, a_j) = 1 - d(a_i, a_j)$ is the support degree for a_i from a_j, in which $d(a_i, a_j)$ is the Minkowski distance between a_i and a_j in this study.

Combined with the operations of q-ROFNs, we can obtain the following result.

Theorem 1 [27]. *Let $a_i(i = 1, 2, \ldots, n)$ be a collection of q-ROFNs; their aggregated value by using the q-ROFPWA operator is also a q-ROFN, and:*

$$q - ROFPWA(a_1, a_2, \ldots, a_n) = \left(\sqrt[q]{1 - \prod_{i=1}^{n}\left(1 - \mu_i^q\right)^{w_i(1+T(a_i))/\sum_{j=1}^{n}(w_j(1+T(a_j)))}}, \prod_{i=1}^{n}(v_i)^{w_i(1+T(a_i))/\sum_{j=1}^{n}(w_j(1+T(a_j)))} \right). \tag{12}$$

4. Green Supplier Selection Method under q-ROF Environment

In this section, we defined the q-ROF consensus measures on three levels, namely, criteria, alternative, and evaluation matrix levels to construct the consensus model. Then, the q-ROFPWA operator was investigated to fuse the q-ROF evaluation information. Finally, combined with the comprehensive weighting method and q-ROF-TODIM method, a novel green supplier selection approach under q-ROF environment was developed. The flowchart of the proposed approach is presented in Figure 2.

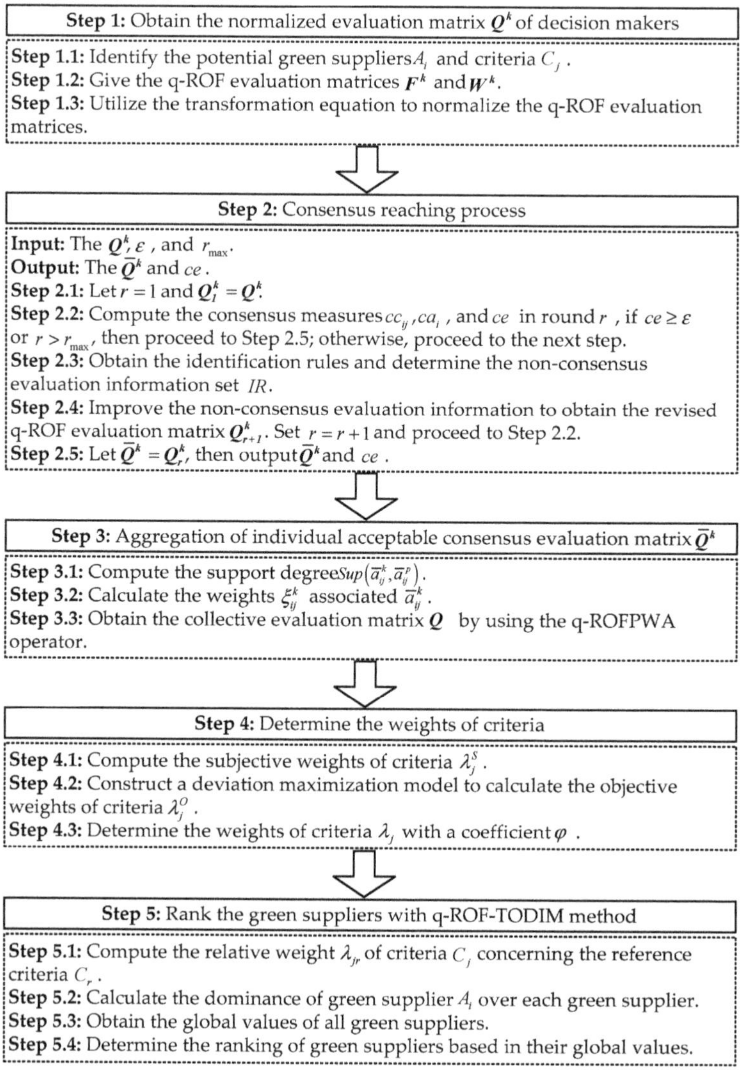

Step 1: Obtain the normalized evaluation matrix Q^k of decision makers

Step 1.1: Identify the potential green suppliers A_i and criteria C_j.
Step 1.2: Give the q-ROF evaluation matrices F^k and W^k.
Step 1.3: Utilize the transformation equation to normalize the q-ROF evaluation matrices.

Step 2: Consensus reaching process

Input: The Q^k, ε, and r_{max}.
Output: The $\bar{Q}^k$ and ce.
Step 2.1: Let $r = 1$ and $Q_1^k = Q^k$.
Step 2.2: Compute the consensus measures cc_{ij}, ca_i, and ce in round r, if $ce \geq \varepsilon$ or $r > r_{max}$, then proceed to Step 2.5; otherwise, proceed to the next step.
Step 2.3: Obtain the identification rules and determine the non-consensus evaluation information set IR.
Step 2.4: Improve the non-consensus evaluation information to obtain the revised q-ROF evaluation matrix Q_{r+1}^k. Set $r = r + 1$ and proceed to Step 2.2.
Step 2.5: Let $\bar{Q}^k = Q_r^k$, then output $\bar{Q}^k$ and ce.

Step 3: Aggregation of individual acceptable consensus evaluation matrix $\bar{Q}^k$

Step 3.1: Compute the support degree $Sup\left(\bar{a}_{ij}^k, \bar{a}_{ij}^p\right)$.
Step 3.2: Calculate the weights ξ_{ij}^k associated $\bar{a}_{ij}^k$.
Step 3.3: Obtain the collective evaluation matrix Q by using the q-ROFPWA operator.

Step 4: Determine the weights of criteria

Step 4.1: Compute the subjective weights of criteria λ_j^S.
Step 4.2: Construct a deviation maximization model to calculate the objective weights of criteria λ_j^O.
Step 4.3: Determine the weights of criteria λ_j with a coefficient φ.

Step 5: Rank the green suppliers with q-ROF-TODIM method

Step 5.1: Compute the relative weight λ_{jr} of criteria C_j concerning the reference criteria C_r.
Step 5.2: Calculate the dominance of green supplier A_i over each green supplier.
Step 5.3: Obtain the global values of all green suppliers.
Step 5.4: Determine the ranking of green suppliers based in their global values.

Figure 2. The flowchart of the proposed approach. Q-ROF: q-rung orthopair fuzzy. Q-ROFPWA: q-rung orthopair fuzzy power weighted average. Q-ROF-TODIM: q-rung orthopair fuzzy TOmada de Decisao Interativa e Multicritevio

4.1. Obtain the Normalized Evaluation Matrices of Decision Makers

For a green supplier selection problem, suppose that a group of decision makers $D_k (k = 1, 2, \ldots, l)$ is assembled to evaluate the green suppliers for an enterprise, in which decision makers may come from different backgrounds of GSCM. Then, the normalized q-ROF evaluation matrices of decision makers can be obtained by the steps as follows:

Step 1.1: After the primary evaluation of the green supplier selection problem, decision makers can identify the potential green supplier $A_i(i = 1, 2, \ldots, m)$ and a collection of criteria $C_j(j = 1, 2, \ldots, n)$.

Step 1.2: Combined with the q-ROFS, the evaluation information of green suppliers can be expressed by q-ROF evaluation matrix $F^k = \left(\tilde{a}_{ij}^k\right)_{m \times n}$, where $\tilde{a}_{ij}^k = \left(\tilde{\mu}_{ij}^k, \tilde{v}_{ij}^k\right)$ indicates the q-ROF evaluation information of green supplier A_i concerning criteria C_j given by decision maker D_k. Moreover, decision makers also evaluate the weights of criteria using q-ROFNs; subsequently, the q-ROF evaluation matrix $W^k = \left(a_j^k\right)_{1 \times n}$ was obtained, where $a_j^k = \left(\mu_j^k, v_j^k\right)$ represents the importance degree of criteria C_j given by decision maker D_k.

Step 1.3: Generally, the criteria of green supplier selection can be divided into two types, namely, cost type and benefit type; thus, we should transform the information with respect to cost type criteria into the information with respect to benefit type criteria to determine the normalized q-ROF evaluation matrix $Q^k = \left(a_{ij}^k\right)_{m \times n}$ as:

$$a_{ij}^k = \left(\mu_{ij}^k, v_{ij}^k\right) = \begin{cases} \tilde{a}_{ij}^k & \text{if } C_j \text{ is the benefit type criteria;} \\ \left(\tilde{a}_{ij}^k\right)^c & \text{if } C_j \text{ is the cost type criteria.} \end{cases} \tag{13}$$

4.2. Consensus-Reaching Process

Most research focused on the consensus model with preference relations that were obtained by pairwise comparison; Wu and Xu [55] proposed an iteration-based consensus model to solve the MCGDM problems based on a hesitant fuzzy linguistic set. Motivated by the literature, we develop the similarity matrix between different q-ROF evaluation matrices.

Definition 8. *Suppose that decision maker $D_k(k = 1, 2, \ldots, l)$ evaluated the alternative $A_i(i = 1, 2, \ldots, m)$ concerning the criteria $C_j(j = 1, 2, \ldots, n)$ using q-ROFNs. For each pair of decision makers, $(D_k, D_p)(k = 1, 2, \ldots, l-1; p = k+1, k+2, \ldots, l)$, the similarity matrix SM^{kp} between the q-ROF evaluation matrices $Q^k = \left(a_{ij}^k\right)_{m \times n}$ and $Q^p = \left(a_{ij}^p\right)_{m \times n}$ is defined by:*

$$SM^{kp} = \left(sm_{ij}^{kp}\right)_{m \times n} = \left(1 - d\left(a_{ij}^k, a_{ij}^p\right)\right)_{m \times n}, \tag{14}$$

where $d\left(a_{ij}^k, a_{ij}^p\right)$ is the Minkowski distance between the q-ROF evaluation information a_{ij}^k and a_{ij}^p. Furthermore, the consensus matrix CM is determined as:

$$CM = \left(cm_{ij}\right)_{m \times n} = \left(\psi\left(sm_{ij}^{kp}\right)\right)_{m \times n}, \tag{15}$$

where ψ is the arithmetic average operator.

The three consensus measures on criteria, alternative, and evaluation matrix levels could then be defined according to the consensus matrix CM, which will be used to complete the consensus-reaching process.

Definition 9. *Criteria level: the consensus measure cc_{ij} for alternative A_i with respect to criteria C_j can be represented by the element of consensus matrix CM as:*

$$cc_{ij} = cm_{ij}. \tag{16}$$

Alternative level: the consensus measure ca_i on alternative A_i can be obtained by:

$$ca_i = \frac{\sum_{j=1}^{n} cc_{ij}}{n}. \tag{17}$$

Evaluation matrix level: the consensus measure ce on the evaluation matrix, i.e., the global consensus measure, can be defined by:

$$ce = \min_{i}\{ca_i\}. \tag{18}$$

Once the q-ROF consensus measures on three levels in Definition 9 were computed, we could check whether the consensus was achieved by comparing the consensus measure ce with the predefined ideal consensus threshold $\varepsilon \in (0,1]$. If $ce \geq \varepsilon$, the consensus was reached; thus, the normalized q-ROF evaluation matrix Q^k was the acceptable consensus evaluation matrix. Otherwise, several identification and direction rules could be obtained according to the aforementioned three consensus measures; identification rules were utilized to determine the non-consensus evaluation information set that contributed less to reach a high consensus level for each iteration round, and direction rules could guide decision makers to revise the non-consensus evaluation information in this round. An iteration-based consensus model under q-ROF environment was constructed to reach consensus as follows.

Input: The original individual evaluation matrix Q^k, the ideal consensus threshold ε, and the maximum permission iterative number of times $r_{\max}$.

Output: The revised individual q-ROF evaluation matrix $\overline{Q}^k$ and the global consensus measure ce.

Step 2.1: Let the initial iterative number be $r = 1$, and the individual evaluation matrix in the first round be $Q_1^k = \left(a_{ij,1}^k\right)_{m \times n} = \left(a_{ij}^k\right)_{m \times n}$.

Step 2.2: Calculate the similarity matrix SM^{kp} $(k = 1, 2, \ldots, l-1; p = k+1, k+2, \ldots, l)$ and aggregate them to obtain the consensus matrix CM; thus, the consensus measures cc_{ij}, ca_i, and ce in round r are computed. If $ce \geq \varepsilon$ or $r > r_{\max}$, then proceed to Step 1.5; otherwise, proceed to the next step.

Step 2.3: Obtain the identification rules as in the following:

(1) Identification rule 1. The non-consensus alternative set $IRA = \{A_i | ca_i < \varepsilon, i = 1, 2, \ldots, m\}$ identifies the rows of the evaluation matrices that should be revised.

(2) Identification rule 2. The non-consensus criteria set $IRC_i = \{C_j | A_i \in IRA \wedge cc_{ij} < \varepsilon, j = 1, 2, \ldots, n\}$ identifies the columns that should be revised for the rows distinguished in the non-consensus alternative set IRA.

(3) Identification rule 3. The non-consensus decision maker set $IRD_{ij} = \left\{D_p \middle| A_i \in IRA \wedge C_j \in IRC_i \wedge d_{ij}^{(p)} = \max_{k}\left\{d_{ij}^{(k)}\right\}\right\}$ identifies the decision makers that should revise the evaluation information at the position (i,j) in evaluation matrices, where $d_{ij}^{(p)}$ is the distance between the similarity measures of D_p and other decision makers, i.e., $d_{ij}^{(p)} = \sum_{k=1, k \neq p}^{l}\left(1 - sm_{ij}^{kp}\right)$.

Subsequently, combined with the identification rules 1~3, the non-consensus evaluation information set IR that should be revised in round r can be determined as:

$$IR = \{(p, (i,j)) | D_p \in IRD_{ij} \wedge A_i \in IRA \wedge C_j \in IRC_i\}. \tag{19}$$

Step 2.4: Aggregate the individual evaluation matrix Q_r^k using the q-ROFAA operator that is reduced by the q-ROFWA operator [18], then, the collective evaluation matrix $Q_r = \left(a_{ij,r}\right)_{m \times n}$ can be obtained as:

$$a_{ij,r} = q - ROFAA\left(a_{ij,r}^1, a_{ij,r}^2, \ldots, a_{ij,r}^l\right) = \left(\sqrt[q]{1 - \prod_{k=1}^{l}\left(1 - \left(\mu_{ij,r}^k\right)^q\right)^{1/l}}, \prod_{k=1}^{l}\left(v_{ij,r}^k\right)^{1/l}\right). \tag{20}$$

Both the collective evaluation information $a_{ij,r}$ and the non-consensus evaluation information set IR show that the direction rules, which suggest decision makers how to change their non-consensus evaluation information as in the following:

(1) If $a_{ij,r} > a_{ij,r}^k$, then the decision maker D_k should decrease the evaluation on alternative A_i concerning criteria C_j when C_j is the benefit type criteria, and the decision maker D_k should increase the evaluation on alternative A_i concerning criteria C_j when C_j is the cost type criteria.

(2) If $a_{ij,r} < a_{ij,r}^k$, the decision maker D_k should increase the evaluation on alternative A_i concerning criteria C_j when C_j is the benefit type criteria, and the decision maker D_k should decrease the evaluation on alternative A_i concerning criteria C_j when C_j is the cost type criteria.

Then, the revised individual q-ROF evaluation matrix Q_{r+1}^k can be obtained. Set $r = r + 1$ and proceed to **Step 1.2**.

Step 2.5: Let $\overline{Q}^k = Q_r^k = \left(\overline{a}_{ij}^k\right)_{m \times n} = \left(\overline{\mu}_{ij}^k, \overline{v}_{ij}^k\right)_{m \times n}$. Output $\overline{Q}^k$ and ce in this round.

4.3. Aggregation of Individual Acceptable Consensus Evaluation Matrices

According to the individual acceptable consensus evaluation matrices, we can use the q-ROFPWA operator to fuse them; then, the weights of decision makers can be determined by both the subjective weights and support degrees between individual evaluation information. Thus, the collective evaluation matrix $Q = \left(a_{ij}\right)_{m \times n}$ is obtained by the steps as below.

Step 3.1: Compute the support degree:

$$Sup\left(\overline{a}_{ij}^k, \overline{a}_{ij}^p\right) = 1 - d\left(\overline{a}_{ij}^k, \overline{a}_{ij}^p\right), k, p = 1, 2, \ldots, l, \tag{21}$$

where $d\left(\overline{a}_{ij}^k, \overline{a}_{ij}^p\right)$ is the Minkowski distance between the evaluation information $\overline{a}_{ij}^k$ and $\overline{a}_{ij}^p$.

Step 3.2: Combined with the subjective weight vector of decision makers $w = (w_1, w_2, \ldots, w_l)^T$ that is provided by the enterprise, the weighted support degree of $\overline{a}_{ij}^k$ can be calculated as:

$$T\left(\overline{a}_{ij}^k\right) = \sum_{p=1, p \neq k}^{l} w_p Sup\left(\overline{a}_{ij}^k, \overline{a}_{ij}^p\right), \tag{22}$$

Then, the weights associated with $\overline{a}_{ij}^k$ can be determined as:

$$\xi_{ij}^k = \frac{w_k\left(1 + T\left(\overline{a}_{ij}^k\right)\right)}{\sum_{k=1}^{l}\left(w_k\left(1 + T\left(\overline{a}_{ij}^k\right)\right)\right)}, \xi_{ij}^k \geq 0, \sum_{k=1}^{l} \xi_{ij}^k = 1. \tag{23}$$

Step 3.3: Use the q-ROFPWA operator to fuse the evaluation matrix $\overline{Q}^k$ to obtain the collective evaluation matrix Q as:

$$a_{ij} = q - ROFPWA\left(\overline{a}_{ij}^1, \overline{a}_{ij}^2, \ldots, \overline{a}_{ij}^l\right) = \left(\sqrt[q]{1 - \prod_{k=1}^{l}\left(1 - \left(\overline{\mu}_{ij}^k\right)^q\right)^{\xi_{ij}^k}}, \prod_{i=1}^{n}\left(\overline{v}_{ij}^k\right)^{\xi_{ij}^k}\right) = (\mu_{ij}, v_{ij}). \tag{24}$$

4.4. Determine the Weights of Criteria

In practice, it is sometimes unreasonable to determine the criteria weights only considering the views of decision makers. To investigate both the subjective and objective factors, we constructed a comprehensive weighting method that consists of a subjective weighting method and a deviation maximization model to calculate the weights of criteria as follows.

Step 4.1: Combined with the evaluation matrix W^k and the similar steps in Sections 4.3 and 4.4, we can obtain the collective evaluation matrix $W = \left(a_j\right)_{1 \times n}$. The larger the score value of a_j, which means the criteria C_j is more important, the higher the weight of criteria C_j and vice versa. Then, the subjective weight vector of criteria $\lambda^S = \left(\lambda_1^S, \lambda_2^S, \ldots, \lambda_n^S\right)^T$ can be determined as:

$$\lambda_j^S = \frac{s\left(a_j\right)}{\sum_{j=1}^n s\left(a_j\right)}. \tag{25}$$

where $s\left(a_j\right)$ is the score value of a_j.

Step 4.2: Let $\sum_{h=1,h\neq i}^m d\left(a_{ij}, a_{hj}\right)\lambda_j^O$ be the deviation between the collective evaluation information on green supplier A_i and other green suppliers concerning C_j, where $d\left(a_{ij}, a_{hj}\right)$ is the Minkowski distance between a_{ij} and a_{hj}; then, the total deviation is obtained as $\sum_{j=1}^n \sum_{i=1}^m \sum_{h=1,h\neq i}^m d\left(a_{ij}, a_{hj}\right)\lambda_j^O$. According to the information theory, if all green suppliers have similar evaluation information concerning one of criteria, a small weight value should be assigned to the criteria as it contributes less to differentiate green suppliers [73]. Subsequently, a deviation maximization model can be developed as:

$$\max \sum_{j=1}^n \sum_{i=1}^m \sum_{h=1,h\neq i}^m d\left(a_{ij}, a_{hj}\right)\lambda_j^O$$
$$s.t. \sum_{j=1}^n \left(\lambda_j^O\right)^2 = 1, \lambda_j^O \geq 0. \tag{26}$$

Solve the model according to the Lagrange function:

$$L\left(\lambda^O, \wp\right) = \sum_{j=1}^n \sum_{i=1}^m \sum_{h=1,h\neq i}^m d\left(a_{ij}, a_{hj}\right)\lambda_j^O + \frac{\wp}{2}\left(\sum_{j=1}^n \left(\lambda_j^O\right)^2 - 1\right). \tag{27}$$

where $\wp$ is the Lagrange multiplier. Differentiate Equation (27) concerning λ_j^O and $\wp$, and let these partial derivatives be equal to zero:

$$\begin{cases} \frac{\partial L\left(\lambda^O, \wp\right)}{\partial \lambda_j^O} = \sum_{i=1}^m \sum_{h=1,h\neq i}^m d\left(a_{ij}, a_{hj}\right) + \wp\lambda_j^O = 0; \\ \frac{\partial L\left(\lambda^O, \wp\right)}{\partial \wp} = \sum_{j=1}^n \left(\lambda_j^O\right)^2 - 1 = 0. \end{cases} \tag{28}$$

By solving Equation (28), the normalized weights of criteria, i.e., objective weight vector of criteria $\lambda^O = \left(\lambda_1^O, \lambda_2^O, \ldots, \lambda_n^O\right)^T$ can be obtained:

$$\lambda_j^O = \frac{\sum_{i=1}^m \sum_{h=1,h\neq i}^m d\left(a_{ij}, a_{hj}\right)}{\sum_{j=1}^n \sum_{i=1}^m \sum_{h=1,h\neq i}^m d\left(a_{ij}, a_{hj}\right)}. \tag{29}$$

Step 4.3: Determine the comprehensive weight vector of criteria $\lambda = \left(\lambda_1, \lambda_2, \ldots, \lambda_n\right)^T$ as:

$$\lambda_j = \varphi\lambda_j^S + (1 - \varphi)\lambda_j^O. \tag{30}$$

where $\varphi \in [0, 1]$ is the importance coefficient of subjective weights, and $1 - \varphi$ is the importance coefficient of objective weights.

4.5. Rank the Green Suppliers Using the TODIM Method under q-ROF Environment

Based on the collective q-ROF evaluation matrix Q and weight vector of criteria λ, we construct the q-ROF-TODIM method that can deal with the multiple criteria decision making problems with q-ROFS to obtain the ranking indices of green suppliers and determine the best green supplier.

Step 5.1: Compute the relative weight λ_{jr} of criteria C_j with respect to the reference criteria C_r as:

$$\lambda_{jr} = \lambda_j / \lambda_r, \tag{31}$$

where λ_j is the weight of criteria C_j and $\lambda_r = \max_j\{\lambda_j\}$ is the weight of reference criteria C_r.

Step 5.2: Calculate the dominance degree of green supplier A_i over each green supplier $A_h (h = 1, 2, \ldots, m)$ by the following equation:

$$\delta(A_i, A_h) = \sum_{j=1}^{n} \phi_j(A_i, A_h), \tag{32}$$

where:

$$\phi_j(A_i, A_h) = \begin{cases} \sqrt{\lambda_{jr} d\left(a_{ij}, a_{hj}\right) / \sum_{j=1}^{n} \lambda_{jr}} & \text{if } a_{ij} > a_{hj}; \\ 0 & \text{if } a_{ij} = a_{hj}; \\ -\frac{1}{\theta} \sqrt{\left(\sum_{j=1}^{n} \lambda_{jr}\right) d\left(a_{ij}, a_{hj}\right) / \lambda_{jr}} & \text{if } a_{ij} < a_{hj}. \end{cases} \tag{33}$$

The parameter $\theta > 0$ indicates the attenuation factor of the losses, and $d\left(a_{ij}, a_{hj}\right)$ is the Minkowski distance between a_{ij} and a_{hj}.

Step 5.3: Compute the global value of green supplier A_i by:

$$\Phi(A_i) = \frac{\sum_{h=1}^{m} \delta(A_i, A_h) - \min_i\{\sum_{h=1}^{m} \delta(A_i, A_h)\}}{\max_i\{\sum_{h=1}^{m} \delta(A_i, A_h)\} - \min_i\{\sum_{h=1}^{m} \delta(A_i, A_h)\}}. \tag{34}$$

Step 5.4: Determine the ranking of potential green suppliers based on their global values; the larger the global value $\Phi(A_i)$, the higher the ranking of green supplier A_i.

5. Numerical Example

In this section, a numerical example in the literature [16] was applied to show the feasibility and advantages of the proposed approach. An electric automobile enterprise plans to purchase a key component of a manufacturing procedure from the green suppliers market; the ranking of green suppliers can be determined by the following steps in the next subsection.

5.1. Implementation

Step 1: Obtain the normalized evaluation matrices of decision makers.

Step 1.1: After a preliminary evaluation, four potential green suppliers $A_i(i = 1, 2, 3, 4)$ are determined by a group of decision makers $D_k(k = 1, 2, 3)$. Decision makers evaluate the four green suppliers concerning six criteria $C_j(j = 1, 2, 3, 4, 5, 6)$, namely, environmental costs (C_1), remanufacturing activity (C_2), energy assumption (C_3), reverse logistics program (C_4), hazardous waste management (C_5), and environmental certification (C_6), where C_1 and C_3 are the cost type criteria, and the others are the benefit type criteria.

Step 1.2: According to the relationships between the linguistic terms and interval-valued Pythagorean fuzzy numbers [74], we can construct the transformation between the linguistic terms and the corresponding q-ROFNs ($q = 3$) as shown in Table 1. Then, decision makers use the linguistic terms to assess the green suppliers as shown in Table 2; thus, the q-ROF evaluation matrix $F^k = \left(\tilde{a}_{ij}^k\right)_{4 \times 6}$ is obtained. It is noteworthy that we adopt the subjective weights of criteria obtained in the literature [16], i.e., $\lambda^S = (0.180, 0.090, 0.130, 0.130, 0.310, 0.160)^T$.

Table 1. Linguistic terms and the corresponding q-rung orthopair fuzzy numbers (q-ROFNs).

Linguistic Terms	Corresponding q-ROFNs
Extremely High (EH)	(0.95,0.15)
Very High (VH)	(0.85,0.25)
High (H)	(0.75,0.35)
Medium High (MH)	(0.65,0.45)
Medium (M)	(0.55,0.55)
Medium Low (ML)	(0.45,0.65)
Low (L)	(0.35,0.75)
Very Low (VL)	(0.25,0.85)
Extremely Low (EL)	(0.15,0.95)

Table 2. Evaluation information of decision makers.

Decision Makers	Alternatives	C_1	C_2	C_3	C_4	C_5	C_6
D_1	A_1	EH	H	L	EL	H	M
	A_2	MH	L	H	H	M	L
	A_3	L	H	M	L	EH	H
	A_4	L	EL	H	MH	MH	H
D_2	A_1	VH	VL	M	L	H	ML
	A_2	H	MH	MH	MH	M	ML
	A_3	L	VH	M	L	EL	H
	A_4	ML	H	H	MH	MH	H
D_3	A_1	ML	MH	ML	VL	VH	M
	A_2	VH	L	H	MH	MH	M
	A_3	ML	MH	MH	ML	EL	MH
	A_4	M	VH	EH	M	M	EL

Step 1.3: After the normalization step according to the different types of criteria, the normalized q-ROF evaluation matrix $Q^k = \left(a_{ij}^k \right)_{4\times6}$ can be obtained by using Equation (13).

Step 2: Consensus-reaching process ($\varepsilon = 0.85, r_{\max} = 5$).

Step 2.1: Let the initial iterative number be $r = 1$, and $Q_1^k = \left(a_{ij,1}^k \right)_{4\times6} = \left(a_{ij}^k \right)_{4\times6}$.

Step 2.2: Calculate the similarity matrix SM^{kp} between the q-ROF evaluation matrices Q_1^k and Q_1^p as

$$
SM^{12} = \begin{pmatrix}
0.9 & 0.5 & 0.8 & 0.8 & 1.0 & 0.9 \\
0.9 & 0.7 & 0.9 & 0.9 & 1.0 & 0.9 \\
1.0 & 0.9 & 1.0 & 1.0 & 0.2 & 1.0 \\
0.9 & 0.4 & 1.0 & 1.0 & 1.0 & 1.0
\end{pmatrix}; SM^{13} = \begin{pmatrix}
0.5 & 0.9 & 0.9 & 0.9 & 0.9 & 1.0 \\
0.8 & 1.0 & 1.0 & 0.9 & 0.9 & 0.8 \\
0.9 & 0.9 & 0.9 & 0.9 & 0.2 & 0.9 \\
0.8 & 0.3 & 0.8 & 0.9 & 0.9 & 0.4
\end{pmatrix}; SM^{23} = \begin{pmatrix}
0.6 & 0.6 & 0.9 & 0.9 & 0.9 & 0.9 \\
0.9 & 0.7 & 0.9 & 1.0 & 0.9 & 0.9 \\
0.9 & 0.8 & 0.9 & 0.9 & 1.0 & 0.9 \\
0.9 & 0.9 & 0.8 & 0.9 & 0.9 & 0.4
\end{pmatrix}.
$$

Then, aggregate them to obtain the consensus matrix CM in round one:

$$
CM = \begin{pmatrix}
0.6667 & 0.6667 & 0.8667 & 0.8667 & 0.9333 & 0.9333 \\
0.8667 & 0.8000 & 0.9333 & 0.9333 & 0.9333 & 0.8667 \\
0.9333 & 0.8667 & 0.9333 & 0.9333 & 0.4667 & 0.9333 \\
0.8667 & 0.5333 & 0.8667 & 0.9333 & 0.9333 & 0.6000
\end{pmatrix}.
$$

Thus, we can calculate the consensus measures cc_{ij}, ca_i, and ce based on Equations (16)~(18); the global consensus measure in round one $ce = 0.7889$. It can be seen that $ce < \varepsilon$ after which we can proceed to the next step.

Step 2.3: Obtain the identification rules as in the following:

(1) Identification rule 1. The non-consensus alternative set: $IRA = \{A_i|ca_i < 0.85\} = \{A_1, A_3, A_4\}$.

(2) Identification rule 2. The non-consensus criteria set:

$$IRC_1 = \{C_j | A_1 \in IRA \wedge cc_{1j} < 0.85\} = \{C_1, C_2\};$$

$$IRC_3 = \{C_j | A_3 \in IRA \wedge cc_{3j} < 0.85\} = \{C_5\};$$

$$IRC_4 = \{C_j | A_4 \in IRA \wedge cc_{4j} < 0.85\} = \{C_2, C_6\}.$$

(3) Identification rule 3. Combined with the distances between the similarity measures of decision maker D_p and the other decision makers at the positions $\{(1,1),(1,2),(3,5),(4,2),(4,6)\}$ in evaluation matrix Q_1^k, we can obtain the non-consensus decision maker set:

$$IRD_{11} = \left\{ D_p \middle| A_1 \in IRA \wedge C_1 \in IRC_1 \wedge d_{11}^{(p)} = \max_k \left\{ d_{11}^{(k)} \right\} \right\} = \{D_3\};$$

$$IRD_{12} = \left\{ D_p \middle| A_1 \in IRA \wedge C_2 \in IRC_1 \wedge d_{12}^{(p)} = \max_k \left\{ d_{12}^{(k)} \right\} \right\} = \{D_2\};$$

$$IRD_{35} = \left\{ D_p \middle| A_3 \in IRA \wedge C_5 \in IRC_3 \wedge d_{35}^{(p)} = \max_k \left\{ d_{35}^{(k)} \right\} \right\} = \{D_1\};$$

$$IRD_{42} = \left\{ D_p \middle| A_4 \in IRA \wedge C_2 \in IRC_4 \wedge d_{42}^{(p)} = \max_k \left\{ d_{42}^{(k)} \right\} \right\} = \{D_1\};$$

$$IRD_{46} = \left\{ D_p \middle| A_4 \in IRA \wedge C_6 \in IRC_4 \wedge d_{46}^{(p)} = \max_k \left\{ d_{46}^{(k)} \right\} \right\} = \{D_3\}.$$

Finally, based on the identification rules 1~3, the non-consensus evaluation information set IR that should be revised in round one can be determined as:

$$IR = \{(p,(i,j)) | D_p \in IRD_{ij} \wedge A_i \in IRA \wedge C_j \in IRC_i\} = \{(3,(1,1)),(2,(1,2)),(1,(3,5)),(1,(4,2)),(3,(4,6))\}.$$

Step 2.4: Aggregate the individual evaluation matrix Q_1^k in round one using the q-ROFAA operator to obtain collective evaluation matrix $Q_1 = (a_{ij,1})_{4\times6}$; then, the direction rules can be put forward to revise the non-consensus evaluation information in set IR as shown in Table 3. Set $r = 2$ and proceed to Step 1.2.

Table 3. Direction rules in round one.

IR	Individual Evaluation Information	Collective Evaluation Information	Direction Rules
(3,(1,1))	ML, (0.45,0.65)	(0.4750,0.7136)	ML→M
(2,(1,2))	VL, (0.25,0.85)	(0.6345,0.5116)	VL→L
(1,(3,5))	EH, (0.95,0.15)	(0.7823,0.5135)	EH→VH
(1,(4,2))	EL, (0.15,0.95)	(0.7332,0.4364)	EL→VL
(3,(4,6))	EL, (0.15,0.95)	(0.6745,0.4882)	EL→VL

Then, combined with the similar **Steps 2.2~2.4**, we can obtain the global consensus measure in round four $ce = 0.8556 > \varepsilon$, which means that a high consensus level between decision makers has been achieved; the individual acceptable consensus q-ROF evaluation matrix $\overline{Q}^k$ are determined as shown in Table 4.

Table 4. Individual acceptable consensus q-rung orthopair fuzzy (q-ROF) evaluation matrices.

Decision Makers	Alternatives	C_1	C_2	C_3	C_4	C_5	C_6
D_1	A_1	(0.15,0.95)	(0.75,0.35)	(0.75,0.35)	(0.15,0.95)	(0.75,0.35)	(0.55,0.55)
	A_2	(0.45,0.65)	(0.35,0.75)	(0.35,0.75)	(0.75,0.35)	(0.55,0.55)	(0.35,0.75)
	A_3	(0.75,0.35)	(0.75,0.35)	(0.55,0.55)	(0.35,0.75)	(0.85,0.25)	(0.75,0.35)
	A_4	(0.75,0.35)	(0.45,0.65)	(0.35,0.75)	(0.65,0.45)	(0.65,0.45)	(0.75,0.35)
D_2	A_1	(0.25,0.85)	(0.45,0.65)	(0.55,0.55)	(0.35,0.75)	(0.75,0.35)	(0.45,0.65)
	A_2	(0.35,0.75)	(0.65,0.45)	(0.45,0.65)	(0.65,0.45)	(0.55,0.55)	(0.45,0.65)
	A_3	(0.75,0.35)	(0.85,0.25)	(0.55,0.55)	(0.35,0.75)	(0.15,0.95)	(0.75,0.35)
	A_4	(0.65,0.45)	(0.75,0.35)	(0.35,0.75)	(0.65,0.45)	(0.65,0.45)	(0.75,0.35)
D_3	A_1	(0.45,0.65)	(0.65,0.45)	(0.65,0.45)	(0.25,0.85)	(0.85,0.25)	(0.55,0.55)
	A_2	(0.25,0.85)	(0.35,0.75)	(0.35,0.75)	(0.65,0.45)	(0.65,0.45)	(0.55,0.55)
	A_3	(0.65,0.45)	(0.65,0.45)	(0.45,0.65)	(0.45,0.65)	(0.15,0.95)	(0.65,0.45)
	A_4	(0.55,0.55)	(0.85,0.25)	(0.15,0.95)	(0.55,0.55)	(0.55,0.55)	(0.45,0.65)

Step 3: Aggregation of individual acceptable consensus evaluation matrices.

Steps 3.1~3.2: Suppose that the subjective weight values of decision makers are equal, i.e., $w = (1/3, 1/3, 1/3)^T$; we can use Equations (21)~(23) to calculate the weighted support degree of $\bar{a}_{ij}^k$ as:

$$T^1 = \begin{pmatrix} 0.5333 & 0.5333 & 0.5667 & 0.5667 & 0.6333 & 0.6333 \\ 0.5667 & 0.5667 & 0.6333 & 0.6000 & 0.6333 & 0.5667 \\ 0.6333 & 0.6000 & 0.6333 & 0.6333 & 0.2000 & 0.6333 \\ 0.5667 & 0.4333 & 0.6000 & 0.6333 & 0.6333 & 0.5667 \end{pmatrix};$$

$$T^2 = \begin{pmatrix} 0.5667 & 0.5000 & 0.5667 & 0.5667 & 0.6333 & 0.6000 \\ 0.6000 & 0.4667 & 0.6000 & 0.6333 & 0.6333 & 0.6000 \\ 0.6333 & 0.5667 & 0.6333 & 0.6333 & 0.4333 & 0.6333 \\ 0.6000 & 0.5333 & 0.6000 & 0.6333 & 0.6333 & 0.5667 \end{pmatrix};$$

$$T^3 = \begin{pmatrix} 0.5000 & 0.5667 & 0.6000 & 0.6000 & 0.6000 & 0.6333 \\ 0.5667 & 0.5667 & 0.6333 & 0.6333 & 0.6000 & 0.5667 \\ 0.6000 & 0.5667 & 0.6000 & 0.6000 & 0.4333 & 0.6000 \\ 0.5667 & 0.5000 & 0.5333 & 0.6000 & 0.6000 & 0.4667 \end{pmatrix}.$$

Then, the weights associated with $\bar{a}_{ij}^k$ can be determined as:

$$\xi^1 = \begin{pmatrix} 0.3333 & 0.3333 & 0.3310 & 0.3310 & 0.3356 & 0.3356 \\ 0.3310 & 0.3406 & 0.3356 & 0.3288 & 0.3356 & 0.3310 \\ 0.3356 & 0.3380 & 0.3356 & 0.3356 & 0.2951 & 0.3356 \\ 0.3310 & 0.3209 & 0.3380 & 0.3356 & 0.3356 & 0.3406 \end{pmatrix};$$

$$\xi^2 = \begin{pmatrix} 0.3406 & 0.3261 & 0.3310 & 0.3310 & 0.3356 & 0.3288 \\ 0.3380 & 0.3188 & 0.3288 & 0.3356 & 0.3356 & 0.3380 \\ 0.3356 & 0.3310 & 0.3356 & 0.3356 & 0.3308 & 0.3356 \\ 0.3380 & 0.3333 & 0.3380 & 0.3356 & 0.3356 & 0.3406 \end{pmatrix};$$

$$\xi^3 = \begin{pmatrix} 0.3261 & 0.3406 & 0.3380 & 0.3380 & 0.3288 & 0.3356 \\ 0.3310 & 0.3406 & 0.3356 & 0.3356 & 0.3288 & 0.3310 \\ 0.3288 & 0.3310 & 0.3288 & 0.3288 & 0.3308 & 0.3288 \\ 0.3310 & 0.3261 & 0.3239 & 0.3288 & 0.3288 & 0.3188 \end{pmatrix}.$$

Step 3.3: Use the q-ROFPWA operator to fuse the evaluation matrix $\overline{Q}^k$ to obtain the collective evaluation matrix Q as shown in Table 5.

Table 5. Collective evaluation matrix.

Alternatives	C_1	C_2	C_3	C_4	C_5	C_6
A_1	(0.3331,0.8082)	(0.6512,0.4677)	(0.6660,0.4425)	(0.2747,0.8461)	(0.7904,0.3133)	(0.5221,0.5811)
A_2	(0.3692,0.7456)	(0.4985,0.6341)	(0.3893,0.7155)	(0.6888,0.4143)	(0.5883,0.5149)	(0.4665,0.6449)
A_3	(0.7225,0.3801)	(0.7690,0.3396)	(0.5221,0.5811)	(0.3893,0.7155)	(0.6271,0.6421)	(0.7225,0.3801)
A_4	(0.6660,0.4425)	(0.7415,0.3869)	(0.3116,0.8097)	(0.6220,0.4807)	(0.6220,0.4807)	(0.6926,0.4264)

Step 4: Determine the weights of the criteria.

Step 4.1: Because the subjective weights of criteria were determined in the literature [16], we adopt the subjective weight vector of criteria as $\lambda^S = (0.180, 0.090, 0.130, 0.130, 0.310, 0.160)^T$.

Step 4.2: Based on the collective evaluation matrix Q, we can construct the programming model, i.e., Equation (26); then, the objective weight vector of criteria can be determined as $\lambda^O = (0.201, 0.160, 0.150, 0.182, 0.151, 0.156)^T$.

Step 4.3: Set the importance coefficient of subjective weights to $\varphi = 0.5$; we can obtain the comprehensive weights of criteria as $\lambda = (0.191, 0.125, 0.140, 0.156, 0.230, 0.158)^T$.

Step 5: Rank the green suppliers using the TODIM method under a q-ROF environment ($\theta = 1$).

Step 5.1: Utilize Equation (31) to compute the relative weight λ_{jr} of criteria C_j concerning the reference criteria C_r as:

$$\lambda_{1r} = 0.8304, \lambda_{2r} = 0.5435, \lambda_{3r} = 0.6087, \lambda_{4r} = 0.6783, \lambda_{5r} = 1.0000, \lambda_{6r} = 0.6870.$$

Step 5.2: Compute the dominance degree of green supplier A_i over each green supplier:

$$\delta = \begin{pmatrix} 0 & -1.5220 & -4.0821 & -4.2886 \\ -3.7277 & 0 & -4.7679 & -4.0119 \\ -1.3699 & -1.2635 & 0 & -1.5385 \\ -0.6062 & -0.7501 & -2.5158 & 0 \end{pmatrix}.$$

Step 5.3: Compute the global value of green supplier A_i by Equation (34):

$$\Phi(A_1) = 0.3028, \Phi(A_2) = 0, \Phi(A_3) = 0.9653, \Phi(A_4) = 1.$$

Step 5.4: Based on the global values of green suppliers, the ranking of potential green suppliers can be determined as $A_4 > A_3 > A_1 > A_2$. The green supplier A_4 is the best choice for the electric automobile company.

5.2. Comparison and Sensitivity Analysis

To investigate the influence of the consensus-reaching process on the ranking result and further verify the effectiveness of the proposed approach, we compared the ranking result of the green suppliers in Section 5.1 with the results that were obtained by the proposed approach without the consensus-reaching process, the green supplier selection approach based on the intuitionistic fuzzy TOPSIS (IF-TOPSIS) method [16], and the green supplier selection approach based on the fuzzy TODIM method [75]. The ranking results of the three green supplier selection approaches are shown in Figure 3; the detailed computation procedures of the proposed approach without consensus-reaching process, IF-TOPSIS method, and fuzzy TODIM method are presented in Appendices A–C, respectively.

The inconsistent ranking results between the proposed approach and the proposed approach without consensus-reaching process, i.e., the different ranking orders of green suppliers A_3 and A_4, can be explained by ignoring the consensus level of q-ROF evaluation information of decision makers. For instance, the linguistic term of decision maker D_1 for green supplier A_3 with respect to criteria C_5 was extremely low (EL); by contrast, the linguistic evaluation information of decision makers D_2 and D_3 for green supplier A_3 with respect to criteria C_5 were both extremely high (EH). Similarly, the linguistic terms of decision makers D_1 and D_2 for green supplier A_4 concerning criteria C_6 were both high (H); however, the linguistic evaluation information of decision maker D_3 for green supplier A_4 concerning criteria C_6 was EL. The aforementioned non-consensus evaluation information led to a change in the ranking orders of green suppliers A_3 and A_4 without a consensus-reaching process. In the procedures of the proposed approach, an iteration-based consensus model under a q-ROF environment was utilized to revise this non-consensus evaluation information until an acceptable consensus level between decision makers was achieved. Thus, we can obtain the ranking of green suppliers that was accepted by decision makers or enterprise; furthermore, the possible extreme evaluation information of individual decision maker was also revised to avoid affecting the accuracy of ranking result.

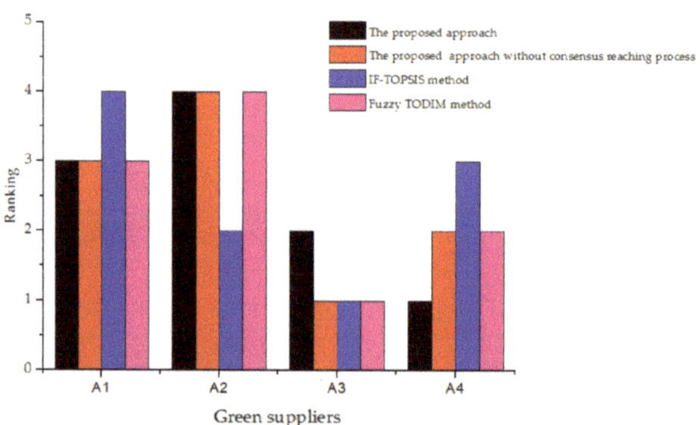

Figure 3. Ranking results of different approaches.

A notable difference existed between the rankings of the proposed method and IF-TOPSIS method. With the exception of ignoring the consensus problem between decision makers, the inconsistent result was caused by several other reasons. First, the evaluation information was represented by the q-ROFS in the proposed method, which is a generalized form of IFS that is used in the IF-TOPSIS method. Different basic data of green suppliers will lead to different result by aggregation tools. Second, combined with the q-ROFPWA operator, the decision maker weights were obtained by the subjective weights and support degrees between the evaluation information in the proposed approach; nevertheless, the determination of decision maker weights was omitted in the IF-TOPSIS method. Third, instead of the TOPSIS method, we utilized the q-ROF-TODIM method to determine the ranking of green suppliers. The q-ROF-TODIM method can consider the bounded rationality behavior of decision makers, which cannot be achieved by the TOPSIS method; consequently, the ranking result of green suppliers may differ.

From Figure 3, we can see that the ranking orders of green suppliers A_3 and A_4 were different between the rankings obtained by the proposed approach and fuzzy TODIM method. The main reason for this result is that the consensus-reaching process was omitted in the fuzzy TODIM method; the non-consensus evaluation information of decision makers made green supplier A_3 rank first in the ranking results determined by the proposed approach without a consensus-reaching process, IF-TOPSIS method, and fuzzy TODIM method. Moreover, the fuzzy TODIM method utilizes the triangular fuzzy numbers to express the evaluation information of decision makers, in which the non-membership and indeterminacy membership levels were ignored. The weights of decision makers in fuzzy TODIM method were assumed to be equal, which was inconsistent with the actual situation.

Furthermore, a sensitivity analysis was implemented by changing the weights of criteria as shown in Table 6. The rankings under different situations of the proposed approach, IF-TOPSIS method, and fuzzy TODIM method are illustrated in Figures 4–6, respectively. Example 0 showed the weights of criteria that were determined by the proposed method, and Examples 1~7 showed the other possible weight values. From Table 6 and Figure 4, we can see that when the weight values of criteria C_4 and C_5 were relatively large, the best green supplier changed from A_4 to A_3, which means that the criteria weights play a crucial role in determining the ranking of green suppliers. Therefore, we should select the appropriate weighting method in practice. The comprehensive weighting approach in the proposed method considered the subjective and objective factors to obtain the more accurate weights of criteria. Once decision makers were confident for the evaluation information of criteria weights, the coefficient φ could be assigned a large value; otherwise, the coefficient φ could be assigned a small value. On the other hand, in addition to Examples 5 and 6, the rankings remain the same as $A_4 > A_3 > A_1 > A_2$ under other situations; the proposed method is proven to be relatively insensitive to the weights of criteria.

Table 6. Different weights of criteria in the sensitivity analysis.

Examples	C_1	C_2	C_3	C_4	C_5	C_6
Example 0	0.191	0.125	0.140	0.156	0.230	0.158
Example 1	1/6	1/6	1/6	1/6	1/6	1/6
Example 2	0.750	0.050	0.050	0.050	0.050	0.050
Example 3	0.050	0.750	0.050	0.050	0.050	0.050
Example 4	0.050	0.050	0.750	0.050	0.050	0.050
Example 5	0.050	0.050	0.050	0.750	0.050	0.050
Example 6	0.050	0.050	0.050	0.050	0.750	0.050
Example 7	0.050	0.050	0.050	0.050	0.050	0.750

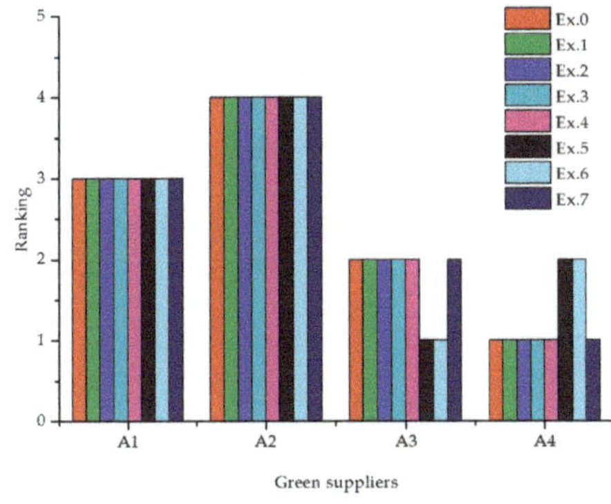

Figure 4. Ranking results of the proposed approach with different weights of criteria.

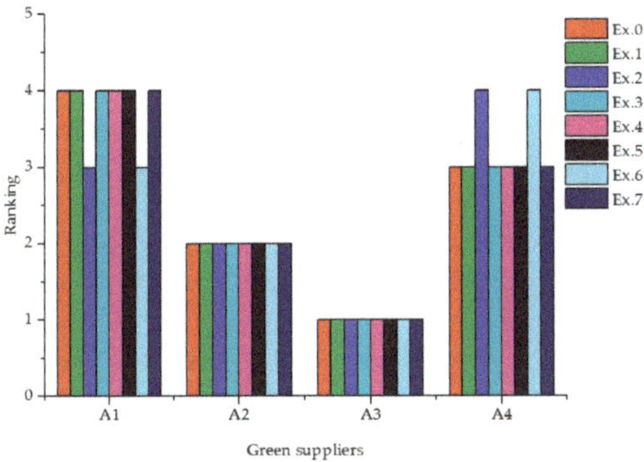

Figure 5. Ranking results of intuitionistic fuzzy (IF)-technique for order performance by similarity to ideal solution (TOPSIS) method with different weights of criteria.

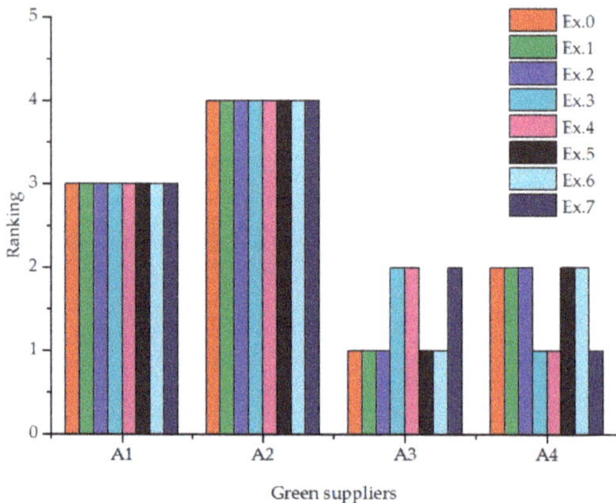

Figure 6. Ranking results of fuzzy TOmada de Decisao Interativa e Multicritevio (TODIM) method with different weights of criteria.

The Spearman's rank correlation coefficient is a powerful tool for measuring the similarity between rankings obtained by MCGDM methods [76]. To investigate the robustness of different green supplier selection approaches, combined with the rankings in Figures 4–6, we can calculate the Spearman's rank correlation coefficients between the ranking of Example 0 and the rankings of other possible weights of criteria, respectively. Thus, the average of these Spearman's rank correlation coefficients can be utilized to measure the robustness of each green supplier selection approach, which are presented in Table 7. The larger the average of Spearman's rank correlation coefficients, which means that the smaller the rankings change with different criteria weights, the stronger the robustness of this green supplier selection approach and vice versa. From Table 7, we can see that the robustness levels of all three green supplier selection approaches were relatively high, and the robustness of the proposed method and IF-TOPSIS was is slightly stronger than that of fuzzy TODIM method.

Table 7. Average of Spearman's rank correlation coefficients of different approaches.

Methods	Average of Spearman's Rank Correlation Coefficients
The proposed approach	0.9429
IF-TOPSIS method	0.9429
Fuzzy TODIM method	0.9143

Based on the analysis above, the advantages of determining the best green supplier by using the proposed approach can be summarized as follows.

(1) The q-ROFS is utilized to represent the evaluation information of decision makers, which can express the membership, non-membership, and indeterminacy membership degrees, simultaneously. Furthermore, with the increasing rung q, the space of acceptable orthopairs of q-ROFS is larger than IFS and PFS; as a generalized form of IFS and PFS, the proposed approach can also be transformed into other green supplier selection approaches under an IF and PF environment if necessary.

(2) In practice, decision makers always differentiate from research fields and domain experiences; the non-consensus evaluation information of green suppliers will inevitably be given. Combined with an iteration-based consensus model under q-ROF environment, the non-consensus evaluation information of all the decision makers can be revised in each round. Therefore, a ranking of green suppliers accepted by decision makers or enterprises can be obtained using the proposed approach, and the efficiency of the consensus-reaching process is relatively high.

(3)   The q-ROFPWA operator is introduced to fuse the individual evaluation matrices; the weight vectors of decision makers can be determined by two aspects, namely, the subjective aspect and the objective aspect. Consequently, we can obtain a ranking of green suppliers that is closer to reality. Additionally, the determination of weights of decision makers is solved, which has been ignored by most existing approaches.

(4)   The weights of criteria are determined by a comprehensive weighting approach, which is composed of the subjective evaluation method and a deviation maximization model. Through changing the valve of coefficient φ, the weights of the criteria can be determined; whether they are closer to subjective weights or objective weights depends on the choice of the decision makers or enterprises. Thus, the proposed approach is more able to cope with different scenarios.

(5)   During the green supplier evaluation process, the bounded rationality behavior of decision makers cannot be avoided. The TODIM method is a powerful tool to solve these MCGDM problems; in the proposed approach, the TODIM method is extended to the q-ROF environment to compute the ranking of green suppliers, which makes the evaluation result more realistic and accurate. In addition, the robustness of the proposed method is relatively strong.

The proposed approach also presents several limitations. With respect to the complicated green supplier selection issues, in which the number of evaluation criteria is relatively large; the interactions or dependencies between the criteria will inevitably exist. These situations cannot be solved combined with the proposed green supplier selection approach. Furthermore, decision makers may have difficulty determining the accurate value of a membership degree or linguistic term in real life. The proposed approach cannot deal with the issue of allowing decision makers to provide several possible values of different membership degrees or linguistic terms, which will be the focus of future research.

6. Conclusions

To deal with the complexity of green supplier selection problems in practice, this paper proposed a novel approach for green supplier selection under q-ROF environment. The q-ROFNs were utilized to express the evaluation information of decision makers; the uncertainty and incompleteness of the evaluation information were effectively addressed. Combined with the consensus measures on three levels, a q-ROF consensus model was developed to revise the non-consensus evaluation information of decision makers to improve the accuracy of the ranking results. To aggregate the q-ROF evaluation information of decision makers, the q-ROFPWA operator that considers both subjective and objective factors of decision maker weights was applied. Furthermore, a comprehensive weighting method was constructed to determine the weights of criteria, which consisted of the subjective weighting method and a deviation maximization model. Finally, the TODIM method under an q-ROF environment was proposed to obtain a ranking of potential green suppliers. An example of a green supplier selection problem in an electric automobile company was used to demonstrate the feasibility of the proposed method; subsequently, the effectiveness of the proposed method was illustrated by the sensitivity analysis and comparative analysis. In the case of increasingly complex green supplier selection issues, the proposed approach can deal with several aspects effectively, such as providing a relaxed evaluation environment for decision makers, promoting a relatively high consensus level between decision makers, and determining the weights of decision makers comprehensively. Thus, this paper provides a more reasonable and effective approach for enterprises to choose green suppliers in practice.

In future research, we will introduce the Choquet integral or Bonferroni mean operator to aggregate the evaluation information, which takes into account the relationships between the criteria. Furthermore, we can extend the proposed method into the q-rung orthopair hesitant fuzzy environment, in which decision makers have difficulty in determining the accurate membership and non-membership degrees.

Author Contributions: Concept of this paper, R.W.; methodology, R.W.; writing—original draft preparation, R.W.; writing—review and editing, Y.L.; funding acquisition, R.W. and Y.L.

Funding: This study was supported by the National Natural Science Foundation of China (no. 71371156) and the Doctoral Innovation Fund Program of Southwest Jiaotong University (D-CX201727).

Conflicts of Interest: The authors declare no conflict of interest.

Appendix A

The ranking of potential green suppliers can be obtained by the proposed approach without the consensus-reaching process as below.

Step 1: Obtain the normalized evaluation matrices of decision makers

Combined with the Steps 1.1~1.3 in Section 5.1, we can obtain the normalized q-ROF evaluation matrix $Q^k = \left(a_{ij}^k\right)_{4\times 6}$.

Step 2: Aggregation of individual evaluation matrices

Steps 2.1~2.2: According to the subjective weight of decision makers $w = (1/3, 1/3, 1/3)^T$, we can utilize Equations (21)~(23) to calculate the weighted support degree of a_{ij}^k as:

$$T^1 = \begin{pmatrix} 0.4667 & 0.4667 & 0.5667 & 0.5667 & 0.6333 & 0.6333 \\ 0.5667 & 0.5667 & 0.6333 & 0.6000 & 0.6333 & 0.5667 \\ 0.6333 & 0.6000 & 0.6333 & 0.6333 & 0.1333 & 0.6333 \\ 0.5667 & 0.2333 & 0.6000 & 0.6333 & 0.6333 & 0.4667 \end{pmatrix};$$

$$T^2 = \begin{pmatrix} 0.5000 & 0.3667 & 0.5667 & 0.5667 & 0.6333 & 0.6000 \\ 0.6000 & 0.4667 & 0.6000 & 0.6333 & 0.6333 & 0.6000 \\ 0.6333 & 0.5667 & 0.6333 & 0.6333 & 0.4000 & 0.6333 \\ 0.6000 & 0.4333 & 0.6000 & 0.6333 & 0.6333 & 0.4667 \end{pmatrix};$$

$$T^3 = \begin{pmatrix} 0.3667 & 0.5000 & 0.6000 & 0.6000 & 0.6000 & 0.6333 \\ 0.5667 & 0.5667 & 0.6333 & 0.6333 & 0.6000 & 0.5667 \\ 0.6000 & 0.5667 & 0.6000 & 0.6000 & 0.4000 & 0.6000 \\ 0.5667 & 0.4000 & 0.5333 & 0.6000 & 0.6000 & 0.2667 \end{pmatrix}.$$

Then, the weights associated with a_{ij}^k can be determined as:

$$\zeta^1 = \begin{pmatrix} 0.3385 & 0.3385 & 0.3310 & 0.3310 & 0.3356 & 0.3356 \\ 0.3310 & 0.3406 & 0.3356 & 0.3288 & 0.3356 & 0.3310 \\ 0.3356 & 0.3380 & 0.3356 & 0.3356 & 0.2881 & 0.3356 \\ 0.3310 & 0.3033 & 0.3380 & 0.3356 & 0.3356 & 0.3492 \end{pmatrix};$$

$$\zeta^2 = \begin{pmatrix} 0.3261 & 0.2971 & 0.3310 & 0.3310 & 0.3356 & 0.3288 \\ 0.3380 & 0.3188 & 0.3288 & 0.3356 & 0.3356 & 0.3380 \\ 0.3356 & 0.3310 & 0.3356 & 0.3356 & 0.3231 & 0.3356 \\ 0.3380 & 0.3116 & 0.3380 & 0.3356 & 0.3356 & 0.3188 \end{pmatrix};$$

$$\zeta^3 = \begin{pmatrix} 0.2971 & 0.3261 & 0.3380 & 0.3380 & 0.3288 & 0.3356 \\ 0.3310 & 0.3406 & 0.3356 & 0.3356 & 0.3288 & 0.3310 \\ 0.3288 & 0.3310 & 0.3288 & 0.3288 & 0.3231 & 0.3288 \\ 0.3310 & 0.3043 & 0.3239 & 0.3288 & 0.3288 & 0.3754 \end{pmatrix}.$$

Step 2.3: Use the q-ROFPWA operator to fuse the evaluation matrix Q^k to obtain the collective evaluation matrix Q as shown in Table A1.

Table A1. Collective evaluation matrix.

Alternatives	C_1	C_2	C_3	C_4	C_5	C_6
A_1	(0.4590,0.7352)	(0.6344,0.5111)	(0.6660,0.4425)	(0.2747,0.8461)	(0.7904,0.3133)	(0.5221,0.5811)
A_2	(0.3692,0.7456)	(0.4985,0.6341)	(0.3893,0.7155)	(0.6888,0.4143)	(0.5883,0.5149)	(0.4665,0.6449)
A_3	(0.7225,0.3801)	(0.7690,0.3396)	(0.5221,0.5811)	(0.3893,0.7155)	(0.7552,0.5600)	(0.7225,0.3801)
A_4	(0.6660,0.4425)	(0.7177,0.4399)	(0.3116,0.8097)	(0.6220,0.4807)	(0.6220,0.4807)	(0.6747,0.4890)

Step 3: Determine the weights of criteria.

Step 3.1: We adopt the subjective weights of criteria in the literature [16] as $\lambda^S = (0.180, 0.090, 0.130, 0.130, 0.310, 0.160)^T$.

Step 3.2: Based on the collective evaluation matrix Q, we construct the programming model, i.e., Equation (26), then, the objective weights of criteria can be determined as $\lambda^O = (0.187, 0.157, 0.157, 0.186, 0.152, 0.161)^T$.

Step 3.3: Set the importance coefficient of subjective weights $\varphi = 0.5$; we can obtain the comprehensive weights of criteria as $\lambda = (0.183, 0.124, 0.143, 0.158, 0.231, 0.161)^T$.

Step 4: Rank the green suppliers using the TODIM method under the q-ROF environment ($\theta = 1$).

Step 4.1: Utilize Equation (31) to compute the relative weight λ_{jr} of criteria C_j concerning the reference criteria C_r as:

$$\lambda_{1r} = 0.7922, \lambda_{2r} = 0.5368, \lambda_{3r} = 0.6190, \lambda_{4r} = 0.6840, \lambda_{5r} = 1.0000, \lambda_{6r} = 0.6970.$$

Step 4.2: Compute the dominance degree of green supplier A_i over each green supplier as:

$$\delta = \begin{pmatrix} 0 & -0.8829 & -4.1540 & -3.8594 \\ -4.3201 & 0 & -5.7749 & -3.8519 \\ -0.8835 & -0.3974 & 0 & -0.2696 \\ -0.7944 & -1.1371 & -3.7778 & 0 \end{pmatrix}.$$

Step 4.3: Compute the global value of green supplier A_i by Equation (34):

$$\Phi(A_1) = 0.4074, \Phi(A_2) = 0, \Phi(A_3) = 1, \Phi(A_4) = 0.6645.$$

Step 4.4: Based on the global values of green suppliers, the ranking of potential green suppliers can be determined as $A_3 > A_4 > A_1 > A_2$. The green supplier A_3 is the best choice for the electric automobile company.

Appendix B

The ranking of potential green suppliers can be obtained by the IF-TOPSIS method [16] as below.

Step 1: According to the linguistic terms of decision makers in Table 2 and the relationships between linguistic terms and intuitionistic fuzzy numbers in the literature [16], we transform the linguistic terms into IF evaluation matri ces of decision makers; then, the intuitionistic fuzzy weighted average operator [77] is utilized to fuse the individual evaluation information to determine the collective evaluation matrix as presented in Table A2.

Table A2. Collective evaluation matrix.

Alternatives	C_1	C_2	C_3	C_4	C_5	C_6
A_1	(0.4590,0.7352)	(0.6344,0.5111)	(0.6660,0.4425)	(0.2747,0.8461)	(0.7904,0.3133)	(0.5221,0.5811)
A_2	(0.3692,0.7456)	(0.4985,0.6341)	(0.3893,0.7155)	(0.6888,0.4143)	(0.5883,0.5149)	(0.4665,0.6449)
A_3	(0.7225,0.3801)	(0.7690,0.3396)	(0.5221,0.5811)	(0.3893,0.7155)	(0.7552,0.5600)	(0.7225,0.3801)
A_4	(0.6660,0.4425)	(0.7177,0.4399)	(0.3116,0.8097)	(0.6220,0.4807)	(0.6220,0.4807)	(0.6747,0.4890)

Step 2: According to the type of criteria, we can obtain the IF positive ideal solution a^+ and IF negative ideal solution a^- as:

$$a^+ = ((0.2348, 0.6649), (0.7116, 0.1817), (0.3458, 0.5944), (0.6366, 0.2621), (1.0000, 0.0000), (0.6698, 0.2289)),$$

$$a^- = ((1.0000, 0.0000), (0.3650, 0.5278), (1.0000, 0.0000), (0.1037, 0.8243), (0.5358, 0.4217), (0.3458, 0.5944)).$$

Step 3: Utilize the maximum average weighted distance method to construct a programming model as:

$$\max \sum_{i=1}^{m} \sum_{j=1}^{n} \lambda_j^O d\left(a_{ij}, a^-\right) \tag{A1}$$
$$s.t. \sum_{j=1}^{n} \left(\lambda_j^O\right)^2 = 1, 0 \le \lambda_j^O \le 1.$$

Then, we can use the Lagrange function to solve this model, and the objective weights of criteria are obtained as $\lambda^O = (0.253, 0.122, 0.217, 0.186, 0.117, 0.105)^T$.

Step 4: Set the importance coefficient of subjective weights $\varphi = 0.5$, combined with the subjective weight vector of criteria $\lambda^S = (0.180, 0.090, 0.130, 0.130, 0.310, 0.160)^T$, we can obtain the comprehensive weights of criteria as $\lambda = (0.217, 0.106, 0.173, 0.158, 0.213, 0.133)^T$. Furthermore, the weighted IF evaluation matrix can be determined as presented in Table A3.

Table A3. Weighted IF evaluation matrix.

Alternatives	C_1	C_2	C_3	C_4	C_5	C_6
A_1	(1.0000,0.0000)	(0.0756,0.8983)	(0.0708,0.9139)	(0.0172,0.9699)	(0.2482,0.6757)	(0.0744,0.9193)
A_2	(0.2365,0.6907)	(0.0470,0.9345)	(0.1744,0.7749)	(0.1478,0.8093)	(0.1508,0.8320)	(0.0549,0.9332)
A_3	(0.0564,0.9153)	(0.1235,0.8346)	(0.1243,0.8612)	(0.0414,0.9376)	(1.0000,0.0000)	(0.1370,0.8219)
A_4	(0.0880,0.8933)	(0.0946,0.8709)	(1.0000,0.0000)	(0.1246,0.8493)	(0.1642,0.8024)	(0.1013,0.8670)

Step 5: Utilize the following equations to calculate the distances between each green supplier and the IF positive ideal solution a^+ and IF negative ideal solution a^-, respectively.

$$S_i^+ = \sum_{j=1}^{n} \left(\left|\mu_{ij} - \mu_j^+\right| + \left|v_{ij} - v_j^+\right|\right), \tag{A2}$$

$$S_i^- = \sum_{j=1}^{n} \left(\left|\mu_{ij} - \mu_j^-\right| + \left|v_{ij} - v_j^-\right|\right). \tag{A3}$$

Subsequently, the relative closeness coefficient of each green supplier concerning the positive ideal solution can be computed by:

$$CC_i = \frac{S_i^-}{S_i^- + S_i^+}. \tag{A4}$$

Thus, the result can be obtained as $CC_1 = 0.3430, CC_2 = 0.4743, CC_3 = 0.5533, CC_4 = 0.3520$.

Step 6: According to the relative closeness coefficient value of each green supplier, we can determine the ranking of the green supplier as $A_3 > A_2 > A_4 > A_1$; the green supplier A_3 is the best choice for the electric automobile company.

Appendix C

The ranking of potential green suppliers can be obtained by the fuzzy TODIM method [75] as below.

Step 1: Because of the linguistic terms utilized in the literature [75] are divided into five grades, we reconstruct the relationships between linguistic terms and triangular fuzzy numbers as presented in Table A4 to implement the numerical example in this paper.

Table A4. Linguistic terms and the corresponding triangular fuzzy numbers.

Linguistic Terms	Corresponding Triangular Fuzzy Numbers
Extremely High (EH)	(0.8,0.9,1.0)
Very High (VH)	(0.6,0.7,0.8)
High (H)	(0.5,0.6,0.7)
Medium High (MH)	(0.4,0.5,0.6)
Medium (M)	(0.3,0.4,0.5)
Medium Low (ML)	(0.2,0.3,0.4)
Low (L)	(0.1,0.2,0.3)
Very Low (VL)	(0.0,0.1,0.2)
Extremely Low (EL)	(0.0,0.0,0.1)

Step 2: According to Tables 2 and A4, we can transform the linguistic evaluation information of decision makers into the corresponding triangular fuzzy numbers. The weights of decision makers are considered equal in the literature [75]; thus, the collective evaluation matrix can be obtained as shown in Table A5.

Table A5. Collective evaluation matrix.

Alternatives	C_1	C_2	C_3	C_4	C_5	C_6
A_1	(0.53,0.63,0.73)	(0.30,0.40,0.50)	(0.20,0.30,0.40)	(0.03,0.10,0.20)	(0.53,0.63,0.73)	(0.27,0.37,0.47)
A_2	(0.50,0.60,0.70)	(0.20,0.30,0.40)	(0.47,0.57,0.67)	(0.43,0.53,0.63)	(0.33,0.43,0.53)	(0.20,0.30,0.40)
A_3	(0.13,0.23,0.33)	(0.50,0.60,0.70)	(0.33,0.43,0.53)	(0.13,0.23,0.33)	(0.27,0.30,0.40)	(0.47,0.57,0.67)
A_4	(0.20,0.30,0.40)	(0.37,0.43,0.53)	(0.60,0.70,0.80)	(0.37,0.47,0.57)	(0.37,0.47,0.57)	(0.33,0.40,0.50)

Step 3: To obtain a more objective comparison result, we adopt the weights of criteria in the Section 5.1 as $\lambda = (0.191, 0.125, 0.140, 0.156, 0.230, 0.158)^T$.

Step 4: Rank the green suppliers using the fuzzy TODIM method ($\theta = 1$); similar to the improved TODIM method in this paper, compute the relative weight λ_{jr} of criteria C_j concerning the reference criteria C_r as

$$\lambda_{1r} = 0.8304, \lambda_{2r} = 0.5435, \lambda_{3r} = 0.6087, \lambda_{4r} = 0.6783, \lambda_{5r} = 1.0000, \lambda_{6r} = 0.6870.$$

Step 5: Compute the dominance degree of green supplier A_i over each green supplier:

$$\delta = \begin{pmatrix} 0 & -1.4414 & -4.3211 & -4.5594 \\ -3.5201 & 0 & -4.8305 & -3.3162 \\ -1.3910 & -1.2951 & 0 & -1.4217 \\ -0.8218 & -1.0348 & -3.7089 & 0 \end{pmatrix}.$$

Step 6: Compute the global value of green supplier A_i:

$$\Phi(A_1) = 0.3102, \Phi(A_2) = 0, \Phi(A_3) = 1, \Phi(A_4) = 0.8072.$$

Step 7: Based on the global values of green suppliers, the ranking of potential green suppliers can be determined as $A_3 > A_4 > A_1 > A_2$. The green supplier A_3 is the best choice for the electric automobile company.

References

1. Sahu, N.K.; Datta, S.; Mahapatra, S.S. Establishing green supplier appraisement platform using grey concepts. *Grey Syst.* **2012**, *2*, 395–418. [CrossRef]
2. Rostamzadeh, R.; Govindan, K.; Esmaeili, A.; Sabaghi, M. Application of fuzzy VIKOR for evaluation of green supply chain management practices. *Ecol. Indic.* **2015**, *49*, 188–203. [CrossRef]

3. Vachon, S. Green supply chain practices and the selection of environmental technologies. *Int. J. Prod. Res.* **2007**, *45*, 4357–4379. [CrossRef]

4. Cabral, I.; Grilo, A.; Cruz-Machado, V. A decision-making model for lean, agile, resilient and green supply chain management. *Int. J. Prod. Res.* **2012**, *50*, 4830–4845. [CrossRef]

5. Beamon, B.M. Designing the green supply chain. *Logist. Inf. Manag.* **2013**, *12*, 332–342. [CrossRef]

6. Bai, C.; Sarkis, J. Green supplier development: Analytical evaluation using rough set theory. *J. Clean. Prod.* **2010**, *18*, 1200–1210. [CrossRef]

7. Wang, K.Q.; Liu, H.C.; Liu, L.; Huang, J. Green supplier evaluation and selection using cloud model theory and the QUALIFLEX method. *Sustainability* **2017**, *9*, 688. [CrossRef]

8. Qin, J.D.; Liu, X.W.; Pedrycz, W. An extended TODIM multi-criteria group decision making method for green supplier selection in interval type-2 fuzzy environment. *Eur. J. Oper. Res.* **2017**, *258*, 626–638. [CrossRef]

9. Kannan, D.; Govindan, K.; Rajendran, S. Fuzzy axiomatic design approach based green supplier selection: Acase study from singapore. *J. Clean. Prod.* **2015**, *96*, 194–208. [CrossRef]

10. Blome, C.; Hollos, D.; Paulraj, A. Green procurement and green supplier development: Antecedents and effects on supplier performance. *Int. J. Prod. Res.* **2014**, *52*, 32–49. [CrossRef]

11. Govindan, K.; Rajendran, S.; Sarkis, J.; Murugesan, P. Multi criteria decision making approaches for green supplier evaluation and selection: A literature review. *J. Clean. Prod.* **2015**, *98*, 66–83. [CrossRef]

12. Zhu, J.H.; Li, Y.L. Green supplier selection based on consensus process and integrating prioritized operator and Choquet integral. *Sustainability* **2018**, *10*, 2744. [CrossRef]

13. Wang, J.; Wei, G.W.; Wei, Y. Models for green supplier selection with some 2-tuple linguistic neutrosophic number Bonferroni mean operators. *Symmetry* **2018**, *10*, 131. [CrossRef]

14. Banaeian, N.; Mobli, H.; Fahimnia, B.; Nielsen, I.E.; Omid, M. Green supplier selection using fuzzy group decision making methods: A case study from the agri-food industry. *Comput. Oper. Res.* **2017**, *89*, 337–347. [CrossRef]

15. Ghorabaee, M.K.; Zavadskas, E.K.; Amiri, M.; Esmaeili, A. Multi-criteria evaluation of green suppliers using an extended WASPAS method with interval type-2 fuzzy sets. *J. Clean. Prod.* **2016**, *137*, 213–229. [CrossRef]

16. Cao, Q.W.; Wu, J.; Liang, C.Y. An intuitionsitic fuzzy judgement matrix and TOPSIS integrated multi-criteria decision making method for green supplier selection. *J. Intell. Fuzzy Syst.* **2015**, *28*, 117–126. [CrossRef]

17. Yager, R.R. Generalized orthopair fuzzy sets. *IEEE Trans. Fuzzy Syst.* **2017**, *25*, 1222–1230. [CrossRef]

18. Liu, P.D.; Wang, P. Some q-rung orthopair fuzzy aggregation operators and their applications to multiple-attribute decision making. *Int. J. Intell. Syst.* **2018**, *33*, 259–280. [CrossRef]

19. Wei, G.W.; Gao, H.; Wei, Y. Some q-rung orthopair fuzzy Heronian mean operators in multiple attribute decision making. *Int. J. Intell. Syst.* **2018**, *33*, 1426–1458. [CrossRef]

20. Liu, P.D.; Liu, J.L. Some q-rung orthopair fuzzy Bonferroni mean operators and their application to multi-attribute group decision making. *Int. J. Intell. Syst.* **2018**, *33*, 315–347. [CrossRef]

21. Alonso, S. A web based consensus support system for group decision making problems and incomplete preferences. *Inf. Sci.* **2010**, *180*, 4477–4495. [CrossRef]

22. Chiclana, F.; Mata, F.; Martinez, L.; Herrera-Viedma, E.; Alonso, S. Integration of a consistency control module within a consensus model. *Int. J. Uncertain. Fuzz. Knowl.-Based Syst.* **2008**, *16*, 35–53. [CrossRef]

23. Herrera-Viedma, E.; Cabrerizo, F.J.; Kacprzyk, J.; Pedrycz, W. A review of soft consensus models in a fuzzy environment. *Inf. Fusion* **2014**, *17*, 4–13. [CrossRef]

24. Alcantud, J.C.R.; Calle, R.D.A.; Cascón, J.M. On measures of cohesiveness under dichotomous opinions: Some characterizations of approval consensus measures. *Inf. Sci.* **2013**, *240*, 45–55. [CrossRef]

25. Alcantud, J.C.R.; Calle, R.D.A.; Cascón, J.M. A unifying model to measure consensus solutions in a society. *Math. Comput. Model.* **2013**, *57*, 1876–1883. [CrossRef]

26. Yager, R.R. The power average operator. *IEEE Trans. Syst. Man Cybern Part A Syst. Hum.* **2001**, *31*, 724–731. [CrossRef]

27. Liu, P.; Chen, S.M.; Wang, P. Multiple-attribute group decision-making based on q-rung orthopair fuzzy power maclaurin symmetric mean operators. *IEEE Trans. Syst. Man Cybern. Syst.* **2018**, 1–16. [CrossRef]

28. Gomes, L.F.A.M.; Lima, M.M.P.P. TODIM: Basic and application to multicriteria ranking of projects with environmental impacts. *Found. Comput. Decis. Sci.* **1991**, *16*, 113–127.

29. Kahneman, D.; Tversky, A. Prospect theory: An analysis of decision under risk. *Econometrica* **1979**, *47*, 263–291. [CrossRef]

30. Morente-Molinera, J.A.; Kou, G.; Crespo, R.G.; Corchado, J.M.; Herrera-Viedma, E. Solving multi-criteria group decision making problems under environments with a high number of alternatives using fuzzy ontologies and multi-granular linguistic modelling methods. *Knowl.-Based Syst.* **2017**, *137*, 54–64. [CrossRef]

31. García-Díaz, V.; Espada, J.P.; Crespo, R.G.; G-Bustelo, B.C.P.; Lovelle, J.M.C. An approach to improve the accuracy of probabilistic classifiers fordecision support systems in sentiment analysis. *Appl. Soft Comput.* **2018**, *67*, 822–833. [CrossRef]

32. Taibi, A.; Atmani, B. Combining fuzzy AHP with GIS and decision rules for industrial site selection. *Int. J. Interact. Multimedia Artif. Intell.* **2017**, *4*, 60–69. [CrossRef]

33. Lee, A.H.I.; Kang, H.Y.; Hsu, C.F.; Hung, H.C. A green supplier selection model for high-tech industry. *Expert Syst. Appl.* **2009**, *36*, 7917–7927. [CrossRef]

34. Chen, H.M.W.; Chou, S.Y.; Luu, Q.D.; Yu, H.K. A fuzzy MCDM approach for green supplier selection from the economic and environmental aspects. *Math. Probl. Eng.* **2016**. [CrossRef]

35. Yazdani, M. An integrated MCDM approach to green supplier selection. *Int. J. Ind. Eng. Comput.* **2014**, *5*, 443–458. [CrossRef]

36. Tsui, C.W.; Wen, U.P. A hybrid multiple criteria group decision-making approach for green supplier selection in the TFT-LCD industry. *Math. Probl. Eng.* **2014**. [CrossRef]

37. Dobos, I.; Vörösmarty, G. Green supplier selection and evaluation using DEA-type composite indicators. *Int. J. Prod. Econ.* **2014**, *157*, 273–278. [CrossRef]

38. Hashemi, S.H.; Karimi, A.; Tavana, M. An integrated green supplier selection approach with analytic network process and improved grey relational analysis. *Int. J. Prod. Econ.* **2015**, *159*, 178–191. [CrossRef]

39. Kuo, R.J.; Wang, Y.C.; Tien, F.C. Integration of artificial neural network and MADA methods for green supplier selection. *J. Clean. Prod.* **2010**, *18*, 1161–1170. [CrossRef]

40. Kuo, T.C.; Hsu, C.W.; Li, J.Y. Developing a green supplier selection model by using the DANP with VIKOR. *Sustainability* **2015**, *7*, 1661–1689. [CrossRef]

41. Sang, X.Z.; Liu, X.W. An interval type-2 fuzzy sets-based TODIM method and its application to green supplier selection. *J. Oper. Res. Soc.* **2016**, *67*, 722–734. [CrossRef]

42. Govindan, K.; Kadziński, M.; Sivakumar, R. Application of a novel PROMETHEE-based method for construction of a group compromise ranking to prioritization of green suppliers in food supply chain. *Omega* **2016**, *71*, 129–145. [CrossRef]

43. Quan, M.Y.; Wang, Z.L.; Liu, H.C.; Shi, H. A hybrid MCDM approach for large group green supplier selection with uncertain linguistic information. *IEEE Access* **2018**, *6*, 50372–50383. [CrossRef]

44. Zadeh, L.A. Fuzzy sets. *Inf. Control* **1965**, *8*, 338–353. [CrossRef]

45. Atanassov, K.T. Intuitionistic fuzzy sets. *Fuzzy Sets Syst.* **1986**, *20*, 87–96. [CrossRef]

46. Bali, O.; Kose, E.; Gumus, S. Green supplier selection based on IFS and GRA. *Grey Syst.* **2013**, *3*, 158–176. [CrossRef]

47. Yager, R.R. Pythagorean membership grades in multicriteria decision making. *IEEE Trans. Fuzzy Syst.* **2014**, *22*, 958–965. [CrossRef]

48. Li, L.; Zhang, R.T.; Wang, J.; Shang, X.P.; Bai, K.Y. A novel approach to multi-attribute group decision-making with q-rung picture linguistic information. *Symmetry* **2018**, *10*, 172. [CrossRef]

49. Cường, B.C. Picture fuzzy sets. *J. Comput. Sci. Cybern.* **2014**, *30*, 409. [CrossRef]

50. Ullah, K.; Mahmood, T.; Jan, N. Similarity measures for T-spherical fuzzy sets with applications in pattern recognition. *Symmetry* **2018**, *10*, 193. [CrossRef]

51. Herrera-Viedma, E.; Herrera, F.; Chiclana, F. A consensus model for multiperson decision making with different preference structures. *IEEE Trans. Syst. Man Cybern. Part A Syst. Hum.* **2002**, *32*, 394–402. [CrossRef]

52. Herrera-Viedma, E.; Alonso, S.; Chiclana, F.; Herrera, F. A consensus model for group decision making with incomplete fuzzy preference relations. *IEEE Trans. Fuzzy Syst.* **2007**, *15*, 863–877. [CrossRef]

53. Chu, J.F.; Liu, X.W.; Wang, Y.M.; Chin, K.S. A group decision making model considering both the additive consistency and group consensus of intuitionistic fuzzy preference relations. *Comput. Ind. Eng.* **2016**, *101*, 227–242. [CrossRef]

54. Wu, J.; Chiclana, F.; Herrera-Viedma, E. Trust based consensus model for social network in an incomplete linguistic information context. *Appl. Soft Comput.* **2015**, *35*, 827–839. [CrossRef]

55. Wu, Z.B.; Xu, J.P. Possibility distribution-based approach for MAGDM with hesitant fuzzy linguistic information. *IEEE Trans. Cybern.* **2016**, *46*, 694–705. [CrossRef] [PubMed]

56. Dong, Y.C.; Chen, X.; Herrera, F. Minimizing adjusted simple terms in the consensus reaching process with hesitant linguistic assessments in group decision making. *Inf. Sci.* **2015**, *297*, 95–117. [CrossRef]
57. Gong, Z.W.; Zhang, H.H.; Forrest, J.; Li, L.S.; Xu, X.X. Two consensus models based on the minimum cost and maximum return regarding either all individuals or one individual. *Eur. J. Oper. Res.* **2015**, *240*, 183–192. [CrossRef]
58. Gong, Z.W.; Xu, X.X.; Li, L.S.; Xu, C. Consensus modeling with nonlinear utility and cost constraints: A case study. *Knowl.-Based Syst.* **2015**, *88*, 210–222. [CrossRef]
59. Xu, G.L.; Wan, S.P.; Wang, F.; Dong, J.Y.; Zeng, Y.F. Mathematical programming methods for consistency and consensus in group decision making with intuitionistic fuzzy preference relations. *Knowl.-Based Syst.* **2016**, *98*, 30–43. [CrossRef]
60. Zhang, Z.M.; Pedrycz, W. Goal programming approaches to managing consistency and consensus for intuitionistic multiplicative preference relations in group decision-making. *IEEE Trans. Fuzzy Syst.* **2018**, In press. [CrossRef]
61. Zhou, L.G.; Chen, H.Y.; Liu, J.P. Generalized power aggregation operators and their applications in group decision making. *Comput. Ind. Eng.* **2012**, *62*, 989–999. [CrossRef]
62. Xu, Z.S. Approaches to multiple attribute group decision making based on intuitionistic fuzzy power aggregation operators. *Knowl.-Based Syst.* **2011**, *24*, 749–760. [CrossRef]
63. Wan, S.P. Power average operators of trapezoidal intuitionistic fuzzy numbers and application to multi-attribute group decision making. *Appl. Math. Model.* **2013**, *37*, 4112–4126. [CrossRef]
64. Liu, P.D.; Liu, Y. An approach to multiple attribute group decision making based on intuitionistic trapezoidal fuzzy power generalized aggregation operator. *Int. J. Comput. Intell. Syst.* **2014**, *7*, 291–304. [CrossRef]
65. He, Y.D.; Chen, H.Y.; Zhou, L.G.; Liu, J.P.; Tao, Z.F. Generalized interval-valued Atanassov's intuitionistic fuzzy power operators and their application to group decision making. *Int. J. Fuzzy Syst.* **2013**, *15*, 401–411.
66. Zhang, X.; Liu, P.D.; Wang, Y.M. Multiple attribute group decision making methods based on intuitionistic fuzzy Frank power aggregation operators. *J. Intell. Fuzzy Syst.* **2015**, *29*, 2235–2246. [CrossRef]
67. Wei, G.W.; Lu, M. Pythagorean fuzzy power aggregation operators in multiple attribute decision making. *Int. J. Intell. Syst.* **2018**, *33*, 169–186. [CrossRef]
68. Song, M.X.; Jiang, W.; Xie, C.H.; Zhou, D.Y. A new interval numbers power average operator in multiple attribute decision making. *Int. J. Intell. Syst.* **2017**, *32*, 631–644. [CrossRef]
69. Liu, P.D.; Shi, L.L. The generalized hybrid weighted average operator based on interval neutrosophic hesitant set and its application to multiple attribute decision making. *Neural Comput. Appl.* **2015**, *26*, 457–471. [CrossRef]
70. Xu, Y.J.; Merigó, J.M.; Wang, H.M. Linguistic power aggregation operators and their application to multiple attribute group decision making. *Appl. Math. Model.* **2012**, *36*, 5427–5444. [CrossRef]
71. Wu, X.H.; Qian, J.; Peng, J.J.; Xue, C.C. A multi-criteria group decision-making method with possibility degree and power aggregation operators of single trapezoidal neutrosophic numbers. *Symmetry* **2018**, *10*, 590. [CrossRef]
72. Du, W.S. Minkowski-type distance measures for generalized orthopair fuzzy sets. *Int. J. Intell. Syst.* **2018**, *33*, 802–817. [CrossRef]
73. Xu, Z.S. A deviation-based approach to intuitionistic fuzzy multiple attribute group decision making. *Group Decis. Negot.* **2010**, *19*, 57–76. [CrossRef]
74. Karaşan, A.; Ilbahar, E.; Cebi, S.; Kahraman, C. A new risk assessment approach: Safety and critical effect analysis (SCEA) and its extension with Pythagorean fuzzy sets. *Saf. Sci.* **2018**, *108*, 173–187. [CrossRef]
75. Tosun, Ö.; Akyüz, G. A fuzzy TODIM approach for the supplier selection problem. *Int. J. Comput. Intell. Syst.* **2015**, *8*, 317–329. [CrossRef]
76. Kou, G.; Lu, Y.Q.; Peng, Y.; Shi, Y. Evaluation of classification algorithms using MCDM and rank correlation. *Int. J. Inf. Technol. Decis. Mak.* **2012**, *11*, 197–225. [CrossRef]
77. Xu, Z.S. Intuitionistic fuzzy aggregation operators. *IEEE Trans. Fuzzy Syst.* **2013**, *14*, 108–116. [CrossRef]

© 2018 by the authors. Licensee MDPI, Basel, Switzerland. This article is an open access article distributed under the terms and conditions of the Creative Commons Attribution (CC BY) license (http://creativecommons.org/licenses/by/4.0/).

Review

Application of MCDM Methods in Sustainability Engineering: A Literature Review 2008–2018

Mirko Stojčić [1], Edmundas Kazimieras Zavadskas [2,*], Dragan Pamučar [3], Željko Stević [1] and Abbas Mardani [4]

[1] Faculty of Transport and Traffic Engineering, University of East Sarajevo, Vojvode Mišića 52, 74000 Doboj, Republic of Srpska, Bosnia and Herzegovina; mirkostojcic1@hotmail.com (M.S.); zeljkostevic88@yahoo.com or zeljko.stevic@sf.ues.rs.ba (Ž.S.)

[2] Institute of Sustainable Construction, Vilnius Gediminas Technical University, Sauletekio al. 11, LT-10223 Vilnius, Lithuania

[3] Department of Logistics, University of Defence in Belgrade, Pavla Jurišića Šturma 33, 11000 Belgrade, Serbia; dpamucar@gmail.com

[4] Azman Hashim International Business School, Universiti Teknologi Malaysia (UTM), Skudai Johor 81310, Malaysia; mabbas3@liveutm.onmicrosoft.com

* Correspondence: edmundas.zavadskas@vgtu.lt

Received: 16 February 2019; Accepted: 5 March 2019; Published: 8 March 2019

Abstract: Sustainability is one of the main challenges of the recent decades. In this regard, several prior studies have used different techniques and approaches for solving this problem in the field of sustainability engineering. Multiple criteria decision making (MCDM) is an important technique that presents a systematic approach for helping decisionmakers in this field. The main goal of this paper is to review the literature concerning the application of MCDM methods in the field of sustainable engineering. The Web of Science (WoS) Core Collection Database was chosen to identify 108 papers in the period of 2008–2018. The selected papers were classified into five categories, including construction and infrastructure, supply chains, transport and logistics, energy, and other. In addition, the articles were classified based on author, year, application area, study objective and problem, applied methods, number of published papers, and name of the journal. The results of this paper show that sustainable engineering is an area that is quite suitable for the use of MCDM. It can be concluded that most of the methods used in sustainable engineering are based on traditional approaches with a noticeable trend towards applying the theory of uncertainty, such as fuzzy, grey, rough, and neutrosophic theory.

Keywords: sustainability; engineering; multi-criteria decision-making

---

1. Introduction

The emergence of the concept of sustainability has been motivated by natural catastrophes, environmental contamination, depletion of natural resources, and other incidents. According to *Our Common Future* (Brundtland Report) adopted by the World Commission on Environment and Development in 1987 [1], sustainability implies an integrative concept that includes environmental, economic, and social aspects. These three aspects are often referred to as the three pillars of sustainability. In this way, sustainability has become a modern principle that explains the long-term relationship between the present and future generations [2]. At the same time, the term "sustainable development' has emerged, which implies "meet[ing] the needs of the present generation without compromising the ability of future generations to meet their own needs" [3]. Although there are many definitions of sustainable development [4], this is one of the most frequently quoted. In order to achieve the balance between the three pillars of sustainability, it is necessary to define the links and interactions between them, i.e., it is necessary to know how they influence each other [5].

In order to achieve sustainability, sustainable engineering is proposed as a potential solution that implies the application of different methods. Examples may include the construction of facilities made of materials that provide energy efficiency, finding energy forms that do not release carbon dioxide into the atmosphere, designing electric vehicles, etc. According to some authors, sustainable engineering implies significantly more serious considerations of environmental and social aspects [6]. Sustainable engineering thereby observes the system as part of a global ecosystem. According to Abraham [7], the following basic principles of sustainable engineering can be set out:

- Using system analysis and integrating environmental impact assessment tools;
- Improving natural ecosystems;
- Using life cycle thinking;
- Using only material and energy inputs and outputs that are clean and safe;
- Minimizing the depletion of the natural resources;
- Preventing waste;
- Applying engineering solutions having in mind geographic area, culture, and aspirations;
- Creating innovation-based solutions;
- Involving all stakeholders and community in the process of developing solutions.

Engineering is the application of scientific and mathematical principles for practical objectives, such as the processes, manufacture, design, and operation of products, while accounting for constraints invoked by environmental, economic, and social factors. There are various factors needing to be considered in order to address engineering sustainability, which is critical for the overall sustainability of human development and activity. In recent decades, decision-making theory has been a subject of intense research activity [8], due to its wide applications in different areas, such as sustainable engineering and environmental sustainability. The decision-making theory approach has become an important means of providing real-time solutions to uncertainty problems, especially for sustainable engineering and environmental sustainability problems in engineering processes. In the recent decades, several techniques and methods have been used for solving problems in the areas of environmental sustainability and sustainable engineering. Multiple criteria decision making (MCDM) is an important method that has been applied in various areas of sustainable engineering. Several prior studies have employed MCDM techniques in different areas of sustainable engineering [9–18]. In addition, several prior papers have reviewed the application MCDM and fuzzy sets theory in different areas of engineering and sustainability [11,19–26].

The main goal of the paper is to review the literature regarding the application of MCDM methods in the field of sustainable engineering. Another goal is to synthesize different areas of engineering and show effective ways of solving various problems in the field by applying various MCDM methods in various forms of uncertainty. Moreover, this review can be very useful for other studies in various areas of sustainable engineering by showing how MCDM methods can be adequate tools for decision-making processes in sustainable engineering. Furthermore, this paper highlights new, important information for all the participants in MCDM processes in sustainable engineering. In addition, this paper, to the authors' knowledge, is the first review of the literature in the area of sustainable engineering from the perspective of the application of MCDM methods.

The remainder of the paper is structured as follows. Section 2 presents the methodology, in which our algorithm for collecting and processing the articles is presented and explained in detail. Section 3 discusses the primary results of the review, i.e., the total number of MCDM articles in the field of science and technology, with an emphasis on the field of sustainable engineering. The results have been presented by various areas and the structure of the published articles has been presented by journal. Section 4 provides a detailed review of various engineering fields including, construction and infrastructure, supply chains, transport and logistics, energy, and other. In this section, the application of MCDM methods in each of the above areas is explained in detail. Section 5 presents our conclusions.

2. Methodology

This paper reviews the collected literature on the topic of MCDM methods in sustainable engineering. In addition to searching in the Web of Science (WoS) Core Collection Database, articles were searched in online journal databases, using Google Scholar and the Google search engine using the following keywords: MCDM, sustainability, and sustainable engineering. Their combinations were also used when searching as follows: MCDM + sustainability + engineering, MCDM + sustainable engineering, MCDM + sustainability, and MCDM + engineering. All the collected articles were published in the period of 2008–2018. The research methodology is shown in Figure 1.

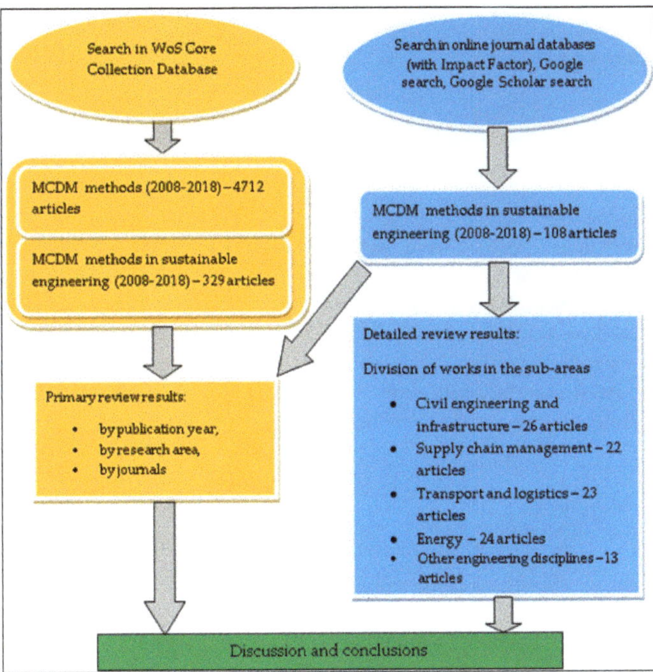

Figure 1. Brief research procedure.

By searching the WoS Core Collection Database, 4712 articles related to the application of MCDM methods in various fields of science and technology have been identified, of which 329 articles deal with the application of MCDM methods in sustainable engineering. In parallel, in the search of online journal databases with impact factors, 108 articles were found, and they were divided into five sub-areas. Based on this, the results of the primary review of articles (by publication year, by area, and by journal) are provided, while a detailed analysis and review of these articles are presented in Section 4.

3. Primary Review Results

By searching the Web of Science Core Collection database, 4712 articles (November 2018) dealing with the application of MCDM methods in various fields of science and technology were found, as shown in Figure 2.

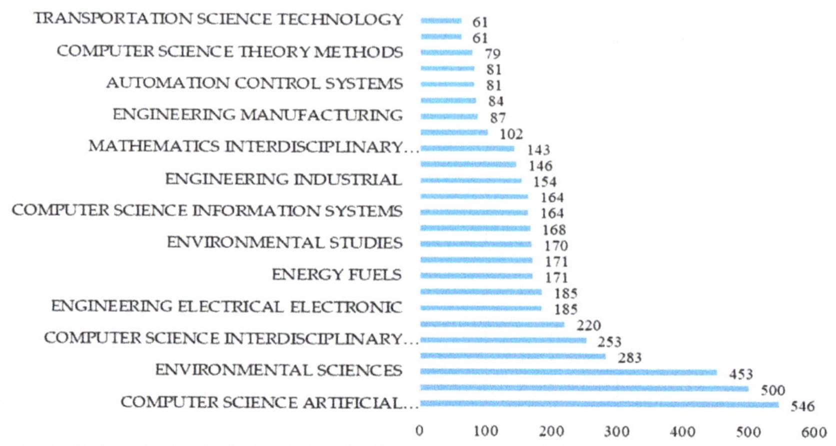

Figure 2. Number of articles on the application of multiple criteria decision making (MCDM) methods in various fields of science and technology.

Figure 2 shows the top 25 areas in which studies applying MCDM methods can be categorized, indicating the number of articles for each area. It appears that the largest number of articles belong to the field of computer science and artificial intelligence (546 articles), while the application of MCDM methods in operational research occupies the second position (500 articles). The smallest number of articles has been published in the field of transport technology (61). It can be concluded that these areas are currently up to date.

In terms of the articles on the application of MCDM methods in sustainable engineering, the Web of Science Core Collection database contains 329 articles, as shown in Figure 3.

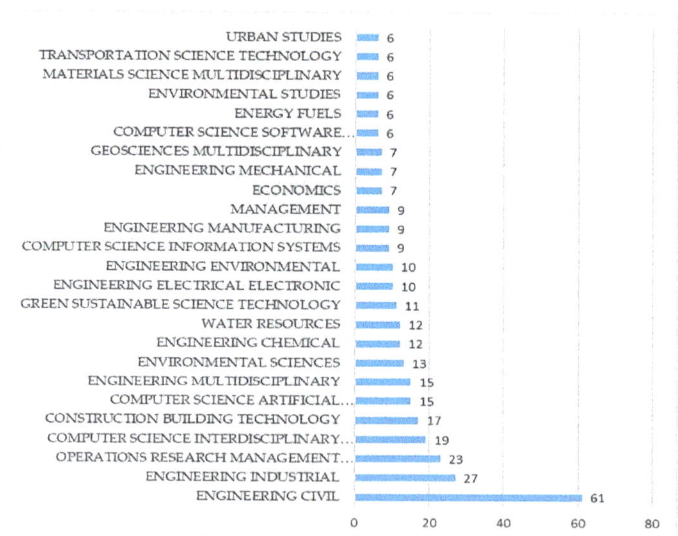

Figure 3. Number of articles on the application of MCDM methods in sustainable engineering.

Figure 3 shows the top 25 areas in which studies applying MCDM methods in sustainable engineering can be categorized. The largest number of articles belongs to the field of civil engineering

(61), and the smallest number belongs to the fields of urban studies (6). It can be observed that the field of transportation science technology, materials science multidisciplinary, environmental studies, energy fuels and computer science software are also at the lower end. In the second position is the area of industrial engineering, followed by operational research, etc.

Figure 4 provides a review of the collected articles by publication years. There is an evident increase in the number of articles in the last few years, because environmental protection, waste minimization, renewable energy sources, energy efficiency, and the concept of sustainability in general have become increasingly frequent and significant subjects of research in many studies in the 21st century [27,28]. In addition, it appears that in 2008, there was not a single published article related to the application of MCDM methods in sustainable engineering.

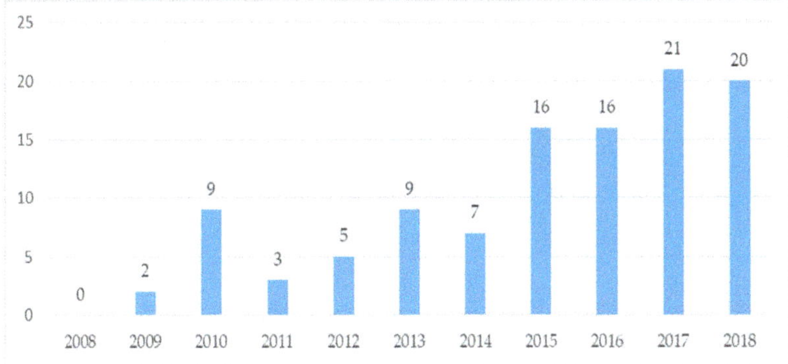

Figure 4. Number of published (collected) articles on MCDM in sustainable engineering by years.

Table 1 provides an overview of the number of articles collected by particular journals.

Table 1. Number of articles by journals.

Title of the Journal	Number of Articles	Percent
Sustainability	14	12.96
Journal of Cleaner Production	10	9.26
Energy	9	8.33
Transport	5	4.63
Journal of Civil Engineering and Management	5	4.63
Applied Energy	5	4.63
International Journal of Production Research	4	3.70
Energies	4	3.70
Construction and Building Materials	4	3.70
Clean Technologies and Environmental Policy	4	3.70
Energy Policy	3	2.78
Transportation Research Part D: Transport and Environment	2	1.85
Renewable and Sustainable Energy Reviews	2	1.85
Land Use Policy	2	1.85
International Journal of Information Technology & Decision Making	2	1.85
Ecological Economics	2	1.85
Water Resources Management	1	0.93
Water	1	0.93
Tunnelling and Underground Space Technology	1	0.93
Transportation Research Part A: Policy and Practice	1	0.93
Transportation Planning and Technology	1	0.93
The International Journal of Advanced Manufacturing Technology	1	0.93
Technological Forecasting and Social Change	1	0.93
Sustainable cities and society	1	0.93
Science of the Total Environment	1	0.93

Table 1. *Cont.*

Title of the Journal	Number of Articles	Percent
Journal of Manufacturing Systems	1	0.93
Journal of Infrastructure Systems	1	0.93
Journal of Environmental Planning and Management	1	0.93
International Journal of Uncertainty, Fuzziness and Knowledge-Based Systems	1	0.93
International Journal of Sustainable Transportation	1	0.93
International Journal of Production Economics	1	0.93
Expert systems with Applications	1	0.93
European Journal of Operational Research	1	0.93
European Journal of Environmental and Civil Engineering	1	0.93
Environmental Modelling & Software	1	0.93
Economic Research	1	0.93
Decision support systems	1	0.93
Computers & Structures	1	0.93
Computers & Industrial Engineering	1	0.93
Computer-Aided Civil and Infrastructure Engineering	1	0.93
Civil Engineering and Environmental Systems	1	0.93
Cities	1	0.93
Building and environment	1	0.93
Automation in Construction	1	0.93
Applied Sciences	1	0.93
Applied Mathematical Modelling	1	0.93
AIChE Journal	1	0.93

Based on Table 1, it can be concluded that most of the collected articles have been published in the journal *Sustainability* (14 articles), which represents 12.96% of the total number. The *Journal of Cleaner Production* can be ranked second with 10 articles or 9.26% of the total articles. Out of a total of 47 journals, 31 have published one article related to MCDM methods in sustainable engineering. It is important to note that all these journals have impact factors.

4. Detailed Review Results

All the collected articles (108 articles) on the topic of applications of MCDM methods in sustainability engineering have been classified into 5 categories: construction and infrastructure, supply chains, transport and logistics, energy, and other. It is important to mention that some areas of engineering, such as mechanical engineering, have not been taken into consideration because of the lack of articles regarding such topics. For each of the above categories, a detailed analysis of the aim and importance of the application of the individual MCDM method has been provided, and the results of the review have also been given in a table. Figure 5 shows the subdivision into the 5 subcategories with the number of articles in each subcategory.

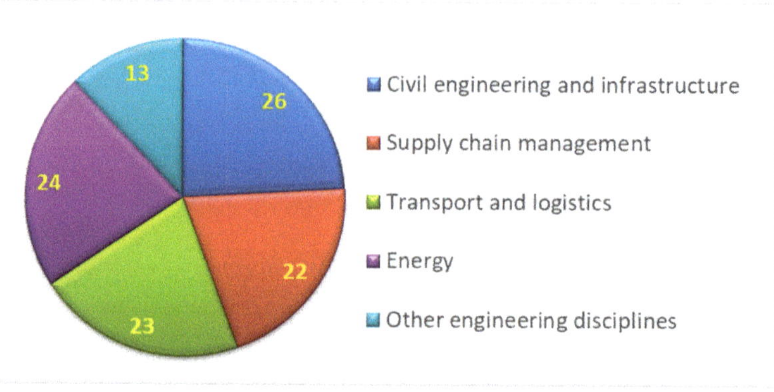

Figure 5. Division of research subjects into five sub-areas.

4.1. Civil Engineering and Infrastructure

In the domain of architecture and construction, increasing attention is being paid to energy efficiency and smart buildings, and therefore, it is necessary to go towards sustainability in the design and construction of facilities and infrastructure. Consequently, it is required to select adequate materials as well. In this section, a detailed analysis of the 26 collected articles in the field of construction and infrastructure is presented.

In their work, Birgani and Yazdandoost [29] provided a framework for a new approach to addressing flood problems in urban areas. In many cases, due to unforeseen and abundant precipitation, the existing drainage network cannot receive large amounts of precipitation. For this reason, for the selection between several alternatives of the sewer system, an integrated approach that implies the sustainability and application of multi-criteria decision-making methods, i.e. the adaptive analytical hierarchy process (AHP), entropy and Technique for Order of Preference by Similarity to Ideal Solution (TOPSIS) was proposed. The framework was applied to the case study for a part of the city of Tehran, Iran. The problem of floods in urban areas due to abundant precipitation was also discussed in [30]. Based on the sustainability criteria, using the AHP method for determining the weights and the Preference Ranking Organization Method for the Enrichment of Evaluations II (PROMETHEE II) for the final ranking of the alternatives, a framework for the selection of an optimum drainage system was proposed. The implementation of the framework was carried out using the example of Buraydah City, Qassim, Saudi Arabia.

Construction is an area that interacts enormously with the natural environment. A large percentage of raw materials are obtained from the earth, and in their treatment and processing and the construction of buildings, certain environmental pollution is inevitable. Lombera and Rojo [31] use the Spanish MIVES (Integrated Value Model for Sustainable Assessment) methodology to define the criteria for the sustainability of industrial buildings and to select the optimum solution with regards to them. Generally speaking, the MIVES methodology combines multi-criteria decision-making and multi-attribute utility theory (MAUT), including a value function concept and weight assignment by the AHP method [32]. A similar study was presented in a study by del Cano et al. [33], in which authors also used the MIVES method but in combination with Monte Carlo simulation in order to assess the sustainability of concrete structures. For the same purpose, de la Fuente et al. [34] applied fuzzy-MIVES. Moreover, de la Fuente et al. [34] also applied the MIVES methodology together with the AHP method in order to reduce the subjective human impact on the selection of sewage pipe material. Akhtar et al. [35] solved the same problem using only the AHP method. The MIVES methodology was also used in a study by de la Fuente et al. [36], assessing the sustainability of alternatives—the types of concrete and their reinforcement for their application in tunnels, depending on environmental, social, and economic criteria. The case study was carried out for the city of Barcelona. Pons and de la Fuente [37] used MIVES to select the most suitable concrete pillars as structural components of buildings, while Pujadas et al. [32] constructed a framework for the evaluation of heterogeneous public investments using this methodology, which is a step towards sustainable urban planning. Different economic, environmental, and social aspects were considered, with five criteria and eight indicators.

The problem of monitoring, repairing, and the returning to function of steel bridge structures is a major challenge for engineers, especially because it is necessary to make key decisions, and wrongly made decisions can be very costly. In order to exclude subjectivity in selecting alternatives in this case, Rashidi et al. [38] presented the decision support system (DSS), in which the simplified AHP (S-AHP) method is used. S-AHP combines the simple multi-attribute rating technique (SMART) and the AHP method. The aim is to help engineers in planning the safety, functionality, and sustainability of steel bridge structures. Jia et al. [39] presented a framework for the selection of bridge construction between the accelerated bridge construction (ABC) method and conventional alternatives, using TOPSIS and fuzzy TOPSIS methods.

In their work, Formisano and Mazzolani [40] presented a new procedure for the selection of the optimum solution for seismic retrofitting of existing reinforced concrete (RC) buildings, as well as optimum solutions for vertical upgrading of existing masonry constructions. The procedure involved the application of three MCDM methods, namely TOPSIS, elimination and choice expressing reality (ELECTRE), and VIseKriterijumska Optimizacija i Kompromisno Rešenje (VIKOR). In two case studies, these methods showed the same results. In their work, Terracciano et al. [41] selected cold-formed, thin-walled steel structures for vertical reinforcement and energy retrofitting systems of existing masonry constructions. The TOPSIS method for selecting alternatives based on structural, economic, environmental, and energy criteria was used.

Improving traditional buildings into modern ones must comply with technical regulations, energy requirements, comfort requirements, and the preservation of existing architecture. Siozinyte et al. [42] applied the AHP and TOPSIS grey MCDM methods to select the optimum solution for modernizing traditional buildings.

Khoshnava et al. [43] applied MCDM methods to select energy efficient, ecological, recyclable materials for building with respect to the three pillars of sustainability. In order to evaluate 23 criteria in the selection of materials, they used the decision-making trial and evaluation laboratory (DEMATEL) hybrid MCDM method together with the fuzzy analytic network process (FANP). Akadiri et al. [44] used fuzzy extended AHP (FEAHP) in order to select sustainable building materials.

In a study by Ozcan-Deniz and Zhu [45], the analytic network process (ANP) method was used to select the most environmentally friendly method for the construction of a highway, because such construction can have a great impact on the environment. Possible alternatives included different types of materials, operations, and project conditions. Constructing traffic infrastructure can greatly increase the level of safety for participants, but also reduce traffic jams. In their work, Stevic et al. [46] selected the locations for the construction of roundabouts using the rough best–worst method (BWM) and the rough weighted aggregated sum product assessment (WASPAS) approach based on the New Rough Hamy Aggregator.

In their work, Rashid et al. [47] used MCDM methods to select sustainable concrete, which implies a mixture of conventional coarse aggregate and ceramic waste aggregate. The AHP and TOPSIS methods were used to select the best performing concrete in terms of the pressure it can endure and its impact on the environment.

During and even after the construction of facilities, a large amount of natural resources is used, which adversely affects the environment. Most systems for evaluating the sustainability of facilities take into account only the environmental aspect and the environmental impact. However, it is necessary to take into account all three basic principles of sustainability, and thus Raslanas et al. [48], in their work, developed a system for evaluating the sustainability of recreational facilities, using the AHP method. Because so-called "green buildings" are environmentally friendly, attention is increasingly being given to the selection of methods for their construction. Taking into account that this is a very complex task, the application of MCDM methods is indispensable, and in the study by Tsai et al. [49], DEMATEL, ANP, and zero–one goal programming (ZOGP) methods were applied.

The selection of construction project managers plays a key role for the entire construction process. Zavadskas et al. [50] used the MCDM approach to this problem and applied AHP and additive ratio assessment (ARAS) methods. The alternatives were selected based on the criteria of education, experience, and personal abilities and skills.

When building larger facilities, i.e., implementing capital projects, it is very important to select a proper transport route for the procurement of raw materials and materials. Marzouk and Elmesteckawi [51] selected the best alternative for the construction of a power plant using the SMART method.

Because the number of vehicles on the roads is increasing every day, the number of parking spaces can hardly follow this trend. Using the MCDM method, Palevicius et al. [52] indicated the worst parking conditions in Vilnius, Lithuania, with all three aspects of sustainability, using simple additive weighting (SAW), TOPSIS, complex proportional assessment (COPRAS), and AHP method. Table 2 summarizes the applied MCDM methods in the sub-area of civil engineering and infrastructure.

Table 2. MCDM methods in the sub-area of civil engineering and infrastructure.

A Problem that Is Solved Using the MCDM Method	Applied Methods	Reference
Evaluating urban drainage plans in terms of their sustainability and resilience	Adaptive AHP, Entropy, TOPSIS	[29]
Assessment of sustainability criteria of industrial buildings	AHP, MIVES	[31]
Assessing sustainability adopted by the Spanish Structural Concrete Code (EHE)	AHP, MIVES	[33]
Assessing sustainability adopted by the Spanish Structural Concrete Code (EHE)	Fuzzy MIVES	[34]
Sustainability analysis of different constituent materials for sewerage pipes	AHP, MIVES	[34]
Evaluating and comparing of four typical sewer pipe materials and identifying a sustainable solution	AHP	[35]
Analyzing the sustainability of different concrete and reinforcement configurations for segmental linings	AHP, MIVES	[36]
Evaluating different stormwater drainage options for urban areas of arid regions	AHP, PROMETHEE II	[30]
Developing the decision support system for the asset management of steel bridges	S-AHP (SMART, AHP)	[38]
Estimating the components of the total cost of ABC versus conventional construction methods	TOPSIS, Fuzzy TOPSIS	[39]
Selection of the optimum solution for the seismic retrofitting of existing RC buildings and for the super-elevation of existing masonry constructions	TOPSIS, ELECTRE, VIKOR	[40]
Establishing a cost–benefit approach for the most suitable vertical addition solution	TOPSIS	[41]
Finding the best compromise solution for effective vernacular architecture change	AHP, TOPSIS Grey	[42]
Developing a methodological and systematic approach for building material selection	DEMATEL, FANP	[43]
Selection of sustainable materials for building projects	FEAHP	[44]
Selecting the most feasible highway construction method	ANP	[45]
Location selection for roundabout construction	Rough BWM, Rough WASPAS	[46]
Sustainability assessment tool for analyzing reinforced concrete structural columns of residential buildings	MIVES	[37]
Finding strategies for the prioritization and selection of heterogeneous investments projects	MIVES	[32]
Assessment of properties of a fresh and hardened concrete by incorporating various amounts of ceramic waste.	AHP, TOPSIS	[47]
Creating a recreational building sustainability assessment model	AHP	[48]
Building an effective evaluation model for green building construction methods	DEMATEL, ANP, ZOGP	[49]
Development of the methodology that serves as a decision support aid in assessing project managers	AHP, ARAS	[50]
Selecting the most efficient procurement/delivery system for multiple contracts power plants	SMART	[51]
Indicating the worst passenger car parking conditions in residential areas	SAW, TOPSIS, COPRAS, AHP	[52]

Based on Table 2, it can be concluded that the AHP method is one of the most frequently applied. In addition, it appears that AHP, as well as other methods, can be synthesized with other MCDM methods, but also with other theories such as fuzzy and grey numbers.

4.2. Supply Chain Management

Supply chains present a very complex field involving many participants. The aim of the complete supply chain is to find an optimum from the perspective of all the participants, which is a rather complex task [53–55].

Supply chain management in terms of sustainability in a number of industries is an increasingly frequent topic of research. Therefore, this section provides an analysis of 22 articles on this topic. In the review by Seuring [56], MCDM, and particularly AHP, was listed as one of the quantitative methods for improving the supply chain management. Additionally, based on the review, it can be concluded that the social component of sustainability is paid the least attention. In their review paper, Zimmer et al. [57] analyzed the use of various models to support decision-making on sustainable supplier selection. The models that were stated as the most commonly used were the mathematical/analytic ones, which include MCDM. Significantly, the biggest percentage of application belongs to the AHP method, followed by ANP, etc. The selection of suppliers, according to many authors, is one of the most demanding problems of sustainable supply chain management. Therefore, numerous methods for ranking suppliers have been developed to date, and Fallahpour et al. [58] used the fuzzy modifications of the AHP and TOPSIS methods. The abovementioned authors used the fuzzy preferences programming (FPP) method to reach the relative weights of criteria, while the fuzzy TOPSIS method was used to rank suppliers. In order to validate the methods, a case study was conducted on a real system. The fuzzy approach in combination with the TOPSIS MCDM method was used by Govindan et al. [59] to assess the sustainable performance of suppliers. In order to perform the selection of suppliers in terms of sustainability and at the same time to take into account the business goals of the company, Dai and Blackhurst [60] presented an integrated approach based on AHP and the quality function deployment (QFD) method with four hierarchical phases. Rezaei et al. [61] presented a new methodology for the selection of suppliers consisting of three phases, where the central phase is the application of the BWM method of multi-criteria decision-making. The methodology presented can be particularly useful for companies that are looking for new markets. For the selection of suppliers, Azadnia et al. [62] proposed an integrated approach that, in addition to the fuzzy AHP method (FAHP), is based on multi-objective mathematical programming, as well as on rule-based weighted fuzzy method. According to Su et al. [63], the assessment of sustainable supply chain management and the selection of suppliers are performed using grey theory in combination with the DEMATEL method. Luthra et al. [64] presented an integrated approach to selecting suppliers consisting of a combination of AHP and VIKOR methods based on 22 criteria for all three aspects of sustainability. Because thermal power plants are the main source of electricity in China, it is necessary to make a selection of sustainable suppliers of raw materials in order to achieve sustainable development of the company. According to Zhao and Guo [65], an integrated approach is based on the fuzzy entropy–TOPSIS method. MCDM methods can be used to assess the degree of organizational sustainability of a company, as presented in [66]. Hsu et al. [67] presented a hybrid approach based on several MCDM methods in order to select suppliers in terms of carbon emissions. The observed framework for the selection of suppliers has been applied to the case of a hotel in Taiwan. A similar study was carried out by Kuo et al. [68] on the example of electronic industry. The evaluation of the supplier performance in the field of electronic industry in order to implement green supply chains is a topic of research in the study by Chatterjee et al. [17]. The authors used rough DEMATEL–ANP (R'AMATEL) in combination with rough multi-attribute ideal real comparative analysis (R'MAIRCA) method. Liu et al. [69] selected the suppliers of fresh products using the BWM and multi-objective optimization on the basis of the ratio analysis (MULTIMOORA) method.

Because innovation plays a very important role in sustainability, Gupta and Sarkis [70] presented a framework for ranking and selecting the criteria for sustainable innovations in supply chain management. This framework is based on the BWM method, and its applicability and efficiency were tested on several manufacturing companies in India. In their work, Validi et al. [71] dealt with the sustainability of the food supply chain. The TOPSIS method was used for the purpose of ranking the traffic routes, taking into account CO_2 emissions and total transport costs.

A quantitative assessment of the performance of a sustainable supply chain was presented in Erol et al. [72] with regard to all three aspects of sustainability. Due to the presence of indeterminacy, it is very difficult to estimate certain criteria, which is why the authors used fuzzy techniques in addition to MCDM. More precisely, the fuzzy entropy and fuzzy MAUT methods were used. Das and Shaw [73] proposed a methodology based on the AHP and Fuzzy TOPSIS methods for selecting a sustainable supply chain, taking into account carbon emissions and various social factors. In the study by Entezaminia et al. [74], the AHP method was used to evaluate products in the supply chain according to environmental criteria such as recyclability, biodegradability, energy consumption, and product risk.

The application of information and communication technologies in supply chains can bring numerous benefits to an organization, and among the most important is sustainability. Luthra et al. [75] proposed the application of delphi and fuzzy DEMATEL methods for identifying and evaluating the guidelines for the application of these technologies in sustainable initiatives in supply chains. In Padhi et al. [76], a framework that identifies sustainable processes in supply chains for individual industries in India was presented. The ranking of industry branches was carried out using six fuzzy MCDM methods. Table 3 provides a summary of the applied MCDM methods for the sub-area of supply chain management. The decision-making process requires the prior definition and fulfillment of certain factors, especially when it comes to complex areas, such as supply chain management [77].

Table 3. MCDM methods in the sub-area of supply chain management.

A Problem that Is Solved Using the MCDM Method	Applied Methods	Reference
Sustainable supplier selection through a questionnaire-based survey	FPP, Fuzzy TOPSIS	[58]
Investigating sustainable supply chains in manufacturing companies	BWM	[70]
Provision of optimized distribution routes based on carbon output and costs for the demand side of a dairy supply chain producing milk products	TOPSIS	[71]
Development of the sustainability-focused supplier assessment methodology that will be able to capture the 'voice of customer' at multiple stages in the supply chain and translate the needs of the end customer back through the supply chain	AHP–QFD	[60]
Proposing an innovative three-phase supplier selection methodology	BWM	[61]
Integrated approach of rule-based weighted fuzzy method, fuzzy analytical hierarchy process, and multi-objective mathematical programming for sustainable supplier selection	FAHP	[62]
Identifying and analyzing criteria and alternatives in incomplete information	Grey-DEMATEL	[63]
Evaluating the sustainable supplier selection	AHP, VIKOR	[64]
Selecting the proper green supplier of thermal power equipment	Fuzzy Entropy–TOPSIS	[65],
Classification of the degree of sustainability of organizations that work in providing supplies to the oil and gas industry	ELECTRE TRI	[66]
Evaluating the carbon and energy management performance of suppliers by using multiple-criteria decision-making	Fuzzy Delphi, DEMATEL, DEMATEL–ANP (DANP), VIKOR	[67]
Novel hybrid multiple-criteria decision-making method for evaluating green suppliers in an electronics company	DANP, DEMATEL, VIKOR	[68]
Evaluating the performance of suppliers in the electronics industry	R'AMATEL, R'MAIRCA	[17]
Novel two-stage fuzzy integrated MCDM method for the selection of suitable suppliers	BWM, MULTIMOORA	[69]
Evaluating the sustainability performance of a supplier	Fuzzy TOPSIS	[59]
Evaluating the sustainability performance of a supply chain	Fuzzy MAUT	[72]
Proposing an uncertain supply chain network design (SCND) model by considering various carbon emissions and social factors	AHP, Fuzzy TOPSIS	[73]
Development of a new comprehensive multi-objective aggregate production planning model in a green supply chain considering a reverse logistic network to be used in many industries	AHP	[74]
Identification and evaluation of key drivers relevant to information and communication technologies for sustainability initiatives in a supply chain	Delphi, Fuzzy DEMATEL	[75]
Identifying the significance of various sustainable supply chain processes on firm performance	Fuzzy TOPSIS, Fuzzy ELECTRE, Fuzzy AHP, Fuzzy Multiplicative AHP, Fuzzy SMART, Fuzzy VIKOR	[76]

In the sub-area of supply chain management, based on the table, it is apparent that most authors apply the AHP and TOPSIS methods. As mentioned in the previous section, their applications can be combined with other methods.

4.3. Transport and Logistics

As in other engineering disciplines, MCDM methods are also applied in the field of transport and logistics. This section provides a review of 23 articles dealing with the above issues. In Mardani et al. [78], a review of the methods used to solve problems in transport systems was provided. The articles were systematically categorized into 10 groups, one of which was sustainability. The authors stated that according to the number of articles published on MCDM in combination with sustainability, this category could be ranked sixth.

In Jeon et al. [79], the application of MCDM methods in selecting sustainable transport plans based on the sustainability index is examined. The weighted sum model (WSM) method was used. In their work, Cadena and Magro [80] presented a new methodology for assigning weight coefficients to sustainability criteria in transport projects. In order to solve the problem of inaccuracy and subjectivity, the REMBRANDT and Delphi methods were applied.

Because the traffic system is the lifeblood of every country and one of the basis for its economic development, Baric et al. [81] proposed the application of the AHP method in selecting the best road section design in urban conditions. The tested model on the real system showed reliable results. One of the disadvantages of the AHP method is that it requires a large number of inputs. In order to solve this problem, Inti and Tandon [82] presented a modified AHP method with the characteristics of the additive transitivity of fuzzy relations. The model was tested in the selection of contractors for the construction of transport infrastructure.

In order to improve transport sustainability, one of the solutions is the application of various alternative fuels and vehicle drives. Mitropoulos and Prevedouros [83], in this way, assessed the characteristics of vehicles using the sustainability index. The identified indicators were classified into five categories of sustainability—environment, technology, energy, economy, and users—followed by the application of the WSM method for their aggregation. Additionally, Mohamadabadi et al. [84] selected the type of fuel for vehicles based on three basic aspects of sustainability. The PROMETHEE method was used for the ranking of alternatives based on five criteria. Intermodal transport can greatly improve the sustainability of the transport system. It is necessary to select the optimum location of terminals in terms of different requirements of different participants in a transport process. Therefore, Zecevic et al. [85] proposed a new hybrid MCDM model for the location selection. Sustainable transport systems have become necessary nowadays, primarily in large cities due to various adverse environmental impacts. An approach to selecting the best alternative of transport systems based on 24 criteria, classified into three categories, was defined in a study by Awasthi et al. [86]. The approach consists of three steps, and the TOPSIS method is applied in combination with fuzzy theory in order to evaluate the criteria and the selection of an alternative. Castillo and Pitfield [87] proposed the evaluative and logical approach to sustainable transport indicator compilation (ELASTIC) framework for selecting the sustainability indicators of the transport system using the AHP and SAW methods. Although, in recent years, improvements have been evidently made to methods of transport planning, according to Lopez and Monzon [88], it is necessary to apply a multidisciplinary approach based on Geographic Information System (GIS) in order to increase the level of sustainability in transport. In addition, it is necessary to integrate multi-criteria decision-making methods within the proposed approach. In his work, Simongati [89] presented a model for the selection of FREIGHT INTEGRATOR with MCDM methods and sustainability indicators. The aforementioned term represents a provider of door-to-door transport services, using different modes of transport in an efficient and sustainable way. The selection of alternatives is based on SAW and PROMETHEE methods. The assessment of transport system sustainability of some European countries based on selected economic, environmental, and social indicators is presented in the work of Bojkovic et al. [90]. The ELECTRE I method has been used together with its modification based on the absolute significance threshold (AST). A framework for the selection of sustainable transport projects in urban areas of developing countries was proposed in the work of Jones et al. [91].

The selection of alternatives is based on the localized sustainability score index using the AHP method. In addition to the AHP method, in order to assess the sustainability of various transport solutions, such as mode sharing, multimodal transport, and intelligent transport systems, Awasthi and Chauhan [92] used the Dempster–Shafer theory in the proposed hybrid approach. While the AHP method serves primarily to rank the criteria based on the weights, the Dempster–Shafer theory allows the synthesis of multiple sources of information. Dimić et al. [93] developed a model for strategic transport management based on Strengths, Weakness, Opportunities, Treats (SWOT) analysis, fuzzy Delphi, and DEMATEL–ANP methods.

Sustainability is a very important concept in logistics, and reverse logistics as one of its subgroups can greatly improve efficiency and the environmental aspect of business. Wang et al. [94] presented a method for identifying the collection mode for used components. A hybrid approach based on AHP and entropy weight (AHP–EW) method was used to estimate the weights of particular criteria, while the multi-attributive border approximation area comparison (MABAC) method was used to rank the alternatives. Different initiatives for city logistics (e.g., the proper location of distribution centers) can significantly contribute to raising the level of sustainability in the city. That is precisely the subject of research in Awasthi and Chauhan [95]. The MCDM methods used in the work were AHP and Fuzzy TOPSIS. Mavi et al. [96], using the fuzzy step-wise weight assessment ratio analysis (SWARA) and fuzzy MOORA methods, selected a third-party provider of reverse logistics service in the plastics industry.

One of the most current problems in logistics and supply chains is the selection of the location of the logistics center in terms of sustainability. Rao et al. [97] used the fuzzy multi-attribute group decision-making (MAGDM) approach to address the problem. Turskis and Zavadskas [98] approached the problem of selecting the location of the logistics center with the fuzzy ARAS (ARAS–F) method, while Pamucar et al. [99] used the DEMATEL–MAIRCA method for the same purpose.

Logistics are closely linked to the processing industry. Therefore, it is necessary to identify the factors that influence their interaction. For this purpose, Jiang et al. [100] applied the grey DEMATEL-based ANP method (DANP). Table 4 provides a summary of the applied MCDM methods for the sub-area of transport and logistics.

Table 4. MCDM methods in the sub-area of transport and logistics.

A Problem that Is Solved Using the MCDM Method	Applied Methods	Reference
Selecting sustainable transport plans based on the sustainability index	WSM	[79]
Assigning weight coefficients to the sustainability criteria in transport projects	REMBRANDT, Delphi	[80]
Evaluating road section design in an urban environment	AHP	[81]
Selection of the contractor for the construction of transport infrastructure	AHP, FAHP	[82]
Assessment of transportation vehicle characteristics.	WSM	[83]
Ranking different renewable- and non-renewable-fuel-based vehicles	PROMETHEE	[84]
Selection of the intermodal transport terminal location	Fuzzy Delphi, Fuzzy Delphi ANP, Fuzzy Delphi VIKOR	[85]
Evaluation and selection of sustainable transportation systems under uncertain (fuzzy) environments	Fuzzy TOPSIS	[86]
Identifying and selecting a small subset of sustainable transport indicators	AHP, SAW	[87]
Assessment model for transport infrastructure plans	REMBRANDT	[88]
Valuation and comparison model adjusted to the decision-making tasks of the freight integrator	SAW, PROMETHEE	[89]
Evaluation of transport–sustainability performance in some European countries.	ELECTRE I, Modified ELECTRE I	[90]
Screening urban transport projects in developing countries to reflect locally derived sustainability criteria	AHP	[91]
Evaluating the impact of environmentally friendly transport measures on city sustainability	AHP	[92]
Developing the model for strategic transport management	Fuzzy Delphi, DEMATEL–ANP	[93]
Identifying the best collection mode for used components	AHP-EW, MABAC	[94]
Hybrid approach for evaluating city logistics initiatives	AHP, Fuzzy TOPSIS	[95]
Evaluation of third-party reverse logistic provider considering sustainability and risk factors	Fuzzy SWARA, Fuzzy MOORA	[96]
Location selection of a city logistics center from a sustainability perspective	Fuzzy MAGDM	[97]
Selecting the most suitable site for a logistics center among a set of alternatives	Fuzzy ARAS, AHP	[98]
Sustainable selection of a location for the development of a multimodal logistics center	DEMATEL–MAIRCA	[99]
Identifying interactions between manufacturing and logistics industries	Grey DANP	[100]

Table 4 indicates which MCDM methods are used in the field of transport and logistics. In this case, the AHP method is also the most applied MCDM method.

4.4. Energy

This section provides a review of the application of MCDM methods in the field of energy. 24 articles were analyzed, and the results have been given in textual and tabular formats. Developing renewable energy sources is a growing trend in the world on a day-to-day basis, especially when it comes to solar energy. The selection of an optimum location for the installation of photovoltaic systems is of great importance, because it can reduce the cost of the project and also ensure the maximum production of electricity. It sufficiently proves the high sustainability of such sources. Al Garni and Awasthi [101] selected the location of solar systems based on MCDM methods and GIS. The AHP method was used to evaluate the weights of feasibility criteria that directly affect the performance of the solar system. A similar study was also presented in the work of Diaz-Cuevas et al. [102], where spatial information instead of GIS was provided with the PostgreSQL-PostGIS database, which was based on Structured Query Language (SQL). The AHP method is used to determine the weights of the criteria. GIS is also necessary in selecting the location of wind farms that are also a very significant alternative source of energy. According to Sanchez-Lozano et al. [103], fuzzy MCDM methods are used to determine the weights of the criteria and the selection of an optimum alternative in solving this problem.

The selection of the optimum type of renewable energy sources using MCDM methods based on the hesitant fuzzy linguistic (HFL) term set was presented in the work of Buyukozkan and

Karabulut [104]. The proposed methodology was tested using the example of the selection between several alternatives in the territory of Turkey. A similar study was presented in the work of Wu et al. [105], where a case study for China was conducted. Based on the AHP and TOPSIS methods, it was found that solar systems are the best solution. Yazdani et al. [18] presented a new hybrid approach for the selection of renewable energy technology, using DEMATEL, ANP, COPRAS, and WASPAS methods. Zhang et al. [106] used the improved MCDM method based on fuzzy measure and integral to select the "pure" form of energy between several alternatives. In their research, Troldborg et al. [107] dealt with the same issue and applied the PROMETHEE method. In their work, Klein and Whalley [108] selected between 13 renewable and non-renewable energy sources based on eight criteria. According to Tsoutsos et al. [109], the selection of an optimum renewable energy source in Crete, Greece was carried out with the PROMETHEE I and PROMETHEE II methods. Countries rich in fossil fuels are forced to seek alternative energy sources in order to reduce CO_2 emissions. Pamucar et al. [110] applied the linguistic neutrosophic numbers pairwise–combinative distance-based assessment (LNN PW–CODAS) to select the optimum energy production technology in Libya. In Pamucar et al. [111], a model for the selection of a location for the construction of wind farms based on GIS in combination with two MCDM methods, BWM and MAIRCA, was presented.

Generated electricity planning is of great importance for the electric power system of a country. Mirjat et al. [112] proposed the application of the AHP method for assessing the sustainability of four types of energy models. A case study was carried out for Pakistan. The European Union is developing its energy plans, and MCDM methods find their application in the ranking of plans. According to Balezentis and Streimikiene [113], for this purpose, WASPAS, ARAS, and TOPSIS methods should be used. The selection of the best energy project between several alternatives, using MCDM methods, was considered in the work of Buyukozkan and Karabulut [104].

Because electric vehicles are becoming increasingly common on roads in the world, it is necessary to provide stations for charging them at optimum locations. Zhao and Li [114] presented a methodology based on MCDM methods. The criteria of the expanded concept of sustainability, which in addition to the traditional three aspects also includes technology, were selected based on fuzzy delphi, while the selection of the best alternative was performed using the fuzzy grey relation analysis (GRA)–VIKOR method. Guo and Zhao [115] dealt with the same issues. In order to eliminate subjectivity when selecting the location of charging stations, in addition to the basic criteria of sustainability, 11 sub-criteria were defined, in which the weights were determined on the basis of literature research, opinion of experts, and feasibility studies. The specific location selection was completed using the fuzzy TOPSIS method.

Nuclear energy implies low values of CO_2 emissions into the atmosphere, which is necessary in terms of the concept of sustainability. Gao et al. [116] presented a framework for selecting the best option for a nuclear fuel cycle at a plant. In order to determine the weights of the criteria, fuzzy AHP and criteria importance through intercriteria correlation (CRITIC) were used, and the selection of alternatives was performed using the TOPSIS and PROMETHEE II methods. The selection of the optimum energy option for a thermal power plant was the subject of research in the work of Skobalj et al. [117]. The selection between seven alternatives, including revitalization and additional production by alternative energy sources, was performed on the basis of the sustainability index, which was determined by the analysis and synthesis of parameters under information deficiency (ASPID) method. The application of MCDM methods in order to select a sustainable energy solution has not been omitted even when it comes to hydroelectric power plants in the work of Vucijak et al. [118]. According to Streimikiene et al. [119], the selection between several alternative technologies for the sustainable production of electricity can be performed with the MULTIMOORA and TOPSIS methods (Barros et al. [120]). Maxim [121] also deals with the same issues in his work. He used a modified SWING method for ranking technologies. Energy is the key to the economic and social development of a particular area. In their work, Jovanovic et al. [122] proposed a new approach based on the predictions of different energy scenarios in urban areas and the application of MCDM methods for evaluating them. Biomass implies a multitude of resources, such as plant waste, animal waste, food waste, etc. Ioannou et al. [123] used MCDM methods in their research to select the location of a biomass power plant. Table 5 provides a summary of the applied MCDM methods for the sub-area of energy.

Table 5. MCDM methods in the sub-area of energy.

A Problem that Is Solved Using the MCDM Method	Applied Methods	Reference
Evaluating and selecting the best location for utility-scale solar photovoltaic (PV) projects	AHP	[101]
Identifying optimum locations for solar plants	AHP	[102]
Approach for the evaluation of available sites to implant onshore wind farms	FAHP, Fuzzy TOPSIS	[103]
Numerical decision support method for identifying the most suitable renewable energy source	HFL–AHP, HFL–COPRAS	[104]
Evaluating the renewable energy sources and selecting the most appropriate one from the perspective of public investors	AHP, TOPSIS	[105]
Multi-criteria assessment of renewable energy systems	DEMATEL, ANP, COPRAS, WASPAS	[18]
Multi-criteria analysis for a national-scale sustainability assessment and ranking of renewable energy technologies in Scotland	PROMETHEE	[107]
Comparing a wide range of conventional and alternative electricity generation technologies across several criteria	Weighted sum method	[108]
Assessing the performance of different renewable energy source alternatives	PROMETHEE I, PROMETHEE II	[109]
Selection of power-generation technology using a linguistic neutrosophic CODAS Method	LNN PW–CODAS	[110]
Location selection for wind farms	Rough BWM, Rough MAIRCA	[111]
Sustainability assessment of energy modeling results for long-term electricity planning	AHP	[112]
Ranking energy development scenarios for the EU by employing MCDM techniques.	WASPAS, ARAS, TOPSIS	[113]
Novel method with a sustainability perspective for selecting energy projects	AHP, VIKOR	[104]
Multi-criteria decision-making framework to address the issue of electric vehicle charging stations siting	Fuzzy Delphi, GRA–VIKOR	[114]
Employing various MCDM techniques to select the optimum electric vehicle charging station	Fuzzy TOPSIS	[115]
Evaluation of decision-making for China's future nuclear fuel cycle options from a sustainability perspective	AHP, CRITIC, TOPSIS, PROMETHEE II	[116]
Estimating the quality of the considered thermal power plant "Kolubara"-A Unit No. 2	ASPID	[117]
Assessing applicability potentials of a specific multi-criteria decision support method to sustainable hydropower design	VIKOR	[118]
Multi-criteria decision support framework for choosing the most sustainable electricity production technologies	MULTIMOORA, TOPSIS	[117]
Comprehensively ranking a large number of electricity-generation technologies	SWING	[121]
Measurement the sustainability of an urban energy system	ASPID	[122]

As can be seen from Table 5, in the field of energy, MCDM methods are mainly used to solve problems of selecting the optimum type of renewable and non-renewable energy sources. In most cases, the AHP method is applied.

4.5. Other Engineering Disciplines

In addition to four previously analyzed areas of the application of MCDM methods in sustainable engineering, uncategorized works are discussed in this section. This includes 13 articles from various fields of engineering, and their detailed analysis is given below. Creating a sustainable environmental management system is of great importance for reducing environmental pollution. Khalili and Duecker [124] created a system for selecting the best solution using the ELECTRE III method. In their research, Egilmez et al. [125] applied the intuitionistic fuzzy decision making (IFDM) approach, which is the integration of fuzzy logic and MCDM theory, in order to rank and select a city (in the US and Canada) with the highest degree of environmental sustainability. According to Alwaer and Clements-Croome [126], the model for the assessment of smart household sustainability includes the application of the AHP method. The level of sustainability of temporary housing units for the accommodation of persons after natural disasters can be assessed using the MIVES method according to Hosseini et al. [127]. A review of the MCDM methods used in assessing the sustainability of the system is shown in Diaz-Balteiro et al. [128], and it can be concluded that the AHP method takes the leading position in a number of applications. In Rosen et al. [129], a new method for assessing the sustainability of renewed contaminated surfaces was developed. The proposed sustainable choice of remediation (SCORE) method is a tool for selecting between several alternatives for possible land remediation. Ren et al. [130] developed a generic framework for the selection of sustainable technology for the treatment of sewage sludge in urban areas, using three MCDM methods: sum weighted method (SWM), digraph model, and TOPSIS. Using a MCDM method, Ren et al. [131] selected industrial systems from the aspect of sustainability. Because fossil fuel reserves are limited and atmospheric pollution is increasing, it is necessary to stimulate bio-diesel consumption. Sivaraja and Sakthivel [132] applied FAHP–TOPSIS, FAHP–VIKOR, and FAHP–ELECTRE to select the best blend of the specified fuel. In their research, Zavadskas et al. [133] selected the site for the incineration of waste, taking into account all the sustainability criteria using the new extension of the WASPAS method, the WASPAS single-valued neutrosophic Set (WASPAS–SVNS). It is known that the global population is growing each year, and it is necessary to provide an adequate amount of food. For this reason, Debnath et al. [134] selected the project portfolio for agricultural production by applying grey DEMATEL and MABAC methods. Huang et al. [135] presented a hybrid MCDM approach for the selection of materials for the production of particulate matter sensors. In their paper, Zhang et al. [136] dealt with the problem of evaluating the supply of rare minerals. For this purpose, fuzzy AHP and PROMETHEE methods were used. The application of the MCDM method within the decision support system can be of great importance to assist in emergency situations such as forest fires. The development of such a system is described in Ioannou et al. [137]. Table 6 provides a summary of the applied MCDM methods for the sub-area of other engineering disciplines.

Table 6. MCDM methods in the sub-area of other engineering disciplines.

A Problem that Is Solved Using the MCDM Method	Applied Methods	Reference
Proposing the design of the sustainable environmental management system	ELECTRE III	[124]
Sustainability performance assessment of 27 US and Canada metropolises	IFDM	[125]
Measuring the level of sustainability for sustainable intelligent buildings	AHP	[126]
Assessing the sustainability of post-disaster temporary housing units Technologies	MIVES	[127]
Assessing the sustainability of contaminated land remediation	SCORE	[128]
Sustainability assessment of the technologies for the treatment of urban sewage sludge	SWM, Digraph model, TOPSIS	[130]
Sustainability prioritization of industrial systems	Fuzzy AHP, Fuzzy ANP, PROMETHEE	[131]
Selection of optimum biodiesel blend in internal combustion engines	Fuzzy AHP–TOPSIS, Fuzzy AHP–VIKOR, Fuzzy AHP–ELECTRE	[132]
Sustainable assessment of alternative sites for the construction of a waste incineration plant	WASPAS–SVNS	[133]
Strategic project portfolio selection of agricultural byproducts	DEMATEL, MABAC	[134]
Evaluation and selection of materials for particulate matter sensors	DEMATEL, VIKOR	[135]
Enrichment evaluation to assess the security performance for China's several critical minerals	Fuzzy AHP, PROMETHEE	[136]

The application of MCDM methods in other engineering disciplines is reduced to the environmental aspect of sustainability. Problems such as environmental pollution, soil contamination, air pollution, and the selection of the best fossil fuel are just a few that are solved by applying MCDM methods, of which AHP is most frequently used, according to Table 6.

5. Conclusions

In this paper, representative studies that include the application of multi-criteria decision-making models in the field of sustainability engineering have been presented. A review of about 108 studies related to the application of multi-criteria decision-making methods in the field of civil engineering and infrastructure, supply chain management, transport and logistics, energy, and other engineering disciplines provides interesting conclusions that can be useful for researchers who deal with the application of MCDM models in different engineering areas.

This literature review has shown that sustainable engineering is an area that is quite suitable for the use of MCDM. It is not surprising that the number of publications related to environmental protection, waste minimization, renewable energy sources, energy efficiency, and the concept of sustainability have tripled in the last decade. Switching to the concept of renewable energy has influenced researchers to try to exploit and improve available knowledge in decision-making.

Most of the methods used in sustainable engineering are based on traditional approaches with a noticeable trend of applying the theory of uncertainty, such as fuzzy, grey, rough, and neutrosophic theory. It can be said that the selection between existing MCDM methods is also a multi-criteria problem. Each of the methods has its advantages and disadvantages, and it is not possible to claim that any method is more suitable than others. The same applies to the selection of uncertainty theory for a considered multi-criteria problem. The choice of the method depends largely on the preferences of decision-makers and analysts. It is therefore important to consider the convenience, validity, and accessibility of methods for a problem considered. Mukhametzyanov and Pamucar [138] emphasize that the choice of method can significantly influence the decision-making process. They also emphasized that several methods should be used in a decision-making process in order to obtain a sustainable and high-quality decision. This is also an explanation of the observed trend of using a large number of methods in the literature. By analyzing the prior research presented in this review, it can be concluded that in the field of civil engineering and infrastructure, MCDM methods in most cases help to solve the problems that arise when selecting methods of building structures and roads. These problems attach great importance as objects require reinforcements in the case of seismic activities, and also, in the trend of the construction of green, ecological houses. In the sub-section of civil engineering and infrastructure there are 6 papers dealing with this topic, which amounts to 23.07% of the total. The most common method is AHP, which is used in 13 papers or 50% of the total. MCDM methods are most commonly used in this field in combination with fuzzy theory. A total of three papers (11.54%) have been analyzed that integrate fuzzy principles along with other MCDM methods. In the field of supply chain management, the selection of the supplier is the most common problem that is solved using the MCDM method. It is necessary to select the most modern supplier in terms of sustainability, but also from the point of view of the customer or user of the service. Out of the total number of analyzed papers in this field, 11 or 50% deal with this problem, and the most commonly used methods are TOPSIS and AHP with 6 papers or 27.27% each. The combination of the MCDM method with the most common fuzzy theory is represented in 8 papers or in 36.36% of the total. The analysis has shown that the selection of the location of terminals and logistics centers from the aspect of sustainability is the most important problem that is solved by MCDM methods in the field of transport and logistics. Of the total number of articles in this field, 4 or 17.39% include a subject of research that is related to the choice of location. In this case, the AHP method is the most commonly applied. More precisely, it was applied in 7 papers or 30.43% of the total number. In this field, it is often a combination of MCDM methods with fuzzy theory. It is the same number as in the previous sub-area: 8 papers or 34.78% of the total. MCDM methods in the field of energy are used to a large extent for the choice of a certain type or mode of energy production. Alternatives most often include renewable but also conventional sources of energy. In this area there are 9 such papers, or 37.50% of the total, while the number of papers in which the AHP method is applied is 8 or 33.33%. Fuzzy theory is most commonly combined with MCDM methods in this field and is present in 3 papers or 12.5% of the total. When it comes to other engineering disciplines, the application of the MCDM method is mainly to assess the sustainability of buildings, land, waste treatment technologies, and cities. Within the mentioned area, 6 papers dealing with this topic were analyzed, or 46.15% of the total number. The AHP method is also the most commonly applied in this field and is applied in 4 papers or 30.77% of the total. In addition, the application of fuzzy theory is used, along with the other methods. In this sub-area, fuzzy principles were applied in 3 papers or in 23.08% of the total.

Based on the analysis of papers and problems that are solved using the MCDM method, it can be concluded that the AHP method has the broadest application when it comes to sustainable engineering. Generally, the total number of papers involving the use of the AHP method was 38 or 35.19% of the total. Among the other theories that integrate with MCDM, fuzzy theory stands out in cases of uncertainty and imprecision with a total of 25 papers or 23.15% of the total.

It can be concluded that there has been a significant increase in the application of MCDM models in all engineering areas in the last decade. The complexity of synchronous problems forces researchers to search for more flexible and simpler methods. Therefore, it is expected that there will be a further increase in works that consider the application of existing MCDM models and the development of new models for multi-criteria decision-making. It is also expected that the validation of results using multiple methods, the development of interactive systems to support the decision-making, and the improvement of fuzzy, grey, rough, and neutrosophic theory for the consideration of uncertainty will encourage researchers in the field of sustainable engineering to expand further research towards the creation of hybrid models, upgrading the existing MCDM models.

Author Contributions: Each author has participated and contributed sufficiently to take public responsibility for appropriate portions of the content.

Funding: This research received no external funding.

Conflicts of Interest: The authors declare no conflicts of interest.

References

1. Special Working Session. *World Commission on Environment and Development*; Oxford University Press: London, UK, 1987.
2. Hansmann, R.; Mieg, H.A.; Frischknecht, P. Principal sustainability components: Empirical analysis of synergies between the three pillars of sustainability. *Int. J. Sustain. Dev. World Ecol.* **2012**, *19*, 451–459. [CrossRef]
3. Moldan, B.; Janoušková, S.; Hák, T. How to understand and measure environmental sustainability: Indicators and targets. *Ecol. Indic.* **2012**, *17*, 4–13. [CrossRef]
4. Lozano, R. Envisioning sustainability three-dimensionally. *J. Clean. Prod.* **2008**, *16*, 1838–1846. [CrossRef]
5. Hutchins, M.J.; Sutherland, J.W. An exploration of measures of social sustainability and their application to supply chain decisions. *J. Clean. Prod.* **2008**, *16*, 1688–1698. [CrossRef]
6. Allenby, B.; Allen, D.; Davidson, C. Sustainable engineering: From myth to mechanism. *Environ. Qual. Manag.* **2007**, *17*, 17–26. [CrossRef]
7. Abraham, M.A. Principles of sustainable engineering. *Sustain. Sci. Eng. Defin. Princ.* **2006**, 3–10. [CrossRef]
8. Pamucar, D.; Bozanic, D.; Lukovac, V.; Komazec, N. Normalized weighted geometric bonferroni mean operator of interval rough numbers—Application in interval rough DEMATEL-COPRAS. *Facta Univ. Ser. Mech. Eng.* **2018**, *16*, 171–191. [CrossRef]
9. Zolfani, S.H.; Pourhossein, M.; Yazdani, M.; Zavadskas, E.K. Evaluating construction projects of hotels based on environmental sustainability with MCDM framework. *Alex. Eng. J.* **2018**, *57*, 357–365. [CrossRef]
10. Zavadskas, E.K.; Antucheviciene, J.; Vilutiene, T.; Adeli, H. Sustainable Decision-Making in Civil Engineering, Construction and Building Technology. *Sustainability* **2018**, *10*, 14. [CrossRef]
11. Zavadskas, E.K.; Saparauskas, J.; Antucheviciene, J. Sustainability in Construction Engineering. *Sustainability* **2018**, *10*, 2236. [CrossRef]
12. Liu, F.; Aiwu, G.; Lukovac, V.; Vukic, M. A multicriteria model for the selection of the transport service provider: A single valued neutrosophic DEMATEL multicriteria model. *Decis. Mak. Appl. Manag. Eng.* **2018**, *1*, 121–130. [CrossRef]
13. Kianpour, K.; Jusoh, A.; Mardani, A.; Streimikiene, D.; Cavallaro, F.; Nor, K.M.; Zavadskas, E.K. Factors Influencing Consumers' Intention to Return the End of Life Electronic Products through Reverse Supply Chain Management for Reuse, Repair and Recycling. *Sustainability* **2017**, *9*, 1657. [CrossRef]
14. Mardani, A.; Zavadskas, E.K.; Streimikiene, D.; Jusoh, A.; Nor, K.M.D.; Khoshnoudi, M. Using fuzzy multiple criteria decision making approaches for evaluating energy saving technologies and solutions in five star hotels: A new hierarchical framework. *Energy* **2016**, *117*, 131–148. [CrossRef]
15. Chatterjee, P.; Mondal, S.; Boral, S.; Banerjee, A.; Chakraborty, S. A novel hybrid method for non-traditional machining process selection using factor relationship and Multi-Attributive Border Approximation Method. *Facta Univ. Ser. Mech. Eng.* **2017**, *15*, 439–456. [CrossRef]
16. Zolfani, S.H.; Zavadskas, E.K.; Khazaelpour, P.; Cavallaro, F. The Multi-Aspect Criterion in the PMADM Outline and Its Possible Application to Sustainability Assessment. *Sustainability* **2018**, *10*, 4451. [CrossRef]

17. Chatterjee, K.; Pamucar, D.; Zavadskas, E.K. Evaluating the performance of suppliers based on using the R'AMATEL-MAIRCA method for green supply chain implementation in electronics industry. *J. Clean. Prod.* **2018**, *184*, 101–129. [CrossRef]

18. Yazdani, M.; Chatterjee, P.; Zavadskas, E.K.; Streimikiene, D. A novel integrated decision-making approach for the evaluation and selection of renewable energy technologies. *Clean Technol. Environ. Policy* **2018**, *20*, 403–420. [CrossRef]

19. Banasik, A.; Bloemhof-Ruwaard, J.M.; Kanellopoulos, A.; Claassen, G.D.H.; van der Vorst, J. Multi-criteria decision making approaches for green supply chains: A review. *Flex. Serv. Manuf. J.* **2018**, *30*, 366–396. [CrossRef]

20. Mardani, A.; Streimikiene, D.; Zavadskas, E.K.; Cavallaro, F.; Nilashi, M.; Jusoh, A.; Zare, H. Application of Structural Equation Modeling (SEM) to Solve Environmental Sustainability Problems: A Comprehensive Review and Meta-Analysis. *Sustainability* **2017**, *9*, 1814. [CrossRef]

21. Mardani, A.; Zavadskas, E.K.; Govindan, K.; Senin, A.A.; Jusoh, A. VIKOR Technique: A Systematic Review of the State of the Art Literature on Methodologies and Applications. *Sustainability* **2016**, *8*, 37. [CrossRef]

22. Mardani, A.; Zavadskas, E.K.; Khalifah, Z.; Zakuan, N.; Jusoh, A.; Nor, K.M.; Khoshnoudi, M. A review of multi-criteria decision-making applications to solve energy management problems: Two decades from 1995 to 2015. *Renew. Sustain. Energy Rev.* **2017**, *71*, 216–256. [CrossRef]

23. Mardani, A.; Zavadskas, E.K.; Streimikiene, D.; Jusoh, A.; Khoshnoudi, M. A comprehensive review of data envelopment analysis (DEA) approach in energy efficiency. *Renew. Sustain. Energy Rev.* **2017**, *70*, 1298–1322. [CrossRef]

24. Sierra, L.A.; Yepes, V.; Pellicer, E. A review of multi-criteria assessment of the social sustainability of infrastructures. *J. Clean. Prod.* **2018**, *187*, 496–513. [CrossRef]

25. Siksnelyte, I.; Zavadskas, E.K.; Streimikiene, D.; Sharma, D. An Overview of Multi-Criteria Decision-Making Methods in Dealing with Sustainable Energy Development Issues. *Energies* **2018**, *11*, 2754. [CrossRef]

26. Mardani, A.; Jusoh, A.; Zavadskas, E.K.; Cavallaro, F.; Khalifah, Z. Sustainable and Renewable Energy: An Overview of the Application of Multiple Criteria Decision Making Techniques and Approaches. *Sustainability* **2015**, *7*, 13947–13984. [CrossRef]

27. Allouhi, A.; El Fouih, Y.; Kousksou, T.; Jamil, A.; Zeraouli, Y.; Mourad, Y. Energy consumption and efficiency in buildings: Current status and future trends. *J. Clean. Prod.* **2015**, *109*, 118–130. [CrossRef]

28. Davidson, C.I.; Hendrickson, C.T.; Matthews, H.S.; Bridges, M.W.; Allen, D.T.; Murphy, C.F.; Allenby, B.R.; Crittenden, J.C.; Austin, S. Preparing future engineers for challenges of the 21st century: Sustainable engineering. *J. Clean. Prod.* **2010**, *18*, 698–701. [CrossRef]

29. Birgani, Y.T.; Yazdandoost, F. An Integrated Framework to Evaluate Resilient-Sustainable Urban Drainage Management Plans Using a Combined-adaptive MCDM Technique. *Water Resour. Manag.* **2018**, *32*, 2817–2835. [CrossRef]

30. Alhumaid, M.; Ghumman, A.R.; Haider, H.; Al-Salamah, I.S.; Ghazaw, Y.M. Sustainability Evaluation Framework of Urban Stormwater Drainage Options for Arid Environments Using Hydraulic Modeling and Multicriteria Decision-Making. *Water* **2018**, *10*, 581. [CrossRef]

31. Lombera, J.T.S.; Rojo, J.C. Industrial building design stage based on a system approach to their environmental sustainability. *Constr. Build. Mater.* **2010**, *24*, 438–447. [CrossRef]

32. Pujadas, P.; Pardo-Bosch, F.; Aguado-Renter, A.; Aguado, A. MIVES multi-criteria approach for the evaluation, prioritization, and selection of public investment projects. A case study in the city of Barcelona. *Land Use Policy* **2017**, *64*, 29–37. [CrossRef]

33. Del Cano, A.; Gomez, D.; de la Cruz, M.P. Uncertainty analysis in the sustainable design of concrete structures: A probabilistic method. *Constr. Build. Mater.* **2012**, *37*, 865–873. [CrossRef]

34. De la Fuente, A.; Pons, O.; Josa, A.; Aguado, A. Multi-Criteria Decision Making in the sustainability assessment of sewerage pipe systems. *J. Clean. Prod.* **2016**, *112*, 4762–4770. [CrossRef]

35. Akhtar, S.; Reza, B.; Hewage, K.; Shahriar, A.; Zargar, A.; Sadiq, R. Life cycle sustainability assessment (LCSA) for selection of sewer pipe materials. *Clean Technol. Environ. Policy* **2015**, *17*, 973–992. [CrossRef]

36. De la Fuente, A.; Blanco, A.; Armengou, J.; Aguado, A. Sustainability based-approach to determine the concrete type and reinforcement configuration of TBM tunnels linings. Case study: Extension line to Barcelona Airport T1. *Tunn. Undergr. Space Technol.* **2017**, *61*, 179–188. [CrossRef]

37. Pons, O.; de la Fuente, A. Integrated sustainability assessment method applied to structural concrete columns. *Constr. Build. Mater.* **2013**, *49*, 882–893. [CrossRef]

38. Rashidi, M.; Ghodrat, M.; Samali, B.; Kendall, B.; Zhang, C.W. Remedial Modelling of Steel Bridges through Application of Analytical Hierarchy Process (AHP). *Appl. Sci.* **2017**, *7*, 168. [CrossRef]

39. Jia, J.M.; Ibrahim, M.; Hadi, M.; Orabi, W.; Xiao, Y. Multi-Criteria Evaluation Framework in Selection of Accelerated Bridge Construction (ABC) Method. *Sustainability* **2018**, *10*, 4059. [CrossRef]

40. Formisano, A.; Mazzolani, F.M. On the selection by MCDM methods of the optimal system for seismic retrofitting and vertical addition of existing buildings. *Comput. Struct.* **2015**, *159*, 1–13. [CrossRef]

41. Terracciano, G.; Di Lorenzo, G.; Formisano, A.; Landolfo, R. Cold-formed thin-walled steel structures as vertical addition and energetic retrofitting systems of existing masonry buildings. *Eur. J. Environ. Civ. Eng.* **2015**, *19*, 850–866. [CrossRef]

42. Siozinyte, E.; Antucheviciene, J.; Kutut, V. Upgrading the old vernacular building to contemporary norms: Multiple criteria approach. *J. Civ. Eng. Manag.* **2014**, *20*, 291–298. [CrossRef]

43. Khoshnava, S.M.; Rostami, R.; Valipour, A.; Ismail, M.; Rahmat, A.R. Rank of green building material criteria based on the three pillars of sustainability using the hybrid multi criteria decision making method. *J. Clean. Prod.* **2018**, *173*, 82–99. [CrossRef]

44. Akadiri, P.O.; Olomolaiye, P.O.; Chinyio, E.A. Multi-criteria evaluation model for the selection of sustainable materials for building projects. *Autom. Constr.* **2013**, *30*, 113–125. [CrossRef]

45. Ozcan-Deniz, G.; Zhu, Y.M. A multi-objective decision-support model for selecting environmentally conscious highway construction methods. *J. Civ. Eng. Manag.* **2015**, *21*, 733–747. [CrossRef]

46. Stevic, Z.; Pamucar, D.; Subotic, M.; Antucheviciene, J.; Zavadskas, E.K. The Location Selection for Roundabout Construction Using Rough BWM-Rough WASPAS Approach Based on a New Rough Hamy Aggregator. *Sustainability* **2018**, *10*, 2817. [CrossRef]

47. Rashid, K.; Razzaq, A.; Ahmad, M.; Rashid, T.; Tariq, S. Experimental and analytical selection of sustainable recycled concrete with ceramic waste aggregate. *Constr. Build. Mater.* **2017**, *154*, 829–840. [CrossRef]

48. Raslanas, S.; Kliukas, R.; Stasiukynas, A. Sustainability assessment for recreational buildings. *Civ. Eng. Environ. Syst.* **2016**, *33*, 286–312. [CrossRef]

49. Tsai, W.H.; Lin, S.J.; Lee, Y.F.; Chang, Y.C.; Hsu, J.L. Construction method selection for green building projects to improve environmental sustainability by using an MCDM approach. *J. Environ. Plan. Manag.* **2013**, *56*, 1487–1510. [CrossRef]

50. Zavadskas, E.K.; Vainiunas, P.; Turskis, Z.; Tamosaitiene, J. Multiple criteria decision support system for assessment of projects managers in construction. *Int. J. Inf. Technol. Decis. Mak.* **2012**, *11*, 501–520. [CrossRef]

51. Marzouk, M.; Elmesteckawi, L. Analyzing procurement route selection for electric power plants projects using SMART. *J. Civ. Eng. Manag.* **2015**, *21*, 912–922. [CrossRef]

52. Palevicius, V.; Paliulis, G.M.; Venckauskaite, J.; Vengrys, B. Evaluation of the requirement for passenger car parking spaces using multi-criteria methods. *J. Civ. Eng. Manag.* **2013**, *19*, 49–58. [CrossRef]

53. Stevic, Z.; Pamucar, D.; Vasiljevic, M.; Stojic, G.; Korica, S. Novel Integrated Multi-Criteria Model for Supplier Selection: Case Study Construction Company. *Symmetry* **2017**, *9*, 279. [CrossRef]

54. Puška, A.; Maksimović, A.; Stojanović, I. Improving organizational learning by sharing information through innovative supply chain in agro-food companies from Bosnia and Herzegovina. *Oper. Res. Eng. Sci. Theory Appl.* **2018**, *1*, 76–90. [CrossRef]

55. Fazlollahtabar, H. Operations and inspection Cost minimization for a reverse supply chain. *Oper. Res. Eng. Sci. Theory Appl.* **2018**, *1*, 91–107. [CrossRef]

56. Seuring, S. A review of modeling approaches for sustainable supply chain management. *Decis. Support Syst.* **2013**, *54*, 1513–1520. [CrossRef]

57. Zimmer, K.; Frohling, M.; Schultmann, F. Sustainable supplier management—A review of models supporting sustainable supplier selection, monitoring and development. *Int. J. Prod. Res.* **2016**, *54*, 1412–1442. [CrossRef]

58. Fallahpour, A.; Olugu, E.U.; Musa, S.N.; Wong, K.Y.; Noori, S. A decision support model for sustainable supplier selection in sustainable supply chain management. *Comput. Ind. Eng.* **2017**, *105*, 391–410. [CrossRef]

59. Govindan, K.; Khodaverdi, R.; Jafarian, A. A fuzzy multi criteria approach for measuring sustainability performance of a supplier based on triple bottom line approach. *J. Clean. Prod.* **2013**, *47*, 345–354. [CrossRef]

60. Dai, J.; Blackhurst, J. A four-phase AHP-QFD approach for supplier assessment: A sustainability perspective. *Int. J. Prod. Res.* **2012**, *50*, 5474–5490. [CrossRef]

61. Rezaei, J.; Nispeling, T.; Sarkis, J.; Tavasszy, L. A supplier selection life cycle approach integrating traditional and environmental criteria using the best worst method. *J. Clean. Prod.* **2016**, *135*, 577–588. [CrossRef]
62. Azadnia, A.H.; Saman, M.Z.M.; Wong, K.Y. Sustainable supplier selection and order lot-sizing: An integrated multi-objective decision-making process. *Int. J. Prod. Res.* **2015**, *53*, 383–408. [CrossRef]
63. Su, C.M.; Horng, D.J.; Tseng, M.L.; Chiu, A.S.F.; Wu, K.J.; Chen, H.P. Improving sustainable supply chain management using a novel hierarchical grey-DEMATEL approach. *J. Clean. Prod.* **2016**, *134*, 469–481. [CrossRef]
64. Luthra, S.; Govindan, K.; Kannan, D.; Mangla, S.K.; Garg, C.P. An integrated framework for sustainable supplier selection and evaluation in supply chains. *J. Clean. Prod.* **2017**, *140*, 1686–1698. [CrossRef]
65. Zhao, H.R.; Guo, S. Selecting Green Supplier of Thermal Power Equipment by Using a Hybrid MCDM Method for Sustainability. *Sustainability* **2014**, *6*, 217–235. [CrossRef]
66. Barata, J.F.F.; Quelhas, O.L.G.; Costa, H.G.; Gutierrez, R.H.; Lameira, V.D.; Meirino, M.J. Multi-Criteria Indicator for Sustainability Rating in Suppliers of the Oil and Gas Industries in Brazil. *Sustainability* **2014**, *6*, 1107–1128. [CrossRef]
67. Hsu, C.W.; Kuo, T.C.; Shyu, G.S.; Chen, P.S. Low Carbon Supplier Selection in the Hotel Industry. *Sustainability* **2014**, *6*, 2658–2684. [CrossRef]
68. Kuo, T.C.; Hsu, C.W.; Li, J.Y. Developing a Green Supplier Selection Model by Using the DANP with VIKOR. *Sustainability* **2015**, *7*, 1661–1689. [CrossRef]
69. Liu, A.J.; Xiao, Y.X.; Ji, X.H.; Wang, K.; Tsai, S.B.; Lu, H.; Cheng, J.S.; Lai, X.J.; Wang, J.T. A Novel Two-Stage Integrated Model for Supplier Selection of Green Fresh Product. *Sustainability* **2018**, *10*, 2371. [CrossRef]
70. Gupta, H.; Sarkis, J. A supply chain sustainability innovation framework and evaluation methodology AU—Kusi-Sarpong, Simonov. *Int. J. Prod. Res.* **2018**, 1–19. [CrossRef]
71. Validi, S.; Bhattacharya, A.; Byrne, P.J. A case analysis of a sustainable food supply chain distribution system-A multi-objective approach. *Int. J. Prod. Econ.* **2014**, *152*, 71–87. [CrossRef]
72. Erol, I.; Sencer, S.; Sari, R. A new fuzzy multi-criteria framework for measuring sustainability performance of a supply chain. *Ecol. Econ.* **2011**, *70*, 1088–1100. [CrossRef]
73. Das, R.; Shaw, K. Uncertain supply chain network design considering carbon footprint and social factors using two-stage approach. *Clean Technol. Environ. Policy* **2017**, *19*, 2491–2519. [CrossRef]
74. Entezaminia, A.; Heydari, M.; Rahmani, D. A multi-objective model for multi-product multi-site aggregate production planning in a green supply chain: Considering collection and recycling centers. *J. Manuf. Syst.* **2016**, *40*, 63–75. [CrossRef]
75. Luthra, S.; Mangla, S.K.; Chan, F.T.S.; Venkatesh, V.G. Evaluating the Drivers to Information and Communication Technology for Effective Sustainability Initiatives in Supply Chains. *Int. J. Inf. Technol. Decis. Mak.* **2018**, *17*, 311–338. [CrossRef]
76. Padhi, S.S.; Pati, R.K.; Rajeev, A. Framework for selecting sustainable supply chain processes and industries using an integrated approach. *J. Clean. Prod.* **2018**, *184*, 969–984. [CrossRef]
77. Stojic, G.; Stevic, Z.; Antucheviciene, J.; Pamucar, D.; Vasiljevic, M. A Novel Rough WASPAS Approach for Supplier Selection in a Company Manufacturing PVC Carpentry Products. *Information* **2018**, *9*, 121. [CrossRef]
78. Mardani, A.; Zavadskas, E.K.; Khalifah, Z.; Jusoh, A.; Nor, K.M.D. Multiple criteria decision-making techniques in transportation systems: A systematic review of the state of the art literature. *Transport* **2016**, *31*, 359–385. [CrossRef]
79. Jeon, C.M.; Amekudzi, A.A.; Guensler, R.L. Evaluating Plan Alternatives for Transportation System Sustainability: Atlanta Metropolitan Region. *Int. J. Sustain. Transp.* **2010**, *4*, 227–247. [CrossRef]
80. Cadena, P.C.B.; Magro, J.M.V. Setting the weights of sustainability criteria for the appraisal of transport projects. *Transport* **2015**, *30*, 298–306. [CrossRef]
81. Baric, D.; Pilko, H.; Strujic, J. An analytic hierarchy process model to evaluate road section design. *Transport* **2016**, *31*, 312–321. [CrossRef]
82. Inti, S.; Tandon, V. Application of Fuzzy Preference-Analytic Hierarchy Process Logic in Evaluating Sustainability of Transportation Infrastructure Requiring Multicriteria Decision Making. *J. Infrastruct. Syst.* **2017**, *23*, 04017014. [CrossRef]
83. Mitropoulos, L.K.; Prevedouros, P.D. Incorporating sustainability assessment in transportation planning: An urban transportation vehicle-based approach. *Transp. Plan. Technol.* **2016**, *39*, 439–463. [CrossRef]

84. Mohamadabadi, H.S.; Tichkowsky, G.; Kumar, A. Development of a multi-criteria assessment model for ranking of renewable and non-renewable transportation fuel vehicles. *Energy* **2009**, *34*, 112–125. [CrossRef]

85. Zecevic, S.; Tadic, S.; Krstic, M. Intermodal Transport Terminal Location Selection Using a Novel Hybrid MCDM Model. *Int. J. Uncertain. Fuzziness Knowl.-Based Syst.* **2017**, *25*, 853–876. [CrossRef]

86. Awasthi, A.; Chauhan, S.S.; Omrani, H. Application of fuzzy TOPSIS in evaluating sustainable transportation systems. *Expert Syst. Appl.* **2011**, *38*, 12270–12280. [CrossRef]

87. Castillo, H.; Pitfield, D.E. ELASTIC—A methodological framework for identifying and selecting sustainable transport indicators. *Transp. Res. Part D Transp. Environ.* **2010**, *15*, 179–188. [CrossRef]

88. Lopez, E.; Monzon, A. Integration of Sustainability Issues in Strategic Transportation Planning: A Multi-criteria Model for the Assessment of Transport Infrastructure Plans. *Comput.-Aided Civ. Infrastruct. Eng.* **2010**, *25*, 440–451. [CrossRef]

89. Simongati, G. Multi-criteria decision making support tool for freight integrators: Selecting the most sustainable alternative. *Transport* **2010**, *25*, 89–97. [CrossRef]

90. Bojkovic, N.; Anic, I.; Pejcic-Tarle, S. One solution for cross-country transport-sustainability evaluation using a modified ELECTRE method. *Ecol. Econ.* **2010**, *69*, 1176–1186. [CrossRef]

91. Jones, S.; Tefe, M.; Appiah-Opoku, S. Proposed framework for sustainability screening of urban transport projects in developing countries: A case study of Accra, Ghana. *Transp. Res. Part A Policy Pract.* **2013**, *49*, 21–34. [CrossRef]

92. Awasthi, A.; Chauhan, S.S. Using AHP and Dempster-Shafer theory for evaluating sustainable transport solutions. *Environ. Model. Softw.* **2011**, *26*, 787–796. [CrossRef]

93. Dimić, S.; Pamučar, D.; Ljubojević, S.; Đorović, B. Strategic Transport Management Models—The Case Study of an Oil Industry. *Sustainability* **2016**, *8*, 954. [CrossRef]

94. Wang, H.; Jiang, Z.G.; Zhang, H.; Wang, Y.; Yang, Y.H.; Li, Y. An integrated MCDM approach considering demands-matching for reverse logistics. *J. Clean. Prod.* **2019**, *208*, 199–210. [CrossRef]

95. Awasthi, A.; Chauhan, S.S. A hybrid approach integrating Affinity Diagram, AHP and fuzzy TOPSIS for sustainable city logistics planning. *Appl. Math. Model.* **2012**, *36*, 573–584. [CrossRef]

96. Mavi, R.K.; Goh, M.; Zarbakhshnia, N. Sustainable third-party reverse logistic provider selection with fuzzy SWARA and fuzzy MOORA in plastic industry. *Int. J. Adv. Manuf. Technol.* **2017**, *91*, 2401–2418. [CrossRef]

97. Rao, C.J.; Goh, M.; Zhao, Y.; Zheng, J.J. Location selection of city logistics centers under sustainability. *Transp. Res. Part D Transp. Environ.* **2015**, *36*, 29–44. [CrossRef]

98. Turskis, Z.; Zavadskas, E.K. A new fuzzy additive ratio assessment method (aras-f). Case study: The analysis of fuzzy multiple criteria in order to select the logistic centers location. *Transport* **2010**, *25*, 423–432. [CrossRef]

99. Pamucar, D.S.; Tarle, S.P.; Parezanovic, T. New hybrid multi-criteria decision-making DEMATEL-MAIRCA model: Sustainable selection of a location for the development of multimodal logistics centre. *Econ. Res.-Ekon. Istraživanja* **2018**, *31*, 1641–1665. [CrossRef]

100. Jiang, P.; Hu, Y.C.; Yen, G.F.; Jiang, H.; Chiu, Y.J. Using a Novel Grey DANP Model to Identify Interactions between Manufacturing and Logistics Industries in China. *Sustainability* **2018**, *10*, 3456. [CrossRef]

101. Al Garni, H.Z.; Awasthi, A. Solar PV power plant site selection using a GIS-AHP based approach with application in Saudi Arabia. *Appl. Energy* **2017**, *206*, 1225–1240. [CrossRef]

102. Diaz-Cuevas, P.; Camarillo-Naranjo, J.M.; Perez-Alcantara, J.P. Relational spatial database and multi-criteria decision methods for selecting optimum locations for photovoltaic power plants in the province of Seville (southern Spain). *Clean Technol. Environ. Policy* **2018**, *20*, 1889–1902. [CrossRef]

103. Sanchez-Lozano, J.M.; Garcia-Cascales, M.S.; Lamata, M.T. GIS-based onshore wind farm site selection using Fuzzy Multi-Criteria Decision Making methods. Evaluating the case of Southeastern Spain. *Appl. Energy* **2016**, *171*, 86–102. [CrossRef]

104. Buyukozkan, G.; Karabulut, Y. Energy project performance evaluation with sustainability perspective. *Energy* **2017**, *119*, 549–560. [CrossRef]

105. Wu, Y.N.; Xu, C.B.; Zhang, T. Evaluation of renewable power sources using a fuzzy MCDM based on cumulative prospect theory: A case in China. *Energy* **2018**, *147*, 1227–1239. [CrossRef]

106. Zhang, L.; Zhou, P.; Newton, S.; Fang, J.X.; Zhou, D.Q.; Zhang, L.P. Evaluating clean energy alternatives for Jiangsu, China: An improved multi-criteria decision making method. *Energy* **2015**, *90*, 953–964. [CrossRef]

107. Troldborg, M.; Heslop, S.; Hough, R.L. Assessing the sustainability of renewable energy technologies using multi-criteria analysis: Suitability of approach for national-scale assessments and associated uncertainties. *Renew. Sustain. Energy Rev.* **2014**, *39*, 1173–1184. [CrossRef]
108. Klein, S.J.W.; Whalley, S. Comparing the sustainability of US electricity options through multi-criteria decision analysis. *Energy Policy* **2015**, *79*, 127–149. [CrossRef]
109. Tsoutsos, T.; Drandaki, M.; Frantzeskaki, N.; Iosifidis, E.; Kiosses, I. Sustainable energy planning by using multi-criteria analysis application in the island of Crete. *Energy Policy* **2009**, *37*, 1587–1600. [CrossRef]
110. Pamucar, D.; Badi, I.; Sanja, K.; Obradovic, R. A Novel Approach for the Selection of Power-Generation Technology Using a Linguistic Neutrosophic CODAS Method: A Case Study in Libya. *Energies* **2018**, *11*, 2489. [CrossRef]
111. Pamucar, D.; Gigovic, L.; Bajic, Z.; Janosevic, M. Location Selection for Wind Farms Using GIS Multi-Criteria Hybrid Model: An Approach Based on Fuzzy and Rough Numbers. *Sustainability* **2017**, *9*, 1315. [CrossRef]
112. Mirjat, N.H.; Uqaili, M.A.; Harijan, K.; Mustafa, M.W.; Rahman, M.M.; Khan, M.W.A. Multi-Criteria Analysis of Electricity Generation Scenarios for Sustainable Energy Planning in Pakistan. *Energies* **2018**, *11*, 757. [CrossRef]
113. Balezentis, T.; Streimikiene, D. Multi-criteria ranking of energy generation scenarios with Monte Carlo simulation. *Appl. Energy* **2017**, *185*, 862–871. [CrossRef]
114. Zhao, H.R.; Li, N.N. Optimal Siting of Charging Stations for Electric Vehicles Based on Fuzzy Delphi and Hybrid Multi-Criteria Decision Making Approaches from an Extended Sustainability Perspective. *Energies* **2016**, *9*, 270. [CrossRef]
115. Guo, S.; Zhao, H.R. Optimal site selection of electric vehicle charging station by using fuzzy TOPSIS based on sustainability perspective. *Appl. Energy* **2015**, *158*, 390–402. [CrossRef]
116. Gao, R.X.; Nam, H.O.; Ko, W.I.; Jang, H. National Options for a Sustainable Nuclear Energy System: MCDM Evaluation Using an Improved Integrated Weighting Approach. *Energies* **2017**, *10*, 2017. [CrossRef]
117. Skobalj, P.; Kijevcanin, M.; Afgan, N.; Jovanovic, M.; Turanjanin, V.; Vucicevic, B. Multi-criteria sustainability analysis of thermal power plant Kolubara-A Unit 2. *Energy* **2017**, *125*, 837–847. [CrossRef]
118. Vucijak, B.; Kupusovic, T.; Midzic-Kurtagic, S.; Ceric, A. Applicability of multicriteria decision aid to sustainable hydropower. *Appl. Energy* **2013**, *101*, 261–267. [CrossRef]
119. Streimikiene, D.; Balezentis, T.; Krisciukaitiene, I.; Balezentis, A. Prioritizing sustainable electricity production technologies: MCDM approach. *Renew. Sustain. Energy Rev.* **2012**, *16*, 3302–3311. [CrossRef]
120. Barros, J.J.C.; Coira, M.L.; Lopez, M.P.D.; Gochi, A.D. Assessing the global sustainability of different electricity generation systems. *Energy* **2015**, *89*, 473–489. [CrossRef]
121. Maxim, A. Sustainability assessment of electricity generation technologies using weighted multi-criteria decision analysis. *Energy Policy* **2014**, *65*, 284–297. [CrossRef]
122. Jovanovic, M.; Afgan, N.; Bakic, V. An analytical method for the measurement of energy system sustainability in urban areas. *Energy* **2010**, *35*, 3909–3920. [CrossRef]
123. Ioannou, K.; Tsantopoulos, G.; Arabatzis, G.; Andreopoulou, Z.; Zafeiriou, E. A Spatial Decision Support System Framework for the Evaluation of Biomass Energy Production Locations: Case Study in the Regional Unit of Drama, Greece. *Sustainability* **2018**, *10*, 531. [CrossRef]
124. Khalili, N.R.; Duecker, S. Application of multi-criteria decision analysis in design of sustainable environmental management system framework. *J. Clean. Prod.* **2013**, *47*, 188–198. [CrossRef]
125. Egilmez, G.; Gumus, S.; Kucukvar, M. Environmental sustainability benchmarking of the US and Canada metropoles: An expert judgment-based multi-criteria decision making approach. *Cities* **2015**, *42*, 31–41. [CrossRef]
126. Alwaer, H.; Clements-Croome, D.J. Key performance indicators (KPIs) and priority setting in using the multi-attribute approach for assessing sustainable intelligent buildings. *Build. Environ.* **2010**, *45*, 799–807. [CrossRef]
127. Hosseini, S.M.A.; de la Fuente, A.; Pons, O. Multi-criteria decision-making method for assessing the sustainability of post-disaster temporary housing units technologies: A case study in Bam, 2003. *Sustain. Cities Soc.* **2016**, *20*, 38–51. [CrossRef]
128. Diaz-Balteiro, L.; Gonzalez-Pachon, J.; Romero, C. Measuring systems sustainability with multi-criteria methods: A critical review. *Eur. J. Oper. Res.* **2017**, *258*, 607–616. [CrossRef]

129. Rosen, L.; Back, P.E.; Soderqvist, T.; Norrman, J.; Brinkhoff, P.; Norberg, T.; Volchko, Y.; Norin, M.; Bergknut, M.; Doberl, G. SCORE: A novel multi-criteria decision analysis approach to assessing the sustainability of contaminated land remediation. *Sci. Total Environ.* **2015**, *511*, 621–638. [CrossRef]

130. Ren, J.Z.; Liang, H.W.; Chan, F.T.S. Urban sewage sludge, sustainability, and transition for Eco-City: Multi-criteria sustainability assessment of technologies based on best-worst method. *Technol. Forecast. Soc. Chang.* **2017**, *116*, 29–39. [CrossRef]

131. Ren, J.Z.; Xu, D.; Cao, H.; Wei, S.A.; Dong, L.C.; Goodsite, M.E. Sustainability Decision Support Framework for Industrial System Prioritization. *AIChE J.* **2016**, *62*, 108–130. [CrossRef]

132. Sivaraja, C.M.; Sakthivel, G. Compression ignition engine performance modelling using hybrid MCDM techniques for the selection of optimum fish oil biodiesel blend at different injection timings. *Energy* **2017**, *139*, 118–141. [CrossRef]

133. Zavadskas, E.K.; Bausys, R.; Lazauskas, M. Sustainable Assessment of Alternative Sites for the Construction of a Waste Incineration Plant by Applying WASPAS Method with Single-Valued Neutrosophic Set. *Sustainability* **2015**, *7*, 15923–15936. [CrossRef]

134. Debnath, A.; Roy, J.; Kar, S.; Zavadskas, E.K.; Antucheviciene, J. A Hybrid MCDM Approach for Strategic Project Portfolio Selection of Agro By-Products. *Sustainability* **2017**, *9*, 1302. [CrossRef]

135. Huang, C.Y.; Chung, P.H.; Shyu, J.Z.; Ho, Y.H.; Wu, C.H.; Lee, M.C.; Wu, M.J. Evaluation and Selection of Materials for Particulate Matter MEMS Sensors by Using Hybrid MCDM Methods. *Sustainability* **2018**, *10*, 3451. [CrossRef]

136. Zhang, L.; Bai, W.; Yu, J.; Ma, L.M.; Ren, J.Z.; Zhang, W.S.; Cui, Y.Z. Critical Mineral Security in China: An Evaluation Based on Hybrid MCDM Methods. *Sustainability* **2018**, *10*, 4114. [CrossRef]

137. Ioannou, K.; Lefakis, P.; Arabatzis, G. Development of a decision support system for the study of an area after the occurrence of forest fire. *Int. J. Sustain. Soc.* **2011**, *3*, 5–32. [CrossRef]

138. Mukhametzyanov, I.; Pamucar, D. A sensitivity analysis in MCDM problems: A statistical approach. *Decis. Mak. Appl. Manag. Eng.* **2018**, *1*, 51–80. [CrossRef]

 © 2019 by the authors. Licensee MDPI, Basel, Switzerland. This article is an open access article distributed under the terms and conditions of the Creative Commons Attribution (CC BY) license (http://creativecommons.org/licenses/by/4.0/).

MDPI

St. Alban-Anlage 66

4052 Basel

Switzerland

Tel. +41 61 683 77 34

Fax +41 61 302 89 18

www.mdpi.com

Symmetry Editorial Office

E-mail: symmetry@mdpi.com

www.mdpi.com/journal/symmetry

www.ingramcontent.com/pod-product-compliance
Lightning Source LLC
LaVergne TN
LVHW071623170726
843515LV00009B/2469